THEORETICAL PHYSICS:

From Classical Mechanics to
Group Theory of Microparticles

*If every positive theory has to be based
on observations, it is, on the other hand,
also true that our mind needs a
theory in order to make observations.*
—A. Conté, *Cours de philosophie positive.*

*Therefore, if all thought is either practical
or productive or theoretical, physics must be
a theoretical science, but it will theorize
about such being as admits of being
moved, and about substance-as-
defined for the most part
only as not separable
from matter.*

—Aristotle, *Metaphysics* Book VI.

*Philosophy is written in this grand book—I mean
the universe—which stands continually open
to our gaze, but it cannot be understood
unless one first learns to comprehend
the language and interpret the
characters in which it is
written. It is written
in the language of
mathematics, . . .*

Galileo Galilei, *Il Saggiatore (Assayer).*

THEORETICAL PHYSICS:

From Classical Mechanics to
Group Theory of Microparticles

Masataka Mizushima

Professor of Physics
UNIVERSITY OF COLORADO

John Wiley & Sons, Inc.
New York   London   Sydney   Toronto

Library of Congress Cataloging in Publication Data:

Mizushima, Masataka, 1923–
    Theoretical physics.

    1. Physics.   I. Title.

QC21.2.M58   530.1   72-3580
ISBN 0-471-61188-3

Printed in the United States of America.

10 9 8 7 6 5 4 3 2 1

To my wife, Yoneko, and our five daughters

Preface

*... for all enquiring and all
learning is but recollection.*
—Plato, *Meno.*

This book is derived from lecture notes I prepared for a graduate course titled "Theoretical Physics" delivered in the Department of Physics and Astrophysics at the University of Colorado. It was a one-year (two-semester) review course for nonphysics as well as physics graduate students. In the course, and in this book, I intended to cover almost all important subjects of graduate level physics: classical mechanics, electricity magnetism, relativity, quantum mechanics, structure of atoms, molecules, and nuclei, and elementary particle physics. Obviously it is impossible to cover everything in one year—I left statistical mechanics and solid-state physics untouched. Originally I had planned to cover quantum field theory, particularly quantum theory of radiation, but I could not find time to go that far in my lectures. Even in the present form, this book may have more than enough material for a one-year course. Asterisks are used before the sections and problems that might be omitted if time is too short.

Although it is a condensed course, I have tried to start each discussion from the lowest level. This is because most of my students were nonphysicists. The background I have assumed here is that of senior physics students. Therefore, first-year graduate students in the physical sciences or mathematics should be able to follow the discussions.

Many exercise problems are provided, which can be used as homework or as examination problems. In any case, students are encouraged to solve as many of them as possible; I believe education is possible only by Socratic dialogue—a teacher's monologue is far from effective. Thus, some important topics are given as problems, in the expectation that students will figure them our by themselves.

Because of limited space, some of my discussions may be too brief. Side readings will therefore be helpful to students. In particular, I recommend four books, which cover nearly the same material as I have covered in this book.

Hylleraas, E. A., *Mathematical and Theoretical Physics*, Vols. I and II, Wiley-Interscience, New York, 1970; Kompaneyts, A. S., *Theoretical Physics*, Dover, New York, 1961; Levich, B. G., *Theoretical Physics*, Vols. I, II, and III, Wiley-Interscience, 1971; Joos, G., *Theoretical Physics*, Blackie, London, 1958.

Since these books discuss the same topics from different viewpoints, usually at more elementary levels than this book, it would be helpful for students to have all or some of them for outside readings.

Footnotes are provided to introduce important original papers and some recent review papers of the various topics. I have not attempted to give full references in a textbook like the present one. I believe, however, it is an important part of education to let students be exposed to the writings of original thinkers. I selected review papers which are readable and up-to-date. Other footnotes give historical anecdotes.

My justification for a short course like this is that in our present curriculum, physics students have to take separate courses for separate fields of physics. One great disadvantage of that method is that students may lose a unified picture of physics: they may forget that some basic concepts and mathematical methods are common to different fields of physics. Thus, for example, the concept of field is basic in Maxwell's theory, but it is often forgotten that this concept appeared in classical mechanics of continuous bodies. Actually this concept can be introduced more easily in the theory of continuous bodies, and it is even necessary that one understand it in that field in order to understand general relativity. Another example is the concept of spin. Since it appears in the Dirac theory in a conventional curriculum, students very often associate spin with quantum mechanics. However, it is better to associate this concept with relativity. In this way it is easier to understand why a magnetic moment is associated with a spin. An example of common mathematical methods is illustrated by Green's functions. Each Green's function is applicable in each field of physics, but one can appreciate its usefulness when one sees that the basic idea is applicable to almost all fields of physics. Actually the method of Green's functions is quite useful in quantum field theory and statistical mechanics (the two major fields I have not discussed here), but I believe students can learn enough about this method in this book to be able to explore these fields without much trouble.

Eliminating repeated discussions of common concepts and methods saves teaching time because several fields of physics can be combined in one course.

In a short course like this one, I had to omit many topics, of course. The situation reminds me of what Socrates said in Plato's *Cratylus*: "If I had not been poor I might have heard the fifty drachma course of the great Prodicus, which is a complete education in grammar and language, ..., and then I should have been able to answer your question about the correctness of names. But, indeed, I have only heard the single-drachma course, and, therefore, I do not know the truth about such matters ...." This may have to be a single-drachma course in a sense. But who can expect to produce a better student than Socrates, after all? If a single-drachma course can produce Socrates, why should we offer a fifty-drachma course?

The level of this book is somewhat higher than that of average textbooks and students may find it difficult on first reading. I admire Kierkegaard's wit when he wrote (in *Concluding Unscientific Postscript*) "... but inasmuch as with your limited capacities it will be impossible to make anything easier than it has become, you must, with the same humanitarian enthusiasm as the others, undertake to make something harder." In my experience, however, there are always some students who welcome the challenge of difficult physics. It has been said that challenge is the source of civilization.

Boulder, Colorado, 1971   *Masataka Mizushima*

Acknowledgments

Professor Eugen Merzbacher sent me numerous comments on the chapters of the final version of this book. His comments and suggestions were extremely helpful in improving and elucidating vital sections. Professor Ali Omar read and commented on the first version of my lecture notes, and Professor Walter Tanttila commented on Part I of the final version. Thomas Dillon, David Cartwright, and Joe Gallagher corrected the English.

M.M.

Contents

THEORETICAL PHYSICS:

From Classical Mechanics to
Group Theory of Microparticles

Part I Physics of the Macroscopic World

. . . But suddenly, about three hundred years ago, an explosion of mental activity occurred.
—Max Born, *Reflection.*

Chapter 1 **Principles of Classical Mechanics**

Aristotle's *Physica* may be the first systematic attempt at theoretical physics. He proposed some basic concepts such as space, time, motion, and force. His work and most of the works by the Greeks, however, were too theoretical, or speculative, and many of their basic postulates were wrong, because they did not care to prove or disprove their postulates by experiment. Galileo was the first physicist in our sense; he was the first who showed that not just a sensible observation but carefully designed experiments are necessary to find physical principles. He was concerned with the important concepts of acceleration, inertial mass, and force, but he did not have enough mathematics, particularly calculus, to formulate the law of motion in a workable form. Theoretical physics, which formulates physical principles found from experiments, was started by Newton around the end of the 17th century.

Newton's genius was exhibited in two fields: optics and mechanics. In optics he found important phenomena and experimental laws. His theory, however, was not quite satisfactory. An entire century was lost before an acceptable theory was found in this field of physics. His theory of mechanics, on the other hand, was one of the most brilliant achievements of the human brain. The 18th century witnessed the first successful development of theoretical physics following Newton's mechanics.

1-1 NEWTON'S MECHANICS

In the celebrated book *Mathematical Principles of Natural Philosophy*,[1] which was first published in 1686, Isaac Newton proposed three laws of motion. His first law is that a free particle moves with a constant velocity $\mathbf{V}$, and his second law is that the change of $\mathbf{V}$ is proportional to the external force $\mathbf{F}$. These two laws are summarized in *Newton's equation of motion*

$$m \frac{d\mathbf{V}}{dt} \equiv m\dot{\mathbf{V}} \equiv m\mathbf{a} = \mathbf{F} \tag{1-1}$$

where $\mathbf{a}$ is the acceleration, and the proportionality constant m is the mass of the particle.

The mass m was defined as the *quantity of matter* by Newton. If the matter is made of one kind of atoms, the mass will be proportional to the number of atoms in the matter. In this definition, however, it is difficult to find the mass ratio of matter made of different kinds of atoms. Although all nuclei are made of protons and neutrons, their masses are not proportional to the number of such nucleons (protons and neutrons). Leonard Euler proposed to define m by

$$m = \frac{|\mathbf{F}|}{|\mathbf{a}|} \tag{1-2}$$

where $|\mathbf{F}|$ and $|\mathbf{a}|$ are the magnitudes of $\mathbf{F}$ and $\mathbf{a}$, respectively. In order to measure m, Euler would first measure $\mathbf{F}$ and $\mathbf{a}$. The ratio of these two vectors is, in general, expected to be a second rank tensor. It is an experimental result that the ratio happens to be a scalar, i.e., a constant which does not depend on a direction in space.[2]

Newton's third law is that action and reaction are opposite in direction and the same in magnitude. When we have two bodies, the force exerted by one on the other, $\mathbf{F}_{12}$, is equal to $-\mathbf{F}_{21}$, the force exerted by the latter on the former (Fig. 1-1). Thus the equations of motion of these two bodies are

$$m_1 \mathbf{a}_1 = \mathbf{F}_{12} = -\mathbf{F}_{21}$$
$$m_2 \mathbf{a}_2 = \mathbf{F}_{21} = -\mathbf{F}_{12} \tag{1-3}$$

if there are no external forces.

---

[1] English translation can be found in *Great Books of the Western World*, Vol. 34, edited by M. I. Adler, Encyclopaedia Britannica, Inc., Chicago. A modern impression of the original publication *Philosophiae Naturalis Principia Mathematica* is available from Culture Civilisation, Bruxells, 1965.

[2] Mach's principle predicts that if the distribution of the stars in the universe is not spherical, the mass m would not be a scalar (see section 4-1).

FIGURE 1–1. Action and reaction.

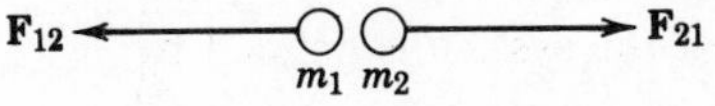

$$\mathbf{F}_{12} \longleftarrow \underset{m_1}{\bigcirc}\ \underset{m_2}{\bigcirc} \longrightarrow \mathbf{F}_{21}$$

Using this third law, Ernst Mach[3] proposed a better, more practical definition of the mass m. Using the situation where (1–3) holds we see

$$\frac{m_1}{m_2} = \frac{|\mathbf{a}_2|}{|\mathbf{a}_1|} \tag{1–4}$$

In this way we do not have to measure $\mathbf{F}$. Since the measurement of $\mathbf{a}$ is much easier than that of $\mathbf{F}$, Mach's definition (1–4) is suitable for a precise measurement of a mass with respect to some standard mass.

The momentum $\mathbf{p}$ is defined by

$$\mathbf{p} = m\mathbf{V} \tag{1–5}$$

Newton's equation (1–1) can be written as

$$\frac{d\mathbf{p}}{dt} \equiv \dot{\mathbf{p}} = \mathbf{F} \tag{1–6}$$

In the relativistic theory (see Chapter 4), equation (1–1) will hold only as an approximation, while (1–6) will be valid with a more general definition of the momentum.

When we have two bodies exerting forces on each other, Newton's equation in the form of (1–6) gives

$$\frac{d}{dt}(\mathbf{p}_1 + \mathbf{p}_2) = \mathbf{F}_{12} + \mathbf{F}_{21} = 0 \tag{1–7}$$

In general, if there are many bodies which are interacting with each other, and there exists no external force, then

$$\frac{d}{dt}\left(\sum_i \mathbf{p}_i\right) = \sum_{i,j} \mathbf{F}_{ij} = 0 \tag{1–8}$$

---

[3] E. Mach, *The Science of Mechanics*, 6th American edition, Open Court Publishing Co., Lasalle, Ill., 1960. Chapter III, section V. The book was originally published in German in 1883.

Thus the total momentum $\sum_i \mathbf{p}_i$ of an isolated system is a constant of the motion, or it is *conserved*.

The center of mass of the system is defined by

$$\mathbf{R} = \frac{\sum_i m_i \mathbf{r}_i}{\sum_i m_i} \tag{1-9}$$

The total momentum of the system is

$$\sum_i \mathbf{p}_i = \left(\sum_i m_i \right) \mathbf{V} \tag{1-10}$$

where $\mathbf{V} = d\mathbf{R}/dt \equiv \dot{\mathbf{R}}$. Thus the velocity of the center of mass $\mathbf{V}$ is also conserved when there is no external force and the total mass $\sum_i m_i$ is conserved.

1–2 TWO-BODY, MANY-BODY PROBLEMS

When two particles of masses m_1 and m_2 are moving under mutual influence but no other external force field, their equations of motion are

$$m_1 \mathbf{a}_1 = \mathbf{F}_{12}, \quad \text{and} \quad m_2 \mathbf{a}_2 = \mathbf{F}_{21} \tag{1-11}$$

where $\mathbf{F}_{ij}$ is the force exerted by j on i.

In (1–7), or from Newton's third law, we see

$$m_1 \mathbf{a}_1 + m_2 \mathbf{a}_2 = 0$$

which means, as shown in the last section, that the center of mass given by (1–9) moves with a constant velocity. Now (1–11) and Newton's third law also give

$$m(\mathbf{a}_1 - \mathbf{a}_2) = \mathbf{F}_{12} \tag{1-12}$$

where

$$\frac{1}{m} = \frac{1}{m_1} + \frac{1}{m_2} \tag{1-13}$$

Since

$$\mathbf{a}_1 - \mathbf{a}_2 = \frac{d^2}{dt^2} (\mathbf{r}_1 - \mathbf{r}_2) = \ddot{\mathbf{r}}_{12} \tag{1-14}$$

where $\mathbf{r}_{12}$ is the directed distance from particle 1 to particle 2 (Fig. 1–2), equation (1–12) reduces to

$$m\ddot{\mathbf{r}}_{12} = \mathbf{F}_{12} \tag{1-15}$$

FIGURE 1–2. Vector $\mathbf{r}_{12}$.

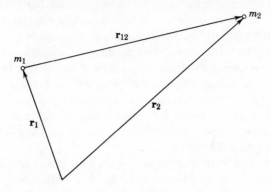

Thus the two-body problem is reduced to two one-body problems: the free motion of the center of mass and the motion of a particle with the reduced mass under the force field $\mathbf{F}_{12}$. The one-body problem in the latter sense will be discussed further in this chapter.

PROBLEM 1–1 When $m_1 \gg m_2$, show that

$$\mathbf{R} = \mathbf{r}_1 + \sum_{n=1}^{\infty} \left(\frac{-m_2}{m_1} \right)^n \mathbf{r}_{12}$$

$$m = m_2 - \left(\frac{m_2^2}{m_1} \right) - O \left(\frac{m_2^3}{m_1^2} \right)$$

where $O(\alpha)$ means small terms of the order of α or higher.

Many-Body Problems

For a three-body problem, we can still find the center of mass whose motion can be solved easily. Since there are nine degrees of freedom in any three-body system and the center of mass has three degrees of freedom, we are left with six degrees of freedom. There is no general way to reduce this number six except for very special cases.[4] In general we have to solve six simultaneous equations.

---

[4] When the force is central, or when $\mathbf{F}_{ij}$ is in the same direction as $\mathbf{r}_{ij}$, three particles, or any number of particles for that matter, will stay on a single line if they started on this line with neither initial velocity nor acceleration perpendicular to the line. When the forces are gravitational, a special case of central forces, a triangle configuration, is also stable in the same sense. In the latter case such stable configurations of a three-body system are called the *Lagrange points*.

In a n-body problem, the degrees of freedom are $3n$, and the isolation of the motion of the center of mass can only reduce the number of unsolved equations to $3n - 3$, which does not help much. Unfortunately, most of the problems of physical interest are such *many-body problems*.

One way to attack the many-body problem is to find as many *conserved quantities*, or *constants of motion*, as possible. The velocity of the center of mass is found to be conserved. We shall find other conserved quantities such as the total energy and the total angular momentum of a system in the following sections. The Hamiltonian formalism which is the main subject of the rest of this chapter is developed mainly to systematize the method of finding these conserved quantities.

Another standard way of approach is to find the *collective coordinates*. These are generalized coordinates (see Section 1–4), which are given by superpositions of coordinates of many particles in such a way that their motions are relatively independent of each other. Common examples of such collective coordinates are the *normal coordinates* in the system of coupled oscillators and the *rotational coordinates* of a rigid body. Normal coordinates will be the main subject of discussion in the next two chapters.

When the number of particles is too large, we may not be interested in following the motion of individual particles, but we may be interested in the discussion of bulk quantities, which specify states of the whole system of particles. Such quantities, sometimes called state variables, are pressure, volume, temperature, etc. This approach of ignoring most of the coordinates and discussing only bulk quantities leads to *statistical mechanics*.

1–3 KINETIC AND POTENTIAL ENERGIES

The *rotation* or the *curl* of a force $\mathbf{F}$ is defined by $\nabla \times \mathbf{F}$, where ∇ is a vector operator, called *del*, whose x, y, and z components are $\partial/\partial x$, $\partial/\partial y$, and $\partial/\partial z$, respectively. Most of the important forces in physics are such that their rotations are zero everywhere. If that is the case we can find a potential U such that

$$\mathbf{F} = -\nabla U \qquad (\text{If} \quad \nabla \times \mathbf{F} = 0) \qquad (1\text{–}16)$$

or the force can be expressed as the gradient of the potential (with a minus sign). This is seen easily since if U is a differentiable scalar function we have the identity

$$\nabla \times (\nabla U) = 0 \qquad (1\text{–}17)$$

This situation can be understood better by using *Stokes' theorem*, which states that

$$\oint \mathbf{A(r)} \cdot d\mathbf{l} = \int (\nabla \times \mathbf{A}) \cdot d\mathbf{a} \qquad (1\text{–}18)$$

for any vector function $\mathbf{A}(\mathbf{r})$. The left-hand side of this equation is the line integral of the tangential component of the vector function $\mathbf{A}$ along a closed curve, while the right-hand side is the area integral of $\mathbf{V} \times \mathbf{A}$ over the whole area enclosed by the closed curve (Fig. 1–3).

FIGURE 1–3.   Line and area integrals in Stokes' Theorem.

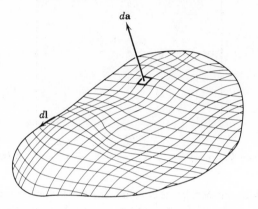

In order to prove this theorem, let us first consider an infinitesimal area in a plane perpendicular to the z axis bounded by the lines $x - \frac{1}{2} dx$, $x + \frac{1}{2} dx$, $y - \frac{1}{2} dy$, and $y + \frac{1}{2} dy$ (Fig. 1–4). The *line integral* around the sides of this area in the direction shown in Fig. 1–4 is

$$A_x(x, y - \tfrac{1}{2} dy)\, dx + A_y(x + \tfrac{1}{2} dx, y)\, dy - A_x(x, y + \tfrac{1}{2} dy)\, dx$$
$$- A_y(x - \tfrac{1}{2} dx, y)\, dy$$

$$= [A_x(x, y - \tfrac{1}{2} dy) - A_x(x, y + \tfrac{1}{2} dy)]\, dx$$
$$+ [A_y(x + \tfrac{1}{2} dx, y) - A_y(x - \tfrac{1}{2} dx, y)]\, dy$$

$$= \left(-\frac{\partial A_x}{\partial y}\, dy\right) dx + \left(\frac{\partial A_y}{\partial x}\, dx\right) dy = (\mathbf{V} \times \mathbf{A})_z\, dx\, dy \qquad (1\text{–}19)$$

Notice that the infinitesimal area $d\mathbf{a}$ in this case is in the plus z direction with magnitude $dx\, dy$ (Figs. 1–4 and 1–5).

When the left-hand side of (1–19) is integrated over a finite area, we see that the inner sides which are shared by adjacent infinitesimal areas contribute nothing, and only the outer sides which are not shared remain without being canceled. The result is just the line integral on the left-hand side of (1–18). Generalizing this result into an arbitrarily orientated curved surface, Stokes' theorem (1–18) is obtained.

FIGURE 1–4. Infinitesimal area *dx dy*.

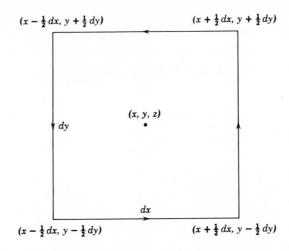

$(x - \tfrac{1}{2}\,dx,\, y + \tfrac{1}{2}\,dy)$ $(x + \tfrac{1}{2}\,dx,\, y + \tfrac{1}{2}\,dy)$

(x, y, z)

dy

dx

$(x - \tfrac{1}{2}\,dx,\, y - \tfrac{1}{2}\,dy)$ $(x + \tfrac{1}{2}\,dx,\, y - \tfrac{1}{2}\,dy)$

Coming back to physics, let us consider the work done in moving a particle under the force field $\mathbf{F}(\mathbf{r})$ from point $\mathbf{r}_a$ to another point $\mathbf{r}_b$. There are infinitely many paths which can be taken between these two points. If 1 and 2 are two such paths, and dl_1 and dl_2 are the line elements of these paths, respectively, the work done in bringing the particle from $\mathbf{r}_a$ to $\mathbf{r}_b$ taking each of these two paths is

$$\int_{\mathbf{r}_a}^{\mathbf{r}_b} \mathbf{F} \cdot d\mathbf{l}_1 \qquad \text{and} \qquad \int_{\mathbf{r}_a}^{\mathbf{r}_b} \mathbf{F} \cdot d\mathbf{l}_2$$

FIGURE 1–5. Line integral.

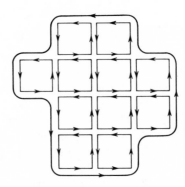

In general they may be different, or the work done may depend on the path one takes. We see, however (Fig. 1–6), using Stokes' theorem, that

$$\int_{\mathbf{r}_a}^{\mathbf{r}_b} \mathbf{F} \cdot d\mathbf{l}_1 - \int_{\mathbf{r}_a}^{\mathbf{r}_b} \mathbf{F} \cdot d\mathbf{l}_2 = \oint_{12} \mathbf{F} \cdot d\mathbf{l} = \int_{a_{12}} (\mathbf{V} \times \mathbf{F}) \cdot d\mathbf{a} \qquad (1\text{--}20)$$

where a_{12} is the area between paths 1 and 2. Therefore, if $\mathbf{V} \times \mathbf{F} = 0$ everywhere in the region of our interest, the work done does not depend on

FIGURE 1–6. $\int_{\mathbf{r}_a}^{\mathbf{r}_b} \mathbf{F} \cdot d\mathbf{l}_1$ and $\int_{\mathbf{r}_a}^{\mathbf{r}_b} \mathbf{F} \cdot d\mathbf{l}_2$.

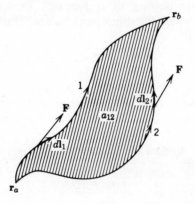

the path, but depends only on the starting and end points. Formula (1–16) is a mathematical expression of this result, because

$$\int_{\mathbf{r}_a}^{\mathbf{r}_b} \mathbf{F} \cdot d\mathbf{l} = - \int_{\mathbf{r}_a}^{\mathbf{r}_b} (\mathbf{V}U) \cdot d\mathbf{l} = - \int_{\mathbf{r}_a}^{\mathbf{r}_b} \left(\frac{\partial U}{\partial x} \, dx + \frac{\partial U}{\partial y} \, dy + \frac{\partial U}{\partial z} \, dz \right)$$

$$= - \int_{\mathbf{r}_a}^{\mathbf{r}_b} dU = U(\mathbf{r}_a) - U(\mathbf{r}_b) \qquad (1\text{--}21)$$

If the potential $U(\mathbf{r})$ exists, Newton's equation for a particle can be written as

$$m\dot{\mathbf{V}} = -\mathbf{V}U \qquad (1\text{--}22)$$

Multiplying this relation by $\mathbf{V}$, we have

$$\mathbf{V} \cdot (m\dot{\mathbf{V}} + \nabla U) = m\mathbf{V} \cdot \dot{\mathbf{V}} + \mathbf{V} \cdot \nabla U$$

$$= m\mathbf{V} \cdot \dot{\mathbf{V}} + \left(\frac{dx}{dt} \frac{\partial}{\partial x} + \frac{dy}{dt} \frac{\partial}{\partial y} + \frac{dz}{dt} \frac{\partial}{\partial z} \right) U$$

$$= \frac{d}{dt} (\tfrac{1}{2}mV^2 + U) = 0 \qquad (1\text{–}23)$$

that is, $\tfrac{1}{2}mV^2 + U$ is conserved. The quantity $\tfrac{1}{2}mV^2$ is called the *kinetic energy*, which we designate as T, while U is called the *potential energy*.

When many particles are interacting with each other the potential energy U is a function of the position of all particles, since the $\mathbf{F}$ exerted on one particle depends on where the other particles are. Thus

$$\frac{dU}{dt} = \sum_i \mathbf{V}_i \cdot \nabla_i U \qquad (1\text{–}24)$$

where the summation is to be taken over all particles. Following the same argument as before, we see that $\sum_i \tfrac{1}{2}m_i V_i^2 + U$ is conserved in this case.

PROBLEM 1–2. When a force is central, that is,

$$\mathbf{F}(\mathbf{r}) = f(r)\mathbf{r}$$

where $\mathbf{r}$ is the vector from the source of the force and $f(r)$ is a function of r, the magnitude of $\mathbf{r}$, show that

$$\mathbf{V} \times \mathbf{F} = 0$$

PROBLEM 1–3. Show that in the two-body system we can express the kinetic energy as

$$T\,(= \tfrac{1}{2}m_1 V_1{}^2 + \tfrac{1}{2}m_2 V_2{}^2) = \tfrac{1}{2}(m_1 + m_2)\dot{\mathbf{R}}^2 + \tfrac{1}{2}\mu\dot{\mathbf{r}}_{12}{}^2$$

The kinetic energy can be separated into that of the center-of-mass motion and that of the internal motion.

1–4 GENERALIZED COORDINATES

In order to represent the position of a particle in space $\mathbf{r}$, we can take a set of three straight lines which are perpendicular to each other and which pass through a single point, called the origin. These three lines are called *X*, *Y*, and *Z* axes. The position of a particle can be given by the three components of **r** along these axes, *x*, *y*, and *z*. This is the *Cartesian coordinate system.*

The Cartesian coordinate system is the most common and often the most convenient one, but this is not the only way to represent the position of a particle. For example one can use *spherical coordinates r*, θ, and φ, which are shown in Fig. 1–7, or *cylindrical coordinates* ρ, *z*, and φ, which are shown in Fig. 1–8. Their relations to the Cartesian coordinates are

$$x = r \sin \theta \cos \varphi$$

$$y = r \sin \theta \sin \varphi \qquad\qquad (1\text{–}25)$$

$$z = r \cos \theta$$

and

$$x = \rho \cos \varphi$$

$$y = \rho \sin \varphi \qquad\qquad (1\text{–}26)$$

$$z = z$$

respectively.

FIGURE 1–7. Spherical coordinates.

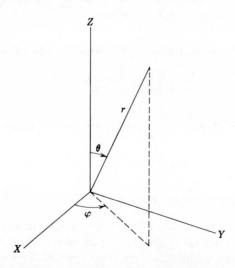

If there is more than one particle, it is often convenient to describe a system by the distance between particles, rather than the position of individual

FIGURE 1–8. Cylindrical coordinates.

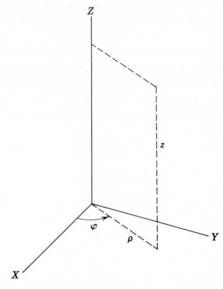

particles. Thus there are many ways to represent the configuration of a many-body system. For a n-body system we can use a set of $3n$ *generalized cordinates* $q_1, q_2, \ldots, q_{3n}$ to express the position of all particles. The relation between such generalized coordinates and the Cartesian coordinates may not be linear, as seen in (1–25) and (1–26).

PROBLEM 1–4. Show that in a cylindrical coordinate system

$$T = \tfrac{1}{2}m(\dot{\rho}^2 + \rho^2\dot{\varphi}^2 + \dot{z}^2)$$

PROBLEM 1–5. Show that in a spherical coordinate system

$$T = \tfrac{1}{2}m(\dot{r}^2 + r^2\dot{\theta}^2 + r^2\dot{\varphi}^2 \sin^2\theta)$$

1–5 LAGRANGE'S EQUATIONS

Lagrange's equations, named after Joseph Louis Lagrange, are equivalent to Newton's equations (1–1), but more convenient when using generalized coordinates. Newton's equations are simple in the Cartesian coordinates, but their formal appearance becomes complicated if we use other coordinates such as the cylindrical or spherical coordinates. Lagrange's equations, on the other hand, have the great advantage that their formal appearance does not change as we go from one coordinate system to another.

Let us consider a one-particle system first. From the expression of the kinetic energy in the Cartesian coordinates

$$T_1 = \tfrac{1}{2}m(\dot{x}^2 + \dot{y}^2 + \dot{z}^2) \tag{1-27}$$

we obtain

$$\frac{\partial T_1}{\partial \dot{q}_i} = m\left(\dot{x}\,\frac{\partial \dot{x}}{\partial \dot{q}_i} + \dot{y}\,\frac{\partial \dot{y}}{\partial \dot{q}_i} + \dot{z}\,\frac{\partial \dot{z}}{\partial \dot{q}_i}\right) \tag{1-28}$$

on the other hand

$$\dot{x} = \sum_i \frac{\partial x}{\partial q_i}\,\dot{q}_i \tag{1-29}$$

which shows that

$$\frac{\partial \dot{x}}{\partial \dot{q}_i} = \frac{\partial x}{\partial q_i} \tag{1-30}$$

Putting this relation into (1–28) and differentiating by t

$$\frac{d}{dt}\frac{\partial T_1}{\partial \dot{q}_i} = m\,\frac{d}{dt}\left(\dot{x}\,\frac{\partial x}{\partial q_i} + \dot{y}\,\frac{\partial y}{\partial q_i} + \dot{z}\,\frac{\partial z}{\partial q_i}\right)$$

$$= m\left(\ddot{x}\,\frac{\partial x}{\partial q_i} + \ddot{y}\,\frac{\partial y}{\partial q_i} + \ddot{z}\,\frac{\partial z}{\partial q_i}\right) + m\left(\dot{x}\,\frac{\partial \dot{x}}{\partial q_i} + \dot{y}\,\frac{\partial \dot{y}}{\partial q_i} + \dot{z}\,\frac{\partial \dot{z}}{\partial q_i}\right) \tag{1-31}$$

where, in obtaining the last expression, we used relations such as

$$\frac{d}{dt}\frac{\partial x}{\partial q_i} = \frac{\partial}{\partial q_i}\frac{dx}{dt} \tag{1-32}$$

In the first terms of (1–31) we can use Newton's equation in the form of (1–22), namely $m\ddot{x} = -\partial U/\partial x$, etc., while the second terms can be expressed by T_1 again. Thus

$$\frac{d}{dt}\frac{\partial T_1}{\partial \dot{q}} = -\frac{\partial U}{\partial q_i} + \frac{\partial T_1}{\partial q_i} \tag{1-33}$$

If U does not depend on $\dot{q}_i$, which is often the case, we can write (1–33) as

$$\frac{d}{dt}\frac{\partial L_1}{\partial \dot{q}_i} - \frac{\partial L_1}{\partial q_i} = 0 \qquad (1\text{–}34)$$

where

$$L_1 = T_1 - U \qquad (1\text{–}35)$$

and is called the Lagrangian function, while (1–34) is called *Lagrange's equation*.

It is easy to generalize the above argument for a many-particle system. We obtain

$$\frac{d}{dt}\frac{\partial L}{\partial \dot{q}_i} - \frac{\partial L}{\partial q_i} = 0, \qquad \text{with} \qquad L = \sum_j T_j - U \qquad (1\text{–}36)$$

for the general case.

PROBLEM 1–6. Prove equation (1–36).

1–6 CANONICAL MOMENTA

Corresponding to a generalized coordinate q_i we define the canonical momentum p_i by

$$p_i \equiv \frac{\partial L}{\partial \dot{q}_i} \qquad (1\text{–}37)$$

For example, when U does not depend on $\dot{x}$, we obtain

$$p_x \equiv \frac{\partial L}{\partial \dot{x}} = \frac{\partial L_1}{\partial \dot{x}} = \frac{\partial T_1}{\partial \dot{x}} = m\dot{x} \qquad (1\text{–}38)$$

which is the ordinary momentum in the x direction.

If we put (1–37) into Lagrange's equation (1–36) we obtain

$$\dot{p}_i = \frac{\partial L}{\partial q_i} \qquad (1\text{–}39)$$

This simple relation shows an important fact that *if q_i does not appear in L explicitly, the corresponding p_i is a constant of motion.*

As a simple example of the above statement, let us take a particle of mass m moving in the xy plane under a central potential, or a potential which does not depend on φ:

$$U = U(\rho) \qquad (1\text{–}40)$$

Using cylindrical coordinates, neglecting z, we see

$$T = \tfrac{1}{2}m(\dot{x}^2 + \dot{y}^2) = \tfrac{1}{2}m(\dot{\rho}^2 + \rho^2\dot{\varphi}^2) \qquad (1\text{–}41)$$

From (1–39) we obtain

$$p_\rho = \frac{\partial L}{\partial \dot{\rho}} = m\dot{\rho}, \qquad p_\varphi = \frac{\partial L}{\partial \dot{\varphi}} = m\rho^2\dot{\varphi} \qquad (1\text{–}42)$$

Since φ does not appear in L explicitly, we see

$$\dot{p}_\varphi = \frac{\partial L}{\partial \varphi} = 0 \qquad (1\text{–}43)$$

or p_φ is a constant of motion.

PROBLEM 1–7. If two particles are interacting with each other but free from the rest of the universe, show that each component of $\mathbf{p}_R$, the momentum conjugate to the center of mass $\mathbf{R}$, is a constant of motion.

1–7 HAMILTONIAN FUNCTION. HAMILTON'S EQUATIONS

The Hamiltonian function is defined in two steps. It is calculated by

$$H = \sum_i^k p_i\dot{q}_i - L(q_1, \ldots, q_k, \dot{q}_1, \ldots, \dot{q}_k, t) \qquad (1\text{–}44)$$

and then $\dot{q}_i$ should be expressed in terms of p_i's to give

$$H = H(q_1, \ldots, q_k, p_1, \ldots, p_k, t) \qquad (1\text{–}45)$$

From the first expression (1–44) we have

$$\begin{aligned} dH &= \sum_i \left(p_i\,d\dot{q}_i + \dot{q}_i\,dp_i - \frac{\partial L}{\partial q_i}\,dq_i - \frac{\partial L}{\partial \dot{q}_i}\,d\dot{q}_i \right) - \frac{\partial L}{\partial t}\,dt \\ &= \sum_i (\dot{q}_i\,dp_i - \dot{p}_i\,dq_i) - \frac{\partial L}{\partial t}\,dt \end{aligned} \qquad (1\text{–}46)$$

where the last expression is obtained by using (1–37) and (1–39). From the second expression (1–45), on the other hand, we see

$$dH = \sum_i \left(\frac{\partial H}{\partial q_i} dq_i + \frac{\partial H}{\partial p_i} dp_i \right) + \frac{\partial H}{\partial t} dt \qquad (1\text{–}47)$$

Comparing (1–46) with (1–47) we obtain Hamilton's equations

$$\frac{\partial H}{\partial q_i} = -\dot{p}_i \qquad (1\text{–}48)$$

$$\frac{\partial H}{\partial p_i} = \dot{q}_i \qquad (1\text{–}49)$$

$$\frac{\partial H}{\partial t} = -\frac{\partial L}{\partial t} \qquad (1\text{–}50)$$

When T is quadratic in $\dot{q}_i$, or of the form

$$T = \sum_{i,j} a_{ij}\dot{q}_i\dot{q}_j, \qquad \text{with} \qquad a_{ij} = a_{ji} \qquad (1\text{–}51)$$

we see

$$p_i = \frac{\partial L}{\partial \dot{q}_i} = \frac{\partial T}{\partial \dot{q}_i} = 2\sum_j a_{ij}\dot{q}_j \qquad (1\text{–}52)$$

so that, from (1–44) and (1–35),

$$H = 2\sum_{ij} a_{ij}\dot{q}_j\dot{q}_i - T + U = T + U \qquad (1\text{–}53)$$

Taking the two-dimensional example in the cylindrical coordinates, using (1–40), (1–41), (1–42), and (1–43), we see

$$H = p_\rho\dot{\rho} + p_\varphi\dot{\varphi} - \tfrac{1}{2}m(\dot{\rho}^2 + \rho^2\dot{\varphi}^2) + U \qquad (1\text{–}54)$$

according to (1–44), but

$$\dot{\rho} = \frac{p_\rho}{m}, \qquad \text{and} \qquad \dot{\varphi} = \frac{p_\varphi}{m\rho^2} \qquad (1\text{–}55)$$

so that

$$H = \frac{1}{2m} p_\rho{}^2 + \frac{1}{2m\rho^2} p_\varphi{}^2 + U \qquad (1\text{--}56)$$

which is of the form of (1–45). Since T of (1–41) is quadratic as (1–51), we see $H = T + U$. Hamilton's equations are

$$-\frac{p_\varphi{}^2}{m\rho^3} + \frac{\partial U}{\partial \rho} = -\dot{p}_\rho, \qquad \frac{\partial U}{\partial \varphi} = -\dot{p}_\varphi$$

$$\frac{p_\rho}{m} = \dot{\rho}, \qquad \frac{p_\varphi}{m\rho^2} = \dot{\varphi}$$

$$\frac{H\partial}{\partial t} = 0 \qquad (1\text{--}57)$$

which are equivalent to Newton's equations for the same problem.

PROBLEM 1–8. Derive all equations in (1–57).

PROBLEM 1–9. Show that the first equation in (1–57) can be written as

$$m(\ddot{\rho} - \rho\dot{\varphi}^2) = F_\rho$$

if

$$F_\rho = -\frac{\partial U}{\partial \rho}$$

PROBLEM 1–10. Show that the second equation in (1–57) can be written as

$$m(\rho\ddot{\varphi} + 2\dot{\rho}\dot{\varphi}) = F_\varphi$$

if

$$F_\varphi = -\frac{\partial U}{\rho\partial \varphi}$$

These are Newton's equations in the cylindrical coordinates.

1–8   POISSON'S BRACKETS

If a function

$$f = f(q_1, \ldots, q_r, p_1, \ldots, p_r, t) \qquad (1\text{--}58)$$

is given, we see

$$\frac{df}{dt} = \sum_i \left(\frac{\partial f}{\partial q_i} \dot{q}_i + \frac{\partial f}{\partial p_i} \dot{p}_i\right) + \frac{\partial f}{\partial t} \tag{1-59}$$

Using Hamilton's equations (1–48) and (1–49), we can write (1–59) as

$$\frac{df}{dt} = \sum_i \left(\frac{\partial f}{\partial q_i} \frac{\partial H}{\partial p_i} - \frac{\partial f}{\partial p_i} \frac{\partial H}{\partial q_i}\right) + \frac{\partial f}{\partial t} \tag{1-60}$$

In 1809 Simeon Denis Poisson introduced Poisson's bracket defined by

$$[f, g]_P \equiv \sum_i \left(\frac{\partial f}{\partial q_i} \frac{\partial g}{\partial p_i} - \frac{\partial f}{\partial p_i} \frac{\partial g}{\partial q_i}\right) \tag{1-61}$$

Using this notation we can write (1–60) as

$$\frac{df}{dt} = [f, H]_P + \frac{\partial f}{\partial t} \tag{1-62}$$

which means that if f does not depend on time explicitly and if its Poisson bracket with H is zero, then f is a constant of motion. This formalism gives us an easy way to find constants of motion.

From the definition (1–61) we can see

$$[f, g]_P = -[g, f]_P \tag{1-63}$$

$$[ef, g]_P = e[f, g]_P + f[e, g]_P \tag{1-64}$$

where e, f, and g are any functions. We can also see easily that

$$[q_i, q_j]_P = 0 \tag{1-65}$$

$$[p_i, p_j]_P = 0 \tag{1-66}$$

$$[q_i, p_j]_P = \delta_{ij} \tag{1-67}^5$$

$$[f, q_i]_P = -\frac{\partial f}{\partial p_i} \tag{1-68}$$

$$[f, p_i]_P = \frac{\partial f}{\partial q_i} \tag{1-69}$$

---

[5] δ_{ij} is Kronecker's delta: $\delta_{ij} = 1$ if $i = j$, and $\delta_{ij} = 0$ if $i \neq j$.

When we use (1–56) we have

$$[p_\varphi, H]_P = \frac{1}{2m} [p_\varphi, p_\rho{}^2]_P + \frac{1}{2m} \left[p_\varphi, \frac{p_\varphi{}^2}{\rho^2} \right]_P + [p_\varphi, U]_P$$

$$= -\frac{\partial U}{\partial \varphi} \qquad (1\text{–}70)$$

by rules (1–64) through (1–69). Thus p_φ is a constant of motion if U does not depend on φ. In the same way

$$[p_\rho, H]_P = \frac{p_\varphi{}^2}{2m} [p_\rho, \rho^{-2}]_P + [p_\rho, U]_P$$

$$= \frac{p_\varphi{}^2}{m\rho^3} - \frac{\partial U}{\partial \rho} \qquad (1\text{–}71)$$

For a particle in the three-dimensional space, the spherical coordinates (1–25) give

$$T = \tfrac{1}{2} m(\dot{r}^2 + r^2 \dot{\theta}^2 + r^2 \sin^2\theta \, \dot{\varphi}^2) \qquad (1\text{–}72)$$

If U does not depend on the velocity we obtain

$$p_r = m\dot{r}, \qquad p_\theta = mr^2\dot{\theta}, \qquad p_\varphi = mr^2 \sin^2\theta \, \dot{\varphi} \qquad (1\text{–}73)$$

from (1–37). By the same procedure as in (1–56) we obtain

$$H = \frac{1}{2m} \left(p_r{}^2 + \frac{p_\theta{}^2}{r^2} + \frac{p_\varphi{}^2}{r^2 \sin^2\theta} \right) + U \qquad (1\text{–}74)$$

If U is central, or depends on r only then

$$[p_\varphi, H]_P = 0 \qquad (1\text{–}75)$$

or p_φ is a constant of motion. We can also see that

$$\left[p_\theta{}^2 + \frac{p_\varphi{}^2}{\sin^2\theta}, H \right]_P$$

$$= \left[p_\theta{}^2 + \frac{p_\varphi{}^2}{\sin^2\theta}, \frac{p_r{}^2}{2m} \right]_P + \left[p_\theta{}^2 + \frac{p_\varphi{}^2}{\sin^2\theta}, \frac{1}{2mr^2} \left(p_\theta{}^2 + \frac{p_\varphi{}^2}{\sin^2\theta} \right) \right]_P$$

$$= 0 + \frac{1}{2mr^2} \left[p_\theta{}^2 + \frac{p_\varphi{}^2}{\sin^2\theta}, p_\theta{}^2 + \frac{p_\varphi{}^2}{\sin^2\theta} \right]_P$$

$$= 0 \qquad (1\text{–}76)$$

which shows that $p_\theta{}^2 + p_\varphi{}^2/\sin^2\theta$ is a constant of motion.

PROBLEM 1–11. Prove relations (1–63) through (1–69).

PROBLEM 1–12. When F and G are functions of $f_1, f_2, \ldots, f_k$, which in turn are functions of $q_1, q_2, \ldots, q_n, p_1, p_2, \ldots, p_n$, show that

$$[F, G]_P = \tfrac{1}{2} \sum_{r,s} \left(\frac{\partial F}{\partial f_r} \frac{\partial G}{\partial f_s} - \frac{\partial F}{\partial f_s} \frac{\partial G}{\partial f_r} \right) [f_r, f_s]_P$$

PROBLEM 1–13. Prove (1–73) and then (1–74).

1–9 CONSTANTS OF MOTION AND SYMMETRY

Formalism so far developed does not produce anything new beyond Newton's equations, but makes constants of motion easier to find as seen in the last example. It also makes the physical meaning of some of the constants of motion clearer.

Suppose we have a set of particles interacting with each other. Their mass and the nature of interaction may be arbitrary, but presumably there exists a Hamiltonian function to describe this physical system. If there is no external field, or if this system is floating in an empty space, we should expect no change when the entire system is moved from one place to another, without changing the internal configuration.

The change of the Hamiltonian function H by the displacement of the ith particle along the x axis, δx_i, is

$$\delta_i H = \frac{\partial H}{\partial x_i} \delta x_i = -\dot{p}_{ix} \, \delta x_i \tag{1–77}$$

Therefore, if we displace all particles by a common amount δx along the x axis, the total change of H is

$$\delta H = \sum_i \frac{\partial H}{\partial x_i} \delta x = \delta x \sum_i \frac{\partial H}{\partial x_i} = -\delta x \sum_i \dot{p}_{ix} = -\delta x \frac{d}{dt} \left(\sum_i p_{ix} \right) \tag{1–78}$$

Since this δH should be zero if there is no external field, and since the displacement δx is arbitrary, we conclude that

$$\frac{d}{dt} \left(\sum_i p_{ix} \right) = 0 \tag{1–79}$$

or the total linear momentum $\sum_i p_{ix}$ is conserved. The same argument is applicable to any other components. The conservation of the total linear momentum is, therefore, a direct result of the homogeneity of the space in each direction.

Sometimes the space may not be homogeneous in all directions. On the earth, for example, the gravity field makes the vertical direction of the space inhomogeneous. In this case the total linear momentum in that direction is not conserved.

The space may often be homogeneous with respect to a rotation around some axis. Suppose

$$\frac{\delta H}{\delta \theta} = 0 \qquad (1\text{--}80)$$

the same argument as above shows immediately that the corresponding total angular momentum $\sum_i p_{i\theta}$ is conserved. For example, our system of particles may be under the influence of a potential of cylindrical symmetry. Then the total angular momentum of the particles around this symmetry axis should be conserved. If, in addition, the field is homogeneous along the symmetry axis, the total linear momentum along that direction is also conserved.

PROBLEM 1–14. Show that symmetry condition (1–80) leads to the conservation of the total angular momentum around the axis $\sum_i p_{i\theta}$.

1–10 HAMILTON'S PRINCIPLE

In 1834, William Rowan Hamilton[6] proposed a teleological interpretation of Lagrange's equation, or Newton's equation, in the following way. Assume Lagrange's function L is a function of $q_1, \ldots, q_i, \ldots, q_r, \dot{q}_1, \ldots, \dot{q}_r$, and t. We take initial and final times t_1 and t_2, and consider $\int_{t_1}^{t_2} L \, dt$. Lagrange's equation can then be obtained by assuming

$$\delta \int_{t_1}^{t_2} L \, dt = 0 \qquad (1\text{--}81)$$

---

[6] Hamilton was promoted to Professor of Trinity College, Dublin at the age of 22 when he was an undergraduate student, and knighted at the age of 30, one year after the publication of Hamilton's principle. A comparable genius was Euler who became the chief mathematician of St. Petersburg Academy at the age of 26.

under the conditions that

$$\delta q_i = 0 \quad \text{for all} \quad i \quad \text{at} \quad t_1 \quad \text{and} \quad t_2 \qquad (1\text{-}82)$$

and

$$\delta t = 0 \qquad (1\text{-}83)$$

where δ means variation.

We see that (1–81), under (1–82) and (1–83), is

$$\delta \int_{t_1}^{t_2} L \, dt = \int_{t_1}^{t_2} (\delta L) \, dt = \int_{t_1}^{t_2} \sum_i \left(\frac{\partial L}{\partial q_i} \delta q_i + \frac{\partial L}{\partial \dot{q}_i} \delta \dot{q}_i \right) dt$$

$$(1\text{-}84)$$

By condition (1–83) we see

$$\delta \dot{q}_i = \delta \frac{dq_i}{dt} = \frac{d}{dt} (\delta q_i) \qquad (1\text{-}85)$$

so that partial integration under condition (1–82) gives

$$\int_{t_1}^{t_2} \frac{\partial L}{\partial \dot{q}_i} \delta \dot{q}_i \, dt = - \int_{t_1}^{t_2} \left(\frac{d}{dt} \frac{\partial L}{\partial \dot{q}_i} \right) \delta q_i \, dt \qquad (1\text{-}86)$$

Putting this expression into (1–84), we see the variational equation (1–81) reduces to

$$\int_{t_1}^{t_2} \sum_i \left(\frac{\partial L}{\partial q_i} - \frac{d}{dt} \frac{\partial L}{\partial \dot{q}_i} \right) \delta q_i \, dt = 0 \qquad (1\text{-}87)$$

In order that this equation holds for an arbitrary variation δq_i, the coefficient of δq_i has to be zero for each time t. Thus, we see that the variational equation (1–81) with (1–82) and (1–83) is equivalent to Lagrange's equations (1–36).

Instead of (1–81), we can use an equivalent formula

$$\delta \int_{t_1}^{t_2} \left(\sum_i p_i \dot{q}_i - H \right) dt = 0 \qquad (1\text{-}88)$$

and take the variations of δp_i as well as those of δp_i. With auxiliary conditions (1–82) and (1–83) we have

$$\delta \int_{t_1}^{t_2} \left(\sum_i p_i \dot{q}_i - H \right) dt$$

$$= \int_{t_1}^{t_2} \sum_i \left\{ p_i \, \delta \dot{q}_i + \dot{q}_i \, \delta p_i - \left(\frac{\partial H}{\partial p_i} \delta p_i + \frac{\partial H}{\partial q_i} \delta q_i \right) \right\} dt \qquad (1\text{-}89)$$

Using (1–85) and (1–82) we see

$$\int_{t_1}^{t_2} p_i \, \delta \dot{q}_i \, dt = - \int_{t_1}^{t_2} \dot{p}_i \, \delta q_i \, dt \qquad (1\text{--}90)$$

so that (1–89) and (1–88) give

$$\int_{t_1}^{t_2} \sum_i \left\{ \left(\dot{q}_i - \frac{\partial H}{\partial p_i} \right) \delta p_i - \left(\dot{p}_i + \frac{\partial H}{\partial q_i} \right) \delta q_i \right\} dt = 0 \qquad (1\text{--}91)$$

Thus the variational equation (1–88) is equivalent to Hamilton's equations (1–48) and (1–49).

In either form (1–81) or (1–88), which are equivalent to each other, our results show that among all possible paths between the final and initial times t_2 and t_1 nature chooses a special one such that the integral

$$\int_{t_1}^{t_2} L \, dt \left(= \int_{t_1}^{t_2} \left(\sum_i p_i \dot{q}_i - H \right) dt \right)$$

takes its extremum value. This is called Hamilton's principle.

PROBLEM 1–15.   When the Lagrangian L depends on the acceleration $\ddot{q}_i$ also, show that the variational equation (1–81) under the conditions (1–82), (1–83), and

$$\delta \dot{q}_i = 0 \qquad \text{at} \qquad t_1 \qquad \text{and} \qquad t_2$$

gives

$$\frac{d^2}{dt^2} \frac{\partial L}{\partial \ddot{q}_i} - \frac{d}{dt} \frac{\partial L}{\partial \dot{q}_i} + \frac{\partial L}{\partial q_i} = 0$$

PROBLEM 1–16.   Show that one can take

$$\delta \int_{t_1}^{t_2} \left(\sum_i \dot{p}_i q_i + H \right) dt = 0$$

instead of (1–88), to obtain Hamilton's equations.

1–11   THE PRINCIPLE OF LEAST ACTION

The integral we discussed in the last section, $\int_{t_1}^{t_2} L \, dt$, is sometimes called *action*; but that name is more commonly associated with another integral

$$\int_{t_1}^{t_2} \sum_i p_i \dot{q}_i \, dt \qquad (1\text{--}92)$$

Maupertuis, in 1744, proposed that this integral, which we call *action*, takes an extremum value when Newton's equation holds. His derivation, however, was more theological than mathematical, and the exact meaning of this variational principle had to be clarified later by Euler and Lagrange. The principle is sometimes called *Maupertuis' principle*, but more often simply called the *least action principle* for this historical reason.

Hamilton's principle is a variational principle under two conditions, (1–82) and (1–83). In the least action principle we say

$$\delta \int_{t_1}^{t_2} \sum_i p_i \dot{q}_i \, dt = 0 \tag{1–93}$$

under two conditions:

$$\delta q_i = 0 \quad \text{for all} \quad i \quad \text{at} \quad t_1 \quad \text{and} \quad t_2 \tag{1–82}$$

$$\delta H = 0, \quad \text{or} \quad H = \mathscr{E}(\text{constant}) \tag{1–94}$$

where the first condition is identical to the previous one in Hamilton's principle, but the second condition (1–94), which means that the total energy is kept constant, replaces (1–83). Because of this change, δt is now allowed to be nonzero. Therefore,

$$\delta \int_{t_1}^{t_2} \sum_i p_i \dot{q}_i \, dt = \delta \int_{t_1}^{t_2} (L + H) \, dt$$

$$= \int_{t_1}^{t_2} (\delta L) \, dt + \int_{t_1}^{t_2} (L + H) \, d(\delta t) \tag{1–95}[7]$$

where condition (1–94) is used. In calculating δL in (1–95), we should pay attention to the fact that under the new conditions

$$\delta \dot{q} \equiv \frac{d(q + \delta q)}{d(t + \delta t)} - \frac{dq}{dt} = \frac{d(q + \delta q)/dt}{d(t + \delta t)/dt} - \dot{q}$$

$$= \frac{\dot{q} + d(\delta q)/dt}{1 + d(\delta t)/dt} - \dot{q} = \frac{d}{dt}(\delta q) - \dot{q}\frac{d}{dt}(\delta t) \tag{1–96}$$

---

[7] $d(\delta t)$ is a shorthand notation which implies that t_1 and t_2 are to be varied. For example, $\int_{t_2}^{t_1} d(\delta t) = \delta t_2 - \delta t_1$.

is obtained in contrast to the previous expression (1–85). Thus we obtain

$$
\begin{aligned}
\int_{t_1}^{t_2} (\delta L)\, dt &= \int_{t_1}^{t_2} \sum_i \left(\frac{\partial L}{\partial q_i} \delta q_i + \frac{\partial L}{\partial \dot{q}_i} \delta \dot{q}_i \right) dt \\
&= \int_{t_1}^{t_2} \sum_i \left(\frac{\partial L}{\partial q_i} \delta q_i + \frac{\partial L}{\partial \dot{q}_i} \frac{d}{dt} (\delta q_i) \right) dt - \int_{t_1}^{t_2} \sum_i \frac{\partial L}{\partial \dot{q}_i} \dot{q}_i \frac{d(\delta t)}{dt}\, dt \\
&= \int_{t_1}^{t_2} \sum_i \left(\frac{\partial L}{\partial q_i} - \frac{d}{dt} \frac{\partial L}{\partial \dot{q}_i} \right) \delta q_i\, dt - \int_{t_1}^{t_2} \sum_i p_i \dot{q}_i\, d(\delta t) \\
&= \int_{t_1}^{t_2} \sum_i \left(\frac{\partial L}{\partial q_i} - \frac{d}{dt} \frac{\partial L}{\partial \dot{q}_i} \right) \delta q_i\, dt - \int_{t_1}^{t_2} (L + H)\, d(\delta t)
\end{aligned}
$$

$$(1\text{–}97)$$

Putting this result into (1–95) we see that (1–93) gives Lagrange's equation. When (1–51) and (1–52) hold we see (1–93) can be written as

$$\delta \int_{t_1}^{t_2} T\, dt = 0 \qquad (1\text{–}98)$$

If, in addition, there is no external force, then the second condition (1–94) reduces to

$$\delta T = 0 \qquad (1\text{–}99)$$

so that (1–98) means

$$\delta(t_2 - t_1) = 0 \qquad (1\text{–}100)$$

which states that a system will choose a path of least transit time between given end points, from all possible paths compatible with a fixed energy. This is actually what Maupertuis attempted to prove.

PROBLEM 1–17. In the case of one particle, show that (1–93) is equivalent to

$$\delta \int_a^b \mathbf{p} \cdot d\mathbf{r} = 0$$

What does this expression mean?

PROBLEM 1–18. If l is the length of a path a particle may take, show that the least action principle can also be expressed as

$$\delta \int_a^b \sqrt{\mathscr{E} - U(\mathbf{r})}\, dl = 0$$

where $\mathscr{E}$ is the total energy introduced in (1–94). This is called *Jacobi's form* of the least action principle.

1–12 CANONICAL TRANSFORMATIONS

The advantage of Lagrange's equations is that their formal appearance is independent of the choice of the generalized coordinates. In this sense Lagrange's equations are invariant under transformations among all possible generalized coordinates. Hamilton's equations are also invariant under the same transformations. Hamilton's equations are, however, invariant under more general transformations, because here we can have transformations not only among generalized coordinates, but also between generalized coordinates and generalized momenta.

Suppose $q_1, \ldots, q_i, \ldots, p_1, \ldots, p_i, \ldots$, are the original set of generalized coordinates and their canonically conjugate momenta, for which Hamilton's equations hold. If under a transformation

$$Q_i = Q_i(q, p, t), \qquad P_i = P_i(q, p, t) \tag{1-101}$$

where q and p stand for all q_i's and p_i's, respectively, there exists a function $K(Q, P, t)$ such that we still have Hamilton's equations

$$\dot{Q}_i = \frac{\partial K}{\partial P_i}, \qquad \dot{P}_i = -\frac{\partial K}{\partial Q_i} \tag{1-102}$$

we say that the transformation is *canonical*. Here K is a new Hamiltonian function, which may be different from the original Hamiltonian function H.

Since Hamilton's equations are obtained from Hamilton's principle, or equation (1–88),

$$\delta \int \left(\sum_i P_i \dot{Q}_i - K \right) dt = 0 \tag{1-103}$$

should hold for the new set of canonical variables. Comparing this equation to the similar one for the original set (1–88), we see

$$\sum_i p_i \dot{q}_i - H = \xi \left(\sum_i P_i \dot{Q}_i - K \right) + \dot{G} \tag{1-104}$$

where ξ is a constant and G is a function called the *generating function* of the canonical transformation, defined by $\delta G = 0$ at t_1 and t_2. In most cases we take $\xi = 1$.

For example the generating function G may be a function of q, Q, and t:

$$G = G(q, Q, t) \tag{1-105}$$

Then

$$\dot{G} = \sum_i \frac{\partial G}{\partial q_i} \dot{q}_i + \sum_i \frac{\partial G}{\partial Q_i} \dot{Q}_i + \frac{\partial G}{\partial t} \tag{1-106}$$

In this case (1–104) holds true when

$$p_i = \frac{\partial G}{\partial q_i} \tag{1-107}$$

$$\xi P_i = -\frac{\partial G}{\partial Q_i} \tag{1-108}$$

$$\xi K = H + \frac{\partial G}{\partial t} \tag{1-109}$$

As an example of such generating function we can take

$$G = \sum_i q_i Q_i \tag{1-110}$$

For this special case, (1–107) through (1–109) reduce to

$$Q_i = p_i, \qquad \xi P_i = -q_i, \qquad \text{and} \qquad \xi K = H \tag{1-111}$$

Putting these relations into (1–102) we obtain the original Hamilton's equations (1–48) and (1–49).

Our example (1–111) shows that coordinates and conjugate momenta are not distinguishable; the nomenclature for them is arbitrary. Thus p and q are to be treated on equal footing, and we simply call them canonically conjugate quantities.

It is important to note that the Poisson bracket defined by (1–61) is actually invariant under a canonical transformation, namely if q, p, and Q, P are two canonically conjugate sets, then

$$\sum_i \left(\frac{\partial f}{\partial q_i} \frac{\partial g}{\partial p_i} - \frac{\partial f}{\partial p_i} \frac{\partial g}{\partial q_i} \right) = \sum_i \left(\frac{\partial f}{\partial Q_i} \frac{\partial g}{\partial P_i} - \frac{\partial f}{\partial P_i} \frac{\partial g}{\partial Q_i} \right) \tag{1-112}$$

for any pair of functions f and g, provided that their variables are transformed from q, p, to Q, P, as we go from the left- to the right-hand side of this equation.

Let us use the notation [ ] for the Poisson bracket by q, p, on the left-hand side of relation (1–112). Then

$$
[f, g]_P = \sum_{i,j} \left\{ \frac{\partial f}{\partial q_i} \left(\frac{\partial g}{\partial Q_j} \frac{\partial Q_j}{\partial p_i} + \frac{\partial g}{\partial P_j} \frac{\partial P_j}{\partial p_i} \right) - \frac{\partial f}{\partial p_i} \left(\frac{\partial g}{\partial Q_j} \frac{\partial Q_j}{\partial p_i} + \frac{\partial g}{\partial P_j} \frac{\partial P_j}{\partial q_i} \right) \right\}
$$

$$
= \sum_{j} \left(\frac{\partial g}{\partial Q_j} [f, Q_j]_P + \frac{\partial g}{\partial P_j} [f, P_j]_P \right) \tag{1–113}
$$

This general relation should hold when $g = H$, in which case it reduces to

$$
[f, H]_P = \sum_{j} \left(\frac{\partial H}{\partial Q_j} [f, Q_j]_P + \frac{\partial H}{\partial P_j} [f, P_j]_P \right)
$$

$$
= \sum_{j} \left(\frac{\partial K}{\partial Q_j} [f, Q_j]_P + \frac{\partial K}{\partial P_j} [f, P_j]_P \right)
$$

$$
= \sum_{j} \left(- \dot{P}_j [f, Q_j]_P + \dot{Q}_j [f, P_j]_P \right) \tag{1–114}
$$

using Hamilton's equations (1–102) and the relations

$$
\frac{\partial H}{\partial Q_j} = \frac{\partial K}{\partial Q_j} - \frac{\partial^2 G}{\partial Q_j \partial t} = \frac{\partial K}{\partial Q_j} + \frac{\partial P_j}{\partial t} = \frac{\partial K}{\partial Q_j} \tag{1–115}
$$

$$
\frac{\partial H}{\partial P_j} = \frac{\partial K}{\partial P_j} - \frac{\partial}{\partial t} \left(\frac{\partial G}{\partial P_j} \right) = \frac{\partial K}{\partial P_j} \tag{1–116}
$$

where G is the generating function which appeared in (1–105), (1–108), and (1–109). Now the left-hand side of (1–114) is

$$
[f, H]_P = \dot{f} - \frac{\partial f}{\partial t}
$$

as given by (1–62). Thus (1–114) requires that

$$
[f, Q_j]_P = - \frac{\partial f}{\partial P_j}, \quad \text{and} \quad [f, P_j]_P = \frac{\partial f}{\partial Q_j} \tag{1–117}
$$

When (1–117) is put into general relation (1–113), we see immediately that (1–112) is proved. Equation (1–117), when compared to (1–68) and (1–69), is seen to give other invariant relations.

PROBLEM 1–19. Show that when

$$G = G(q, P, t)$$

we obtain

$$p_i = \frac{\partial G}{\partial q_i}$$

$$\xi Q_i = \frac{\partial G}{\partial P_i}$$

$$\xi K = H + \frac{\partial G}{\partial t}$$

Use the variational condition given in Problem 1–16.

PROBLEM 1–20. Obtain the similar sets of equations as in the previous problem, assuming that

$$G = G(p, Q, t)$$

and

$$G = G(p, P, t)$$

PROBLEM 1–21. Show that under the canonical transformation generated by the generating function given in (1–105), the equation of motion (1–62) is transformed into

$$\frac{df}{dt} = [f, K]_P - \left[f, \frac{\partial G}{\partial t}\right]_P + \frac{\partial f}{\partial t}$$

PROBLEM 1–22. Show that

$$P_i = \xi p_i, \qquad Q_i = q_i, \qquad \text{and} \qquad K = \xi H$$

is a canonical transformation. Here ξ is a constant.

1–13 HAMILTON'S PRINCIPAL FUNCTION. HAMILTON–JACOBI EQUATIONS

A special case of the canonical transformation is of particular interest. If we take the generating function in such a way that

$$\frac{\partial G}{\partial t} = -H\left(q_i, \frac{\partial G}{\partial q_i}, t\right) \tag{1-118}$$

then, from (1–109), the new Hamiltonian is identically zero,

$$K = 0 \tag{1-119}$$

which means, from (1–102), that

$$\dot{Q}_i = \dot{P}_i = 0 \tag{1-120}$$

or the new canonical variables are all constants of motion. *Thus, this canonical transformation gives us the complete set of constants of motion for a given system.*

The solution of (1–118) is called *Hamilton's principal function,* and denoted as S. If we take the Cartesian coordinates, for example, the equation for this function, (1–118), can be explicitly written as

$$\sum_i^n \frac{1}{2m_i} (\nabla S) \cdot (\nabla S) + U(\mathbf{r}_1, \cdots, \mathbf{r}_n, t) + \frac{\partial S}{\partial t} = 0 \tag{1-121}$$

for a system of n particles. This equation is called the Hamilton–Jacobi equation.

When the Hamiltonian function H, particularly the potential U, does not depend on time explicitly, the Hamiltonian is a constant of motion

$$H = \mathscr{E} \tag{1-122}$$

and we see (1–121), the Hamilton–Jacobi equation, has a solution of the form

$$S = S_0(q_i) - \mathscr{E}t \tag{1-123}$$

Since the Hamilton–Jacobi equation for a system of n particles is a first-order differential equation with $3n + 1$ variables (including t), the solution S should have $3n + 1$ integration constants. One of these constants, however, is just an additive constant to S, since if S is a solution of (1–121) obviously $S + \alpha$ is also a solution, where α is the arbitrary constant. Since S, as a generating function of the canonical transformation in the form of (1–105), is a function of the old variables q_i, and the new variables Q_i, the remaining

$3n$ integration constants should correspond to Q_i's, which are now the constants of motion. Neglecting the additive constant α, we can, therefore, write

$$S = S(q_1, \ldots, q_{3n}, \alpha_1, \ldots, \alpha_{3n}, t) \tag{1-124}$$

with the physical interpretation

$$Q_i = \alpha_i \tag{1-125}$$

With this physical interpretation, using the canonical transformation formula (1–108), we obtain

$$\frac{\partial S}{\partial \alpha_i} = -\xi P_i = \beta_i \tag{1-126}$$

where β_i is another constant, since the new momentum P_i is also a constant of motion as (1–120) shows. This equation (1–126) can be used to find the solution

$$q_i = q_i(\alpha_j, \beta_j, t) \tag{1-127}$$

of the original problem.

An important relation

$$\frac{dS}{dt} = \sum_i \frac{\partial S}{\partial q_i} \dot{q}_i + \frac{\partial S}{\partial t} = \sum_i p_i \dot{q}_i - H = L \tag{1-128}$$

is obtained from (1–107) and (1–118). Therefore,

$$S = \int L \, dt + \alpha \tag{1-129}$$

where α is the additive constant mentioned before.

PROBLEM 1–23. When

$$H = \frac{p_x^2}{2m} + U(x)$$

in a one-dimensional problem, show that

$$S = -\mathscr{E}t + \sqrt{2m} \int \sqrt{\mathscr{E} - U(x)} \, dx$$

HINT. The Hamilton–Jacobi equation is

$$\frac{1}{2m}\left(\frac{\partial S}{\partial x}\right)^2 + U(x) = -\frac{\partial S}{\partial t}$$

Note that in this problem there are two variables, x and t; therefore only one meaningful constant, namely, $\mathscr{E}$.

PROBLEM 1–24. Using (1–126) and the principal function S obtained in Problem 1–23, show that

$$\beta + t = \int \frac{m\,dx}{\sqrt{2m(\mathscr{E} - U(x))}}$$

PROBLEM 1–25. Let

$$U(x) = \tfrac{1}{2}kx^2$$

in the previous problem, and obtain

$$x = \sqrt{\frac{2\mathscr{E}}{k}} \sin{(\omega t + \omega\beta)}$$

where $\omega = \sqrt{k/m}$

PROBLEM 1–26. Expressing the Lagrangian L by means of the constant of motion previously obtained

$$\mathscr{E} = \tfrac{1}{2}mv_x^2 + U(x)$$

obtain the result (1–129) directly.

HINT. $L = -\mathscr{E} + mv_x^2$.

ADDITIONAL PROBLEMS

1–27. Obtain Hamilton's equations in the spherical coordinates.

ANSWER.

$$p_r = m\dot{r}, \qquad p_\theta = mr^2\,\dot{\theta}, \qquad p_\varphi = mr^2\sin^2\theta\,\dot{\varphi}$$

$$\dot{p}_r = \left(\frac{l^2}{mr^3}\right) - \left(\frac{\partial U}{\partial r}\right),$$

l^2 given in Problem 1–34

$$\dot{p}_\theta = \left(\frac{p_\varphi{}^2 \cos\theta}{mr^2 \sin^3\theta} \right) - \left(\frac{\partial U}{\partial\theta} \right)$$

$$\dot{p}_\varphi = - \frac{\partial U}{\partial\varphi}$$

1–28. Obtain Newton's equations in the spherical coordinates.

HINT. Use the result of Problem 1–27.

1–29. The potential may depend on the velocity. Show that Lagrange's equation (1–34) is still valid, if the force is given by

$$F_i = - \frac{\partial U}{\partial q_i} + \frac{d}{dt} \left(\frac{\partial U}{\partial \dot{q}_i} \right)$$

Such a velocity dependent force will appear in Chapter 5.

*1–30.** Show that for any functions e, f, and g

$$[e, [f, g]_P]_P + [f, [g, e]_P]_P + [g, [e, f]_P]_P = 0$$

This relation is called Jacobi's identity. (C. G. J. Jacobi, 1866.)

1–31. Show that if f and g are constants of motion, the $[f, g]_P$ is also a constant of motion.

HINT. Use Jacobi's identity. Do not neglect $\partial f/\partial t$ and $\partial g/\partial t$.

1–32. Show that

$$[l_x, l_y]_P = l_z$$

where $\mathbf{l} \equiv \mathbf{r} \times \mathbf{p}$ is the angular momentum.

1–33. Show that

$$[l^2, l_x]_P = 0$$

1–34. Show that

$$l^2 = p_\theta{}^2 + \frac{p_\varphi{}^2}{\sin^2\theta}$$

Comparing this result to (1–76), we see that l^2 is a constant of motion when U is a function of r only.

1–35. When the motion of a particle is constrained by $f(\mathbf{r}) = 0$, the variation $\delta \int L\, dt = 0$ can be modified, using the Lagrange multiplier method, as $\delta \int (L + \lambda f)\, dt = 0$, where λ is the Lagrange multiplier. When we take the variation $\delta\lambda$ we obtain the subsidiary condition back. Show that Lagrange's equation under the constraint can be written as

$$\frac{\partial L}{\partial q_i} - \frac{d}{dt}\frac{\partial L}{\partial \dot{q}_i} + \lambda \frac{\partial f}{\partial q_i} = 0$$

A constraint which can be expressed as $f(\mathbf{r}) = 0$ is called *holonomic*.

1–36. (a) Obtain the canonical transformation, using

$$G = \tfrac{1}{2}\sqrt{mk}\, x^2 \cot Q$$

as the generator.

(b) Applying this canonical transformation to

$$H = \frac{p_x{}^2}{2m} + \tfrac{1}{2}kx^2$$

obtain

$$K = \omega P \ (= \text{constant})$$

$$Q = \omega t + \alpha, \qquad \text{where} \qquad \omega = \sqrt{k/m}$$

and then

$$x = \sqrt{2K/k}\, \sin(\omega t + \alpha)$$

1–37. In the two-dimensional problem, using the cylindrical coordinates, (a) obtain the Hamilton–Jacobi equation

$$\frac{1}{2m}\left(\frac{\partial S_0}{\partial \rho}\right)^2 + \frac{1}{2m\rho^2}\left(\frac{\partial S_0}{\varphi\partial}\right)^2 + U(\rho) = \mathscr{E}$$

(b) show that

$$S_0 = \alpha\varphi + \int \sqrt{2m(\mathscr{E} - U(\rho)) - (\alpha/\rho)^2}\, dr$$

when U does not depend on φ.

(c) obtain

$$\varphi = \beta + \alpha \int \frac{d\rho}{\rho^2\sqrt{2m(\mathscr{E} - U(\rho)) - (\alpha/\rho)^2}}$$

1–38. A particle of mass m comes from infinity towards the center of a spherical potential of

$$U(r) = \frac{A}{r}$$

where A is a constant (Fig. 1–9).

FIGURE 1–9. Geometry of the classical scattering problem (Problem 1–38).

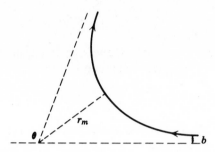

(a) If, at infinity, the energy of the particle is $\mathscr{E}_\infty$, and the path is such that if we draw a line parallel to the path and passing through the center of the potential, the distance between this line and the path is b, then show that α of Problem 1–37(b) is

$$\alpha = b\sqrt{2m\mathscr{E}_\infty}$$

(b) As the particle approaches the center, the path bends due to the potential. Show that if r_m is the shortest distance between the particle and the center during the flight, then

$$\mathscr{E}_\infty r_m{}^2 - Ar_m - \mathscr{E}_\infty b^2 = 0$$

(c) As the particle goes away from the center, the path will gradually become straight again. Show that at infinity the path of getaway makes an angle θ with the direction of the incident path such that

$$2\mathscr{E}_\infty b = A \cot (\tfrac{1}{2}\theta)$$

HINT. Use (1–43) and (1–57).

1–39. If the motion of a system of particles is periodic, so that $\sum_i \mathbf{p}_i \cdot \mathbf{r}_i$ repeat itself with a period τ, show that

$$\overline{\sum_i T_i} = -\tfrac{1}{2} \overline{\sum_i \mathbf{F}_i \cdot \mathbf{r}_i}$$

where the upper bars mean the average over the period τ. This relation is called the virial theorem.

HINT. Show that

$$\frac{d}{dt}\left(\sum_i \mathbf{p}_i \cdot \mathbf{r}_i\right) = \sum_i (m_i v_i^2 + \mathbf{F}_i \cdot \mathbf{r}_i)$$

1–40. If

$$U = \sum_{i,j} \alpha r_{ij}^{\,n}$$

where α and n are constants, and r_{ij} is the distance between particles i and j, show, using the virial theorem, that

$$\overline{T} = \tfrac{1}{2} n \overline{U}$$

where $T = \sum_i T_i$, and the upper bars mean the average introduced in Problem 1–39.

Chapter 2 Simple Harmonic Oscillators Normal Vibrations

The motion of a particle around the stable equilibrium position is always approximately simple harmonic, in the small amplitude limit, whatever the potential. We will see in later chapters that in electromagnetic field theory, or in any other field theories for that matter, not only in particle physics, simple harmonic oscillation gives the dynamical behavior of the field quantities to a high accuracy. Most of the arguments in the present chapter will also be applicable to discussions of field theories.

Most actual problems in physics are many-body problems, and they are usually very difficult to solve. One way of attacking a many-body problem is to find as many constants of motion as possible, and that approach was discussed in the last chapter. The other way is to find collective coordinates, such that the motion of each collective coordinate is independent from the other coordinates. The normal coordinates are the typical and most common collective coordinates.

2–1 SIMPLE HARMONIC OSCILLATOR

The potential $U(x)$, for most cases in physics, can be assumed to be an analytic function of the coordinate x, so that the Taylor expansion

$$U(x) = U(a) + (x - a)\left(\frac{dU}{dx}\right)_{x=a} + \tfrac{1}{2}(x - a)^2 \left(\frac{d^2U}{dx^2}\right)_{x=a} + \cdots$$

(2–1)

exists around an arbitrary point a. If the potential has a minimum at $x = a_0$, then

$$\left(\frac{dU}{dx}\right)_{x=a_0} = 0$$

(2–2)

so that the Taylor expansion around this point is

$$U(x) = U_0 + \tfrac{1}{2}k(x - a_0)^2 + \cdots$$

(2–3)

where

$$U_0 = U(a_0) \quad \text{and} \quad k = \left(\frac{d^2U}{dx^2}\right)_{x=a_0} > 0$$

(2–4)

are constants. Thus the motion of a particle of mass m in the vicinity of the potential minimum can be approximately given by Newton's equation of

$$m\ddot{x} = -\frac{\partial U}{\partial x} \cong -k(x - a_0)$$

(2–5)

which is easily solved to give

$$x - a_0 = A'e^{i\omega_0 t} + B'e^{-i\omega_0 t} = A \cos(\omega_0 t + \delta)$$

(2–6)

where

$$\omega_0 = \sqrt{k/m}$$

(2–7)

and A', B', A, and δ are constants which are to be determined from the initial conditions, x and $\dot{x}$ at $t = 0$. The oscillatory motion given by (2–6) is called simple harmonic oscillation.

As seen in this discussion, any oscillatory motion in the region near a potential minimum is approximately simple harmonic. When the amplitude of the vibration is large, however, the higher-order terms, such as $(x - a_0)^3$ term, in the potential expansion produce appreciable *anharmonicity*, or deviation from the simple harmonic oscillation. The anharmonicity is important in the vibration of atoms in molecules and solids, but it is negligible, at least to the present experimental accuracy, in the case of the oscillations of electromagnetic field strength in vacuum.

Even when the anharmonicity is so large that the oscillatory motion is not simple harmonic at all, we can obtain an expression for the period τ of the oscillation without too much difficulty. If $x = A$ is one of the turning points, the velocity should be zero there, so that, from energy conservation,

$$\tfrac{1}{2}m\dot{x}^2 + U(x) = U(A) \tag{2-8}$$

which gives

$$\dot{x} \equiv \frac{dx}{dt} = \sqrt{[U(A) - U(x)](2/m)} \tag{2-9}$$

Thus $\tfrac{1}{2}\tau$, which is the time the particle takes moving from the turning point A to another turning point B is

$$\tfrac{1}{2}\tau = \int_B^A \frac{dx}{\sqrt{[U(A) - U(x)](2/m)}} \tag{2-10}$$

as can be seen from Problem 1–24 (see also Fig. 2–1).

FIGURE 2–1. General potential. Turning points.

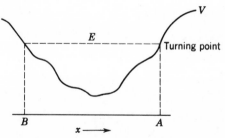

PROBLEM 2–1. A pendulum is made of a mass m suspended at one end of a weightless unstretchable string of length l. The other end is fixed (Fig. 2–2). Show that if the gravity force is F_g, the potential energy is

$$U = lF_g(1 - \cos\varphi) = \tfrac{1}{2}lF_g\varphi^2 + O(\varphi^4)$$

where φ is the angle between the string and the vertical, and $O(\varphi^4)$ means terms of order of φ^4 or higher. Using T of Problem 1–4, obtain the equation of motion

$$ml\ddot{\varphi} = -F_g\varphi + O(\varphi^3)$$

from Lagrange's equation

FIGURE 2–2. Pendulum.

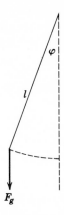

F_g

PROBLEM 2–2. A massless elastic string is stretched between two points A and B. The tension is T_0 and the length is $2l$ in this situation. Attach a mass m at the middle of this string and displace it by a distance x_0 in a direction perpendicular to AB (Fig. 2–3). Find the position of this mass after it is released from the above position with no initial velocity. Neglect gravity, and assume that $x_0 \ll l$.

FIGURE 2–3. Oscillator of Problem 2–2.

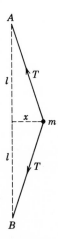

ANSWER. When the displacement is x, the length of each section of the string is $\sqrt{l^2 + x^2}$. The tension T at this length is

$$T = T_0 + k(\sqrt{l^2 + x^2} - l)$$

where k is the elastic constant. Newton's equation is

$$m\ddot{x} = -\frac{2T_x}{\sqrt{l^2 + x^2}} = -2[T_0 + k(\sqrt{l^2 + x^2} - l)]\frac{x}{\sqrt{l^2 + x^2}}$$

$$= -2\left\{T_0 + k\left[\frac{1}{2}\left(\frac{x^2}{l}\right) + \cdots\right]\right\} x \frac{(1 - \frac{1}{2}(x^2/l^2) - \cdots)}{l}$$

$$= -2\left(\frac{T_0}{l}\right)x - \left(\frac{k}{l^2} - \frac{T_0}{l^3}\right)x^3 - \cdots$$

If terms of the order of x^3 and higher are neglected

$$x = x_0 \cos(\omega t)$$

where $\omega = \sqrt{2T_0/lm}$.

PROBLEM 2–3. Solve the equation of motion

$$\ddot{x} + \omega^2 x + \varepsilon x^3 = 0$$

to the first-order approximation, assuming that the εx^3 term can be neglected in the zeroth-order approximation.

ANSWER. If εx^3 is neglected, $x = x_0 \cos(\omega t)$. In the next order approximation

$$\ddot{x} + x_0\omega^2 \cos(\omega t) + \varepsilon x_0{}^3 \cos^3(\omega t) = 0$$

Solving this equation we obtain

$$x = \left(x_0 + \frac{3\varepsilon x_0{}^3}{4\omega^2}\right)\cos(\omega t) + \frac{\varepsilon x_0{}^3}{36\omega^2}\cos(3\omega t)$$

2–2 DAMPED HARMONIC OSCILLATORS

The oscillator may be submerged in some viscous medium which produces a friction force proportional to the velocity. The simple harmonic motion with such a friction force is given by

$$m\ddot{x} = -kx - R\dot{x} \qquad (2\text{–}11)$$

where we shifted the origin of the coordinate to eliminate a_0. If we set

$$x = e^{\lambda t} \tag{2-12}$$

in equation (2–11) we obtain

$$e^{\lambda t} \left(\lambda^2 + \frac{R}{m} \lambda + \omega_0{}^2 \right) = 0 \tag{2-13}$$

where ω_0 is given by (2–7). Since (2–13) gives

$$\lambda = -\frac{R}{2m} \pm \sqrt{(R/2m)^2 - \omega_0{}^2} \tag{2-14}$$

the general solution of (2–11) is

$$x = e^{-(R/2m)t}\{A' \exp \left[\sqrt{(R/2m)^2 - \omega_0{}^2t} \right]$$
$$+ B' \exp \left[-\sqrt{(R/2m)^2 - \omega_0{}^2t} \right]\} \tag{2-15}$$

The solution gives oscillatory or nonoscillatory motion depending on whether $R/2m < \omega_0$ or $R/2m \geq \omega_0$. We call

$$\frac{R}{2m} < \omega_0 \qquad \text{underdamped case}$$

$$\frac{R}{2m} = \omega_0 \qquad \text{critically damped case} \tag{2-16}$$

$$\frac{R}{2m} > \omega_0 \qquad \text{overdamped case}$$

In the underdamped case, the motion is oscillatory and

$$x = e^{-(R/2m)t}A \cos (\omega_0' t + \delta) \tag{2-17}$$

where

$$\omega_0' = \sqrt{\omega_0{}^2 - (R/2m)^2} \tag{2-18}$$

and A and δ are constants. For example if $\dot{x} = 0$ at $t = 0$, Fig. 2–4 illustrates the motion. When $R/2m$ is very small compared to ω_0 we can regard the motion as a simple harmonic oscillation with frequency ω_0 but amplitude decreasing exponentially with time instead of being constant. In (2–17) $R/2m$ is called the *damping constant* for that reason.

FIGURE 2–4.   Underdamped oscillator.

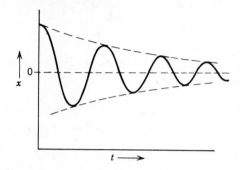

In the critically damped and overdamped cases the motion is not oscillatory, as illustrated in Fig. 2–5.

If we simply set $\omega_0 = R/2m$ in (2–15) we obtain

$$x = (A' + B')e^{-(R/2m)t} \qquad (2\text{–}19)$$

which has only one integration constant $A' + B'$, and, therefore, cannot be a general solution of the second-order differential equation (2–11). In this special case we take

$$x = y(t)e^{\lambda t} \qquad (2\text{–}20)$$

FIGURE 2–5.   Two typical critically damped oscillators.

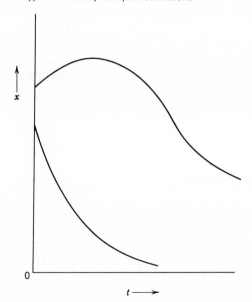

instead of (2–12). We see that the general solution in the critically damped case is

$$x = (A + Bt)e^{-(R/2m)t} \qquad (2\text{–}21)$$

PROBLEM 2–4. Show that (2–21) is the general solution for the critically damped case.

PROBLEM 2–5. If $x = x_0$ and $\dot{x} = v_0$ (> 0) at $t = 0$, show that the displacement x of the critically damped oscillator (2–21) is maximum at

$$t = \frac{v_0}{\gamma(v_0 + \gamma x_0)}$$

where $\gamma = R/2m$.

2–3 FORCED HARMONIC OSCILLATOR. RESONANCE

The simple harmonic oscillator may be under the influence of a time-dependent external field in addition to the previously discussed forces. A typical time-dependent external force is $F \cos \omega t$, where F and ω are constants. Actually more general time-dependent forces can be expressed in the form of a Fourier integral, $\int F(\omega)e^{i\omega t}\,d\omega$, or a linear superposition of the above forms, and the following discussion can be easily generalized. Thus we consider the following Newton's equation:

$$m\ddot{x} = -kx - R\dot{x} + F \cos \omega t \qquad (2\text{–}22)$$

This is an inhomogeneous equation, whose solution can be written as a linear combination of the solution of the corresponding homogeneous equation (which is obtained by setting $F = 0$), and the *particular solution*. The particular solution can be obtained by assuming

$$x_F = C \sin \omega t + D \cos \omega t \qquad (2\text{–}23)$$

in (2–22). A straightforward calculation gives

$$C = \frac{FR\omega}{\omega^2 R^2 + m^2(\omega_0^2 - \omega^2)^2} \qquad (2\text{–}24)$$

$$D = \frac{Fm(\omega_0^2 - \omega^2)}{\omega^2 R^2 + m^2(\omega_0^2 - \omega^2)^2} \qquad (2\text{–}25)$$

The particular solution (2–23) is then

$$x_F = \frac{F}{\sqrt{\omega^2 R^2 + m^2(\omega_0^2 - \omega^2)^2}} \cos(\omega t - \delta_F) \qquad (2\text{–}26)$$

where

$$\tan \delta_F = \frac{R\omega}{m(\omega_0^2 - \omega^2)} \qquad (2\text{–}27)$$

The general solution of (2–22) is given by adding the previously obtained solution of the corresponding homogeneous equation to x_F of (2–26). Thus in the underdamped case, for example,

$$x = Ae^{-(R/2m)t} \cos(\omega_0' t + \delta) + x_F \qquad (2\text{–}28)$$

Notice, however, that, whatever the initial condition may be, the first term in (2–28) decreases as time elapses, so that after some time it will always be negligible compared to the particular solution x_F.

The particular solution x_F oscillates with the same frequency as that of the external field, but there is a phase difference δ_F between them. The amplitude of this oscillation of x_F is proportional to that of the external field F but it depends on the frequency ω in a complicated way.

Since (2–26) can also be written as

$$x_F^2 = \frac{F^2}{(\omega_0' R)^2 + m^2(\omega_R^2 - \omega^2)^2} \cos^2(\omega t - \delta_F) \qquad (2\text{–}26')$$

where ω_0' is given by (2–18), and

$$\omega_R = \sqrt{\omega_0^2 - (R^2/2m^2)} \qquad (2\text{–}29)$$

We see that the amplitude of x_F has a maximum at $\omega = \omega_R$. The frequency ω_R is called the *resonance frequency*; it is near but slightly less than the natural frequency ω_0, when the damping constant $R/2m$ is small. When the frequency ω is not far from the resonance frequency ω_R, we may assume that

$$\omega_R^2 - \omega^2 = (\omega_R + \omega)(\omega_R - \omega) \cong 2\omega_0'(\omega_R - \omega) \qquad (2\text{–}30)$$

and obtain the approximation

$$x_F^2 \cong \frac{(F/2m\omega_0')^2}{(R/2m)^2 + (\omega_R - \omega)^2} \cos^2(\omega t - \delta_F) \qquad (2\text{–}31)$$

If we plot the ω dependence of the amplitude of $x_F{}^2$ according to this expression, we obtain the so-called Lorentzian line shape.[1] The Lorentzian line shape is relatively sharp around the peak, but decreases slowly in the tail regions. At frequencies

$$|\omega - \omega_R| = \frac{R}{2m} \tag{2-32}$$

the Lorentzian line shape formula has a value one-half of the peak value. The frequency difference given by (2–32) is called the half-width for that reason. We see from this expression that the half-width gives the damping constant $R/2m$ (Fig. 2–6).

FIGURE 2–6.   Lorentzian line shape.

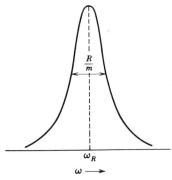

In physics we have electrical and quantum mechanical resonances, in addition to the mechanical one we just considered. In all such resonances, the Lorentzian line shape usually is a good approximation.

PROBLEM 2–6.   Obtain (2–24), (2–25), (2–26), and (2–27).

PROBLEM 2–7.   Obtain (2–26′) from (2–26).

PROBLEM 2–8.   Obtain (2–31).

2–4   TWO COUPLED OSCILLATORS

Suppose we have two identical simple harmonic oscillators with mass m and force constant k. If the displacements are x_1 and x_2 for these oscillators, respectively, their motions are given by

$$m\ddot{x}_1 = -kx_1 \tag{2-33}$$

$$m\ddot{x}_2 = -kx_2 \tag{2-34}$$

---

[1] H. A. Lorentz, *The Theory of Electrons*, Dover Publications, Inc., New York, 1909, Formula (230).

and their solutions are

$$x_1 = a_1 \cos(\omega_0 t + \delta_1) \qquad (2\text{--}35)$$

$$x_2 = a_2 \cos(\omega_0 t + \delta_2) \qquad (2\text{--}36)$$

$$\omega_0 = \sqrt{k/m}$$

where a_1 and a_2 are amplitudes, and δ_1 and δ_2 are phase constants. These constants should be decided by the initial conditions.

When these two oscillators are coupled by a spring with force constant κ and the equilibrium length just equal to the distance between these two particles (Fig. 2–7), we have

$$m\ddot{x}_1 = -kx_1 - \kappa(x_1 - x_2) \qquad (2\text{--}37)$$

$$m\ddot{x}_2 = -kx_2 - \kappa(x_2 - x_1) \qquad (2\text{--}38)$$

instead of (2–33) and (2–34).

FIGURE 2–7. Coupled simple harmonic oscillators.

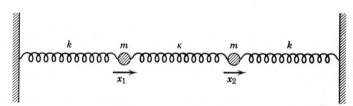

Equations (2–37) and (2–38) can be solved easily. Adding and subtracting these equations we obtain

$$m(\ddot{x}_1 + \ddot{x}_2) = -k(x_1 + x_2) \qquad (2\text{--}39)$$

$$m(\ddot{x}_1 - \ddot{x}_2) = -(k + 2\kappa)(x_1 - x_2) \qquad (2\text{--}40)$$

respectively. The solutions are

$$x_1 + x_2 \equiv X_1 = A_1 \cos(\omega_0 t + \delta_1) \qquad (2\text{--}41)$$

$$x_1 - x_2 \equiv X_2 = A_2 \cos(\omega' t + \delta_2) \qquad (2\text{--}42)$$

where ω_0 is given by (2–7) and

$$\omega' = \sqrt{(k + 2\kappa)/m} \qquad (2\text{--}43)$$

If the original two oscillators are not identical to each other, but still coupled in the same manner, we have

$$m_1\ddot{x}_1 = -k_1x_1 - \kappa(x_1 - x_2) \tag{2-44}$$

$$m_2\ddot{x}_2 = -k_2x_2 - \kappa(x_2 - x_1) \tag{2-45}$$

The solutions of these simultaneous equations must be similar to the previous solutions (2–41) and (2–42), but apparently more complicated since the symmetry is now lost.

In view of the previous solutions (2–41) or (2–42), we may try

$$x_1 = a_1 \cos (\Omega t) \quad \text{and} \quad x_2 = a_2 \cos (\Omega t) \tag{2-46}$$

as a possible solution of (2–44) and (2–45). If we actually put (2–46) into these equations we obtain

$$(\Omega^2 - \omega_1{}^2 - \omega_1'^2)a_1 + \omega_1'^2 a_2 = 0 \tag{2-47}$$

$$\omega_2'^2 a_1 + (\Omega^2 - \omega_2{}^2 - \omega_2'^2)a_2 = 0 \tag{2-48}$$

where

$$
\begin{aligned}
\omega_1{}^2 &= \frac{k_1}{m_1} \\[4pt]
\omega_2{}^2 &= \frac{k_2}{m_2} \\[4pt]
\omega_1'^2 &= \frac{\kappa}{m_1} \\[4pt]
\omega_2'^2 &= \frac{\kappa}{m_2}
\end{aligned}
\tag{2-49}
$$

In (2–47) and (2–48), we took out $\cos (\Omega t)$ because it would not be zero in general.

Equations (2–47) and (2–48) have an obvious solution of

$$a_1 = a_2 = 0 \tag{2-50}$$

but such a solution should be excluded because it is not an interesting one. Suppose a_1 is not zero, we can solve (2–47) and (2–48) for a_2/a_1 and obtain a single equation

$$(\Omega^2 - \omega_1{}^2 - \omega_1'^2)(\Omega^2 - \omega_2{}^2 - \omega_2'^2) - \omega_1'^2\omega_2'^2 = 0 \tag{2-51}$$

from which we obtain two solutions

$$\Omega_\pm{}^2 = \tfrac{1}{2}[\omega_1{}^2 + \omega_2{}^2 + \omega_1'^2 + \omega_2'^2$$

$$\pm \sqrt{(\omega_1{}^2 - \omega_2{}^2 + \omega_1'^2 - \omega_2'^2)^2 + 4\omega_1'^2\omega_2'^2}] \qquad (2\text{-}52)$$

If we put these solutions back into (2–47) we obtain

$$\left(\frac{a_2}{a_1}\right)_\pm \equiv \rho_\pm = - \left(\frac{\Omega_\pm{}^2 - \omega_1{}^2 - \omega_1'^2}{\omega_1'^2}\right) \qquad (2\text{-}53)$$

The corresponding solutions from (2–48) may look different but actually are the same as (2–53) when (2–52) is substituted explicitly.

According to a general theorem, a set of simultaneous linear homogeneous equations, like (2–47) and (2–48), can have nontrivial solutions [a trivial solution is (2–50)], only if a determinant given by the coefficients is equal to zero. In our case this condition is

$$\begin{vmatrix} \Omega^2 - \omega_1{}^2 - \omega_1'^2 & \omega_1'^2 \\ \omega_2'^2 & \Omega^2 - \omega_2{}^2 - \omega_2'^2 \end{vmatrix} = 0 \qquad (2\text{-}54)$$

We can easily see that this is the same condition as (2–51), but this formulation is easier to obtain when more than two oscillators are coupled.

PROBLEM 2–9. Illustrate the motion of the two particles, assuming that $A_1 = 0$. Do the same thing assuming that $A_2 = 0$. A_1 and A_2 are defined by (2–41) and (2–42).

PROBLEM 2–10. Confirm steps (2–51), (2–52), and (2–53).

PROBLEM 2–11. When $\omega_1 = \omega_2$ and $\omega_1' = \omega_2'$, show that (2–52) reduces to the previous solutions (2–7) and (2–43).

2–5 NORMAL COORDINATES

Coordinates X_1 and X_2 introduced by (2–41) and (2–42) are generalized coordinates, which are sometimes called *collective coordinates* because each of them represents the motion of the entire system of coupled oscillators. They are actually a special kind of collective coordinates. The Lagrangian function of two coupled identical oscillators is

$$L = \tfrac{1}{2}m\dot{x}_1{}^2 + \tfrac{1}{2}m\dot{x}_2{}^2 - \tfrac{1}{2}kx_1{}^2 - \tfrac{1}{2}kx_2{}^2 - \tfrac{1}{2}\kappa(x_1 - x_2)^2$$

$$(2\text{-}55)$$

If we use X_1 and X_2 given by (2–41) and (2–42), respectively, we see that this Lagrangian function can be separated into two parts

$$L = L_1 + L_2 \tag{2–56}$$

where

$$L_1 = \tfrac{1}{4}m\dot{X}_1{}^2 - \tfrac{1}{4}kX_1{}^2 \tag{2–57}$$

$$L_2 = \tfrac{1}{4}m\dot{X}_2{}^2 - \tfrac{1}{4}(k + 2\kappa)X_2{}^2 \tag{2–58}$$

Thus we obtain completely separated equations of motion, if Lagrange's equations are calculated using the coordinates X_1 and X_2. Collective coordinates for which such a separation occurs are called *normal coordinates*. The vibration of a normal coordinate is called the *normal vibration*. Thus the normal vibrations are independent of each other in the sense that we can excite one of them without exciting any other normal vibrations.

From (2–41) and (2–42) we see

$$x_1 = \tfrac{1}{2}(X_1 + X_2) \tag{2–59}$$

$$x_2 = \tfrac{1}{2}(X_1 - X_2) \tag{2–60}$$

When $X_2 = 0$, we see that

$$x_1 = x_2 \qquad : X_1\text{-normal vibration} \tag{2–61}$$

which means that in the X_1-normal vibration both particles are displaced by the same amount in the same direction at one time. The X_2-normal vibration is visualized in the same way by letting $X_1 = 0$ in (2–59) and (2–60). Namely,

$$x_1 = -x_2 \qquad : X_2\text{-normal vibration} \tag{2–62}$$

They are illustrated in Fig. 2–8.

If we invert the oscillators at the middle point, the phase of the X_1-normal vibration will be changed by π, that is, X_1 is transformed into $-X_1$. In this sense, this vibration is called an antisymmetric vibration. In a similar way we see that the X_2-normal vibration is a symmetric vibration.

FIGURE 2–8.   Normal vibrations of two coupled identical oscillators.

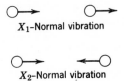

X_1-Normal vibration

X_2-Normal vibration

For the case of coupled nonidentical oscillators, the normal coordinates exist, but they do not have the symmetry properties just mentioned. The result given in the previous section shows that

$$x_1 = A_+ \rho_+ \cos (\Omega_+ t + \delta_+) + A_- \rho_- \cos (\Omega_- t + \delta_-)$$

$$(2\text{--}63)$$

$$x_2 = A_+ \cos (\Omega_+ t + \delta_+) + A_- \cos (\Omega_- t + \delta_-) \qquad (2\text{--}64)$$

is the general solution of this problem. A_+, A_-, δ_+, and δ_- are the constants to be decided from the initial conditions, while Ω_+ and Ω_- are given by (2–52), and ρ_+ and ρ_- are $(a_2/a_1)_+$ and $(a_2/a_1)_-$, respectively, given by (2–53). The amplitudes of the normal vibrations are given by A_+ and A_-, respectively.

PROBLEM 2–12.   Show that Lagrange's equations obtained by using (2–55) are the equations of motion (2–37) and (2–38).

PROBLEM 2–13.   Explain that one can directly understand the result obtained previously in (2–41) and (2–42), particularly the frequency of each normal vibration, using (2–61) and (2–62), or Fig. 2–8.

PROBLEM 2–14.   Show that when $\omega_1 = \omega_2$ and $\omega_1' = \omega_2'$, (2–63) and (2–64) reduce to (2–59) and (2–60).

PROBLEM 2–15.   In the case of two coupled identical oscillators, if the initial conditions are

$$\dot{x}_1 = \dot{x}_2 = x_2 = 0, \qquad x_1 = a \qquad \text{at} \qquad t = 0$$

show that

$$x_1 = a \cos \left[\tfrac{1}{2}(\omega_0 - \omega')t\right] \cos \left[\tfrac{1}{2}(\omega_0 + \omega')t\right]$$

$$x_2 = -a \sin \left[\tfrac{1}{2}(\omega_0 - \omega')t\right] \sin \left[\tfrac{1}{2}(\omega_0 + \omega')t\right]$$

where ω' is given by (2–43).

2–6   SECULAR EQUATION

Let us take a general case of n simple harmonic oscillators which are coupled to each other by a quadratic potential. The kinetic energy of this system is simply

$$T = \tfrac{1}{2} \sum_i m_i \dot{x}_i{}^2 \qquad (2\text{--}65)$$

while the potential energy can be written as

$$U = \tfrac{1}{2} \sum_i k_i x_i^2 - \tfrac{1}{2} \sum_{i \neq j} k_{ij} x_i x_j \qquad (2\text{–}66)$$

where the terms $k_{ij} x_i x_j$ are the terms due to the coupling.

If the normal coordinates are found for this system we should have

$$T = \tfrac{1}{2} \sum_i M_i \dot{X}_i^2 \qquad (2\text{–}67)$$

$$U = \tfrac{1}{2} \sum_i K_i X_i^2 \qquad (2\text{–}68)$$

Thus our problem is to find a transformation by which (2–65) and (2–66) are transformed into (2–67) and (2–68), respectively.

Instead of the original coordinates x_i, it is better to introduce

$$\xi_i = \sqrt{m_i}\, x_i \qquad (2\text{–}69)$$

to simplify the mathematics. In this new "coordinate system" we have

$$T = \tfrac{1}{2} \sum_i \dot{\xi}_i^2 \qquad (2\text{–}70)$$

$$U = \tfrac{1}{2} \sum_i \omega_i^2 \xi_i^2 - \tfrac{1}{2} \sum_{i \neq j} \omega_{ij}^2 \xi_i \xi_j \qquad (2\text{–}71)$$

where

$$\omega_i^2 = \frac{k_i}{m_i} \qquad (2\text{–}72)$$

$$\omega_{ij}^2 = \frac{k_{ij}}{\sqrt{m_i m_j}} \qquad (2\text{–}73)$$

It is always possible to take $\omega_{ij} = \omega_{ji}$, but it is not necessary in the following discussions.[2]

Taking the same unit in the normal coordinates, our problem is to find a transformation by which (2–70) and (2–71) are transformed into

$$T = \tfrac{1}{2} \sum_i \dot{X}_i^2 \qquad (2\text{–}74)$$

$$U = \tfrac{1}{2} \sum_i \Omega_i^2 X_i^2 \qquad (2\text{–}75)$$

[2] The form of (2–71) implies that the coefficients of the coupling terms are all negative, but these coefficients can actually be positive as well. The following discussions do not depend on the sign of these coefficients.

where Ω_i is the frequency of the ith normal vibration. Let us assume the required transformation to be

$$X_i = \sum_j a_{ij}\xi_j \tag{2-76}$$

which can be written in the matrix form as

$$\begin{pmatrix} X_1 \\ X_2 \\ \vdots \\ X_n \end{pmatrix} = \begin{pmatrix} a_{11} & a_{12} & \cdots & a_{1n} \\ a_{21} & a_{22} & \cdots & a_{2n} \\ \vdots & & & \\ a_{n1} & & \cdots & a_{nn} \end{pmatrix} \begin{pmatrix} \xi_1 \\ \xi_2 \\ \vdots \\ \xi_n \end{pmatrix} \tag{2-77}$$

It is convenient to abbreviate this expression as

$$\mathbf{X} = \mathbf{a}\boldsymbol{\xi} \tag{2-78}$$

and, in addition, introduce the transposed matrices as

$$\mathbf{X}^T = (X_1 \quad X_2 \quad \cdots \quad X_n) \tag{2-79}$$

$$\mathbf{a}^T = \begin{pmatrix} a_{11} & a_{21} & \cdots & a_{n1} \\ a_{12} & a_{22} & & a_{n2} \\ \vdots & & & \vdots \\ a_{1n} & & \cdots & a_{nn} \end{pmatrix} \tag{2-80}$$

by which we have, for example,

$$\mathbf{X}^T = \boldsymbol{\xi}^T \mathbf{a}^T \tag{2-81}$$

(see Problem 2–17).

Using this notation, (2–70) and (2–71) can be expressed as

$$T = \tfrac{1}{2}\dot{\boldsymbol{\xi}}^T \dot{\boldsymbol{\xi}} \tag{2-82}$$

$$U = \tfrac{1}{2}\boldsymbol{\xi}^T \omega^2 \boldsymbol{\xi} \tag{2-83}$$

where

$$\omega^2 = \begin{pmatrix} \omega_1{}^2 & -\omega_{12}{}^2 & \cdots & -\omega_{1n}{}^2 \\ -\omega_{21}{}^2 & \omega_2{}^2 & \cdots & -\omega_{2n}{}^2 \\ \vdots & & & \vdots \\ -\omega_{n1}{}^2 & & \cdots & \omega_n{}^2 \end{pmatrix} \tag{2-84}$$

In this notation, using (2–78) and (2–81), the kinetic energy after the transformation is, from (2–74),

$$T = \tfrac{1}{2}\dot{X}^T\dot{X} = \tfrac{1}{2}\dot{\xi}^T\mathbf{a}^T\mathbf{a}\dot{\xi} \tag{2–85}$$

Comparing this result with (2–82), we see that the transformation matrix should be such that

$$\mathbf{a}^T\mathbf{a} = \mathbf{1} \equiv \begin{pmatrix} 1 & 0 & 0 & \cdots & 0 \\ 0 & 1 & 0 & \cdots & 0 \\ \vdots & & & & \vdots \\ 0 & 0 & & \cdots & 1 \end{pmatrix} \tag{2–86}$$

A matrix which satisfies this relation is called an *orthogonal matrix*, and the corresponding transformation is called the *orthogonal transformation*. The determinant of an orthogonal matrix is given by

$$\det \mathbf{a} = \pm 1 \tag{2–87}$$

For the potential energy we have, from (2–75), (2–78), and (2–81)

$$U = \tfrac{1}{2}X^T\Omega^2 X = \tfrac{1}{2}\xi^T\mathbf{a}^T\Omega^2\mathbf{a}\xi \tag{2–88}$$

where Ω^2 is a diagonal matrix

$$\Omega^2 = \begin{pmatrix} \Omega_1{}^2 & 0 & \cdots & 0 \\ 0 & \Omega_2{}^2 & \cdots & 0 \\ \vdots & & & \vdots \\ & & \cdots & \Omega_n{}^2 \end{pmatrix} \tag{2–89}$$

Comparing this result with (2–83) we obtain

$$\mathbf{a}^T\Omega^2\mathbf{a} = \omega^2 \tag{2–90}$$

From (2–86) we easily obtain

$$\mathbf{a}^T\mathbf{1}\mathbf{a} = \mathbf{1} \tag{2–91}$$

Multiplying both sides of this equation by a constant Ω^2, and subtracting the result from equation (2–90), we have

$$\mathbf{a}^T(\Omega^2 - \Omega^2\mathbf{1})\mathbf{a} = \omega^2 - \Omega^2\mathbf{1} \tag{2–92}$$

If we take the corresponding determinant relation and use (2–87), we obtain

$$
\begin{vmatrix}
\omega_1^2 - \Omega^2 & -\omega_{12}^2 & \cdots & -\omega_{1n}^2 \\
-\omega_{21}^2 & \omega_2^2 - \Omega^2 & & \vdots \\
\vdots & & & \\
-\omega_{n1}^2 & & \cdots & \omega_n^2 - \Omega^2
\end{vmatrix}
$$

$$
=
\begin{vmatrix}
\Omega_1^2 - \Omega^2 & 0 & \cdots & 0 \\
0 & \Omega_2^2 - \Omega^2 & \cdots & 0 \\
\vdots & & & \vdots \\
0 & & \cdots & \Omega_n^2 - \Omega^2
\end{vmatrix}
$$

$$
= (\Omega_1^2 - \Omega^2)(\Omega_2^2 - \Omega^2) \cdots (\Omega_n^2 - \Omega^2) \qquad (2\text{–}93)
$$

Thus the frequencies of the normal vibrations $\Omega_1, \ldots, \Omega_n$ are obtained from the *secular equation*[3]

$$
\begin{vmatrix}
\omega_1^2 - \Omega^2 & -\omega_{12}^2 & \cdots & -\omega_{1n}^2 \\
-\omega_{21}^2 & \omega_2^2 - \Omega^2 & \cdots & -\omega_{2n}^2 \\
\vdots & & & \vdots \\
-\omega_{n1}^2 & & \cdots & \omega_n^2 - \Omega^2
\end{vmatrix}
= 0 \qquad (2\text{–}94)
$$

PROBLEM 2–16. Express the potential energy of the two coupled oscillators, given by (2–44) and (2–45), in the form of (2–66). Notice that $k_{ij} = k_{ji}$.

PROBLEM 2–17. Show that if

$$
\mathbf{AB} = \mathbf{C}
$$

where $\mathbf{A}$, $\mathbf{B}$, and $\mathbf{C}$ are matrices, then

$$
\mathbf{C}^T = \mathbf{B}^T \mathbf{A}^T
$$

PROBLEM 2–18. (a) Knowing $\mathbf{X} = \mathbf{a}\boldsymbol{\xi}$ show that $\boldsymbol{\xi} = \mathbf{a}^T\mathbf{X}$. (b) Show that $\mathbf{aa}^T = \mathbf{1}$.

---

[3] The secular (meaning "long period") equation was introduced in astronomy in relation to the disturbances of planetary orbits. A history of this equation is found in M. Jammer, *Conceptual Development of Quantum Mechanics*, McGraw-Hill, New York, 1966, pp. 215–218.

HINT.   Use the result of part (a).

PROBLEM 2–19.   Show that the ith row of the **a** matrix can be obtained from the equation

$$(\omega^{2T} - \Omega_i^2 \mathbf{1}) \begin{pmatrix} a_{i1} \\ a_{i2} \\ \vdots \\ a_{in} \end{pmatrix} = \begin{pmatrix} 0 \\ 0 \\ \vdots \\ 0 \end{pmatrix}$$

PROBLEM 2–20.   Show that a_{ij} gives the relative amplitude of the ξ_j coordinate of the ith normal vibration. The ξ_j coordinate is given by

$$\xi_j = a_{ij} X_i^{(0)} \cos(\Omega_i t + \delta_i)$$

where $X_i^{(0)}$ and δ_i are constants which are independent of j.

2–7   LINEAR MOLECULE

As an example of the normal vibration problem let us take a chain made of n identical atoms which are bound together by $n - 1$ identical springs of the force constant k. We assume that the chain is along the x axis and that each atom can move in the x direction only (Fig. 2–9).

FIGURE 2–9.   Linear molecule.

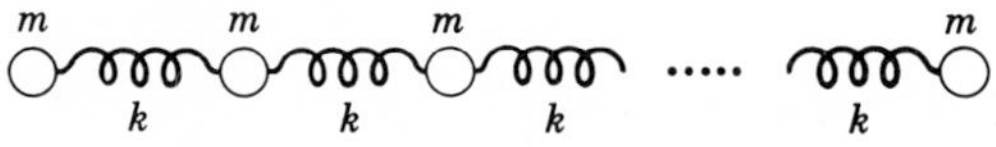

If the displacement of the ith atom along the x axis is x_i, the kinetic energy of the whole system is

$$T = \tfrac{1}{2}m(\dot{x}_0^2 + \dot{x}_1^2 + \cdots + \dot{x}_{n-1}^2) \tag{2-95}$$

where m is the mass of each atom, while the potential energy is

$$U = \tfrac{1}{2}k[(x_0 - x_1)^2 + (x_1 - x_2)^2 + \cdots + (x_{n-2} - x_{n-1})^2]$$

$$= \tfrac{1}{2}k(x_0^2 + 2x_1^2 + \cdots + 2x_{n-2}^2 + x_{n-1}^2$$

$$- 2x_0 x_1 - 2x_1 x_2 - \cdots - 2x_{n-2} x_{n-1}) \tag{2-96}$$

From (2–94) we obtain the secular equation as

$$
\begin{vmatrix}
\omega_0^2 - \Omega^2 & -\omega_0^2 & 0 & \cdots & 0 & 0 \\
-\omega_0^2 & 2\omega_0^2 - \Omega^2 & -\omega_0^2 & \cdots & 0 & 0 \\
0 & -\omega_0^2 & 2\omega_0^2 - \Omega^2 & \cdots & 0 & 0 \\
\vdots & \vdots & & & & \\
0 & 0 & & \cdots & 2\omega_0^2 - \Omega^2 & -\omega_0^2 \\
0 & 0 & & \cdots & -\omega_0^2 & \omega_0^2 - \Omega^2
\end{vmatrix} = 0
$$

$$(2\text{–}97)$$

where

$$\omega_0^2 = \frac{k}{m}$$

A large determinant equation like (2–97) is usually difficult to solve, but our particular one can be solved using its symmetry properties. Let us multiply all members of the $(l + 1)$th column of the determinant by $\cos(vl + \delta)$; the resulting determinant should still be zero. Do the same thing for all columns and add them all to the first column, a procedure which leaves the resulting determinant equal to zero. Now a determinant is zero if every member of the first column is zero. Therefore, we require

$$(\omega_0^2 - \Omega^2) \cos \delta - \omega_0^2 \cos(v + \delta) = 0$$

$$(2\text{–}98)$$

$$-\omega_0^2 \cos \delta + (2\omega_0^2 - \Omega^2) \cos(v + \delta) - \omega_0^2 \cos(2v + \delta) = 0$$

$$(2\text{–}99)$$

. . . .

$$-\omega_0^2 \cos[v(l - 1) + \delta] + (2\omega_0^2 - \Omega^2) \cos(vl + \delta)$$
$$- \omega_0^2 \cos[v(l + 1) + \delta] = 0 \qquad (2\text{–}99')$$

. . . .

$$-\omega_0^2 \cos[v(n - 2) + \delta] + (\omega_0^2 - \Omega^2) \cos[v(n - 1) + \delta] = 0$$

$$(2\text{–}100)$$

Equation (2–99′) is satisfied for all l except for 0 and $n - 1$, i.e., except for (2–98) and (2–100), when

$$\Omega^2 = 2\omega_0^2 (1 - \cos v)$$

$$(2\text{–}101)$$

where v is still arbitrary. On the other hand, noticing that (2–100) can be written as

$$[(\omega_0{}^2 - \Omega^2) \cos \delta - \omega_0{}^2 \cos (v + \delta)] \cos [v(n - 1) + 2\delta]$$
$$- [(\omega_0{}^2 - \Omega^2) \sin \delta + \omega_0{}^2 \sin (v + \delta)] \sin [v(n - 1) + 2\delta] = 0$$

$$(2\text{–}102)$$

we see that (2–100) and (2–98) agree if

$$\sin [v(n - 1) + 2\delta] = 0 \qquad (2\text{–}103)$$

that is, if

$$v(n - 1) + 2\delta = 0, \quad \pm\pi, \quad \pm 2\pi, \ldots \qquad (2\text{–}104)$$

Now if we put (2–101) into (2–98) we obtain

$$\cos (v - \delta) = \cos \delta \qquad (2\text{–}105)$$

which means that

$$\delta = \tfrac{1}{2}v, \quad \tfrac{1}{2}v + \pi, \quad \tfrac{1}{2}v + 2\pi, \ldots \qquad (2\text{–}106)$$

Combining (2–106) with (2–104) we see

$$v = \frac{k\pi}{n} \qquad \text{where} \qquad k = 0, \pm 1, \pm 2, \ldots \qquad (2\text{–}107)$$

which gives the required solution

$$\Omega_k{}^2 = 2\omega_0{}^2 \left[1 - \cos \left(\frac{k\pi}{n}\right)\right] \qquad (2\text{–}108)$$

The result is plotted in Fig. 2–10.

FIGURE 2–10. $\Omega_k{}^2$ as a function of k.

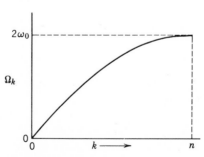

The remaining problem is to find the normal coordinates, or the orthogonal transformation matrix $\mathbf{a}$ (see Problem 2–19). For that purpose, take (2–92), or its equivalent expression

$$(\Omega^2 - \Omega^2 \mathbf{1})\mathbf{a} = \mathbf{a}(\omega^2 - \Omega^2 \mathbf{1}) \qquad (2\text{–}92')$$

and let $\Omega^2 = \Omega_k^2$; which should give us the kth normal coordinate. Since every member of the k row of $\Omega^2 - \Omega_k^2 \mathbf{1}$ is zero, the members of the k row of $(\Omega^2 - \Omega_k^2 \mathbf{1})\mathbf{a}$ are also zero. The corresponding part of the matrix equation (2–92'), therefore, gives

$$(\omega_0^2 - \Omega_k^2)a_{k0} - \omega_0^2 a_{k1} = 0 \qquad (2\text{–}109)$$

$$\cdots\cdots\cdots\cdots\cdots\cdots\cdots\cdots\cdots\cdots\cdots\cdots\cdots$$

$$-\omega_0^2 a_{kl-1} + (2\omega_0^2 - \Omega_k^2)a_{kl} - \omega_0^2 a_{kl+1} = 0 \qquad (2\text{–}110)$$

$$\cdots\cdots\cdots\cdots\cdots\cdots\cdots\cdots\cdots\cdots\cdots\cdots\cdots$$

$$-\omega_0^2 a_{kn-2} + (\omega_0^2 - \Omega_k^2)a_{kn-1} = 0 \qquad (2\text{–}111)$$

Comparing these equations to (2–98), (2–99), and (2–100), we see immediately that

$$a_{kl} = \alpha_k \cos\left[(l + \tfrac{1}{2})\frac{k\pi}{n}\right] \qquad (2\text{–}112)$$

where the constant α_k should be decided from one of the orthogonality conditions

$$\sum_{l=0}^{n-1} a_{kl}^2 = 1 \qquad (2\text{–}113)$$

Thus

$$\alpha_k^2 = \left\{\sum_{l=0}^{n-1} \cos^2\left[(l + \tfrac{1}{2})\frac{k\pi}{n}\right]\right\}^{-1} = \left(\frac{2}{n+1}\right) \qquad (2\text{–}114)$$

From (2–78) and (2–86) we see

$$\sqrt{m}\,x_l = \sum_{k=0}^{n-1} a_{lk}X_k$$

$$= \sqrt{2/(n+1)} \sum_{k=0}^{n-1} \cos\left[(l + \tfrac{1}{2})\frac{k\pi}{n}\right] X_k^{(0)} \cos(\Omega_k t + \delta_k) \qquad (2\text{–}115)$$

where $X_k^{(0)}$ and δ_k are the amplitude and the phase constant of the kth normal vibration, respectively, while Ω_k is given by (2–108). The jth normal coordinate is obtained if we let all $X_k^{(0)}$, except for $X_j^{(0)}$, be zero. For the jth normal coordinate we see from (2–115) that

$$x_{n-1-l} = \sqrt{2/(n+1)m} \, \cos \left[(n - l - \tfrac{1}{2}) \frac{j\pi}{n} \right] X_j$$

$$= \sqrt{2/(n+1)m} \, \cos(j\pi) \cos \left[(l + \tfrac{1}{2}) \frac{j\pi}{n} \right] X_j$$

$$= \cos(j\pi) x_l = (-1)^j x_l \tag{2–116}$$

Therefore, the normal coordinate X_j is symmetric or antisymmetric with respect to the inversion at the center of the molecule, depending on whether j is even or odd.

The normal coordinate X_0, obtained from (2–115) in the same way, is such that $x_0 = x_1 = \cdots = x_{n-1}$. This is just the translational motion of the molecule as a whole.

PROBLEM 2–21. Confirm the results of (2–98) through (2–100), and then (2–101).

PROBLEM 2–22. Show directly that the matrix **a** given by (2–112) and (2–114) is an orthogonal matrix, or $\mathbf{aa}^T = \mathbf{a}^T\mathbf{a} = \mathbf{1}$.

PROBLEM 2–23. Illustrate the normal vibrations for $k = 0, \tfrac{1}{2}n$, and n; and find an intuitive explanation why $\Omega_k = 0, \sqrt{2}\omega_0$, and $2\omega_0$, respectively, in these cases.

ADDITIONAL PROBLEMS

2–24. Show directly that in a simple harmonic oscillator $\overline{T} = \overline{U}$, where the upper bars denote a time average over the period of motion. Show also that this is a special case of the virial theorem shown in Problem 1–40.

2-25. Given an isotropic three-dimensional simple harmonic oscillator

$$\ddot{\mathbf{r}} = -\omega^2 \mathbf{r}$$

show that

$$\left(\frac{d}{dt}\right)^{2n} \mathbf{r} = (-\omega^2)^n \mathbf{r}$$

$$\left(\frac{d}{dt}\right)^{2n+1} \mathbf{r} = (-\omega^2)^n \dot{\mathbf{r}}$$

and obtain

$$\mathbf{r}(t) = \mathbf{r}_0 \cos \omega t + \frac{1}{\omega} \mathbf{v}_0 \sin \omega t$$

from the Taylor expansion

$$\mathbf{r}(t) = \mathbf{r}_0 + \mathbf{v}_0 t + \frac{1}{2}\left(\frac{d^2 \mathbf{r}}{dt^2}\right)_0 t^2 + \frac{1}{6}\left(\frac{d^3 \mathbf{r}}{dt^3}\right)_0 t^3 + \cdots$$

FIGURE 2–11.   A triatomic molecule.

2-26. Obtain all normal vibrations of the system shown in Fig. 2–11, assuming that the motions are restricted to the direction parallel to the springs. Particularly, what is the normal coordinate with zero frequency in this problem?

FIGURE 2–12.   The solution of Problem 2–27:

$$\Omega^2 = \omega_1{}^2 + \omega_2{}^2 \pm \sqrt{(\omega_1{}^2 - \omega_2{}^2) + 4\omega_0{}^2 \cos^2 \nu}$$

where

$$\omega_1{}^2 = k/m, \qquad \omega_2{}^2 = k/M, \qquad \text{and} \qquad \omega_0{}^2 = k/\sqrt{mM}$$

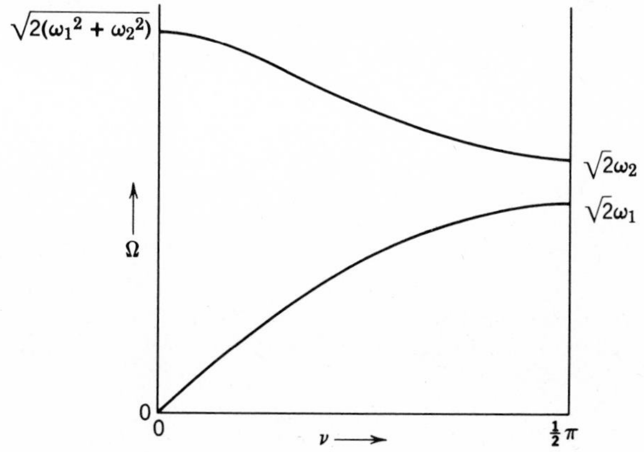

*2–27.** A long linear molecule is made of two kinds of atom the masses of which are m and M, respectively. Suppose they are placed alternatively as $mMmMmM \cdots mMm$ and the force constant between neighboring atoms is k, obtain all frequencies for the longitudinal vibrations of this molecule. Show particularly that they can be divided into high- and low-frequency groups in this case.

ANSWER. See Fig. 2–12.

2–28. In the previous problem, 2–27, the vibrational modes of high- and low-frequency groups are called optical and acoustical modes, respectively. Obtain the normal coordinates of problem 2–27, and show that the basic difference between the optical and acoustical modes is whether the neighboring atoms vibrate with phase difference of larger or less than $\frac{1}{2}\pi$.

Chapter 3 Continuous Body, Rigid Body, and Rotating Frame

Point particles and distributed continuous media, or fields, are the two basic concepts in physics. As an introduction to the latter concept we shall discuss classical mechanics of an elastic continuous body in this chapter. The continuous body is, actually, a macroscopic approximation of physical objects, which are known to be made of atoms. By smoothing out the atomic structure the mass density is introduced as a continuous function, and the stress tensor and the elastic constants are introduced to represent the combined effects of all interatomic forces.

The important step from the mechanics of particles to that of continuous bodies is the parametrization of the space coordinates. Instead of describing the motion for each particle, we fix our attention on a given volume element fixed in space. In this approach we treat the space coordinates as parameters just like the time coordinate while the space coordinates are dynamical variables and time is a unique parameter in the classical mechanics of particles. This conceptual development will find deeper meaning in the next chapter where we shall discuss the special theory of relativity.

A rigid body is, conceptually, a step back to a particle, except that it has a finite extension. Here, however, all internal degrees of freedom are assumed to be frozen; the remaining difference between a point particle and a rigid body is that the latter has the freedom of rotational motion.

3–1 VIBRATING STRING

Let us take an elastic string with mass μ per unit length and tension T. In this section we assume that μ and T are constant, or the string is homogeneous. Suppose that the string is stretched along an x axis. A small displacement of this string from the x axis at x is designated as $u(x)$. Notice that x is a parameter now. It is not x but $u(x)$ that gives the motion of matter.

FIGURE 3–1. Infinitesimal segment of string. The perpendicular component of tension T at one end is $T\,du/\sqrt{(dx)^2 + (du)^2} \cong T\,du/dx$ assuming that $|du/dx| \ll 1$.

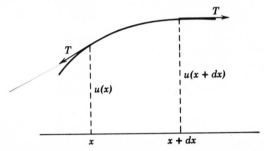

A small segment of the string which is between x and $x + dx$ has mass $\mu\,dx$. The tension T pulls this segment from both ends along the string. If the segment is straight these two tensions will exactly cancel each other, but if it is curved there will be a nonzero restoring force. If u is small enough, the perpendicular component of the tension force at one end of the segment is approximately

$$T\,\frac{du}{dx} \tag{3–1}$$

as can be seen in Fig. 3–1. The difference of this force at x and $x + dx$ is the net restoring force which pulls the segment back towards the x axis. Thus the equation of motion of this segment is

$$\mu\,dx\,\frac{d^2u}{dt^2} = \frac{d}{dx}\left(T\,\frac{du}{dx}\right)dx \tag{3–2}$$

or

$$\mu\,\frac{\partial^2 u}{\partial t^2} = T\,\frac{\partial^2 u}{\partial x^2} \tag{3–3}$$

when T is a constant. We used the partial derivative notation in (3–3), because x and t are independent parameters in our formalism now.

A solution[1] of (3-3) can be written as

$$u = u_0 e^{i(kx - \omega t)} \tag{3-4}$$

where

$$\mu \omega^2 = T k^2 \tag{3-5}$$

This solution gives a wave traveling in the $+x$ direction with the phase velocity

$$V = \frac{\omega}{k} = \sqrt{T/\mu} \tag{3-6}$$

The general solution of (3-3) can be expressed as

$$u(x, t) = \int_{-\infty}^{\infty} f(k) e^{i(kx - \omega t)} \, dk \tag{3-7}$$

where $f(k)$ is a function of k.

The transformation from $f(k)$ to $u(x)$ given in (3-7) is the *Fourier transformation*. Using the *Dirac delta function*,[2] which is expressed as

$$\int_{-\infty}^{\infty} e^{ikx} \, dk = 2\pi \, \delta(x) \tag{3-8}$$

and has the property of

$$\int F(x) \, \delta(x) \, dx = F(0) \tag{3-9}$$

for any function $F(x)$, provided the range of integration includes the point $x = 0$, we can find the inverse Fourier transformation for (3-7) as

$$f(k) = \frac{1}{2\pi} e^{i\omega t} \int_{-\infty}^{\infty} u(x, t) e^{-ikx} \, dx \tag{3-10}$$

PROBLEM 3-1. Show the equivalence between (3-7) and (3-10).

---

[1] Another solution $u_0 e^{-i(kx + \omega t)}$, or equivalently $u_0 e^{i(kx + \omega t)}$, gives a wave traveling in the $-x$ direction.

[2] An interesting history of the δ function can be found on p. 301 of M. Jammer, *The Conceptual Development of Quantum Mechanics*, McGraw-Hill, New York, 1966.

PROBLEM 3–2. A wave packet is a relatively localized wave. A typical wave packet is

$$u(x, t) = e^{-\alpha(x - Vt)^2} \qquad \alpha > 0$$

where α is a constant which gives the width of this wave packet, while V is the velocity. Taking $V = \omega/k$, show that

$$f(k) = \frac{1}{2\sqrt{\pi\alpha}} e^{-k^2/4\alpha}$$

in this case.

PROBLEM 3–3. Show that
(a) $(x - x_0) \delta(x) = -x_0 \delta(x)$, or $x \delta(x) = 0$.
(b) $\delta(cx) = \delta(x)/|c|$; c is a constant; $(c \neq 0)$.
(c) $\delta[(x - c)(x - d)] = [\delta(x - c) + \delta(x - d)]/|c - d|$; c and d are constants; $(c \neq d)$.

Boundary and Initial Conditions

If the string has *boundary conditions* such that

$$u = 0 \quad \text{at} \quad x = 0 \quad \text{and} \quad x = L \qquad (3\text{--}11)$$

then the solution is limited to be a linear combination of

$$U_n = U_n^0 e^{-i\omega_n t} \sin\left(\frac{n\pi x}{L}\right) \qquad (3\text{--}12)$$

where

$$n = 1, 2, 3, \ldots$$

and U_n^0 is a constant. The relation (3–5) gives

$$\omega_n = \left(\frac{n\pi}{L}\right) \sqrt{T/\mu} \qquad (3\text{--}13)$$

The general solution under the boundary conditions of (3–11) is, therefore,

$$u(x, t) = \sum_n [A_n \cos(\omega_n t) + B_n \sin(\omega_n t)] \sin\left(\frac{n\pi x}{L}\right) \qquad (3\text{--}14)$$

The coefficients A_n and B_n in (3–14) are determined from the *initial conditions*. If, at $t = 0$,

$$u(x, 0) = \eta(x) \tag{3-15}$$

and

$$\dot{u}(x, 0) = \xi(x) \tag{3-16}$$

then, from (3–14),

$$\eta(x) = \sum_n A_n \sin\left(\frac{n\pi x}{L}\right) \tag{3-17}$$

and

$$\xi(x) = \sum_n \omega_n B_n \sin\left(\frac{n\pi x}{L}\right) \tag{3-18}$$

Since

$$\int_0^L \sin\left(\frac{n\pi x}{L}\right) \sin\left(\frac{m\pi x}{L}\right) dx$$

$$= -\frac{1}{2}\int_0^L \left\{ \cos\left[\frac{(n+m)\pi x}{L}\right] - \cos\left[\frac{(n-m)\pi x}{L}\right] \right\} dx$$

$$= \tfrac{1}{2}L\,\delta_{n,m} \tag{3-19}$$

we see

$$A_n = \frac{2}{L}\int_0^L \eta(x) \sin\left(\frac{n\pi x}{L}\right) dx \tag{3-20}$$

$$B_n = \frac{2}{L\omega_n}\int_0^L \xi(x) \sin\left(\frac{n\pi x}{L}\right) dx \tag{3-21}$$

Each term like (3–12) is a normal coordinate, and it vibrates with a definite frequency ω_n. We see from this expression that the normal coordinate is symmetric or antisymmetric with respect to the inversion, $x \to L - x$, depending on whether n is even or odd. One may notice, however, that the relation is opposite to the case of the linear molecule we discussed in section 2–6. This is because in the previous section we considered the motion along the x axis, while here we consider the motion perpendicular to the x axis. The corresponding vibrations are called *longitudinal* and *transverse* vibrations, respectively.

PROBLEM 3–4. If the initial conditions are (Fig. 3–2)

$$\eta(x) = ax \quad \text{for} \quad 0 \le x \le \tfrac{1}{2}L$$
$$\quad = a(L - x) \quad \text{for} \quad \tfrac{1}{2}L \le x \le L$$
$$\xi(x) = 0$$

show that

$$A_n = 0 \qquad \text{when} \qquad n = 2, 4, 6, \ldots$$
$$\quad = \frac{4aL}{(n\pi)^2} \qquad \text{when} \qquad n = 1, 5, 9, \ldots$$
$$\quad = -\frac{4aL}{(n\pi)^2} \qquad \text{when} \qquad n = 3, 7, 11, \ldots$$
$$B_n = 0$$

FIGURE 3–2. The initial condition of Problem 3–4.

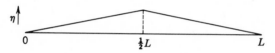

3–2 STRING WITH VARIABLE TENSION

If the tension T is not a constant along the string, we have

$$\mu \frac{\partial^2 u}{\partial t^2} = \frac{\partial}{\partial x} T \frac{\partial u}{\partial x} \tag{3–22}$$

instead of (3–3). In this case it is difficult to obtain the normal coordinates U_n, which correspond to (3–12) for the simpler case. One may assume, however, that a normal coordinate U_n satisfies

$$\frac{\partial^2 U_n}{\partial t^2} = -\omega_n{}^2 U_n \tag{3–23}$$

where ω_n is a constant. If the string is bound, all normal coordinates should satisfy the boundary condition

$$U_n = 0, \quad \text{at} \quad x = 0 \quad \text{and} \quad x = L \tag{3–24}$$

From (3–22) and (3–23) we have

$$-\mu\omega_n^2 U_n = \frac{\partial}{\partial x}\left(T\,\frac{\partial}{\partial x}\,U_n\right) \tag{3-25}$$

$$-\mu\omega_m^2 U_m = \frac{\partial}{\partial x}\left(T\,\frac{\partial}{\partial x}\,U_m\right) \tag{3-26}$$

where U_m is another normal coordinate. From (3–25) and (3–26) we obtain, if μ is constant,

$$\mu(\omega_n^2 - \omega_m^2)\int_0^L U_n U_m\,dx$$

$$= \int_0^L U_n\,\frac{\partial}{\partial x}\left(T\,\frac{\partial}{\partial x}\,U_m\right)dx - \int_0^L U_m\,\frac{\partial}{\partial x}\left(T\,\frac{\partial}{\partial x}\,U_n\right)dx$$

$$= \int_0^L T\left(\frac{\partial U_n}{\partial x}\right)\left(\frac{\partial U_m}{\partial x}\right)dx - \int_0^L T\left(\frac{\partial U_m}{\partial x}\right)\left(\frac{\partial U_n}{\partial x}\right)dx = 0 \tag{3-27}$$

where we used (3–24) in the partial integrations. Equation (3–27) shows the orthogonality between the normal coordinates

$$\mu\int_0^L U_n U_m\,dx = 0, \quad \text{if} \quad \omega_n^2 \neq \omega_m^2 \tag{3-28}$$

We saw in (3–19) that such an orthogonality relation is satisfied with the normal coordinates of the simpler problem (3–12), but this discussion shows that orthogonality is a more general property of normal coordinates. When the mass density μ is not constant we obtain

$$\int_0^L \mu(x)U_n U_m\,dx = 0, \quad \text{if} \quad \omega_n^2 \neq \omega_m^2 \tag{3-29}$$

as a generalized orthogonality condition.

PROBLEM 3–5. Show that (3–29) holds true even when μ is a function of t and x. [Note that (3–22) should be modified in this case.]

3–3 CONTINUOUS BODY. EQUATION OF CONTINUITY

Let us consider a three-dimensional body in which the mass is continuously distributed. We know that these bodies, solid, liquids and gases, are actually made of molecules, which exclude each other through some quantum mechanical interactions (section 11–16), and carry most of their mass in their nuclei. In this sense the mass distribution is actually rather discontinuous since each nucleus is about 10^{-15} m in diameter while the separation between them is about 10^{-10} m. In this chapter, however, we shall consider dimensions much larger than atomic or nuclear dimensions, something like 10^{-3} m. In the latter viewpoint, that is, the *macroscopic* viewpoint, we neglect the atomic or *microscopic* structure of matter mentioned above, and smooth out the mass distribution as an approximation. This viewpoint is admittedly an approximation for matter, but it will turn out to be a more exact, or at least a more useful viewpoint in treating fields, such as electromagnetic fields, gravity fields, and meson fields.

In treating continuous bodies we take a coordinate system which is fixed in space, but not on the body. Just like in the treatment of the motion of a string, the coordinates x, y, and z are not the dynamical variables, but parameters as the time t. Let us take a mathematically infinitesimal lattice of volume $dx \, dy \, dz$, located at a point $\mathbf{r}$, whose components are x, y, and z (Fig. 3–3). The infinitesimal lengths dx, dy, and dz are assumed to be much larger than the atomic dimension mentioned above, so that the mass distribution throughout this infinitesimal volume can be regarded as continuous. We introduce the mass density μ such that the mass inside this infinitesimal volume at time t is

$$\mu(\mathbf{r}, t) \, dv \qquad (3\text{–}30)$$

FIGURE 3–3. An infinitesimal volume. Coordinates of each corner are given in a parenthesis.

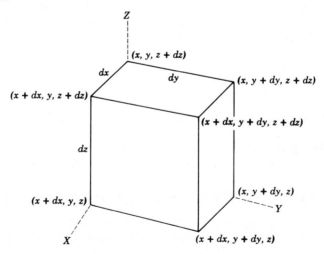

and assume that μ is a continuous function of $\mathbf{r}$ and t. Note that our infinitesimal volume is fixed in space and does not move with the matter, so that the coordinate $\mathbf{r}$ does not depend on t; $\mathbf{r}$ and t should be regarded as independent parameters.

If the mass is moving through this infinitesimal volume, the motion can be described by the mass current density

$$\mathbf{j}(\mathbf{r}, t) \equiv \mu \mathbf{V}(\mathbf{r}, t) \tag{3-31}$$

such that if $d\mathbf{a}$ is an infinitesimal area, the mass which passes through this area per unit time is

$$\mathbf{j} \cdot d\mathbf{a} \tag{3-32}$$

Thus the mass which flows per unit time into the infinitesimal volume shown in Fig. 3–3 through the lower $dx\,dy$ plane located at z is $j_z(x, y, z, t)\,dx\,dy$. On the other hand, the mass which flows per unit time out of the infinitesimal volume through the upper $dx\,dy$ plane located at $x + dz$ is

$$j_z(x, y, z + dz, t)\,dx\,dy = \left[j_z(x, y, z, t) + \frac{\partial j_z(x, y, z, t)}{\partial z}\,dz\right]dx\,dy \tag{3-33}$$

Taking into account all six surfaces of dv in the same manner, we see that the net change of mass per unit time for this volume is

$$\frac{\partial}{\partial t}(\mu\,dx\,dy\,dz) = -\left(\frac{\partial j_x}{\partial x} + \frac{\partial j_y}{\partial y} + \frac{\partial j_z}{dz}\right)dx\,dy\,dz$$

$$\equiv -\mathbf{\nabla} \cdot \mathbf{j}\,dv \tag{3-34}$$

which can be written as

$$\frac{\partial \mu}{\partial t} + \mathbf{\nabla} \cdot \mathbf{j} = 0 \tag{3-35}$$

This is the *equation of continuity* for mass.

$\mathbf{\nabla} \cdot \mathbf{j}$ is called the *divergence* of $\mathbf{j}$. As clearly seen from (3–33) the total current through the surface of a given volume v is given by

$$\oint \mathbf{j} \cdot d\mathbf{a} = \int_v \mathbf{\nabla} \cdot \mathbf{j}\,dv \tag{3-36}$$

This relation is called *Gauss' theorem*.

The equation of continuity (3–35) gives the conservation of mass. If the mass is not conserved but created (or annihilated) at the rate of $n(r, t)$ per unit volume in the infinitesimal volume dv, then (3–35) should be generalized as

$$\frac{\partial \mu}{\partial t} + \mathbf{V} \cdot \mathbf{j} = n \qquad (3\text{--}37)$$

The mass current density is actually the momentum density because of relation (3–31). In the following discussions on the dynamics of continuous body we shall be very much interested in the change of the momentum density per unit time. When the mass is conserved, we see, using (3–35), that

$$\frac{\partial \mathbf{j}}{\partial t} = \mu \frac{\partial \mathbf{V}}{\partial t} + \mathbf{V} \frac{\partial \mu}{\partial t} = \mu \frac{\partial \mathbf{V}}{\partial t} - \mathbf{V}(\mathbf{V} \cdot (\mu \mathbf{V})) \qquad (3\text{--}38)$$

On the other hand, if we integrate over the whole body so that μ is zero at the boundaries of the integral, we obtain

$$- \int \mathbf{V}(\mathbf{V} \cdot (\mu \mathbf{V})) \, dv = \int (\mu \mathbf{V} \cdot \mathbf{V}) \mathbf{V} \, dv \qquad (3\text{--}39)$$

by the partial integration. Expecting such integration we see that (3–38) can be written as

$$\frac{\partial \mathbf{j}}{\partial t} = \mu \frac{\partial \mathbf{V}}{\partial t} + \mu (\mathbf{V} \cdot \mathbf{V}) \mathbf{V} \qquad (3\text{--}40)$$

PROBLEM 3–6. When the body is homogeneous, show directly, that is, without assuming (3–31), that for a given infinitesimal volume

$$d\mu = -\mu \mathbf{V} \cdot \mathbf{u}$$

holds true. Note that this implies

$$\frac{\partial \mu}{\partial t} + \mu \mathbf{V} \cdot \mathbf{V} = 0$$

$\mathbf{u}(\mathbf{r})$ is a small displacement at $\mathbf{r}$.

3–4 STRESS TENSOR

There are two kinds of forces acting on matter in the infinitesimal volume, immersed in the continuous body; *volume* or *body forces* and *surface forces*. The gravity force and the electrostatic Coulomb force may be regarded as volume forces, at least in the present chapter, since they are exerted directly on matter in the infinitesimal volume from sources which are at some distance or may even be completely outside of the continuous body. (Note, however, that we will reformulate the gravity and electromagnetic forces in later chapters in another way.) Since the body is continuous through the infinitesimal volume which we are considering now, matter inside this volume interacts with that in the neighboring infinitesimal volume at the surface between them, and this interaction gives the surface forces.

Let us take an infinitesimal area $dx\,dy$ located at z, and imagine, as illustrated in Fig. 3–4, that we can consider separately the upper and lower parts of the continuous body, with respect to this plane. For the matter

FIGURE 3–4.   The surface forces. Two infinitesimal volumes which are next to each other are drawn as separated for the purpose of illustration.

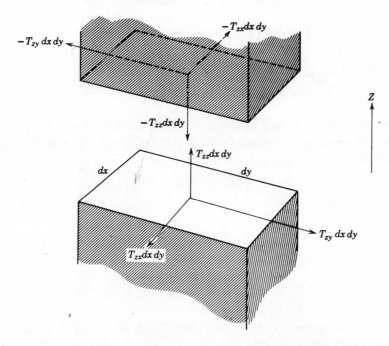

just below this mathematical surface, we express the surface force exerted on this area as

$$F_x = T_{zx}(\mathbf{r}) \, dx \, dy, \qquad F_y = T_{zy}(\mathbf{r}) \, dx \, dy, \qquad F_z = T_{zz}(\mathbf{r}) \, dx \, dy$$

$$(3\text{--}41)$$

The proportionality constant for each component $T_{\alpha\beta}(\mathbf{r})$ is a function of the position of the surface $dx \, dy$. Since the body is continuous there should be reaction forces exerted on the matter on the other side of this plane $dx \, dy$, with the same magnitudes but opposite signs as the corresponding components of (3–41). If we consider the infinitesimal volume dv again, we see that the surface force exerted on the upper $dx \, dy$ plane is given by (3–41), except that z is replaced by $z + dz$, while the surface force exerted on the lower $dx \, dy$ plane is the reaction force. These two forces, however, may not cancel each other since they are at slightly different locations, i.e., $z + dz$ and z. The net force on dv through these two $dx \, dy$ surfaces is

$$F_x = -T_{zx}(x, y, z) \, dx \, dy + T_{zx}(x, y, z + dz) \, dx \, dy$$

$$= \left(\frac{\partial T_{zx}}{\partial z} \, dz \right) dx \, dy = \frac{\partial T_{zx}}{\partial z} \, dv \qquad\qquad (3\text{--}42)$$

$$F_y = \frac{\partial T_{zy}}{\partial z} \, dv \qquad F_z = \frac{\partial T_{zz}}{\partial z} \, dv$$

The total force exerted on the matter in this infinitesimal volume is obtained by summing up the results similar to (3–42) over all six surfaces. The result is

$$F_x = \left(\frac{\partial T_{xx}}{\partial x} + \frac{\partial T_{yx}}{\partial y} + \frac{\partial T_{zx}}{\partial z} \right) dv$$

$$F_y = \left(\frac{\partial T_{xy}}{\partial x} + \frac{\partial T_{yy}}{\partial y} + \frac{\partial T_{zy}}{\partial z} \right) dv \qquad\qquad (3\text{--}43)$$

$$F_z = \left(\frac{\partial T_{xz}}{\partial x} + \frac{\partial T_{yz}}{\partial y} + \frac{\partial T_{zz}}{\partial z} \right) dv$$

These nine quantities $T_{\alpha\beta}$ form a second rank tensor which is called the *stress tensor*. The diagonal terms, T_{xx}, T_{yy}, and T_{zz}, give the *normal stress*, or *pressure*, and the other terms give the *shearing*, or *tangential stress*.

The stress tensor is a symmetric tensor, or

$$T_{\alpha\beta} = T_{\beta\alpha} \tag{3-44}$$

This can be seen when we consider the torque, that is, the moment of the surface forces. In Fig. 3–5 all surface forces exerted on dv which lie in the xy plane are shown. These forces produce a torque around an axis which passes through the center of the $dx\,dy$ plane and parallel to the z axis, which tends to rotate the matter inside dv. The torque is defined as $\mathbf{r} \times \mathbf{F}$, and in our case the four forces produce a torque in the z direction with magnitude

$$dxT_{xy}\,dy\,dz - dyT_{yx}\,dz\,dx = (T_{xy} - T_{yx})\,dv \tag{3-45}$$

FIGURE 3–5. Torque around an axis which passes through the center and parallel to the z-axis.

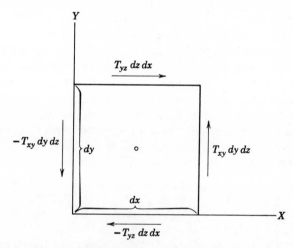

The torque should be equal to the time derivative of the angular momentum, but the latter is of the order of $(dl)^5/dt$ where dl is the linear infinitesimal dimension [Formulas (3–121) and (3–129)]. Since the torque given by (3–45) is of the order of $T(dl)^3$, if it is not zero, we should have

$$T_{xy}(\mathbf{r}) = T_{yx}(\mathbf{r}) \tag{3-46}$$

to the order of $(dl)^2$. It is easy to generalize this result to other directions to obtain (3–44).

PROBLEM 3–7. Obtain the result of (3–43).

3–5 STRAIN. ELASTIC CONSTANTS

Since tensor equations like (3–43) tend to be long and complicated, it is customary to abbreviate these formulas by introducing the notation

$$
\begin{aligned}
x^1 &\equiv x \\
x^2 &\equiv y \\
x^3 &\equiv z
\end{aligned}
\tag{3–47}
$$

and then introducing the convention that when a superscript or subscript is repeated in a single term we mean the summation over three components. Thus, for example, we can write equations (3–43) as

$$
F_\alpha = \left(\frac{\partial T_{\beta\alpha}}{\partial x^\beta} \right) dv
\tag{3–43'}
$$

using this convention.

The displacement of matter in dv at $\mathbf{r}$ of the continuous body is given by a three-dimensional vector $\mathbf{u}(\mathbf{r})$. The displacement itself, however, does not produce any stress; but the difference of displacements between neighboring points in the continuous body does. Thus important quantities in investigating the dynamics of a continuous body are the derivatives of u_x, u_y, and u_z with respect to x, y, and z. These nine derivatives form a second rank tensor since $\mathbf{u}$ and $\mathbf{r}$ are vectors. The antisymmetric part of this second rank tensor, such as $\partial u_x / \partial y - \partial u_y / \partial x$, corresponds to rotation, so that the real strain should be given by the remaining symmetric part

$$
S_{xy} = S_{yx} = \frac{1}{2} \left(\frac{\partial u_x}{\partial y} + \frac{\partial u_y}{\partial x} \right) \qquad S_{xx} = \frac{\partial u_x}{\partial x}
\tag{3–48}
$$

and so forth. There are six independent components of the strain tensor.

If the strain is small, Hooke's law is expected to hold, namely, that the stress is a linear function of strain:

$$
T_{\alpha\beta} = C_{\alpha\beta,\gamma\delta} S_{\gamma\delta} \equiv \sum_{\gamma,\delta} C_{\alpha\beta,\gamma\delta} S_{\gamma\delta}
\tag{3–49}
$$

where the proportionality constant $C_{\alpha\beta,\gamma\delta}$, called the *elastic constant*, forms a fourth rank tensor.

There are 81 elastic constants which appear in (3–49), but the number can be reduced to 21 for a general elastic body. First of all we obtain

$$
C_{\alpha\beta,\gamma\delta} = C_{\beta\alpha,\gamma\delta} = C_{\beta\alpha,\delta\gamma} = C_{\alpha\beta,\delta\gamma}
\tag{3–50}
$$

because $T_{\alpha\beta}$ and $S_{\gamma\delta}$ are both symmetric tensors. This relation reduces the number of independent constants from 81 to 36. It can be shown that the potential energy density of an elastic body is expressed as

$$\mathscr{U} = \tfrac{1}{2}C_{\alpha\beta,\gamma\delta}S_{\alpha\beta}S_{\gamma\delta} \qquad (3\text{–}51)$$

as expected. This form implies that

$$C_{\alpha\beta,\gamma\delta} = C_{\gamma\delta,\alpha\beta} \qquad (3\text{–}52)$$

and this symmetry reduces the number of independent constants to 21.

If the elastic body is a crystal, some additional symmetry of structure may exist (see section 11–14). For example, if the crystal is symmetric under the transformation $(x, y, z) \rightarrow (x, y, -z)$, then S_{yz} and $-S_{yz}$ should result in the same stress, e.g., T_{xx}, and the only way such a symmetry property can be satisfied is

$$C_{xx,xz} = -C_{xx,xz} = 0 \qquad (3\text{–}53)$$

In the same way we find

$$C_{xx,xz} = C_{yy,yz} = C_{yy,xz} = C_{xx,yz} = C_{xy,xz} = C_{xy,yz} = C_{xz,zz} = C_{yz,zz} = 0 \qquad (3\text{–}54)$$

Thus only 13 elastic constants remain to be determined in such symmetric crystals.

PROBLEM 3–8. Show that the number of independent elastic constants is reduced from 81 to 36 by (3–50), and then to 21 by (3–52).

PROBLEM 3–9. If a crystal is symmetric under the rotation by $\tfrac{1}{2}\pi$ around the z axis, that is, invariant under the transformations $(x, y, z) \rightarrow (y, -x, z) \rightarrow (-x, -y, z)$, show that only 7 elastic constants remain to be determined.

3–6 ISOTROPIC ELASTIC BODY

When the continuous body is completely isotropic, that is, x, y, and z directions are completely equivalent to each other, then

$$\mathscr{U} = \tfrac{1}{2}\Lambda(S_{xx} + S_{yy} + S_{zz})^2$$
$$+ M(S_{xx}{}^2 + S_{yy}{}^2 + S_{zz}{}^2 + 2S_{xy}{}^2 + 2S_{yz}{}^2 + 2S_{zx}{}^2) \qquad (3\text{–}55)$$

is the most general form for the energy density $\mathcal{U}$, a scalar function. Two elastic constants, Λ and M, are called Lamé's constants. Comparing (3–55) with (3–51) we obtain

$$
\begin{aligned}
T_{xx} &= C_{xx,\alpha\beta}S_{\alpha\beta} = (\Lambda + 2M)S_{xx} + \Lambda(S_{yy} + S_{zz}) \\
T_{yy} &= (\Lambda + 2M)S_{yy} + \Lambda(S_{zz} + S_{xx}) \\
T_{zz} &= (\Lambda + 2M)S_{zz} + \Lambda(S_{xx} + S_{yy}) \\
T_{xy} &= 2MS_{xy} \\
T_{yz} &= 2MS_{yz} \\
T_{zx} &= 2MS_{zx}
\end{aligned}
\tag{3–56}
$$

Let us take a simple case of the elastic body under a hydrostatic pressure P, so that

$$
\begin{aligned}
T_{xx} &= T_{yy} = T_{zz} = -P \\
T_{xy} &= T_{yz} = T_{zx} = 0
\end{aligned}
\tag{3–57}
$$

According to (3–56), we see a similar symmetry should exist for $S_{\alpha\beta}$ also. In this case we notice that

$$
S_{xx} = S_{yy} = S_{zz} = \frac{1}{3}\frac{\partial u_\alpha}{\partial x^\alpha} = \tfrac{1}{3}\nabla \cdot \mathbf{u} = \frac{v - v_0}{3v_0}
\tag{3–58}
$$

where v_0 is the natural volume of the elastic body, while v is the volume under the pressure. [Fig. 3–6 illustrates the relation used in obtaining the last expression of (3–58).] Therefore, from (3–56), we can write

$$
\frac{v_0 - v}{v_0} = \frac{1}{k}P
\tag{3–59}
$$

where

$$
k = \Lambda + \tfrac{2}{3}M
\tag{3–60}
$$

is called the *bulk modulus*.

Another simple case is that of an isotropic elastic body (perhaps in the form of a wire) stretched along the x direction. Then

$$
T_{xx} = T
\tag{3–61}
$$

FIGURE 3–6. Illustration of relation $V_0(\nabla \cdot \mathbf{u}) = V - V_0$. Take an infinitesimal volume $dv = dx\,dy\,dz$ and consider u_z. Because of the displacement matter which occupied dv before the displacement now occupies a volume of $dv - u_z\,dx\,dy + (u_z + (\partial u_z/\partial z)\,dz)\,dx\,dy = dv(1 + \partial u_z/\partial z)$.

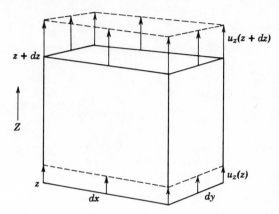

is the only nonzero component of the stress tensor. In this case, from (3–56) we obtain

$$S_{yy} = S_{zz} = -\sigma S_{xx} \tag{3–62}$$

where

$$\sigma = \frac{\Lambda}{2(\Lambda + M)} \tag{3–63}$$

is called *Poisson's ratio*, and gives the ratio of the sidewise contraction to the lengthwise stretch in this deformation. We also obtain

$$T_{xx} = T = ES_{xx} \tag{3–64}$$

where

$$E = \frac{M(3\Lambda + 2M)}{\Lambda + M} \tag{3–65}$$

is called *Young's modulus*.

An experiment can be designed in such a way that T_{xy} is the only nonzero component of the stress tensor. In this case

$$T_{xy} = MS_{xy} \tag{3–66}$$

is the only nontrivial equation. In this sense M is called the *rigidity*.

The elastic bodies are made of atoms, and the elastic constants should be related to the interatomic forces between them. When the atomic force between two atoms is along the line joining them, and when each atom is at a center of symmetry, Cauchy showed that the elastic constants resulting from such interatomic forces should have the symmetry of

$$C_{\alpha\beta,\gamma\delta} = C_{\gamma\beta,\alpha\delta} \qquad (3\text{-}67)$$

If Cauchy's relation holds for the isotropic elastic body, we obtain

$$\Lambda = M \qquad (3\text{-}68)$$

and

$$\sigma = \tfrac{1}{4} \qquad (3\text{-}69)$$

PROBLEM 3-10. Obtain (3–60), (3–63), and (3–65) from (3–56).

PROBLEM 3-11. When Cauchy's relation (3–67) holds in an isotropic body, show that the elastic constant C is completely symmetric with respect to all permutations of the four indices.

PROBLEM 3-12. Obtain (3–68) and (3–69).

3-7 ELASTIC WAVES

The equation of motion of a homogeneous deformable elastic body is given, from (3–40), (3–43'), (3–49), and (3–48), as

$$\mu \left[\frac{\partial^2 u_\alpha}{\partial t^2} + \left(\frac{\partial \mathbf{u}}{\partial t} \cdot \mathbf{V} \right) \frac{\partial u_\alpha}{\partial t} \right] = \frac{\partial T_{\beta\alpha}}{\partial x^\beta} = C_{\alpha\beta,\gamma\delta} \frac{\partial S_{\gamma\delta}}{\partial x^\beta}$$

$$= C_{\alpha\beta,\gamma\delta} \frac{\partial^2 u_\gamma}{\partial x^\beta \, \partial x^\delta} \qquad (3\text{-}70)$$

where the summation convention is used. When the amplitude of the displacement u stays small enough, we can neglect the second term on the left-hand side of (3–70), since that is the only term of the order of u^2. In this approximation (3–70) reduces to a linear differential equation, for which we can assume a solution of the form of

$$u_\alpha = a_\alpha \exp (ik_\beta x^\beta - i\omega t) \qquad (3\text{-}71)$$

for each component α. Here the constants a_α, k_1, k_2, k_3, and ω should satisfy

$$-\mu\omega^2 a_\alpha = -C_{\alpha\beta,\gamma\delta} a_\gamma k_\beta k_\delta \qquad (3\text{-}72)$$

This is the same kind of equation we discussed in the last chapter, and in order to have nontrivial values for a_1, a_2, and a_3 the secular equation

$$\begin{vmatrix} C_{1\beta,1\delta}k_\beta k_\delta - \mu\omega^2 & C_{1\beta,2\delta}k_\beta k_\delta & C_{1\beta,3\delta}k_\beta k_\delta \\ C_{2\beta,1\delta}k_\beta k_\delta & C_{2\beta,2\delta}k_\beta k_\delta - \mu\omega^2 & C_{2\beta,3\delta}k_\beta k_\delta \\ C_{3\beta,1\delta}k_\beta k_\delta & C_{3\beta,2\delta}k_\beta k_\delta & C_{3\beta,3\delta}k_\beta k_\delta - \mu\omega^2 \end{vmatrix} = 0 \qquad (3\text{-}73)$$

should be satisfied by ω and k. Since this secular equation is of the third order, there are three solutions.

When the elastic body is isotropic, equation (3-72) reduces to

$$(\mu\omega^2 - Mk^2)\mathbf{a} - (\Lambda + M)(\mathbf{k} \cdot \mathbf{a})\mathbf{k} = 0 \qquad (3\text{-}74)$$

We obtain two kinds of solutions from (3-74). One solution is

$$\mathbf{k} \cdot \mathbf{a} = 0 \qquad \mu\omega^2 = Mk^2 \qquad (3\text{-}75)$$

Since, as seen from (3-71), $\mathbf{k}$ gives the direction of propagation of this wave, the solution (3-75) gives transverse waves. The phase velocity of these waves is obtained from (3-71) and (3-75) to be

$$\text{phase velocity} = \omega/k = \sqrt{M/\mu} \qquad (3\text{-}76)$$

The other solution of (3-74) is obtained by assuming $\mathbf{k} \| \mathbf{a}$, and is

$$\mu\omega^2 = (\Lambda + 2M)k^2 \qquad (3\text{-}77)$$

or

$$\text{phase velocity} = \sqrt{(\Lambda + 2M)/\mu} \qquad (3\text{-}78)$$

This is the longitudinal wave. We see from (3-76) and (3-78) that

phase velocity of longitudinal wave $>$ phase velocity of transverse wave

$$(3\text{-}79)$$

for a given elastic body. In seismology the faster wave or the longitudinal wave is called the *P wave*, and the slower one, or the transverse wave is called the *S wave*. The terminology is unfortunately confusing for general physicists, since *P* and *S* waves in the scattering theory in quantum mechanics refer to entirely different concepts. (See section 9-7.)

PROBLEM 3-13. Obtain (3-70) in the way suggested.

PROBLEM 3-14. Obtain (3-74).

PROBLEM 3-15. Find the relation between the phase velocities of the longitudinal and transversal waves when Cauchy's relation holds true.

PROBLEM 3-16. Show that the equation of motion for an isotropic elastic body can be written as

$$\mu \frac{\partial^2 \mathbf{u}}{\partial t^2} = (\Lambda + M)\nabla(\nabla \cdot \mathbf{u}) + M\nabla^2 \mathbf{u}$$

3-8 IDEAL FLUIDS

The ideal fluid is characterized by zero shearing stresses, or

$$T_{\alpha\beta} = -P\,\delta_{\alpha\beta} \tag{3-80}$$

where P is the pressure. The equation of motion is thus

$$\mu \left[\frac{\partial \mathbf{V}}{\partial t} + (\mathbf{V} \cdot \nabla)\mathbf{V}\right] = -\nabla P + \mathbf{F}_v \tag{3-81}$$

where $\mathbf{F}_v$ is a possible volume force (density), such as the gravitational force. This equation is called *Euler's equation of motion for ideal fluids.*

The flow of fluid is either laminar or turbulent. More than one whirlpool exists in the turbulent flow. Since the integral $\oint \mathbf{V} \cdot d\mathbf{l}$ is not zero for each whirlpool, the laminar and the turbulent flows can be distinguished from each other by the *vorticity*

$$\mathbf{w} = \nabla \times \mathbf{V} \tag{3-82}$$

If the vorticity is zero throughout the flow, it is laminar.

Since the theory of turbulent flow is quite complicated, although of practical importance, we will not discuss it in this book.[3]

---

[3] The circulation theorem, which states that if a flow originally has no vorticity then it will stay that way, is useful. The theorem is based on Helmholz' equation

$$\frac{\partial w_\alpha}{\partial t} = -V_\beta \frac{\partial w_\alpha}{\partial x^\beta} - w_\alpha \frac{\partial V_\beta}{\partial x^\beta} + w_\beta \frac{\partial V_\beta}{\partial x^\alpha}$$

Since a stationary fluid, for which $V = 0$ everywhere, does not have any vorticity, any flow which starts from this stationary state is irrotational, according to this theorem.

When the vorticity is zero everywhere, one can express the velocity **V** by means of the velocity potential ϕ as

$$\mathbf{V} = -\nabla\phi \tag{3-83}$$

(See section 1-3.) If in addition, the fluid is incompressible, the equation of continuity (3-35) gives

$$\nabla \cdot \mathbf{V} = 0 \tag{3-84}$$

In this case, combining (3-83) and (3-84), we see that the velocity potential satisfies *Laplace's equation*

$$\nabla^2\phi = 0 \tag{3-85}$$

This equation will be discussed further in sections 5-4 and 5-5.

When the flow is irrotational, we see that

$$\nabla(V^2) = 2(\mathbf{V} \cdot \nabla)\mathbf{V} + 2\mathbf{V} \times (\nabla \times \mathbf{V}) = 2(\mathbf{V} \cdot \nabla)\mathbf{V} \tag{3-86}$$

using one of the formulas in Appendix I(a). If, in addition, the volume force density $\mathbf{F}_v$ is given by a potential

$$\mathbf{F}_v = -\nabla U \tag{3-87}$$

then Euler's equation (3-81) gives

$$\mu\frac{\partial \mathbf{V}}{\partial t} = -\nabla(\tfrac{1}{2}\mu V^2 + P + U) \tag{3-88}$$

When the flow is steady, i.e., $\partial V/\partial t = 0$, we obtain *Bernoulli's equation*

$$\tfrac{1}{2}\mu V^2 + P + U = \text{constant} \tag{3-89}$$

Readers may see that this is a form of the energy conservation law.

PROBLEM 3-17. Show that

$$\phi = -\phi_0 \left(r + \frac{a^3}{2r^2}\right) \cos\theta$$

is the velocity potential for an irrotational flow of the ideal incompressible fluid around a rigid sphere of radius a. Here ϕ_0 is a constant, r is the distance from the center of the sphere, and θ is the angle between the direction of the flow at infinity and the position vector $\mathbf{r}$. Notice that the boundary condition at the surface of the sphere is $\partial\phi/\partial r = 0$ (Fig. 3–7).

FIGURE 3–7. Flow around a sphere.

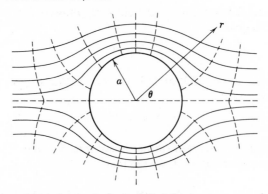

HINT. Show that it is a solution of (3–85).

PROBLEM 3–18. Calculate V for Problem 3–17 on the surface of the sphere, and show that no net force is exerted on the sphere by the flow. Assume that $U = 0$. This is called D'Alembert's paradox.

HINT. V is symmetric under the transformation $\theta \to \pi - \theta$, so that P is also symmetric under the same transformation according to (3–89).

PROBLEM 3–19. When the vorticity $\mathbf{w}$ is not zero, obtain

$$\mu \left(\frac{\partial \mathbf{V}}{\partial t} + \mathbf{w} \times \mathbf{V} \right) = -\nabla(\tfrac{1}{2}\mu V^2 + P + U)$$

3–9 VISCOUS FLOW

A real fluid has a nonzero shearing stress because of the viscosity η;

$$T_{\alpha\beta} = \eta \left(\frac{\partial V_\alpha}{\partial x^\beta} + \frac{\partial V_\beta}{\partial x^\alpha} \right) \tag{3–90}$$

The equation of motion is accordingly corrected to

$$\mu \left[\frac{\partial \mathbf{V}}{\partial t} + (\mathbf{V} \cdot \nabla)\mathbf{V} \right] = -\nabla P + \mathbf{F}_v + \eta[\nabla^2 \mathbf{V} + \nabla(\nabla \cdot \mathbf{V})] \tag{3–91}$$

This equation is called *Navier–Stokes' equation.*

FIGURE 3–8. Viscous flow in a cylinder.

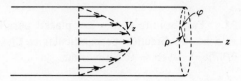

Let us consider a steady viscous flow inside a straight cylinder of radius a (Fig. 3–8). If $\mathbf{F}_v = 0$ and the flow is symmetric around the center axis (z axis) of the cylinder, Navier–Stokes' formula with $\partial \mathbf{V}/\partial t = 0$ and $\partial \mathbf{V}/\partial z = 0$ gives

$$\frac{\partial P}{\partial z} = \eta \nabla^2 V_z = \eta \left(\frac{d^2 V_z}{d\rho^2} + \frac{1}{\rho} \frac{d V_z}{d\rho} \right) \tag{3-92}$$

$$\frac{\partial P}{\partial \rho} = 0 \qquad \frac{\partial P}{\partial \varphi} = 0$$

using cylindrical coordinates (1–26). Assuming that

$$\frac{\partial P}{\partial z} = k \tag{3-93}$$

is a constant, we obtain from (3–92)

$$\frac{d}{d\rho} \left(\rho \frac{d V_z}{d\rho} \right) = \frac{k}{\eta} \rho \tag{3-94}$$

If the boundary condition at the wall of the cylinder is

$$V_z = 0 \qquad \text{at} \qquad \rho = a \tag{3-95}$$

the solution of (3–94) is

$$V_z = -\frac{k}{4\eta} (a^2 - \rho^2) \tag{3-96}$$

The amount of flow which passes through a cross section of the tube per unit time is

$$Q = 2\pi \int_0^a \rho V_z \, d\rho = \frac{\pi a^4 k}{8\eta} \tag{3-97}$$

This is called *Poiseuille's law*, and gives a way to measure η.

PROBLEM 3–20. Obtain (3–91).

PROBLEM 3–21. Two infinite planes are placed parallel to each other at distance $2d$. Let z be the direction perpendicular to these planes, and let $\partial P/\partial x = k$, but $\partial P/\partial y = \partial P/\partial z = 0$. Show that

$$V_x = - \frac{k}{2\eta} (d^2 - z^2)$$

3–10 LAGRANGIAN DENSITY

The potential energy density of an elastic body was introduced in (3–51) and (3–55). When the displacement $\mathbf{u}$ remains small in the elastic body, the kinetic energy density can be defined as $\frac{1}{2}\mu\dot{\mathbf{u}}^2$. The Lagrangian density $\mathscr{L}$, defined such that the Lagrangian of the elastic body L is written as

$$L = \int \mathscr{L} \, dv \qquad (3\text{–}98)$$

can be introduced as

$$\mathscr{L} = \tfrac{1}{2}\mu\dot{\mathbf{u}}^2 - \mathscr{U} \qquad (3\text{–}99)$$

We will be discussing many continuous media other than the elastic body in the future, and it will be useful to generalize the concept of the Lagrangian density in discussing these continuous media. In our discussion on Hamilton's principle in section 1–10, we assumed

$$L = L(\mathbf{r}, \dot{\mathbf{r}}) \qquad (3\text{–}100)$$

that is, the Lagrangian function does not depend on higher time derivatives of $\mathbf{r}$ than $\dot{\mathbf{r}}$. In going from the Lagrangian function to the Lagrangian density, we should remember that in the continuous body we have four parameters t, x, y, and z, instead of just one t. The dynamical variables, which were $\mathbf{r}$ for the case of a particle, are $\mathbf{u}$ for the continuous body, and we need three additional parameters x, y, and z, to specify the location of the infinitesimal volume described by the Lagrangian density. With this modification in our minds, we may assume, in the same spirit as (3–100), that

$$\mathscr{L} = \mathscr{L}\left(\mathbf{u}, \frac{\partial \mathbf{u}}{\partial x}, \frac{\partial \mathbf{u}}{\partial y}, \frac{\partial \mathbf{u}}{\partial z}, \dot{\mathbf{u}}\right) \qquad (3\text{–}101)$$

i.e., that the Lagrangian density depends on **u** and the first derivatives of **u** with respect to x, y, z, and t, but not on any higher-order derivatives.

Let us apply Hamilton's principle to obtain an equation of motion for the Lagrangian density. As in section 1–10 we require

$$\delta \int_{t_1}^{t_2} L \, dt = \delta \int_{t_1}^{t_2} \int_{z_1}^{z_2} \int_{y_1}^{y_2} \int_{x_1}^{x_2} \mathscr{L} \, dx \, dy \, dz \, dt = 0 \qquad (3\text{--}102)$$

under the conditions that

$$\delta t = 0 \qquad (3\text{--}103a)$$

and

$$\delta \mathbf{u} = 0 \qquad \text{at} \qquad t = t_1 \qquad \text{and} \qquad t = t_2 \qquad (3\text{--}103b)$$

Since x, y, and z are parameters, like t, the first condition $\delta t = 0$ should now be generalized as

$$\delta t = \delta x = \delta y = \delta z = 0 \qquad (3\text{--}104a)$$

Also if the boundary of our continuous body is fixed, we can add the boundary conditions

$$\delta \mathbf{u} = 0 \qquad \text{at} \qquad x = x_1, \qquad x = x_2, \qquad y = y_1, \qquad y = y_2,$$
$$z = z_1 \qquad \text{and} \qquad z = z_2 \qquad (3\text{--}104b)$$

We then see that

$$\delta \iiiint \mathscr{L} \, dv \, dt$$

$$= \iiiint \delta \mathscr{L} \left(\mathbf{u}, \frac{\partial \mathbf{u}}{\partial x}, \frac{\partial \mathbf{u}}{\partial y}, \frac{\partial \mathbf{u}}{\partial z}, \frac{\partial \mathbf{u}}{\partial t} \right) dv \, dt$$

$$= \iiiint \left[\frac{\partial \mathscr{L}}{\partial \mathbf{u}} \delta \mathbf{u} + \frac{\partial \mathscr{L}}{\partial (\partial \mathbf{u}/\partial x)} \delta \frac{\partial \mathbf{u}}{\partial x} + \frac{\partial \mathscr{L}}{\partial (\partial \mathbf{u}/\partial y)} \delta \frac{\partial \mathbf{u}}{\partial y} \right.$$

$$\left. + \frac{\partial \mathscr{L}}{\partial (\partial \mathbf{u}/\partial z)} \delta \frac{\partial \mathbf{u}}{\partial z} + \frac{\partial \mathscr{L}}{\partial (\partial \mathbf{u}/\partial t)} \delta \frac{\partial \mathbf{u}}{\partial t} \right] dv \, dt$$

$$= \iiiint \left[\frac{\partial \mathscr{L}}{\partial \mathbf{u}} - \frac{\partial}{\partial x} \frac{\partial \mathscr{L}}{\partial (\partial \mathbf{u}/\partial x)} - \frac{\partial}{\partial y} \frac{\partial \mathscr{L}}{\partial (\partial \mathbf{u}/\partial y)} - \frac{\partial}{\partial z} \frac{\partial \mathscr{L}}{\partial (\partial \mathbf{u}/\partial z)} \right.$$

$$\left. - \frac{\partial}{\partial t} \frac{\partial \mathscr{L}}{\partial (\partial \mathbf{u}/\partial t)} \right] \delta \mathbf{u} \, dv \, dt \qquad (3\text{--}105)$$

We used (3–103b), (3–104a) and (3–104b) in the integration by parts in the above calculation. We thus obtain *Lagrange–Euler's equation*:

$$\frac{\partial \mathscr{L}}{\partial u} - \frac{\partial}{\partial x}\frac{\partial \mathscr{L}}{\partial(\partial u/\partial x)} - \frac{\partial}{\partial y}\frac{\partial \mathscr{L}}{\partial(\partial u/\partial y)} - \frac{\partial}{\partial z}\frac{\partial \mathscr{L}}{\partial(\partial u/\partial z)} - \frac{\partial}{\partial t}\frac{\partial \mathscr{L}}{\partial(\partial u/\partial t)} = 0$$

$$(3\text{–}106)$$

which corresponds to Lagrange's equation (1–36).

PROBLEM 3–22. Show that if

$$\mathscr{L} = \tfrac{1}{2}\mu \left(\frac{\partial u}{\partial t}\right)^2 - \tfrac{1}{2}T \left(\frac{\partial u}{\partial x}\right)^2$$

Lagrange–Euler's equation gives the equation of motion of a string (3–3).

PROBLEM 3–23. For the transverse displacement $u(x, y, t)$ of a membrane, we may assume that

$$\mathscr{L} = \tfrac{1}{2}\mu \left(\frac{\partial u}{\partial t}\right)^2 - \tfrac{1}{2}T \left(\frac{\partial u}{\partial x}\right)^2 - \tfrac{1}{2}T' \left(\frac{\partial u}{\partial y}\right)^2$$

Obtain the equation of motion for u, and show that the solution can be written as

$$u(x, y, t) = u_1(x, t)u_2(y, t)$$

Hamiltonian Density

In the present formalism we have infinitely many dynamic variables, since $\mathbf{u}(\mathbf{r})$ at any one point $\mathbf{r}$ is independent from $\mathbf{u}(\mathbf{r}')$ at any other point, and there are infinitely many such points in the continuous space. The Hamiltonian formalism developed in the first chapter, however, can be generalized to the present case.

The momentum canonically conjugate to $\mathbf{u}(\mathbf{r})$ is

$$\pi'(\mathbf{r}) = \frac{\partial L}{\partial \dot{\mathbf{u}}(\mathbf{r})} = \frac{\partial \mathscr{L}(\mathbf{r})}{\partial \dot{\mathbf{u}}(\mathbf{r})}\, dv \qquad (3\text{–}107)$$

It is more convenient to introduce the canonical momentum density defined by

$$\pi = \frac{\pi'}{dv} \qquad (3\text{–}108)$$

The Hamiltonian function H of the continuous medium is

$$H = \sum \boldsymbol{\pi}' \cdot \dot{\mathbf{u}} - L = \int (\boldsymbol{\pi} \cdot \dot{\mathbf{u}} - \mathscr{L}) \, dv \qquad (3\text{-}109)$$

from which we can introduce the Hamiltonian density

$$\mathscr{H} = \boldsymbol{\pi} \cdot \dot{\mathbf{u}} - \mathscr{L} \qquad (3\text{-}110)$$

If the kinetic energy is quadratic in $\dot{\mathbf{u}}$, $\mathscr{H}$ is the energy density.

PROBLEM 3–24. Starting from the Lagrangian density of Problem 3–22, obtain π and $\mathscr{H}$ for the vibrating string.

ANSWER.

$$\pi = \mu\dot{u} \qquad \mathscr{H} = \left(\frac{\pi^2}{2\mu}\right) + \tfrac{1}{2}T\left(\frac{\partial u}{\partial x}\right)^2$$

PROBLEM 3–25. Since $\mathbf{u(r)}$ and $\boldsymbol{\pi}'\mathbf{(r)}$ are canonically conjugate, Hamilton's equations should be

$$\frac{\partial H}{\partial \boldsymbol{\pi}'\mathbf{(r)}} = \dot{\mathbf{u}}\mathbf{(r)}$$

$$\frac{\partial H}{\partial \mathbf{u(r)}} = -\dot{\boldsymbol{\pi}}'\mathbf{(r)}$$

Applying these formulas for the one-dimensional case where the Hamiltonian density is given by the result of Problem 3–24, and assuming the boundary condition that u is zero at both boundaries, obtain the original equation of motion (3–3).

ANSWER.

$$\frac{\partial H}{\partial u} = \tfrac{1}{2}T\frac{\partial}{\partial u}\int\left(\frac{\partial u'}{\partial x'}\right)^2 dx' = 2T\int\left(\frac{\partial}{\partial x'}\frac{\partial u'}{\partial u}\right)\frac{\partial u'}{\partial x'}\,dx'$$

$$= 2T\int\left(\frac{\partial}{\partial x'}\,\delta(x - x')\right)\frac{\partial u'}{\partial x'}\,dx' = -T\frac{\partial^2 u}{\partial x^2}\,dx$$

$$\dot{\pi}' = \dot{\pi}\,dx = \mu\ddot{u}\,dx$$

Poisson Bracket Density

If dynamical quantities F and G can be expressed by their densities as

$$F = \int \mathscr{F} \, dv \quad \text{and} \quad G = \int \mathscr{G} \, dv \qquad (3\text{-}111)$$

and if $\mathscr{F}(\mathbf{r})$ and $\mathscr{G}(\mathbf{r})$ are both functions of u and π at the same point in the space, then

$$[F, G]_P = \int \int \left[\frac{\partial \mathscr{F}(\mathbf{r})}{\partial u(\mathbf{r})} \frac{\partial \mathscr{G}(\mathbf{r}')}{\partial \pi'(\mathbf{r})} - \frac{\partial \mathscr{F}(\mathbf{r})}{\partial \pi'(\mathbf{r})} \frac{\partial \mathscr{G}(\mathbf{r}')}{\partial u(\mathbf{r})} \right] dv \, dv'$$

$$= \int \left[\frac{\partial \mathscr{F}(\mathbf{r})}{\partial u(\mathbf{r})} \frac{\partial \mathscr{G}(\mathbf{r})}{\partial \pi(\mathbf{r})} - \frac{\partial \mathscr{F}(\mathbf{r})}{\partial \pi(\mathbf{r})} \frac{\partial \mathscr{G}(\mathbf{r})}{\partial u(\mathbf{r})} \right] dv \qquad (3\text{-}112)$$

Thus

$$\frac{\partial \mathscr{F}}{\partial u} \frac{\partial \mathscr{G}}{\partial \pi} - \frac{\partial \mathscr{F}}{\partial \pi} \frac{\partial \mathscr{G}}{\partial u} \equiv [\mathscr{F}, \mathscr{G}]_P \qquad (3\text{-}113)$$

can be regarded as the Poisson bracket density. In this way we obtain

$$\frac{d\mathscr{F}}{dt} = [\mathscr{F}, \mathscr{H}]_P + \frac{\partial \mathscr{F}}{\partial t} \qquad (3\text{-}114)$$

corresponding to (1–62).

PROBLEM 3–26. Obtain (3–114) from (1–62).

PROBLEM 3–27. In the one-dimensional string discussed in Problem 3–24, show that (3–114) gives

$$\dot{u} = \frac{\pi}{\mu} \quad \text{and} \quad \dot{\pi}(= \mu \ddot{u}) = \frac{T \partial^2 u}{\partial x^2}$$

RIGID BODY AND ROTATING FRAME

3–11 KINETIC ENERGY OF A RIGID BODY. INERTIA TENSOR

We have seen that coupled simple harmonic oscillators give a typical case where the many-body problem can be solved exactly. Another solvable example of a many-body problem is a rigid body. A rigid body is made of many particles (atoms), but the force between these particles is so strong that the distances between them do not change. The only motion possible is motion of the rigid body as a whole, its translational and rotational motion.

Let us designate the center of mass of a rigid body as 0, whose position in space is $\mathbf{R}$. If $d\phi$ is an infinitesimal rotation of the rigid body around 0, the infinitesimal displacement of the portion of the rigid body at $\mathbf{r}$ from the center of mass 0 due to this rotation is $d\phi \times \mathbf{r}$. The total displacement of the portion with respect to the space is due to the displacement of the center of mass and the rotation around the center of mass, or $d\mathbf{R} + d\phi \times \mathbf{r}$ as seen in Fig. 3–9. Thus the velocity of this portion in space is

$$\mathbf{v} = \mathbf{V} + \mathbf{\Omega} \times \mathbf{r} \qquad (3\text{–}115)$$

where $\mathbf{V}(= d\mathbf{R}/dt)$ is the velocity of the center of mass, and $\mathbf{\Omega} = d\phi/dt$ is the angular velocity of the rigid body, common to all portions.

FIGURE 3–9.   Coordinates of a rigid body.

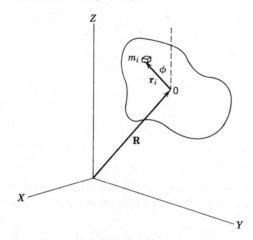

The total kinetic energy of the rigid body is thus

$$T = \sum_i \tfrac{1}{2} m_i (\mathbf{V} + \mathbf{\Omega} \times \mathbf{r}_i)^2 = \sum_i \tfrac{1}{2} m_i V^2 + \sum_i m_i \mathbf{V} \cdot (\mathbf{\Omega} \times \mathbf{r}_i)$$

$$+ \sum_i \tfrac{1}{2} m_i (\mathbf{\Omega} \times \mathbf{r}_i)^2 \qquad (3\text{–}116)$$

The first terms are the kinetic energy of the translational motion of the center of mass, and are equal to that of a particle with total mass $\sum_i m_i$ moving with velocity $\mathbf{V}$. Since

$$\sum_i m_i \mathbf{V} \cdot (\mathbf{\Omega} \times \mathbf{r}_i) = \mathbf{V} \cdot \left(\mathbf{\Omega} \times \left(\sum_i m_i \mathbf{r}_i \right) \right) \qquad (3\text{–}117)$$

and $\sum_i m_i \mathbf{r}_i = 0$ by the definition of the center of mass, the second terms in (3–116) disappear. The last terms in (3–116) give the kinetic energy of the rotation of the rigid body T_{rot}. Using an easily proved identity we obtain

$$T_{\text{rot},} = \sum_i \tfrac{1}{2} m_i (\boldsymbol{\Omega} \times \mathbf{r}_i)^2 = \tfrac{1}{2} \sum_i m_i [\Omega^2 r_i^2 - (\boldsymbol{\Omega} \cdot \mathbf{r}_i)^2] \qquad (3\text{–}118)$$

It is convenient to introduce the body-fixed Cartesian coordinates ξ, η, and ζ fixed to the rigid body (Fig. 3–10). If the origin of such a coordinate system is 0, it rotates with angular velocity $\boldsymbol{\Omega}$ in space. The components of $\mathbf{r}_i$ with respect to this coordinate system, ξ_i, η_i, and ζ_i are constants because of the rigid-body assumption.

FIGURE 3–10.   Body-fixed axes ξ, η, and ζ.

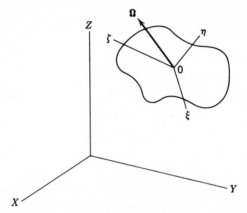

Using the components in this body-fixed coordinate system, the rotational part of the kinetic energy (3–118) can be written as

$$T_{\text{rot}} = \tfrac{1}{2}(I_{\xi\xi}\Omega_\xi^2 + I_{\eta\eta}\Omega_\eta^2 + I_{\zeta\zeta}\Omega_\zeta^2 + 2I_{\xi\eta}\Omega_\xi\Omega_\eta + 2I_{\eta\zeta}\Omega_\eta\Omega_\zeta + 2I_{\zeta\xi}\Omega_\zeta\Omega_\xi)$$
$$(3\text{–}119)$$

where

$$I_{\xi\xi} = \sum_i m_i(\eta_i^2 + \zeta_i^2) \qquad I_{\eta\eta} = \sum_i m_i(\zeta_i^2 + \xi_i^2)$$

$$I_{\zeta\zeta} = \sum_i m_i(\eta_i^2 + \xi_i^2) \qquad I_{\xi\eta} = -\sum_i m_i \xi_i \eta_i \qquad (3\text{–}120)$$

$$I_{\eta\zeta} = -\sum_i m_i \eta_i \zeta_i \qquad I_{\zeta\xi} = -\sum_i m_i \zeta_i \xi_i$$

Quantities in (3–120) are components of a second rank tensor called the inertia tensor.

We can express T_{rot} by matrices. If we define

$$\mathbf{\Omega} = \begin{pmatrix} \Omega_\xi \\ \Omega_\eta \\ \Omega_\zeta \end{pmatrix} \tag{3-121}$$

and

$$\mathbf{I} = \begin{pmatrix} I_{\xi\xi} & I_{\xi\eta} & I_{\zeta\xi} \\ I_{\xi\eta} & I_{\eta\eta} & I_{\eta\zeta} \\ I_{\zeta\xi} & I_{\eta\zeta} & I_{\zeta\zeta} \end{pmatrix} \tag{3-122}$$

we see

$$T_{\text{rot}} = \tfrac{1}{2}\mathbf{\Omega}^T\mathbf{I}\mathbf{\Omega} \tag{3-123}$$

From the similarity between (3–123) and (2–83) we can see immediately that by applying an orthogonal transformation we can express T_{rot} as

$$T_{\text{rot}} = \tfrac{1}{2}(I_a\Omega_a{}^2 + I_b\Omega_b{}^2 + I_c\Omega_c{}^2) \tag{3-124}$$

where I_a, I_b, and I_c are solutions of

$$\begin{vmatrix} I_{\xi\xi} - I & I_{\xi\eta} & I_{\zeta\xi} \\ I_{\eta\xi} & I_{\eta\eta} - I & I_{\eta\zeta} \\ I_{\zeta\xi} & I_{\zeta\eta} & I_{\zeta\zeta} - I \end{vmatrix} = 0 \tag{3-125}$$

I_a, I_b, and I_c are called the principal moments of inertia of the rigid body. The axes, a, b, and c, which are obtained from the original axes, ξ, η, and ζ by the orthogonal transformation, are called the *principal axes of inertia*.

A continuous set of points given by sets of values (ξ, η, ζ), which satisfy

$$1 = I_{\xi\xi}\xi^2 + I_{\eta\eta}\eta^2 + I_{\zeta\zeta}\zeta^2 + 2I_{\xi\eta}\xi\eta + 2I_{\eta\zeta}\eta\zeta + 2I_{\zeta\xi}\zeta\xi$$

$$\tag{3-126}$$

give the surface of an ellipsoid called the *inertia ellipsoid* which is illustrated in Fig. 3–11. The ellipsoid is fixed to the rigid body, and represents the rotational motion of a rigid body. For the principal axes of inertia the corresponding inertia tensor is diagonal. Thus the principal axes of inertia are actually the principal axes of the inertia ellipsoid.

PROBLEM 3–28. Obtain the final expression of (3–118) and (3–119).

FIGURE 3–11. The inertia ellipsoid. The lengths of major axes are $2/\sqrt{I_a}$, $2/\sqrt{I_b}$, and $2/\sqrt{I_c}$, respectively.

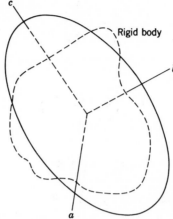

PROBLEM 3–29. Show that for a very thin planer disk, one of the principal axes (a axis) of inertia is perpendicular to the disk, and that $I_a = I_b + I_c$.

PROBLEM 3–30. When the very thin planer disk has the form of an equilateral triangle, show that any axis in the plane of the disk is a principal axis of inertia as far as it passes through the center of the triangle.

3–12 ANGULAR MOMENTUM OF A RIGID BODY

The angular momentum of the ith particle around the center of mass 0, is $m_i \mathbf{r}_i \times \mathbf{v}_i$, but $\mathbf{v}_i = \boldsymbol{\Omega} \times \mathbf{r}_i$ using the angular velocity $\boldsymbol{\Omega}$. Thus the total angular momentum of the rigid body is

$$\mathbf{L} = \sum_i m_i \mathbf{r}_i \times (\boldsymbol{\Omega} \times \mathbf{r}_i) = \sum_i m_i [r_i^2 \boldsymbol{\Omega} - \mathbf{r}_i (\mathbf{r}_i \cdot \boldsymbol{\Omega})] \qquad (3\text{–}127)$$

For arbitrary body-fixed axes we obtain from (3–120) that

$$L_\xi = I_{\xi\xi}\Omega_\xi + I_{\xi\eta}\Omega_\eta + I_{\zeta\xi}\Omega_\zeta$$
$$L_\eta = I_{\eta\eta}\Omega_\eta + I_{\eta\zeta}\Omega_\zeta + I_{\xi\eta}\Omega_\xi \qquad (3\text{–}128)$$
$$L_\zeta = I_{\zeta\zeta}\Omega_\zeta + I_{\zeta\xi}\Omega_\xi + I_{\eta\zeta}\Omega_\eta$$

or using the matrix notations

$$\mathbf{L} = \mathbf{I}\boldsymbol{\Omega} \qquad (3\text{–}129)$$

If we take the principal axes of inertia for the body fixed axes, we see (3–128) reduce to

$$L_a = I_a\Omega_a \qquad L_b = I_b\Omega_b \qquad L_c = I_c\Omega_c \qquad (3\text{–}130)$$

Comparing this result with (3–124) we obtain

$$T_{\text{rot}} = \frac{1}{2}\left(\frac{L_a^{\,2}}{I_a} + \frac{L_b^{\,2}}{I_b} + \frac{L_c^{\,2}}{I_c}\right) \qquad (3\text{–}131)$$

PROBLEM 3–31. When two of the principal moments of inertia are equal to each other, for example $I_a = I_b$, the rotator is called a symmetric top rotator. Show that

$$T_{\text{rot}} = \frac{L^2}{2I_a} + \frac{1}{2}\left(\frac{1}{I_c} - \frac{1}{I_a}\right)L_c^{\,2}$$

in such symmetric top rotator. Here

$$L^2 = L_a^{\,2} + L_b^{\,2} + L_c^{\,2}$$

3–13   FREE ROTATION OF A RIGID BODY

When the rigid body is free of any external forces the above rotational energy (3–131) is conserved. The total angular momentum **L** is also conserved when the rigid body is free, as the arguments in section 1–11 clearly show. As seen from (3–130) $\boldsymbol{\Omega}$ and **L** are not parallel to each other in general. The vector $\boldsymbol{\Omega}$ itself is, therefore, not conserved in general, but the component of $\boldsymbol{\Omega}$ along **L** is conserved, because

$$\frac{\boldsymbol{\Omega}\cdot\mathbf{L}}{|\mathbf{L}|} = \frac{2T_{\text{rot}}}{|\mathbf{L}|} \qquad (3\text{–}132)$$

An ingenious way to visualize the free rotational motion of a rigid body was given by Poinsot (Fig. 3–12). We represent the rigid body by the inertia ellipsoid whose surface is given by equation (3–126). The point on this surface through which the vector $\boldsymbol{\Omega}$ passes is given by $\rho\boldsymbol{\Omega}/|\boldsymbol{\Omega}|$, where ρ is the distance of that point from the center of mass 0. From (3–126), in the principal axes representation, we see ρ should satisfy

$$1 = \rho^2 \frac{I_a\Omega_a^{\,2} + I_b\Omega_b^{\,2} + I_c\Omega_c^{\,2}}{|\boldsymbol{\Omega}|^2}$$

$$= \rho^2 \frac{2T_{\text{rot}}}{|\boldsymbol{\Omega}|^2} \qquad (3\text{–}133)$$

FIGURE 3–12. Poinsot's representation of free rotational motion of a rigid body.

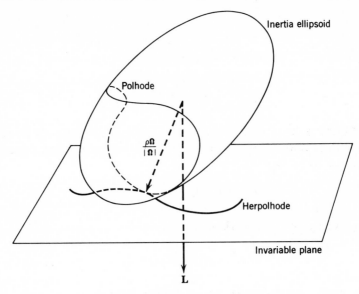

or

$$\rho = \frac{|\Omega|}{\sqrt{2T_{\text{rot}}}} \tag{3-134}$$

The component of the vector $\rho\Omega/|\Omega|$ along $\mathbf{L}$ is thus

$$\rho\frac{\Omega \cdot \mathbf{L}}{|\Omega|\,|\mathbf{L}|} = \frac{\sqrt{2T_{\text{rot}}}}{|\mathbf{L}|} \tag{3-135}$$

which is another constant of motion.

An amazing fact is that $\mathbf{L}$ is perpendicular to the surface of the inertia ellipsoid at the point $\rho\Omega/|\Omega|$. In order to see this we first let

$$f = I_{\xi\xi}\xi^2 + I_{\eta\eta}\eta^2 + I_{\zeta\zeta}\zeta^2 + 2I_{\xi\eta}\xi\eta + 2I_{\eta\zeta}\eta\zeta + 2I_{\zeta\xi}\zeta\xi \tag{3-136}$$

instead of (3–126). Then a vector perpendicular to the surface of the inertia ellipsoid is given by ∇f calculated at $f = 1$. Using the rotating coordinate and putting $\xi = \rho\Omega_\xi/|\mathbf{\Omega}|$, $\eta = \rho\Omega_\eta/|\mathbf{\Omega}|$, and $\zeta = \rho\Omega_\zeta/|\mathbf{\Omega}|$, the components of the perpendicular vector ∇f are

$$\left.\frac{\partial f}{\partial \xi}\right|_{f=1} = 2(I_{\xi\xi}\xi + I_{\xi\eta}\eta + I_{\zeta\xi}\zeta) = 2\rho\,\frac{I_{\xi\xi}\Omega_\xi + I_{\xi\eta}\Omega_\eta + I_{\zeta\xi}\Omega_\zeta}{|\mathbf{\Omega}|}$$

$$= \frac{2\rho L_\xi}{|\mathbf{\Omega}|} = \frac{2L_\xi}{\sqrt{2T_{\text{rot}}}}$$

$$\left.\frac{\partial f}{\partial \eta}\right|_{f=1} = \frac{2L_\eta}{\sqrt{2T_{\text{rot}}}} \qquad \left.\frac{\partial f}{\partial \zeta}\right|_{f=1} = \frac{2L_\zeta}{\sqrt{2T_{\text{rot}}}} \qquad\qquad (3\text{--}137)$$

where (3–128) and (3–134) are used. Thus the above statement that $\mathbf{L}$ is perpendicular to the inertia ellipsoid at $\rho\mathbf{\Omega}/|\mathbf{\Omega}|$, is proved.

If we take a plane perpendicular to $\mathbf{L}$ which touches the inertia ellipsoid, the point where it touches the inertia ellipsoid is $\rho\mathbf{\Omega}/|\mathbf{\Omega}|$ according to the above statement. The distance from the center of the inertia ellipsoid to this plane is given by (3–135) and is conserved. Since $\mathbf{L}$ and the center of the inertia ellipsoid, the center of mass, are also conserved, the above plane is invariant during free rotational motion of the rigid body. The plane is called the invariable plane. Since the point at which the inertia ellipsoid touches this invariable plane is on the rotation vector $\mathbf{\Omega}$ around which the rigid body rotates, the inertia ellipsoid rolls on the invariable plane without slipping.

The trace of the contact point on the invariable plane by such nonslipping rolling motion of the inertia ellipsoid is called the *herpolhode*. It gives the precession of the rotation vector $\mathbf{\Omega}$ in space during free rotation. The trace of the contact point on the surface of the inertia ellipsoid is called the *polhode*, which gives the precession of $\mathbf{\Omega}$ in the body-fixed frame (Fig. 3–12).

PROBLEM 3–32. Obtain (3–132).

HINT. Use (3–124) and (3–130).

3–14 EULER'S EQUATIONS OF MOTION FOR A RIGID BODY

Let us consider the time derivative of the total angular momentum $\mathbf{L}$ of a system of many particles. Using Newton's equation (1–6) we obtain

$$\dot{\mathbf{L}} = \sum_i \frac{d}{dt}(\mathbf{r}_i \times \mathbf{p}_i) = \sum_i \{\dot{\mathbf{r}}_i \times \mathbf{p}_i) + (\mathbf{r}_i \times \dot{\mathbf{p}}_i)\} = \sum_i (\mathbf{r}_i \times \dot{\mathbf{p}}_i)$$

$$= \sum_i (\mathbf{r}_i \times \mathbf{F}_i) \equiv \mathbf{N} \qquad\qquad (3\text{--}138)$$

Here **N** is the total moment of force exerted on the system, and is called the *torque* when the system is a rigid body.

It is important to note that each component of the inertia tensor in the space-fixed coordinate system changes with time in general, but the component is invariant with respect to the body-fixed coordinate system for a rigid body. The body-fixed coordinates are, therefore, much more convenient than the space-fixed coordinates for expressing the equation of motion, $\dot{\mathbf{L}} = \mathbf{N}$, explicitly. In calculating the time derivative $\dot{\mathbf{L}}$ in the body-fixed coordinates, we observe that

$$\left(\frac{d}{dt}\right)_{\text{space}} = \left(\frac{d}{dt}\right)_{\text{body}} + \mathbf{\Omega} \times \qquad (3\text{-}139)$$

holds true as an operator equation which can be applied to any vector quantity. Thus if ξ, η, and ζ are body-fixed coordinates we obtain from (3–138) and (3–139)

$$\dot{\mathbf{L}}_\xi + \Omega_\eta L_\zeta - \Omega_\zeta L_\eta = N_\xi \qquad (3\text{-}140)$$

and so on. It is most convenient to take the principal axes of inertia as the body-fixed coordinates. In that case using (3–130), we see (3–140) can be written as

$$I_a \dot{\Omega}_a - \Omega_b \Omega_c (I_b - I_c) = N_a$$

and, in the same way,

$$I_b \dot{\Omega}_b - \Omega_c \Omega_a (I_c - I_a) = N_b$$
$$I_c \dot{\Omega}_c - \Omega_a \Omega_b (I_a - I_b) = N_c \qquad (3\text{-}141)$$

These are Euler's equations of motion, of a rigid body.

PROBLEM 3-33. When $\mathbf{N} = 0$, show from Euler's equations (3–141) that the rotational kinetic energy $I_a \Omega_a{}^2 + I_b \Omega_b{}^2 + I_c \Omega_c{}^2$ is a constant of motion.

PROBLEM 3-34. When $\mathbf{N} = 0$ and $I_b = I_c$, show from Euler's equations that

$$\dot{\Omega}_a = 0$$
$$\Omega_b = \Omega_0 \cos (kt + \delta)$$
$$\Omega_b{}^2 + \Omega_c{}^2 = \Omega_0{}^2$$

where Ω_0 is a constant and

$$k = (I_a - I_b) \frac{\Omega_a}{I_b}$$

Also give an intuitive explanation of these results using Poinsot's representation.

PROBLEM 3–35. When $I_b = I_c$ and $\mathbf{N}$ is along the a axis, show that $\Omega_b{}^2 + \Omega_c{}^2$ is a constant of motion.

3–15 EQUATIONS OF MOTION IN A ROTATING FRAME

As seen in (3–139) the time derivative of a vector is different depending on whether we measure it with respect to a space-fixed frame or a body-fixed frame if the body is rotating. An observer on the earth, for example, sees the velocity of a particle differently from that seen by another observer in outer space, since the earth is rotating with respect to outer space. Operating (3–139) on the position vector of the particle $\mathbf{r}$, we see

$$(\dot{\mathbf{r}})_{\text{space}} = (\dot{\mathbf{r}})_{\text{body}} + \mathbf{\Omega} \times \mathbf{r} \tag{3–142}$$

or

$$(\dot{\mathbf{R}} + \dot{\mathbf{r}})_{\text{space}} = (\dot{\mathbf{R}})_{\text{space}} + (\dot{\mathbf{r}})_{\text{body}} + \mathbf{\Omega} \times \mathbf{r} \tag{3–143}$$

If $\mathbf{R}$ is the distance from the origin of the space-fixed coordinate to that of the body-fixed coordinate, and if $(\dot{\mathbf{r}})_{\text{body}} = 0$, we obtain the previous equation (3–115).

The second time derivative of the position vector $\mathbf{R} + \mathbf{r}$ can be obtained by applying (3–140) twice. Since

$$\left(\frac{d^2}{dt^2}\right)_{\text{space}} = \left[\left(\frac{d}{dt}\right)_{\text{body}} + \mathbf{\Omega} \times\right]\left[\left(\frac{d}{dt}\right)_{\text{body}} + \mathbf{\Omega} \times\right]$$

$$= \left(\frac{d^2}{dt^2}\right)_{\text{body}} + \dot{\mathbf{\Omega}} \times + 2\mathbf{\Omega} \times \left(\frac{d}{dt}\right)_{\text{body}} + \mathbf{\Omega} \times (\mathbf{\Omega} \times) \tag{3–144}$$

we obtain

$$(\ddot{\mathbf{r}})_{\text{space}} = (\ddot{\mathbf{r}})_{\text{body}} + \dot{\mathbf{\Omega}} \times \mathbf{r} + 2\mathbf{\Omega} \times (\dot{\mathbf{r}})_{\text{body}} + \mathbf{\Omega} \times (\mathbf{\Omega} \times \mathbf{r})$$

$$\tag{3–145}$$

from which the acceleration $(\ddot{\mathbf{R}} + \ddot{\mathbf{r}})_{\text{space}}$ is obtained by just adding $(\ddot{\mathbf{R}})_{\text{space}}$, the acceleration of the moving frame with respect to space.

Our result (3–145) shows that the acceleration of a particle depends on whether the observer is rotating or not. Even when the particle is stationary in space, that is $(\ddot{\mathbf{r}})_{\text{space}} = 0$, it seems to be accelerated in a rotating frame.

If Newton's equations hold true in space as

$$m(\ddot{\mathbf{r}})_{\text{space}} = \mathbf{F} \qquad (3\text{--}146)$$

we have

$$m(\ddot{\mathbf{r}})_{\text{body}} = \mathbf{F} - m\,\dot{\boldsymbol{\Omega}} \times \mathbf{r} - 2m\boldsymbol{\Omega} \times (\dot{\mathbf{r}})_{\text{body}} - m\boldsymbol{\Omega} \times (\boldsymbol{\Omega} \times \mathbf{r})$$

$$(3\text{--}147)$$

for the observer on the rotating frame. The additional three terms which appear on the right-hand side of (3–147) are fictitious forces due to the acceleration of the reference frame. The last term $-m\boldsymbol{\Omega} \times (\boldsymbol{\Omega} \times \mathbf{r})$ is called the *centrifugal force* while the second term $-2m\boldsymbol{\Omega} \times (\dot{\mathbf{r}})_{\text{body}}$ is called the *Coriolis force*. Note that all such fictitious forces are proportional to m.

PROBLEM 3–36. A bullet is fired horizontally from a gun which is fixed on the equator of the earth. Show that in an idealized situation, the bullet hits the ground sooner when it is fired west than when it is fired east.

HINT. Consider the Coriolis force.

ADDITIONAL PROBLEMS

3–37. Show that the total energy of the vibrating string is $\frac{1}{4}L\mu \sum_n \omega_n^2 \times (A_n^2 + B_n^2)$, where A_n and B_n are the amplitude of the nth normal coordinate as defined by (3–14).

HINT. Use the orthogonality relation (3–19).

3–38. A planar circular membrane can vibrate in the direction perpendicular to its own plane. Assuming that the circumference is fixed and the displacement is small, obtain the normal coordinates.

3–39. Show that the energy flux of a homogeneous elastic string is

$$S = -T \frac{\partial u}{\partial t} \frac{\partial u}{\partial x}$$

where S is defined by the continuity equation

$$\frac{\partial}{\partial t} (\text{energy density}) + \frac{\partial}{\partial x} S = 0$$

This can be generalized to $\mathbf{S} = -T\dot{u}\nabla u$.

3–40. If $d\mathbf{a}$ is an infinitesimal surface area of an elastic body, an external force exerted on this surface can be written in terms of the reaction force

$$F_\alpha = T_{\alpha\beta} \, da_\beta$$

so that, if δu_α is the change in the displacement u_α on this surface, then the work done by this external force is

$$F_\alpha \, \delta u_\alpha = \delta u_\alpha T_{\alpha\beta} \, da_\beta$$

The total work done on the elastic body is obtained by integrating the above expression over the total surface of this body. Show that this total work done is equal to

$$\left(\oint \delta u_\alpha T_{\alpha\beta} \, da_\beta \right) = \int T_{\alpha\beta} \, \delta S_{\alpha\beta} \, dv$$

In this way $T_{\alpha\beta} \, \delta S_{\alpha\beta}$ can be interpreted as the change in the potential energy density of the elastic body $\delta \mathscr{U}$.

HINT. Use Gauss' theorem (3–36), and then (3–43) with $\mathbf{F} = 0$ for the bulk of the body.

3–41. Obtain (3–51) from the result of Problem 3–40.

3–42. Assuming that μ, P, and $\mathbf{V}$ of an ideal fluid do not deviate much from their equilibrium values μ_0, P_0, and 0, obtain the wave equation

$$\frac{\partial^2 \mu}{\partial t^2} = \left(\frac{\partial P}{\partial \mu} \right)_0 \nabla^2 \mu$$

where $(\partial P/\partial \mu)_0$ is $\partial P/\partial \mu$ at the equilibrium. Note that this is a longitudinal (acoustic) wave with velocity $\sqrt{(\partial P/\partial \mu)_0}$.

HINTS. From (3–35) $\partial \mu/\partial t \cong -\mu_0 \mathbf{V} \cdot \mathbf{V}$, and from (3–81) $\mu_0(\partial \mathbf{V}/\partial t) \cong -\nabla P$. Also $P - P_0 \cong (\partial P/\partial \mu)_0(\mu - \mu_0)$.

3–43. As a special case of Navier–Stokes' equation (3–91), Korteweg and deVries in 1895 proposed a nonlinear equation

$$\frac{\partial V}{\partial t} + \alpha V \frac{\partial V}{\partial x} + \delta^2 \frac{\partial^3 V}{\partial x^3} = 0$$

Show that

$$V = 3c\, \text{sech}^2 \left(\frac{x - ct}{\Delta} \right) \qquad \text{where} \qquad \Delta^2 = \frac{4\delta^2}{c}$$

is a solution of this equation (often called *KdV* equation), and that this solution represents a pulse of velocity c and width Δ. This pulse is called a soliton (solitary wave pulse).[4]

3–44. A simple harmonic oscillator

$$\ddot{\mathbf{r}} = -\omega_0^2 \mathbf{r}$$

is restricted to oscillate in a plane. This plane (xy plane) is fixed on a shaft which passes through the origin and makes an angle θ with the z axis. The shaft rotates with a circular frequency ω. Assuming that ω is so small that the centrifugal force is negligible, show that the equations of motion of the simple harmonic oscillator as given for an observer moving with the xy plane are

$$\ddot{x} = -\omega_0^2 x + 2\omega \cos \theta \dot{y}$$
$$\ddot{y} = -\omega_0^2 y - 2\omega \cos \theta \dot{x}$$

and that their approximate solution when $\omega_0 \gg \omega$ is

$$x = A \cos \omega_0 t \cos \omega' t, \qquad y = -A \cos \omega_0 t \sin \omega' t$$

where $\omega' = \omega \cos \theta$ (see Fig. 3–13).

---

[4] N. J. Zabusky, *Nonlinear Partial Differential Equations*, (edited by W. F. Ames), Academic Press, New York, 1967, p. 234. See also H. Ikezi *et al.*, *Phys. Rev. Letters*, **25**, 11 (1970) for recent experimental and theoretical works on solitons.

FIGURE 3–13.   The rotating frame of Problem 3–45.

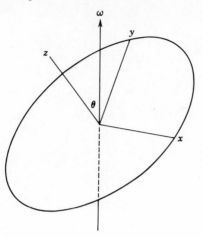

3–45. Show that

$$\tfrac{1}{2}mv^2 + V - \mathbf{l} \cdot \boldsymbol{\Omega} \qquad (\mathbf{l} = \mathbf{r} \times m\mathbf{v})$$

is conserved in a frame which is rotating with constant angular velocity $\boldsymbol{\Omega}$. This quantity is called the *Jacobi integral.*

Chapter 4   Special Theory of Relativity

Galileo Galilei first proposed the concept of relativity in *Risota* (Reply) which he distributed privately to defend Copernicanism. He said he had performed several experiments on board moving ships and concluded "... nor will you be able to determine whether the ship moves or stands still by events pertaining to your person." This statement was repeated in his famous book, *Dialogue Concerning Two Chief Systems of the World*, which was published in 1632. Newton must have agreed with Galileo on this point, but, from his theological standpoint, asserted the existence of an absolute space, one and only one coordinate system, in which his equation of motion held true. Bishop Berkeley criticized Newton in his book, *A Treatise Concerning the Principle of Human Knowledge*, which was published in 1710. He said, "..., it does not appear to me that there can be any motion other than relative; ... A man in a ship may be said to be quiescent with relation to the side of the vessel, and yet move with relation to the land." It is probable that Berkeley read Galileo.

It took almost 300 years, however, for physics to realize the deeper meaning of the proposition of Galileo. Albert A. Michelson, the first American physicist to receive the Nobel prize in 1907, and E. W. Morley showed that the velocity of light measurement cannot be used to determine an absolute space. Albert Einstein,[1] in 1905, abandoned the concept of absolute space completely, and pushed the concept of relativity of physical measurements to include the velocity of light measurement. He showed that this new viewpoint

---

[1] A. Einstein, *Ann. Physik* **17**, 891 (1905).

gives us a deeper understanding of such physical quantities as momentum, energy, and fields.

History would have been more logical if Einstein's theory was proposed to explain Michelson Morley's experiment, but the fact is[2] that Einstein did not know Michelson Morley's experiment when he published the famous paper.

[2] R. S. Shankland, *Amer. J. Phys.* **31**, 47 (1963).

4-1 ABSOLUTE SPACE. MACH'S PRINCIPLE

In a two-dimensional space we can take (1–41) for the kinetic energy and obtain the Lagrangian function as

$$L = \tfrac{1}{2}m(\dot{\rho}^2 + \rho^2\dot{\phi}^2) - U \tag{4-1}$$

Therefore, the equation of motion in the radial direction is

$$\frac{d}{dt}\frac{\partial L}{\partial \dot{\rho}} - \frac{\partial L}{\partial \rho} = m\ddot{\rho} - m\rho\dot{\phi}^2 + \frac{\partial U}{\partial \rho} = 0$$

or

$$m\ddot{\rho} = -\frac{\partial U}{\partial \rho} + m\rho\dot{\phi}^2 \tag{4-2}$$

An observer sitting on such a rotating frame would feel a fictitious force $m\rho\dot{\phi}^2$ pulling him towards increasing r. This is the centrifugal force which we obtained in (3–149). For a given force $-\partial U/\partial \rho$ on a given object with mass m, the observer in this frame will see that the acceleration $\ddot{\rho}$ is different at different positions r. Such an apparent deviation from Newton's equation can be attributed to the fictitious force (centrifugal force), and by correcting that, Newton's equation can be confirmed. Since the earth is rotating around its axis, an experiment in the laboratory should be corrected for this effect. After such correction one would still feel a deviation due to the rotation of the earth around the sun. That is still not the end of the story since the sun may be rotating around the center of our galaxy. The question is raised where we can find a laboratory such that Newton's equation holds true without any further corrections due to such fictitious forces. Newton assumed that an absolute space exists fixed among all stars or galaxies, in which his equation holds true without any corrections due to fictitious forces.

Such an absolute space should contain some physical entities besides the object one measures, because otherwise one could not establish a frame to measure the acceleration. The existence of stars or the universe is thus essential to an absolute space and Newton's equation. Mach proposed[3] that the mass of an object must somehow reflect the entire universe. This is called *Mach's principle*. Since our milky way is a flat disk this principle predicts that the mass of an object may be different when it is accelerated in that disk

---

[3] The statement here is a modern interpretation of Mach's intention in his discussions on the absolute space presented on and around p. 290 of his book *The Science of Mechanics*, 6th American Edition, Open Court Publishing Co., Lasalle, Ill., 1960.

and perpendicular to that disk. Experiments[4] so far, however, have not shown any such effect. It is an amazing feature of Newton's theory that the mass is a scalar and not a tensor, i.e., the direction of the acceleration and the force coincide exactly.

4–2 INERTIAL SYSTEM

Suppose Newton's equation holds in a coordinate system X, Y, Z. One can take another coordinate system X', Y', Z' which is moving with respect to the first one with a constant velocity V along the X axis. An object whose position along the X axis is x will have a position

$$x' = x - Vt \qquad (4\text{–}3)$$

in the second coordinate system. If V is constant, or $\dot{V} = 0$,

$$\ddot{x}' = \ddot{x} \qquad (4\text{–}4)$$

that is, the acceleration of the object is the same in both coordinate systems. Thus if Newton's equation holds true in the first system, it holds true in the second system with the same mass, if the force is independent of the velocity of the coordinate system.

A coordinate system in which Newton's equation holds true is called an *inertial system*. When XYZ is an inertial system, the $X'Y'Z'$ mentioned above is also an inertial system.

Suppose an absolute space exists, then another coordinate system moving with a constant velocity with respect to the absolute space will be equivalent as far as Newton's equation is concerned. Since such velocity can have any direction and magnitude, an infinite number of inertial systems exist. Only one of them is the absolute space. The question would then be how can we pick out the absolute space among infinitely many inertial systems. Apparently Newton's equation cannot serve for that purpose. One possible candidate is the velocity of light measurement.[5] Light propagates through empty space. One would expect that the velocity of light c is the same for all directions in the absolute space. In an inertial system which is moving with an appreciable

---

[4] V. W. Hughes, H. G. Robinson, and V. Beltran-Lopez, *Phys. Rev. Letters* **4**, 342 (1960).
[5] Galileo, who noticed the relativity between inertial systems, also tried to measure the velocity of light. In "First Day" of his book, *Dialogues of the Two New Sciences*, which was published in 1638, he stated: "Let each of two persons take a light contained in a lantern, or other receptacle, such that by the interposition of the hand, the one can shut off or admit the light to the vision of the other. Next let them stand opposite each other at a distance of a few cubits and practice until they acquire such skill in uncovering and occulting their light that the instant one sees the light of his companion he will uncover his own. After a few trials the response will be so prompt that without sensible error the uncovering

velocity V with respect to that, the velocity of light must be between $c - V$ and $c + V$ depending on the direction of the light propagation. The famous Michelson–Morley experiment tested this expectation. The result, however, showed that the velocity of light is the same for all directions in a laboratory fixed on the earth, which is apparently moving with respect to the absolute space.

Einstein proposed that the velocity of light is the same in all inertial systems. Thus, hope to determine an absolute space has been given up.

PROBLEM 4–1. If a particle has velocity **v** in the XYZ system, show that the same particle is seen to be moving with velocity **v′** such that

$$v'_{X'} = v_X - V$$

$$v'_{Y'} = v_Y \qquad v'_{Z'} = v_Z$$

This is called the Galileo transformation.

4–3 LORENTZ TRANSFORMATIONS[6]

The wave front of light which is emitted at the origin when $t = 0$ is given by

$$c^2 t^2 - x^2 - y^2 - z^2 = 0 \qquad (4\text{–}5)$$

---

of one light is immediately followed by the uncovering of the other, so that as soon as one exposes his light he will instantly see that of the other. Having acquired skill at this short distance let the two experimenters, equipped as before, take up positions separated by a distance of two or three miles and let them perform the same experiment at night, noting carefully whether the exposures and occultations occur in the same manner as at short distances; if they do, we may safely conclude that the propagation of light is instantaneous; but if time is required at a distance of three miles which, considering the going of one light and the coming of the other, really amounts to six, then the delay ought to be easily observable."

[6] H. A. Lorentz, *Proc. Acad. Sci. Amsterdam* **6**, 809 (1904). Einstein gave the same transformation in his 1905 paper without knowing Lorentz's paper. Some people like E. Whittaker (*A History of the Theories of Aether and Electricity*, Harper, New York, 1960, originally published by Thomas Nelson, London, 1910) gave the credit of the discovery of relativity to Lorentz and Poincaré rather than to Einstein. It is true that Poincaré enunciated the principle of relativity in 1899, 1900, and in 1904, but his talks were philosophical rather than physical, similar to the enunciation by Bishop Berkeley. Lorentz proposed the Lorentz transformation to explain Michelson Morley's experiment, and that transformation showed that the velocity of light is independent of the velocity of the observer with respect to an absolute space. Einstein, who was 25 years old and not even a bona fide physicist in 1905, started from the assumption of the constancy of the velocity of light and obtained, not only the Lorentz transformation, but many other revolutionary results as seen in this chapter. The relation between Lorentz and Einstein may be compared to that between Ptolemy and Copernicus. See G. Holton, *Amer. J. Phys.* **28**, 627 (1960) and *American Scholar* **37**, 59 (1967–68), for Einstein's background. Lorentz also served his country as the head of the Lorentz Commission which laid a plan of reclaiming the Zuiderzee polder, which, when completed in the year 2000, will increase the land area of the Netherlands by 10%.

in one inertial system. The same wave front will have different coordinates x', y', and z' in another inertial system which is moving with respect to the first one. The velocity of light, however, must also be c for the second system according to Einstein's proposition. Thus if these two systems coincide with each other at $t = 0$, time for the second system must advance differently than t in order to satisfy

$$c^2 t'^2 - x'^2 - y'^2 - z'^2 = 0 \tag{4-6}$$

Take a four-dimensional space made of t, x, y, and z, which is often called the Minkowski space (Fig. 4–1). Each point in this space is called an *event*. The wave front of the light we are talking about will give a cone around the t axis. This is called the *light cone*. For a given event we define s as

$$s^2 = c^2 t^2 - x^2 - y^2 - z^2 \tag{4-7}$$

If $s^2 = 0$, the event is on the light cone. Events such that

$$s^2 > 0 \tag{4-8}$$

are called *timelike*, and events such that

$$s^2 < 0 \tag{4-9}$$

are called *spacelike*.

FIGURE 4–1. *txyz* space (Minkowski space).

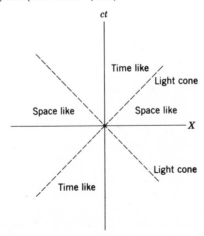

Formulas (4–5) and (4–6) give the transformation between two events on the light cone. We should find a similar transformation for general events. For that purpose we consider an event infinitesimally close to the origin. From (4–7) we have

$$(ds)^2 = (c\,dt)^2 - (dx)^2 - (dy)^2 - (dz)^2 \qquad (4\text{--}10)$$

For another inertial system which has the same origin as ours the same event will also have an infinitesimal quantity $(ds')^2$, which should be related to $(ds)^2$ as

$$(ds')^2 = \alpha(ds)^2 \qquad (4\text{--}11)$$

since when ds' is zero, ds must also be zero, according to (4–5) and (4–6).

The coefficient α may be a function of $\mathbf{V}$, the relative velocity of the second system with respect to the first one. However, since the space is isotropic we expect that α is a function of $|\mathbf{V}|$, the magnitude of the relative velocity, at most. Now the transformation from the second system back to the first one is the same transformation as the previous one except that the sign of the velocity is the opposite. But, as we just said, we expect the coefficient α to be independent of the sign of the velocity. Therefore we obtain

$$(ds)^2 = \alpha(ds')^2 = \alpha^2(ds)^2 \qquad (4\text{--}12)$$

from which we see that

$$\alpha = \pm 1 \qquad (4\text{--}13)$$

However, α should be a continuous function of V. Therefore it should be either 1 or -1, not both. It should actually be 1 because when $V \to 0$, $ds' \to ds$, or in this limit $\alpha = +1$. We thus have

$$(ds')^2 = (ds)^2 \qquad (4\text{--}14)$$

or integrating

$$s'^2 = s^2 \qquad (4\text{--}15)$$

4–4 COVARIANT AND CONTRAVARIANT COMPONENTS

In (4–7) we see ct, x, y, and z appear in an equivalent form. It will be convenient to define a vector **x** whose components are

$$x^0 = ct, \qquad x^1 = x, \qquad x^2 = y, \qquad \text{and} \qquad x^3 = z \qquad (4\text{--}16)$$

We see that (4–7) can be written as

$$s^2 = \sum_i \sum_j x^i g_{ij} x^j \qquad (4\text{–}17)$$

if

$$g_{00} = 1, \qquad g_{11} = g_{22} = g_{33} = -1, \quad \text{and} \quad g_{ij} = 0 \quad \text{if} \quad i \neq j \qquad (4\text{–}18)$$

This 4×4 tensor is called the *metric tensor*. In special relativity the metric tensor is a constant tensor and is purely a mathematical quantity. However, it will have very important physical meanings in general relativity, which will be discussed in Chapter 7.

In the following we use the convention that when an index is repeated that index is to be summed up over all four components.[7] Thus (4–17) can be simply written as

$$s^2 = x^i g_{ij} x^j \qquad (4\text{–}19)$$

When the index is given as a superscript like x^i, we call it a *contravariant component*. It will be convenient to introduce the corresponding *covariant components* which are defined as

$$x_i = g_{ij} x^j \left(= \sum_{j=0}^{3} g_{ij} x^j \right) \qquad (4\text{–}20)$$

Thus

$$x_0 = ct, \qquad x_1 = -x, \qquad x_2 = -y, \quad \text{and} \quad x_3 = -z \qquad (4\text{–}21)$$

and

$$s^2 = x^i x_i = x_i x^i \qquad (4\text{–}22)$$

The covariant components in special relativity are so simply related to the contravariant components that they do not seem to serve any purpose except to introduce unnecessary complications. That is true as long as we stay in special relativity. However, when we go to general relativity in Chapter 7 we shall find that the notion of covariant components helps simplify formulas a

[7] Greek letters are used for indices to be summed up over three space components (see section 3–5), and Latin letters are used for indices to be summed up over four components.

great deal. Actually we will find that many of the formulas we are going to present in this chapter in terms of contravariant and covariant components are valid even in general relativity only if definition formula (4–20) is given in terms of general metric tensor g_{ij}.

4–5 LORENTZ TRANSFORMATION MATRICES

The transformation from one inertial system to another can be expressed by a 4 × 4 matrix **L** called a *Lorentz transformation matrix*

$$\mathbf{Lx} = \mathbf{x}' \tag{4–23}$$

where **x** and **x**' are vectors defined by (4–16). If **g** is a matrix defined as

$$\mathbf{g} = \begin{pmatrix} 1 & 0 & 0 & 0 \\ 0 & -1 & 0 & 0 \\ 0 & 0 & -1 & 0 \\ 0 & 0 & 0 & -1 \end{pmatrix} \tag{4–24}$$

(4–15) and (4–17) can be expressed as

$$\mathbf{x}'^T \mathbf{g} \mathbf{x}' = \mathbf{x}^T \mathbf{L}^T \mathbf{g} \mathbf{L} \mathbf{x} = \mathbf{x}^T \mathbf{g} \mathbf{x} \tag{4–25}$$

which shows that **L** should satisfy

$$L^T \mathbf{g} \mathbf{L} = \mathbf{g} \tag{4–26}$$

The corresponding determinant equation gives

$$\{\det \mathbf{L}\}^2 = 1 \tag{4–27}$$

or

$$\det \mathbf{L} = \pm 1 \tag{4–28}$$

The Lorentz transformations with det **L** = 1 and det **L** = −1 are called the *proper* and *improper* Lorentz transformations, respectively. The *time reversal* where

$$t' = -t, \quad x' = x, \quad y' = y, \quad z' = z \tag{4–29}$$

is given by

$$\mathbf{L}_T = \begin{pmatrix} -1 & 0 & 0 & 0 \\ 0 & 1 & 0 & 0 \\ 0 & 0 & 1 & 0 \\ 0 & 0 & 0 & 1 \end{pmatrix} \tag{4-30}$$

and the *space inversion* where

$$t' = t, \qquad x' = -x, \qquad y' = -y, \qquad z' = -z \tag{4-31}$$

is given by

$$\mathbf{L}_P = \begin{pmatrix} 1 & 0 & 0 & 0 \\ 0 & -1 & 0 & 0 \\ 0 & 0 & -1 & 0 \\ 0 & 0 & 0 & -1 \end{pmatrix} \tag{4-32}$$

These two transformations are typical improper Lorentz transformations. In the following, however, we shall be concerned with proper Lorentz transformations.

4-6 SPECIAL LORENTZ TRANSFORMATIONS

Suppose two inertial systems coincide at $t = 0$ but one moves with velocity V with respect to the other along the X axis (Fig. 4–2). Let us consider an event $(ct', 0, y', z')$, which is in the $Y'Z'$ plane of the moving system. The

FIGURE 4–2. Two inertial systems.

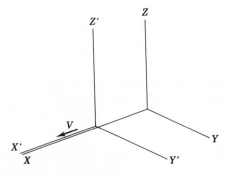

same event in the stationary system is (ct, Vt, y, z), but we may assume that $y' = y$ and $z' = z$. The Lorentz transformation matrix is of the form

$$\mathbf{Lx} = \begin{pmatrix} L_0{}^0 & L_1{}^0 & 0 & 0 \\ L_0{}^1 & L_1{}^1 & 0 & 0 \\ 0 & 0 & 1 & 0 \\ 0 & 0 & 0 & 1 \end{pmatrix} \begin{pmatrix} ct \\ Vt \\ y \\ z \end{pmatrix} = \begin{pmatrix} ct' \\ 0 \\ y \\ z \end{pmatrix} \tag{4-33}$$

Thus

$$L_0{}^1 ct + L_1{}^1 Vt = 0 \tag{4-34}$$

but the general requirement (4–26) gives

$$(L_0{}^0)^2 - (L_0{}^1)^2 = 1$$
$$(L_0{}^1)^2 - (L_1{}^1)^2 = -1$$
$$L_0{}^0 L_1{}^0 - L_0{}^1 L_1{}^1 = 0 \tag{4-35}$$

From (4–34) and (4–35) we obtain four solutions

$$\begin{pmatrix} L_0{}^0 & L_1{}^0 \\ L_0{}^1 & L_1{}^1 \end{pmatrix} = \begin{cases} \begin{pmatrix} \gamma & -\beta\gamma \\ -\beta\gamma & \gamma \end{pmatrix} & (4\text{-}36a) \\[2mm] \begin{pmatrix} -\gamma & \beta\gamma \\ \beta\gamma & -\gamma \end{pmatrix} & (4\text{-}36b) \\[2mm] \begin{pmatrix} -\gamma & \beta\gamma \\ -\beta\gamma & \gamma \end{pmatrix} & (4\text{-}36c) \\[2mm] \begin{pmatrix} \gamma & -\beta\gamma \\ \beta\gamma & -\gamma \end{pmatrix} & (4\text{-}36d) \end{cases}$$

where

$$\beta = \frac{V}{c} \tag{4-37}$$

and

$$\gamma = \frac{1}{\sqrt{1 - \beta^2}} \tag{4-38}$$

When we take the limit $\beta \rightarrow 0$, only the first solution (4–36a) reduces to the proper limit, i.e., unit matrix. Therefore, the Lorentz transformation matrix we are looking for is

$$\mathbf{L}_V = \begin{pmatrix} \gamma & -\beta\gamma & 0 & 0 \\ -\beta\gamma & \gamma & 0 & 0 \\ 0 & 0 & 1 & 0 \\ 0 & 0 & 0 & 1 \end{pmatrix} \qquad (4\text{–}39)$$

The other solutions in (4–36) are obtained by combining $\mathbf{L}_V$ with the time reversal $\mathbf{L}_T$, or the space reflection, $t \rightarrow t, x \rightarrow -x, y \rightarrow y, z \rightarrow z$, or both. Thus, for example

$$\begin{pmatrix} -\gamma & \beta\gamma & 0 & 0 \\ -\beta\gamma & \gamma & 0 & 0 \\ 0 & 0 & 1 & 0 \\ 0 & 0 & 0 & 1 \end{pmatrix} = \mathbf{L}_T\mathbf{L}_V \qquad (4\text{–}40)$$

We will consider $\mathbf{L}_V$ only in the following discussions.

Fig. 4–3 gives a graphical derivation of the Lorentz transformation (4–39).

FIGURE 4–3. The special Lorentz transformation $\mathbf{L}_V$ can be obtained graphically in three steps (Born): (1) Start with a rectilinear system $ct\text{-}X$. The origin ($x' = 0$) of a moving system (velocity V) is given by straight line $x = Vt = \beta ct$ in this diagram, and this is the ct' axis of the moving system. Since the light cone, which is given by $x = ct$ in the original system, must be given by $x' = ct'$ in the moving system, the X' axis is given by $ct = \beta x$. (2) Let E and E' be unit lengths in the $ct\text{-}X$ and $ct'\text{-}X'$ systems, respectively. Then their lengths in the $ct'\text{-}X'$ and $ct\text{-}X$ systems are e'/E' and e/E, respectively. Since these two inertial systems must be equivalent to each other we set $e'/E' = e/E$. Combining this equation with the result obtained in the first step, we obtain $e = E\sqrt{1 - \beta^2}$. [$e'^2 = E^2 + E^2\beta^2$ and $E'^2(1 - \beta^2)^2 = e^2(1 + \beta^2)$.] (3) Let $E = 1$ in the last result to obtain $e^{-1} = \gamma$. Finally $x' = \xi'/E' = (x - Vt)/e = -\beta\gamma ct + \gamma x$.

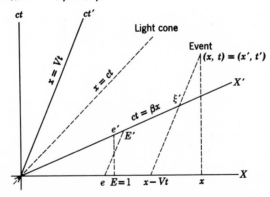

PROBLEM 4–2. Check that all four solutions in (4–36) satisfy (4–34) and (4–35).

PROBLEM 4–3. Show that

$$
L_V' = \begin{pmatrix}
\gamma & -\sqrt{\tfrac{1}{2}}\,\beta\gamma & -\sqrt{\tfrac{1}{2}}\,\beta\gamma & 0 \\
-\sqrt{\tfrac{1}{2}}\,\beta\gamma & \tfrac{1}{2}(\gamma + 1) & \tfrac{1}{2}(\gamma - 1) & 0 \\
-\sqrt{\tfrac{1}{2}}\,\beta\gamma & \tfrac{1}{2}(\gamma - 1) & \tfrac{1}{2}(\gamma + 1) & 0 \\
0 & 0 & 0 & 1
\end{pmatrix}
$$

is a Lorentz transformation matrix when V is in the XY plane making 45° to both X and Y axes (Fig. 4–4).

FIGURE 4–4. Two inertial systems.

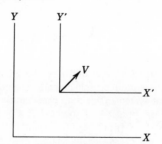

PROBLEM 4–4. If L_R is a Lorentz transformation matrix for the rotation of the frame by 45° around the Z axis, that is,

$$
L_R = \begin{pmatrix}
1 & 0 & 0 & 0 \\
0 & \sqrt{\tfrac{1}{2}} & \sqrt{\tfrac{1}{2}} & 0 \\
0 & -\sqrt{\tfrac{1}{2}} & \sqrt{\tfrac{1}{2}} & 0 \\
0 & 0 & 0 & 1
\end{pmatrix}
$$

then show that L_V' of Problem 4–3 can be expressed as

$$
L_V' = L_R{}^T L_V L_R
$$

where L_V is given by (4–41). Interpret this expression graphically.

Time Dilation and Lorentz Contraction

For a general event L_V gives

$$x' = -\gamma Vt + \gamma x \qquad (4\text{-}41)$$

$$ct' = \gamma ct - \beta \gamma x \qquad (4\text{-}42)$$

The distance between two such events has the components

$$\Delta x' = -\gamma V \Delta t + \gamma \Delta x \qquad (4\text{-}43)$$

$$c\Delta t' = \gamma c \Delta t - \beta \gamma \Delta x \qquad (4\text{-}44)$$

For an observer moving with the $X'Y'Z'$ frame the time interval should be given by setting $\Delta x' = 0$, since the position of his clock remains the same for him. In this case (4–43) gives $\Delta x = V \Delta t$ so that (4–44) gives

$$\Delta t' = \sqrt{1 - \beta^2}\ \Delta t \leq \Delta t \qquad (4\text{-}45)$$

FIGURE 4–5.   Time dilation. The time unit is defined by means of a reproducible physical phenomenon, such as atomic oscillations or meson decays, taking place at a given spatial position in that system. Thus if an oscillating atom staying at $x = 0$ is in phase 0 at $t = 0$ and in phase 2π at $t = T$ then OT is a unit time in the ct-X system. If another atom staying at $x' = 0$ is in phase O and 2π at $t' = O$ and T', respectively, then OT' is a unit time in the ct'-X' system. All events which take place on the $T\tau'$ line are simultaneous in the ct-X system, while all events which take place on the $\tau T'$ line are simultaneous in the ct'-X' system (showing the non-absoluteness of simultaneity). The time interval $O\tau$ is the unit time of the ct'-X' system measured in the ct-X system, and $O\tau = OT/\gamma$ if the same physical phenomenon used in defining the time unit in both systems. The event T', however, takes place at $t = \tau''$ in the ct-X system and $O\tau'' > OT$.

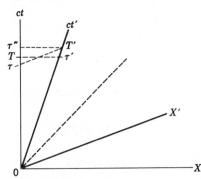

Thus a moving clock beats slower than a stationary clock in this sense (Fig. 4–5). For the moving observer the spatial distance $\Delta x'$ should be measured with $\Delta t' = 0$. In this case (4–43) and (4–44) give

$$\Delta x' = \sqrt{1 - \beta^2}\, \Delta x \le \Delta x \qquad (4\text{–}46)$$

which is called *Lorentz contraction* (Fig. 4–6).

FIGURE 4–6.   Lorentz contraction. The length of an object is defined as the difference of the spatial positions of two ends measured "simultaneously." Thus the length of a rod is *OL* in the *ct-X* system at $t = 0$, and that of another rod is *OL'* in the *ct'-X'* system at $t' = 0$. If the first rod is stationary with respect to the *ct-X* system, then one end follows the *ct* axis and the other end follows the *LI'* line as time elapses. If the second rod is stationary with respect to the *ct'-X'* system their ends follow the *ct'* axis and the *IL'* line. The length of the second rod measured in the *ct-X* system (at $t = 0$) is *Ol* by the above definition. If these two rods are physically identical, i.e., made of the same number of a given kind of atoms arranged in the same way, the principle of relativity requires that $Ol/OL = Ol'/OL'$, and the contraction results as shown in relation to Fig. 4–3.

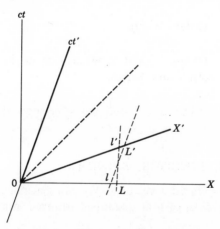

PROBLEM 4–5.   A circular disk is rotating around an axis perpendicular to the disk passing through the center, with angular velocity ω with respect to an inertial system fixed to the rotation axis. The circumference of the disk l experiences Lorentz contraction but the radius r does not. Show that an observer moving with the disk will find

$$l = 2\pi r \sqrt{1 - (\omega r/c)^2}$$

Velocity Addition Rule

The X component of the velocity of an object is

$$v_x' = \frac{dx'}{dt'} \qquad (4\text{–}47)$$

for the moving observer, while it is

$$v_x = \frac{dx}{dt} \tag{4-48}$$

for the stationary observer. Letting $\Delta \to d$ in (4–43) and (4–44) we see that

$$v'_x = \frac{-\gamma V\, dt + \gamma\, dx}{\gamma\, dt - \beta\gamma\, dx/c} = \frac{v_x - c\beta}{1 - (\beta v_x/c)} \tag{4-49}$$

For the other components the same definition gives

$$v'_y = \frac{v_y}{\gamma[1 - (\beta v_x/c)]} \qquad v'_z = \frac{v_z}{\gamma[1 - (\beta v_x/c)]} \tag{4-50}$$

PROBLEM 4-6. Obtain (4–50).

PROBLEM 4-7. Obtain v'_x and v'_y when V is in the XY plane in the direction 45° from both X and Y axes.

PROBLEM 4-8. When $\beta < 1$ and $|v_x| < c$, show that $|v'_x|$ can never exceed c.

4-7 VECTORS, TENSORS, SPINORS

In (4–16) we introduced a vector **x**. We can generalize this concept and say that if there is a set of four quantities arranged in a column matrix **A** such that

$$\mathbf{A}' = \mathbf{LA} \qquad \text{when} \qquad \mathbf{x}' = \mathbf{Lx} \tag{4-51}$$

we call such a set a vector. Sometimes it is called a *4-vector* to distinguish it from a three-component vector defined with respect to rotation in the ordinary three-dimensional space. These vectors can have both contravariant and covariant components A^i and A_i, respectively.[8]

If we have two vectors **A** and **B** we see

$$A_i B^i = \text{scalar} \tag{4-52}$$

[8] $A_i = g_{ij} A^j$

which is invariant under any proper Lorentz transformation. Two vectors can also form a tensor of second rank:

$$T_{ij} = A_i B_j, \qquad T_j{}^i = A^i B_j, \qquad T^{ij} = A^i B^j \qquad (4\text{-}53)$$

Using (4–52) we see

$$T_{ij} A^j = B_i, \qquad T_j{}^i A^j = B^i, \qquad T^{ij} A_j = B^i \qquad (4\text{-}54)$$

where **A** and **B** are vectors not necessarily the same as those in (4–53). Thus

$$T_{ij} U^{ij} = T_j{}^i U_i{}^j = T^{ij} U_{ij} = \text{scalar} \qquad (4\text{-}55)$$

These processes of reducing the rank of tensors such as (4–52), (4–54), and (4–55) are called *contractions*.

The metric tensor g_{ij} defined by (4–18) is a second rank tensor. Another special second rank tensor is the Kronecker tensor $\delta_i{}^j$ defined by

$$\delta_i{}^j A^i = A^j \qquad (4\text{-}56)$$

for any vector **A**. This means $\delta_i{}^j = 1$ if $i = j$, but $\delta_i{}^j = 0$ if $i \neq j$. Note that

$$
\begin{aligned}
\delta_{ji} = \delta_{ij} = g_{jk}\delta_i{}^k = \quad & 1 \text{ when } i = j = 0 \\
= \,& -1 \text{ when } i = j \neq 0 \qquad (4\text{-}57) \\
= \,& 0 \text{ when } i \neq j
\end{aligned}
$$

In general T_{ij} and T_{ji} may be different. When $T_{ij} = T_{ji}$, it is called a symmetric tensor, while if $T_{ij} = -T_{ji}$, it is called an antisymmetric tensor.

If f is a scalar function of t, x, y, and z, then

$$df = dt\,\frac{\partial f}{\partial t} + dx\,\frac{\partial f}{\partial x} + dy\,\frac{\partial f}{\partial y} + dz\,\frac{\partial f}{dz} \qquad (4\text{-}58)$$

is also scalar. Therefore, the set of differential operators

$$\left(\frac{\partial}{c\,\partial t}, \, -\frac{\partial}{\partial x}, \, -\frac{\partial}{\partial y}, \, -\frac{\partial}{\partial z} \right) \equiv \left(\frac{\partial}{\partial x_0}, \, \frac{\partial}{\partial x_1}, \, \frac{\partial}{\partial x_2}, \, \frac{\partial}{\partial x_3} \right) \qquad (4\text{-}59)$$

form a 4-vector (contravariant components), as seen by comparing (4–58) with (4–52).

Using the *Pauli spin matrices*

$$\sigma_0 = \begin{pmatrix} 1 & 0 \\ 0 & 1 \end{pmatrix}, \qquad \sigma_1 = \begin{pmatrix} 0 & 1 \\ 1 & 0 \end{pmatrix}, \qquad \sigma_2 = \begin{pmatrix} 0 & -i \\ i & 0 \end{pmatrix}, \qquad \sigma_3 = \begin{pmatrix} 1 & 0 \\ 0 & -1 \end{pmatrix}$$

$$(4\text{--}60)$$

one can express a vector in a 2×2 matrix form

$$A^i \sigma_i = \begin{pmatrix} A^0 + A^3 & A^1 - iA^2 \\ A^1 + iA^2 & A^0 - A^3 \end{pmatrix} \tag{4--61}$$

Corresponding to a Lorentz transformation $\mathbf{LA} = \mathbf{A}'$ we can find a 2×2 matrix $\mathbf{X}$ such that

$$\mathbf{X} A^i \sigma_i \mathbf{X}^\dagger = A'^i \sigma_i \tag{4--62}$$

where $\mathbf{X}^\dagger$ is the Hermitian conjugate of $\mathbf{X}$, or the complex conjugate of $\mathbf{X}^T$. For example we can easily see that

$$\mathbf{X}_V = \begin{pmatrix} \sqrt{\gamma(1 - \beta)} & 0 \\ 0 & \sqrt{\gamma(1 + \beta)} \end{pmatrix} \tag{4--63}$$

corresponds to $\mathbf{L}_V$ of (4–39) except that $\mathbf{V}$ is along the Z axis rather than the X axis. If we have a two-component quantity $\mathbf{S}$

$$\mathbf{S} \equiv \begin{pmatrix} S^1 \\ S^2 \end{pmatrix} \tag{4--64}$$

which is transformed as

$$\mathbf{XS} = \mathbf{S}' \tag{4--65}$$

under the Lorentz transformation, then such quantity $\mathbf{S}$ is called a *spinor*.
The space–time element

$$d^4x \equiv dx^0 \, dx^1 \, dx^2 \, dx^3 = c \, dt \, dv \tag{4--66}$$

appears often in our calculations. The transformation of a d^4x between two inertial systems is given by the formula

$$
d^4x' = \begin{vmatrix} \dfrac{\partial x^{0'}}{\partial x^0} & \dfrac{\partial x^{0'}}{\partial x^1} & \dfrac{\partial x^{0'}}{\partial x^2} & \dfrac{\partial x^{0'}}{\partial x^3} \\[2ex] \dfrac{\partial x^{1'}}{\partial x^0} & \dfrac{\partial x^{1'}}{\partial x^1} & \dfrac{\partial x^{1'}}{\partial x^2} & \dfrac{\partial x^{1'}}{\partial x^3} \\[2ex] \dfrac{\partial x^{2'}}{\partial x^0} & \dfrac{\partial x^{2'}}{\partial x^1} & \dfrac{\partial x^{2'}}{\partial x^2} & \dfrac{\partial x^{2'}}{\partial x^3} \\[2ex] \dfrac{\partial x^{3'}}{\partial x^0} & \dfrac{\partial x^{3'}}{\partial x^1} & \dfrac{\partial x^{3'}}{\partial x^2} & \dfrac{\partial x^{3'}}{\partial x^3} \end{vmatrix} d^4x \qquad (4\text{-}67)
$$

but this determinant, which is called the Jacobian is just the determinant of **L**, the Lorentz transformation matrix. From (4–28) we know that the latter is 1 or -1, depending on whether the transformation is proper or improper. For all proper Lorentz transformations we therefore have

$$
d^4x' = \det \mathbf{L}\, d^4x = d^4x \qquad (4\text{-}68)
$$

i.e., the space–time element is invariant. Since an integration does not change this property, we can say that a space–time volume is also invariant.

Some important tensor quantities are tabulated in Table 4–1 at the end of this chapter.

PROBLEM 4–9. Obtain

$$
T^{00'} = \gamma^2 T^{00} - \beta\gamma^2(T^{01} + T^{10}) + (\beta\gamma)^2 T^{11}
$$
$$
T^{01'} = -\beta\gamma^2 T^{00} + \gamma^2 T^{01} - (\beta\gamma)^2 T^{10} - \beta\gamma^2 T^{11}
$$
$$
T^{11'} = (\beta\gamma)^2 T^{00} - \beta\gamma^2(T^{01} + T^{10}) + \gamma^2 T^{11}
$$

when $\mathbf{L}_V$ is given by (4–39).

PROBLEM 4–10. Show that

$$
\square = \nabla^2 - \left(\frac{\partial}{c\,\partial t}\right)^2
$$

is a scalar operator. The operator $\square$ is called the *D'Alembertian*.[9]

---

[9] Other definitions one may find are $\square = (\partial/c\,\partial t)^2 - \nabla^2$ and $\square^2 = (\partial/c\,\partial t)^2 - \nabla^2$.

PROBLEM 4-11. Show that

$$\frac{\partial T_i^{\ j}}{\partial x^j}$$

is a vector.

PROBLEM 4-12. Show that (4–63) corresponds to (4–39) except that **V** is along the Z axis, that is,

$$\begin{pmatrix} \gamma & 0 & 0 & -\beta\gamma \\ 0 & 1 & 0 & 0 \\ 0 & 0 & 1 & 0 \\ -\beta\gamma & 0 & 0 & \gamma \end{pmatrix}$$

PROBLEM 4-13. Show that

$$\begin{pmatrix} e^{i\varphi/2} & 0 \\ 0 & e^{-i\varphi/2} \end{pmatrix}$$

corresponds to the rotation around the Z axis by φ, or the Lorentz transformation of

$$\begin{pmatrix} 1 & 0 & 0 & 0 \\ 0 & \cos\varphi & \sin\varphi & 0 \\ 0 & -\sin\varphi & \cos\varphi & 0 \\ 0 & 0 & 0 & 1 \end{pmatrix}$$

Note that this result means that both

$$\begin{pmatrix} 1 & 0 \\ 0 & 1 \end{pmatrix} \quad \text{and} \quad \begin{pmatrix} -1 & 0 \\ 0 & -1 \end{pmatrix}$$

correspond to the identity Lorentz transformation (unit matrix).

4-8 RELATIVISTIC EQUATION OF MOTION

From (4–10) we see

$$ds = c\sqrt{1 - \beta^2}\, dt \qquad (4\text{–}69)$$

and this quantity is invariant under the proper Lorentz transformations as shown in (4–14). If we compare this formula with (4–45) we see that

$$d\tau = \frac{ds}{c} = \sqrt{1 - \beta^2}\, dt \qquad (4\text{–}70)$$

is the time interval measured by a clock fixed on the object, which is moving with velocity $c\beta$ with respect to us, when the same time interval is dt by our clock. This time τ is called the *proper time*, and it is invariant under proper Lorentz transformations.

We define the 4-velocity

$$u^i = \frac{dx^i}{d\tau} \tag{4-71}$$

This 4-velocity obviously is a vector under Lorentz transformations. We see that

$$u^0 = c\,\frac{dt}{d\tau} = c\gamma, \qquad u^1 = V_x\gamma, \qquad u^2 = V_y\gamma, \qquad u^3 = V_z\gamma \tag{4-72}$$

where γ is defined by (4–38) and **V** is the velocity of the object measured by us. Since

$$u^i u_i = \left(\frac{ds}{d\tau}\right)^2 = c^2 \tag{4-73}$$

we see from (4–8) that it is a timelike vector.

Another 4-vector can be defined as

$$p^i = mu^i \tag{4-74}$$

with a scalar quantity m which is called the rest mass. This 4-vector is called the 4-momentum, and its components are

$$p^0 = mc\gamma, \qquad p^1 = mV_x\gamma, \qquad p^2 = mV_y\gamma, \qquad p^3 = mV_z\gamma \tag{4-75}$$

In the limit of $\beta \to 0$, we see the last three components (space components) approach those of the ordinary momentum.

Corresponding to Newton's equation (1–1) one may propose four equations

$$\frac{dp^i}{d\tau} = K^i \tag{4-76}$$

where K^i is a 4-vector called the *Minkowski force*. This type of formula which equates one vector to another vector is a *covariant equation*. By covariant here we mean that if we go to a different inertial system each component will be changed but the same form of the equation will hold for the new components. Readers should not confuse this covariant with the covariant components we introduced before.

If p^1, p^2, and p^3 are identified to the momentum components p_x, p_y, and p_z in Newton's equation, the Minkowski force should be related to the ordinary force **F** as

$$\mathbf{F} = \frac{d\mathbf{p}}{dt} = \mathbf{K}\frac{d\tau}{dt} = \mathbf{K}\sqrt{1 - \beta^2} \qquad (4\text{-}77)$$

where **p** and **K** represent the space components (1, 2, 3 components) of each vector. In order to calculate the time component K^0 we note first that

$$K^i u_i = m\frac{du^i}{d\tau}u_i = m\frac{du_i}{d\tau}u^i = \tfrac{1}{2}m\frac{d(u^i u_i)}{d\tau} = 0 \qquad (4\text{-}78)$$

due to (4–73). Thus

$$K^0 = \frac{\mathbf{K} \cdot \mathbf{u}}{u_0} = \frac{\mathbf{F} \cdot \mathbf{V}}{c}\gamma \qquad (4\text{-}79)$$

using (4–72), (4–77), and (4–78).

From the time component of (4–76) we obtain

$$p^0 = \int K^0\,d\tau = \frac{1}{c}\int \mathbf{F}\cdot\mathbf{V}\,dt = \frac{1}{c}\int \mathbf{F}\cdot d\mathbf{r} \qquad (4\text{-}80)$$

Since the last expression gives the work done on the object divided by c, $p^0 c$ must be the relativistic kinetic energy. Actually from (4–75)

$$T \equiv p^0 c = mc^2\gamma = mc^2(1 - \beta^2)^{-1/2} = mc^2 + \tfrac{1}{2}mV^2 + \cdots \qquad (4\text{-}81)$$

where the second term is the familiar classical kinetic energy. The first term mc^2 is called the rest-mass energy. Since the relativistic momentum is

$$\mathbf{p} = m\mathbf{V}\gamma \qquad (4\text{-}82)$$

we obtain the useful relativistic relation

$$\frac{\mathbf{p}c^2}{T} = \mathbf{V} \qquad (4\text{-}83)$$

Another important relation

$$T^2 = (pc)^2 + (mc^2)^2 \qquad (4\text{--}84)$$

can be obtained from (4–81) and (4–82), if p is the magnitude of the relativistic momentum (Fig. 4–7).

FIGURE 4–7. Illustration of (4–84).

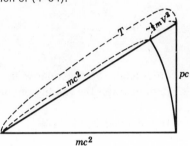

PROBLEM 4–14. Obtain (4–72) and then (4–73).

PROBLEM 4–15. Show how T and $\mathbf{p}$ are transformed when we go from one inertial frame to another which is moving with respect to the first frame with a constant velocity $\mathbf{V}$ along the X axis.

PROBLEM 4–16. Obtain the relativistic Hamilton–Jacobi equation of a free particle $(U = 0)$

$$\frac{1}{c^2}\left(\frac{\partial S}{\partial t}\right)^2 - \left(\frac{\partial S}{\partial x}\right)^2 - \left(\frac{\partial S}{\partial y}\right)^2 - \left(\frac{\partial S}{\partial z}\right)^2 = m^2 c^2$$

Then assuming

$$S = S' - mc^2 t$$

obtain the nonrelativistic Hamilton–Jacobi equation (1–121) in the limit of $c \to \infty$.

4–9 RELATIVISTIC LAGRANGIAN AND HAMILTONIAN FUNCTIONS

Let us propose a Lagrangian function

$$L = -mc^2 \sqrt{1 - \beta^2} - U \qquad (4\text{--}85)$$

This function is not equal to $T - U$ as one might expect from the classical definition (1–35). We see, however, that

$$\frac{\partial L}{\partial \mathbf{V}} = m\mathbf{V}\gamma = \mathbf{p} \tag{4-86}$$

which is the relativistic extension of (1–37), and Lagrange's equations (1–34) are

$$\frac{d\mathbf{p}}{dt} = -\nabla U = \mathbf{F} \tag{4-87}$$

which give the space components of the relativistic equation of motion (4–76). This result (4–87) demonstrates the validity of our Lagrangian function (4–85).

The Hamiltonian function defined by (1–44) is

$$\begin{aligned} H = \mathbf{p} \cdot \mathbf{V} - L &= mV^2\gamma + mc^2\sqrt{1 - \beta^2} + U \\ &= (mV^2 + mc^2 - mV^2)\gamma + U \\ &= mc^2\gamma + U = T + U \end{aligned} \tag{4-88}$$

It is amazing that although L is not $T - U$, H is still $T + U$, or equal to the total energy.

PROBLEM 4–17. Using the L of (4–85) show that Hamilton's principle can be expressed as

$$\delta \int ds = 0$$

for a free particle.

HINT. $L\,dt = -mc\,ds$.

PROBLEM 4–18. The potential U may depend on the velocity $\mathbf{V}$. If

$$U = \mathbf{A} \cdot \mathbf{V}$$

where $\mathbf{A}$ is a constant vector, show that

$$\mathbf{p} = m\mathbf{V}\gamma - \mathbf{A}$$

$$H = T = \sqrt{(\mathbf{p} + \mathbf{A})^2 c^2 + (mc^2)^2}$$

PROBLEM 4-19. Using the Hamiltonian (4–88), show that Hamilton's equations give the correct relativistic equations of motion.

4-10 LAGRANGIAN DENSITY. STRESS-ENERGY TENSOR[10]

When matter is continuously distributed in ordinary space, we consider the Lagrangian density $\mathscr{L}$ which is introduced by

$$L = \int \mathscr{L} \, dv \qquad (3\text{--}98)$$

Hamilton's principle is written, as in (3–102), in the form of

$$\delta \int \mathscr{L} \, dv \, dt = \delta \int \mathscr{L} \, \frac{d^4 x}{c} = 0 \qquad (4\text{--}89)$$

where $d^4 x$ is the space–time element introduced in (4–66). Since $d^4 x$ is invariant under the Lorentz transformation, as shown in (4–68), we see that (4–89) is covariant *if the Lagrangian density $\mathscr{L}$ is invariant*. The boundary conditions we took in (3–103) and (3–104) can be written in the covariant form as

$$\delta q = 0 \qquad \text{at both ends of} \qquad x^0, \, x^1, \, x^2, \quad \text{and} \quad x^3$$
$$(4\text{--}90)$$

where q is the field variable. (We switched the notation from **u** of (3–104) to q here.) If $\mathscr{L}$ is invariant, or scalar, the resulting Lagrange–Euler's equation (3–106) is, therefore, covariant, as can be seen better by writing it as

$$\frac{\partial \mathscr{L}}{\partial q} - \frac{\partial}{\partial x^i} \frac{\partial \mathscr{L}}{\partial (\partial q / \partial x^i)} = 0 \qquad (4\text{--}91)$$

The field variable q is implicitly assumed to be a scalar quantity, but in general it may have more than one component. In that case (4–91) should be generalized to

$$\frac{\partial \mathscr{L}}{\partial q^{(l)}} - \frac{\partial}{\partial x^i} \frac{\partial \mathscr{L}}{\partial (\partial q^{(l)} / \partial x^i)} = 0 \qquad (4\text{--}92)$$

In the rest of this chapter we shall treat the case of a scalar field, to simplify the formulas. The case of a vector field will be discussed in Chapter 6.

---

[10] Also called energy-momentum tensor.

The stress-energy tensor, whose components are defined by

$$T_i{}^j = \mathscr{L}\, \delta_i{}^j - \frac{\partial q}{\partial x^i}\, \frac{\partial \mathscr{L}}{\partial(\partial q/\partial x^j)} \tag{4-93}$$

will be useful. Here $\delta_i{}^j$ is the Kronecker tensor introduced in (4–56). Since

$$\frac{\partial \mathscr{L}}{\partial x^i} = \frac{\partial \mathscr{L}}{\partial q}\, \frac{\partial q}{\partial x^i} + \frac{\partial \mathscr{L}}{\partial(\partial q/\partial x^k)}\, \frac{\partial^2 q}{\partial x^i\, \partial x^k} \tag{4-94}$$

and because of (4–91), we obtain

$$\frac{\partial T_i{}^j}{\partial x^j} = \frac{\partial \mathscr{L}}{\partial x^i} - \frac{\partial}{\partial x^j}\left[\frac{\partial q}{\partial x^i}\, \frac{\partial \mathscr{L}}{\partial(\partial q/\partial x^j)}\right]$$

$$= 0 \tag{4-95}$$

From (4–93) we see that

$$T_0{}^0 = \mathscr{L} - \frac{\partial q}{\partial t}\, \frac{\partial \mathscr{L}}{\partial(\partial q/\partial t)} = \mathscr{L} - \dot{q}\pi = -\mathscr{H} \tag{4-96}$$

using (3–110). $\mathscr{H}$ is the Hamiltonian density, which is the energy density. Since (4–95) gives

$$\frac{\partial \mathscr{H}}{c\, \partial t} - \frac{\partial T_0{}^1}{\partial x} - \frac{\partial T_0{}^2}{\partial y} - \frac{\partial T_0{}^3}{\partial z} = 0 \tag{4-97}$$

and since energy conservation in the continuous medium implies the continuity equation

$$\frac{\partial \mathscr{H}}{\partial t} + \mathbf{V} \cdot \mathbf{S} = 0 \tag{4-98}$$

where $\mathbf{S}$ is the energy flux, we see that $-cT_0{}^1$, $-cT_0{}^2$, and $-cT_0{}^3$ components of the stress-energy tensor are the x, y, and z components of the energy flux, respectively.

Since $dt\, dv$ is invariant, as shown in (4–68), $\int T_i{}^0\, dv = -cp_i$ should be the ith covariant component of a 4-vector. We know, on the other hand, that $-\int T_0{}^0\, dv = cp_0$ is the energy, according to (4–96), so that the other components $-\int T_\alpha{}^0 \, dv/c$ should be the αth covariant component of the momentum. Using the metric tensor g_{ij} of (4–18), we obtain

$$\frac{1}{c} \int T_\alpha{}^0 \, dv = -p_\alpha = p^\alpha \tag{4–99}$$

Thus $T_\alpha{}^0/c$ is the density of the αth component of the field momentum, which we designate $\mathscr{G}$. From (4–93) we obtain an explicit expression

$$\mathscr{G}_x = \frac{1}{c} T_1{}^0 = -\pi \frac{\partial q}{\partial x} \tag{4–100}$$

or

$$\mathscr{G} = -\pi \nabla q \tag{4–101}$$

Another way of showing that (4–101) is the momentum density is to calculate the Poisson bracket of this quantity with a function of q. Since $\pi(\mathbf{r}) \, dv$ is the canonical momentum conjugate to $q(\mathbf{r})$, as seen in (3–107), we obtain

$$\left[\{q(\mathbf{r})\}^n, \int \mathscr{G}(\mathbf{r}') \, dv' \right]_P = - \left[\{q(\mathbf{r})\}^n, \int \pi \nabla q(\mathbf{r}') \, dv' \right]_P$$
$$= -n\{q(\mathbf{r})\}^{n-1} \nabla q(\mathbf{r}) = -\nabla \{q(\mathbf{r})\}^n \tag{4–102}$$

for an arbitrary power n. If a function $f(q(\mathbf{r}))$ is defined as a power series of q, then

$$\left[f(q(\mathbf{r})), \int \mathscr{G}(\mathbf{r}') \, dv' \right]_P = -\nabla f(q(\mathbf{r})) \tag{4–103}$$

This result is similar to (1–117) for the particle momentum, except that the sign on the right-hand side is the opposite of that in (1–117). This difference in sign is due to the different meaning of ∇: In particle physics we take the position of the particle $\mathbf{r}$ for the dynamical variable, so that the momentum is $m\dot{\mathbf{r}}$, for example, but in the field theory we are discussing now, the coordinate $\mathbf{r}$ is a parameter which designates the part of the space we are considering, and the dynamical variable by which the motion of matter is described is q, not $\mathbf{r}$. As a matter of fact, the displacement of matter in one direction is equivalent to displacement of the coordinate system in the opposite direction by the same amount.

From equation (4–95) we obtain

$$-\frac{\partial \mathscr{G}_x}{\partial t} = \frac{\partial T_1{}^1}{\partial x} + \frac{\partial T_1{}^2}{\partial y} + \frac{\partial T_1{}^3}{\partial z} \qquad (4\text{–}104)$$

(and two other equations for $\mathscr{G}_y$ and $\mathscr{G}_z$.) Since the right-hand side of this equation should be the force exerted on the matter in the volume element, we see from (3–43), that $T_1{}^1$, $T_1{}^2$, and $T_1{}^3$ are the $-T_{xx}$, $-T_{yx}$, and $-T_{zx}$ components of the stress tensor, respectively. Thus if we write the stress-energy tensor in matrix form, we obtain

$$\begin{pmatrix} T_0{}^0 & T_0{}^1 & T_0{}^2 & T_0{}^3 \\ T_1{}^0 & T_1{}^1 & T_1{}^2 & T_1{}^3 \\ T_2{}^0 & T_2{}^1 & T_2{}^2 & T_2{}^3 \\ T_3{}^0 & T_3{}^1 & T_3{}^2 & T_3{}^3 \end{pmatrix} = \begin{pmatrix} -\mathscr{H} & -S_x/c & -S_y/c & -S_z/c \\ c\mathscr{G}_x & -T_{xx} & -T_{yx} & -T_{zx} \\ c\mathscr{G}_y & -T_{xy} & -T_{yy} & -T_{zy} \\ c\mathscr{G}_z & -T_{xz} & -T_{yz} & -T_{zz} \end{pmatrix} \qquad (4\text{–}105)$$

Instead of the mixed components of the stress-energy tensor, we can take the pure contravariant components T^{ij} $(= g^{ik}T_k{}^j)$ as

$$\begin{pmatrix} T^{00} & T^{01} & T^{02} & T^{03} \\ T^{10} & T^{11} & T^{12} & T^{13} \\ T^{20} & T^{21} & T^{22} & T^{23} \\ T^{30} & T^{31} & T^{32} & T^{33} \end{pmatrix} = \begin{pmatrix} -\mathscr{H} & -S_x/c & -S_y/c & -S_z/c \\ -c\mathscr{G}_x & T_{xx} & T_{yx} & T_{zx} \\ -c\mathscr{G}_y & T_{xy} & T_{yy} & T_{zy} \\ -c\mathscr{G}_z & T_{xz} & T_{yz} & T_{zz} \end{pmatrix} \qquad (4\text{–}106)$$

It was shown in (3–46) that the stress tensor should be a symmetric tensor. Since the same argument is still applicable, and since the symmetry property should hold for any inertial system, the entire stress-energy tensor should be symmetric

$$T^{ij} = T^{ji} \qquad (4\text{–}107)$$

which means, that

$$\mathbf{S} = c^2 \mathscr{G} \qquad (4\text{–}108)$$

PROBLEM 4–20. Verify (4–95).

PROBLEM 4–21. Confirm each step in (4–102).

PROBLEM 4–22. Using partial integration, show that we can express

$$\mathscr{G} = q\nabla\pi$$

and then show that if $f(\pi(\mathbf{r}))$ is a power series of $\pi(\mathbf{r})$

$$\left[f(\pi(\mathbf{r})), \int \mathcal{G}(\mathbf{r}') \, dv' \right]_P = -\nabla f(\pi(\mathbf{r}))$$

PROBLEM 4–23. Show the relation between (4–108) and (4–83).

PROBLEM 4–24. It is shown in Problem 3–22 that

$$\mathcal{L} = \tfrac{1}{2}\mu \left(\frac{\partial u}{\partial t} \right)^2 - \tfrac{1}{2}T \left(\frac{\partial u}{\partial x} \right)^2$$

for an homogeneous elastic string. Using (4–93), show that the energy flux is

$$S = -T \frac{\partial u}{\partial t} \frac{\partial u}{\partial x}$$

in this case. (This is another way to solve Problem 3–39.)

4–11 MASS DENSITY. STRESS-ENERGY TENSOR OF DUST, IDEAL FLUID, AND IDEAL GAS

The mass density μ, introduced in (3–30), is related to the rest mass of a particle as

$$m = \int \mu \, dv \tag{4–109}$$

Since the rest mass m is scalar, and since $d^4x = c \, dt \, dv$ is also scalar, according to (4–68), (4–109) shows that μ should transform as the time component of a 4-vector. This 4-vector can actually be found using the displacement vector $\mathbf{q}$ of matter in the infinitesimal volume we are considering. Thus it is easy to see that $\mathbf{J}$ defined by

$$J^0 = c\mu, \qquad J^1 = \mu \frac{\partial q_x}{\partial t}, \qquad J^2 = \mu \frac{\partial q_y}{\partial t}, \qquad J^3 = \mu \frac{\partial q_z}{\partial t} \tag{4–110}$$

is the 4-vector we are looking for. Using the mass current density introduced in (3–31) the 4-vector can be written as

$$J^0 = c\mu, \qquad J^1 = j_x, \qquad J^2 = j_y, \qquad J^3 = j_z \tag{4–111}$$

The relativistic Lagrangian function L given by (4–85) implies that the Lagrangian density can be written as

$$\mathscr{L} = -\mu c^2 \sqrt{1 - \beta^2} - \mathscr{U} \tag{4-112}$$

where $\mathscr{U}$ is the potential energy density. Notice that the first term is scalar since

$$-\mu c \sqrt{1 - \beta^2} = -\sqrt{J^i J_i} \tag{4-113}$$

The Hamiltonian density is expected to be

$$\mathscr{H} = \mu c^2 \gamma + \mathscr{U} \tag{4-114}$$

from (4–88). This can be confirmed from (4–96).

The stress-energy tensor can be calculated from (4–93), but can also be obtained easily, if $\mathscr{U} = 0$, by observing that

$$\mu c^2 \gamma = J^0 u_0 \tag{4-115}$$

where u_0 can be obtained from (4–71) or (4–72). In the case of $\mathscr{U} = 0$, we therefore obtain

$$T_i{}^j = -u_i J^j \tag{4-116}$$

This is the stress-energy tensor of dust, or aggregate of noninteracting particles.

The symmetry of this tensor can be seen better if we introduce the rest-mass density

$$\mu_r = \mu \sqrt{1 - \beta^2} \tag{4-117}$$

by which we can write

$$J^i = \mu_r u^i \tag{4-118}$$

Equation (4–116) is now

$$T_i{}^j = -\mu_r u_i u^j \tag{4-119}$$

An ideal fluid was introduced in section 3–8 by means of the stress tensor, as given by (3–80),

$$T_\alpha{}^\beta = P\, \delta_\alpha{}^\beta \tag{4-120}$$

where P is the pressure and α and β stand for 1, 2, and 3. Since the discussions in the previous chapter are all nonrelativistic, we expect that (4–120) is valid in an inertial system for which the matter in the volume element is stationary. In such an inertial system the only other nonzero element of the stress-energy tensor is

$$T_0{}^0 = -\mathcal{H}_0 \qquad (4\text{--}121)$$

where $\mathcal{H}_0$ is the energy density in this special inertial system. The covariant expression of the stress-energy tensor which reduces to (4–120) and (4–121) for the special inertial system is

$$T_i{}^j = -\left(\frac{\mathcal{H}_0 + P}{c^2}\right) u_i u^j + P\, \delta_i{}^j \qquad (4\text{--}122)$$

This is the stress-energy tensor of an ideal fluid.

An ideal gas is made of molecules, each of which moves independently from the others. When there are n molecules in a unit volume and each molecule has mass m, the stress-energy tensor of the ideal gas is

$$T_i{}^j = -\overline{nmu_i u^j \sqrt{1 - \beta^2}} \qquad (4\text{--}123)$$

where the upper bar means average. In an inertial system where the distribution of molecular velocity is completely isotropic, the stress-energy is diagonal and has the form of (4–120) and (4–121). Thus

$$\mathcal{H}_0 = nmc^2 \overline{\gamma} \qquad (4\text{--}124)$$

$$P = \tfrac{1}{3}nm\overline{V^2 \gamma} \qquad (4\text{--}125)$$

This expression of the pressure P of an ideal gas is the relativistic extension of the well-known formula obtained from entirely different considerations of kinetic theory.

PROBLEM 4–25. Express the equation of continuity (3–35) in a covariant form.

PROBLEM 4–26. Show that the **J** defined by (4–110) forms a 4-vector, and then show (4–113).

PROBLEM 4–27. In (4–112) assuming that $\mathcal{U}$ does not depend on $\partial \mathbf{q}/\partial t$, obtain π and then show that $-T_0{}^0$ is given by (4–114) itself.

PROBLEM 4–28. Obtain T_α^β, T_α^0, and T_0^0 of an ideal fluid in a general inertial system.

ANSWER.

$$T_\alpha^\beta = \frac{(\mathscr{H}_0 + P)V^\alpha V^\beta}{c^2(1 - \beta^2)} + P\,\delta_\alpha^\beta$$

$$T_\alpha^0 = \frac{(\mathscr{H}_0 + P)V^\alpha}{c(1 - \beta^2)}$$

$$T_0^0 = -\frac{\mathscr{H}_0 + \beta^2 P}{1 - \beta^2}$$

TABLE 4–1 *Important Tensor Quantities in (Proper) Lorentz Transformations*

Scalars (Invariants)					
s, $\quad d\tau \equiv dt\sqrt{1-\beta^2}$, $\quad d^4x \equiv c\,dt\,dx\,dy\,dz$					

4-Vectors (Contravariant Components)					
ct	x	y	z		
$c\gamma$	$V_x\gamma$	$V_y\gamma$	$V_z\gamma$	4-velocity	(4–72)
$\dfrac{T}{c}$	p_x	p_y	p_z	energy-momentum	(4–75)
$\dfrac{\partial}{c\,\partial t}$	$-\dfrac{\partial}{\partial x}$	$-\dfrac{\partial}{\partial y}$	$-\dfrac{\partial}{\partial z}$		(4–59)
$\dfrac{\mathbf{F}\cdot\mathbf{V}\gamma}{c}$	$F_x\gamma$	$F_y\gamma$	$F_z\gamma$	Minkowski force	(4–77), (4–79)
$c\mu$	j_x	j_y	j_z	mass density current	(4–111)
$c\rho$	J_x	J_y	J_z	charge density current	(6–12)
$\dfrac{\phi}{c}$	A_x	A_y	A_z	electromagnetic potentials	(6–14)
$\dfrac{\omega}{c}$	k_x	k_y	k_z	4–wave vector	(6–66)

Second-Rank Tensors	
Metric tensor $g^{ij} = \delta^{ij}$ (Kronecker tensor $= \delta_i{}^j$)	
Stress-energy tensors. General expression	(4–106)
$\quad$ Of dust: $\qquad T^{ij} = -\mu u^i u^j$	(4–119)
$\quad$ Of ideal fluid: $\quad T^{ij} = -((\mathscr{H}_0 + P)/c^2)u^i u^j + P\delta^{ij}$	(4–122)
$\quad$ Of ideal gas: $\quad T^{ij} = -nm(u^i u^j \sqrt{1-\beta^2})_{\text{average}}$	(4–123)
Electromagnetic field tensor	
$\qquad F^{ij} = (\partial A^j/\partial x_i{}^i) - (\partial A^i/\partial x_j)$	(6–16), (6–17)

ADDITIONAL PROBLEMS

4-29. Obtain the Lorentz transformation matrix **L** for the spinor transformation matrice **X**:

$$\begin{pmatrix} \sqrt{(1 + \gamma)/2} & -\sqrt{(1 - \gamma)/2} \\ -\sqrt{(1 - \gamma)/2} & \sqrt{(1 + \gamma)/2} \end{pmatrix}$$

4-30. (a) Obtain the relativistic equation of motion

$$m(\gamma\mathbf{a} + \gamma^3(\boldsymbol{\beta} \cdot \mathbf{a})\boldsymbol{\beta}) = \mathbf{F}$$

where **a** is the second derivative of **r** by t.
(b) Calculate the effective mass defined by **F/a** when **V** is parallel and perpendicular to **a**.

ANSWER. $m\gamma^3$, $m\gamma$.

4-31. Covariant angular momentum can be defined as

$$J^{ij} = x^i p^j - x^j p^i$$

Show that the conservation of $\sum J^{0j}$ for a set of particles under no external field is equivalent to the conservation of velocity of the *center of inertia* $\mathbf{R} \equiv \sum T\mathbf{r}/\sum T$. The center of inertia reduces to the center of mass in the nonrelativistic limit. Here T is the relativistic kinetic energy of a particle whose position is **r**, and the summations are over all particles.

HINT. $\sum \mathbf{p}$ and $\sum T$ are conserved.

4-32. The total angular momentum of a set of particles **M** is defined as $M_x = \sum J^{23}$, $M_y = \sum J^{31}$, $M_z = \sum J^{12}$, where J^{ij} is defined in problem 4-31. Suppose **M** is measured in a system for which the center of inertia of the set of particles is stationary at the origin, and **M'** is measured in another system moving with respect to the first system with velocity **V** along the X direction. Find the relation between **M** and **M'**.

ANSWER.

$$M_x = M_x', \qquad M_y = \gamma M_y', \qquad M_z = \gamma M_z'$$

4-33. Two particles with masses m_1 and m_2 are moving in the same direction with velocities v_1 and v_2, respectively. If they collide and move together after the collision, find the final velocity.

ANSWER.

$$v = \frac{pc^2}{T} = \frac{m_1 \gamma_1 v_1 + m_2 \gamma_2 v_2}{m_1 \gamma_1 + m_2 \gamma_2}$$

4-34.* Show that

$$\frac{dp_x \, dp_y \, dp_z}{T}$$

is invariant under the proper Lorentz transformation.

HINT. Using the result of Problem 4-15 obtain

$$\frac{\partial p'_x}{\partial p_x} = \frac{p^{0'}}{p^0}$$

etc. Follow the argument in (4-67).

4-35. Show that $p \sin \theta \, d\theta \, d\varphi \, dT$ is invariant under the proper Lorentz transformation. Here p, θ, and φ are magnitude and direction angles of **p**.

4-36.* Entropy S is defined as $dS = dQ/T$, where Q is heat and T is temperature. Knowing that S does not change during contraction of matter unless the arrangement of atoms in that matter is changed show that

(a) the Lorentz transformation of the temperature[11] is

$$T' = T\sqrt{1 - \beta^2}$$

(b) the gas law

$$pV = NkT$$

is covariant in the Lorentz transformation. V is the volume, P is the pressure, N is the total number of molecules, and k is the Boltzmann constant.

HINT. P is scalar since it is isotropic.

[11] M. Planck, *Ann. Phys.* [4] **26**, 1 (1908). However H. Ott [*Zeits. Phys.* **175**, 70 (1963)] proposed that another temperature which transforms as $T_v = T/\sqrt{1 - \beta^2}$ is more useful. For a review, see N. G. von Kampen, *J. Phys. Soc. Japan*, Suppl. **26**, 316 (1969).

Chapter 5 Electromagnetic Fields

The concept of forces in classical mechanics was rather abstract until, in the 19th century, the investigation of the electromagnetic field started to clarify the physical nature of forces. Human beings had been worshipping lightning before civilization, and had been using compasses for centuries after the birth of civilization, but they did not realize, until about a century ago, that electricity and magnetism constitute the fundamental nature of force, along with the gravity force. The complete understanding of the nature of force, however, had to await the development of quantum mechanics.

Light has been a favorite subject of physics from the earliest time, but no satisfactory theory was found until Maxwell showed that the electromagnetic field can propagate as a wave, and that light is an electromagnetic wave of a particular frequency region.

When the frequency is high, a wave can behave like a particle. Because of this relation the corpuscular theory of light and geometrical optics had been partially successful, although it was later replaced by Maxwell's theory for more exact treatment. This same relation suggested that matter, which had been regarded as an aggregate of particles, may exhibit wave properties under more exact treatments. This suggestion led to quantum mechanics in the 20th century.

5–1 COULOMB'S LAW OF ELECTROSTATIC FORCE

Charles Augustus Coulomb, a French engineer in the 18th century, measured a force $\mathbf{F}$ between two charges q_1 and q_2 as a function of the distance between the charges $\mathbf{r}$, and found that

$$\mathbf{F} = \frac{kq_1q_2}{r^3}\,\mathbf{r} \qquad (5\text{--}1)$$

where k is a constant.[1] This is called *Coulomb's law*. His experiment, which used a very accurate device of his own invention, a torsion balance, showed that F was proportional to r^{-n} where n was 2 ± 0.20, while a recent result[2] showed that n was $2 + (2.7 \pm 3.1) \times 10^{-16}$. This, undoubtedly, pleased him and other physicists very much, because this form is the same as that of *Newton's law of gravity force* already known:

$$\mathbf{F} = \frac{-Gm_1m_2}{r^3}\,\mathbf{r} \qquad (5\text{--}2)$$

Einstein, who proposed the theory of general relativity (Chapter 7) to explain Newton's law of gravity, was naturally tempted to find a similar explanation of Coulomb's law. He proposed a so-called *unified field theory* to explain both laws from a single viewpoint. The theory, however, was not completed in an acceptable form before he died. Quite a few attempts have been made by other physicists, but this parallelism between Coulomb's and Newton's laws is still a mystery today.[3]

Although these two laws have the same r dependence, more accurately confirmed by later experiments, large differences between the charge and the mass should be noticed. First of all, we know of no repulsive gravitational force, but the electrostatic force is known to be either attractive or repulsive depending on the relative sign of the interacting charges. The concept of negative mass has been proposed in relation to antiparticles (Chapter 10), but, as will be pointed out in section 7–2, even such antiparticles are attracted by the earth, that is, there exists no repulsive gravity force. On the other

[1] $k = 8.98755 \times 10^9$ Newton m²/C².
[2] E. R. Williams, J. E. Faller, and H. A. Hill, *Phys. Rev. Letters*, **26**, 721 (1971).
[3] A review of unified field theories is given in Part III of P. G. Bergmann, *Introduction to the Theory of Relativity*, Prentice-Hall, Englewood Cliffs, New Jersey, 1942. See also M. A. Tonnelat, *Einstein's Unified Field Theory*, English transl., Gordon and Breach, New York, 1966; and V. Hlavaty, *Geometry of Einstein's Unified Field Theory*, P. Nordhoff, Groningen, Netherlands, 1957.

hand, as du Fay discovered during the 18th century, there are two kinds of charges, negative and positive, and charges of the same sign repel each other, while those of opposite sign attract each other.

It is known now that the electron, one of the fundamental elementary particles, has the charge of $-e$, where

$$e = 1.6021917 \times 10^{-19} \text{ coulomb} \tag{5-3}$$

and the sign (negative) is defined by the convention. The same convention defines the charge of the proton, another fundamental elementary particle, to be positive. An amazing fact[4] observed is that the charge of the proton is exactly $+e$, the same magnitude as that for the electron. A large number of elementary particles are known, but we know of only three charges carried by them, $+e$, 0, and $-e$. Elementary particles of charge $\frac{2}{3}e$ and $\frac{1}{3}e$ are proposed theoretically,[5] but they have not been found. This situation is in sharp contrast to that for the mass. As seen in Table 11–19 there are all sorts of masses, and it is not easy to find a formula by which the masses of elementary particles are given. (See section 11–18 for further discussions.)

An even more important fact is that the mass is not conserved, but the charge is strictly conserved. We have seen in the last chapter that if the mass is defined by $\mathbf{p}/\mathbf{V}$, it depends on the velocity as $m_0/\sqrt{1 - (V/c)^2}$, while the charge is independent of the velocity, or is invariant under the Lorentz transformation. We have also seen that mc^2 contains the energy, so that in a reaction the mass is not conserved but only $mc^2 +$ (energy) is. Thus the mass of the proton and neutron are 1.007596 and 1.008986, respectively, in the atomic mass unit (amu), but that of the deuteron is 2.014194 amu. The charge, on the other hand, is exactly conserved, that is, the charge of the deuteron is exactly $+e$. At least at present we do not know of any reaction where the charge is not conserved.

The electrostatic and gravity forces are two forces which with nuclear forces and weak interactions are fundamental to modern physics. If we take the ratio between them for a pair of electrons, it is a dimensionless constant

$$\frac{ke^2}{Gm^2} = 4.17 \times 10^{42} \tag{5-4}$$

---

[4] V. W. Hughes, *Phys. Rev.* **97**, 380 (1957); and an article by V. W. Hughes in *Gravitation and Relativity* (edited by M. Y. Chiu and W. F. Hoffmann), Benjamin, New York, 1964.
[5] The quarks (see section 11–17). For the present experimental results see R. W. Holcomb, *Science*, **165**, 340 (1969), and D. C. Rahn and R. I. Louttit, *Phys. Rev. Letters*, **24**, 276 (1970). An "elementary particle" Δ^{++} has charge of $2e$, but decays into $\pi^+ + p$ in 10^{-23} sec.

Dirac[6] noticed that this huge number is obtained approximately if the present age of the universe, 3×10^{17} sec, is divided by an atomic time unit, $ke^2/mc^3 \cong 10^{-23}$ sec. Dirac, therefore, suggested that the gravitational constant G is inversely proportional to the age of the universe. Gamow[7] showed that this assumption contradicts the known age of the sun, and he proposed that e, instead of G, may be changing in proportion to the fourth root of the age of the universe. Gamow's proposition stirred up a big controversy, and some experimental evidences in atomic and nuclear spectra have been presented to disprove Gamow's idea. As Gamow said,[7] "it would be too bad to abandon an idea so elegant and so attractive as Dirac's proposal." But contemporary physics, like the contemporary world situation, does not seem to be as elegant as it used to be in the days of Newton or Coulomb.

5-2 ELECTRIC FIELD

The concept of the electric field was introduced by Michael Faraday. He assumed that to each point of space a vector quantity **E**, called the *electric field strength*, could be assigned. If a small amount of charge q is placed at **r**, the electrostatic force

$$\mathbf{F} = q\mathbf{E}(\mathbf{r}) \tag{5-5}$$

will be exerted on this test charge. Measuring the force **F**, we can measure the field strength **E** at this point **r**.

According to Coulomb's law (5-1) we obtain

$$\mathbf{E}(\mathbf{r}) = k \sum_i \frac{q_i(\mathbf{r} - \mathbf{r}_i)}{|\mathbf{r} - \mathbf{r}_i|^3} \tag{5-6}$$

where $\mathbf{r}_i$ is the position of charge q_i, and the summation is over all charges excepting the test charge mentioned above.

The *divergence* of **E** is

$$\mathbf{V} \cdot \mathbf{E} \equiv \frac{\partial E_x}{\partial x} + \frac{\partial E_y}{\partial y} + \frac{\partial E_z}{\partial z} \tag{5-7}$$

---

[6] P. A. M. Dirac, *Nature* **139**, 323 (1937); *Proc. Roy. Soc.* **A165**, 198 (1938).

[7] G. Gamow, *Phys. Rev. Letters* **19**, 759 (1967). Also see I. I. Shapiro, *et al., Phys. Rev. Letters*, **26**, 27 (1971) for a further evidence that G is constant within 3 parts in 10^{11} per year.

just as $\mathbf{V} \cdot \mathbf{j}$, introduced in equation (3–34). Since $\mathbf{E}$ is explicitly given by (5–6) we can calculate $\mathbf{V} \cdot \mathbf{E}$ directly, except at the points $\mathbf{r}_i$ where $\mathbf{E}$ diverges. If we assume that $\mathbf{r}$ does not coincide with any of $\mathbf{r}_i$'s, namely no charge exists at the point we are interested in, then direct calculation shows

$$\mathbf{V} \cdot \mathbf{E} = 0 \qquad (5–8)$$

The divergence of $\mathbf{E}$ at a singular point, one of $\mathbf{r}_i$'s, can be obtained from an entirely different viewpoint. Let us take an infinitesimal segment of a plane $d\mathbf{a}$ located at $\mathbf{r}$. The direction of $d\mathbf{a}$ is perpendicular to the area, with a magnitude $|d\mathbf{a}|$. Then, using (5–6) again, we see

$$\mathbf{E} \cdot d\mathbf{a} = k \sum_i q_i \frac{(\mathbf{r} - \mathbf{r}_i) \cdot d\mathbf{a}}{|\mathbf{r} - \mathbf{r}_i|^3} = k \left(\sum_j q_j \, d\Omega_j - \sum_h q_h \, d\Omega_h \right) \qquad (5–9)$$

where $d\Omega_i$ is the infinitesimal solid angle subtended at q_i by this area $d\mathbf{a}$. The sign of each term in the last expression of (5–9) is positive or negative depending on whether the charge is on the back side or the front side of the plane of $d\mathbf{a}$. (See Fig. 5–1.) If we take a volume v enclosed by a surface S and integrate (5–9) over the entire surface S, we obtain

$$\oint_S \mathbf{E} \cdot d\mathbf{a} = 4\pi k \sum_j^v q_j \qquad (5–10)$$

where the summation is over all charges contained in the volume v. Notice that when a charge is outside of the volume v, an infinitesimal solid angle

FIGURE 5–1. $\oint \mathbf{E} \cdot d\mathbf{a} = 4\pi k \sum_j^v q_j$.

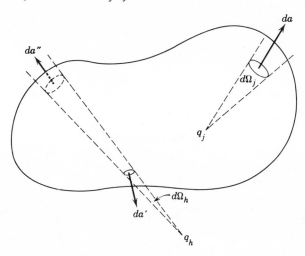

subtended at the charge by any infinitesimal area on the surface S will pass through the surface S at another infinitesimal area, and the contributions from these two infinitesimal areas to the integral (5–10) exactly cancel each other. Thus, on the right-hand side of (5–10), those charges outside of volume v do not appear.

Gauss' theorem, which we proved in section 3–2, states that

$$\int_v \mathbf{V} \cdot \mathbf{E} \, dv = \oint_S \mathbf{E} \cdot d\mathbf{a} \tag{5–11}$$

when applied to our case. Combining (5–10) and (5–11) we obtain

$$\int_v \mathbf{V} \cdot \mathbf{E} \, dv = 4\pi k \sum_j^v q_j \tag{5–12}$$

If we introduce the three-dimensional Dirac delta function by

$$\int_v \delta(\mathbf{r} - \mathbf{r}_i) \, dv = 1 \quad \text{if} \quad \mathbf{r}_i \quad \text{is inside } v$$
$$= 0 \quad \text{if} \quad \mathbf{r}_i \quad \text{is outside of } v \tag{5–13}$$

and write

$$\rho(\mathbf{r}) = \sum_i q_i \, \delta(\mathbf{r} - \mathbf{r}_i) \tag{5–14}$$

we can express (5–12) as

$$\int \mathbf{V} \cdot \mathbf{E} \, dv = 4\pi k \int \rho(\mathbf{r}) \, dv \tag{5–15}$$

We see $\rho(\mathbf{r})$ is the charge density at $\mathbf{r}$. Since (5–15) holds true for any volume v, we obtain

$$\mathbf{V} \cdot \mathbf{E} = 4\pi k \rho(\mathbf{r}) \tag{5–16}$$

This is the generalization of (5–8).

PROBLEM 5–1. From (5–6) and (5–7), show that (5–8) is obtained if $\mathbf{r}$ does not coincide to any of $\mathbf{r}_i$'s.

PROBLEM 5–2. Show

$$\delta(\mathbf{r} - \mathbf{r}_0) = \delta(x - x_0)\,\delta(y - y_0)\,\delta(z - z_0)$$

$$= \frac{\delta(r - r_0)\,\delta(\theta - \theta_0)\,\delta(\varphi - \varphi_0)}{r^2 \sin\theta}$$

$$= \frac{\delta(\rho - \rho_0)\,\delta(z - z_0)\,\delta(\varphi - \varphi_0)}{\rho}$$

5–3 SCALAR POTENTIAL. POISSON'S EQUATION. GREEN'S FUNCTION

Since the electrostatic force is central, as seen by (5–6), we obtain

$$\mathbf{V} \times \mathbf{E} = 0 \tag{5-17}$$

(See Problem 1–2.) We can, therefore, find a potential ϕ such that

$$\mathbf{E} = -\mathbf{V}\phi \tag{5-18}$$

We call ϕ the electrostatic scalar potential, or simply, the *scalar potential*. From (5–6) we obtain

$$\phi(\mathbf{r}) = k \sum_i \frac{q_i}{|\mathbf{r} - \mathbf{r}_i|} \tag{5-19}$$

From (5–16) and (5–18) we obtain

$$\mathbf{V}^2\phi = -4\pi k\rho(\mathbf{r}) \left(= -4\pi k \sum_i q_i\,\delta(\mathbf{r} - \mathbf{r}_i) \right) \tag{5-20}$$

This is called *Poisson's equation*.

In general, if we have an inhomogeneous differential equation

$$\hat{D}(\mathbf{V})\Phi(\mathbf{r}) = S(\mathbf{r}) \tag{5-21}$$

where $\hat{D}(\mathbf{V})$ is a differential operator which is a function of $\mathbf{V}$, and $S(\mathbf{r})$ is a source function, a general solution of this equation can be expressed as

$$\Phi(\mathbf{r}) = F(\mathbf{r}) + \int S(\mathbf{r}')G(\mathbf{r}, \mathbf{r}')\,dv' \tag{5-22}$$

where $F(\mathbf{r})$ is a solution of the corresponding homogeneous differential equation

$$\hat{D}(\mathbf{V})F(\mathbf{r}) = 0 \qquad (5\text{--}23)$$

and $G(\mathbf{r}, \mathbf{r}')$ is the solution of[8]

$$\hat{D}(\mathbf{V})G(\mathbf{r}, \mathbf{r}') = \delta(\mathbf{r} - \mathbf{r}') \qquad (5\text{--}24)$$

The function $G(\mathbf{r}, \mathbf{r}')$ is called *Green's function* of the differential operator $\hat{D}(\mathbf{V})$.

Looking at Poisson's equation (5–20) from this viewpoint, we see that the differential operator in this case is ∇^2, which is called the *Laplacian*, and the corresponding Green's function defined by (5–24) is

$$G(\mathbf{r}, \mathbf{r}') = -\frac{1}{4\pi} \frac{1}{|\mathbf{r} - \mathbf{r}'|} \qquad (5\text{--}25)$$

since (5–19) should be a solution of (5–20). Formula (5–22), however, shows that (5–19) is not the general solution of Poisson's equation (5–20), but that the solution can be written as

$$\phi(\mathbf{r}) = F(\mathbf{r}) + k \int \frac{\rho(\mathbf{r}')}{|\mathbf{r} - \mathbf{r}'|} \, dv' \qquad (5\text{--}26)$$

where $F(r)$ is a solution of

$$\nabla^2 F(\mathbf{r}) = 0 \qquad (5\text{--}27)$$

which is Laplace's equation which we have seen in (3–85). Comparing the present case with Poisson's equation (5–20), we see Laplace's equation is the equation for the scalar potential in a region free from charges.

Formula (5–26) shows that the scalar potential $\phi(\mathbf{r})$ cannot be decided uniquely from the charge distribution in a finite volume alone. Some boundary conditions are necessary to decide $F(\mathbf{r})$ and $\phi(\mathbf{r})$ completely.

PROBLEM 5–3. Derive (5–19) from (5–6).

PROBLEM 5–4. Using the Fourier transformation, show that (5–19) is a solution of (5–20).

---

[8] Some authors define G by $\hat{D}G = -\delta(\mathbf{r} - \mathbf{r}')$.

ANSWER. Let

$$\phi'(\mathbf{r}) = \int F(\mathbf{k}) e^{i\mathbf{k}\cdot\mathbf{r}} \, d\mathbf{k}$$

where

$$d\mathbf{k} = k^2 \sin\theta \, dk \, d\theta \, d\varphi$$

If

$$\nabla^2 \phi' = -\delta(\mathbf{r})$$

then, from (3–8),

$$\int F(\mathbf{k}) k^2 e^{i\mathbf{k}\cdot\mathbf{r}} \, d\mathbf{k} = \frac{\int e^{i\mathbf{k}\cdot\mathbf{r}} \, d\mathbf{k}}{(2\pi)^3}$$

so that

$$F(\mathbf{k}) = \frac{1}{k^2 (2\pi)^3}$$

Putting this result back into the original expression

$$\phi'(\mathbf{r}) = \frac{1}{(2\pi)^3} \int \frac{e^{i\mathbf{k}\cdot\mathbf{r}}}{k^2} k^2 \sin\theta \, dk \, d\theta \, d\varphi$$

$$= \frac{1}{(2\pi)^2} \int_0^\infty \int_0^\pi e^{ikr\cos\theta} \sin\theta \, d\theta \, dk$$

$$= \frac{1}{2\pi^2} \int_0^\infty \frac{\sin(kr)}{kr} \, dk = \frac{1}{4\pi r}$$

PROBLEM 5–5. Prove that (5–22) is a solution of (5–21).

HINT. Operate $\hat{D}(\mathbf{V})$ on (5–22).

PROBLEM 5–6. Knowing the Green's function of the Laplacian as (5–25), show that the Green's function of $\nabla^2 + k^2$, that is, a solution of $(\nabla^2 + k^2) G_k(r, r') = \delta(r - r')$, is[9]

$$G_k(\mathbf{r}, \mathbf{r}') = -\frac{1}{4\pi} \frac{\exp(ik|\mathbf{r} - \mathbf{r}'|)}{|\mathbf{r} - \mathbf{r}'|}$$

---

[9] Actually k in this Green's function can be either positive or negative, but we will use this function with positive k in the following.

HINT.

$$\nabla^2 \left(\frac{r}{r}\right) = r^{-1}\nabla^2 r + 2(\nabla r) \cdot (\nabla r^{-1}) + r\nabla^2 r^{-1}$$

but

$$\nabla^2 \left(\frac{r}{r}\right) = \nabla^2 1 = 0, \qquad \nabla^2 r = 0, \qquad \text{and} \qquad r\nabla^2 r^{-1} = -4\pi r\, \delta(r) = 0$$

so that

$$(\nabla r) \cdot (\nabla r^{-1}) = 0$$

5–4 SOLUTIONS OF LAPLACE'S EQUATION I

The Laplacian is

$$\nabla^2 = \frac{\partial^2}{\partial x^2} + \frac{\partial^2}{\partial y^2} + \frac{\partial^2}{\partial z^2} \tag{5–28}$$

in the Cartesian coordinate system. It is convenient to have this expression in the other coordinate systems. In the cylindrical coordinate system

$$\nabla^2 = \frac{1}{\rho}\frac{\partial}{\partial \rho}\left(\rho\,\frac{\partial}{\partial \rho}\right) + \frac{1}{\rho^2}\frac{\partial^2}{\partial \varphi^2} + \frac{\partial^2}{\partial z^2} \tag{5–29}$$

while in the spherical coordinate system

$$\nabla^2 = \frac{1}{r^2}\frac{\partial}{\partial r}\left(r^2\,\frac{\partial}{\partial r}\right) + \frac{1}{r^2 \sin\theta}\frac{\partial}{\partial \theta}\left(\sin\theta\,\frac{\partial}{\partial \theta}\right) + \frac{1}{r^2 \sin^2\theta}\frac{\partial^2}{\partial \varphi^2} \tag{5–30}$$

The solution of Laplace's equation can be written in several ways. In the Cartesian coordinate system, using (5–28), we see a solution

$$F_{\mathbf{k}}(\mathbf{r}) = e^{\mathbf{k}\cdot\mathbf{r}} = \exp\,(k_x x + k_y y + k_z z) \tag{5–31}$$

where

$$k_x^2 + k_y^2 + k_z^2 = 0 \tag{5–32}$$

Equation (5–32) implies that one or two components of **k** are imaginary. The equation has two other kinds of solutions:

$$F_{ak}(\mathbf{r}) = (ax + b)\exp\,(k_y y + k_z z) \tag{5–33}$$

where a and b are arbitrary constants and

$$k_y^2 + k_z^2 = 0 \qquad (5\text{–}34)$$

and

$$F_a(\mathbf{r}) = (a_x x + b_x)(a_y y + b_y)(a_z z + b_z) \qquad (5\text{–}35)$$

Any linear combinations of these solutions are also solutions of Laplace's equation.

Since $\mathbf{E}$ is zero everywhere in the ideal conductor, the whole body of an ideal conductor must be at a constant potential. To a good approximation metals can be regarded as ideal conductors.

If we have two infinite flat plate conductors which are placed parallel to each other at a distance d, and if the potential of one plate is ϕ_1 and that of the other plate is ϕ_2, the potential at distance x from the first plate in the space between these plates is

$$\phi = \phi_1 + (\phi_2 - \phi_1)\left(\frac{x}{d}\right) \qquad (5\text{–}36)$$

This is a special case of solution (5–35) (Fig. 5–2).

PROBLEM 5–7. Obtain (5–29) and (5–30) from (5–28).

FIGURE 5–2. Parallel plate conductors.

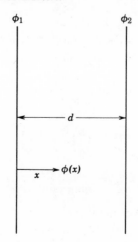

HINT. This is easier to do backward using

$$\frac{\partial}{\partial \rho} = \frac{\partial x}{\partial \rho} \frac{\partial}{\partial x} + \frac{\partial y}{\partial \rho} \frac{\partial}{\partial y} + \frac{\partial z}{\partial \rho} \frac{\partial}{\partial z} , \quad \text{etc.}$$

and (1–25) and (1–26).

PROBLEM 5–8. A rectangular box of dimensions a, b, and c in the x, y, and z directions, respectively, is made of a conductor (Fig. 5–3). If the

FIGURE 5–3. Rectangular box.

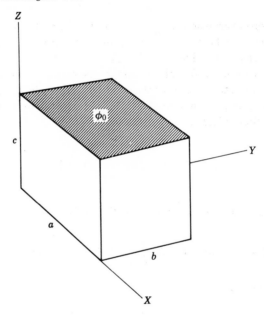

potential at one surface $z = c$ is ϕ_0, but that of the other five surfaces is zero, and if no charge exists inside the box, show that the potential inside the box is

$$\phi(r) = \sum_{n,m} A_{nm} \sin\left(\frac{n\pi x}{a}\right) \sin\left(\frac{m\pi y}{b}\right) \sinh(l\pi z)$$

where

$$l^2 = \left(\frac{n}{a}\right)^2 + \left(\frac{m}{b}\right)^2$$

$$A_{nm} = \frac{16\phi_0}{nm\pi^2 \sinh(l\pi c)}$$

and n and m are odd integer numbers.

PROBLEM 5–9. When the potential is given by (5–36), show that the charge of

$$|\sigma| = \frac{|\phi_2 - \phi_1|}{4\pi k\ d}$$

per unit area should exist on the surface of each conductor plate.

HINT. Take a "pill box" whose two flat surfaces are parallel to the surface of the plates, but one is inside a conductor and the other is in the space between the plates. The curved surface of the pill box is perpendicular to the surface of the plate as shown in Fig. 5–4. Apply Gauss' theorem to this pill box.

FIGURE 5–4.   Pill box on a conducting plate.

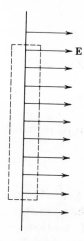

5–5 SOLUTIONS OF LAPLACE'S EQUATION II

Cylindrical Functions

Laplace's equation in the cylindrical coordinate system, given by the Laplacian of (5–29), gives a solution of the form

$$F = P(\rho)Z(z)\Phi(\varphi) \tag{5–37}$$

In this case Laplace's equation is reduced to

$$\frac{P''}{P} + \frac{1}{\rho}\frac{P'}{P} + \frac{Z''}{Z} + \frac{1}{\rho^2}\frac{\Phi''}{\Phi} = 0 \tag{5–38}$$

where

$$P'' = \frac{d^2P}{d\rho^2} \quad P' = \frac{dP}{d\rho} \quad Z'' = \frac{d^2Z}{dz^2} \quad \Phi'' = \frac{d^2\Phi}{d\varphi^2} \quad (5\text{-}39)$$

Equation (5–38) can be separated into three equations

$$\frac{Z''}{Z} = k^2 \qquad (5\text{-}40)$$

$$\frac{\Phi''}{\Phi} = -m^2 \qquad (5\text{-}41)$$

and

$$2P'' + \rho P' + (k^2\rho^2 - m^2)P = 0 \qquad (5\text{-}42)$$

The first two equations, (5–40) and (5–41), are solved easily to obtain

$$Z = Z_0 e^{\pm kz} \qquad (5\text{-}43)$$

$$\Phi = \Phi_0 e^{\pm im\varphi} \qquad (5\text{-}44)$$

and the last equation (5–42) can be solved by the Bessel function Z_m as

$$P = P_0 Z_m(k\rho) \qquad (5\text{-}45)$$

Z_0, Φ_0, and P_0 are constants. Some Bessel functions are tabulated in Appendix I(c). In (5–44) m has to be an integer number for $\Phi(\varphi) = \Phi(\varphi + 2\pi)$. There are two kinds of Bessel functions, the Bessel functions of the first kind J_m, and the Neumann functions N_m. A boundary condition in the ρ direction is necessary to choose the solution.

There are other kinds of solutions: When $m = 0$

$$\Phi = a\varphi + b \qquad (5\text{-}46)$$

and when $k = 0$

$$Z = az + b \quad P = \rho^{\pm m} \qquad (5\text{-}47)$$

In these solutions, a and b are arbitrary constants.

PROBLEM 5-10. Obtain solutions (5–46) and (5–47).

PROBLEM 5–11.   Show that when $k = m = 0$

$$F = (a \ln \rho + b)(cz + d)(e\varphi + f)$$

where a, b, c, d, e, and f are constants.

PROBLEM 5–12.   If two coaxial cylinders of radii ρ_1 and ρ_2 are kept at potentials ϕ_1 and ϕ_2, respectively, as shown in Fig. 5–5, and if no charge

FIGURE 5–5.   Coaxial cylinders.

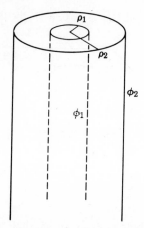

exists in the space between them, show that the potential at ρ, where $\rho_1 < \rho < \rho_2$, is

$$\phi = a \ln \rho + b$$

where

$$a = \frac{(\phi_1 - \phi_2)}{\ln (\rho_1/\rho_2)} \qquad b = \frac{(\phi_1 \ln \rho_2 - \phi_2 \ln \rho_1)}{\ln (\rho_2/\rho_1)}$$

PROBLEM 5–13.   An infinite circular cylindrical shell of radius a is at potential $\phi_0 \cos (m\varphi)$, where m is an integer number and φ is the azimuthal angle around the axis of the cylinder. Find the potential outside and inside the shell.

ANSWER.   Using (5–47)

$$\phi = \phi_0 a^m \cos (m\varphi)\rho^{-m} \qquad \text{for outside}$$

$$\phi = \phi_0 a^{-m} \cos (m\varphi)\rho^{m} \qquad \text{for inside}$$

PROBLEM 5–14. A circular thin disk is placed on an infinite plane. When the infinite plane is at a zero potential, but the disk is kept at potential ϕ_0, find the potential outside the plane (Fig. 5–6).

ANSWER. We find the formula

$$a \int_0^\infty J_0(k\rho)J_1(ka)\,dk = 1 \quad \text{if} \quad a > \rho > 0$$
$$= 0 \quad \text{if} \quad \rho > a > 0$$

FIGURE 5–6. The infinite plate is at $\phi = 0$, while the thin disk placed on that is at $\phi = \phi_0$.

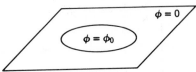

in a table. Thus, if a is the radius of the disk

$$\phi = a\phi_0 \int_0^\infty e^{-k|z|}J_0(k\rho)J_1(ka)\,dk$$

Spherical Functions

In the spherical coordinate system, we see that

$$F = R(r)\Theta(\theta)\Phi(\varphi) \tag{5–48}$$

separates Laplace's equation, expressed using the Laplacian of (5–30), into

$$\Phi'' = -m^2\Phi \tag{5–41}$$

$$\frac{1}{\sin\theta}\frac{d}{d\theta}\left(\sin\theta\frac{d\Theta}{d\theta}\right) = \left(\frac{m^2}{\sin^2\theta} - l'\right)\Theta \tag{5–49}$$

$$\frac{d}{dr}\left(r^2\frac{dR}{dr}\right) = l'R \tag{5–50}$$

The solution of the equation for Φ is the same as before, namely (5–44). If we require that the solution (5–48) be unique for the entire space, that is,

$$F(r, \theta, \varphi) = F(r, \theta, \varphi + 2\pi) = F(r, \theta + 2\pi, \varphi) \tag{5–51}$$

then m should be an integer, and

$$l' = l(l + 1) \qquad l: \text{zero or positive integer} \geq |m| \qquad (5\text{-}52)$$

$$\Theta = P_l^m (\cos \theta) \qquad (5\text{-}53)$$

which is called the *associated Legendre polynomial*.[10] Accordingly we obtain

$$R_l = ar^l + br^{-l-1} \qquad (5\text{-}54)$$

where a and b are constants.

Instead of the associated Legendre polynomials we often use the *spherical harmonics*, defined as

$$Y_{lm}(\theta, \varphi) = \sqrt{\frac{(2l + 1)}{4\pi} \frac{(l - m)!}{(l + m)!}} \, P_l^m (\cos \theta) e^{im\varphi} \qquad (5\text{-}55)$$

and

$$Y_{l-m}(\theta, \varphi) = (-1)^m Y_{lm}^*(\theta, \varphi) \qquad m \geq 0 \qquad (5\text{-}56)$$

The spherical harmonics defined in this way satisfy the orthonormality condition

$$\int_0^{2\pi} d\varphi \int_0^{\pi} \sin \theta \, d\theta Y_{lm}^*(\theta, \varphi) Y_{l'm'}(\theta, \varphi) = \delta_{l'l} \delta_{m'm} \qquad (5\text{-}57)$$

An important property of the spherical harmonics is the parity. When we invert the position vector $\mathbf{r}$ into $-\mathbf{r}$, we see r, θ, and φ are transformed into $r, \pi - \theta,$ and $\varphi + \pi$, respectively. The spherical harmonics have the property that

$$Y_{lm}(\pi - \theta, \varphi + \pi) = (-1)^l Y_{lm}(\theta, \varphi) \qquad (5\text{-}58)$$

The factor $(-1)^l$ is called the *parity* of the spherical harmonics. Using the spherical harmonics, the solution (5–48) of Laplace's equation can be written as

$$F = R_l(r) Y_{lm}(\theta, \varphi) \qquad (5\text{-}59)$$

Some Legendre polynomials and spherical harmonics are tabulated in Appendix I(c).

---

[10] When $m = 0$, $P_l^0 (\cos \theta)$ is designated simply as $P_l (\cos \theta)$ and is called the *Legendre polynomial*.

PROBLEM 5–15. A spherical shell of radius a is at a constant potential ϕ_0. Obtain the potential inside and outside the shell which approaches zero at a large distance from the shell.

HINT. Since $l = m = 0$ from the symmetry

$$\phi = A + \frac{B}{r}$$

where A and B are constants to be determined.

PROBLEM 5–16. When $\rho = \rho(r)$, or the charge distribution is spherical, show that

$$\phi = \phi(r) = \frac{4\pi k \int_0^r \rho(r')r'^2 \, dr'}{r}$$

PROBLEM 5–17. A point charge q is on the z axis at $z = a$. Find the potential at a point (r, θ, φ) (Fig. 5–7).

FIGURE 5–7. Charge q is at $z = a$.

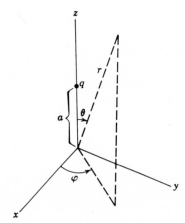

ANSWER.

$$\phi = kq \sum_{l=0}^{\infty} \frac{a^l}{r^{l+1}} P_l (\cos \theta) \qquad \text{when} \qquad r > a$$

$$\phi = kq \sum_{l=0}^{\infty} \frac{r^l}{a^{l+1}} P_l (\cos \theta) \qquad \text{when} \qquad r < a$$

PROBLEM 5–18. Two point charges q and $-q$ are on the z axis at $z = \frac{1}{2}a$ and $z = -\frac{1}{2}a$, respectively. Find the potential at $r \gg a$.

ANSWER.

$$\phi = kq \left[\frac{a}{r^2} P_1 (\cos \theta) + \frac{a^3}{4r^4} P_3 (\cos \theta) + \cdots \right]$$

qa is called the *electric dipole moment*, and qa^3 may be called the *electric octupole moment* of this system.

PROBLEM 5–19. Show from the result of Problem 5–18 that the field due to an electric dipole moment can be expressed as

$$\phi = k \frac{\boldsymbol{\mu} \cdot \mathbf{r}}{r^3}$$

$$\mathbf{E} = -k \left[\frac{\boldsymbol{\mu}}{r^3} - \frac{3\mathbf{r}(\boldsymbol{\mu} \cdot \mathbf{r})}{r^5} \right]$$

where $\boldsymbol{\mu}$ is the dipole moment vector with magnitude qa, and the direction is from $-q$ to $+q$.

PROBLEM 5–20. Obtain the potential due to the same system as that in Problem 5–18 except that the two charges are both $+q$. qa^2 is called the electric quadrupole moment of the system.

PROBLEM 5–21. A conducting sphere, whose potential is zero, is in a space of homogeneous electric field $\mathbf{E}_0$, as shown in Fig. 5–8. Find the potential around the sphere.

HINT. Boundary conditions are $\phi = 0$ on the surface of the sphere and $\phi = |\mathbf{E}_0|z$ when $r \to \infty$. Notice that this problem is mathematically equivalent to Problem 3–17.

PROBLEM 5–22. Show that a charge per unit area of $3|E_0| \cos \theta / 4\pi k$ is induced on the surface of the conducting sphere when it is placed in a homogeneous external field $\mathbf{E}_0$ (Problem 5–21). If the polarizability α is defined as (induced electric dipole moment)$/|\mathbf{E}_0|$, show that α of the conducting sphere is equal to a^3/k, where a is the radius of the sphere.

FIGURE 5–8. Potential ϕ around the sphere of $\phi = 0$. Arrows are **E**, which approaches **E**$_0$ at $z \to \pm\infty$.

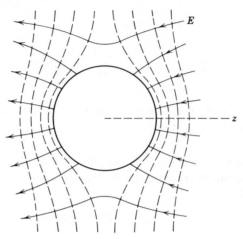

5–6 MAGNETIC FIELD

Magnets, which attract iron, were mentioned by Plato,[11] Lucretius,[12] and even the *Arabian Nights*.[13] A compass was described by Shen Kua, a Chinese mathematician, during the 11th century, and had become an essential navigation instrument in Europe by the 13th century. Because of its strange but useful properties the magnet attracted many speculations, most of which were mythical and superstitious.[14]

In 1600 William Gilbert published a book, *De Magnete*,[15] which was the first scientific attempt to investigate magnetism. Gilbert, and later René Descartes, a famous French philosopher for whom Cartesian coordinates are named, conceived the idea of a magnetic field produced by a large magnet such as the earth, aligning a smaller magnet such as a compass.[16] In modern

[11] *Ion*, [533], in the *Dialogues of Plato*.
[12] Luretius Carus, *De Rerum Natura*, v. 906.
[13] The third mendicant of the *Arabian Nights*.
[14] Mariners, particularly steermen, were forbidden to eat onions and garlic or to wear diamonds in order to keep the attention of their compasses to their own business.
[15] William Gilbert, *De Magnete Magneticisque et de Magno Magnete Tellure Physiologia Nova*. English translation can be found in *Great Books*, Vol. 28, Encyclopaedia Brittanica Inc., Chicago.
[16] A brief history of magnetism can be found in Chapter 1 of D. C. Mattis, *The Theory of Magnetism*, Harper & Row, New York, 1965.

terminology we say that a magnetic field **B** can be measured by the torque **N** exerted on a magnetic moment μ as

$$\mathbf{N} = \mu \times \mathbf{B} \qquad (5\text{-}60)$$

More precisely, **B** is called the *magnetic flux density* or *magnetic induction*. This fundamental formula, which corresponds to (5–5) for the electric field, implies an important difference between the electric and magnetic properties of matter: There is no magnetic charge or monopole, corresponding to the electric charge. Simon Denis Poisson, a French mathematical physicist for whom Poisson's equation is named, correctly assumed that a magnet can be divided into smaller magnets, but can never be separated into isolated poles. Some elementary particles are known to carry an elementary charge of e or $-e$, but no elementary particle is found which carries a magnetic charge.

Dirac[17] suggested that a magnetic charge of the magnitude $(137/2)e$ may exist. About 2.4 BeV would be required to isolate this magnetic charge. Attempts to find them using modern accelerators have failed so far.[18]

FIGURE 5–9a. Change of the earth's magnetic field during the past 9000 years, observed in Europe and Japan. BP stands for "before present," and F_0 is the field strength at present. [K. Kitazawa, *Butsuri* **24**, 825 (1969).]

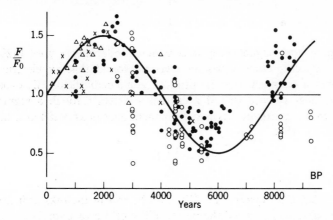

[17] P. A. M. Dirac, *Proc. Roy. Soc.* **133**, 60 (1931); *Phys. Rev.* **74**, 817 (1948); also, H. A. Wilson, *Phys. Rev.* **75**, 1968 (1949).
[18] H. Brandner and W. M. Isbell, *Phys. Rev.* **114**, 603 (1959). M. Fidecarco *et al.*, *Nuovo Cimento* **22**, 657 (1961). E. M. Purcell *et al.*, *Phys. Rev.* **129**, 2326 (1963).

Cosmic rays may have enough energy to produce Dirac's magnetic charge but so far attempts in this field have not been successful.[19]

There exists decisive evidence that the direction and magnitude of the earth's magnetic field have changed during the past (Fig. 5–9).

FIGURE 5–9b. Estimated positions of the magnetic North pole during historic time. Numbers are years in AD, and the broken circles indicate the regions of 95% confidence. [N. Kawai, K. Hirooka, and S. Sasajima, *Proc. Japan Acad.* **41**, 398 (1965); M. Keimatsu, N. Fukushima, and T. Nagata, *J. Geomag. Geoel.* **20**, 45 (1968).]

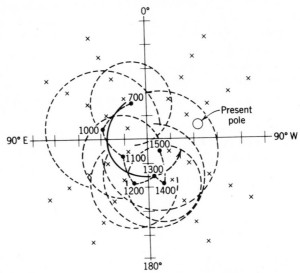

Electricity and magnetism were considered independent until 1820 when a Danish physicist, Hans Christian Oersted, found that an electric current can deflect a compass, that is, produce a magnetic field. His experiments and later ones showed that if a current I passes through a segment of a wire $d\mathbf{l}$, that segment produces the magnetic flux density $d\mathbf{B}$ at a distance $\mathbf{r}$ in such a way that

$$d\mathbf{B} = \frac{\mu_0}{4\pi} I \frac{d\mathbf{l} \times \mathbf{r}}{r^3} \qquad (5\text{–}61)$$

This is called *Biot-Savart's law* (Fig. 5–10), and in the mks unit (see Table 5–1) the constant μ_0, called the permeability of the vacuum, is

$$\mu_0 = 4\pi \times 10^{-7} \text{ weber/ampere meter} \qquad (5\text{–}62)$$

[19] W. V. R. Malkus, *Phys. Rev.* **83**, 899 (1951); E. Goto, *J. Phys. Soc. Japan* **10**, 1413 (1958); K. H. Shatten, *Phys. Rev. D* **1**, 2245 (1970). For reviews see L. W. Alvarez *et al.*, *Science*, **167**, 703 (1970); and H. H. Kolm, *Physics Today*, **20**, No. 10, 69 (1967).

FIGURE 5–10. Biot–Savart's law.

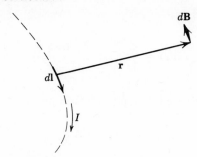

TABLE 5-1 *mks Unit System (the Unit System Used in this Book)*

Quantity	Symbol	Dimensions	Unit	Abbreviation
Length	l	l	Meter	m (= 10^2 cm)
Mass	m	m	Kilogram	kg (= 10^3 gm)
Time	t	t	Second	sec
Force	F	mlt^{-2}	Newton	(= 10^5 dyne)
Energy	$T, U, \mathcal{E}$	ml^2t^{-2}	Joule	J (= 10^7 erg)
Power		ml^2t^{-3}	Watt	
Frequency	$\omega/2\pi$	t^{-1}	Herz	Hz (= 1 cps)
Charge	q	q	Coulomb	C
Current	I	qt^{-1}	Ampere	A
Electric potential	ϕ	$q^{-1}ml^2t^{-2}$	Volt	V
Resistance	R	$q^{-2}ml^2t^{-1}$	Ohm	
Capacitance	C	$q^2m^{-1}l^{-2}t^2$	Farad	F
Magnetic flux	Φ	$q^{-1}ml^2t^{-1}$	Weber	(= 10^8 maxwell)
Charge density	ρ	ql^{-3}	C/m³	
Current density	J	$ql^{-2}t^{-1}$	A/m²	
Conductivity	σ	$q^2m^{-1}l^{-3}t$	1/m Ohm	
Electric field intensity	E	$q^{-1}mlt^{-2}$	V/m	
Displacement	D	ql^{-2}	C/m²	
Magnetic flux density	B	$q^{-1}mt^{-1}$	Weber/m²	(= 10^4 gauss)
Magnetic field intensity	H	$ql^{-1}t^{-1}$	A-turn/m	(= $4\pi10^{-3}$ oersted)
Capacitivity	ε	$q^2m^{-1}l^{-3}t^2$	F/m	
Permeability	μ	$q^{-2}ml$	Henry/m	

When the electric current density $\mathbf{J}(\mathbf{r}')$ is distributed in a conducting volume, the magnetic flux density $\mathbf{B}(\mathbf{r})$ produced at $\mathbf{r}$ is

$$\mathbf{B}(\mathbf{r}) = \frac{\mu_0}{4\pi} \int \mathbf{J}(\mathbf{r}') \times \frac{\mathbf{r} - \mathbf{r}'}{|\mathbf{r} - \mathbf{r}'|^3}\, dv' \qquad (5\text{–}63)$$

following Biot-Savart's law. If $\mathbf{V}$ is the differentiation with respect to $\mathbf{r}$, we can express (5–63) as

$$\mathbf{B}(\mathbf{r}) = \frac{\mu_0}{4\pi} \mathbf{V} \times \int \frac{\mathbf{J}(\mathbf{r}')}{|\mathbf{r} - \mathbf{r}'|} \, dv' \qquad (5\text{–}64)$$

from which we can obtain the important formulas

$$\mathbf{V} \cdot \mathbf{B} = 0 \qquad (5\text{–}65)$$

$$\mathbf{V} \times \mathbf{B} = \mu_0 \mathbf{J} \qquad (5\text{–}66)$$

when $\mathbf{V} \cdot \mathbf{J} = 0$. (5–66) is often called *Ampère's law*.

PROBLEM 5–23. A current I runs through a straight infinite wire. Find the magnetic flux density $\mathbf{B}$ at a distance d from the wire.

ANSWER. $B = \mu_0 I/(2\pi d)$.

PROBLEM 5–24. A current I runs through a circular wire of radius a. Find the magnetic flux density at a point on the line perpendicular to the plane of the wire passing through the center of the circle.

ANSWER. If R is the distance of the point from the center of the circle,

$$B = \frac{\mu_0 a^2 I}{2(a^2 + R^2)^{3/2}}$$

in the direction shown in Fig. 5–11.

PROBLEM 5–25. Obtain (5–65) from (5–64).

PROBLEM 5–26. Obtain (5–66) from (5–64) assuming $\mathbf{V} \cdot \mathbf{J} = 0$.

HINT. For a vector $\mathbf{A}$, $\mathbf{V} \times (\mathbf{V} \times \mathbf{A}) = \mathbf{V}(\mathbf{V} \cdot \mathbf{A}) - \nabla^2 \mathbf{A}$. Remember also that the Green's function of ∇^2 is given by (5–25).

PROBLEM 5–27. We have a straight infinitely long cylindrical shell conductor. When a total current I is homogeneously distributed over the surface of the shell running in the direction parallel to the symmetry axis of

FIGURE 5–11. **B** due to a circular current.

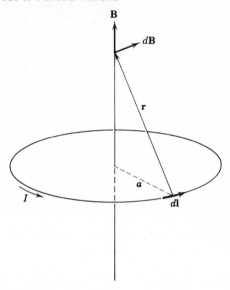

the cylinder, calculate the magnetic flux density **B** at distance R from the center of the cylinder.

HINT. From Stokes' theorem (1–20) and Ampère's law (5–66)

$$\mu_0 I = \oint \mathbf{B} \cdot d\mathbf{l} = 2\pi R B$$

if we take a circle with radius R, with its center on the symmetry axis of the cylinder, and situated perpendicular to the symmetry axis.

PROBLEM 5–28. A conducting wire is wound a large number of turns on a cylindrical form of radius a. This device is called a *solenoid*. Assuming that the length of the solenoid is infinite and the number of turns per unit length n is constant, find the magnetic flux density **B** inside and outside the solenoid as a function of the current I.

ANSWER. From the symmetry we see that **B** should be in a direction parallel to the axis of the solenoid. In a plane which contains the axis of the solenoid, take a rectangular loop *abcda* (see Fig. 5–12), such that the sides *ad* and *bc*, both with length l, are parallel to the axis and outside the solenoid. Integrate both sides of (5–69) over the area enclosed by this loop. From

FIGURE 5–12. Application of Stokes' theorem in finding **B** due to a solenoid.

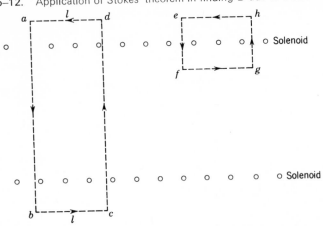

Stokes' theorem (1–20) and the expectation mentioned above that B should be in the direction of the axis, we obtain

$$(B_a - B_b)l = \mu_0 \int \mathbf{J} \cdot d\mathbf{a} = 0$$

where B_a and B_b are the magnitude of **B** along the sides ad and bc, respectively. $\int \mathbf{J} \cdot d\mathbf{a} = 0$, because the area cuts the wire at two points and the total current going into the area at one line is equal to that going out of the area at another line. Let one side ad be very far from the solenoid; then we expect that B_a will approach zero in the limit, and see that

$$B_b = 0$$

Therefore, **B** = 0 everywhere outside the solenoid.

Take another loop $efghe$ (see Fig. 5–12), such that the side eh is outside the solenoid, but the side fg is inside. The same procedure as before gives

$$B_f l = \mu_0 I n l \quad \text{or} \quad B_f = \mu_0 n I$$

if I is the current. Notice that **B** is independent of the position as far as it is inside the solenoid, that is, the magnetic field is homogeneous.

5–7 LORENTZ FORCE

A few weeks after Oersted's discovery (mentioned in the last section) was published a French physicist, André Marie Ampère, discovered[20] that

---

[20] Both Oersted's and Ampère's papers were published in *Annales de Chemie et de Physique* **15** (1820). Ampère may be one of the most colorful characters in the history of physics. He ignored Napoleon, perhaps because his father was beheaded during the French revolution.

two currents I_1 and I_2 interact with each other giving

$$\mathbf{F}_{12} = \frac{\mu_0}{4\pi} I_1 I_2 \oint \oint \frac{d\mathbf{l}_1 \times (d\mathbf{l}_2 \times \mathbf{r}_{12})}{r_{12}^{\ 3}} \qquad (5\text{-}67)$$

where $\mathbf{F}_{12}$ is the total force exerted on the current loop #1 by the current loop #2, $d\mathbf{l}_1$ and $d\mathbf{l}_2$ are segments of these loops, respectively, and $\mathbf{r}_{12}$ is the distance from $d\mathbf{l}_1$ to $d\mathbf{l}_2$.

Expression (5-67) looks asymmetric with respect to loops #1 and #2, but using

$$d\mathbf{l}_1 \times (d\mathbf{l}_2 \times \mathbf{r}_{12}) = -(d\mathbf{l}_1 \cdot d\mathbf{l}_2)\mathbf{r}_{12} + d\mathbf{l}_2(d\mathbf{l}_1 \cdot \mathbf{r}_{12}) \qquad (5\text{-}68)$$

we can reduce (5-67) to

$$\mathbf{F}_{12} = -\frac{\mu_0}{4\pi} I_1 I_2 \oint \oint \frac{(d\mathbf{l}_1 \cdot d\mathbf{l}_2)\mathbf{r}_{12}}{r_{12}^{\ 3}} \qquad (5\text{-}69)$$

which clearly shows that Newton's third law is satisfied.

Combining (5-61) and (5-67) together, we see that the magnetic field exerts a force on a circuit of

$$\mathbf{F} = I \oint d\mathbf{l} \times \mathbf{B} \qquad (5\text{-}70)$$

which can be generalized as

$$\mathbf{F} = \int (\mathbf{J} \times \mathbf{B}) \, dv \qquad (5\text{-}71)$$

for a distributed current.

If a current is made of moving charged particles, the current density can be expressed as

$$\mathbf{J}(\mathbf{r}) = \sum_i q_i \mathbf{V}_i \, \delta(\mathbf{r} - \mathbf{r}_i) \qquad (5\text{-}72)$$

When we have only one particle interacting with the magnetic field, (5-71) and (5-72) give

$$\mathbf{F} = q\mathbf{V} \times \mathbf{B} \qquad (5\text{-}73)$$

Combining this result with (5-5) we obtain the *Lorentz force*

$$\mathbf{F} = q(\mathbf{E} + \mathbf{V} \times \mathbf{B}) \qquad (5\text{-}74)$$

PROBLEM 5–29. Obtain (5–69) from (5–67).

HINT.

$$\oint d\mathbf{l}_1 \cdot \frac{\mathbf{r}_{12}}{r_{12}{}^3} = 0$$

PROBLEM 5–30. Two parallel infinite straight wires carry currents of I_1 and I_2, respectively. Calculate the force per unit length as a function of the distance d between them.

ANSWER. $F = \mu_0 I_1 I_2 / (4\pi\, d)$. Attractive when the currents are parallel, but repulsive when they are antiparallel.

PROBLEM 5–31. Calculate the magnetic interaction between two moving charges.

ANSWER. From (5–63) and (5–72) we obtain

$$\mathbf{B}(\mathbf{r}_2) = \frac{\mu_0}{4\pi}\, q_1\, \frac{\mathbf{V}_1 \times \mathbf{r}_{21}}{r_{21}{}^3}$$

Thus

$$\mathbf{F}_{21} = \frac{\mu_0}{4\pi}\, q_1 q_2\, \frac{\mathbf{V}_2 \times (\mathbf{V}_1 \times \mathbf{r}_{21})}{r_{21}{}^3} = \frac{\mu_0 q_1 q_2}{4\pi r_{21}{}^3}\, [\mathbf{V}_1(\mathbf{V}_2 \cdot \mathbf{r}_{21}) - \mathbf{r}_{21}(\mathbf{V}_1 \cdot \mathbf{V}_2)]$$

The resulting force is not in the direction of $\mathbf{r}_{21}$. We will discuss this point again in the next chapter.

PROBLEM 5–32. Show that if T is the relativistic kinetic energy

$$\frac{dT}{dt} = q\mathbf{V} \cdot \mathbf{E}$$

and the magnetic field does not contribute.

HINT. From (4–83) and (4–84) we see

$$dT = mV\gamma^3\, dV = \mathbf{V} \cdot d\mathbf{p}$$

Use (5–74) with $d\mathbf{p}/dt = \mathbf{F}$.

PROBLEM 5–33. Find the motion of a charge in a constant magnetic field.

ANSWER. If **B** is in the z direction

$$V_z = \text{constant} \qquad V_x = V_0 \sin \omega t \qquad V_y = V_0 \cos \omega t$$

where V_0 is a constant, and

$$\omega = \frac{Bqc^2}{T} = \frac{Bq}{m\gamma}$$

is called the *Larmor frequency* or *cyclotron frequency*.

5–8 MAGNETIC MOMENT

From (5–70) the torque exerted on a circuit by a magnetic field is

$$\mathbf{N} = I \oint \mathbf{r} \times (d\mathbf{l} \times \mathbf{B}) \tag{5–75}$$

where **r** is the position of $d\mathbf{l}$ measured from an origin; which means

$$d\mathbf{l} = d\mathbf{r} \tag{5–76}$$

Using the formulas in Appendix I(a), we obtain

$$\begin{aligned}
\mathbf{B} \times (\mathbf{r} \times d\mathbf{l}) &= -\mathbf{r} \times (d\mathbf{l} \times \mathbf{B}) - d\mathbf{l} \times (\mathbf{B} \times \mathbf{r}) \\
&= -2\mathbf{r} \times (d\mathbf{l} \times \mathbf{B}) + \mathbf{r} \times (d\mathbf{r} \times \mathbf{B}) - d\mathbf{r} \times (\mathbf{B} \times \mathbf{r}) \\
&= -2\mathbf{r} \times (d\mathbf{l} \times \mathbf{B}) + d(\mathbf{r} \times (\mathbf{r} \times \mathbf{B})) \\
&= -2\mathbf{r} \times (d\mathbf{l} \times \mathbf{B}) + d(\mathbf{r}(\mathbf{B} \cdot \mathbf{r})) - \mathbf{B} \, d(r^2)
\end{aligned} \tag{5–77}$$

where we used (5–76) and assumed that **B** is constant in the region of interest. Integrating both sides of (5–77) over the loop, we obtain

$$\begin{aligned}
\mathbf{B} \times \oint (\mathbf{r} \times d\mathbf{l}) &= -2 \oint \mathbf{r} \times (d\mathbf{l} \times \mathbf{B}) + \oint d(\mathbf{r}(\mathbf{B} \cdot \mathbf{r})) - \mathbf{B} \oint d(r^2) \\
&= -2 \oint \mathbf{r} \times (d\mathbf{l} \times \mathbf{B})
\end{aligned} \tag{5–78}$$

Therefore (5–75) can be written as

$$\mathbf{N} = \left(\tfrac{1}{2}I \oint \mathbf{r} \times d\mathbf{l} \right) \times \mathbf{B} \tag{5–79}$$

Comparing this expression to the fundamental formula (5–60) we see that the magnetic moment μ can be expressed as

$$\mu = \tfrac{1}{2}I \oint \mathbf{r} \times d\mathbf{l} = I\mathbf{a} \tag{5–80}$$

where $\mathbf{a}$ is the total area enclosed by the circuit.

Our result (5–80) provides a physical model of a magnet. Obviously we do not observe any gross current in a magnet, but if each atom in the magnet has a current loop around itself, and if they all rotate in the same direction, they can add up to produce a large magnetic moment. This picture, which was first proposed by Ampère, turns out to be about right, except that in most magnetic materials the atomic current is due to the electronic spin rather than the orbital motion. In any case, we need quantum mechanics to see why such atomic currents can be sustained and aligned together.

Formula (5–80) can be generalized to a distributed current giving

$$\mu = \frac{1}{2} \int (\mathbf{r} \times \mathbf{J}) \, dv \tag{5–81}$$

When the distributed current is made of moving charged particles, (5–72) and (5–81) give

$$\mu = \frac{1}{2} \sum_i q_i(\mathbf{r}_i \times \mathbf{V}_i) = \sum_i \frac{q_i}{2m_i} (\mathbf{r}_i \times \mathbf{p}_i) = \sum_i \frac{q_i}{2m_i} \mathbf{l}_i \tag{5–82}$$

in the nonrelativistic approximation. Thus for each charge, the magnetic moment is proportional to the angular momentum $\mathbf{l}$. The proportionality constant $q/(2m)$ is called the *magneto-mechanical ratio*. The relation (5–82) will be important in quantum mechanics.

PROBLEM 5–34. Using (5–63) and (5–81), show that the magnetic flux density produced by a magnetic moment μ is

$$\mathbf{B(r)} = \frac{\mu_0}{4\pi} \left[\frac{3\mathbf{r}(\mu \cdot \mathbf{r})}{r^5} - \frac{\mu}{r^3} \right]$$

Since this result is similar to that of the electric dipole moment (Problem 5–19), the interaction between two magnetic monopoles should be in the same form as Coulomb's law (5–1). This was experimentally verified by Coulomb himself.

5–9 FARADAY'S LAW OF INDUCTION. MAXWELL'S EQUATIONS

On August 29, 1831, Faraday discovered[21] that when the magnetic flux crossing a circuit

$$\Phi = \int \mathbf{B} \cdot d\mathbf{a} \tag{5–83}$$

changes in time, an electromotive force

$$E = \oint \mathbf{E} \cdot d\mathbf{l} \tag{5–84}$$

is induced in the circuit in such a way that

$$E = -\frac{\partial \Phi}{\partial t} \tag{5–85}$$

Using Stokes' theorem (Fig. 5–13), the corresponding differential form is obtained as

$$\mathbf{\nabla} \times \mathbf{E} = -\frac{\partial \mathbf{B}}{\partial t} \tag{5–86}$$

FIGURE 5–13.   Faraday's law of induction.

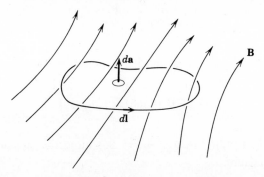

---

[21] M. Faraday, *Diary*, Royal Institution, London, 1932. This and relevant parts of Faraday's diary are quoted in G. Gamow, *Biography of Physics*, Harpers & Row, New York, 1961.

This is the time-dependent generalization of the previous law (5–17), and shows that the electric and magnetic fields are dynamically coupled.

Summarizing the field equations obtained so far:

$$\mathbf{V} \cdot \mathbf{E} = 4\pi k \rho \qquad (5\text{–}16)$$

$$\mathbf{V} \cdot \mathbf{B} = 0 \qquad (5\text{–}65)$$

$$\mathbf{V} \times \mathbf{B} = \mu_0 \mathbf{J} \qquad (5\text{–}66)$$

and (5–86), James Clark Maxwell noticed that a modification was necessary, and with that modification a symmetry between the electric and magnetic fields could be attained. He realized that

$$\mathbf{V} \cdot (\mathbf{V} \times \mathbf{B}) = 0 \qquad (5\text{–}87)$$

should hold identically, even if (5–65) was not valid; but looking at (5–66), $\mathbf{V} \cdot \mathbf{J}$ is not identically zero. On the other hand,

$$\frac{\partial \rho}{\partial t} + \mathbf{V} \cdot \mathbf{J} = 0 \qquad (5\text{–}88)$$

holds true due to conservation of charge. Using (5–16) we see

$$\mathbf{V} \cdot \mathbf{J} + \frac{\partial \rho}{\partial t} = \mathbf{V} \cdot \mathbf{J} + \frac{\partial}{\partial t} \frac{\mathbf{V} \cdot \mathbf{E}}{4\pi k} = \mathbf{V} \cdot \left(\mathbf{J} + \varepsilon_0 \frac{\partial \mathbf{E}}{\partial t} \right) = 0 \qquad (5\text{–}89)$$

where we defined

$$\varepsilon_0 \equiv \frac{1}{4\pi k} \qquad (5\text{–}90)$$

Therefore, Maxwell proposed that (5–66) should be replaced by

$$\mathbf{V} \times \mathbf{B} = \mu_0 \left(\mathbf{J} + \varepsilon_0 \frac{\partial \mathbf{E}}{\partial t} \right) \qquad (5\text{–}91)$$

This equation and (5–16), (5–65), and (5–86), altogether are called Maxwell's equations in vacuum.

PROBLEM 5–35. Show that (5–86) and (5–65) are consistent.

PROBLEM 5–36. In Faraday's law of induction (5–86), show that the sign of the induced electric field (or the electromotive force) is such that if it produces a current in the circuit, the current, in turn, produces additional magnetic flux to oppose the original change. This is called *Lenz's law*.

5–10 MICROSCOPIC AND MACROSCOPIC FIELDS

In the microscopic view, space is mostly vacuum with point particles, electrons and nuclei, scattered around. The electromagnetic fields are carried by the vacuum, and the point-charged particles give singularities of the fields. The fields, with strengths defined by the Lorentz force (5–74), change appreciably within the atomic dimension in substances such as crystals.

While the microscopic view is correct and necessary in many calculations, the macroscopic view, in which we neglect the atomic structure of a substance regarding it as a continuous medium, is often useful.

Let us denote the microscopic fields as **e** and **b**, corresponding to **E** and **B** of previous sections, respectively. If we place a subatomic test charge q with velocity **V**, we can define **e** and **b** by the Lorentz force exerted on this subatomic particle moving among atoms of a given medium as

$$\mathbf{F} = q(\mathbf{e} + \mathbf{V} \times \mathbf{b}) \qquad (5\text{–}74')$$

Maxwell's equations are assumed to hold true for **e** and **b**:

$$\nabla \cdot \mathbf{e} = 4\pi k\rho \qquad (5\text{–}92a)$$

$$\nabla \cdot \mathbf{b} = 0 \qquad (5\text{–}92b)$$

$$\nabla \times \mathbf{e} = -\frac{\partial \mathbf{b}}{\partial t} \qquad (5\text{–}92c)$$

$$\nabla \times \mathbf{b} = \mu_0 \left(\mathbf{J} + \varepsilon_0 \frac{\partial \mathbf{e}}{\partial t} \right) \qquad (5\text{–}92d)$$

Now take a volume much larger than the atomic volume (Å^3) but still much smaller than a macroscopic volume (say mm^3), and let the subatomic test particle move in this volume with constant **V**. The Lorentz force exerted on this particle averaged over this volume defines the macroscopic fields **E** and **B** as

$$\bar{\mathbf{F}} = q(\mathbf{E} + \mathbf{V} \times \mathbf{B}) \qquad (5\text{–}74'')$$

Obviously

$$\mathbf{E} = \bar{\mathbf{e}} \tag{5-93}$$

$$\mathbf{B} = \bar{\mathbf{b}} \tag{5-94}$$

or the macroscopic fields are average of the microscopic fields. The variation of the microscopic fields within the atomic distance is smoothed out in the macroscopic fields.

Maxwell's equations for the macroscopic fields can be obtained by averaging equations (5–92a)–(5–92d):

$$\nabla \cdot \mathbf{E} = 4\pi k \bar{\rho} \tag{5-95a}$$

$$\nabla \cdot \mathbf{B} = 0 \tag{5-95b}$$

$$\nabla \times \mathbf{E} = -\frac{\partial \mathbf{B}}{\partial t} \tag{5-95c}$$

$$\nabla \times \mathbf{B} = \mu_0 \left(\bar{\mathbf{J}} + \varepsilon_0 \frac{\partial \mathbf{E}}{\partial t} \right) \tag{5-95d}$$

which are about the same as the basic equations (5–92a)–(5–92d) except that the average charge density $\bar{\rho}$ and the average current density $\bar{\mathbf{J}}$ appear.

5–11 MACROSCOPIC SUSCEPTIBLE MEDIA

Electric Susceptibility

Macroscopic substances are either *conductors* or *dielectrics*, as far as their electric properties are concerned.[22] The conductors were introduced in section 5–4; these are bodies where macroscopic current will be induced when an electric field exists inside. The current will continue until the electric field inside the conductor disappears. In a dielectric, on the other hand, an applied electric field does not produce any macroscopic current, but can polarize the medium producing a local charge density ρ_P. If the substance is neutral as a whole, we have

$$\int \rho_P \, dv = 0 \tag{5-96}$$

---

[22] In a few other crystals, such as Rochelle salt or barium titanate, the polarization can exist without any external electric field; they are called *ferroelectrics*. In still other crystals, an external pressure can produce the polarization; they are called *piezoelectrics*.

where the integration is over the entire volume of the dielectric substance. In order to discuss the electric field inside such substance, it is convenient to introduce the polarization vector $\mathbf{P}$ defined by

$$\mathbf{V} \cdot \mathbf{P} = -\rho_P \qquad (5\text{-}97)$$

which, when combined with (5–96), means that $\mathbf{P}$ can be taken to be zero outside the given substance. It is easy to prove that

$$\int \mathbf{r}\rho_P(\mathbf{r}) \, dv = \int \mathbf{P}(\mathbf{r}) \, dv \qquad (5\text{-}98)$$

which shows that $\mathbf{P}$ can be interpreted as the density of the electric dipole moment due to the polarization.

A dielectric can have extra net charge density ρ_{ex} in addition to ρ_P; thus

$$\bar{\rho} = \rho_P + \rho_{ex} \qquad (5\text{-}99)$$

If we define the displacement $\mathbf{D}$ as

$$\mathbf{V} \cdot \mathbf{D} = \rho_{ex} \qquad (5\text{-}100)$$

then we can express $\mathbf{E}$ as

$$\mathbf{E} = 4\pi k(\mathbf{D} - \mathbf{P}) \qquad (5\text{-}101)$$

Note that any external field that is produced by charges outside of the dielectric but penetrating into the volume element of interest is included in $\mathbf{D}$, as well as the field produced by the local extra charge density ρ_{ex}.

The polarization in most of the dielectrics is due to the deformation of the charge distribution produced by the average Lorentz force $\bar{\mathbf{F}}$. Therefore, $\mathbf{P}$ is expected to be linear in $\mathbf{E}$, if the latter is not too strong. If the medium is isotropic

$$\mathbf{P} = \varepsilon_0 \chi_e \mathbf{E} \qquad (5\text{-}102)$$

where χ_e is called the *electric susceptibility* of the substance. In this case

$$\mathbf{D} = \varepsilon_0 \mathbf{E} + \mathbf{P} = \varepsilon_0 (1 + \chi_e)\mathbf{E} \equiv \varepsilon \mathbf{E} \qquad (5\text{-}103)$$

and ε is called the *capacitivity*: Thus ε_0 ($= 1/4\pi k$) is the capacitivity of the vacuum. The ratio

$$\kappa \equiv \frac{\varepsilon}{\varepsilon_0} = 1 + \chi_e \qquad (5\text{-}104)$$

is called the *dielectric constant*.

PROBLEM 5–37. Prove (5–98).

HINT. Use (5–97) and the partial integration, remembering that $P = 0$ outside a given substance.

PROBLEM 5–38. Extra charge of σ per unit area is distributed on a flat interface between dielectrics a and b. If n_{ab} is a unit vector perpendicular to the interface and point from a to b dielectrics, show that

$$(D_a - D_b) \cdot n_{ab} = -\sigma$$

where D_a and D_b are the displacement vectors at the a and b sides, respectively, of the interface. Thus if $\sigma = 0$, the perpendicular component of D is continuous through the interface.

PROBLEM 5–39. At the same interface as in Problem 5–38, show that

$$(E_a - E_b) \times n_{ab} = 0$$

Thus the tangential component of E is continuous through the interface.

HINT. Take a rectangular area the two sides of which are perpendicular to the interface and infinitesimal in length, while the other two sides are both parallel to the interface but one is in the dielectric a, and the other is in b. Integrate (5–95c) over this area.

PROBLEM 5–40. A flat board of infinite area but a finite thickness is made of a neutral dielectric of susceptibility χ_e. An external homogeneous field D_0 is applied in the direction θ with respect to the surface of the dielectric. Find D inside the dielectric (Fig. 5–14).

FIGURE 5–14. Dielectric board in a vacuum. D_n and E_t are continuous. **P** is zero in a vacuum.

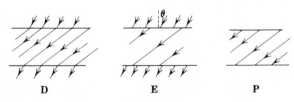

D	E	P

ANSWER.

$$D_n \text{ (outside)} = D_0 \cos \theta = D_n \text{ (inside)}$$

$$E_t \text{ (outside)} = \left(\frac{D_0}{\varepsilon_0}\right) \sin \theta = E_t \text{ (inside)} = \frac{D_t}{\varepsilon}$$

$$D \text{ (inside)} = \varepsilon E \text{ (inside)} = D_0\sqrt{\cos^2 \theta + (1 + \chi_e)^2 \sin^2 \theta}$$

PROBLEM 5–41. A spherical cavity of radius a is cut in a large homogeneous isotropic dielectric of capacitivity ε. If the electric field is homogeneously $\mathbf{E}_0$ in the dielectric at large distance from the cavity, calculate $\mathbf{E}$ in and around the cavity (Fig. 5–15).

FIGURE 5–15.   Lorentz field.

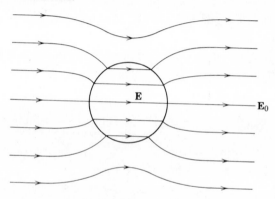

ANSWER.

$$\phi_+ = \left(-E_0 r + \frac{\alpha}{r^2}\right) \cos \theta \qquad \text{for} \qquad r > a$$

$$\phi_- = -E r \cos \theta \qquad \text{for} \qquad r < a$$

taking $\mathbf{E}_0$ along the z axis.

$$\mathbf{E} = \frac{3\varepsilon}{\varepsilon_0 + 2\varepsilon} \mathbf{E}_0$$

$$\alpha = (E_0 - E)a^3$$

It is interesting to note that at a large distance from the cavity

$$\mathbf{P} = \mathbf{D} - \varepsilon_0 \mathbf{E}_0 = (\varepsilon - \varepsilon_0)\mathbf{E}_0$$

so that

$$\mathbf{E} = \mathbf{E}_0 + \frac{\mathbf{P}}{\varepsilon_0 + 2\varepsilon}$$

This field in a spherical cavity is often called the *Lorentz field.*

Magnetic Susceptibility

Corresponding to the polarization **P** due to an electric field, we define the magnetization of a substance by

$$\mathbf{V} \times \mathbf{M} = \mathbf{J}_i \qquad (5\text{–}105)$$

where $\mathbf{J}_i$ is the density of the atomic current loops, each of which is localized in each atom. $\mathbf{J}_i$ does not contribute to the macroscopic current, but can contribute to the magnetic flux density. For an arbitrary cross section of a substance

$$\int \mathbf{J}_i \cdot d\mathbf{a} = 0 \qquad (5\text{–}106)$$

because $\mathbf{J}_i$ is made of atomic current loops. Thus **M**, defined by (5–105) can be taken to be zero outside of the substance. Definition (5–105) also gives

$$\int \mathbf{M} \, dv = \int \boldsymbol{\mu} \, dv \qquad (5\text{–}107)$$

or **M** can be regarded as the density of the atomic magnetic moments.

Any external magnetic field, and the field produced by macroscopic current, which may exist in a given substance if the substance is a conductor, are put together into the *magnetic field intensity* **H**, defined by

$$\mathbf{V} \times \mathbf{H} = \mathbf{J}_{ex} \qquad (5\text{–}108)$$

where $\mathbf{J}_{ex}$ is the density of the macroscopic current. Since

$$\bar{\mathbf{J}} = \mathbf{J}_i + \mathbf{J}_{ex} \qquad (5\text{–}109)$$

we see

$$\mathbf{B} = \mu_0(\mathbf{H} + \mathbf{M}) \qquad (5\text{–}110)$$

holds.

Most of the substances are either *ferromagnetic, paramagnetic,* or *diamagnetic,* with respect to their magnetic properties. In the ferromagnetic substances, the magnetization **M** can exist without any external magnetic field. Thus permanent magnets are ferromagnetic. In a ferromagnetic substance **M** is a complicated function of **H**: There is a maximum value that |**M**| can take for a given sample, called the saturation magnetization, and **M** at a given **H** depends on the history of the sample, *hysteresis* phenomena (Fig. 5–16).

FIGURE 5–16.   Hysteresis curve.

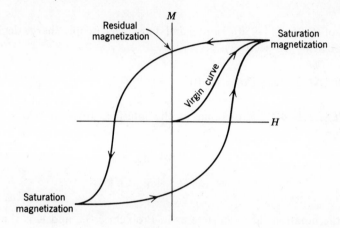

In the para- and diamagnetic substances, **M** and **H** are conventionally[23] related by

$$\mathbf{M} = \chi_m \mathbf{H} \tag{5-111}$$

where χ_m is called the magnetic susceptibility. In the paramagnetic substances $\chi_m > 0$, while in the diamagnetic substances $\chi_m < 0$. In either case

$$\mathbf{B} = \mu_0(\mathbf{H} + \mathbf{M}) = \mu_0(1 + \chi_m)\mathbf{H} = \mu\mathbf{H} \tag{5-112}$$

and μ is called the permeability of the substance.

---

[23] In the atomic viewpoint, the magnetization **M** is an expression of atomic deformations produced by the Lorentz force, or by **B** field. It would be more natural, in this viewpoint, to assume a proportionality between **M** and **B**.

From this viewpoint, Maxwell's equations in a macroscopic medium can be written as

$$\mathbf{\nabla} \cdot \mathbf{D} = \rho \tag{5-113a}$$

$$\mathbf{\nabla} \cdot \mathbf{B} = 0 \tag{5-113b}$$

$$\mathbf{\nabla} \times \mathbf{E} = - \frac{\partial B}{\partial t} \tag{5-113c}$$

$$\mathbf{\nabla} \times \mathbf{H} = \mathbf{J} + \frac{\partial \mathbf{D}}{\partial t} \tag{5-113d}$$

instead of (5–95a)–(5–95d). Here ρ and $\mathbf{J}$ are macroscopic charge density and current density, respectively.

PROBLEM 5–42. Prove (5–107).

PROBLEM 5–43. Show that at the interface between two substances a and b,

$$(\mathbf{B}_a - \mathbf{B}_b) \cdot \mathbf{n}_{ab} = 0$$

$$(\mathbf{H}_a - \mathbf{H}_b) \times \mathbf{n}_{ab} = \mathbf{K}$$

where the notations are the same as in Problem 5–38, and $\mathbf{K}$ is the current density per unit length passing on the interface. (Table 5–2 summarizes the boundary conditions obtained here and in Problems 5–38 and 5–39.)

TABLE 5–2 *Boundary Conditions of Electromagnetic Fields*

Tangential Components	$\mathbf{E}_a \times \mathbf{n}_{ab} = \mathbf{E}_b \times \mathbf{n}_{ab}$	$\mathbf{H}_a \times \mathbf{n}_{ab} = \mathbf{H}_b \times \mathbf{n}_{ab} + \mathbf{K}$
Perpendicular Components	$\mathbf{D}_a \cdot \mathbf{n}_{ab} = \mathbf{D}_b \cdot \mathbf{n}_{ab} + \sigma$	$\mathbf{B}_a \cdot \mathbf{n}_{ab} = \mathbf{B}_b \cdot \mathbf{n}_{ab}$

$\mathbf{n}_{ab}$ *is a unit vector perpendicular to the interface between media a and b, in the direction from a to b.*
σ *is the interface charge density (charge/unit area).*
$\mathbf{K}$ *is the interface current density (current/unit length).*

PROBLEM 5–44. The work required to increase the charge density by $\delta \rho$ is

$$\delta U = \int \phi \, \delta \rho \, dv$$

Show that

$$\delta U = \int \mathbf{E} \cdot \delta \mathbf{D} \, dv$$

when the integral is taken over a volume on whose surface $\phi = 0$.

PROBLEM 5–45. The work done by the field $\mathbf{E}$ on the current $\mathbf{J}$ during a time δt is

$$-\delta U = \delta t \int \mathbf{E} \cdot \mathbf{J} \, dv$$

in other words, δU is the increase of the field energy in time δt. Show that

$$\delta U = \delta t \int (\mathbf{E} \times \mathbf{H}) \cdot d\mathbf{a} + \int (\mathbf{H} \cdot \delta \mathbf{B} + \mathbf{E} \cdot \delta \mathbf{D}) \, dv$$

where

$$\delta \mathbf{B} = \frac{\partial \mathbf{B}}{\partial t} \delta t \qquad \delta \mathbf{D} = \frac{\partial \mathbf{D}}{\partial t} \delta t$$

and the first integral is over the surface of the volume on which the second integral is performed.

HINT. Use (5–113c) and (5–113d).

5–12 ELECTROMAGNETIC PLANE WAVE

One of the important results of Maxwell's modification (5–91) of (5–66) is that the existence of the electromagnetic wave is predicted, and this gives a complete physical explanation of light and associated optical phenomena.[24]

In an isotropic homogeneous medium ε and μ are constants independent of time and position, and we may assume that

$$\rho = 0 \qquad (5\text{--}114)$$

$$\mathbf{J} = \sigma \mathbf{E} \qquad (5\text{--}115)$$

---

[24] A brief but good historical review of theories of light before Maxwell can be found in the Introduction of M. Born and E. Wolf, *Principles of Optics*, Pergamon, Oxford, 1965. Also see the introduction of M. Klein and I. W. Kay, *Electromagnetic Theory and Geometrical Optics*, Interscience, New York, 1965.

where σ is the conductivity. When σ is constant equation (5–115) expresses *Ohm's law*. In this case (5–113c) and (5–113d) give

$$\mathbf{V} \times (\mathbf{V} \times \mathbf{E}) = -\frac{\partial}{\partial t}(\mathbf{V} \times \mathbf{B}) = -\mu\left(\sigma\frac{\partial \mathbf{E}}{\partial t} + \varepsilon\frac{\partial^2 \mathbf{E}}{\partial t^2}\right) \qquad (5\text{–}116)$$

but

$$\mathbf{V} \times (\mathbf{V} \times \mathbf{E}) = \mathbf{V}(\mathbf{V} \cdot \mathbf{E}) - \mathbf{V}^2\mathbf{E} = -\mathbf{V}^2\mathbf{E} \qquad (5\text{–}117)$$

because of (5–113a) and (5–114). Thus we obtain

$$\mathbf{V}^2\mathbf{E} - \varepsilon\mu\frac{\partial^2 \mathbf{E}}{\partial t^2} - \sigma\mu\frac{\partial \mathbf{E}}{\partial t} = 0 \qquad (5\text{–}118)$$

In the same way we obtain

$$\mathbf{V}^2\mathbf{B} - \varepsilon\mu\frac{\partial^2 \mathbf{B}}{\partial t^2} - \sigma\mu\frac{\partial \mathbf{B}}{\partial t} = 0 \qquad (5\text{–}119)$$

When $\sigma = 0$, or in a dielectric medium, (5–118) has a solution

$$\mathbf{E} = \mathbf{E}_0 \cos(\mathbf{k} \cdot \mathbf{r} - \omega t) \qquad (5\text{–}120)$$

which is a plane wave of the amplitude $\mathbf{E}_0$ and the propagation direction $\mathbf{k}$. The phase velocity of this wave is given by

$$\frac{\omega}{|\mathbf{k}|} = \frac{1}{\sqrt{\varepsilon\mu}} \qquad (5\text{–}121)$$

If the medium is the vacuum, this velocity is

$$c = \frac{1}{\sqrt{\varepsilon_0\mu_0}} = 2.9979250 \times 10^8 \text{ m/sec} \qquad (5\text{–}122)$$

using the value of ε_0 and μ_0 given before. The resulting number agrees exactly with that of the velocity of light! The velocity of an electromagnetic wave in a medium, as given by (5–121), is usually smaller than c. If the index of refraction is n, the velocity of the electromagnetic wave in that medium can be written (Problem 5–49) as

$$V = \frac{1}{\sqrt{\varepsilon\mu}} = \frac{c}{n} \qquad (5\text{–}123)$$

It will be shown later that n is a function of the frequency ω, and the phenomenon that the velocity of a wave depends on its frequency will be called *dispersion*.

Because of (5–113c) the E wave given by (5–120) is associated with a $\mathbf{B}$ wave given by

$$\mathbf{B} = \mathbf{k} \times \frac{\mathbf{E}}{\omega} = \left(\frac{\mathbf{k}}{|\mathbf{k}|}\right) \times \frac{\mathbf{E}n}{c} \tag{5–124}$$

The whole wave, called the *electromagnetic wave*, is made of $\mathbf{E}$ and $\mathbf{B}$ fields oscillating in the same frequency and phase but perpendicular to each other (Fig. 5–17).

FIGURE 5–17. Electromagnetic wave (linearly polarized).

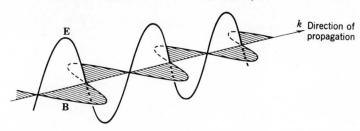

Since $\mathbf{E}$ and $\mathbf{B}$ fields oscillate in a plane perpendicular to the direction of the propagation, given by $\mathbf{k}$, the electromagnetic wave is a transverse wave.[25] The plane in which the $\mathbf{E}$ field oscillates is called the polarization plane of the electromagnetic wave. Since there are two mutually perpendicular polarization planes for a given propagation direction, there are two independent plane waves. If $\mathbf{k}$ is taken along the z axis

$$E_x = E_0' \cos (kz - \omega t + \delta_x) \tag{5–125}$$

and

$$E_y = E_0'' \cos (kz - \omega t + \delta_y) \tag{5–125'}$$

can be the electric components of these two independent plane waves. Each of them is called the linear, or plane polarized plane wave. A linear combination of these plane polarized waves is also a solution of the wave equation (5–118). If we superimpose two waves with the phase relation

$$\delta_x - \delta_y = \pm\tfrac{1}{2}\pi \tag{5–126}$$

[25] In 1817 Thomas Young suggested that light must be a transverse wave, to explain the interference properties observed by Augustin Fresnel and Dominique F. Arago. It was another triumph of Maxwell's theory that this property of light was explained without further assumptions.

we see

$$\left(\frac{E_x}{E_0'}\right)^2 + \left(\frac{E_y}{E_0''}\right)^2 = 1 \tag{5-127}$$

which shows that in this superimposed wave, the **E** vector moves along an ellipse. This is called an elliptically polarized wave. In a special case of $E_0' = E_0''$, we obtain a circularly polarized wave. It is easy to see that the double sign in (5–126) corresponds to the two possible senses of rotation of the **E** vector with respect to the propagation direction (Fig. 5–18).

FIGURE 5–18. Two circularly polarized lights.

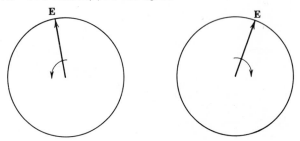

When the conductivity σ is finite, we substitute

$$\mathbf{E} = \mathbf{E}_0 e^{i(\mathbf{k}\cdot\mathbf{r}-\omega t)} \tag{5-128}$$

in (5–118) and obtain

$$k^2 = \varepsilon\mu\omega^2 \left(1 + i\left(\frac{\sigma}{\varepsilon\omega}\right)\right) \tag{5-129}$$

Therefore, we have to have a complex **k** vector in (5–128). When

$$\sigma \ll \varepsilon\omega \tag{5-130}$$

we obtain

$$k \cong \sqrt{\varepsilon\mu}\,\omega + i\tfrac{1}{2}\sigma\sqrt{\mu/\varepsilon} \tag{5-131}$$

and

$$\mathbf{E} \cong \mathbf{E}_0 \exp\left(-\tfrac{1}{2}z\sigma\sqrt{\mu/\varepsilon}\right)e^{i(\sqrt{\varepsilon\mu}z-t)\omega} \tag{5-132}$$

which shows that the electromagnetic wave decays as it propagates into a conducting medium. The amplitude decreases to $1/e$ of the original value at a distance of

$$\delta \equiv \frac{2}{\sigma}\sqrt{\frac{\varepsilon}{\mu}} \tag{5-133}$$

This distance is called the *skin depth*.

PROBLEM 5–46.   Obtain (5–119).

PROBLEM 5–47.   An electromagnetic wave propagating along the z axis is linearly polarized in the plane making a 45° angle to both x and y axes. When it comes into a dielectric such that $\varepsilon_x \neq \varepsilon_y$, show that the polarization character changes from linear to elliptic, and then back to a linear polarization at a certain depth as the wave propagates through the dielectric. (In most dielectrics $\mu_x = \mu_y = \mu_0$.)

PROBLEM 5–48.   Show that for a good conductor, where $\sigma \gg \omega$, we obtain

$$\mathbf{E} = \mathbf{E}_0 e^{-\alpha z} e^{i(\alpha z - \omega t)}$$

where

$$\alpha = \sqrt{\omega \mu \sigma / 2}$$

HINT.   From (5–129) $k \cong \alpha(1 + i)$.

PROBLEM 5–49.   When an electromagnetic wave hits an interface between two dielectrics, it is partially reflected back to the original medium a, and partially refracted into the second medium b. Show that if the incident angle is θ_a with respect to the interface, the reflection angle is also θ_a, but the refraction angle is θ_b such that $\sin \theta_a / \sin \theta_b = n_b/n_a$, where n_a and n_b are the indices of refraction of the media defined by (5–123). (See Fig. 5–19.)

FIGURE 5–19.   Reflection and refraction.

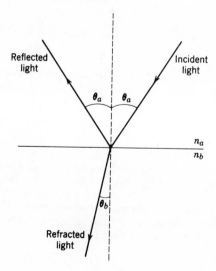

PROBLEM 5–50. An electromagnetic plane wave propagating in a medium with an index of refraction n_a is incident on the interface between the original medium and another with an index of refraction of n_b in a direction perpendicular to the interface. Calculate the amplitude of the refracted and reflected waves in units of the incident wave amplitude. Assume that $\mu_a = \mu_b$.

ANSWER.

$$E_r = \frac{2n_a}{n_a + n_b} E_i \qquad E_l = \frac{n_a - n_b}{n_a + n_b} E_i$$

where E_i, E_r, and E_l are the amplitudes of incident, refracted, and reflected waves, respectively.

5–13 HUYGENS' PRINCIPLE

In 1678, Christian Huygens enunciated the principle that every point upon which a wave falls at one time can be regarded as the center of a new disturbance propagated in the form of spherical waves; and the envelope of these secondary waves determines the wave front at any later time (Fig. 5–20). However, this principle, called *Huygens' principle*, was rejected by Newton, and was forgotten for over a century until, at the beginning of the 19th century, Fresnel and Young applied it successfully in their theory of diffraction.[26]

Each component of **E** and **B** obeys the wave equation

$$\left(\nabla^2 - \frac{\partial^2}{c^2 \, \partial t^2}\right) Q \equiv \Box \, Q = 0 \tag{5-134}$$

where $\Box$ is the *D'Alembertian*, which was introduced in Problem 4–10. (We will take the vacuum as a typical medium for the rest of this chapter, but all resulting formulas can be easily modified for more general media.) If we set

$$Q = qe^{i\omega t} \tag{5-135}$$

[26] Even at the beginning of the 19th century, the authority of Newton was still so strong in England that Young's theory was strongly criticized by an authority of his time, one Lord Brougham, who declared "we searched [in Young's paper] without success for some traces of learning, acuteness, and ingenuity, that might compensate his evident deficiency in the power of development of the laws of nature, by steady and modest observation of his operation." Young prepared his reply as a pamphlet, but could only sell one copy. Fresnel, being a French physicist, was much more fortunate.

FIGURE 5–20. Huygens' principle.

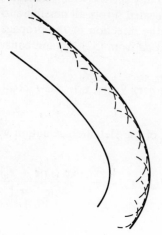

the wave equation (5–134) gives

$$\nabla^2 q + k^2 q = 0 \qquad \left(k = \frac{\omega}{c}\right) \tag{5-136}$$

which is the Helmholz equation.

Now if we have two arbitrary functions u and v, we see

$$\nabla \cdot (u\nabla v - v\nabla u) = u\nabla^2 v - v\nabla^2 u \tag{5-137}$$

so that

$$\int (u\nabla^2 v - v\nabla^2 u)\, dv = \oint (u\nabla v - v\nabla u) \cdot d\mathbf{a} \tag{5-138}$$

where the second integral is over the entire surface of the volume of the first integral. This relation is called *Green's theorem.*

Let us take the wave function q and the Green's function $G_k(\mathbf{r}, \mathbf{r}_0)$ of the Helmholz equation, for which

$$(\nabla^2 + k^2)G_k(\mathbf{r}, \mathbf{r}_0) = \delta(\mathbf{r} - \mathbf{r}_0) \tag{5-139}$$

and apply them to Green's theorem (5–138). We obtain

$$q(\mathbf{r}_0) = \oint (q(\mathbf{r})\nabla G_k(\mathbf{r}, \mathbf{r}_0) - G_k(\mathbf{r}, \mathbf{r}_0)\nabla q(\mathbf{r})) \cdot d\mathbf{a} \tag{5-140}$$

where the integral is over any surface which encloses r_0, and r is the position where the element $d a$ is located. From all possible surfaces we take one which is perpendicular to ∇q, the direction of the propagation of the q wave. In this case[27] $\nabla q \cdot d\mathbf{a} = ikq\, da$ from (5–136), and equation (5–140) reduces to

$$q(\mathbf{r}_0) = \oint q(\mathbf{r})[\nabla G_k(\mathbf{r}, \mathbf{r}_0) \cdot d\mathbf{a} - ikG_k(\mathbf{r}, \mathbf{r}_0)\, da] \qquad (5\text{--}140')$$

The Green's function of the Helmholz equation was obtained in Problem 5–6 as

$$G_k(\mathbf{r}, \mathbf{r}_0) = - \frac{\exp(ik|\mathbf{r} - \mathbf{r}_0|)}{4\pi|\mathbf{r} - \mathbf{r}_0|} \qquad (5\text{--}141)$$

so that

$$\nabla G_k(\mathbf{r}, \mathbf{r}_0) = - \left(\frac{ik}{|\mathbf{r} - \mathbf{r}_0|} - \frac{1}{|\mathbf{r} - \mathbf{r}_0|^2} \right) (\mathbf{r} - \mathbf{r}_0)G_k(\mathbf{r}, \mathbf{r}_0)$$

$$(5\text{--}142)$$

If we consider the region such that

$$k|r - r_0| \gg 1 \qquad (5\text{--}143)$$

that is, if the distance is much larger than the wavelength, we can neglect the second term of (5–142), and, from (5–140′), we obtain the final formula

$$q(\mathbf{r}_0) = - \frac{ik}{4\pi} \oint q(\mathbf{r}) \frac{\exp(ik|\mathbf{r} - \mathbf{r}_0|)}{|\mathbf{r} - \mathbf{r}_0|^2} [(\mathbf{r} - \mathbf{r}_0) \cdot d\mathbf{a} - |\mathbf{r} - \mathbf{r}_0|\, da]$$

$$(5\text{--}144)$$

This is Kirchhoff's derivation of *Huygens' principle*. Fig. 5–21 shows the closed integral area and other quantities which appear in (5–144).

PROBLEM 5–51. Obtain (5–140).

PROBLEM 5–52. Generalize Kirchhoff's formula (5–144) for waves which are not necessarily monochromatic.

---

[27] When $\nabla q \cdot d\mathbf{a} = ikq\, da$ the wave is propagating into the volume enclosed by the surface integration da, while we should take $\nabla q \cdot d\mathbf{a} = -ikq\, da$ if the wave is propagating out.

FIGURE 5–21. Integral (5–144) is taken over a surface perpendicular to the direction of propagation of q at everywhere ,and encloses r_0.

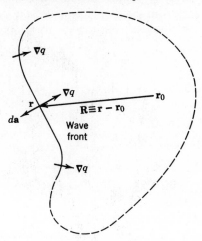

ANSWER. A general wave can be expressed as the Fourier integral:

$$Q(\mathbf{r}_0, t) = \frac{1}{\sqrt{2\pi}} \int_{-\infty}^{\infty} q_\omega(\mathbf{r}_0) e^{-i\omega t}\, d\omega$$

$$= \frac{1}{4\pi\sqrt{2\pi}} \int_{-\infty}^{\infty} d\omega \oint e^{-i\omega t} q_\omega(\mathbf{r}) \left(\frac{de^{ikR}}{dR}\right) \left(\frac{\mathbf{R}}{R^2} \cdot d\mathbf{a} + \frac{da}{R}\right)$$

$$= \frac{1}{4\pi\sqrt{2\pi}} \oint \left(\frac{d}{dR} \int_{-\infty}^{\infty} q_\omega(\mathbf{r}) e^{ikR - i\omega t}\, d\omega\right) \left(\frac{\mathbf{R}}{R^2} \cdot d\mathbf{a} + \frac{da}{R}\right)$$

$$= \frac{1}{4\pi} \oint \left[\frac{d}{dR} Q\left(\mathbf{r}, t - \frac{R}{c}\right)\right] \left(\frac{\mathbf{R}}{R^2} \cdot d\mathbf{a} + \frac{da}{R}\right)$$

since $k = \omega/c$. Here $\mathbf{R} = \mathbf{r} - \mathbf{r}_0$.

5–14 DIFFRACTION

When a plane electromagnetic (or any) wave passes through a slit, a diffraction pattern of the slit is obtained.[28] If the plane of the slit is perpendicular to the propagation direction of the plane wave, the amplitude q and ∇q are both constant over the area of the slit and zero outside the area.

[28] First observed by Leonardo da Vinci.

The amplitude of a component of $\mathbf{E}$ or $\mathbf{B}$ at a position $\mathbf{r}_0$ in the other side of the screen can be given by (5–144), if the surface integral is taken over the area of the slit. If the slit is in the xy plane, and if the dimensions of the slit are much less than the distance from the slit to the position $\mathbf{r}_0$, which we designate R, then

$$q(\mathbf{r}_0) = - \frac{ikq}{4\pi R} \left(\cos \theta + 1 \right) \int_s \int \exp \left(ik\sqrt{r_0^2 - 2\mathbf{r} \cdot \mathbf{r}_0 + r^2} \right) dx \, dy$$

(5–145)

where θ is the angle between the z axis and a vector from the position of the slit to the observation point $\mathbf{r}_0$.

Fraunhofer Diffraction

If we take the origin of $\mathbf{r}$ and $\mathbf{r}_0$ at a center of the slit, and assume that R (which is $|\mathbf{r}_0|$ in this case) is much larger than the dimensions of the slit, then

$$\exp \left(ik\sqrt{r_0^2 - 2\mathbf{r} \cdot \mathbf{r}_0 + r^2} \right) \cong e^{ikR} e^{-ik(lx + my)}$$

(5–146)

where

$$l = \frac{x_0}{R} \qquad m = \frac{y_0}{R}$$

if x_0 and y_0 are the x and y components of $\mathbf{r}_0$, respectively. Using this expression in (5–145) we obtain

$$q(\mathbf{r}_0) = A \int_s \int e^{-ik(lx + my)} \, dx \, dy$$

(5–147)

where A is a constant, assuming that we stay in the region of very small θ. The diffraction pattern calculated by this formula is called the *Fraunhofer diffraction*. For example, if the slit is rectangular with dimensions of $a \times b$,

$$q(\mathbf{r}_0) = A \int_{-\frac{1}{2}a}^{\frac{1}{2}a} e^{-iklx} \, dx \int_{-\frac{1}{2}b}^{\frac{1}{2}b} e^{-ikmy} \, dy$$

$$= A a b \left(\frac{\sin \left(\frac{1}{2} kal \right)}{\frac{1}{2} kal} \right) \left(\frac{\sin \left(\frac{1}{2} kbm \right)}{\frac{1}{2} kbm} \right)$$

(5–148)

The intensity is given by $|q(\mathbf{r}_0)|^2$ (Fig. 5–22).

FIGURE 5–22.

$$y = \left(\frac{\sin x}{x}\right)^2.$$

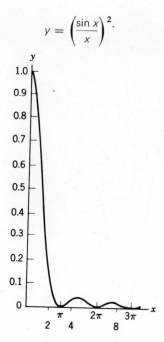

PROBLEM 5–53. Obtain the Fraunhofer diffraction of a circular slit of radius a.

ANSWER. Since the diffraction pattern should be cylindrically symmetric, we can set $m = 0$, without losing generality.

$$q(\mathbf{r}_0) = A \int \int e^{-ikl\rho \cos\varphi} \rho \, d\rho \, d\varphi$$

$$= 2\pi A \int_0^a J_0(kl\rho)\rho \, d\rho$$

$$= 2\pi A a^2 \frac{J_1(kla)}{kla}$$

where J_0 and J_1 are the Bessel functions. Fig. 5–23 gives the resulting intensity.

PROBLEM 5–54. Show that when the wavelength $(= 2\pi/k)$ is very small, light which was originally a plane wave forms a sharp ray after it passes through a slit.

FIGURE 5–23. $y = \left(\dfrac{2J_1(x)}{x}\right)^2.$

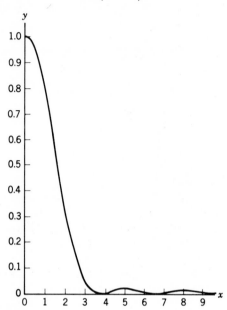

Fresnel Diffraction

When the observation point $\mathbf{r}_0$ is closer to the slit, we have to take higher-order terms in expansion (5–146). In considering second-order terms such as x^2 and y^2, it is more convenient to use a coordinate system in which the first-order terms disappear. For that purpose we still take the slit to be in the xy plane, and take the z axis to be perpendicular to the slit plane but passing through the observation point $\mathbf{r}_0$. In this coordinate system

$$l = m = \mathbf{r} \cdot \mathbf{r}_0 = 0 \tag{5–149}$$

so that

$$\exp\left(ik\sqrt{r_0^2 - 2\mathbf{r} \cdot \mathbf{r}_0 + r^2}\right) \cong e^{ikR} \exp\left[\frac{ik(x^2 + y^2)}{2R}\right] \tag{5–150}$$

Therefore

$$q(\mathbf{r}_0) = A \int_s \exp\left(\frac{ikx^2}{2R}\right) dx \int_s \exp\left(\frac{iky^2}{2R}\right) dy \tag{5–151}$$

The diffraction pattern obtained in this approximation is called the *Fresnel diffraction*.

Using the Fresnel integrals

$$C(\sigma) = \int_0^\sigma \cos\left(\tfrac{1}{2}\pi s^2\right) ds \quad \text{and} \quad S(\sigma) = \int_0^\sigma \sin\left(\tfrac{1}{2}\pi s^2\right) ds \quad (5\text{–}152)$$

we obtain

$$\int_{x_1}^{x_2} \exp\left(\frac{ikx^2}{2R}\right) dx = \sqrt{\pi R/k}\, [C(\sigma_2) - C(\sigma_1) + iS(\sigma_2) - iS(\sigma_1)]$$

$$(5\text{–}153)$$

if

$$\sigma = x\sqrt{k/\pi R} \qquad (5\text{–}154)$$

The intensity, being proportional to $|q(\mathbf{r}_0)|^2$, is proportional to

$$[C(\sigma_2) - C(\sigma_1)]^2 + [S(\sigma_2) - S(\sigma_1)]^2 \qquad (5\text{–}155)$$

Equations

$$x = C(\sigma) \qquad \text{and} \qquad y = S(\sigma) \qquad (5\text{–}156)$$

FIGURE 5–24. Cornu spiral. Numbers on the spiral are σ.

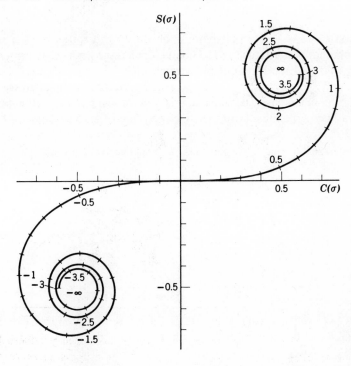

give a curve called the *Cornu spiral* (Fig. 5–24). If we know σ_1 and σ_2 from given dimensions of the slit, the quantity given by (5–155) is obtained from the distance between the corresponding two points on the Cornu spiral.

As an example, let us consider a straight edge of a semi-infinite screen. The edge may be taken to be parallel to the y axis, and the slit, or the open area may be $\xi \leq x < \infty$; this means that the observation point r_0 is on the dark side from the edge by distance ξ (Fig. 5–25). First find σ from $x = \xi$ from

FIGURE 5–25.   Straight edge.

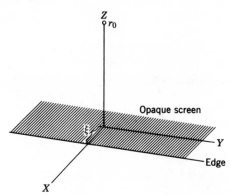

equation (5–154), locate that point σ on the Cornu spiral, draw a straight line from that point to the $\sigma = \infty$ point (which is the upper right center of the spiral), and find the length of this straight line. The square of this length is the quantity (5–155) of this case and is proportional to the intensity at this r_0. Plotting the intensity obtained in this way for several observation points r_0 along a line perpendicular to the edge, we obtain Fig. 5–26.

FIGURE 5–26.   Fresnel diffraction due to a straight edge (Born and Wolf).

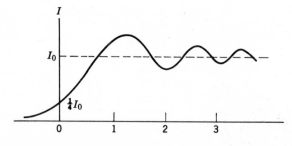

Fig. 5–27 schematically shows that a given screen produces Fresnel and Fraunhoffer diffraction patterns at short and long distances, respectively.

FIGURE 5–27. Diffraction pattern due to a slit changes from Fraunhofer diffraction to Fresnel diffraction as a screen gets closer to the slit.

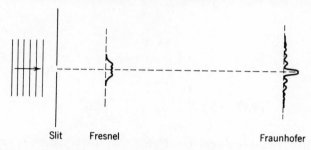

Slit   Fresnel   Fraunhofer

PROBLEM 5–55. Knowing that $S(\pm\infty) = C(\pm\infty) = \pm\frac{1}{2}$, show that the intensity at a point directly behind the edge ($x = 0$), is $\frac{1}{4}$ of that at a point far from the edge on the clear side ($x = -\infty$), in the above example. (See Figs. 5–24 and 5–26.)

PROBLEM 5–56. Show that if the diffraction field due to an opaque screen with a given shape of slit is $q_a(\mathbf{r}_0)$, and the field due to a screen of the same shape as the slit of the first setup is $q_b(\mathbf{r}_0)$, then $q_a(\mathbf{r}_0) + q_b(\mathbf{r}_0) = q(\mathbf{r}_0)$ is the field when no screen is present. (This is called *Babinet's principle.*)

HINT. If the area of the slits in these complementary setups are S_a and S_b, respectively, then $S_a + S_b$ = total area.

5–15 EIKONAL EQUATION. FERMAT'S PRINCIPLE

A general monochromatic wave may be expressed as

$$Q(\mathbf{r}, t) = Ae^{i\omega(S/c - t)} \tag{5–157}$$

where in general A and S are both real functions of $\mathbf{r}$. A plane wave such as (5–128) is a special case where A is constant and $S = \mathbf{k} \cdot \mathbf{r}(c/\omega)$. For this general wave (5–157) we see the phase is common to all points on a surface given by S = constant. Therefore, this function S, called the *eikonal*, gives the wave surface, and ∇S gives the propagation direction of the wave at that point.

If we put expression (5–157) into the wave equation we obtain

$$-\left(\frac{\omega}{c}\right)^2 [(\nabla S) \cdot (\nabla S) - n^2] + 2i \left(\frac{\omega}{c}\right) [\tfrac{1}{2}\nabla^2 S + \nabla(lnA) \cdot \nabla S]$$

$$+ \frac{(\nabla^2 A)}{A} = 0 \tag{5–158}$$

where n is the index of refraction of the medium, which can be a function of $\mathbf{r}$. In the limit of very high frequency

$$\frac{\omega}{c} \to \infty \tag{5-159}$$

(5–158) gives

$$(\boldsymbol{\nabla}S) \cdot (\boldsymbol{\nabla}S) = n^2 \quad \text{or} \quad |\boldsymbol{\nabla}S| = n \tag{5-160}$$

which is called the *eikonal equation*. Notice that this equation is very similar to the Hamilton–Jacobi equation introduced by (1–121) and reduces to

$$(\boldsymbol{\nabla}S') \cdot (\boldsymbol{\nabla}S') = 2m(\mathscr{E} - U(\mathbf{r})) \tag{5-161}$$

for a particle with constant energy $\mathscr{E}$. Here S' is Hamilton's principal function instead of the eikonal. This comparison strongly suggests that in this limit of high frequency, a wave can behave like a particle. Actually, $\boldsymbol{\nabla}S'$ gives the momentum of a particle while we have just seen that $\boldsymbol{\nabla}S$ gives the direction of propagation of the wave, so that the wave can propagate in a medium of given index of refraction in the same way as a particle in a potential $U(\mathbf{r})$. This explains why the particle theory of light, supported by many people including Newton, had been partially successful.[29]

The branch of optics which corresponds to the limit of short wavelength, or $k \to \infty$, is known as *geometrical optics*. The eikonal equation can give all formulas of that branch of optics. Among them, *Fermat's principle* is a famous and useful one. Fermat's principle states that in a medium with the index of refraction $n(\mathbf{r})$, which may depend on $\mathbf{r}$, the ray of light chooses its path l between given end points in such a way that $\int n \, dl$ is minimum; or

$$\delta \int_a^b n \, dl = 0 \tag{5-162}$$

We can justify this variational principle immediately by comparing the eikonal equation (5–160) with the Hamilton–Jacobi equation (5–161) and remembering Jacobi's form of the least action principle (Problem 1–18 of section 1–11).

---

[29] Hamilton himself intended to present a unified theory which covered both mechanics and optics. Historically, however, the latter part of his theory has been forgotten for about a century.

PROBLEM 5–57. Obtain (5–158).

PROBLEM 5–58. Show that Fermat's principle (5–162) can also be written as

$$\delta(t_2 - t_1) = 0$$

where $t_2 - t_1$ is the transit time. Notice that this is the same as (1–100).

·   HINT.   Use (5–123).

PROBLEM 5–59. Derive the law of refraction (Problem 5–49) using Fermat's principle.

ADDITIONAL PROBLEMS

5–60. Show that a system of charged particles, interacting with each other through electrostatic forces alone, cannot be maintained in stable equilibrium (Earnshaw's theorem.)

HINT.   If **r** is a stable equilibrium position for a particle, one should be able to find a small volume around **r** such that $\int \nabla^2 U(r) \, dv > 0$, where $U (= q\phi(\mathbf{r}))$ is the potential energy.

5–61. The nuclear potential, in contrast to the electrostatic potential, is approximately of the form of

$$\phi_Y(\mathbf{r}) = -g \int \frac{e^{-\kappa|\mathbf{r}-\mathbf{r}'|}}{|\mathbf{r} - \mathbf{r}'|} \rho(\mathbf{r}') \, dv'$$

where ρ is the density of nucleons and κ and g are constants.
(a) Show that such a field obeys the equation

$$\nabla^2 \phi_Y - \kappa^2 \phi_Y = 4\pi g\rho$$

instead of Poisson's equation. (Yukawa's theory; see section 11–12.)
(b) Show that a system made of a point charge $+q$ at the center and a
negative charge $-q$ distributed around the center as

$$\rho = -q \, \frac{\kappa^2}{4\pi} \, \frac{e^{-\kappa r}}{r}$$

gives rise to the electrostatic potential of

$$\phi = \frac{q}{4\pi\varepsilon} \, \frac{e^{-\kappa r}}{r}$$

or the same form as the Yukawa potential of (a).

5–62. Show that

$$\phi(\mathbf{r}) = \int \frac{1}{4\pi\varepsilon} \frac{\rho(\mathbf{r}')}{|\mathbf{r} - \mathbf{r}'|} \, dv' + \frac{1}{4\pi} \oint \frac{1}{|\mathbf{r} - \mathbf{r}_s'|} \, \nabla\phi(\mathbf{r}_s) \cdot d\mathbf{a}$$

$$- \frac{1}{4\pi} \oint \phi(\mathbf{r}_s) \nabla \frac{1}{|\mathbf{r} - \mathbf{r}_s|} \cdot d\mathbf{a}$$

where the area integrals are over the entire surface of the volume of
the first integral and $\mathbf{r}_s$ is the position of $d\mathbf{a}$. This shows that the
potential is uniquely fixed when, in addition to the charge distribution,
ϕ and $\nabla\phi$ at the boundary are given.

HINT. Use Green's theorem (5–138).

5–63. A conducting sphere of radius a is at potential $\phi = 0$. Place a point
charge q at distance R from the center of the sphere, assuming $R > a$.
Find the potential at $\mathbf{r}$, where $r > a$.

ANSWER. According to (5–26) $\phi = \phi_q + F$, where ϕ_q is the potential
due to the point charge and F is a solution of Laplace's equation.
From the result of Problem 5–17, we see that

$$\phi_q = kq \sum_n \frac{r^n}{R^{n+1}} \, P_n(\cos\theta) \qquad \text{when} \qquad r < R$$

while from (5–59) we obtain

$$F = \sum_n \frac{\alpha_n}{r^{n+1}} \, P_n(\cos\theta)$$

where α_n are constants. Since $\phi = 0$ when $r = a$, we see that

$$\alpha_n = -\frac{a^{2n+1}}{R^{n+1}} kq$$

so that

$$F = -kq\,\frac{a}{R}\sum_n \frac{(a^2/R)^n}{r^{n+1}}\,P_n\,(\cos\theta)$$

which is the potential due to $-q(a/R)$ placed at a^2/R from the center of sphere on the same line as the charge q. This effective charge $-q(a/R)$ is called the *image* of the charge q (Fig. 5–28).

FIGURE 5–28.   Image of q at R is $-qa/R$ at a^2/R.

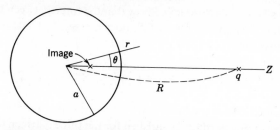

5–64.   A nonconducting sphere of radius a, capacitivity ε_2 (homogeneous) is immersed in an infinite homogeneous dielectric of capacitivity ε_1. A point charge q is at distance R from the center of the sphere ($R > a$). Find the potential inside and outside the sphere.

ANSWER.   Outside,

$$\phi_+ = \frac{q}{4\pi\varepsilon_1}\sum_n \frac{r^n}{R^{n+1}}\,P_n\,(\cos\theta) + \sum_n \frac{\alpha_n}{r^{n+1}}\,P_n\,(\cos\theta)$$

inside,

$$\phi_- = \sum_n \beta_n r^n P_n\,(\cos\theta)$$

where

$$\alpha_n = \frac{q}{4\pi\varepsilon_1}\,\frac{n(\varepsilon_2 - \varepsilon_1)}{(n+1)\varepsilon_1 + n\varepsilon_2}\,\frac{a^{2n+1}}{R^{n+1}}.$$

$$\beta_n = \frac{q(2n+1)}{4\pi[(n+1)\varepsilon_1 + n\varepsilon_2]}\,\frac{1}{R^{n+1}}$$

Notice that this result cannot be interpreted by an "image."

5-65. An infinite circular cylindrical sheet of radius a is divided longitudinally into two parts by a plane which contains the axis of the cylinder. If these two parts are at constant potentials ϕ_0 and $-\phi_0$, respectively, show that the potential at distance ρ from the cylinder is

$$\phi \cong \phi_0 \frac{4}{\pi} \sum_{n=1}^{\infty} \frac{\sin [(2n-1)\varphi]}{2n-1} \left(\frac{a}{\rho}\right)^{2n-1}$$

where φ is the angle measured from the dividing plane. Note that

$$\frac{4}{\pi} \sum_{n=1}^{\infty} \frac{\sin [(2n-1)\varphi]}{2n-1} = 1 \qquad \text{when} \qquad 0 < \varphi < \pi$$
$$= -1 \qquad \text{when} \qquad -\pi < \varphi < 0$$

5-66. (a) Equation (5–69) shows that the total force between two circuits satisfies Newton's third law. Show that the force between line elements dl_1 and dl_2, as given from (5–67), does not satisfy Newton's third law. (b) Show that if

$$d^2\mathbf{F}_{12} = -\frac{\mu_0}{4\pi} \frac{\mathbf{r}_{12}}{r_{12}{}^3} I_1 I_2 \left(\frac{\partial r_{12}}{\partial l_1} \frac{\partial r_{12}}{\partial l_2} - r_{12} \frac{\partial^2 r_{12}}{\partial l_1 \partial l_2}\right) dl_1 \, dl_2$$

is assumed to be the force exerted on the line element dl_1 by dl_2, then Newton's third law is satisfied for the force between line elements, and still the total force between two circuits is given by (5–69). (This form was proposed by Ampère himself.)

5-67. An infinite straight wire has a constant charge density σ per unit length. A point charge q is placed without velocity at distance a from the wire.
(a) Obtain the force $\mathbf{F}$ exerted on the point charge.
(b) Let both q and the wire (with σ) move along the direction of the wire with velocity V. Calculate the force exerted on the moving point charge by the moving wire.

ANSWER. (a) $F = 2kq\sigma/a$ in the direction away from the wire.
(b) $F/\sqrt{1 - \beta^2}$ in the same direction as before. Consider the Lorentz contraction, and see Problem 4–30 in interpreting the result.

5-68. When no macroscopic currents exist, $\mathbf{V} \times \mathbf{H} = 0$, and we can define the magnetic potential ϕ_m such that

$$\mathbf{H} = -\mathbf{V}\phi_m$$

(a) Show that

$$\phi_m(\mathbf{r}) = -\int \frac{\mathbf{V}' \cdot \mathbf{M}(\mathbf{r}')}{4\pi|\mathbf{r} - \mathbf{r}'|} \, dv'$$

(b) Calculate ϕ_m due to a homogeneously magnetized cylindrical magnet of radius a and length l. The magnetization is along the length of the cylinder.

ANSWER. for (b).

$$\phi_m = \rho_m \int_0^\infty (e^{-k|z-(1/2)l|} - e^{-k|z+(1/2)l|}) J_0(k\rho) J_1(ka)(2k)^{-1}\, dk$$

See Problem 5–14.

5–69. A grating is made of N parallel straight slits, each of which has the width a and is separated from neighboring slits by d. An incident plane wave has the propagation vector **k** perpendicular to the plane of the grating. Show that the intensity of the diffracted wave at angle θ is

$$I = I_0 \left(\frac{\sin\left(\frac{1}{2}ka\alpha\right)}{\frac{1}{2}ka\alpha} \right)^2 \left(\frac{\sin\left(\frac{1}{2}Nkd\alpha\right)}{\sin\left(\frac{1}{2}kd\alpha\right)} \right)^2$$

where

$$\alpha = \sin\theta$$

5–70. Using the Cornu spiral, calculate the Fresnel diffraction due to an infinitely long straight slit of width s, where $s\sqrt{k/\pi R} = 1$.

ANSWER. See Fig. 5–29.

FIGURE 5–29. Fresnel diffraction due to a slit of width $s = \sqrt{\pi R/k}$. (Problem 5–70).

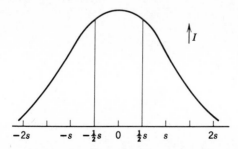

5–71. The index of refraction of a medium depends on the y coordinate as $n = ay + b$, where a and b are constants. Find the path of light between points $(x = d, y = 0)$ and $(x = -d, y = 0)$.
HINT. Compare (5–160) with (5–161), and solve (5–161) for $U = -(ay + b)^2/2m$.

Chapter 6   Classical Theory of Radiation

Following the spectacular discoveries of the laws of electricity and magnetism during the 19th century, more sophisticated theoretical formulations of field theory have been attempted. Particularly the incorporation of field theory with the special theory of relativity, which we discussed in Chapter 4, has been the most important problem of this part of modern physics.

The potentials including the vector potentials, in addition to the scalar potential we discussed in Chapter 5, are found to be more useful than the field strengths $\mathbf{E}$ and $\mathbf{B}$, themselves, in the relativistic formalism. The Lagrangian density, stress-energy tensor, and Hamiltonian density are given following the procedure we developed in Chapters 3 and 4 for continuous bodies.

As a field theory, the electromagnetic field theory is more complicated than most others, because this field is a vector field and the corresponding particle (photon) does not have mass. A few meson fields are investigated in modern physics in relation to nuclear and other forces. They are formulated analogously to the procedure we are using in this chapter on electromagnetic fields, but most of them are actually simpler than the present one.

The present theory of electromagnetic fields interacting with an electron is actually not in a satisfactory form, although radiation theory, which explains emission, absorption, and scattering of electromagnetic waves by a system of charged particles, has been quite successful. The trouble with the existing theory is that an electron, which is supposed to have no structure, is found to have a diverging self-energy, and the field carried by the electron does not behave like a particle under Lorentz transformations. These conclusions apparently contradict our experience. Some improvements are obtained when we apply quantum theory, but the contradictions are still essentially unresolved.

6–1   VECTOR POTENTIAL. LORENTZ GAUGE

We have already discussed scalar potentials (sections 1–3, 3–8, and 5–3). We now introduce a vector potential **A** such that a vector quantity **B** is expressed as

$$\mathbf{B} = \mathbf{\nabla} \times \mathbf{A} \qquad (6\text{–}1)$$

In order to express a vector quantity **E** with a scalar potential, we required $\mathbf{\nabla} \times \mathbf{E} = 0$. The corresponding condition that (6–1) holds is that $\mathbf{\nabla} \cdot \mathbf{B} = 0$, because

$$\mathbf{\nabla} \cdot \mathbf{B} = \mathbf{\nabla} \cdot (\mathbf{\nabla} \times \mathbf{A}) = \frac{\partial}{\partial x}\left(\frac{\partial A_z}{\partial y} - \frac{\partial A_y}{\partial z}\right) + \frac{\partial}{\partial y}\left(\frac{\partial A_x}{\partial z} - \frac{\partial A_z}{\partial x}\right)$$

$$+ \frac{\partial}{\partial z}\left(\frac{\partial A_y}{\partial x} - \frac{\partial A_x}{\partial y}\right) = 0 \qquad (6\text{–}2)$$

holds identically. Since for the magnetic flux density $\mathbf{\nabla} \cdot \mathbf{B}$, not $\mathbf{\nabla} \times \mathbf{B}$, is always zero, it is expressed by a vector potential rather than a scalar potential. Therefore, from now on we take (6–1) as the equation for the magnetic flux density.

Inserting (6–1) into one of Maxwell's equations (5–113c) we obtain

$$\mathbf{\nabla} \times \left(\mathbf{E} + \frac{\partial \mathbf{A}}{\partial t}\right) = 0 \qquad (6\text{–}3)$$

Therefore, the scalar potential ϕ, which we called the electrostatic potential, is related to **E** as

$$\mathbf{E} = -\mathbf{\nabla}\phi - \frac{\partial \mathbf{A}}{\partial t} \qquad (6\text{–}4)$$

rather than the previous relation (5–18).

It can easily be seen that when the set of potentials ϕ and **A** are transformed

$$\phi' = \phi + \frac{\partial \chi}{\partial t} \qquad (6\text{–}5)$$

$$\mathbf{A}' = \mathbf{A} - \mathbf{\nabla}\chi \qquad (6\text{–}6)$$

by an arbitrary function χ, the corresponding vectors **E** and **B** remain invariant. In this sense the original set, ϕ and **A**, is equivalent to the transformed set, ϕ' and $\mathbf{A}'$. This transformation is called the *gauge transformation*; physical quantities should be invariant under the gauge transformation.

Using this arbitrariness we can impose an auxiliary condition on the potentials. One typical auxiliary condition, which we are going to consider in this book, is

$$\mathbf{V} \cdot \mathbf{A} + \frac{1}{c^2} \frac{\partial \phi}{\partial t} = 0 \qquad (6\text{--}7)$$

This condition is called the *Lorentz condition,* and the potentials which satisfy this condition are said to be in the *Lorentz gauge.*

Let us consider Maxwell's equations in a vacuum. Using the scalar and vector potentials we see two of them, (5–113b) and (5–113c), are automatically satisfied. The potentials, therefore, should be derived from the other two; (5–113a) and (5–113d). In the case of a vacuum, we find that the latter two equations give

$$\nabla^2 \phi + \frac{\partial}{\partial t} (\mathbf{V} \cdot \mathbf{A}) = -\frac{\rho}{\varepsilon_0} \qquad (6\text{--}8)$$

and

$$\nabla^2 \mathbf{A} - \frac{1}{c^2} \frac{\partial^2 \mathbf{A}}{\partial t^2} - \mathbf{V} \left(\mathbf{V} \cdot \mathbf{A} + \frac{1}{c^2} \frac{\partial \phi}{\partial t} \right) = -\mu_0 \mathbf{J} \qquad (6\text{--}9)$$

When the Lorentz condition (6–7) is assumed, we see that (6–8) and (6–9) reduce to

$$\Box \, \phi = -\frac{\rho}{\varepsilon_0} \qquad (6\text{--}10)$$

$$\Box \, \mathbf{A} = -\mu_0 \mathbf{J} \qquad (6\text{--}11)$$

where $\Box$ is the D'Alembertian which appeared in (5–134).

PROBLEM 6–1. Show that $\mathbf{E}$ and $\mathbf{B}$ are invariant under the gauge transformation.

PROBLEM 6–2. If the initially chosen potentials do not satisfy the Lorentz condition, or if

$$\mathbf{V} \cdot \mathbf{A}' + \frac{1}{c^2} \frac{\partial \phi'}{\partial t} = C$$

where C is a constant, find a gauge transformation such that the transformed potentials satisfy the Lorentz condition.

ANSWER.

$$\nabla^2 \chi - \frac{1}{c^2} \frac{\partial^2 \chi}{\partial t^2} = C$$

PROBLEM 6–3. Show that the Lorentz condition does not fix the potentials uniquely, that is, we can always find gauge transformations to transform one set of potentials in the Lorentz gauge into another set of potentials also in the Lorentz gauge.

ANSWER.

$$\nabla^2 \chi - \frac{1}{c^2} \frac{\partial^2 \chi}{\partial t^2} = 0$$

PROBLEM 6–4. Obtain (6–8) and (6–9).

6–2 COVARIANT FORMALISM OF ELECTROMAGNETIC FIELDS

If we assume that charge q of a particle is invariant under Lorentz transformations, then we see immediately that the charge density ρ should be transformed like the time component of a 4-vector, and that the 4-vector is

$$J^0 = c\rho, \qquad J^1 = J_x, \qquad J^2 = J_y, \qquad J^3 = J_z \qquad (6\text{–}12)$$

where **J** is the current density we introduced in the last chapter. This argument is identical to that of section 4–11, on the mass density and mass current.

Since the D'Alembertian $\square$ is a scalar operator (see Problem 4–10), we see that the wave equations (6–10) and (6–11) can be written as a covariant equation

$$\square A^i = -\mu_0 J^i \qquad (6\text{–}13)$$

if

$$A^0 = \frac{\phi}{c}, \qquad A^1 = A_x, \qquad A^2 = A_y, \qquad A^3 = A_z \qquad (6\text{–}14)$$

form a 4-vector. Notice that the Lorentz condition (6–7) can be written in a covariant form as

$$\frac{\partial A^i}{\partial x^i} = 0 \qquad (6\text{–}15)$$

This is a great advantage of the Lorentz gauge.

In this formalism, **E** and **B** are seen to be the components of an antisymmetric tensor

$$F^{ij} = \frac{\partial A^j}{\partial x_i} - \frac{\partial A^i}{\partial x_j} \tag{6-16}$$

From (6–1) and (6–4) we see that

$$(F^{ij}) = \begin{pmatrix} 0 & -\dfrac{E_x}{c} & -\dfrac{E_y}{c} & -\dfrac{E_z}{c} \\[2ex] \dfrac{E_x}{c} & 0 & -B_z & B_y \\[2ex] \dfrac{E_y}{c} & B_z & 0 & -B_x \\[2ex] \dfrac{E_z}{c} & -B_y & B_x & 0 \end{pmatrix} \begin{matrix} \downarrow \\ i \end{matrix} \tag{6-17}$$

and that Maxwell's equations can be expressed as

$$\frac{\partial F^{ij}}{\partial x^j} = -\mu_0 J^i \tag{6-18}$$

and

$$\frac{\partial F^{ik}}{\partial x_j} + \frac{\partial F^{kj}}{\partial x_i} + \frac{\partial F^{ji}}{\partial x_k} = 0 \tag{6-19}$$

We can obtain two invariants from **E** and **B**. They are

$$F^{ij}F_{ij} = g_{ik}g_{jl}F^{ij}F^{kl} = 2\left(B^2 - \left(\frac{E}{c}\right)^2\right) \tag{6-20}$$

and

$$e_{ijkl}F^{ij}F^{kl} = \frac{8\mathbf{B} \cdot \mathbf{E}}{c} \tag{6-21}$$

where e_{ijkl} is a completely antisymmetric tensor of fourth rank, namely, $e_{1234} = 1$, $e_{1243} = -1$, $e_{2143} = 1$, $e_{1123} = 0$, etc.

Finally, the Lorentz force (5–74) can be expressed as a Minkowski force

$$K^i = qF^{ij}u_j \tag{6-22}$$

because, when (6–22) is substituted into (4–76) we obtain

$$\frac{d\mathbf{p}'}{dt} = q(\mathbf{E} + \mathbf{V} \times \mathbf{B}) \qquad (6\text{–}23)$$

for the space components.

PROBLEM 6–5. Obtain (6–17).

PROBLEM 6–6. Show that (6–18) and (6–19) give Maxwell's equations.

PROBLEM 6–7. Calculate (6–20) and (6–21) from (6–17).

PROBLEM 6–8. Show that (6–23) is obtained from (6–22) and (4–76).

PROBLEM 6–9. Show that (6–19) can be written as

$$e_{ijkl} \frac{\partial F^{ij}}{\partial x_k} = 0$$

PROBLEM 6–10. An infinite straight wire carries a homogeneously distributed stationary charge of σ per unit length. Calculate $\mathbf{E}$ at distance d from the wire. Now let an observer move by velocity V in a direction parallel to the wire. Calculate $\mathbf{E}$ and $\mathbf{B}$ at the same distance d from the wire for the moving observer, and show that (6–20) and (6–21) are the same between stationary and moving observers.

ANSWER.

$$E = \frac{\sigma\gamma}{(2\pi\varepsilon_0\, d)}, \qquad B = \frac{\mu_0\sigma\gamma V}{(2\pi\, d)}$$

PROBLEM 6–11. We have seen that when (6–22) is put ìnto (4–76) the space components give the right equation (6–23). Show that the corresponding time component gives

$$\frac{dT}{dt} = q\mathbf{E} \cdot \mathbf{V}$$

which is previously obtained in Problem 5–32.

HINT. p^0 is given by (4–81), while $K^0 = qF^{0j}u_j$ can be obtained from (6–17) and (4–72).

6–3 LAGRANGIAN DENSITY OF ELECTROMAGNETIC FIELDS

The Lagrangian density should be chosen in such a way that Lagrange–Euler's equation (4–92) gives the equations of motion, i.e., Maxwell's equations. Actually, we see that

$$\mathscr{L}_{(1)} = - \frac{1}{4\mu_0} F^{ij} F_{ij} - A_i J^i \tag{6–24}$$

gives Maxwell's equation (6–18), from the Lagrange–Euler's equation

$$\frac{\partial \mathscr{L}}{\partial A_i} - \frac{\partial}{\partial x^j} \frac{\partial \mathscr{L}}{\partial (\partial A_i / \partial x^j)} = 0 \tag{6–25}$$

Since we chose the Lorentz condition (6–15), another Lagrangian density

$$\mathscr{L}_{(2)} = - \frac{1}{4\mu_0} \left[F^{ij} F_{ij} + 2 \left(\frac{\partial A^i}{\partial x^i} \right)^2 \right] - A_i J^i \tag{6–26}$$

is also acceptable.

The canonical momentum density, defined by (3–107) and (3–108) for a scalar field, can be generalized to a tensor

$$\pi^{ij} = \frac{1}{c} \frac{\partial \mathscr{L}}{\partial (\partial A_i / \partial x^j)} \tag{6–27}$$

in the case of a vector field. We find

$$\pi^{\alpha 0} = \frac{1}{c\mu_0} F^{0\alpha} = -\varepsilon_0 E^\alpha \qquad \alpha = 1, 2, 3 \tag{6–28}$$

from both (6–24) and (6–26). (We used an unconventional notation $E^1 = E_x$, etc. in (6–28), just for abbreviation.) We see, however, that

$$\pi^{00} = 0 \tag{6–29}$$

when $\mathscr{L}_{(1)}$ of (6–24) is used, but that

$$\pi^{00} = \frac{1}{c\mu_0} \frac{\partial A^i}{\partial x^i} \tag{6–30}$$

when $\mathscr{L}_{(2)}$ of (6–26) is used. Of course, under the Lorentz condition (6–15), these two expressions are equivalent; but we will find in quantum electro-dynamics that it is necessary to have π^{00} not *identically* zero, thus $\mathscr{L}_{(2)}$ is preferable to $\mathscr{L}_{(1)}$ in quantum electrodynamics. However, in this chapter we will use $\mathscr{L}_{(1)}$, because the difference is not as essential in classical electro-dynamics, and $\mathscr{L}_{(1)}$ is simpler than $\mathscr{L}_{(2)}$.

Note that the canonical momentum conjugate to A^i is $\pi^{i0}\, dv$, as given by (3–108). Since dv transforms as a time component of a 4-vector, the canonical momenta form a 4-vector, in contrast to the canonical momentum densities, which form a tensor.

PROBLEM 6–12. Show that the Lagrangian density, either (6–24) or (6–26), reduces to $\frac{1}{2}(\mathbf{E}\cdot\mathbf{D} - \mathbf{B}\cdot\mathbf{H}) - \rho\phi + \mathbf{A}\cdot\mathbf{J}$.

HINT. Use (6–20).

PROBLEM 6–13. Show that (6–24) and (6–25) give (6–18).

PROBLEM 6–14. Verify (6–28).

6–4 STRESS-ENERGY TENSOR OF ELECTROMAGNETIC FIELDS

The stress-energy tensor of a scalar field is defined by (4–93). We now generalize that definition for a vector field as

$$T_i^{\ j} = \mathscr{L}\, \delta_i^{\ j} - \frac{\partial A_k}{\partial x^i}\, \frac{\partial \mathscr{L}}{\partial(\partial A_k/\partial x^j)} \tag{6–31}$$

This quantity is still a component of a second rank tensor, and we can show that the fundamental equation

$$\frac{\partial T_i^{\ j}}{\partial x^j} = 0 \tag{4–95}$$

still holds, when the J^i's are fixed during the differentiation, i.e., if

$$\frac{\partial \mathscr{L}}{\partial x^j} = \frac{\partial \mathscr{L}}{\partial A^k}\frac{\partial A^k}{\partial x} + \frac{\partial \mathscr{L}}{\partial(\partial A_k/\partial x^l)}\frac{\partial}{\partial x^j}\left(\frac{\partial A_k}{\partial x^l}\right) \tag{6–32}$$

Using the Lagrangian density (6–24) in (6–31), we obtain

$$T_i^j = \mathcal{L}_{(1)} \delta_i^j - \frac{1}{\mu_0} \frac{\partial A_k}{\partial x^i} F^{kj} \qquad (6\text{–}33)$$

Since

$$\frac{\partial A_i}{\partial x^k} F^{kj} = \frac{\partial}{\partial x^k} (A_i F^{kj}) - A_i \frac{\partial F^{kj}}{\partial x^k}$$

$$= \frac{\partial}{\partial x^k} (A_i F^{kj}) - \mu_0 A_i J^j \qquad (6\text{–}34)$$

using equation (6–18), and since the first term in the last expression of (6–34) disappears when integrated over a volume large enough so that field is zero on its surface, we can express (6–33) as

$$T_i^j = - \left(\frac{1}{4\mu_0} F^{kl} F_{kl} + A_k J^k \right) \delta_i^j + \frac{1}{\mu_0} F_{ki} F^{kj} + A_i J^j$$

$$= t_i^j - A_k J^k \delta_i^j + A_i J^j \qquad (6\text{–}35)$$

where

$$t_i^j = - \frac{1}{4\mu_0} (F^{kl} F_{kl} \delta_i^j - 4 F_{ki} F^{kj}) \qquad (6\text{–}36)$$

is the stress-energy tensor of the free field. ($J^i = 0$.)

If we use (6–35) in (4–95) with (6–32) we obtain

$$\frac{\partial t_i^j}{\partial x^j} = \frac{\partial A_k}{\partial x^i} J^k - \frac{\partial A_i}{\partial x^j} J^j = F_{ij} J^j \qquad (6\text{–}37)$$

When the current density is due to a charged particle, we can use (5–14) and (5–72) to obtain

$$\frac{\partial \int t_i^j \, dv}{\partial x^j} = q F_{ij} u^j = K_i \qquad (6\text{–}38)$$

where K_i is the covariant expression of the Lorentz force given by (6–22). This relation (6–38) allows the same interpretation of the stress-energy tensor which we obtained in section 4–10, for a scalar field. Thus,

$$\mathcal{H} = - t_0^{\ 0} = \frac{1}{4\mu_0} (F^{kl} F_{kl} - 4 F_{k0} F^{k0}) = \tfrac{1}{2} (\mathbf{E} \cdot \mathbf{D} + \mathbf{B} \cdot \mathbf{H}) \qquad (6\text{–}39)$$

is the energy density of the field;

$$-ct_0{}^\alpha = \frac{cF_{k0}F^{k\alpha}}{\mu_0} = (\mathbf{E} \times \mathbf{H})^\alpha \tag{6-40}$$

is the αth component of the energy current density

$$\mathbf{S} = c^2\mathscr{G} = \mathbf{E} \times \mathbf{H} \tag{6-41}$$

where $\mathscr{G}$ is the momentum density of the field, while the remaining space components $t_\alpha{}^\beta$ give the stress tensor of the field. $\mathbf{S}$ of (6–41) is called the *Poynting vector*. When t^{ij} is calculated from (6–36), we see this tensor is symmetric as required.

PROBLEM 6–15. Show that (4–95) is obtained from the Lagrange–Euler's equation (6–25) under (6–32).

PROBLEM 6–16. Express the stress-energy tensor of the field t^{ij} by means of $\mathbf{E}$ and $\mathbf{B}$.

ANSWER.

$$t^{00} = -\frac{1}{2}\left(\varepsilon_0 E^2 + \frac{1}{\mu_0} B^2\right)$$

$$t^{01} = t^{10} = -\frac{E_y B_z - E_z B_y}{c\mu_0}$$

$$t^{11} = -\frac{1}{2}\left(\varepsilon_0 E^2 + \frac{1}{\mu_0} B^2\right) + \varepsilon_0 E_x{}^2 + \frac{1}{\mu_0} B_x{}^2$$

$$t^{12} = \varepsilon_0 E_x E_y + \frac{1}{\mu_0} B_x B_y$$

etc.

PROBLEM 6–17. Compare the results of (6–39) and (6–40) to the result of Problem 5–45.

PROBLEM 6–18. A spherical shell of radius r_0 has a homogeneously distributed total charge of q. Show that the field due to this charged spherical shell contains energy of

$$H = \frac{kq^2}{2r_0}$$

HINT.

$$\mathbf{E} = \frac{\mathbf{D}}{\varepsilon_0} = \frac{kq\mathbf{r}}{r^3}$$

PROBLEM 6–19. An infinite straight wire of resistivity R carries a current I. Show that $\mathbf{S}$ of the field at distance a from the wire has the magnitude of $RI^2/2\pi a$ directed towards the wire, or the energy of the field flows into the wire at the rate of RI^2 per unit time per unit length. This is called the *Joule heat*.

HINT.

$$|\mathbf{E}| = RI$$

PROBLEM 6–20. (a) Show that if we do not neglect the first term in the last expression of (6–34),

$$-T_0{}^\alpha = \frac{1}{\mu_0} F_{k0}F^{k\alpha} - A_0 J^\alpha + \frac{\partial}{\partial x^k}(A_0 F^{k\alpha})$$

where $\alpha = 1, 2,$ or 3.
(b) Show that using the above expression we still obtain

$$-\frac{\partial}{\partial x^\alpha} T_0{}^\alpha = \mathbf{V} \cdot \left[(\mathbf{E} \times \mathbf{H}) - \frac{\phi \mathbf{J}}{c} \right]$$

$$= \mathbf{V} \cdot (\mathbf{E} \times \mathbf{H}) + (\mathbf{E} + \dot{\mathbf{A}}) \cdot \frac{\mathbf{J}}{c}$$

which is the same expression obtained when (6–35) is used.

HINT. The last term in the formula of (a) gives

$$\mathbf{V} \cdot \left(\mathbf{V} \times \frac{\phi \mathbf{B}}{c} \right)$$

in the formula of (b).

6–5 HAMILTONIAN OF A PARTICLE-FIELD SYSTEM

When a point particle of mass m and charge q is interacting with an electromagnetic field, the Lagrangian of the whole system may be assumed as

$$L = \int \mathscr{L}_{(1)} \, dv - mc^2\sqrt{1 - \beta^2} \qquad (6\text{–}42)$$

where $\mathscr{L}_{(1)}$ is the Lagrangian density of the field with the interaction density given by (6–24), and the last term is the Lagrangian of a free particle given by (4–85). Using (5–14) and (5–72) we see that

$$L = - \frac{1}{4\mu_0} \int F^{ij}F_{ij} \, dv - q(\phi_0 - \mathbf{V} \cdot \mathbf{A}_0) - mc^2\sqrt{1 - \beta^2} \qquad (6\text{–}43)$$

where ϕ_0 and $\mathbf{A}_0$ are the potentials at the position of the particle.
The canonical momentum of the particle is

$$\mathbf{p} = \nabla_V L = q\mathbf{A}_0 + m\mathbf{V}\gamma \qquad (6\text{–}44)$$

and Lagrange's equation (1–34) gives

$$\frac{d\mathbf{p}}{dt} + q\nabla(\phi_0 - \mathbf{V} \cdot \mathbf{A}_0) = 0 \qquad (6\text{–}45)$$

Substituting (6–44), and using the relation

$$\nabla(\mathbf{V} \cdot \mathbf{A}_0) = (\mathbf{V} \cdot \nabla)\mathbf{A}_0 + \mathbf{V} \times (\nabla \times \mathbf{A}_0) \qquad (6\text{–}46)$$

which can be obtained from a general formula given in Appendix I(a), we see that (6–45) gives

$$\frac{d(m\mathbf{V}\gamma)}{dt} = q(\mathbf{E} + \mathbf{V} \times \mathbf{B}) \qquad (6\text{–}47)$$

which is the explicit expression of (6–23). Note that $\mathbf{p}'$ in (6–23) is not the canonical momentum, which is given by (6–44). This result and the results we obtained in the last section justify the Lagrangian we assumed in (6–42).
The Hamiltonian of the system is

$$H = \mathbf{p} \cdot \mathbf{V} + mc^2\sqrt{1 - \beta^2} - \int T_0{}^0 \, dv \qquad (6\text{–}48)$$

from the definition (1–44). The first two terms of (6–48) can be expressed as

$$\mathbf{p} \cdot \mathbf{V} + mc^2\sqrt{1 - \beta^2} = q\mathbf{V} \cdot \mathbf{A}_0 + mV^2\gamma + mc^2\sqrt{1 - \beta^2}$$
$$= q\mathbf{V} \cdot \mathbf{A}_0 + mc^2\gamma$$
$$= q\mathbf{V} \cdot \mathbf{A}_0 + c\sqrt{(\mathbf{p} - q\mathbf{A}_0)^2 + (mc)^2}$$

$$(6\text{–}49)$$

using the same procedure employed in obtaining (4–88).

The last term of (6–48) has been calculated in the last section. From (6–35) and (6–36) we see that

$$- \int T_0{}^0 \, dv = - \int t_0{}^0 \, dv + q(\phi_0 - \mathbf{V} \cdot \mathbf{A}_0) - q\phi_0$$

$$= \int \frac{1}{2} \left[\frac{\pi^2}{\varepsilon_0} + \frac{(\mathbf{V} \times \mathbf{A})^2}{\mu_0} \right] dv - q\mathbf{V} \cdot \mathbf{A}_0 \qquad (6\text{–}50)$$

where we used the abbreviated notation

$$\boldsymbol{\pi} = (\pi^{10}, \pi^{20}, \pi^{30}) = -\varepsilon_0 \mathbf{E} \qquad (6\text{–}51)$$

From (6–48), (6–49), and (6–50), we obtain

$$H = mc^2\gamma + \int \tfrac{1}{2}(\mathbf{E} \cdot \mathbf{D} + \mathbf{B} \cdot \mathbf{H}) \, dv$$

$$= c\sqrt{(\mathbf{p} - q\mathbf{A}_0)^2 + (mc)^2} + \int \frac{1}{2} \left[\frac{\pi^2}{\varepsilon_0} + \frac{(\mathbf{V} \times \mathbf{A})^2}{\mu_0} \right] dv \qquad (6\text{–}52)$$

PROBLEM 6–21. Obtain (6–47) as suggested.

HINT.

$$\frac{d\mathbf{A}_0}{dt} = \frac{\partial \mathbf{A}_0}{\partial t} + (\mathbf{V} \cdot \boldsymbol{\nabla})\mathbf{A}$$

PROBLEM 6–22. (a) Show that if we integrate over a finite volume whose surface element is $d\mathbf{a}$, we obtain

$$H = mc^2\gamma + \int \tfrac{1}{2}(\mathbf{E} \cdot \mathbf{D} + \mathbf{B} \cdot \mathbf{H}) \, dv + c^{-2} \oint \phi \mathbf{E} \cdot d\mathbf{a} \qquad (6\text{–}52')$$

instead of (6–52).

(b) When ϕ and $\mathbf{E}$ are due to the charged particle at the center of the space, show that the surface integral term in (6–52′) can be neglected in the limit of an infinite volume.

HINT. See (6–34). Note that if we take a sphere whose center is at the charged particle, and whose radius is r, then

$$\oint \phi \mathbf{E} \cdot d\mathbf{a} = \frac{4\pi k^2 q^2}{r}$$

PROBLEM 6–23. Microscopic charged particles are distributed in space with the rest-mass density μ_r and 4-velocity (u^i). Show that the stress-energy tensor can be written as

$$T_i{}^j = -\mu_r u_i u^j + t_i{}^j \qquad \text{`}$$

when we expect an integration over an infinitely large volume, so that the first term in the last expression of (6–34) can be neglected.

HINT. See (4–119).

6–6 ENERGY TRANSFER BETWEEN PARTICLE AND FIELD

According to the general formula (1–62), the kinetic energy of a particle changes at the rate of

$$\frac{d(mc^2\gamma)}{dt} = [mc^2\gamma, H]_P \qquad (6\text{–}53)$$

where $[\ ,\ ]_P$ is the Poisson bracket. Using the canonical expression of the Hamiltonian given by (6–52), we see the above formula reduces to

$$\frac{d(mc^2\gamma)}{dt} = \left[c\sqrt{(\mathbf{p} - q\mathbf{A}_0)^2 + (mc)^2}, \int \frac{1}{2}\left\{ \frac{\pi^2}{\varepsilon_0} + \frac{(\mathbf{\nabla} \times \mathbf{A})^2}{\mu_0} \right\} dv \right]_P$$

$$(6\text{–}54)$$

In our case, the dynamical variables are $\mathbf{r}$ (position of the particle) and A^i at each point of space, and their canonical momenta are $\mathbf{p}$ and $\pi^i\, dv$ at the same point of space as A^i. We see that the only surviving terms in the Poisson bracket of (6–54) are

$$\frac{d(mc^2\gamma)}{dt} = \frac{c\, \partial\sqrt{(\mathbf{p} - q\mathbf{A}_0)^2 + (mc)^2}}{\partial \mathbf{A}_0} \frac{\partial \int \frac{1}{2}\{(\pi^2/\varepsilon_0) + ((\mathbf{\nabla} \times \mathbf{A})^2/\mu_0)\}\, dv}{\partial \pi_0\, dv}$$

$$(6\text{–}55)$$

where $\pi_0\, dv$ is the canonical momentum at the position of the particle. We obtain

$$\frac{c\, \partial\sqrt{(\mathbf{p} - q\mathbf{A}_0)^2 + (mc)^2}}{\partial \mathbf{A}_0} = \frac{-cq(\mathbf{p} - q\mathbf{A}_0)}{\sqrt{(\mathbf{p} - q\mathbf{A}_0)^2 + (mc)^2}} = -\frac{cqm\mathbf{V}\gamma}{mc\gamma} = -q\mathbf{V}$$

$$(6\text{–}56)$$

and

$$\frac{\partial \int \frac{1}{2}\{(\pi^2/\varepsilon_0) + ((\nabla \times \mathbf{A})^2/\mu_0)\}\ dv}{\partial \pi_0\ dv} = \frac{\pi_0}{\varepsilon_0} = -\mathbf{E}_0 \qquad (6\text{--}57)$$

Therefore

$$\frac{d(mc^2\gamma)}{dt} = q\mathbf{V} \cdot \mathbf{E}_0 \qquad (6\text{--}58)$$

This is the result we obtained in Problems 5–32 and 6–11, showing the consistency of the canonical formalism.

The particle may have potential energy in addition to kinetic energy. The potential energy of the particle in the field is $q\phi_0$, if ϕ_0 is the scalar potential at the position of the particle. Looking at the expression of the total energy given by (6–52), we see that this potential energy must be included in the field energy $\int t_0{}^0\ dv$. As a matter of fact, if we do not use transformation (6–34), the original expression (6–33) gives us

$$-\int T_0{}^0\ dv = \int \{\tfrac{1}{2}(\mathbf{E} \cdot \mathbf{D} + \mathbf{B} \cdot \mathbf{H}) + \mathbf{D} \cdot \nabla\phi\}\ dv - q\mathbf{V} \cdot \mathbf{A}_0 + q\phi_0$$

$$(6\text{--}59)$$

which, when combined with (6–48) and (6–49) shows explicitly that

$$H = mc^2\gamma + q\phi_0 + \int \{\tfrac{1}{2}(\mathbf{E} \cdot \mathbf{D} + \mathbf{B} \cdot \mathbf{H}) + \mathbf{D} \cdot \nabla\phi\}\ dv \qquad (6\text{--}60)$$

Since the total energy H is obviously conserved, we can say that the loss of the particle energy at the rate of $-q\mathbf{V} \cdot \mathbf{E}_0$, as given by (6–58), corresponds to an increase of the field energy. However, if we want to talk about radiation, the energy taken out of the system in the form of an electromagnetic wave propagating into infinity, then the expression $-q\mathbf{V} \cdot \mathbf{E}_0$ is not adequate, because the potential energy $q\phi_0$ is not treated in the derivation. The potential energy may belong to the field, but that part of the field energy cannot be dissipated from the system, since this is the part which the particle can re-claim at any later time. This part $q\phi_0$ of the field energy is deposited but reserved by the particle, and cannot be used for radiation. Thus the rate of the radiation energy loss out of the system is[1]

$$
\mathbb{R} = -\frac{d(mc^2\gamma + q\phi_0)}{dt} = -q\left(\mathbf{V}\cdot\mathbf{E}_0 + \frac{d\phi_0}{dt}\right)
$$

$$
= -q\left(\mathbf{V}\cdot\mathbf{E}_0 + \frac{\partial\phi_0}{\partial t} + \mathbf{V}\cdot\nabla\phi_0\right)
$$

$$
= -q(\mathbf{V}\cdot\mathbf{E}_0 + \dot{\phi}_0 - \mathbf{V}\cdot(\mathbf{E}_0 + \dot{\mathbf{A}}_0)) = q(\mathbf{V}\cdot\dot{\mathbf{A}}_0 - \dot{\phi}_0)
$$

$$(6\text{--}61)$$

where $d\phi_0/dt$ is the change of ϕ_0 measured while following the motion of the particle, but $\partial\phi_0/\partial t$, or $\dot{\phi}_0$, is that measured at a fixed point in space.

PROBLEM 6–24. Obtain (6–59), and then (6–60) starting from (6–33). Also show that (6–60) reduces to (6–52) when the integration volume is such that the corresponding surface integral $\oint \phi\mathbf{D}\cdot d\mathbf{a} = 0$.

6–7 PLANE WAVE

Let us consider a free electromagnetic field, that is, a field in regions where $J^i = 0$. If, in addition, we assume that

$$
A^i(x) = A^i(\mathbf{r})e^{-i\omega t} \tag{6--62}
$$

the wave equation (6–13) reduces to

$$
\left(\nabla^2 + \left(\frac{\omega}{c}\right)^2\right) A^i(\mathbf{r}) = 0 \tag{6--63}
$$

which is the Helmholtz equation. In (6–62) x stands for the four coordinates.

We have seen in section 5–4 that the Laplacian ∇^2 can be expressed in several coordinate systems and has its simplest form in the Cartesian coordinate system, as given by (5–28). The solution of the Helmholtz equation (6–63) in the Cartesian coordinate system is easily obtained as

$$
A^i(\mathbf{r}) = a^i e^{i\mathbf{k}\cdot\mathbf{r}} \tag{6--64}
$$

---

[1] The same result is obtained by the author in a slightly different way: M. Mizushima, *Quantum Mechanics of Atomic Spectra and Atomic Structure*, W. A. Benjamin, New York, 1970, Formula (1–89).

where a^i is a constant and

$$k^2 = \left(\frac{\omega}{c}\right)^2 \tag{6-65}$$

Introducing a 4-vector whose components are

$$(k^0, k^1, k^2, k^3) = \left(\frac{\omega}{c}, k_x, k_y, k_z\right)$$

$$(k_0, k_1, k_2, k_3) = \left(\frac{\omega}{c}, -k_x, -k_y, -k_z\right) \tag{6-66}$$

the solutions (6–62) and (6–64) can be written in the covariant form

$$A^i(x) = a^i \exp\left(-ik_j x^j\right) \tag{6-67}$$

or equivalently

$$A^i(x) = a^i \sin\left(-k_j x^j\right) \tag{6-67'}$$

where the origin of the coordinates is suitably chosen to take care of the phase factor.

Since the Lorentz condition (6–7) should hold for any value of x^j, we obtain

$$k_i a^i = k_0 a^0 - \mathbf{k} \cdot \mathbf{a} = 0 \tag{6-68}$$

where $\mathbf{a}$ is a vector whose components are a^1, a^2, and a^3. Since the choice of the X, Y, and Z axes is still arbitrary, we may choose them in such a way that one of them is parallel to the direction of $\mathbf{k}$-vector, and the others are perpendicular to $\mathbf{k}$. The component of $\mathbf{a}$, or $\mathbf{A}$, in the parallel direction is called the longitudinal component, and the other two space components are called the transverse components, while the time component, namely a^0, of A^0, is called the scalar component of the plane wave. If the longitudinal component is designated as a^l, (6–68) reduces to

$$0 = k_0 a^0 - |\mathbf{k}| a^l = \frac{\omega}{c}(a^0 - a^l)$$

or

$$a^0 = a^l \tag{6-69}$$

using (6–65). Since this result implies

$$a_0 = -a_l \tag{6-70}$$

for the covariant components, we obtain an invariant relation

$$a_i a^i = a_t a^t + a_{t'} a^{t'} \tag{6-71}$$

where t and t' designate two transverse components, assuming that the a^i's are all real. This relation (6–71) shows that we can find an inertial system in which the longitudinal and scalar components disappear simultaneously. In such an inertial system

$$\mathbf{V} \cdot \mathbf{A} = 0, \quad \text{or} \quad \mathbf{k} \cdot \mathbf{a} = 0$$

and

$$\phi = 0 \tag{6-72}$$

In this case (6–67') gives potentials which give electromagnetic fields of

$$\mathbf{E} = -\dot{\mathbf{A}} = \omega \mathbf{a} \cos (k_j x^j)$$
$$\mathbf{B} = \mathbf{V} \times \mathbf{A} = \mathbf{k} \times \mathbf{a} \cos (k_j x^j) \tag{6-73}$$

which correspond exactly to those previously obtained in section 5–12. The waves in this case are illustrated in Fig. 6–1.

FIGURE 6–1. Linear polarized plane wave. — or solid line is **A** wave, $\cdots$ dotted lines are **E** and **B** waves. See Fig. 5–17

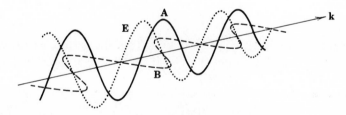

PROBLEM 6–25. For the plane wave (6–73), show that

$$\mathcal{H} = \tfrac{1}{2}(\mathbf{E} \cdot \mathbf{D} + \mathbf{B} \cdot \mathbf{H}) = \mathbf{E} \cdot \mathbf{D} = \mathbf{B} \cdot \mathbf{H}$$

$$\mathcal{H} = \varepsilon_0 (\omega \mathbf{a})^2 \cos^2 (k_j x^j)$$

$$\mathbf{S} = \mathbf{E} \times \mathbf{H} = \varepsilon_0 \omega (c \mathbf{a})^2 \mathbf{k} \cos^2 (k_j x^j)$$

6–8 INTERACTION OF A PLANE WAVE WITH FREE AND BOUND CHARGED PARTICLES

The motion of a free particle under the influence of an electromagnetic plane wave, which is given by the potentials

$$A_x(z) = a \sin [k^0 (z - ct)]$$
$$A_y = A_z = \phi = 0 \tag{6-74}$$

is obtained from the Hamiltonian

$$H_p = c\sqrt{(p_x - qA_x(z_0))^2 + p_y{}^2 + p_z{}^2 + (mc)^2} \qquad (6\text{–}75)$$

where z_0 is the position of the particle.

From Hamilton's equation (1–48) we obtain

$$\frac{dp_x}{dt} = -\frac{\partial H_p}{\partial x} = 0 \qquad (6\text{–}76)$$

Therefore p_x is a constant of motion, which we designate as α, and

$$mV_x\gamma = p_x - qA_x(z_0) = \alpha - qa \sin\left[k^0(z_0 - ct)\right] \qquad (6\text{–}77)$$

In the nonrelativistic approximation, where $\gamma \to 1$, equation (6–58) gives the rate of change of the particle energy as

$$\frac{d(mc^2\gamma)}{dt} = q\mathbf{V} \cdot \mathbf{E} = -qV_x\dot{A}_x$$

$$\cong \frac{q\omega a}{m}\{\alpha - qa \sin\left[k^0(z_0 - ct)\right]\}\left[\cos k^0(z_0 - ct)\right] \qquad (6\text{–}78)$$

assuming that z_0 is nearly constant in time. Thus the particle energy changes in time with a period of $2\pi k_0 c$ or $2\pi/\omega$. If we take an average over a period much larger than $2\pi/\omega$, we obtain

$$\overline{\frac{d(mc^2\gamma)}{dt}} = 0 \qquad (6\text{–}79)$$

or on the average the particle energy stays constant. The equation of motion of the particle can be solved without the nonrelativistic approximation,[2] but this conclusion is still valid (Fig. 6–2).

When the particle is bound, or under the influence of a static scalar potential ϕ, in addition to that of the plane wave (6–74), $q\phi_0$ should be added to the Hamiltonian H_p and equation (6–76) is replaced by

$$\frac{dp_x}{dt} = -q\frac{\partial \phi}{\partial x} \qquad (6\text{–}80)$$

---

[2] See L. Landau and E. Lifschitz, *Classical Theory of Fields*, Addison-Wesley, Reading, Mass., 1951, p. 120.

FIGURE 6–2. The relativistic equation of motion of a particle under the influence of a linear polarized plane wave will give an 8-shape trajectory. (Landau and Lifschitz).

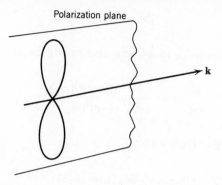

If we take the nonrelativistic approximation ($\gamma \to 1$) and assume that the position of the particle is $z = 0$, and if

$$\phi = \frac{1}{2}\frac{k}{q}x^2 = \frac{1}{2}\frac{m\omega_0^2}{q}x^2 \qquad (6\text{–}81)$$

we obtain the previous equation (2–22) for a forced harmonic oscillator, except for the damping term $R\dot{x}$. The damping term will be obtained in section 6–15; meanwhile we may assume its existence phenomenologically. Using solution (2–26), we see that the radiation rate $\mathbb{R}$ of (6–61) is (with $\dot\phi = 0$)

$$\mathbb{R} = q\mathbf{V}\cdot\dot{\mathbf{A}}_0 = \frac{(aq)^2\omega^3}{\sqrt{\omega^2 R^2 + m^2(\omega_0^2 - \omega^2)^2}}\sin(\omega t - \delta_F)\cos(\omega t)$$

$$(6\text{–}82)$$

Since

$$\sin(\omega t - \delta_F)\cos(\omega t) = \tfrac{1}{2}\left[\sin(2\omega t - \delta_F) - \sin\delta_F\right] \qquad (6\text{–}83)$$

and since δ_F is given by (2–27), we see that, when averaged over a period much larger than $2\pi/\omega$, (6–82) reduces to

$$\bar{\mathbb{R}} = -\frac{\tfrac{1}{2}(aq)^2 R\omega^4}{\omega^2 R^2 + m^2(\omega_0^2 - \omega^2)^2} \qquad (6\text{–}84)$$

Thus, on the average, the energy of the plane wave is absorbed by the oscillator, and the rate is the largest when the resonance condition $\omega_0 \cong \omega$ is satisfied.

Since the plane wave carries an average energy current density

$$|\bar{S}| = \tfrac{1}{2}(a\omega)^2 c\varepsilon_0 \tag{6-85}$$

as given by Problem 6–25, (6–84) can also be written as

$$\frac{\bar{\mathbb{R}}}{|\bar{S}|} = -\frac{(q\omega)^2 R/(c\varepsilon_0)}{\omega^2 R^2 + m^2(\omega_0{}^2 - \omega^2)^2} \tag{6-86}$$

PROBLEM 6–26.   Obtain (6–82).

PROBLEM 6–27.   Obtain (6–84) from (6–82).

6–9   RETARDED AND ADVANCED GREEN'S FUNCTIONS

Our discussions in the last section are not exact, because we neglected the field produced by the moving charge itself. The approximation may be all right when the external field, or the plane wave, is much larger than the field produced by the charge itself, but otherwise it will not be valid. We shall consider the field produced by the moving charge in this and the following few sections.

If we introduce the Green's function of the D'Alembertian, by

$$\Box \, G(x, x') = \delta(x - x') \tag{6-87}$$

where

$$\delta(x - x') = \delta(ct - ct')\,\delta(\mathbf{r} - \mathbf{r}')$$

and $x(= x^0, x^1, x^2, x^3)$ is the event for the D'Alembertian, the solution of the wave equation

$$\Box \, A^i = -\mu_0 J^i \tag{6-13}$$

can be written as

$$A^i(x) = A_0{}^i(x) - \mu_0 \int G(x - x') J^i(x')\, d^4x' \tag{6-88}$$

where $A_0{}^i$ is a solution of the homogeneous equation

$$\Box \, A_0{}^i = 0 \tag{6-89}$$

Equation (6–87) can be solved by introducing $G(k)$ as

$$G(x, x') = \frac{1}{(2\pi)^4} \int e^{-ik_j(x^j - x'^j)} G(k)\, d^4 k \qquad (6\text{–}90)$$

Putting this expression into (6–87) and noting that

$$\delta(x - x') = \frac{1}{(2\pi)^4} \int e^{-ik_j(x^j - x'^j)}\, d^4 k \qquad (6\text{–}91)$$

we obtain

$$G(k) = \frac{1}{(k^0)^2 - \mathbf{k}^2} \qquad (6\text{–}92)$$

If we substitute this expression into (6–90) we obtain

$$
\begin{aligned}
G(x, x') &= \frac{1}{(2\pi)^4} \int e^{-ik_j(x^j - x'^j)} \frac{1}{(k^0)^2 - \mathbf{k}^2}\, d^4 k \\
&= \frac{1}{(2\pi)^4} \int e^{i\mathbf{k}\cdot(\mathbf{r} - \mathbf{r}')} \left[\int_{-\infty}^{\infty} \frac{e^{-ik^0(x^0 - x'^0)}}{(k^0)^2 - \mathbf{k}^2}\, dk^0 \right] d\mathbf{k} \qquad (6\text{–}93)
\end{aligned}
$$

The integral over k^0 is not straightforward because of the poles of the integrand at

$$k^0 = +|\mathbf{k}| \qquad \text{and} \qquad k^0 = -|\mathbf{k}| \qquad (6\text{–}94)$$

In order to calculate the integral over k^0 in (6–93), we first consider a complex integral over a closed contour

$$\oint \frac{e^{-iz\xi}}{z^2 - k^2}\, dz \qquad (6\text{–}95)$$

where ξ is a real number. Let us take a semicircle, as shown in Fig. 6–3, for the contour. When z is on the real axis, the integrand of (6–95) equals that of our integral with replacements

$$z \to k^0 \qquad \xi \to x^0 - x'^0 \qquad (6\text{–}96)$$

while when z is on the circumference of the semicircle

$$z = re^{i\varphi} = r \cos \varphi + ir \sin \varphi$$

and

$$e^{-iz\xi} = \exp(\xi r \sin \varphi) \exp(-i\xi r \cos \varphi) \qquad (6\text{–}97)$$

FIGURE 6–3.   Integral contours in the complex z plane. Take the semicircle in the upper or lower half-plane depending on whether $\xi < 0$ or $\xi > 0$.

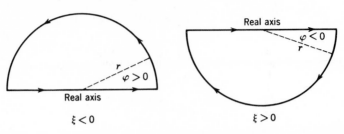

$$\xi < 0 \qquad\qquad\qquad \xi > 0$$

where r and φ are the radial and angular coordinates in the complex plane, respectively. When $\xi \sin \varphi < 0$, the first factor of (6–97) approaches zero exponentially as the radius of the contour r approaches infinity even for an infinitesimally small φ. Thus in the limit of the infinite radius this contour gives

$$\oint_C \frac{e^{-iz\xi}}{z^2 - \mathbf{k}^2}\, dz = \int_{-\infty}^{\infty} \frac{e^{-ik^0\xi}}{(k^0)^2 - \mathbf{k}^2}\, dk^0 \qquad (6\text{–}98)$$

According to Cauchy's integration theorem,[3] if $f(z)$ is analytic in the region of integration and if the contour C encircles a point a once in the counterclockwise direction, then

$$\oint_C \frac{f(z)}{z - a}\, dz = 2\pi i f(a) \qquad (6\text{–}99)$$

but if C encircles a in the clockwise direction, the integral should give $-2\pi i f(a)$.

In applying Cauchy's theorem to the present integral (6–98), an ambiguity occurs since the contour we just established goes right over the poles given by (6–94). In order to obtain a definite answer we can bend the contour infinitesimally around each pole. Since there are two poles, we have four ways of doing this; they are illustrated in Fig. 6–4.

If we take the contour C_R of Fig. 6–4, and if $\xi > 0$, then we see that this contour encircles both poles clockwise, since the contour closes with the semicircle in the lower plane. On the other hand, if the contour goes around the poles in the manner of C_R but if $\xi < 0$, then the contour does not contain

---

[3] For Cauchy's theorem and complex integrals see Morse and Feshbach, *Methods of Theoretical Physics*, McGraw-Hill, New York, 1953, Chapter 4.

FIGURE 6–4. Four contours along the real axis.

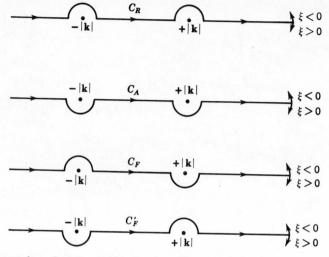

any of the poles, since it closes with the semicircle in the upper plane. Thus, according to Cauchy's theorem, we obtain

$$\oint_{C_R} \frac{e^{-iz\xi}}{z^2 - \mathbf{k}^2} \, dz = -\frac{\pi i}{|\mathbf{k}|} (e^{-i|\mathbf{k}|\xi} - e^{i|\mathbf{k}|\xi}) \qquad \text{when} \qquad \zeta > 0$$

$$= 0 \qquad\qquad\qquad \text{when} \qquad \zeta < 0$$

(6–100)

Introducing the step function

$$\theta(\xi) = 1, \quad \text{if} \quad \xi > 0$$

$$= 0, \quad \text{if} \quad \xi < 0$$

(6–101)

in the result (6–100) we see (6–98) gives

$$\int_{-\infty}^{\infty} \frac{\exp\left[-ik^0(x^0 - x'^0)\right]}{(k^0)^2 - \mathbf{k}^2} \, dk^0$$

$$= -\frac{\pi i}{|\mathbf{k}|} \{\exp\left[-i|\mathbf{k}|(x^0 - x'^0)\right] - \exp\left[i|\mathbf{k}|(x^0 - x'^0)\right]\}\theta(x^0 - x'^0)$$

(6–102)

The corresponding Green's function is designated as D_R, and is, from (6–93) and (6–102),

$$D_R(x - x') = -\frac{\pi i}{(2\pi)^4} \int e^{i\mathbf{k} \cdot (\mathbf{r}-\mathbf{r}')} \frac{1}{|\mathbf{k}|}$$

$$\times \{\exp[-i|\mathbf{k}|(x^0 - x'^0)]$$

$$-\exp[i|\mathbf{k}|(x^0 - x'^0)]\}\theta(x^0 - x'^0)\, d\mathbf{k}$$

$$= -\frac{1}{8\pi^2 R} \int_0^\infty (e^{i|\mathbf{k}|R} - e^{-i|\mathbf{k}|R})$$

$$\times (e^{-i|\mathbf{k}|(x^0-x'^0)} - e^{i|\mathbf{k}|(x^0-x'^0)})\, d|\mathbf{k}|\theta(x^0 - x'^0)$$

$$= -\frac{1}{8\pi^2 R} \int_{-\infty}^\infty (e^{-i|\mathbf{k}|(x^0-x'^0-R)} - e^{i|\mathbf{k}|(x^0-x'^0+R)})\, d|\mathbf{k}|\theta(x^0 - x'^0)$$

$$= -\frac{1}{4\pi R}[\delta(x^0 - x'^0 - R) - \delta(x^0 - x'^0 + R)]\theta(x^0 - x'^0)$$

$$= -\frac{1}{4\pi R}\delta(x^0 - x'^0 - R) \tag{6–103}$$

where

$$R = |\mathbf{r} - \mathbf{r}'| > 0 \tag{6–104}$$

This Green's function D_R is called the *retarded Green's function*.
The second contour C_A of Fig. 6–4 gives

$$\oint_{C_A} \frac{e^{-iz(x^0-x'^0)}}{z^2 - \mathbf{k}^2}\, dz = \int_{-\infty}^\infty \frac{e^{-i|\mathbf{k}|(x^0-x'^0)}}{(k^0)^2 - \mathbf{k}^2}\, dk^0$$

$$= -\frac{\pi i}{|\mathbf{k}|}[e^{-i|\mathbf{k}|(x^0-x'^0)} - e^{i|\mathbf{k}|(x^0-x'^0)}]\theta(x'^0 - x^0) \tag{6–105}$$

instead of (6–102). A calculation similar to (6–103) gives the *advanced Green's function* D_A

$$D_A(x, x') = -\frac{1}{4\pi R}\delta(x^0 - x'^0 + R) \tag{6–106}$$

The other Green's functions D_F and $D_{F'}$, obtained by taking the integration contours C_F and $C_{F'}$, appear in quantum electrodynamics, but not in classical electrodynamics. Therefore, we shall not discuss D_F and $D_{F'}$ here.

If we define

$$D(x, x') \equiv D_R(x, x') - D_A(x, x') \tag{6-107}$$

this function satisfies

$$\Box\, D(x, x') = 0 \tag{6-108}$$

Differentiating (6–102) and (6–105) by $x^0 - x'^0$, subtracting one result from the other, and letting $x^0 = x'^0$, we obtain

$$\left. \frac{\partial D(x, x')}{\partial(x^0 - x'^0)} \right|_{x^0 = x'^0} = -\delta(\mathbf{r} - \mathbf{r}') \tag{6-109}$$

PROBLEM 6–28. Obtain (6–109).

HINT. First show that

$$D(x, x') = -\frac{i}{(2\pi)^3} \int e^{-ik_j(x^j - x'^j)}\, \delta(k_j k^j)\varepsilon(k^0)\, d^4k$$

where

$$\varepsilon(k^0) = 1 \qquad \text{if} \qquad k^0 > 0$$
$$= -1 \qquad \text{if} \qquad k^0 < 0$$

PROBLEM 6–29. When the free fields $A_0{}^i$ satisfy the Lorentz condition (6–15), show that the general fields A^i, given by

$$A^i(x) = A_0{}^i(x) - \mu_0 \int D_R(x, x')J^i(x')\, d^4x'$$

also satisfy the Lorentz condition.

PROBLEM 6–30. The Green's function $G_k(\mathbf{r}, \mathbf{r}')$ is defined by

$$(\nabla^2 + k^2)G_k(\mathbf{r}, \mathbf{r}') = \delta(\mathbf{r} - \mathbf{r}')$$

(a) Show that

$$G_k(\mathbf{r}, \mathbf{r}') = -\int e^{-i\boldsymbol{\kappa} \cdot (\mathbf{r} - \mathbf{r}')} \frac{1}{k^2 - \kappa^2}\, d\kappa$$

(b) Using the Cauchy theorem, evaluate the above integral to obtain

$$G_k(\mathbf{r}, \mathbf{r}') = -\frac{e^{ik|\mathbf{r}-\mathbf{r}'|}}{4\pi|\mathbf{r} - \mathbf{r}'|}$$

which is the same result we obtained before in Problem 5–6.

6–10 LIÉNARD–WIECHERT POTENTIALS

If we use the retarded Green's function (6–103) in (6–88), with $A_0{}^i(x) = 0$, we obtain

$$A^i(x) = \frac{\mu_0}{4\pi} \int \delta(x^0 - x'^0 - R) \frac{J^i(x')}{R} dx' \qquad (6\text{–}110)$$

which is called the retarded potential. This potential satisfies the causality principle, which states that the signal (potential) we receive can tell the past of the source only. Actually the potential propagates from the source with velocity c.

For a point particle, we have

$$J^0(x') = cq\,\delta(\mathbf{r}' - \mathbf{r}_q(t')) \qquad (6\text{–}111)$$

$$\mathbf{J}(x') = q\mathbf{V}(t')\,\delta(\mathbf{r}' - \mathbf{r}_q(t')) \qquad (6\text{–}112)$$

from (5–14) and (5–72). Therefore (6–110) gives

$$\phi(t, \mathbf{r}) = \frac{\mu_0 c^2 q}{4\pi} \int \frac{\delta(\mathbf{r}' - \mathbf{r}_q(t'))\,\delta(ct - ct' - |\mathbf{r} - \mathbf{r}'|)}{|\mathbf{r} - \mathbf{r}'|} dv'c\,dt'$$

$$= \frac{q}{4\pi\varepsilon_0} \int \frac{\delta(ct - ct' - |\mathbf{r} - \mathbf{r}_q(t')|)}{|\mathbf{r} - \mathbf{r}_q(t')|} c\,dt' \qquad (6\text{–}113)$$

In order to calculate the last integral we introduce t'' by

$$ct'' = ct' + |\mathbf{r} - \mathbf{r}_q(t')| \qquad (6\text{–}114)$$

which gives

$$c\,dt'' = c\,dt' + \frac{\partial}{\partial \mathbf{r}_q} |\mathbf{r} - \mathbf{r}_q(t')| \frac{\partial \mathbf{r}_q}{\partial t'} dt'$$

$$= c\,dt' \left[1 - \frac{\mathbf{R}_q \cdot \mathbf{V}(t')}{cR_q} \right] \qquad (6\text{–}115)$$

where

$$\mathbf{R}_q = \mathbf{r} - \mathbf{r}_q(t') \qquad R_q = R_q \qquad (6\text{--}116)$$

Thus (6–113) yields

$$\phi(t, \mathbf{r}) = \frac{q}{4\pi\varepsilon_0} \int \frac{\delta(t - t'')}{R_q - \mathbf{R}_q \cdot \mathbf{V}(t')/c} \, dt''$$

$$= \frac{q}{4\pi\varepsilon_0} \frac{1}{R_q - \mathbf{R}_q \cdot \mathbf{V}/c} \bigg|_{t\mathrm{ret}} \qquad (6\text{--}117)$$

In the same way we also obtain

$$\mathbf{A}(t, \mathbf{r}) = \frac{\mu_0 q}{4\pi} \frac{\mathbf{V}}{R_q - \mathbf{R}_q \cdot \mathbf{V}/c} \bigg|_{t\mathrm{ret}} \qquad (6\text{--}118)$$

In (6–117) and (6–118) the expressions should be evaluated at

$$t_{\mathrm{ret}} = t - \frac{R_q(t_{\mathrm{ret}})}{c} \qquad (6\text{--}119)$$

These potentials (6–117) and (6–118) are called the Liénard–Wiechert potentials. Fig. 6–5 shows $\mathbf{V}$ and $\mathbf{R}_q$ used in the Liénard–Wiechert potentials.

PROBLEM 6–31. Show that the Liénard–Wiechert potentials reduce to the Coulomb potential (5–19) and $\mathbf{A}$ which gives $\mathbf{B}$ of Problem 5–23, when $V \ll c$.

FIGURE 6–5. Liénard–Wiechert potential.

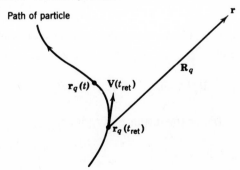

PROBLEM 6–32. Show that the Liénard–Wiechert potentials can be expressed in a covariant form of

$$A^i(x) = \frac{q}{4\pi\varepsilon_0 c} \frac{\dot{z}^i}{(x - z)^j \dot{z}_j}\bigg|_{\tau=\tau_0}$$

where z^i are the coordinates of the source particle,

$$\dot{z}^i = \frac{dz^i}{d\tau}$$

and τ_0 is the proper time when the light cone which originates at x meets the source particle.

HINT.

$$(x - z)^0|_{\tau=\tau_0} = ct - ct_{ret} = R_q(t_{ret})$$

PROBLEM 6–33. Calculate the potentials due to a moving charged particle using the advanced Green's function D_A.

ANSWER.

$$\phi(t, \mathbf{r}) = \frac{q}{4\pi\varepsilon_0} \frac{1}{R_q + \mathbf{R}_q \cdot \mathbf{V}/c}\bigg|_{t_{ad}}$$

$$\mathbf{A}(t, \mathbf{r}) = \frac{\mu_0 q}{4\pi} \frac{\mathbf{V}}{R_q + \mathbf{R}_q \cdot \mathbf{V}/c}\bigg|_{t_{ad}}$$

where

$$t_{ad} = t + R_q \frac{(t_{ad})}{c}$$

PROBLEM 6–34. When the radiating particle is moving at a constant velocity $\mathbf{V}$, show that

$$(R_q - \mathbf{R}_q \cdot \mathbf{V}/c)_{t_{ret}} = \sqrt{R_{q0}^2 - (\mathbf{R}_q \times \mathbf{V}/c)^2}$$

where $\mathbf{R}_{q0} = \mathbf{r} - \mathbf{r}_q(t)$, or the distance at t, not t_{ret}.

HINT. See Fig. 6–6.

FIGURE 6–6.   If **V** is constant $R_q - \mathbf{R}_q \cdot \mathbf{V}/c = \sqrt{R_{q0}{}^2 - (\mathbf{R}_q \times \mathbf{V}/c)^2}$.

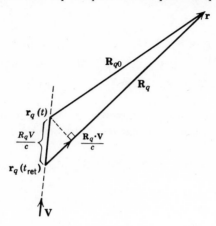

6–11   ELECTROMAGNETIC FIELDS DUE TO A MOVING CHARGE

From the Liénard–Wiechert potentials we can calculate **E** and **B** due to a moving charge. In this calculation, however, we should remember that the fields at a time t are expressed in terms of the position and velocity of the particle at time t_{ret}, in the Liénard–Wiechert potentials. Thus, for example, in calculating $\dot{\mathbf{A}}$ using (6–118) we have to calculate

$$\frac{\partial \mathbf{A}(t, \mathbf{r})}{\partial t} = \frac{\partial \mathbf{A}(t_{\text{ret}})}{\partial t_{\text{ret}}} \frac{\partial t_{\text{ret}}}{\partial t} \tag{6–120}$$

where $\mathbf{A}(t_{\text{ret}})$ is the expression on the right-hand side of (6–118). The factor $\partial t_{\text{ret}}/\partial t$ which appears in (6–120) can be calculated as follows: From (6–119)

$$\frac{\partial R_q}{\partial t} = c \left(1 - \frac{\partial t_{\text{ret}}}{\partial t} \right) \tag{6–121}$$

while

$$\frac{\partial R_q}{\partial t} = \frac{\partial R_q}{\partial t_{\text{ret}}} \frac{\partial t_{\text{ret}}}{\partial t} = -\frac{\mathbf{R}_q \cdot \mathbf{V}(t_{\text{ret}})}{R_q} \frac{\partial t_{\text{ret}}}{\partial t} \tag{6–122}$$

Combining (6–121) and (6–122) we obtain

$$\frac{\partial t_{\text{ret}}}{\partial t} = \frac{R_q}{R_q - \mathbf{R}_q \cdot \mathbf{V}/c} \bigg|_{t_{\text{ret}}} \tag{6–123}$$

Therefore (6–120) and (6–123) give

$$\frac{\partial \mathbf{A}(t, \mathbf{r})}{\partial t} = \frac{\mu_0 q}{4\pi} \left[\frac{R_q \dot{\mathbf{V}}}{(R_q - \mathbf{R}_q \cdot \mathbf{V}/c)^2} \right.$$

$$\left. + \frac{R_q \mathbf{V}}{(R_q - \mathbf{R}_q \cdot \mathbf{V}/c)^3} \left(\frac{\mathbf{R}_q \cdot \mathbf{V}}{R_q} - \frac{V^2}{c} + \frac{\mathbf{R}_q \cdot \dot{\mathbf{V}}}{c} \right) \right] \Bigg|_{t_{\text{ret}}} \quad (6\text{–}124)$$

In calculating $\nabla \phi$ we should remember that by changing the observation point $\mathbf{r}$, the retarded time also changes. Thus from (6–119)

$$\nabla R_q = -c \nabla t_{\text{ret}} \quad (6\text{–}125)$$

but

$$\nabla R_q = \mathbf{V}' R_q + \frac{\partial R_q}{\partial t_{\text{ret}}} \nabla t_{\text{ret}} = \frac{\mathbf{R}_q - (\mathbf{R}_q \cdot \mathbf{V}) \nabla t_{\text{ret}}}{R_q} \quad (6\text{–}126)$$

where $\mathbf{V}'$ is the gradient calculated with fixed t_{ret}; therefore

$$\nabla t_{\text{ret}} = - \frac{R_q}{cR_q - \mathbf{R}_q \cdot \mathbf{V}} \Bigg|_{t_{\text{ret}}} \quad (6\text{–}127)$$

Now

$$\nabla \phi = \mathbf{V}' \phi + (\nabla t_{\text{ret}}) \frac{\partial \phi}{\partial t_{\text{ret}}} \quad (6\text{–}128)$$

and (6–117) give

$$\nabla \phi = - \frac{q}{4\pi \varepsilon_0} \left[\frac{(\mathbf{R}_q/R_q) - (\mathbf{V}/c)}{(R_q - \mathbf{R}_q \cdot \mathbf{V}/c)^2} \right.$$

$$\left. + \frac{R_q}{c(R_q - \mathbf{R}_q \cdot \mathbf{V}/c)^3} \left(\frac{\mathbf{R}_q \cdot \mathbf{V}}{R_q} - \frac{V^2}{c} + \frac{\mathbf{R}_q \cdot \dot{\mathbf{V}}}{c} \right) \right] \Bigg|_{t_{\text{ret}}}$$

$$(6\text{–}129)$$

From (6–124) and (6–129) we finally obtain

$$\mathbf{E} = \frac{q}{4\pi \varepsilon_0} \left\{ \frac{(1 - \beta^2)(\mathbf{n} - \boldsymbol{\beta})}{R_q^2 (1 - \mathbf{n} \cdot \boldsymbol{\beta})^3} + \frac{\mathbf{n} \times [(\mathbf{n} - \boldsymbol{\beta}) \times \dot{\boldsymbol{\beta}}]}{cR_q (1 - \mathbf{n} \cdot \boldsymbol{\beta})^3} \right\} \Bigg|_{t_{\text{ret}}} \quad (6\text{–}130)$$

where

$$\mathbf{n} = \frac{\mathbf{R}_q}{R_q} \qquad \boldsymbol{\beta} = \frac{\mathbf{V}}{c} \quad (6\text{–}131)$$

The corresponding calculation of the magnetic flux density **B** shows that

$$\mathbf{B} = \mathbf{n} \times \frac{\mathbf{E}}{c} \tag{6-132}$$

where **n** is the unit vector defined by (6–131) and measured at t_{ret}.

PROBLEM 6–35. Obtain (6–124).

PROBLEM 6–36. Confirm (6–129) and (6–130).

PROBLEM 6–37. Show that **E** is in the direction of $\mathbf{R}_{q0}$ ($\equiv \mathbf{r} - \mathbf{r}_q(t)$), when the velocity **V** is constant.

6-12 EMISSION AND SCATTERING OF RADIATION

In (6–130), the first term is proportional to R_q^{-2}, while the second term is proportional to R_q^{-1}. Therefore, at a large distance R_q from the particle, the first term becomes negligible compared to the second term. Since this predominant (second) term has a direction perpendicular to **n**, or $\mathbf{R}_q$, the **E** field at a large distance R_q is nearly perpendicular to **n**. The **B** field, given by (6–132), is always perpendicular to both **n** and **E**, and thus the Poynting vector **S** will be parallel to **n** at a large distance from the particle (Fig. 6–7): Using the second term of (6–130) and the corresponding term in **B**, we obtain

$$\mathbf{S} = \mathbf{E} \times \mathbf{H} \cong \frac{q^2}{16\pi^2 \varepsilon_0 c} \frac{[(\mathbf{n} - \boldsymbol{\beta}) \times \dot{\boldsymbol{\beta}}]^2}{R_q^2(1 - \mathbf{n} \cdot \boldsymbol{\beta})^6} \mathbf{n} \Bigg|_{t_{\text{ret}}} \tag{6-133}$$

Bremsstrahlung

If a charged particle enters a dense material at a high speed it may be stopped in a short time interval: The deceleration during this process produces *Bremsstrahlung*. From energy conservation we expect that the amount

FIGURE 6–7. Electromagnetic fields and the Poynting vector due to acceleration $\dot{\boldsymbol{\beta}}$.

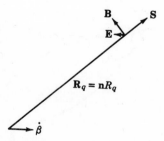

of power carried as radiation into a *unit solid angle*, $dP/d\Omega$, must be equal to that lost by the particle at the retarded time t_{ret}:

$$\frac{dP}{d\Omega} = -\frac{d^2W}{d\Omega\, dt_{ret}} = |S|R_q^2\, \frac{dt}{dt_{ret}} \qquad (6\text{-}134)$$

Using (6–123) for dt/dt_{ret} and assuming that β and $\dot{\beta}$ are in the same direction in (6–133), we obtain

$$\frac{dP}{d\Omega} = \frac{(q\dot{V}\sin\theta)^2}{16\pi^2\varepsilon_0 c^3(1 - \beta\cos\theta)^5} \qquad (6\text{-}135)$$

for the intensity of Bremsstrahlung, where θ is the angle between β and **n**. When $\beta \ll 1$, the radiation is mostly in the direction perpendicular to β (and $\dot{\beta}$), but when the velocity is relativistic, or β is comparable to 1, the intensity is stronger in the forward direction than in the backward direction. In any case the pattern is symmetric around the axis given by β (and $\dot{\beta}$). The Bremsstrahlung by each charged particle is a pulse of a short duration, since the radiation stops when the particle is stopped. If we assume that the particle velocity V changes from V_0 to 0 with constant $\dot{V}$ ($= \dot{V}_0$) and that **V** and $\dot{\mathbf{V}}$ stay in the same direction during the braking process, we obtain

$$\int |S|\, \frac{dt}{dt_{ret}}\, R_q^2\, dt_{ret} = \frac{q^2\dot{V}_0 \sin^2\theta}{64\pi^2\varepsilon_0 c^2 \cos\theta} \left[\frac{1}{(1 - \beta_0\cos\theta)^4} - 1 \right] \qquad (6\text{-}136)$$

for the energy of the pulse radiation into the solid angle $d\Omega$. In this derivation θ is assumed to be constant during the braking process, since the distance covered by the particle before it stops is very short compared to R_q. Fig. 6–8 shows the intensity distribution given by (6–136).

FIGURE 6–8. $dP/d\Omega$ in the Bremsstrahlung; equation (6–134).

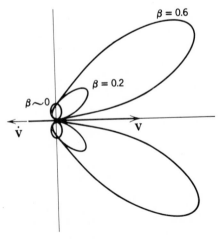

PROBLEM 6–38. Obtain (6–136).

Radiation by a Simple Harmonic Oscillator

If a charge is making a simple harmonic motion

$$\mathbf{r}_q = \mathbf{r}_0 \cos \omega t$$

then

$$\ddot{\mathbf{r}}_q = -\omega^2 \mathbf{r}_0 \cos \omega t \tag{6–137}$$

From (6–133) and (6–137), the intensity of radiation (emission) by the simple harmonic oscillator is obtained as

$$\frac{dP}{d\Omega} = \frac{(qr_0\omega^2 \sin \theta \cos (\omega t))^2}{16\pi^2\varepsilon_0 c^3} \tag{6–138}$$

when $\beta \ll 1$, and the amplitude r_0 is so small that the variation of θ during the oscillation is negligible. This formula gives *electric dipole radiation*.[4]

Thomson Scattering

The acceleration of a charged particle may be due to an incoming electromagnetic wave. Thus, if $\mathbf{E}$ is the electric component of the incoming electromagnetic wave, a charged particle of charge q and mass m is accelerated with

$$\ddot{\mathbf{r}} = \frac{q}{m} \mathbf{E} \tag{6–139}$$

to the first approximation. This acceleration, in turn, produces a radiation of intensity

$$\frac{dP}{d\Omega} = \frac{(q^2 E \sin \theta)^2}{16\pi^2\varepsilon_0 c^2 m^2} \tag{6–140}$$

into the solid angle $d\Omega$ which is at an angle θ from the polarization direction of the incident light. This scattering of a plane wave by a free charge is called *Thomson scattering* (Fig. 6–9). Since the intensity of the incident light is

$$I = \frac{E^2}{(c\mu_0)} \tag{6–141}$$

---

[4] For classical discussions on higher-order radiations, see Chapter 16 of J. D. Jackson, *Classical Electrodynamics*, John Wiley & Sons, New York, 1965.

FIGURE 6-9. Scattering of light by an electron.

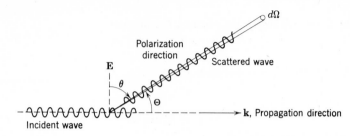

the *differential scattering cross section*, which gives the efficiency for scattering, is

$$\frac{d\sigma}{d\Omega} = \frac{1}{I}\frac{dP}{d\Omega} = \frac{(\mu_0 q^2 \sin \theta)^2}{16\pi^2 m^2} \qquad (6\text{-}142)$$

in the case of Thomson scattering.

PROBLEM 6-39. If the incident light is not polarized, or is an incoherent mixture of every polarization with a fixed propagation direction **k**, show that the differential cross section of the Thomson scattering is

$$\left(\frac{d\sigma}{d\Omega}\right)_{\text{unpolarized}} = \frac{\mu_0^2 q^4}{32\pi^2 m^2} (1 + \cos^2 \Theta)$$

where Θ is the angle between the directions of incident to scattered waves.

HINT. Average $\sin^2 \theta$ over the polarization direction.

Scattering by a Bound Charge

We have seen in section 6-8 that a charged particle bound to an origin with a simple harmonic potential absorbs energy from an incident plane wave; but the forced vibration of the particle should, in turn, produce radiation. Thus, unless an additional mechanism which removes absorbed energy from the system exists, the bound particle will scatter light instead of absorbing it. Using the previously obtained solution (2-26) we obtain

$$\ddot{\mathbf{r}} = \frac{-\omega^2 q \mathbf{E}_0}{\sqrt{\omega^2 R^2 + m^2(\omega_0^2 - \omega^2)^2}} \cos (\omega t - \delta_F) \qquad (6\text{-}143)$$

where E_0 is the amplitude of the electric field of the incident wave. This formula implies that the incident and the scattered waves have a phase difference of δ_F, although their frequencies are the same. In the nonrelativistic approximation, the differential cross section of the scattering is

$$\frac{d\sigma}{d\Omega} = \frac{\mu_0^2 q^4}{16\pi^2} \frac{\omega^4}{\omega^2 R^2 + m^2(\omega_0^2 - \omega^2)^2} \sin^2\theta \qquad (6\text{--}144)$$

When

$$(m\omega)^2 \gg (m\omega_0)^2 \qquad \text{and} \qquad (\omega R)^2 \qquad (6\text{--}145)$$

this formula reduces to (6–142), while if

$$(m\omega_0)^2 \gg (m\omega)^2 \qquad \text{and} \qquad (\omega R)^2 \qquad (6\text{--}146)$$

it reduces to

$$\frac{d\sigma}{d\Omega} = \frac{\mu_0^2 q^4 \omega^4}{16\pi^2 m^2 \omega_0^2} \sin^2\theta \qquad (6\text{--}147)$$

which is the cross section for the *Rayleigh scattering*. Since for most molecules ω_0 is in the ultraviolet region, this formula is applicable to the scattering of visible light by molecules, and shows that blue light is scattered very much more than red light, explaining why the sky is blue and why the rising sun is red.

PROBLEM 6–40. The complex polarizability α may be defined as

$$q\mathbf{r} = \alpha\mathbf{E} = |\alpha|e^{-i\delta_F}\mathbf{E}$$

where $\mathbf{E} = \mathbf{E}_0 e^{i\omega t}$, and δ_F is the phase difference between $\mathbf{E}$ and $\mathbf{r}$. (a) Show that in the model we took

$$\alpha(\omega) = \frac{q^2 \exp(-i\,\delta_F)}{\sqrt{\omega^2 R^2 + m^2(\omega_0^2 - \omega^2)^2}}$$

(b) Show that the differential scattering cross section can be written as

$$\frac{d\sigma}{d\Omega} = \frac{\mu_0^2}{16\pi^2} \omega^4 |\alpha(\omega)|^2 \sin^2\theta$$

(c) Show from (6–86) that

$$-\frac{\overline{\mathbb{R}}}{|\overline{\mathbf{S}}|} = \frac{\omega}{c\varepsilon_0} \mathscr{I}m\,\alpha$$

6–13* RETARDED POTENTIALS AT SHORT DISTANCES

So far we have been discussing the Liénard–Wiechert potentials at a large distance from a moving source. We now turn our attention to the field at close distances, where the retardation effects are relatively small.

We shall use the abbreviated notations

$$\mathbf{R}' \equiv \mathbf{R}_q(t_{\text{ret}})$$
$$\mathbf{R} \equiv \mathbf{R}_q(t) \tag{6–148}$$

in this section. Since t_{ret} is related to t by (6–119), we can expand $\mathbf{R}'$ in a Taylor series as

$$\mathbf{R}' \left(= \mathbf{R}_q\left(t - \frac{R'}{c} \right) \right)$$

$$= \mathbf{R} - \frac{R'}{c} \dot{\mathbf{R}} + \frac{1}{2} \left(\frac{R'}{c} \right)^2 \ddot{\mathbf{R}} - \frac{1}{6} \left(\frac{R'}{c} \right)^3 \dddot{\mathbf{R}} + \cdots$$

$$= \mathbf{R} + \frac{R'}{c} \mathbf{V} - \frac{1}{2} \left(\frac{R'}{c} \right)^2 \dot{\mathbf{V}} + \frac{1}{6} \left(\frac{R'}{c} \right)^3 \ddot{\mathbf{V}} - \cdots \tag{6–149}$$

where $\mathbf{V}$ is the velocity of the source particle at time t. This expression for $\mathbf{R}'$ includes R', the magnitude of $\mathbf{R}'$, or R_q at t_{ret}. In order to express $\mathbf{R}'$ by quantities at t only, we assume an expansion of R' as

$$R' = R + c^{-1}a + c^{-2}b + c^{-3}d + \cdots \tag{6–150}$$

Inserting this expression into (6–149) we obtain an expression $(R')^2 = \mathbf{R}' \cdot \mathbf{R}'$. Comparing that expression with another obtained directly from (6–150), that is,

$$(R')^2 = R^2 + c^{-1}(2aR) + c^{-2}(a^2 + 2bR) + \cdots \tag{6–151}$$

we obtain

$$a = \mathbf{R} \cdot \mathbf{V}$$
$$b = \tfrac{1}{2}[RV^2 + R(\mathbf{n} \cdot \mathbf{V})^2 - R^2(\mathbf{n} \cdot \dot{\mathbf{V}})] \tag{6–152}$$
$$d = R(\mathbf{n} \cdot \mathbf{V})V^2 - R^2(\mathbf{n} \cdot \mathbf{V})(\mathbf{n} \cdot \dot{\mathbf{V}}) - \tfrac{1}{2}R^2(\mathbf{V} \cdot \dot{\mathbf{V}}) + \tfrac{1}{6}R^3(\mathbf{n} \cdot \ddot{\mathbf{V}})$$

where $\mathbf{n} = \mathbf{R}_q/R_q$, as defined in (6–131). Inserting this result into (6–150) and (6–149), we now have an expression for R' and thus $\mathbf{R}'$ in terms of quantities at t only.

Using an expansion similar to (6–148), we see that

$$\mathbf{V'} \ (\equiv \mathbf{V}(t_{\text{ret}})) = \mathbf{V} - c^{-1}(R\dot{\mathbf{V}}) - c^{-2}[(\mathbf{R} \cdot \mathbf{V})\dot{\mathbf{V}} - \tfrac{1}{2}R^2\ddot{\mathbf{V}}] + \cdots \tag{6–153}$$

and we obtain expressions for the retarded potentials as a function of quantities at t as

$$\phi(t, \mathbf{r}) = \frac{q}{4\pi\varepsilon_0} \frac{1}{R' - \mathbf{R'} \cdot \mathbf{V'}/c}$$

$$= \frac{q}{4\pi\varepsilon_0} \left\{ \frac{1}{R} + \frac{1}{2c^2R} \left[V^2 - (\mathbf{n} \cdot \mathbf{V})^2 - R(\mathbf{n} \cdot \dot{\mathbf{V}}) \right] \right.$$

$$\left. - \frac{1}{c^3} \left[(\mathbf{V} \cdot \dot{\mathbf{V}}) - \tfrac{1}{3}(\mathbf{R} \cdot \ddot{\mathbf{V}}) \right] + \cdots \right\} \tag{6–154}$$

$$\mathbf{A}(t, \mathbf{r}) = \frac{q}{4\pi\varepsilon_0} \left(\frac{\mathbf{V}}{c^2R} - \frac{\dot{\mathbf{V}}}{c^3} + \cdots \right) \tag{6–155}$$

PROBLEM 6–41. Obtain (6–152).

PROBLEM 6–42. When $\mathbf{V}$ is constant, or $\dot{\mathbf{V}} = \ddot{\mathbf{V}} = \cdots = 0$, show that

$$R' = \frac{R[(\mathbf{n} \cdot \boldsymbol{\beta}) + \sqrt{1 - \beta^2 + (\mathbf{n} \cdot \boldsymbol{\beta})^2}]}{(1 - \beta^2)}$$

where

$$\boldsymbol{\beta} = \frac{\mathbf{V}}{c}$$

PROBLEM 6–43. Obtain (6–154) and (6–155).

PROBLEM 6–44. A total charge q is homogeneously distributed on a spherical shell of radius R_0.
(a) If the shell is moving with a constant velocity $\mathbf{V}$, show from (6–154) that the scalar potential at the center is

$$\phi = \frac{q}{4\pi\varepsilon_0 R_0} \left(1 + \frac{V^2}{3c^2} + \cdots \right)$$

(b) If the shell was spherical, it is deformed from that shape when it is moving, because of the relativistic contraction effect given by formula (4–46). Show that this contraction effect produces the scalar potential

$$\phi_{\text{cont}} = \frac{q}{4\pi\varepsilon_0 R_0} \left[1 + \frac{1}{2} \frac{\overline{(\mathbf{n} \cdot \mathbf{V})^2}}{c^2} + \cdots \right] = \frac{q}{4\pi\varepsilon_0 R_0} \left(1 + \frac{V^2}{6c^2} + \cdots \right)$$

6–14* SELF-ENERGY OF AN ELECTRON

If we assume that an electron is made of infinitesimal charges $-de$, assembled in a small volume, the potential energy of the electron itself due to the interaction among these elements may be calculated with formulas (6–154) and (6–155). This energy is called the *self-energy* of an electron.

When the electron is at rest, we may assume that each part is also at rest. The self-energy in this case is

$$U(\mathbf{V} = 0) = \frac{1}{4\pi\varepsilon_0} \int \int \frac{de\, de'}{R} \tag{6–156}$$

If the electron is a spherical shell of radius r_0, on which the total charge e is homogeneously distributed (Fig. 6–10), we see (6–156) gives

$$U(\mathbf{V} = 0) = \frac{e^2}{8\pi\varepsilon_0 r_0} \tag{6–157}$$

while if it is a solid sphere of radius r_0, in which the charge is homogeneously distributed, we obtain

$$U(\mathbf{V} = 0) = \frac{3e^2}{40\pi\varepsilon_0 r_0} \tag{6–158}$$

FIGURE 6–10. Classical electron theory.

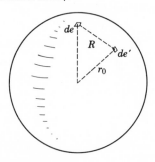

The self-energy can be calculated from the field viewpoint as well. Since a field of

$$E = \frac{e^2}{4\pi\varepsilon_0 R^3} R \qquad (6\text{-}159)$$

exists around an electron at rest, the total field energy, given by (6–39), is

$$U(\mathbf{V} = 0) = \tfrac{1}{2}\varepsilon_0 \int E^2 \, dv = \frac{e^2}{8\pi\varepsilon_0 r_0} \qquad (6\text{-}160)$$

if the electron has the shell structure; this result agrees with (6–157).

Since the self-energy cannot be taken out of the electron, it must be included in the rest-mass energy mc^2 of the electron. Thus the electron mass consists of an electromagnetic mass and a purely mechanical mass. Lorentz assumed that the electron mass was purely electromagnetic: The self-energy depends on the structure of the electron, as seen in (6–157) and (6–158), but to an order of magnitude, the Lorentz model gives

$$mc^2 = U(V = 0) \cong \frac{e^2}{4\pi\varepsilon_0 r_0} \qquad (6\text{-}161)$$

defining a *classical electron radius* r_0 of

$$r_0 = \frac{e^2}{4\pi\varepsilon_0 mc^2} = 2.817939 \times 10^{-15} \text{ m} \qquad (6\text{-}162)$$

While such a simple model of the electron is not acceptable, since it gives contradictory results as will be shown below, and since quantum mechanical properties such as spin are not considered, the number given in (6–162) gives a limit below which some unknown subatomic physical phenomena may exist; that is, the present laws of electricity and magnetism are expected to be violated for phenomena taking place within dimensions smaller than this classical electron radius.

This Lorentz model, sometimes called the Lorentz–Abraham model of the electron, confronts severe difficulty when a moving electron is considered. Since the stress-energy tensor $t_i{}^j$ of the field due to an electron at rest is

$$\begin{pmatrix} -\tfrac{1}{2}\varepsilon_0 E^2 & 0 & 0 & 0 \\ 0 & \tfrac{1}{2}\varepsilon_0(E^2 - 2E_x{}^2) & -\varepsilon_0 E_x E_y & -\varepsilon_0 E_z E_x \\ 0 & -\varepsilon_0 E_x E_y & \tfrac{1}{2}\varepsilon_0(E^2 - 2E_y{}^2) & -\varepsilon_0 E_y E_z \\ 0 & -\varepsilon_0 E_z E_x & -\varepsilon_0 E_y E_z & \tfrac{1}{2}\varepsilon_0(E^2 - 2E_z{}^2) \end{pmatrix}$$

$$(6\text{-}163)$$

where $\mathbf{E}$ is given by (6–159), the energy density of the field due to a moving electron, which is given by the 00 component of the stress-energy tensor obtained from a Lorentz transformation of the above tensor (6–163), is

$$-t_0{}^0(\mathbf{V}) = \tfrac{1}{2}\gamma^2\varepsilon_0 \left[E^2 + \beta^2 \left(E^2 - 2\left(\frac{\mathbf{E}\cdot\mathbf{V}}{V}\right)^2 \right) \right] \qquad (6\text{--}164)$$

where β and γ are defined by (4–37) and (4–38). The total field energy after integrating (6–164) and using (6–160), is given as

$$U(\mathbf{V}) = -\frac{1}{\gamma} \int t_0{}^0 \, dv = \gamma(1 + \tfrac{1}{3}\beta^2)U(\mathbf{V} = 0)$$

$$\cong (1 + \tfrac{5}{6}\beta^2)U(\mathbf{V} = 0) \qquad (6\text{--}165)$$

or $mc^2 + \tfrac{5}{6}mV^2$, for the Lorentz model. However, the electron, being a particle, should have an energy of $mc^2 + \tfrac{1}{2}mV^2$. In the same way, the momentum of the field due to the moving electron is obtained from (6–163), with a Lorentz transformation, as

$$G = -\frac{1}{c\gamma} \int t_0{}^1 \, dv = \frac{\gamma\beta\varepsilon_0}{c} \int \left(E^2 - \left(\frac{\mathbf{E}\cdot\mathbf{V}}{V}\right)^2 \right) dv$$

$$= \frac{4}{3}\frac{\gamma\beta}{c} U(V = 0) \qquad (6\text{--}166)$$

which gives $\tfrac{4}{3}mV$ in the nonrelativistic limit, for the Lorentz model. These are obviously the symptoms of the well-known fact that the field (tensor) and a particle (vector) behave differently under Lorentz transformations.

There are two difficulties in the present theory. One is the self-energy, and the other is its transformation property. Even in high-energy scattering experiments we have not found any finite radius, nor any structure, for the electron, but the electron cannot be a point charge since the self-energy would diverge and the electron mass would be infinite if it were structureless. The Lorentz model saves this divergence, but the behavior under the Lorentz transformations is contradictory. Many serious attempts have been made to reconcile these inconsistencies, but no convincing theory has been found. Poincaré[5] assumed that there is an additional stress field which pulls infinitesimal charges together to form a stable electron, and compensates for

---

[5] A discussion on the Poincaré stress is given in section 17.5 of J. D. Jackson, *Classical Electrodynamics*, John Wiley & Sons, New York, 1965.

the undesired parts of the electromagnetic field which produce the extra terms in (6–165) and (6–166). Quantum electrodynamics saves the situation to some extent; the so-called renormalization procedure is found to give a consistent and meaningful solution to many problems related to the self-energy, replacing it with the observed electron mass. However, the self-energy of an electron itself is still infinite even in quantum electrodynamics. The divergence in quantum electrodynamics is not $1/r$, but $ln(r)$, showing an improvement over the classical theory discussed here.

PROBLEM 6–45. Obtain (6–157) and (6–158).

PROBLEM 6–46. Show that the total cross section of the Thomson scattering is

$$\sigma \equiv \int \frac{d\sigma}{d\Omega}\, d\Omega = \frac{1}{6\pi} \left(\frac{\mu_0 q^2}{m}\right)^2 = \frac{8\pi}{3} r_0^{\,2} \left(\frac{q}{e}\right)^4$$

PROBLEM 6–47. Obtain (6–163).

PROBLEM 6–48. Obtain the results of (6–165) and (6–166).

PROBLEM 6–49. Show that when the self-energy is calculated as $\frac{1}{2} \int\!\int de\, d\phi$, using the potential given by (6–154) with constant $\mathbf{V}$, we obtain

$$U'(V) = (1 + \tfrac{1}{3}\beta^2)U(V = 0)$$

6–15* SELF-FORCE

It sometimes happens in physics that although a large lower-order term is difficult to interpret, higher-order terms in the same expansion provide clear physical pictures. The expansions (6–154) and (6–155) are such examples.

Formula (6–61) shows that the rate of radiation $\mathbb{R}$ from a charged particle can be obtained if $\dot\phi$ and $\dot{\mathbf{A}}$ at the position of the particle are known. On the other hand, in sections 6–11 and 6–12 we obtained formulas for the radiation rate from an accelerated charge from a rather different viewpoint: Liénard–Wiechert potentials. Now let us try to obtain formula (6–138) of section 6–12 for electric dipole radiation using formula (6–61).

Differentiating (6–154) with respect to t, for fixed $\mathbf{r}$, we obtain

$$\dot{\phi}(t, \mathbf{r}) = \frac{q}{4\pi\varepsilon_0} \left\{ \frac{\mathbf{n} \cdot \mathbf{V}}{R^2} \left[1 + \frac{1}{2c^2} (V^2 - (\mathbf{n} \cdot \mathbf{V})^2 - R(\mathbf{n} \cdot \dot{\mathbf{V}})) \right] \right.$$

$$+ \frac{1}{c^2 R} \left[2\mathbf{V} \cdot \dot{\mathbf{V}} - (\mathbf{n} \cdot \mathbf{V})(\mathbf{n} \cdot \dot{\mathbf{V}}) + \frac{V^2(\mathbf{n} \cdot \mathbf{V})}{R} \right.$$

$$\left. - \frac{(\mathbf{n} \cdot \mathbf{V})^3}{R} - R(\mathbf{n} \cdot \ddot{\mathbf{V}}) \right]$$

$$\left. - \frac{1}{c^3} (\dot{V}^2 + \tfrac{4}{3}\mathbf{V} \cdot \ddot{\mathbf{V}} - \tfrac{1}{3}\mathbf{R} \cdot \dddot{\mathbf{V}}) + \cdots \right\} \qquad (6\text{–}167)$$

and differentiating (6–155) in the same way, we obtain

$$\dot{\mathbf{A}}(t, \mathbf{r}) = \frac{q}{4\pi\varepsilon_0} \left[\frac{1}{c^2 R^2} (R\dot{\mathbf{V}} - (\mathbf{n} \cdot \mathbf{V})\mathbf{V}) - \frac{\ddot{\mathbf{V}}}{c^3} + \cdots \right] \qquad (6\text{–}168)$$

Let us take a charged particle, which may be an electron, and assume that it is composed of infinitesimal charges de, de', etc., and that they are distributed with spherical symmetry in a small volume moving with the common velocity $\mathbf{V}$. If $d\dot{\phi}$ and $d\dot{\mathbf{A}}$ are the time derivatives of potentials due to an infinitesimal charge de, as obtained from (6–167) and (6–168) replacing q by de, then the rate of radiation $\mathbb{R}$ due to this particle is, from (6–61),

$$\mathbb{R} = \int\!\!\int de'(\mathbf{V} \cdot d\dot{\mathbf{A}} - d\dot{\phi}) \qquad (6\text{–}169)$$

Since the particle is assumed to be spherical, we see that terms such as $(\mathbf{n} \cdot \mathbf{V})$ or $(\mathbf{n} \cdot \mathbf{V})^3$ disappear when the integration of (6–169) is performed. Thus we obtain

$$\int\!\!\int de'(\mathbf{V} \cdot d\dot{\mathbf{A}}) = \frac{e^2}{4\pi\varepsilon_0} \left(\frac{\mathbf{V} \cdot \dot{\mathbf{V}}}{c^2 r} - \frac{\mathbf{V} \cdot \ddot{\mathbf{V}}}{c^3} + \cdots \right) \qquad (6\text{–}170)$$

$$\int\!\!\int de' \, d\dot{\phi} = \frac{e^2}{4\pi\varepsilon_0} \left[\frac{3\mathbf{V} \cdot \dot{\mathbf{V}}}{2c^2 r} - \frac{1}{c^3} (\dot{V}^2 + \tfrac{4}{3}\mathbf{V} \cdot \ddot{\mathbf{V}}) + \cdots \right] \qquad (6\text{–}171)$$

where e is the total charge and r^{-1} is the average of R^{-1}. Finally

$$\mathbb{R} = \frac{e^2}{4\pi\varepsilon_0} \left[\frac{\mathbf{V} \cdot \dot{\mathbf{V}}}{2c^2 r} + \frac{1}{3c^3} (3\dot{V}^2 + \mathbf{V} \cdot \ddot{\mathbf{V}}) + \cdots \right] \qquad (6\text{–}172)$$

In (6–61) $\mathbb{R}$ is defined as the time rate at which the energy of the particle is lost. Therefore, if we define $\mathbf{F}_s$ by

$$\mathbb{R} \, dt = \mathbf{F}_s \cdot d\mathbf{l} \tag{6-173}$$

where $\mathbf{l}$ is the distance covered by the particle, $\mathbf{F}_s$ can be interpreted as a force responsible for the radiation loss. In the present case of (6–172), the force which will be obtained in this way is due to the particle itself; so we call this force the *self-force*. The self-force is a kind of viscous force due to the interaction between the particle and the field it produces. When the particle tries to move it has to drag the field with it, but the field has its own inertia, or laziness, and the self-force results.

The first term of (6–172) can be easily expressed in the form of (6–173) because $\dot{\mathbf{V}} \cdot \mathbf{V} \, dt = \dot{\mathbf{V}} \cdot d\mathbf{l}$. The coefficient, particularly r^{-1}, depends on the structure of the particle. If we take the Lorentz model, which we discussed in the last section, we see that the coefficient $e^2/(8\pi\varepsilon_0 c^2 r)$ is exactly the electromagnetic mass m. Therefore, if all higher-order terms are neglected

$$\mathbf{F}_s = m\dot{\mathbf{V}} + \cdots \tag{6-174}$$

In the second term of (6–172), $\dot{V}^2 \, dt$ cannot be expressed in the form of (6–173); but, since

$$\int \dot{V}^2 \, dt = \dot{V}V \bigg| - \int \ddot{\mathbf{V}} \cdot \mathbf{V} \, dt = \dot{V}V \bigg| - \int \ddot{\mathbf{V}} \cdot d\mathbf{l} \tag{6-175}$$

we see that if the motion of the particle is periodic, the average of this term can also be expressed in the form of (6–173). Thus (6–172) corresponds to an average self-force of

$$\overline{\mathbf{F}}_s = m\dot{\mathbf{V}} - \frac{e^2}{6\pi\varepsilon_0 c^3} \ddot{\mathbf{V}} + \cdots \tag{6-176}$$

From the form of the original expansion (6–149), we can see easily that all other higher-order terms, neglected in (6–176), disappear when the radius of the particle goes to zero.

When the motion of the particle is nearly simple harmonic, or

$$\mathbf{V} = -\omega\mathbf{r}_0 \sin \omega t \tag{6-177}$$

then

$$\ddot{\mathbf{V}} = \omega^3 \mathbf{r}_0 \sin \omega t = -\omega^2 \mathbf{V} \tag{6-178}$$

and the self-force can be written as

$$\mathbf{F}_s = m\dot{\mathbf{V}} + \frac{(\omega e)^2}{6\pi\varepsilon_0 c^3} \mathbf{V} \tag{6-179}$$

which shows that the second term is the damping term, introduced in (2–11), and the damping constant R is

$$R \text{ (damping constant)} = \frac{(\omega e)^2}{6\pi\varepsilon_0 c^3} \tag{6–180}$$

in this case. When there is no other damping mechanism, this is the expression of R to be used in formulas (6–84) and (6–144). Since the self-force should always exist for a moving charge, this damping is called the natural damping, and the width of spectral lines, given by (6–84) and (6–144), due to the natural damping, is called the *natural width*. Note that when the oscillator is not isolated but interacts with other oscillators or some environment, additional damping mechanisms appear and the spectral linewidth will be larger than the natural width.

Finally, using (6–177) and (6–178) in (6–172), we obtain the rate of radiation from the simple harmonic oscillator as

$$\mathbb{R} = \frac{(er_0\omega^2)^2}{12\pi\varepsilon_0 c^3} \tag{6–181}$$

corresponding to (6–138).

PROBLEM 6–50. Obtain (6–170) and (6–171) using (6–167) and (6–168). Note that

$$\int \int de' \, de (\mathbf{n} \cdot \mathbf{V})(\mathbf{n} \cdot \dot{\mathbf{V}}) = \frac{e^2}{3} \mathbf{V} \cdot \dot{\mathbf{V}}$$

PROBLEM 6–51. Obtain (6–181) and show that (6–138) agrees with (6–181) when averaged over t and integrated over $d\Omega$.

ADDITIONAL PROBLEMS

6–52. Calculate how $\mathbf{E}$ and $\mathbf{B}$ change under the Lorentz transformation.

ANSWER.

$$\mathbf{E}'_{\parallel} = \mathbf{E}_{\parallel}, \qquad \mathbf{E}'_{\perp} = \gamma(\mathbf{E}_{\perp} + \mathbf{V} \times \mathbf{B})$$

$$\mathbf{B}'_{\parallel} = \mathbf{B}_{\parallel}, \qquad \mathbf{B}'_{\perp} = \gamma\left(\mathbf{B}_{\perp} - \frac{\mathbf{V} \times \mathbf{E}}{c^2}\right)$$

6–53. Show that

$$\mathbf{B} = \frac{\mu_0}{2\pi} \frac{\mathbf{I} \times \mathbf{d}}{d^2}$$

which gives $\mathbf{B}$ at a distance $\mathbf{d}$ from a straight wire carrying current $\mathbf{I}$ (see Problem 5–23), can be obtained by a Lorentz transformation of the following formula for $\mathbf{E}$ due to a homogeneously charged straight wire:

$$\mathbf{E} = \frac{\rho \mathbf{d}}{2\pi \varepsilon_0 d^2}$$

HINTS. Use the result of Problem 6–52. Note that ρ goes to $\gamma \rho$ because of the relativistic contraction.

6–54. Show that the following Lagrangian densities are acceptable as well as (6–24) and (6–26):

$$\mathcal{L}_{(3)} = -\frac{1}{2\mu_0} \frac{\partial A^i}{\partial x^j} \frac{\partial A^i}{\partial x^j} - A_i J^i$$

$$\mathcal{L}_{(4)} = \frac{1}{2\mu_0} \left(A_i \frac{\partial F^{ji}}{\partial x^j} - \frac{\partial A_i}{\partial x^j} F^{ji} \right) - \tfrac{1}{2} A_i J^i$$

6–55. We have seen, in Problem 6–23, that

$$T_i{}^j = -\mu_r u_i u^j + \frac{1}{4\mu_0} [4 F_{im} F^{jm} - (F_{kl} F^{kl}) \delta_i{}^j]$$

for a system of charged particles distributed with the rest-mass density μ_r. Assuming that an average value $\overline{T_i{}^j}$ is time independent, or $\partial \overline{T_i{}^j}/\partial t = 0$ for all components;

(a) Show that

$$\int \overline{(T_1{}^1 + T_2{}^2 + T_3{}^3)} \, dv = 0$$

(b) Show that

$$-\int \overline{T_0{}^0} \, dv = \int \overline{H} \, dv = \int \overline{\mu c^2 \sqrt{1 - \beta^2}} \, dv \simeq \int \overline{\mu c^2} \, dv - \frac{1}{2} \int \overline{\mu V^2} \, dv$$

that is, the total energy of the system of charged particles, interacting with each other through the electromagnetic field, is equal to (rest-mass energy) − (kinetic energy) on the average. This is a relativistic extension of the virial theorem (Problem 1–40).

HINT.

$$\sum_{\alpha=1}^{3} \left(\frac{\partial \overline{T_\alpha^i}}{\partial x_\alpha}\right) = -\frac{\partial \overline{T_0^i}}{\partial x_0} = 0$$

so that

$$\sum_{\alpha,\beta=1}^{3} \int x_\beta \frac{\partial \overline{T_\alpha^\beta}}{\partial x_\alpha} \, dv = -\sum_{\alpha,\beta=1}^{3} \int \frac{\partial x_\beta}{\partial x_\alpha} \overline{T_\alpha^\beta} \, dv = 0$$

(b) is solved using the result of (a) and the result that the trace of the tensor is

$$T_i^i = -\mu_r c^2 = -\mu c^2 \sqrt{1-\beta^2}$$

6–56. The wave equation for the Yukawa field ϕ_Y is

$$\Box \, \phi_Y - \kappa^2 \phi_Y = -4\pi g \rho$$

where g is a coupling constant. The static case of this equation was discussed in Problem 5–61. Obtain the Green's function for this general case.

ANSWER. We replace (6–90) and (6–92) by

$$G(x - x') = \frac{1}{(2\pi)^4} \int \frac{\exp -ik_j(x^j - x'^j)}{(k^0)^2 - k^2 - \kappa^2} \, d^4k$$

which gives

$$k^0 = \pm\sqrt{k^2 + \kappa^2}$$

for the poles, instead of (6–94). Let us denote

$$|k| = y \qquad \sqrt{k^2 - \kappa^2} = K$$

Then the retarded Green's function is

$$\Delta_R = -(8\pi^2 R)^{-1} \int_0^\infty (e^{iyR} - e^{-iyR})(e^{-iKc(t-t')} - e^{iKc(t-t')})$$

$$\times \left(\frac{y}{K}\right) dy\theta(t - t')$$

$$= +i(4\pi^2 R)^{-1} \int_{-\infty}^\infty e^{iyR} y \left\{\sin \frac{[Kc(t - t')]}{K}\right\} dy\theta(t - t')$$

$$= \frac{1}{4\pi^2 R} \frac{\partial}{\partial R} \int_{-\infty}^\infty e^{iyR} \frac{\sin [c(t - t')\sqrt{y^2 + \kappa^2}]}{\sqrt{y^2 + \kappa^2}} dy\theta(t - t')$$

Now

$$\int_{-\infty}^{\infty} e^{iyR} \frac{\sin\,(b\sqrt{y^2 + \kappa^2})}{\sqrt{y^2 + \kappa^2}}\, dy = \pi J_0(\kappa\sqrt{b^2 - R^2}) \qquad \text{if} \qquad b > R > 0$$
$$= 0 \qquad\qquad\qquad \text{if} \qquad b < R$$

and

$$\frac{dJ_0(z)}{dz} = -J_1(z)$$

Therefore

$$\Delta_R = -\frac{1}{4\pi R}\,\delta(R - ct + ct') - \frac{\kappa^2}{4\pi}\,\frac{J_1(\kappa\sqrt{c^2(t - t')^2 - R^2})}{\kappa\sqrt{c^2(t - t')^2 - R^2}}$$
$$\text{if} \qquad c(t - t') \geq R$$
$$= 0 \qquad \text{if} \qquad c(t - t') < R$$

The additional part, or its essential part $J_1(x)/x$, is shown in Fig. 6–11. The advanced Green's function is easily obtained:

$$\Delta_A = \frac{1}{4\pi R}\,\delta(R + ct - ct') + \frac{\kappa^2}{4\pi}\,\frac{J_1(\kappa\sqrt{c^2(t - t')^2 - R^2})}{\kappa\sqrt{c^2(t - t')^2 - R^2}}$$
$$\text{if} \qquad c(t - t') \leq R$$
$$= 0 \qquad \text{if} \qquad c(t - t') > R$$

FIGURE 6–11. Function $J_1(x)/x$, which appears in Δ_R and Δ_A.

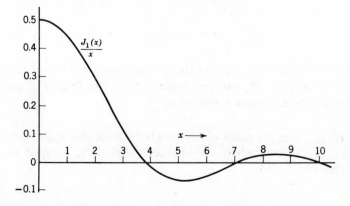

6–57. Show that if a charge q is moving with a constant velocity $c\beta$ the Liénard–Wiechert potentials give

$$\mathbf{E}(t, \mathbf{r}) = \frac{q}{4\pi\varepsilon_0} \frac{(1 - \beta^2)\mathbf{R}_{q0}}{R_{q0}^{*3}}$$

$$\mathbf{B}(t, \mathbf{r}) = \frac{cq\mu_0}{4\pi} \frac{(1 - \beta^2)\boldsymbol{\beta} \times \mathbf{R}_{q0}}{R_{q0}^{*3}}$$

where $\mathbf{R}_{q0}$ is the distance at the time t from the charged particle to the observation point $\mathbf{r}$, and

$$R_{q0}^{*2} = R_{q0}^2(1 + \beta^2) + (\mathbf{R}_{q0} \cdot \boldsymbol{\beta})^2$$

HINT.

$$R_{q0}^* = \mathbf{R}_q(t_{\text{ret}}) \cdot (\mathbf{n}(t_{\text{ret}}) - \boldsymbol{\beta})$$

$$\mathbf{R}_{q0} = R_q(t_{\text{ret}})(\mathbf{n}(t_{\text{ret}}) - \boldsymbol{\beta})$$

See Fig. 6–12.

FIGURE 6–12. R_{q0}^* of Problem 6–57.

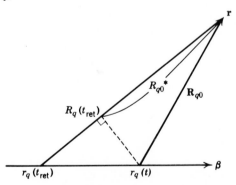

6–58. Show that the field due to a charge with constant velocity, as obtained in Problem 6–57, can be obtained from the field of a stationary charge by a Lorentz transformation.

HINT. Use the result of Problem 6–52. Note that R_{q0} in the original inertial system, where the charge is stationary, is equal to $R_{q0}^*\gamma$. (See Fig. 6–13.)

FIGURE 6–13. R_{q0} and R'_{q0} in the inertial system, in which the charged particle is stationary.

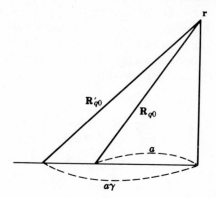

6–59. Obtain *Liénard's formula* for the total radiation power from a charged particle:

$$P \equiv \int \frac{dP}{d\Omega} \, d\Omega = \frac{q^2}{6\pi\varepsilon_0 c^3} \gamma^6 [\dot{V}^2 - \frac{1}{c^2} (\mathbf{V} \times \dot{\mathbf{V}})^2]$$

HINT. From (6–134), taking the nonrelativistic approximation, we obtain *Larmor's formula*

$$P = \frac{q^2 \dot{V}^2}{6\pi\varepsilon_0 c^3}$$

Since P has the dimension of (energy)/(time), it is scalar under Lorentz transformations. Thus we expect

$$P = \frac{q^2}{6\pi\varepsilon_0 c^3} \frac{-1}{m^2} \left(\frac{dp_i}{d\tau} \frac{dp^i}{d\tau} \right) = \frac{q^2}{6\pi\varepsilon_0 c^3 m^2} \left[\left(\frac{dp}{d\tau} \right)^2 - \frac{1}{c^2} \left(\frac{dT}{d\tau} \right)^2 \right]$$

where (4–84) can be used for T.

6–60. When light is scattered by a charge which is not moving on the average, the frequency of the incident light is the same to that of the scattered light. Does the frequency change if the charge is moving?

ANSWER.

$$\omega(1 - \beta \cos \theta) = \omega'(1 - \beta \cos \theta')$$

where $\beta = V/c$ of the charge, and θ and θ' are the angles between $\mathbf{V}$ and $\mathbf{k}$ and $\mathbf{k}'$, respectively.

6-61. A particle of mass m, charge q_1 passes by a very heavy particle of charge q_2. When the velocity of the first particle is very large, and the second particle is not moving before the collision, we may assume, as the zeroth approximation, that the first particle passes by on a straight line, at a distance b from the second particle, with a constant velocity V. Obtain the total radiation intensity produced by this collision in the first approximation.

HINT.

$$a = \frac{F}{m} = \frac{q_1 q_2}{4\pi\varepsilon_0 m} \frac{1}{b^2 + V^2 t^2}$$

$$\int a^2 \, dt = \left(\frac{q_1 q_2}{4\pi\varepsilon_0 m}\right)^2 \frac{\pi}{2V b^3}$$

6-62. A particle with charge q is confined on a circle of radius a and moving with a constant velocity V. Calculate $dP/d\Omega$.

6-63. An electron passes through a plane with a constant velocity V, perpendicular to the plane. Assuming that the electron is a spherical shell of radius r_0, calculate the total energy current of the field which passes through the plane, when the center of the electron is in the plane. Note that the result is the same as the mass current which passes through the plane at the same time, if

$$mc^2 = \frac{e^2}{8\pi\varepsilon_0 r_0}$$

ANSWER. $mc^2 \gamma V / (2r_0)$, where mc^2 is given above.

6-64. An infinite straight hollow cylinder of radius r_0 has charge ρ per unit length.
(a) Calculate the self-energy per unit length.
(b) Let the cylinder move along its length with constant velocity V. Show that the total field energy current which passes through a plane perpendicular to the velocity is $2\gamma^2 \mu V$, where μ is the electromagnetic mass density obtained by equating the result of (a) to μc^2.

ANSWER To (a).

$$\frac{\rho^2}{4\pi\varepsilon_0} \ln \frac{\infty}{r_0} = \infty$$

HINT TO (b). See Problem 6-52.

6–65. Show that

$$K^i = \frac{e^2}{6\pi\varepsilon_0} \left(\frac{d^2 u^i}{ds^2} - u^i u_j \frac{d^2 u^j}{ds^2} \right)$$

reduces to the self-force given by the second term of (6–175) in the nonrelativistic limit, and also satisfies

$$K^i j_i = 0$$

as a Minkowski force should.

Chapter 7 General Theory of Relativity

The special theory of relativity is formulated on the assumption that ds^2 is invariant under the transformations between inertial systems. In general relativity ds^2 is assumed to be invariant under the transformations between any space–time coordinate systems, including accelerated systems.

The theory of general relativity is particularly important because of Einstein's equivalence principle,[1] which states that the gravitational field is equivalent to an accelerated space–time. Thus, general relativity is also a theory of gravity; and has been a successful one. In 1916 Einstein proposed[2] equations which relate the gravitational field to the stress-energy tensor of matter and the electromagnetic field, and give Newton's expression for the gravitational field as its first approximation: zeroth approximation being an inertial system, that is, the Euclidean space–time.

Einstein's equations have been applied to cosmology, which tries to explain over-all features of the universe. Since few data are available in this branch of physics, no theory of cosmology can claim definite superiority over others, but the general relativistic cosmological theories may be regarded as the first scientific approach to this problem, which has been the most fascinating one through the entire history of civilization.

[1] A. Einstein, *Ann. Physik*, **35**, 898 (1911).
[2] A. Einstein, *Ann. Physik*, **49**, 769 (1916). These two and other important papers are translated into English in Einstein *et al.*, *The Principle of Relativity*, Dover, New York, 1923.

7–1   GRAVITATIONAL FORCE

The gravitational force on the earth is seen as an attractive force, which causes everything to fall towards the center of the earth. Aristotle dealt a severe blow to physics when he said,[3] "By absolutely light, then, we mean that which moves upward or to the extremity, and by absolutely heavy that which moves downward or to the center. By lighter or relatively light we mean that one, of two bodies endowed with weight and equal in bulk, which is exceeded by the other in the speed of its natural downward movement." People believed Aristotle's statement for about 2000 years until Galileo Galilei, according to an anecdote,[4] demonstrated at the leaning tower of Pisa that the acceleration by the gravity force is independent of the mass of the object; everything falls at the same speed. With Newton's equation (1–1), it means that the gravity force F_g is proportional to the mass.

The same result can be confirmed by using a pendulum. It was shown in Problem 2–1 that if the pendulum is made of a massless string of length l with a mass m attached at the end, the equation of motion for the angle φ, which the string makes with a vertical line, is

$$lm\ddot{\varphi} = -F_g\varphi \tag{7–1}$$

when φ is small. A solution of this equation is

$$\varphi = \varphi_0 \cos(\omega t) \tag{7–2}$$

where

$$\omega = \sqrt{F_g/(lm)} \tag{7–3}$$

Galileo, who found that the frequency of a pendulum is independent of the amplitude, must have noticed that it is also independent of the mass and depends only on the length of the string. Newton[5] did this experiment and confirmed that ω is independent of m, proving that the gravity force is proportional to m.

When we write

$$F_g = mg \tag{7–4}$$

---

[3] Aristotle, *De Coel* (*On the Heavens*), Book IV, Chapter 1, [30].
[4] Whether the anecdote is true or not is discussed by F. Cajori, *A History of Physics*, Dover, New York, 1962, pp. 37–39. Also see chapter 2 of L. Gaymonat, *Galileo Galilei*, McGraw-Hill, New York, 1965.
[5] I. Newton, *Mathematical Principles of Natural Philosophy*, Book III, Proposition 6, Theorem 6.

g is the gravitational acceleration. It is found that

$$g = 9.80665 \text{ m/sec}^2 \tag{7-5}$$

on the surface of the earth. This number itself depends slightly on the place on the earth. However, the fact that F_g is proportional to m at a given place on the earth, namely (7-4), has been confirmed experimentally[6] to the accuracy of 1 part in 10^{11}.

7-2 EINSTEIN'S EQUIVALENCE PRINCIPLE

Newton's equation holds in any inertial system: As shown in section 4-2, the acceleration of a particle $\ddot{\mathbf{r}}$ is the same for all inertial systems. If, however, we measure the same acceleration in a coordinate system which is accelerated by $\mathbf{a}$ with respect to an inertial system, then we find the acceleration $\ddot{\mathbf{r}}'$ to be

$$\ddot{\mathbf{r}}' = \ddot{\mathbf{r}} - \mathbf{a} \tag{7-6}$$

If the particle is free, Newton's equation holds in an inertial system as $m\ddot{\mathbf{r}} = 0$, but in the accelerated system, the same equation reads

$$m(\ddot{\mathbf{r}}' + \mathbf{a}) = 0 \tag{7-7}$$

or

$$m\ddot{\mathbf{r}}' = -m\mathbf{a} \tag{7-8}$$

Therefore, if we were in this accelerated system looking at the "free particle," and used Newton's equation to describe its motion, we would say that the particle is not free, but under the influence of an apparent force $-m\mathbf{a}$. Such force is called fictitious, because it is not an actual physical force, but it appears simply because of the fact that the coordinate system is not an inertial system but an accelerated one. An example of such a fictitious force is the centrifugal force in a rotating coordinate system, as shown in (4-2).

The fictitious force $-m\mathbf{a}$ is proportional to m due to its own nature. Einstein proposed that the gravity force, which is found to be proportional to m, is a fictitious force; that is, the gravity field around the earth, or any large mass, is equivalent to an acceleration of the space.

According to this view, a falling object is actually free with respect to an inertial system, while we, standing on the earth looking at the falling object, are actually "accelerating" with respect to the inertial system.

---

[6] R. H. Dicke, *Sci. Am.* **205**(6), 84 (1961).

Light propagates through a vacuum on a straight line when observed in an inertial system. If an observer is accelerated in a direction perpendicular to the path of light, he will see that the path of light is bent in a direction opposite the acceleration. According to the equivalence principle, the space around the sun is accelerated outwards; thus the light path around the sun must be bent towards the sun. This bending of about 1.75 sec has been observed and agrees with the theory. (See section 7–12.)

The equivalence principle predicts that all elementary particles including their antiparticles would experience the same gravitational acceleration g of the earth. This prediction is confirmed by experiment for the K mesons and the electron and positron.[7]

7–3 FREE PARTICLE IN GENERAL SPACE–TIME

Hamilton's principle, introduced in section 1–10, states that the motion of a free particle is given by

$$\delta \int_{s_1}^{s_2} ds = 0 \qquad (7\text{–}9)$$

as shown in Problem 4–17. Mathematically, this is the equation to find geodesics, paths which give the maximum or minimum distances between two fixed end points s_1 and s_2. In the Euclidean space the geodesic is a straight line, but in a more general space (Riemann space) the situation may be different (Fig. 7–1).

FIGURE 7–1. A geodesic between points a and b.

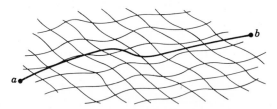

A space is specified by the metric tensor g_{ij} giving ds^2 as

$$ds^2 = g_{ij} \, dx^i \, dx^j \qquad (7\text{–}10)$$

Taking $x^0 = ct$, $x^1 = x$, etc. as in (4–16), we found that the metric tensor of an inertial system was given by (4–18). In a more general space–time, each

[7] M. L. Good, *Phys. Rev.* **121**, 311 (1961); F. C. Witteborn and W. M. Fairbank, *Phys. Rev. Letters* **19**, 1049 (1967).

component of the metric tensor may depend on the event (x) and not be a constant. Hamilton's principle, in that case, is

$$\delta \int ds = \delta \int \sqrt{g_{ij} \, dx^i \, dx^j} = \delta \int \sqrt{g_{ij} \frac{dx^i}{ds} \frac{dx^j}{ds}} \, ds = 0 \qquad (7\text{--}11)$$

which gives

$$\left(\frac{\partial}{\partial x^k} - \frac{d}{ds} \frac{\partial}{\partial (dx^k/ds)} \right) \sqrt{g_{ij} \frac{dx^i}{ds} \frac{dx^j}{ds}} = 0 \qquad (7\text{--}12)$$

assuming that g_{ij} is a function of x, but not a function of dx/ds, or any higher derivatives.

Now

$$\frac{\partial}{\partial x^k} \sqrt{g_{ij} \frac{dx^i}{ds} \frac{dx^j}{ds}} = \frac{1}{2} \frac{\partial g_{ij}}{\partial x^k} \frac{dx^i}{ds} \frac{dx^j}{ds} \qquad (7\text{--}13)$$

and

$$\frac{d}{ds} \frac{\partial}{\partial (dx^k/ds)} \sqrt{g_{ij} \frac{dx^i}{ds} \frac{dx^j}{ds}} = \frac{d}{2 \, ds} \left(g_{ik} \frac{dx^i}{ds} + g_{kj} \frac{dx^j}{ds} \right)$$

$$= \frac{d}{ds} \left(g_{ik} \frac{dx^i}{ds} \right) = \left(\frac{\partial g_{ik}}{\partial x^l} \frac{dx^l}{ds} \frac{dx^i}{ds} + g_{ik} \frac{d^2 x^i}{ds^2} \right)$$

$$= \left[\frac{1}{2} \left(\frac{\partial g_{ik}}{\partial x^j} + \frac{\partial g_{jk}}{\partial x^i} \right) \frac{dx^i}{ds} \frac{dx^j}{ds} + g_{ik} \frac{d^2 x^i}{ds^2} \right]$$

$$(7\text{--}14)$$

where we assume

$$g_{ij} = g_{ji} \qquad (7\text{--}15)$$

Putting these results into (7–12), we obtain the equation of motion of a "free particle" in this general space–time as

$$g_{ik} \frac{d^2 x^i}{ds^2} = -\frac{1}{2} \left(\frac{\partial g_{ik}}{\partial x^j} + \frac{\partial g_{jk}}{\partial x^i} - \frac{\partial g_{ij}}{\partial x^k} \right) \frac{dx^i}{ds} \frac{dx^j}{ds} \qquad (7\text{--}16)$$

If we use the formula

$$g^{kl} g_{ki} = \delta^l{}_i \qquad (7\text{--}17)$$

formula (7–16) can also be written as

$$\frac{d^2x^l}{ds^2} = -\Gamma_{ij}{}^l \frac{dx^i}{ds} \frac{dx^j}{ds} \qquad (7\text{–}18)$$

where

$$\Gamma_{ij}{}^l = \tfrac{1}{2}g^{lk}\left(\frac{\partial g_{ik}}{\partial x^j} + \frac{\partial g_{jk}}{\partial x^i} - \frac{\partial g_{ij}}{\partial x^k}\right) \qquad (7\text{–}19)$$

is called the *Christoffel symbol* (second kind).[8] Using (7–15) we can easily show that

$$\Gamma_{ij}{}^l = \Gamma_{ji}{}^l \qquad (7\text{–}20)$$

Our result (7–18) implies that in a general space–time, where g_{ij} depends on x, or the Christoffel symbol is not zero, a fictitious force appears.

PROBLEM 7–1. Obtain (7–12). Note the analogy to Lagrange's equation.

PROBLEM 7–2. Prove relation (7–17).

HINT. $ds^2 = g_{ij}\, dx^i\, dx^j = g^{ij}\, dx_i\, dx_j = g^{ij}(g_{ik}\, dx^k)(g_{jl}\, dx^l)$

PROBLEM 7–3. Obtain (7–20).

PROBLEM 7–4. Show that

$$g_{ij}\frac{\partial g^{jk}}{\partial x^m} = -g^{jk}\frac{\partial g_{ij}}{\partial x^m}$$

7–4 TRANSFORMATIONS IN GENERAL RELATIVITY

Einstein proposed that the laws of physics should be equivalent not only among inertial systems but among all systems including accelerated systems, in the sense that ds^2 is invariant under transformations between them. Thus

$$g_{ij}\, dx^i\, dx^j = g'_{ij}\, dx'^i\, dx'^j \qquad (7\text{–}21)$$

This is the generalization of (4–14) which now includes cases where g_{ij} may depend on x, and the theory resulting from (7–21) is called the general theory of relativity. One immediate difference between general relativity and special relativity is that the integrated relation $s^2 = s'^2$ of special relativity [Eq.

[8] Another notation $\left\{\begin{matrix} l \\ i \quad j \end{matrix}\right\}$ is often used for $\Gamma_{ij}{}^l$. $g_{lk}\Gamma_{ij}{}^l \equiv g_{lk}\left\{\begin{matrix} l \\ i \quad j \end{matrix}\right\} \equiv [ij, k]$ is called the Christoffel symbol of the first kind.

(4–15)] does not hold in general relativity. Since g_{ij} depends on x, (7–21) does *not* give

$$g_{ij}x^ix^j = g'_{ij}x'^ix'^j \qquad (7\text{–}22)$$

The contravariant components x^0, x^1, etc. still have the same meaning as before [Eq. (4–16)], namely $x^0 = ct$, $x^1 = x$, etc., but the meaning of the corresponding covariant components is not simple anymore; x_0 might not be ct; x_1 might not be $-x$; that is, (4–21) does not hold, in general.

Any four-component quantity A^i, which transforms in the same way as dx^i at a given event, is called a vector. Thus a vector is defined by the transformation property of

$$A'^i = \frac{\partial x'^i}{\partial x^j} A^j \qquad (7\text{–}23)$$

The corresponding covariant components of this vector (Fig. 7–2 and Fig. 7–3) are given by the same formula as in special relativity:

$$A_i = g_{ij}A^j \qquad (7\text{–}24)$$

FIGURE 7–2. Covariant and contravariant components of vector **A**. (Illustration in a flat two-dimensional space.) $\mathbf{e}_1$ and $\mathbf{e}_2$ (unit vectors) form a coordinate system in which **A** has contravariant components A^1 ($= \mathbf{A} \cdot \mathbf{e}_1$) and A^2 ($= \mathbf{A} \cdot \mathbf{e}_2$). Another set of unit vectors $\mathbf{e}^1$ and $\mathbf{e}^2$, which are defined as $\mathbf{e}^1 \cdot \mathbf{e}_2 = \mathbf{e}^2 \cdot \mathbf{e}_1 = 0$ and $|\mathbf{e}_1| = |\mathbf{e}^2|$, $|\mathbf{e}_2| = |\mathbf{e}^1|$, form a coordinate system in which **A** has covariant components A_1 ($= \mathbf{A} \cdot \mathbf{e}^1$) and A_2 ($= \mathbf{A} \cdot \mathbf{e}^2$).

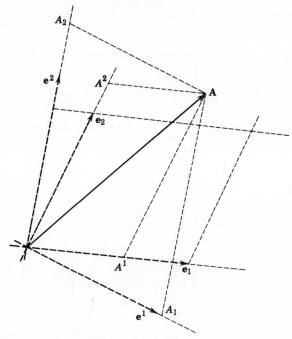

FIGURE 7-3. A simple example of metric tensor. On a flat two-dimensional plane, $ds^2 = (dx^1)^2 + (dx^2)^2 + 2 \cos \theta \, dx^1 \, dx^2$. Therefore

$$\mathbf{g} = \begin{pmatrix} 1 & \cos \theta \\ \cos \theta & 1 \end{pmatrix}$$

In this case $g = 1 - \cos^2 \theta = \sin^2 \theta$, and volume element $= \sin \theta \, dx^1 \, dx^2 = \sqrt{g} \, dx^1 \, dx^2$.

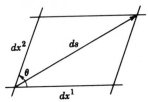

From (7–23) and (7–24) we see that the contraction formula

$$A'_i B'^i = A_i B^i \tag{4-52'}$$

(which means that if A_i and B^i are vectors then $A_i B^i$ is a scalar, invariant under the relativistic transformation) still holds.

If A^i and B^j are both vectors, then $A^i B^j$ is a T^{ij} component (contravariant) of a second rank tensor, which transforms as

$$T'^{ij} = \frac{\partial x'^i}{\partial x^k} \frac{\partial x'^j}{\partial x^l} T^{kl} \tag{7-25}$$

The mixed and covariant components of a second rank tensor can be obtained by

$$T^i{}_j = g_{kj} T^{ik} \quad \text{and} \quad T_{ij} = g_{ik} T^k{}_j \tag{7-26}$$

and the contraction formulas (4–54) and (4–55) are still valid for them.

The basic equation (7–21) may be expressed in a matrix form as

$$d\mathbf{x}'^T \mathbf{g}' \, d\mathbf{x}' = d\mathbf{x}^T \mathbf{g} \, d\mathbf{x} \tag{7-27}$$

in the same way as (4–25), except that

$$\mathbf{g} = \begin{pmatrix} g_{00} & g_{10} & g_{20} & g_{30} \\ g_{01} & g_{11} & g_{21} & g_{31} \\ g_{02} & g_{12} & g_{22} & g_{32} \\ g_{03} & g_{13} & g_{23} & g_{33} \end{pmatrix} \tag{7-28}$$

Since the transformation from $d\mathbf{x}$ to $d\mathbf{x}'$ is given by (7–23) we see that

$$d\mathbf{x}' = \mathbf{J}\, d\mathbf{x} \qquad (7\text{–}29)$$

where $\mathbf{J}$ is the Jacobian matrix, whose determinant form is given in (4–67). Thus (7–27) gives

$$\mathbf{J}^T\mathbf{g}'\mathbf{J} = \mathbf{g} \qquad (7\text{–}30)$$

If we introduce an abbreviated notation for the determinant of $\mathbf{g}$ as

$$g \equiv \det \mathbf{g} \quad \text{and} \quad g' \equiv \det \mathbf{g}' \qquad (7\text{–}31)$$

then (7–30) shows that the Jacobian of the transformation is

$$\det \mathbf{J} = \sqrt{g/g'} \qquad (7\text{–}32)$$

Relation (4–68) should, therefore, be generalized as

$$\sqrt{-g}\, d^4x = \sqrt{-g'}\, d^4x' \ (= \text{scalar}) \qquad (7\text{–}33)$$

Here $-g$ is taken inside the square root sign since g is negative at least in inertial systems.

PROBLEM 7–5. Prove (4–52').

HINT.

$$A_i' = g_{ij}'A'^j = g_{ij}'\,\frac{\partial x'^j}{\partial x^k}\, A^k$$

and use (7–21).

PROBLEM 7–6. Obtain (7–29).

7–5 COVARIANT DERIVATIVE

When $A^i(\mathbf{x})$ is a vector at $\mathbf{x}$, then $A^i(\mathbf{x}) - A^i(\mathbf{x}')$ is not a vector in general since $g_{ij}(\mathbf{x})$ is different from $g_{ij}(\mathbf{x}')$ in general space–time. Thus, in a curved space, dA^i may not be a vector, and $\partial A^i/\partial x^j$ may not be a tensor. It is necessary to introduce something which corresponds to the derivatives and still behaves in a definite manner under the transformations of general relativity.

If A^i is a vector, then $A^i g_{ij} (dx^j/ds)$ is a scalar. Let us calculate the derivative of the latter by ds along the geodesics, that is, under the condition of (7–18):

$$\frac{d}{ds}\left(A^i g_{ij} \frac{dx^j}{ds}\right) = \frac{\partial A^i}{\partial x^k} g_{ij} \frac{dx^j}{ds}\frac{dx^k}{ds} + A^i \frac{\partial g_{ij}}{\partial x^k}\frac{dx^j}{ds}\frac{dx^k}{ds} + A^i g_{ij}\frac{d^2 x^j}{ds^2}$$

$$= \frac{\partial A^i}{\partial x^k}\frac{dx_i}{ds}\frac{dx^k}{ds} + A^i \frac{\partial g_{ij}}{\partial x^k} g^{lj}\frac{dx_l}{ds}\frac{dx^k}{ds} - A^i g_{ij}\Gamma_{kl}{}^j \frac{dx^k}{ds}\frac{dx^l}{ds}$$

$$= \left[\frac{\partial A^j}{\partial x^k} + A^l \left(g^{ij}\frac{\partial g_{il}}{\partial x^k} - g_{il}\Gamma_{km}{}^i g^{mj}\right)\right]\frac{dx^k}{ds}\frac{dx_j}{ds} \qquad (7\text{–}34)$$

Using (7–19) we find

$$g_{il}\Gamma_{km}{}^i g^{mj} = \tfrac{1}{2} g_{il} g^{in}\left(\frac{\partial g_{kn}}{\partial x^m} + \frac{\partial g_{mn}}{dx^k} - \frac{\partial g_{km}}{\partial x^n}\right) g^{mj}$$

$$= \tfrac{1}{2}\delta_l{}^n \left(\frac{\partial g_{kn}}{\partial x^m} + \frac{\partial g_{mn}}{\partial x^k} - \frac{\partial g_{km}}{\partial x^n}\right) g^{mj}$$

$$= \tfrac{1}{2} g^{mj}\left(\frac{\partial g_{kl}}{\partial x^m} + \frac{\partial g_{ml}}{\partial x^k} - \frac{\partial g_{km}}{\partial x^l}\right) \qquad (7\text{–}35)$$

so that

$$g^{ij}\frac{\partial g_{il}}{\partial x^k} - g_{il}\Gamma_{km}{}^i g^{mj} = \Gamma_{kl}{}^j \qquad (7\text{–}36)$$

Putting the result into (7–34) we see

$$\frac{d}{ds}\left(A^i g_{ij}\frac{dx^j}{ds}\right) = A^j{}_{;k}\frac{dx_j}{ds}\frac{dx^k}{ds} \qquad (7\text{–}37)$$

where

$$A^j{}_{;k} = \frac{\partial A^j}{\partial x^k} + \Gamma_{kl}{}^j A^l \qquad (7\text{–}38)$$

and is called the *covariant derivative* of A^j by x^k. From (7–37) this is clearly a second rank tensor. The second term on the right-hand side of (7–38) comes from the parallel displacement of vector A^j. Since the jth coordinate itself changes its direction from x^k to $x^k + dx^k$, the component A^j changes even when the vector is displaced without changing its direction (Fig. 7–4).

In the same way we obtain

$$A_{j;k} = \frac{\partial A_j}{\partial x^k} - \Gamma_{jk}{}^l A_l \qquad (7\text{–}39)$$

FIGURE 7–4. Covariant derivative of a vector **A**. During the parallel displacement of a given vector **A**, its component $\mathbf{A}^x$ changes to $\mathbf{A}^{x\prime}$ ($= \mathbf{A}^x + \delta\mathbf{A}^x$) simply because the direction of the x-axis changes. As we go from point (x, y) to point $(x, y + dy)$ the vector itself changes from **A** to $\mathbf{A}(x, y + dy)$, and the x component changes from $\mathbf{A}^x$ to $\mathbf{A}^x + d\mathbf{A}^x$; but among the change $d\mathbf{A}^x$, as much as $\delta\mathbf{A}^x$ should be subtracted to find the intrinsic change. *(Shown by a bold arrow.)*

$$dA^i = \frac{\partial A^i}{\partial x^j} \, dx^j, \qquad \delta A^i = -\Gamma_{jk}{}^i A^j \, dx^k$$

and

$$dA^i - \delta A^i = A^i{}_{;j} \, dx^j$$

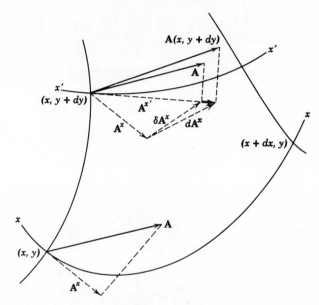

Then

$$(A^i B^j)_{;k} = (A^i{}_{;k})B^j + A^i(B^j{}_{;k}) = \frac{\partial(A^i B^j)}{\partial x^k} + \Gamma_{kl}{}^i A^l B^j + \Gamma_{kl}{}^j A^i B^l$$

$$(7\text{–}40)$$

Thus the covariant derivative of a tensor is

$$T^{ij}{}_{;k} = \frac{\partial T^{ij}}{\partial x^k} + \Gamma_{kl}{}^i T^{lj} + \Gamma_{kl}{}^j T^{il} \tag{7–41}$$

It is easy to obtain

$$T^i{}_{j;k} = \frac{\partial T^i{}_j}{\partial x^k} + \Gamma_{kl}{}^i T^l{}_j - \Gamma_{jk}{}^l \, T^i{}_l \tag{7–42}$$

and

$$T_{ij;k} = \frac{\partial T_{ij}}{\partial x^k} - \Gamma_{ik}{}^l T_{lj} - \Gamma_{jk}{}^l T_{il} \tag{7-43}$$

Quantities in (7–41)–(7–43) are third rank tensors.

The second covariant derivative of a vector can be calculated from (7–43) as

$$A_{i;jk} \equiv (A_{i;j})_{;k} = \frac{\partial(A_{i;j})}{\partial x^k} - \Gamma_{ik}{}^l A_{l;j} - \Gamma_{jk}{}^l A_{i;l}$$

$$= \frac{\partial^2 A_i}{\partial x^j \partial x^k} - \Gamma_{ij}{}^l \frac{\partial A_l}{\partial x^k} - \Gamma_{ik}{}^l \frac{\partial A_l}{\partial x^j} + \Gamma_{ik}{}^l \Gamma_{lj}{}^m A_m$$

$$- \Gamma_{jk}{}^l \frac{\partial A_i}{\partial x^l} + \Gamma_{jk}{}^l \Gamma_{il}{}^m A_m - A_l \frac{\partial \Gamma_{ij}{}^l}{\partial x^k} \tag{7-44}$$

PROBLEM 7–7. Obtain (7–42) and (7–43) from (7–41).

PROBLEM 7–8. Show that

$$g^{ij}{}_{;k} = 0$$

HINT.

$$(A^i A_i)_{;k} = 0$$

PROBLEM 7–9. Show that

$$\frac{\partial A_i}{\partial x^j} - \frac{\partial A_j}{\partial x^i}$$

is a tensor.

PROBLEM 7–10. Show that

$$\frac{\partial T_{ij}}{\partial x^k} + \frac{\partial T_{jk}}{\partial x^i} + \frac{\partial T_{ki}}{\partial x^j}$$

is a tensor if $T_{ij} = -T_{ji}$.

PROBLEM 7–11. When F is a scalar and

$$A_i = \frac{\partial F}{\partial x^i}$$

show that

$$A_{i;j} = A_{j;i}$$

7–6 SOME USEFUL MATHEMATICAL FORMULAS

The determinant g [see (7–31) and (7–28)], can be expanded as

$$g = \sum_i \Delta_{ij} g_{ij} \tag{7–45}$$

where Δ_{ij} is the cofactor of g_{ij}, and the summation is over one index only. According to the theory of determinants

$$\sum_i \Delta_{ik} g_{ij} = 0 \quad \text{if} \quad k \neq j \tag{7–46}$$

Comparing these formulas with (7–17) we obtain

$$\Delta_{ij} = g g^{ij} \tag{7–47}$$

and

$$\frac{\partial g}{\partial g_{ij}} (= \Delta_{ij}) = g g^{ij} \tag{7–48}$$

Note that g is not a scalar, so that Δ_{ij} is not a tensor.

Let us contract the Christoffel symbol as $\Gamma_{ji}{}^{i}$. From its definition (7–19) we see that

$$\Gamma_{ji}{}^{i} = \tfrac{1}{2} g^{ik} \left(\frac{\partial g_{ik}}{\partial x^j} + \frac{\partial g_{jk}}{\partial x^i} - \frac{\partial g_{ij}}{\partial x^k} \right) = \tfrac{1}{2} g^{ik} \frac{\partial g_{ik}}{\partial x^j} \tag{7–49}$$

since the middle and the last terms in the second expression cancel each other. On the other hand,

$$\frac{\partial g}{\partial x^j} = \frac{\partial g}{\partial g_{ik}} \frac{\partial g_{ik}}{\partial x^j} = g g^{ik} \frac{\partial g_{ik}}{\partial x^j} \tag{7–50}$$

from (7–48). Comparing this result with (7–49), we obtain

$$\Gamma_{ji}{}^{i} = \frac{1}{2g} \frac{\partial g}{\partial x^j} = \frac{1}{\sqrt{-g}} \frac{\partial \sqrt{-g}}{\partial x^j} \tag{7–51}$$

The covariant divergence of a vector $A^i_{;i}$ can be expressed as

$$A^i_{;i} = \frac{\partial A^i}{\partial x^i} + \Gamma_{ji}{}^i A^j = \frac{\partial A^i}{\partial x^i} + \frac{1}{\sqrt{-g}} \frac{\partial \sqrt{-g}}{\partial x^j} A^j = \frac{1}{\sqrt{-g}} \frac{\partial (A^i \sqrt{-g})}{dx^i}$$

(7–52)

PROBLEM 7–12. (a) Show that

$$T^{ij}{}_{;i} = \frac{1}{\sqrt{-g}} \frac{\partial (T^{ij}\sqrt{-g})}{\partial x^i} + \Gamma_{ki}{}^j T^{ik}$$

(b) Show that if T^{ij} is antisymmetric, that is, $T^{ij} = -T^{ji}$, then the last term in the above formula disappears.

PROBLEM 7–13. (a) Show that

$$T^i_{j;i} = \frac{1}{\sqrt{-g}} \frac{\partial (T^i_j \sqrt{-g})}{\partial x^i} - \Gamma_{ji}{}^k T^i_k$$

(b) Show that if $T^{ij} = T^{ji}$ (which does not necessarily mean that T^i_j is symmetric), then

$$T^i_{j;i} = \frac{1}{\sqrt{-g}} \frac{\partial (T^i_j \sqrt{-g})}{\partial x^i} - \frac{1}{2} \frac{\partial g_{ik}}{\partial x^j} T^{ik}$$

PROBLEM 7–14. Show that

$$\frac{\partial g}{\partial g^{ij}} = -g g_{ij}$$

PROBLEM 7–15. Show that

$$g^{jk} \Gamma_{jk}{}^i = \frac{1}{\sqrt{-g}} \frac{\partial (\sqrt{-g}\, g^{ij})}{\partial x^j}$$

7–7 RIEMANN TENSOR, RICCI TENSOR, AND EINSTEIN TENSOR

The *Riemann tensor*[9] R^l_{ijk} is introduced by

$$A_{i;jk} - A_{i;kj} = R^l_{ijk} A_l$$

(7–53)

[9] Also called the *curvature tensor*.

From (7–44) we obtain

$$R^l_{ijk} = \frac{\partial \Gamma_{ik}{}^l}{\partial x^j} - \frac{\partial \Gamma_{ij}{}^l}{\partial x^k} + \Gamma_{mj}{}^l \Gamma_{ik}{}^m - \Gamma_{mk}{}^l \Gamma_{ij}{}^m \qquad (7\text{–}54)$$

Obviously this tensor is antisymmetric with respect to j and k:

$$R^l_{ijk} = -R^l_{ikj} \qquad (7\text{–}55)$$

Also from (7–54) we see

$$R^l_{ijk} + R^l_{kij} + R^l_{jki} = 0 \qquad (7\text{–}56)$$

Using (7–53) one can show that

$$R^m_{ijk;l} + R^m_{ilj;k} + R^m_{ikl;j} = 0 \qquad (7\text{–}57)$$

This relation is called the *Bianchi identity*.

The covariant Riemann tensor is defined by

$$R_{ijkl} = g_{im} R^m_{jkl} \qquad (7\text{–}58)$$

From (7–54) and (7–19) we obtain

$$R_{ijkl} = \frac{1}{2} \left(\frac{\partial^2 g_{il}}{\partial x^j\, \partial x^k} + \frac{\partial^2 g_{jk}}{\partial x^i\, \partial x^l} - \frac{\partial^2 g_{ik}}{\partial x^j\, \partial x^l} - \frac{\partial^2 g_{jl}}{\partial x^i\, \partial x^k} \right)$$
$$+ g_{mn}(\Gamma_{jk}{}^m \Gamma_{il}{}^n - \Gamma_{jl}{}^m \Gamma_{ik}{}^n) \qquad (7\text{–}59)$$

Thus we see

$$R_{ijkl} = -R_{jikl} \qquad (7\text{–}60)$$

$$R_{ijkl} = -R_{ijlk} \qquad (7\text{–}61)$$

$$R_{ijkl} = R_{klij} \qquad (7\text{–}62)$$

By a contraction we obtain a second rank tensor

$$R_{ik} = R^j_{ijk} = -R^j_{ikj} \qquad (7\text{–}63)$$

which is called the *Ricci tensor*. Contracting again we obtain the *curvature scalar*

$$R = g^{ik} R_{ik} \qquad (7\text{–}64)$$

The tensor

$$R_{ik} - \tfrac{1}{2}g_{ik}R \tag{7–65}$$

is called the *Einstein tensor.*

Contracting the Bianchi identity (7–57) for m and j we obtain

$$R_{ik;l} - R_{il;k} + R^j_{ikl;j} = 0 \tag{7–66}$$

Using the result of Problem 7–8, we can write (7–66) as

$$R_{;l} - R^k_{l;k} + (g^{ik}R^j_{ikl})_{;j} = 0 \tag{7–67}$$

Using (7–58) and (7–60) we see that

$$
\begin{aligned}
(g^{ik}R^j_{ikl})_{;j} &= (g^{ik}g^{jm}R_{mikl})_{;j} = -(g^{ik}g^{jm}R_{imkl})_{;j} \\
&= -(g^{jm}R^k_{mkl})_{;j} = -(g^{jm}R_{ml})_{;j} = -R^j_{l;j}
\end{aligned} \tag{7–68}
$$

Putting this result into (7–67) we obtain

$$(R^i_j - \tfrac{1}{2}\,\delta^i_{\,j}R)_{;j} = 0 \tag{6–69}$$

or

$$(R_{ij} - \tfrac{1}{2}g_{ij}R)_{;j} = 0 \tag{7–70}$$

Thus the covariant divergence of the Einstein tensor is identically zero.

PROBLEM 7–16.   (a) Confirm (7–54).
(b) Confirm (7–56).

PROBLEM 7–17.   Obtain

$$A^l_{\;;jk} - A^l_{\;;kj} = -R^l_{ijk}A^i$$

PROBLEM 7–18.   Obtain

$$T_{ij;kl} - T_{ij;lk} = -R^m_{ikl}T_{mj} - R^m_{jkl}T_{im}$$

where T_{ij} is a tensor.

PROBLEM 7–19*.   Prove the Bianchi identity (7–57).

HINTS. From Problem 7-18, obtain

$$A_{i;jkl} - A_{i;jlk} = -R^m_{ikl}A_{m;j} - R^m_{jkl}A_{i;m}$$

Take the covariant derivative of (7-53), rotate the indices to obtain two other relations, sum them up, and use the above formula, to obtain

$$A_m(R^m_{ijk;l} + R^m_{ilj;k} + R^m_{ikl;j}) = 0$$

PROBLEM 7-20. Show that

$$R_{ij} = \frac{\partial \Gamma_{ij}{}^k}{\partial x^k} - \frac{\partial \Gamma_{ik}{}^k}{\partial x^j} + \Gamma_{ij}{}^k\Gamma_{lk}{}^l - \Gamma_{ik}{}^l\Gamma_{jl}{}^k$$

7-8 HAMILTON'S PRINCIPLE. STRESS-ENERGY TENSOR

Hamilton's principle, which was generalized as (4-89) in special relativity, should be modified to

$$\delta \int \mathscr{L}\sqrt{-g}\, d^4x = 0 \qquad (7\text{-}71)$$

in general relativity, since we found in (7-33) that, $\sqrt{-g}\, d^4x$, not d^4x, is a scalar quantity. The Lagrangian density $\mathscr{L}$, which is assumed to be scalar, is a function of field variables, such as A^i in the case of the electromagnetic field, and their first derivatives; which, actually, should be the first covariant derivatives, since $\mathscr{L}$ is expected to be scalar:

$$\mathscr{L} = \mathscr{L}(A^i, A^i{}_{;j}) \qquad (7\text{-}72)$$

When we take the variation in (7-71) with respect to δA^i, we obtain, in the same manner as before,

$$\frac{\partial \mathscr{L}}{\partial A^i} - \left(\frac{\partial \mathscr{L}}{\partial A^i{}_{;j}}\right)_{;j} = 0 \qquad (7\text{-}73)$$

which is the covariant form of the Lagrange–Euler equation.

In general relativity, the Lagrangian density $\mathscr{L}$ can be regarded as a function of the metric tensor g_{ij} and its first derivatives, so that the variation in (7-71) can be taken with respect to δg_{ij}. The result can be written as

$$\int T_{ij}\, \delta g^{ij}\sqrt{-g}\, d^4x = -\int T^{ij}\, \delta g_{ij}\sqrt{-g}\, d^4x = 0 \qquad (7\text{-}74)$$

where

$$\tfrac{1}{2} T^{ij} \sqrt{-g} = \frac{\partial}{\partial x^k} \frac{\partial(\mathscr{L}\sqrt{-g})}{\partial(\partial g_{ij}/\partial x^k)} - \frac{\partial(\mathscr{L}\sqrt{-g})}{\partial g_{ij}} \tag{7-75}$$

assuming, again, that δg_{ij} is zero at the boundaries. We take this T^{ij}, defined by (7–75) as the stress-energy tensor of the field given by $\mathscr{L}$. This definition looks quite different from our previous one of (4–93), but turns out to give the proper general relativistic extension. It is superior in the sense that the tensor is already symmetrized, as can be seen immediately from (7–75), since g_{ij} is a symmetric tensor.

One good reason that T^{ij} of (7–75) must be the stress-energy tensor is that it satisfies

$$T^{ij}{}_{;j} = 0 \tag{7-76}$$

which is the proper generalization of (4–95). This relation can be shown by first investigating what is δg_{ij}. If the coordinate is changed from x^i to x'^i as

$$x'^i = x^i - \xi^i \tag{7-77}$$

then

$$dx'^i = dx^i - \frac{\partial \xi^i}{\partial x^j} dx^j \tag{7-78}$$

so that

$$g_{ij} dx^i dx^j = g'_{ij} dx'^i dx'^j = \left(g'_{ij} - g'_{kj} \frac{\partial \xi^k}{\partial x^i} - g'_{ik} \frac{\partial \xi^k}{\partial x^j} \right) dx^i dx^j \tag{7-79}$$

where g'_{ij} is the metric tensor at x'. In order to find δg_{ij} we should transform g'_{ij} back to x, and compare the result with g_{ij}. Thus, expanding g'_{ij} as a power series around x and taking the first-order terms only, we see

$$g'_{ij} = g_{ij} + \delta g_{ij} - \xi^k \frac{\partial g_{ij}}{\partial x^k} \tag{7-80}$$

where δg_{ij} is the variation of the metric tensor. Inserting this expression into (7–79) we obtain

$$\delta g_{ij} = \xi^k \frac{\partial g_{ij}}{\partial x^k} + g_{kj} \frac{\partial \xi^k}{\partial x^i} + g_{ik} \frac{\partial \xi^k}{\partial x^j} \tag{7-81}$$

It is a straightforward calculation to show that this expression can be also written as

$$\delta g_{ij} = \xi_{i;j} + \xi_{j;i} \qquad (7\text{-}82)$$

Substituting (7-82) into (7-74) we obtain

$$\int T^{ij}\xi_{i;j}\sqrt{-g}\ d^4x = \int (T^{ij}\xi_i)_{;j}\sqrt{-g}\ d^4x - \int T^{ij}_{\ ;j}\xi_i\sqrt{-g}\ d^4x = 0 \qquad (7\text{-}83)$$

because T^{ij} is a symmetric tensor and the covariant derivative of g is zero. Now the first term in the middle expression of (7-83) is

$$\int (T^{ij}\xi_i)_{;j}\sqrt{-g}\ d^4x = \int \frac{\partial(T^{ij}\xi_i\sqrt{-g})}{\partial x^j}\ d^4x = \oint T^{ij}\xi_i\sqrt{-g}\ d^3x \qquad (7\text{-}84)$$

where the last term is the surface integral in the space–time; and this integral vanishes if we assume that the variation ξ_i is zero on this surface (hypersurface) which forms the boundary of the d^4x integration. Thus the variation equation (7-74) is equivalent to

$$T^{ij}_{\ ;j} = 0 \qquad (7\text{-}76)$$

PROBLEM 7–21. Obtain (7–73).

HINT. Use (7–52).

PROBLEM 7–22. Obtain (7–81) from (7–82).

PROBLEM 7–23. Show that when the Lagrangian density is given by (4-113), i.e.,

$$\mathscr{L} = -c\sqrt{J^iJ_i} = -c\sqrt{g^{ij}J_jJ_i}$$

the new definition (7–75) gives an expression of T^{ij} which reduces to the previous expression (4–119) in the special relativistic limit.

PROBLEM 7–24. Show that when the Lagrangian density is given by (6–24) with $J^i = 0$, i.e., when the electromagnetic field is free, our new definition (7–75) gives an expression of t^{ij} which reduces to the previous result of (6–36) in the special relativistic limit.

Stress-Energy Pseudotensor
Since the stress-energy tensor is symmetric the result of Problem 7–13(b) can be applied to reduce (7–76) as

$$\frac{\partial(T^i_j\sqrt{-g})}{\partial x^i} - \tfrac{1}{2}\sqrt{-g}\ T^{ik}\frac{\partial g_{ik}}{\partial x^j} = 0 \tag{7-85}$$

If we define t^i_j by

$$\frac{\partial(t^i_j\sqrt{-g})}{\partial x^i} = -\tfrac{1}{2}\sqrt{-g}\ T^{ik}\frac{\partial g_{ik}}{\partial x^j} \tag{7-86}$$

then (7–85) can be written as

$$\frac{\partial((T^i_j + t^i_j)\sqrt{-g})}{\partial x^i} = 0 \tag{7-87}$$

which gives the conservation law: Since $-T^0_0$ is the energy density of the matter or electromagnetic field, or both, $-t^0_0$ is interpreted as the energy density of the gravity field, while $-ct^\alpha_0$ is the αth component of the gravitational energy flux. t^i_j, however, is not a tensor, and is called the stress-energy pseudotensor of the gravity field.[10]

7-9   EINSTEIN'S EQUATIONS AND THEIR LINEAR LIMIT

In view of (7–70) and (7–76) Einstein proposed that

$$R_{ij} - \tfrac{1}{2}g_{ij}(R - 2\Lambda) = KT_{ij} \tag{7-88}$$

or

$$R^i_j - \tfrac{1}{2}\delta^i_j(R - 2\Lambda) = KT^i_j \tag{7-88'}$$

[10] Using Einstein's equation, the stress-energy pseudotensor of the gravity field is calculated as

$$t^{ik} = c^4(16G)^{-1}[(2\Gamma^n_{lm}\Gamma^p_{np} - \Gamma^n_{lp}\Gamma^p_{mn} - \Gamma^n_{ln}\Gamma^p_{mp})(g^{il}g^{km} - g^{ik}g^{lm})$$
$$+ g^{il}g^{mn}(\Gamma^k_{lp}\Gamma^p_{mn} + \Gamma^k_{mn}\Gamma^p_{lp} - \Gamma^k_{np}\Gamma^p_{lm} - \Gamma^k_{lm}\Gamma^p_{np})$$
$$+ g^{kl}g^{mn}(\Gamma^i_{lp}\Gamma^p_{mn} + \Gamma^i_{mn}\Gamma^p_{lp} - \Gamma^i_{np}\Gamma^p_{lm} - \Gamma^i_{lm}\Gamma^p_{np})$$
$$+ g^{lm}g^{np}(\Gamma^i_{ln}\Gamma^k_{mp} - \Gamma^i_{lm}\Gamma^k_{np})]$$

[L. Landau and E. Lifschitz, *The Classical Theory of Fields*, Addison-Wesley, Reading, Mass., 1951, Formula (11–82).]

where K is a constant which we shall presently be deciding in equation (7–111), and Λ is a constant called the *cosmical constant*. Einstein thought the cosmical constant had to be nonzero in his cosmology, as will be seen in equation (7–169), but more recent cosmological data have shown that his original argument is not necessarily true. In any case, Λ is a very small number if not zero.

In Newton's theory of gravity, on the other hand, the gravitational potential ϕ due to a mass of mass density μ is given by

$$\phi(\mathbf{r}) = G \int \frac{\mu(\mathbf{r'})}{|\mathbf{r} - \mathbf{r'}|} \, dv' \qquad (7\text{–}89)$$

where G is the gravitational constant, which is observed to be

$$G = (6.670 \pm 0.005) \times 10^{-11} \text{ newton m}^2/\text{kg}^2 \qquad (7\text{–}90)$$

The motion of a particle of mass m in the gravitational field is given by

$$\delta \int L \, dt = -\delta \int (mc^2 \sqrt{1 - \beta^2} + m\phi) \, dt$$

$$\cong -\delta \int mc^2 \left(1 - \tfrac{1}{2}\beta^2 + \frac{\phi}{c^2}\right) dt = 0 \qquad (7\text{–}91)$$

using the Lagrangian function of (4–85). Since Hamilton's principle in general relativity is given by (7–9), we expect that the space–time around a mass of density μ, should be given as

$$ds \cong c \left(1 - \tfrac{1}{2}\beta^2 + \frac{\phi}{c^2}\right) dt = c \left(1 - \tfrac{1}{2}\beta^2 + \frac{G}{c^2} \int \frac{\mu(\mathbf{r'})}{|\mathbf{r} - \mathbf{r'}|} \, dv'\right) dt \qquad (7\text{–}92)$$

Einstein's equation should give this metric as a first approximation.

We assume that when the space is empty, the curvature of the space–time is zero, that is, we have an inertial system,[11] for which the metric tensor is given by

$$g_{ij} = \delta_{ij} \qquad (7\text{–}93)$$

[11] When the space is empty, $T_{ij} = 0$, and Einstein's equation does have a solution of (7–93). This solution, however, is not the only solution for the empty space.

which is (4–24), since δ_{ij} is given by (4–57). When a mass is present in the space with a small density μ, the stress-energy tensor $T_j{}^i$ of this space becomes nonzero, which, through Einstein's equation, makes the curvature of the space–time also nonzero; that is, the metric tensor should deviate from (7–93).

Let us write

$$g_{ij} = \delta_{ij} + h_{ij} \tag{7–94}$$

and assume that all h_{ij}'s are very small, so that we can neglect all terms except those which are linear in h_{ij}. This is called the *linear limit*.

From (7–19), we see that the Christoffel symbol is

$$\Gamma_{ij}{}^l \cong \tfrac{1}{2}\,\delta^{lk}\left(\frac{\partial h_{ik}}{\partial x^j} + \frac{\partial h_{jk}}{\partial x^i} - \frac{\partial h_{ij}}{\partial x^k}\right) \tag{7–95}$$

in this linear limit. The Riemann tensor, given by (7–59),

$$R_{ijkl} \cong \frac{1}{2}\left(\frac{\partial^2 h_{il}}{\partial x^j\,\partial x^k} + \frac{\partial^2 h_{jk}}{\partial x^i\,\partial x^l} - \frac{\partial^2 h_{ik}}{\partial x^j\,\partial x^l} - \frac{\partial^2 h_{jl}}{\partial x^i\,\partial x^k}\right) \tag{7–96}$$

in the same limit, so that

$$R_{jl} \cong \delta^{ik}R_{ijkl} \cong -\tfrac{1}{2}\,\delta^{ik}\frac{\partial^2 h_{jl}}{\partial x^i\,\partial x^k} + \frac{1}{2}\left(\frac{\partial^2 h_j{}^i}{\partial x^l\,\partial x^i} + \frac{\partial^2 h_l{}^i}{\partial x^j\,\partial x^i} - \frac{\partial^2 h}{\partial x^j\,\partial x^l}\right) \tag{7–97}$$

where

$$h^i{}_j = \delta^{ik}h_{kj} \qquad \text{and} \qquad h = h^i{}_i \tag{7–98}$$

It will be shown in the next section that we can always assume an auxiliary condition, similar to the Lorentz condition of the electromagnetic field,

$$\frac{\partial \varphi^i{}_j}{\partial x^i} = 0 \tag{7–99}$$

where

$$\varphi^i{}_j = h^i{}_j - \tfrac{1}{2}\,\delta^i{}_j h \tag{7–100}$$

Under the condition of (7–99), we see (7–97) gives

$$R_{jl} \cong -\tfrac{1}{2}\,\delta^{ik}\frac{\partial^2 h_{jl}}{\partial x^i\,\partial x^k} = \tfrac{1}{2}\,\square\, h_{jl} \tag{7–101}$$

where $\square$ is the D'Alambertian, which appeared in Problem 4–10 and equation (5–134). Equation (7–64) gives,

$$R = \delta^{jl} R_{jl} = \tfrac{1}{2} \square\, h \tag{7–102}$$

Therefore, Einstein's equation (7–88) reduces to

$$\square\, \varphi_{ij} = 2K T_{ij} \tag{7–103}$$

in the linear limit. ($\Lambda = 0$.)

When the mass density is μ, and is not moving, the stress-energy tensor T_{ij} of the mass system is given by (4–119) as

$$T_{00} = -\mu c^2 \tag{7–104}$$

with all other components zero. Assuming that φ_{00} is time independent, we obtain, from (7–103),

$$\nabla^2 \varphi_{00} = -2K\mu c^2 \tag{7–105}$$

which immediately gives

$$\varphi_{00}(\mathbf{r}) = \frac{Kc^2}{2\pi} \int \frac{\mu(\mathbf{r'})}{|\mathbf{r} - \mathbf{r'}|} \, dv' \tag{7–106}$$

using Green's function (5–25).

Since T_{00} is the only nonzero component of T_{ij} in this case, we may assume that φ_{00} is the only nonzero component of φ_{ij}. Then, from (7–100),

$$h = -\varphi^i{}_i = -\varphi^0{}_0 = -\varphi_{00} \tag{7–107}$$

and

$$h^0{}_0 = \tfrac{1}{2}\varphi_{00} \qquad h^1{}_1 = h^2{}_2 = h^3{}_3 = -\tfrac{1}{2}\varphi_{00} \tag{7–108}$$

or

$$h_{00} = h_{11} = h_{22} = h_{33} = \tfrac{1}{2}\varphi_{00} \tag{7–109}$$

while all other components of h_{ij} or g_{ij} are zero. Putting this result into (7–94) we obtain

$$\begin{aligned} ds^2 &= (1 + \tfrac{1}{2}\varphi_{00})(c^2\,dt^2 - dr^2) \\ &= (1 + \tfrac{1}{2}\varphi_{00})(1 - \beta^2)c^2\,dt^2 \\ &\cong (1 - \beta^2 + \tfrac{1}{2}\varphi_{00})c^2\,dt^2 \cong (1 - \tfrac{1}{2}\beta^2 + \tfrac{1}{4}\varphi_{00})^2 c^2\,dt^2 \end{aligned} \tag{7–110}$$

This result agrees with the prediction (7-92) if

$$\frac{Kc^2}{8\pi} = \frac{G}{c^2}$$

or

$$K = \frac{8\pi}{c^4} G \qquad (7\text{-}111)$$

PROBLEM 7-25. (a) Show that (7-88) gives

$$R = 4\Lambda - KT$$

(b) Show that (7-88) is equivalent to

$$R_{ij} = \Lambda g_{ij} + K(T_{ij} - \tfrac{1}{2}g_{ij}T)$$

where

$$T = g^{ij}T_{ij}$$

PROBLEM 7-26. Obtain (7-95).

PROBLEM 7-27. Obtain (7-107), (7-108), and (7-109).

7-10 GRAVITATIONAL WAVE

Since equation (7-103), which holds under condition (7-99), is quite similar to equation (6-13) for the electromagnetic potentials A^i, we immediately expect that the gravitational potential also propagates as a wave with velocity c. In the present case φ_{ij}, which appears instead of A^i, is a 4×4 symmetric tensor, which has 10 independent components in general. We shall show, however, that only two out of ten components are really independent in the present case.

In the case of the electromagnetic potentials, an ambiguity exists because of the Gauge transformation. The Lorentz condition (6-7) or (6-15) could be imposed employing this arbitrariness. The analogous situation in the general relativity is an ambiguity with respect to the choice of the coordinate system. Since ds^2 must be invariant under the transformation (7-77) between coordinate systems, formula (7-79) results, which, for the linear limit, is

$$h_{ij} = h'_{ij} - \delta_{kj}\frac{\partial \xi^k}{\partial x^i} - \delta_{ik}\frac{\partial \xi^k}{\partial x^j}$$

or

$$h^i_{\ j} = h'^i_{\ j} - \frac{\partial \xi^i}{\partial x^j} - \frac{\partial \xi_j}{\partial x_i} \tag{7-112}$$

From (7–100) we obtain

$$\varphi^i_{\ j} = \varphi'^i_{\ j} - \frac{\partial \xi^i}{\partial x^j} - \frac{\partial \xi_j}{\partial x_i} + \delta^i_j \frac{\partial \xi^k}{\partial x^k} \tag{7-113}$$

from which

$$\frac{\partial \varphi^i_{\ j}}{\partial x^i} = \frac{\partial \varphi'^i_{\ j}}{\partial x^i} - \Box\, \xi_j \tag{7-114}$$

The last expression shows that one can choose ξ_j in a suitable way so that the condition we assumed in the last section (7–99) can always be imposed on $\varphi^i_{\ j}$. This condition, however, still does not fix the transformation function ξ_j's completely, since any further transformation by a ξ_j which satisfies

$$\Box\, \xi_j = 0 \tag{7-115}$$

will not change condition (7–99) as can be seen from (7–114).

A solution for the gravitational wave in a vacuum

$$\Box\, \varphi^i_{\ j} = 0 \tag{7-116}$$

is a plane wave such that all $\varphi^i_{\ j}$ depend on $x - ct$ only. In this case (7–99) reduces to

$$\frac{\partial \varphi^0_{\ i}}{\partial t} = \frac{\partial \varphi^1_{\ i}}{\partial t} \tag{7-117}$$

which essentially means

$$\varphi^0_{\ i} = \varphi^1_{\ i} \tag{7-118}$$

We can make some $\varphi^i_{\ j}$ vanish using the transformation by ξ_j, each of which is also a function of $x - ct$ only. Such ξ_j certainly satisfies (7–115), so that conditions (7–99) or (7–116) are not violated. Since (7–113) gives

$$\varphi^0_{\ \alpha} = \varphi'^0_{\ \alpha} - \frac{\partial \xi_\alpha}{c\, \partial t}, \qquad \alpha = 2, 3 \tag{7-119}$$

$$\varphi^0_{\ 1} = \varphi'^0_{\ 1} - \frac{\partial \xi^0}{\partial x} - \frac{\partial \xi_1}{c\, \partial t} = \varphi'^0_1 + \frac{\partial(\xi_0 - \xi_1)}{c\, \partial t} \tag{7-120}$$

$$\varphi^2_{\ 2} + \varphi^3_{\ 3} = \varphi'^2_{\ 2} + \varphi'^3_{\ 3} + 2\,\frac{\partial(\xi_0 + \xi_1)}{c\, \partial t} \tag{7-121}$$

All these components can be made zero by choosing suitable ζ's.

$$\varphi^0{}_1 = \varphi^0{}_2 = \varphi^0{}_3 = \varphi^2{}_2 + \varphi^3{}_3 = 0 \qquad (7\text{--}122)$$

Because of (7–118), (7–122) also implies

$$\varphi^1{}_1 = \varphi^1{}_2 = \varphi^1{}_3 = 0 \qquad (7\text{--}123)$$

From the symmetry we obtain

$$\varphi^1{}_0 = \varphi^2{}_0 = \varphi^3{}_0 = \varphi^2{}_1 = \varphi^3{}_1 = 0 \qquad (7\text{--}124)$$

also. Finally, from (7–116) and (7–122), we obtain

$$\varphi^0{}_0 \; (= \varphi^1{}_0) = 0 \qquad (7\text{--}125)$$

Thus the only remaining components which cannot be made zero after a suitable coordinate transformation are $\varphi^2{}_3$ and $\varphi^2{}_2 - \varphi^3{}_3$. Therefore, the plane gravitational wave is transverse, has two components, with velocity c. We will see later that this is the same as the wave for massless, spin one-half particles such as neutrinos.

Radiation of the gravitational wave can be implied from (7–103) and (4–119) similar to the interpretation of the radiation of the electromagnetic wave by a moving charge in sections 6–11 and 6–12. We obtain after some calculation that

$$\varphi_{\alpha\beta} = \frac{2G}{c^4 R} \frac{\partial^2}{\partial t^2} \int \rho x_\alpha x_\beta \, dv \qquad (7\text{--}126)$$

where α and β take 1, 2, and 3. We saw in section 7–8 that $-ct_0{}^1$ introduced by (7–86) gives the energy flux of the gravitational field in the x direction. A lengthy calculation which is not presented here[12] gives

$$S_x = -ct_1{}^0 = \frac{G}{36\pi c^4 R^2} \left[\frac{(\ddot{D}_{22} - \ddot{D}_{33})^2}{4} + (\ddot{D}_{23})^2 \right] \qquad (7\text{--}127)$$

where $D_{\alpha\beta}$ are components of the mass quadrupole moment defined as

$$D_{\alpha\beta} = \int \mu (3 x_\alpha x_\beta - \delta_{\alpha\beta} r^2) \, dv \qquad (7\text{--}128)$$

[12] The result is taken from Landau and Lifschitz, *Classical Theory of Fields*, Formula (11–115). Further discussion can be found in J. Weber's book *General Relativity and Gravitational Waves*, Interscience, New York, 1961. (Chapters 7 and 8.)

where

$$\delta_{\alpha\beta} = 1 \qquad \text{when} \qquad \alpha = \beta$$
$$= 0 \qquad \text{when} \qquad \alpha \neq \beta$$

A dot denotes differentiation with respect to time, and R is the distance from the moving mass. The gravitational wave is very weak but recently detected by Weber.[13]

7–11   SCHWARZSCHILD SOLUTION

Einstein's equation is a nonlinear equation, and the solutions we have discussed so far are valid only in its linear limit. Nonlinear equations are difficult to solve in general, and Einstein's equation is not an exception. However, a few exact solutions of Einstein's equation are known, of which Schwarzschild's solution is the most useful one: This is the solution for an empty static field with spherical symmetry around a mass at the center, but approaches the inertial system solution as the distance from the center increases to infinity.

When the space we are considering is empty (free from matter and electromagnetic fields), Einstein's equation (7–88) gives ($\Lambda = 0$)

$$R_{ij} - \tfrac{1}{2}g_{ij}R = 0 \tag{7–129}$$

Multiplying this equation by g^{ij} and contracting, we obtain

$$R = 0 \tag{7–130}$$

which, when put into (7–129), gives us

$$R_{ij} = 0 \tag{7–131}$$

Thus the Ricci tensor should be a null tensor when the space is empty. The inertial system solution, where $g_{ij} = \delta_{ij}$, certainly satisfies (7–131), but that is not the only solution; many other solutions are possible including the Schwarzschild solution.

When the space has spherical symmetry, it is convenient to use spherical coordinates with the space–time metric expressed as

$$ds^2 = e^\nu(c\,dt)^2 - e^\lambda\,dr^2 - r^2\,d\theta^2 - r^2\sin^2\theta\,d\varphi^2 \tag{7–132}$$

[13] J. Weber, *Phys. Rev. Letters*, **24**, 276 (1970). See G. L. Wick, *Science*, **167**, 1237 (1970) for a review.

where v and λ may be functions of r, but not of t, for a static field. From (7–19) we obtain

$$\Gamma_{10}{}^0 = \frac{1}{2}\frac{\partial v}{\partial r}$$

$$\Gamma_{00}{}^1 = \frac{1}{2}e^{v-\lambda}\frac{\partial v}{\partial r} \qquad \Gamma_{11}{}^1 = \frac{1}{2}\frac{\partial \lambda}{\partial r} \qquad \Gamma_{22}{}^1 = -re^{-\lambda} \tag{7–133}$$

$$\Gamma_{33}{}^1 = -r\sin^2\theta\, e^{-\lambda}$$

$$\Gamma_{12}{}^2 = \Gamma_{13}{}^3 = \frac{1}{r} \qquad \Gamma_{33}{}^2 = -\sin\theta\cos\theta \qquad \Gamma_{23}{}^3 = \cot\theta$$

and all other Christoffel symbols are zero. Using the expression of the Ricci tensor given in Problem 7–20, we obtain

$$R_{00} = e^{v-\lambda}\left[-\frac{1}{2}\frac{\partial^2 v}{\partial r^2} - \frac{\partial v}{r\,\partial r} + \frac{1}{4}\frac{\partial v}{\partial r}\frac{\partial \lambda}{\partial r} - \frac{1}{4}\left(\frac{\partial v}{\partial r}\right)^2\right]$$

$$R_{11} = \frac{1}{2}\frac{\partial^2 v}{\partial r^2} - \frac{\partial \lambda}{r\,\partial r} - \frac{1}{4}\frac{\partial v}{\partial r}\frac{\partial \lambda}{\partial r} + \frac{1}{4}\left(\frac{\partial v}{\partial r}\right)^2 \tag{7–134}$$

$$R_{33} = \sin^2\theta\, R_{22} = \sin^2\theta\left[e^{-\lambda} + \frac{1}{2}re^{-\lambda}\left(\frac{\partial v}{\partial r} - \frac{\partial \lambda}{\partial r}\right) - 1\right]$$

and, according to (7–131), all components R_{ij} shown in (7–134) should vanish. A solution is obtained in the form of

$$e^v = e^{-\lambda} = 1 - \frac{\text{constant}}{r} \tag{7–135}$$

It was shown in (7–106) and (7–110), that when a mass of M exists at the origin, we should have

$$g_{00} = e^v = 1 - \frac{2GM}{c^2 r} \tag{7–136}$$

at least in the linear limit. Comparing this expression with (7–135) we may choose the constant and see that the linear limit expression is actually exact in this case. Therefore

$$ds^2 = \left(1 - \frac{2GM}{c^2 r}\right)(c\,dt)^2 - \frac{dr^2}{1 - (2GM/c^2 r)} - r^2\,d\theta^2 - r^2\sin^2\theta\,d\varphi^2 \tag{7–137}$$

is the exact solution of Einstein's equation we were looking for.

Beside the obvious singularity at the origin, the Schwarzschild solution (7–137) has another singularity at $r = 2GM/c^2$, which is 2.956 km if we use the mass M of the sun (Fig. 7–5). It is 0.895 cm for the earth, and 1.32×10^{-55} cm for an electron. What this singularity means physically is not clear. Some speculative theories are proposed,[14] but nothing has been observed yet. When quantum mechanical effects are taken into account this singularity may disappear.[15]

FIGURE 7–5. $-g_{rr}$ of the Schwartzschild solution. A singularity appears at $r = 2GM/c^2$.

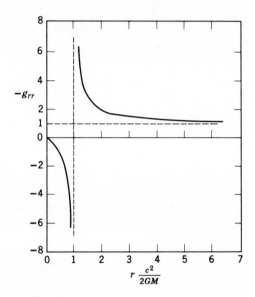

PROBLEM 7–28.   Obtain (7–133).

PROBLEM 7–29.   Obtain (7–134) from (7–133).

PROBLEM 7–30.   Show that (7–135) is a solution of $R_{ij} = 0$, when (7–134) is given.

---

[14] See R. Ruffini and J. A. Wheeler, *Physics Today* **24**, No. 1, 30 (1971).
[15] See, for example, B. K. Harrison *et al.*, *Gravitation Theory and Gravitational Collapse*, University of Chicago Press, Chicago, 1965, p. 142.

7–12 EXPERIMENTAL TESTS OF GENERAL RELATIVITY[16]

The existing experimental tests of general relativity are all based on the Schwarzschild solution. We should note, however, that $2GM/c^2r$ is 4.24 × 10^{-6} and 1.41 × 10^{-9} at the surface of the sun and earth, respectively. Since these numbers are so small compared to one, the experimental verification of the effects of general relativity is extremely difficult, and the best one can achieve is a test of the lowest-order effects.

The gravitational red shift appears because of the r dependence of g_{00}. The period of an atomic oscillation at two distances from the center of the earth are related by

$$\left(1 - \frac{2GM}{c^2r_1}\right)\Delta t_1 = \left(1 - \frac{2GM}{c^2r_2}\right)\Delta t_2 \qquad (7\text{–}138)$$

Taking M for the earth, we see that a given atomic oscillation has a higher frequency when the atom is placed 47 ft above the surface of the earth than at the surface of the earth, by the factor $1 + 2.5 \times 10^{-15}$. This very small correction term was measured and confirmed using the Mössbauer effect.[17] White dwarfs are very dense stars. The red shift of some spectral lines coming from two white dwarfs were successfully explained in terms of the gravitational effect.[18] The red shift might arise from the Doppler effect due to the relative motion of the star with respect to earth. Actually in most of the stars, the Doppler effect is much larger than the gravitational effect. The spectral lines from the sun have been carefully measured, but due to the turbulent motion of the gas on the surface of the sun, the Doppler effect is quite complicated. A careful analysis, however, shows the existence of a gravitational red shift, when Doppler effects are corrected out.[19]

The bending of a light ray due to a gravitational field is a direct result of the equivalence principle, as already mentioned in section 7–2. The bending angle is calculated taking $ds^2 = 0$ in the Schwarzschild solution (7–137). Let us assume that the light path is in the plane for which $\theta = \frac{1}{2}\pi$; then from (7–137), we see that the motion of a light pulse is approximately given by

$$0 = \left(1 - \frac{2GM}{c^2r}\right)c^2 - \left(1 + \frac{2GM}{c^2r}\right)\dot{r}^2 - r^2\dot{\varphi}^2 \qquad (7\text{–}139)$$

[16] P. H. Dicke, *The Theoretical Significance of Experimental Relativity*, Gordon and Breach, New York, 1965. R. W. Holcomb, *Science*, **169**, No. 3940, 40 (1970).

[17] R. V. Pound and G. A. Rebka, Jr., *Phys. Rev. Letters*, **4**, 337 (1960).

[18] D. M. Popper, *Astrophys. J.* **120**, 316 (1954).

[19] J. E. Blaumont and E. Roddier, *Phys. Rev. Letters* **7**, 437 (1961).

Since the gravitational effect is very small, the light path is expected to be nearly straight. Consider a straight line very close to the actual path of the pulse (Fig. 7–6). If the distance between this straight line and the mass M is b, the position of a light pulse, which moves on this straight line passing through the center point (whose distance from M is b) at time $t = 0$, is given by

$$r_0{}^2 = b^2 + (ct)^2 \qquad (7\text{–}140)$$

$$\cos \varphi_0 = \frac{b}{r}$$

which give

$$\dot{r}_0 = \frac{c^2 t}{r_0} \qquad (7\text{–}141)$$

$$\dot{\varphi}_0 = \frac{b\dot{r}_0}{r_0{}^2}\frac{r_0}{ct} = \frac{bc}{r_0{}^2}$$

FIGURE 7–6. Bending of the light path due to the gravitational field of mass M. (Very much exaggerated.)

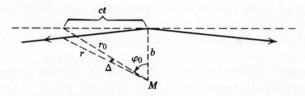

If, at a given time, the position of a light pulse on the bent path is related to that on the straight line path as

$$r = r_0 + \delta \quad \text{and} \quad \varphi = \varphi_0 + \Delta \qquad (7\text{–}142)$$

then

$$\dot{r} = \dot{r}_0 + \dot{\delta} \quad \text{and} \quad \dot{\varphi} = \dot{\varphi}_0 + \dot{\Delta} \qquad (7\text{–}143)$$

where

$$\dot{\delta} = \frac{r_0 ct}{b}\dot{\Delta} \qquad (7\text{–}144)$$

Inserting (7–141), (7–142), (7–143), and (7–144) into (7–139), and neglecting higher-order terms in $\dot{\Delta}$, we obtain

$$0 = c^2 - \dot{r}^2 - r^2\dot{\varphi}^2 - \frac{2GM}{r}\left(1 + \left(\frac{\dot{r}}{c}\right)^2\right)$$

$$\cong c^2 - \dot{r}_0^2 - r_0^2\dot{\varphi}_0^2 - 2\dot{r}_0\,\delta - 2r_0^2\dot{\varphi}_0\,\dot{\Delta}$$

$$- \frac{2GM}{r_0}\left(1 + \left(\frac{\dot{r}_0}{c}\right)^2\right)$$

$$\cong - \frac{2cr_0^2}{b}\dot{\Delta} - \frac{4GM}{r_0} \tag{7–145}$$

which shows that the total deflection angle is

$$\Delta_t = -\int_{-\infty}^{\infty} \frac{2GMb}{cr_0^3}\,dt = -\frac{2GMb}{c}\int_{-\infty}^{\infty} \frac{dt}{(c^2t^2 + b^2)^{3/2}} = -\frac{4GM}{bc^2} \tag{7–146}$$

When we take the mass and radius of the sun for M and b, respectively, we obtain 1.75 sec for the total deflection angle. About 11 observations involving nearly 300 stars have been done, and the results agree fairly well with the theory.

The orbit of a planet around the sun can be calculated using (7–18) and the Schwarzschild solution. The details are given in Appendix II. In Newton's classical theory, with the gravity potential of (7–89) due to the sun, one obtains an elliptic orbit given by

$$\frac{C}{r} = 1 + e\cos\varphi \tag{7–147}$$

where C and e are constants. This orbit is stable in the sense that when $\varphi = 2\pi\,r$ takes the same value as when $\varphi = 0$. In special relativity the orbit is modified to be

$$\frac{C'}{r} = 1 + e\cos(\varphi\xi) \tag{7–148}$$

where

$$\xi = \sqrt{1 - (GMm/cl)^2} \cong 1 - \frac{1}{2}\left(\frac{GMm}{cl}\right)^2 \tag{7–149}$$

M and m are the masses of the sun and the planet, respectively, and l is a constant decided by the eccentricity of the orbit. Since ξ is not one, the orbit is not stable. The perihelion of the orbit, given by $\varphi\xi = \pi$, moves by $2\pi \times (1 - \xi)$ during each cycle (Fig. 7-7). In general relativity, or the Schwarzschild space–time, we obtain the orbit of the same form as (7–148), except that ξ is replaced by ξ_g where

$$\xi_g = \sqrt{1 - 6(GMm/cl)^2} \cong 1 - 3 \left(\frac{GMm}{cl}\right)^2 \qquad (7\text{–}150)$$

FIGURE 7–7.   Precession of the perihelion of a planetary orbit. (Very much exaggerated.)

Thus the precession of the perihelion is much larger than in special relativity. The calculation according to (7–150) agrees quite well with the observed values as shown in Table 7–1. However, Dicke[20] observed a slight oblateness of the sun and argued that a solar mass quadrupole moment might be the main cause of this precession.

TABLE 7–1   *Precession of Perihelion (sec/century)*

Planet	Mercury	Venus	Earth	Mars
Calc.	43.03	8.63	3.84	1.35
Obs.	43.11 ± 0.45	8.4 ± 4.8	5.0 ± 1.2	⋯

When a radar signal from earth to Venus, or any planet, and back (echo) passes near the sun, an extra time delay must appear due to the gravity

[20] R. H. Dicke and H. M. Goldenberg, *Phys. Rev. Letters* **18**, 313 (1967). Also, see R. H. Dicke, *Physics Today* **20**, No. 1, 55 (1967).

field of the sun. Such an extra time delay of about 200 sec was observed and successfully explained by Einstein's theory of general relativity.[21]

We saw in section 7–10 that Weber detected the gravitational wave.

Since the general relativistic effects are all so small in any case, these observations could confirm only first-order effects. Therefore, any other theories of gravity which agree with Einstein's theory in the first approximation can also claim observational verification. Quite a few such alternative theories of gravity exist.[22]

7–13 HOMOGENEOUS MODEL UNIVERSE

The universe is observed to be made of mass points, stars, and galaxies, distributed in a random way. The space–time of the universe must be similar to that of Schwarzschild near each mass point. Since, however, Einstein's equations are not linear, the solution for several mass points cannot be given as a superposition of the Schwarzschild solutions: The solution for more than one mass point is so difficult that it is not known. The present relativistic cosmology, therefore, is based on model universes, which are not given by several mass points, but by a fluid model assuming that mass is continuously distributed, neglecting local nonuniformity.

The gross features of the universe, such as the mass density over a sufficiently large volume, seem to be the same in all directions and at all places.[23] Therefore, a homogeneous fluid model may serve as the model universe: To develop a general relativistic cosmology, we assume the stress-energy tensor T_{ij} of (4–122):

$$T_{ij} = - \frac{(\mathscr{H}_0 + P)}{c^2} u_i u_j + P g_{ij} \qquad (7-151)$$

where $\mathscr{H}_0$ is the energy density and P is the pressure, both of which are assumed to be independent of the spatial coordinates in the homogeneous model universe. When the universe, at some stage of its development, has an extremely high energy density, the stress-energy tensor may be due largely to the electromagnetic field.[24] However, at least at the present stage, the

---

[21] I. Shapiro, M. Ash, R. Ingalls, W. Smith, D. Campbell, R. Dyce, R. Jurgens, and G. Pettengill, *Phys. Rev. Letters*, **26**, 1132 (1971). Similar experiments using Mariner VI and VII as the targets also confirmed Einstein's theory.

[22] A review of theories of gravity is given by G. J. Whitrow and G. E. Morduch, *Relativistic Theories of Gravitation*, pp. 1–67 of *Vistas in Astronomy*, Vol. 6 (edited by A. Beer), Pergamon Press, New York, 1965. Also see ref. 20.

[23] This is often called the cosmical principle.

[24] See G. Gamow, *Phys. Rev.* **70**, 572 (1946).

rest-mass energy of matter seems to be larger than that of the radiation field. In the following discussion, therefore, we neglect the radiation field; in which case

$$\mathscr{H}_0 = \mu c^2 \quad \text{and} \quad |P| < \mathscr{H}_0 \tag{7-152}$$

The average mass density of the universe is observed to be

$$10^{-29} < \mu < 10^{-26} \text{ kg/m}^3 \tag{7-153}$$

The 4-velocity u_i of the matter may depend on the radial distance r, the distance from a point chosen as the origin of the spatial coordinates. It is observed[25] that the wave length λ of a given atomic spectral line emitted from a star at a distance r from us is shifted towards red by $\Delta\lambda$ in such a way that

$$\frac{\Delta\lambda}{\lambda} = \frac{Hr}{c} \tag{7-154}$$

where H is a universal constant called the Hubbele constant, and is

$$H = (1.8 \text{ or } 0.24) \times 10^{-17} \text{ sec}^{-1} \tag{7-155}$$

If this red shift is attributed to the Doppler effect, as we assume in our discussion, (7-154) means that the matter at r is receding from us with velocity proportional to r. Such motion, however, does not destroy the homogeneity of the universe.

The space–time line element of the homogeneous universe must be given by

$$ds^2 = D(c\, dt)^2 - A(dx^2 + dy^2 + dz^2) \tag{7-156}$$

where D and A are the components of the metric tensor and may depend on t and r. When the velocity of the matter is radial and has the magnitude $c\beta$, relation (4-73) gives

$$c^2 = g_{ij}u^i u^j = D(u^0)^2 - Au^2 \tag{7-157}$$

or

$$D = 1 - (1 - A)\beta^2 \tag{7-158}$$

[25] E. Hubbele, *Astrophys. J.*, **84**, 158, 270, 517 (1936). Also see A. R. Sandage, *Astrophys. J.*, **127**, 513 (1958), where the smaller value of H, shown in (7-155), is proposed.

If we choose our coordinate system such that $\beta = 0$, then

$$D = 1 \tag{7-159}$$

This coordinate system is called the *co-moving system*, and the following calculations are much easier in this system.

In the co-moving coordinate system the stress-energy tensor is

$$
\begin{aligned}
T^{00} &= -\mu c^2 \\
T^{11} = T^{22} &= T^{33} = P \\
T^{ij} &= 0 \quad \text{if} \quad i \neq j
\end{aligned}
\tag{7-160}
$$

with the approximation (7–152).

From one of Einstein's equations we obtain

$$T^{10} = 0 = -\frac{A_{10}}{A} + \frac{A_1 A_0}{A^2} \tag{7-161}$$

where

$$A_0 = \frac{\partial A}{c\,\partial t}, \qquad A_1 = \frac{\partial A}{\partial x}, \qquad A_{10} = \frac{\partial^2 A}{c\,\partial t\,\partial x} \tag{7-162}$$

This is satisfied if A is of the form of

$$A = (\mathcal{R}(t)\mathcal{P}(r))^2 \tag{7-163}$$

Also

$$
\begin{aligned}
T^{12} = 0 &= -\frac{1}{2}\frac{A_{12}}{A} + \frac{3}{4}\frac{A_1 A_2}{A^2} \\
&= -\frac{1}{2}\frac{(\mathcal{P}^2)_{12}}{\mathcal{P}^2} + \frac{3}{4}\frac{(\mathcal{P}^2)_1(\mathcal{P}^2)_2}{\mathcal{P}^4}
\end{aligned}
\tag{7-164}
$$

that is,

$$\mathcal{P}^2\,\frac{\partial^2(\mathcal{P}^2)}{\partial x\,\partial y} = \frac{3}{2}\frac{\partial(\mathcal{P}^2)}{\partial x}\frac{\partial(\mathcal{P}^2)}{\partial y} \tag{7-165}$$

This relation is satisfied if

$$\mathcal{P}^2 = \frac{1}{(1 + \alpha r^2)^2} \tag{7-166}$$

where α is an undetermined constant. The other components of Einstein's equation give

$$8\pi G\mu = \frac{3}{\mathscr{R}^2}(4\alpha c^2 + \dot{\mathscr{R}}^2) - \Lambda c^2 \qquad (7\text{-}167)$$

and

$$\frac{8\pi GP}{c^2} = -\frac{2\ddot{\mathscr{R}}}{\mathscr{R}} - \left(\frac{\dot{\mathscr{R}}}{\mathscr{R}}\right)^2 - 4\alpha\left(\frac{c}{\mathscr{R}}\right)^2 + \Lambda c^2 \qquad (7\text{-}168)$$

where a dot denotes a time derivative.

PROBLEM 7–31. Obtain (7–161), (7–164), (7–167), and (7–168).

PROBLEM 7–32. Show that (7–166) satisfies (7–165).

7–14 COSMOLOGICAL THEORIES

Einstein Model

Einstein assumed that the universe is static, therefore, $\dot{\mathscr{R}}$ and $\ddot{\mathscr{R}}$ should be zero. In addition the pressure P must be negligible. In this case (7–167) and (7–168) give

$$\mathscr{R}^2 = \frac{\alpha}{\pi G\mu}$$

$$\Lambda = \frac{4\pi G\mu}{c^2} \cong 10^{-54}\ \text{m}^{-2} \qquad (7\text{-}169)$$

(Einstein model)

This is called Einstein's static model, and this is the reason that Einstein thought the cosmical constant Λ had to be nonzero. Later data, particularly Hubbele's observation (7–154), showed, however, that the universe is not static; Einstein's model, therefore, should be modified, and during that course, it will be seen that Λ does not have to be nonzero.

Before we begin the discussion of other theories, it is worthwhile to note some physical interpretations of the constant α, introduced in (7–166). In Einstein's model (7–169) α is positive since A of (7–163) should be positive; but the relation (7–166) in which α is originally introduced does not require this constant to be positive. This constant can be either positive or negative, or even zero.

The volume of the universe may be defined as

$$v = \int A^{3/2} \, dx \, dy \, dz = \mathscr{R}^3 \int_0^\infty \frac{4\pi r^2 \, dr}{(1 + \alpha r^2)^3} \qquad (7\text{--}170)$$

from which we obtain

$$v = \frac{\frac{1}{4}\pi^2 \mathscr{R}^3}{\alpha^{3/2}} \qquad \text{if} \qquad \alpha > 0 \qquad (7\text{--}171a)$$

$$= \lim_{r \to \infty} \frac{4\pi}{3} (\mathscr{R}r)^3 \qquad \text{if} \qquad \alpha = 0 \qquad (7\text{--}171b)$$

$$= \lim_{r \to \infty} [4\pi\mathscr{R}^3 \sqrt{-\alpha}\, r(1 + \alpha r^2)(1 - \alpha r^2)^{-2} - \tanh^{-1}(\sqrt{-\alpha}\, r)]$$
$$\text{if} \qquad \alpha < 0 \qquad (7\text{--}171c)$$

Since the volume of the universe is finite when $\alpha > 0$, this condition gives a *closed universe*, while in the other cases the universe is open.

It is instructive to consider a three-dimensional hypersurface

$$\bar{x}^2 + \bar{y}^2 + \bar{z}^2 + \zeta^2 = \frac{1}{4\alpha} \qquad (7\text{--}172)$$

embedded in a four-dimensional Euclidean space. (This four-dimensional space has nothing to do with the space–time.) Comparing this formula to the more familiar formula in the three-dimensional space, we may say that the surface given by (7–172) is

$$\begin{array}{lll} \text{hypersphere,} & \text{if} & \alpha > 0 \\[4pt] \text{hyperplane,} & \text{if} & \alpha = 0 \qquad (7\text{--}173) \\[4pt] \text{hyperpseudosphere,} & \text{if} & \alpha < 0 \end{array}$$

If the four-dimensional space is Euclidean, the line element on this hypersurface is

$$dl^2 = d\bar{x}^2 + d\bar{y}^2 + d\bar{z}^2 + \frac{(\bar{x} \, d\bar{x} + \bar{y} \, d\bar{y} + \bar{z} \, d\bar{z})^2}{(4\alpha)^{-1} - (\bar{x}^2 + \bar{y}^2 + \bar{z}^2)} \qquad (7\text{--}174)$$

because (7–172) gives

$$\bar{x} \, d\bar{x} + \bar{y} \, d\bar{y} + \bar{z} \, d\bar{z} + \zeta \, d\zeta = 0 \qquad (7\text{--}175)$$

If we introduce spherical coordinates as

$$\bar{x} = \bar{r} \sin \theta \cos \varphi \qquad \bar{y} = \bar{r} \sin \theta \sin \varphi \qquad \bar{z} = \bar{r} \cos \theta$$

then

$$d\bar{x}^2 + d\bar{y}^2 + d\bar{z}^2 = d\bar{r}^2 + \bar{r}^2 \, d\theta^2 + \bar{r}^2 \sin^2 \theta \, d\varphi^2$$

$$\bar{x} \, d\bar{x} + \bar{y} \, d\bar{y} + \bar{z} \, d\bar{z} = \bar{r} \, d\bar{r} \tag{7-176}$$

so that

$$dl^2 = d\bar{r}^2 + \bar{r}^2(d\theta^2 + \sin^2 \theta \, d\varphi^2) + \frac{\bar{r}^2 \, d\bar{r}^2}{(4\alpha)^{-1} - \bar{r}^2}$$

$$= \frac{d\bar{r}^2}{1 - 4\alpha\bar{r}^2} + \bar{r}^2(d\theta^2 + \sin^2 \theta \, d\varphi^2) \tag{7-177}$$

Changing the scale of the radial coordinate to

$$\bar{r} = \frac{r}{1 + \alpha r} \tag{7-178}$$

we finally obtain

$$dl^2 = (1 + \alpha r^2)^{-2}(dr^2 + r^2 \, d\theta^2 + r^2 \sin^2 \theta \, d\varphi^2)$$

$$= (1 + \alpha r^2)^{-2}(dx^2 + dy^2 + dx^2) \tag{7-179}$$

which is essentially what we have as the space part of (7–156).

The Ricci tensor R_{ij} can be calculated for the three-dimensional space where

$$d\sigma^2 = \mathscr{R}^2(1 + \alpha r^2)^{-2} (dx^2 + dy^2 + dz^2) \tag{7-180}$$

and we obtain

$$^3R_{ii} = 8\alpha(1 + \alpha r^2)^{-2} \qquad ^3R_{ij} = 0 \quad \text{if} \quad i \neq j \tag{7-181}$$

which gives the three-dimensional curvature scalar as

$$^3R = \frac{24\alpha}{\mathscr{R}^2} \tag{7-182}$$

In these formulas the notation ^3R is used to distinguish from the regular R, which is for space–time. Result (7–182) shows again that the curvature of the space and α have the same sign.

De Sitter Model

If the universe is void of anything, that is, if the stress-energy tensor T^{ij} is zero, one would expect that the space–time of such universe would be static and Euclidean. That is certainly a solution of Einstein's equations for the empty universe, but it is not the only solution. It was first pointed out by de Sitter that a strange solution of

$$\mathscr{R} = \exp\left(ct\sqrt{\Lambda/3}\right) \qquad \mathscr{P} = 1 \qquad (7\text{--}183)$$

does exist for Einstein's equations of the empty universe. This can be easily verified from (7–167) and (7–168). This is a flat universe since $\alpha = 0$, but it is an expanding one.

It can be shown in general that the Hubbele constant H, defined by (7–154), can be expressed as

$$H = \frac{\dot{\mathscr{R}}}{\mathscr{R}} \qquad (7\text{--}184)$$

In the de Sitter model, therefore,

$$H = c\sqrt{\Lambda/3} \qquad (7\text{--}185)$$

Using the smaller value of H given in (7–155), we obtain

$$\Lambda \cong 10^{-52} \text{ m}^{-2} \qquad (7\text{--}186)$$

PROBLEM 7–33. When

$$d\sigma^2 = a^2(d\theta^2 + \sin^2\theta \, d\varphi^2)$$

is given, calculate the Ricci tensor of this two-dimensional space (θ and φ), and show that

$$^2R = -\frac{2}{a^2}$$

where ^2R is the curvature scalar in the two-dimensional space.

PROBLEM 7–34. Obtain (7–181) and (7–182).

Dynamical Models with $\Lambda = 0$, $P = 0$

When the radiation field is negligible, we can neglect P compared to μc^2. Equations (7–167) and (7–168), which are equivalent to Einstein's equations for the case of a homogeneous model universe, can be rearranged to give

$$\dot{\mathscr{R}}^2 = \frac{8\pi}{3} G\mu\mathscr{R}^2 - 4\alpha c^2 + \tfrac{5}{3}\Lambda c^2\mathscr{R}^2 \qquad (7\text{–}187)$$

$$\ddot{\mathscr{R}} = -\frac{4\pi}{3} G\mu\mathscr{R} - \tfrac{1}{3}\Lambda c^2\mathscr{R} \qquad (7\text{–}188)$$

With our approximation of $P = 0$, we can easily obtain from the conservation law (7–85) that $\mathscr{R}^3\mu$ is a conserved quantity. Let us introduce m by

$$m = \frac{4\pi}{3} \mathscr{R}^3\mu \qquad (7\text{–}189)$$

then m is a constant during the evolution of the universe. Equations (7–187) and (7–188) are now

$$\dot{\mathscr{R}}^2 = \frac{2m}{\mathscr{R}} G - 4\alpha c^2 + \tfrac{5}{3}\Lambda c^2\mathscr{R}^2 \qquad (7\text{–}190)$$

$$\ddot{\mathscr{R}} = -\frac{m}{\mathscr{R}^2} G - \tfrac{1}{3}\Lambda c^2\mathscr{R} \qquad (7\text{–}191)$$

Equation (7–190) is often called the Friedmann equation.

Integrating these equations we can obtain $\mathscr{R}$ as a function of t. Since the integrations are much simpler if $\Lambda = 0$, we neglect Λ and see what happens. In this case the Friedmann equation gives

$$d\mathscr{R} = \mp \frac{2mG\, d\mathscr{R}}{\mathscr{R}\sqrt{2m\mathscr{R}G - 4\alpha c^2\mathscr{R}^2}} \qquad (7\text{–}192)$$

Putting this result into (7–191) with $\Lambda = 0$ we obtain

$$-t = \mp \int \frac{2\mathscr{R}\, d\mathscr{R}}{\sqrt{2m\mathscr{R}G - 4\alpha c^2\mathscr{R}^2}} \qquad (7\text{–}193)$$

If $\alpha = 0$, this equation gives

$$t = \pm\sqrt{8/9 m G} \; \mathcal{R}^{3/2} \qquad (\alpha = 0) \tag{7-194}$$

Taking the upper sign, we obtain an expanding universe.
If $\alpha > 0$, the same equation gives

$$\pm \mathcal{A}t = \sqrt{\mathcal{R}'(1 - \mathcal{R}')} - \sin^{-1}(1 - \mathcal{R}') \tag{7-195}$$

while if $\alpha < 0$, we obtain

$$\pm \mathcal{A}t = \sqrt{\mathcal{R}'(1 + \mathcal{R}')} + \tfrac{1}{2}ln \left| \frac{\sqrt{1 + \mathcal{R}'} - \sqrt{\mathcal{R}'}}{\sqrt{1 + \mathcal{R}'} + \sqrt{\mathcal{R}'}} \right| \tag{7-196}$$

where

$$\mathcal{A} = \frac{4|\alpha|^{3/2}c^3}{m G} \tag{7-197}$$

$$\mathcal{R}' = \frac{2|\alpha|c^2}{m G} \mathcal{R} \tag{7-198}$$

These three cases are illustrated in Figs. 7–8, 7–9, and 7–10. It is seen that both $\alpha = 0$ and $\alpha < 0$ give an expanding universe, while $\alpha > 0$ gives a cycloydal universe the metric of which oscillates with the period $\pi/\mathcal{A}$.
If $\Lambda = 0$, (7–184) and (7–187) give

$$\frac{4\alpha c^2}{\mathcal{R}^2} = \frac{8\pi G \mu}{3} - H^2 \tag{7-199}$$

FIGURE 7–8. Homogeneous model universe with $P = \Lambda = 0$, and $\alpha = 0$.

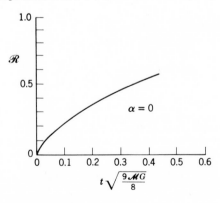

FIGURE 7–9. Homogeneous model universe with $P = \Lambda = 0$, and $\alpha > 0$.

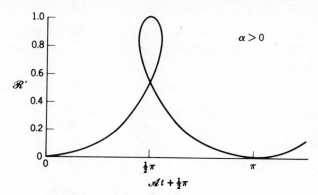

where H is the Hubbele's constant. Using the numerical values of these constants as given before, we obtain

$$5 \times 10^{-39} < \frac{8\pi G\mu}{3} < 5 \times 10^{-36}$$

$$5 \times 10^{-36} < H^2 < 3 \times 10^{-34}$$

(7–200)

Therefore, (7–199) implies that α is either zero or negative.[26]

PROBLEM 7–35. Prove that m of (7–189) is independent of time.

FIGURE 7–10. Homogeneous model universe with $P = \Lambda = 0$, and $\alpha < 0$.

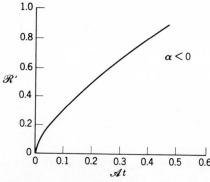

[26] Another experimental support can be seen in R. H. Dicke, P. J. E. Peebles, P. G. Roll, and D. J. Wilkinson, *Astrophys. J.* **142**, 1149 (1965).

ADDITIONAL PROBLEMS

7–36. (a) Show that on a plane that is rotating with the angular velocity ω with respect to an inertial system around an axis perpendicular to itself, we have

$$ds^2 = (c^2 - \omega^2\rho^2)\,dt^2 - d\rho^2 - \rho^2\,d\varphi^2 + 2\rho^2\omega\,dt\,d\varphi$$

where ρ and φ are the cylindrical coordinates.
(b) Calculate the curvature scalar ^3R of this three-dimensional space–time.

7–37. Prove "Gauss's theorem"

$$\oint A^i\sqrt{-g}\,d^3x = \int A^i_{\;;i}\sqrt{-g}\,d^4x$$

in the general relativistic space–time. Here A^i is a vector, and the integral on the left-hand side is over the hypersurface which encloses the four-dimensional volume of the integral on the right-hand side of this equation.

HINT. Use (7–52).

7–38. (a) Show that F^{ij} of the electromagnetic field, as defined by (6–16), is a tensor as it is in general relativity.
(b) Show that four of Maxwell's equations

$$\frac{\partial F^{ik}}{\partial x_j} + \frac{\partial F^{kj}}{\partial x_i} + \frac{\partial F^{ji}}{\partial x_k} = 0 \tag{6–19}$$

are valid tensor equations in general relativity, without further modification.

HINT. See Problems 7–9 and 7–10.

7–39. (a) Show that the equation of continuity of charge, which was given by (5–88), should be modified to

$$\frac{\partial}{\sqrt{-g}\,\partial x^i}(\sqrt{-g}\,J^i) = 0$$

in general relativity.
(b) Knowing that the Lagrangian density for electrodynamics

$$\mathscr{L} = -\frac{1}{4\mu_0}F^{ij}F_{ij} - A_iJ^i \tag{6–24}$$

is scalar in general relativity, show that four of Maxwell's equations, which were given by (6–18) before, have to be modified to

$$\frac{\partial(F^{ij}\sqrt{-g})}{\sqrt{-g}\,\partial x^j} = -\mu_0 J^i$$

HINT. See (7–52) and (7–73).

7-40. If we neglect the time component, the Schwarzschild metric is

$$d\sigma^2 = \frac{dr^2}{1 - (2GM/c^2 r)} + r^2\,d\theta^2 + r^2\sin^2\theta\,d\varphi^2$$

Show that in this three-dimensional space, the components of the Ricci tensor are

$$R_\theta{}^\theta = R_\varphi{}^\varphi = \frac{GM}{c^2 r^3} \qquad R_r{}^r = -\frac{2GM}{c^2 r^3}$$

and that all other components are zero.

7-41. Show that

$$\delta\int R\sqrt{-g}\,d^4x = \int (\delta R_{ij})g^{ij}\sqrt{-g}\,d^4x + \int R_{ij}\,\delta(g^{ij}\sqrt{-g})\,d^4x$$

$$= \int (R_{ij} - \tfrac{1}{2}g_{ij}R)\,\delta g^{ij}\sqrt{-g}\,d^4x$$

Note that when this result is combined with (7–71) and (7–74), Einstein's equations (7–88) can be obtained from

$$\delta\int \left(\frac{c^2}{16\pi G}R + \mathscr{L}\right)\sqrt{-g}\,d^4x = 0$$

7-42. An infinite plane sheet has homogeneous mass distribution of μ per unit area. Show that

$$ds^2 = \left(1 + \frac{8\pi G\mu z}{c^2}\right)c^2\,dt^2 - dx^2 - dy^2 - \frac{dz^2}{1 + (8\pi G\mu/c^2)z}$$

is a solution of Einstein's equations. z is the distance from the sheet. (The corresponding gravitational potential in Newton's theory is $\phi = 4\pi G\mu z$.)

Part II Physics of the Microscopic World

. . . for at the moment that the observer approaches, then they become other and of another nature, so that you cannot get any further in knowing their nature or state, for you cannot know that which has no state.

—Plato, *Cratylus*

Chapter 8 Quantum Mechanics

On December 14, 1900, Max Planck announced his discovery that the concept of photons, or quanta of electromagnetic fields, had to be introduced to explain blackbody radiation. This concept could have been discovered in other, better-known phenomena, such as the specific heats of gases and solids, as Einstein and Debye showed a little later. As will be seen in the following chapters, the quantum effect is essential for understanding the physics of the atomic structure of matter.

Experimental and theoretical work followed Planck's discovery elucidating the real meaning of quanta, and it rapidly became clear to many physicists that this was "a discovery comparable only to the discoveries of Newton." A new mathematical formulation of this new physics was proposed by Niels Bohr[1] in 1913, the so-called Bohr's quantization condition. The approach was semiclassical in the sense that classical concepts and formulas, discussed in Part I of this book, were retained and the quantization condition was imposed on them as an additional restriction. Werner Heisenberg, stimulated by Bohr's works, proposed[2] a fundamentally new approach and published his results in 1925. The burst of theoretical physics that followed his publication was the most spectacular in the history of physics. Most of the formulas in this chapter and the essential parts of the following chapters of Part II of this book were published in about two years, 1925 and 1926.

---

[1] N. Bohr, *Phil. Mag.* **26**, 1, 476, 857 (1913).
[2] W. Heisenberg, *Zeits. Physik*, **33**, 879 (1925). See also M. Jammer, *The Conceptual Development of Quantum Mechanics*, McGraw-Hill, New York, 1966.

8–1 HILBERT SPACE

Vector Space

The ordinary physical space is said to be three dimensional because if we take unit vectors $\mathbf{i}$, $\mathbf{j}$, and $\mathbf{k}$ along X, Y, and Z axes, arbitrary vectors $\mathbf{a}$, $\mathbf{b}, \ldots$, can be expressed as

$$\mathbf{a} = a_x\mathbf{i} + a_y\mathbf{j} + a_z\mathbf{k}, \qquad \mathbf{b} = b_x\mathbf{i} + b_y\mathbf{j} + b_z\mathbf{k}, \ldots \qquad (8\text{–}1)$$

and also all vectors produced by additions $\mathbf{a} + \mathbf{b}, \ldots$, and multiplications by scalars $\alpha, \beta, \ldots$, in such ways that $(\alpha + \beta)\mathbf{a} = \alpha\mathbf{a} + \beta\mathbf{a}$, $\alpha(\mathbf{a} + \mathbf{b}) = \alpha\mathbf{a} + \alpha\mathbf{b}, \ldots$, can be expressed as linear combinations of $\mathbf{i}$, $\mathbf{j}$, and $\mathbf{k}$ in the same way as (8–1).

This concept can be generalized to (functional) *vector spaces*. A set of n linearly independent functions $\psi_1(x), \psi_2(x), \ldots, \psi_n(x)$, where x is a variable, span a n-dimensional complete vector space, if functions $a(x), b(x), \ldots$, and functions produced by additions $a(x) + b(x), \ldots$, and multiplications by scalars $\alpha, \beta, \ldots$, in such ways that $(\alpha + \beta)a(x) = \alpha a(x) + \beta a(x)$, $\alpha(a(x) + b(x)) = \alpha a(x) + \alpha b(x), \ldots$, can all be expressed as linear combinations of $\psi_1(x), \ldots, \psi_n(x)$, that is

$$a(x) = \sum_{i=1}^{n} a_i\psi_i(x), \qquad b(x) = \sum_{i=1}^{n} b_i\psi_i(x), \ldots \qquad (8\text{–}2)$$

The number of variables can be more than one, and they can be anything, not just spacial coordinates.

Norm. Inner Product

In the ordinary physical space, the length of a vector $\mathbf{a}$ is defined by $\sqrt{\mathbf{a} \cdot \mathbf{a}}$. In analogy we define a *norm* $\|a\|$ of a function $a(x)$ in a vector space such that

$$\|a\| = \|b\| \quad \text{if} \quad a(x) = b(x)$$
$$\|a\| \geq 0, \quad \|\alpha a\| = |\alpha| \, \|a\|, \quad \text{and} \quad \|a + b\| \leq \|a\| + \|b\| \qquad (8\text{–}3)$$

A complete normed vector space is called *Banach space*.

In analogy to the scalar product $\mathbf{a} \cdot \mathbf{b}$ of two vectors, the *inner product* of two functions $a(x)$ and $b(x)$ in a vector space is defined as

$$(a, b) = \int a^*(x)b(x) \, dx \qquad (8\text{–}4)$$

where a^* is the complex conjugate of a. The norm in a vector space can be explicitly defined by

$$\|a\| \equiv \sqrt{(a, a)} = \left[\int a^*(x)a(x) \, dx \right]^{1/2} \tag{8-5}$$

It is easy to see that the norm defined in this way satisfies the required relations of (8–3).

Hilbert Space

Almost all functions can be expressed as Fourier series, or linear combinations of the trigonometric functions. The trigonometric functions span a complete vector space in this sense.[3] However, there are infinitely many trigonometric functions, and Fourier series are usually infinite series. Therefore, the vector space we are talking about now has infinite dimension. This is true even when we impose boundary conditions, such as zero at $x = 0$ and L, which allow only $\sin(n\pi x/L)$, where $n = 0, 1, 2, \ldots$.

The inner products of these sine functions, which are the base functions (or base vectors) of the vector space, can be calculated as

$$\left(\sin \frac{n\pi x}{L}, \sin \frac{m\pi x}{L} \right) = \tfrac{1}{2} L \, \delta_{n,m} \tag{8-6}$$

from the definition (8–4). We say that those base functions are orthogonal to each other, because their inner products are zero when $n \neq m$. We can normalize each base function by multiplying it by $\sqrt{2/L}$:

$$\left\| \sqrt{2/L} \sin \left(\frac{n\pi x}{L} \right) \right\| = 1 \tag{8-7}$$

When base functions are orthogonal and normalized (orthonormal)[4]

[3] Completeness in infinite dimensional vector space needs elaborate discussions. This section is a very brief introduction to a big branch of mathematical physics. See J. von Neumann, *Mathematical Foundations of Quantum Mechanics*, Princeton University Press, Princeton, 1955, Chapter 2; E. Butkov, *Mathematical Physics*, Addison-Wesley, Reading, Mass., 1968, Chapter 11, and other books for full discussions on Hilbert space.

[4] Many other special functions form orthonormal sets. See Chapter II of R. Courant and D. Hilbert, *Methods of Mathematical Physics*, Interscience, New York, 1953. This was the most important reference book on mathematics during the development of quantum mechanics from 1925 to now; yet an amazing fact is that the book was published originally in German in 1924, before quantum mechanics appeared.

we have a "complete" analogy to the ordinary physical space, because if

$$(\psi_i(x), \psi_j(x)) = \delta_{i,j} \qquad (8\text{--}8)$$

and functions $a(x)$, $b(x)$, . . . , are expanded as (8–2), where n may be infinite, we obtain

$$(a, b) = \sum_i a_i{}^* b_i \qquad (8\text{--}9)$$

corresponding to

$$\mathbf{a} \cdot \mathbf{b} = a_x b_x + a_y b_y + a_z b_z \qquad (8\text{--}9')$$

of the ordinary physical space. Note, however, that normalization (8–8) is not always possible.

Complete vector spaces of infinite dimension with inner products and norms defined by (8–5) are called *Hilbert spaces*.

8–2 HERMITIAN OPERATORS AND THEIR EIGENFUNCTIONS AND EIGENVALUES

Take an arbitrary pair of functions ϕ and χ of a Hilbert space. If an operator $\hat{f}$ is such that

$$(\phi, \hat{f}\chi) = (\hat{f}\phi, \chi) \qquad (8\text{--}10)$$

for any pair of functions in this Hilbert space, it is called *Hermitian*. More explicitly, (8–10) is

$$\int \phi^*(\hat{f}\chi)\, d\tau = \int (\hat{f}\phi)^* \chi\, d\tau \qquad (8\text{--}10')$$

where τ stands for all appropriate variables.

An operator $\hat{f}^\dagger$ is called Hermitian conjugate or adjoint to $\hat{f}$ if

$$(\phi, \hat{f}^\dagger \chi) = (\hat{f}\phi, \chi) \qquad (8\text{--}11)$$

for all pairs of functions in a Hilbert space. Therefore, a Hermitian operator $\hat{f}$ has the property

$$\hat{f} = \hat{f}^\dagger \qquad (8\text{--}12)$$

or self-adjoint. There are other types of operators such that

$$\hat{f} = -\hat{f}^\dagger \qquad (8\text{-}13)$$

which are called anti-Hermitian.

Consider the differential operator $\partial/\partial x$ as an example. If ϕ and χ are differentiable and satisfy a boundary condition that $\phi^*(a)\chi(a) = \phi^*(b)\chi(b)$ then

$$\int_a^b \phi^* \frac{\partial}{\partial x} \chi \, dx = \phi^*\chi \Big|_a^b - \int_a^b \chi \frac{\partial}{\partial x} \phi^* \, dx = - \int_a^b \left(\frac{\partial}{\partial x} \phi\right)^* \chi \, dx \qquad (8\text{-}14)$$

by partial integration. Therefore $\partial/\partial x$ is anti-Hermitian in this space. One can immediately see that $i \, \partial/\partial x$ is a Hermitian operator.

PROBLEM 8–1. Show that $(\partial/\partial x)^2$ is a Hermitian operator in the same Hilbert space as the one in which $\partial/\partial x$ is anti-Hermitian.

PROBLEM 8–2. The reflection operator $\hat{I}_x$ is defined such that

$$\hat{I}_x\chi(x) = \chi(-x)$$

for any function $\chi(x)$. Show that $\hat{I}_x$ is Hermitian.

PROBLEM 8–3. Show that

$$(\hat{f}\hat{g})^\dagger = \hat{g}^\dagger\hat{f}^\dagger$$

PROBLEM 8–4. If $\hat{f}$ is Hermitian, show that $e^{\hat{f}}$, defined by

$$e^{\hat{f}} = \sum_{n=0}^{\infty} \frac{1}{n!} \hat{f}^n$$

is also Hermitian.

Eigenfunctions, Eigenvalues
If

$$\hat{f}\psi_n = f_n\psi_n \qquad (8\text{-}15)$$

holds for an operator $\hat{f}$, the function ψ_n is called an *eigenfunction* of $\hat{f}$, and the number f_n is called the corresponding *eigenvalue*. For example, we know that

$$\frac{\partial}{\partial x} e^{kx} = k e^{kx} \tag{8-16}$$

Therefore, e^{kx} is an eigenfunction of $\partial/\partial x$ with the eigenvalue k.

An important fact is that eigenvalues of a Hermitian operator are all real. Let us assume that $\hat{f}$ is a Hermitian operator in (8–15). Then

$$\int \psi_n^* \hat{f} \psi_n \, d\tau = \int \psi_n^* f_n \psi_n \, d\tau = f_n \int |\psi_n|^2 \, d\tau \tag{8-17}$$

since the eigenvalue f_n is a number. On the other hand, since $\hat{f}$ is Hermitian, equation (8–10′) holds true for any pair of functions ϕ and χ. Let $\phi = \chi = \psi_n$ in (8–10′). Then we obtain

$$\int \psi_n^* \hat{f} \psi_n \, d\tau = \int (\hat{f} \psi_n)^* \psi_n \, d\tau = f_n^* \int |\psi_n|^2 \, d\tau \tag{8-18}$$

Therefore

$$f_n = f_n^* \tag{8-19}$$

or f_n is real.

Orthogonality

Another important property of a Hermitian operator is that its eigenfunctions are orthogonal to each other. To show this let us take two eigenfunctions of the same Hermitian operator $\hat{f}$:

$$\hat{f} \psi_n = f_n \psi_n \tag{8-20a}$$

and

$$\hat{f} \psi_m = f_m \psi_m \tag{8-20b}$$

Because of (8–19), (8–20) also means

$$\psi_m^* \hat{f}^\dagger = \hat{f}^* \psi_m^* = f_m \psi_m^* \tag{8-21}$$

Now by (8–20a)

$$\int \psi_m^* \hat{f} \psi_n \, d\tau = f_n \int \psi_m^* \psi_n \, d\tau \tag{8-22}$$

but, since $\hat{f}$ is Hermitian

$$\int \psi_m{}^* \hat{f} \psi_n \, d\tau = \int \psi_m{}^* \hat{f}^\dagger \psi_n \, d\tau = f_m \int \psi_m{}^* \psi_n \, d\tau \qquad (8\text{-}23)$$

where (8–21) was used. Therefore, from (8–22) and (8–23), we see

$$\int \psi_m{}^* \psi_n \, d\tau = 0 \qquad \text{if} \qquad f_n \neq f_m \qquad (8\text{-}24)$$

which proves that the eigenfunctions of a Hermitian operator with different eigenvalues are orthogonal to each other.

There are cases where more than one linearly independent function corresponds to a given eigenvalue f_n. Thus we may have k functions, ψ_{n1}, $\psi_{n2}, \ldots, \psi_{nk}$, which are eigenfunctions of an operator $\hat{f}$ with a common eigenvalue f_n, such that

$$a_1 \psi_{n1} + a_2 \psi_{n2} + \cdots + a_k \psi_{nk} = 0 \qquad (8\text{-}25)$$

can be satisfied only when all coefficients a_1 to a_k are zero. In this case we say that ψ_n is k-fold degenerate. If such degeneracy exists, the above argument on the orthogonality, equation (8–24), does not hold. We can, however, construct k mutually orthogonal functions using linear combinations of $\psi_{n1}, \psi_{n2}, \ldots, \psi_{nk}$. For example

$$\psi_{n1} \qquad \text{and} \qquad \psi_{n2} - \psi_{n1} \frac{\int \psi_{n1}{}^* \psi_{n2} \, d\tau}{\int |\psi_{n1}|^2 \, d\tau} \qquad (8\text{-}26)$$

are orthogonal to each other even when ψ_{n1} and ψ_{n2} are not. By extending this same procedure, one can obtain k mutual orthogonal functions, whenever a k-fold degeneracy exists (Gram–Schmidt process).

Since any eigenfunction has an arbitrary multiplication factor, usually we can choose that factor to satisfy the normalization condition

$$\int |\psi_n|^2 \, d\tau = 1 \qquad (8\text{-}27)$$

In this manner we can choose all eigenfunctions of a given operator to form an orthonormal set

$$\int \psi_m{}^* \psi_n \, d\tau = \delta_{m,n} \qquad (8\text{-}28)$$

There are cases where the eigenvalue is distributed continuously. In such cases (8–28) may be replaced by

$$\int \psi_m{}^*\psi_n \, d\tau = \delta(m - n) \qquad (8\text{–}29)$$

where $\delta(m - n)$ is Dirac's delta function, which is defined by (3–9).

PROBLEM 8–5. Following the procedure of (8–26), find a function which contains ψ_{n3} and orthogonal to both of the functions given in (8–26).

PROBLEM 8–6. When $\hat{f}$ is an anti-Hermitian operator,
(a) show that its eigenvalues are all pure imaginary,
(b) show that its eigenfunctions are orthogonal to each other.

PROBLEM 8–7. If there exists a k-fold degeneracy, that is, if $\psi_{n1}, \dots,$ ψ_{nk} are all eigenfunctions of $\hat{f}$ with a common eigenvalue f_n,
(a) show that linear transformed functions

$$\begin{aligned}
\psi_{n1'} &= \langle 1 \mid 1' \rangle \psi_{n1} + \langle 2 \mid 1' \rangle \psi_{n2} + \cdots + \langle k \mid 1' \rangle \psi_{nk} \\
\psi_{n2'} &= \langle 1 \mid 2' \rangle \psi_{n1} + \cdots\cdots\cdots\cdots\cdots\cdots + \langle k \mid 2' \rangle \psi_{nk} \\
&\cdots\cdots\cdots\cdots\cdots\cdots\cdots\cdots\cdots\cdots\cdots\cdots\cdots\cdots \\
\psi_{nk'} &= \langle 1 \mid k' \rangle \psi_{n1} + \cdots\cdots\cdots\cdots\cdots + \langle k \mid k' \rangle \psi_{nk}
\end{aligned}$$

are also eigenfunctions of $\hat{f}$ with the same eigenvalue f_n, where $\langle i \mid j' \rangle$ are the transformation coefficients.
(b) if $\int \psi_{ni}{}^*\psi_{nj} \, d\tau = \delta_{ij}$, show that in order that the transformed functions also satisfy $\int \psi_{ni'}{}^*\psi_{nj'} \, d\tau = \delta_{i'j'}$, the transformation coefficients should be such that

$$\sum_i \langle i \mid j' \rangle^* \langle i \mid k' \rangle = \delta_{j'k'}$$

This equation means that the $k \times k$ matrix formed by the transformation coefficients

$$\mathbf{U} \equiv \begin{pmatrix} \langle 1 \mid 1' \rangle & \langle 1 \mid 2' \rangle \cdots \langle 1 \mid k' \rangle \\ \langle 2 \mid 1' \rangle & \langle 2 \mid 2' \rangle \cdots \langle 2 \mid k' \rangle \\ \cdots\cdots\cdots\cdots\cdots\cdots\cdots\cdots \\ \langle k \mid 1' \rangle & \langle k \mid 2' \rangle \cdots \langle k \mid k' \rangle \end{pmatrix}$$

is a *unitary matrix*, which is defined by

$$\mathbf{U}^\dagger \mathbf{U} = \mathbf{U}^{*T}\mathbf{U} = \mathbf{1}$$

[See Appendix I(b).]

8-3 WAVE FUNCTIONS

In quantum mechanics the dynamical variables such as position, momentum, and Hamiltonian are replaced by operators and a state, in which a particle is, is represented by a *wave function* $\Psi_a(\mathbf{r}, t)$. If $\hat{f}$ is the operator which replaces a classical dynamical variable, then we say that the average value of this dynamical variable which will be observed for a state which is represented by $\Psi_a(\mathbf{r}, t)$, is given by

$$\langle a|\hat{f}|a\rangle \equiv \int \Psi_a^* \hat{f} \Psi_a \, d\tau \qquad (8\text{-}30)$$

This quantity is also called the *expectation value* of $\hat{f}$ in state a.

When the wave function Ψ_a is one of the eigenfunctions of $\hat{f}$,

$$\Psi_a = \psi_n \qquad \text{where} \qquad \hat{f}\psi_n = f_n\psi_n \qquad (8\text{-}31)$$

then (8-30) gives

$$\langle a|\hat{f}|a\rangle = f_n \qquad (8\text{-}32)$$

If we define an operator $(\widehat{\Delta f})$ by

$$(\widehat{\Delta f}) \equiv \hat{f} - \langle a|\hat{f}|a\rangle \qquad (8\text{-}33)$$

then the mean square deviation from the average value $\langle a|\hat{f}|a\rangle$ is given by

$$\langle a|(\widehat{\Delta f})^2|a\rangle = \langle a|\hat{f}^2|a\rangle - \langle a|\hat{f}|a\rangle^2 \qquad (8\text{-}34)$$

Since in our special case of (8-31)

$$\langle a|\hat{f}^2|a\rangle = f_n^2 \qquad (8\text{-}35)$$

we see that the mean square deviation is

$$\langle a|(\widehat{\Delta f})^2|a\rangle = 0 \qquad (8\text{-}36)$$

Thus, the value of a dynamical variable represented by $\hat{f}$ is definitely f_n for a state whose wave function coincides with an eigenfunction ψ_n.

A more general wave function is given by

$$\Psi_a = \sum_n \langle n \mid a\rangle \psi_n \qquad (8\text{-}37)$$

where the expansion coefficient $\langle n \mid a \rangle$ can be expressed as

$$\langle n \mid a \rangle = \int \psi_n^* \Psi_a \, d\tau \tag{8-38}$$

using the orthonormality relation (8–28). The expectation value of $\hat{f}$ in this state is

$$\langle f \rangle_a = \sum_{n,m} \langle m \mid a \rangle^* \langle n \mid a \rangle \int \psi_m^* \hat{f} \psi_n \, d\tau = \sum_n |\langle n \mid a \rangle|^2 f_n \tag{8-39}$$

Since the expectation value is interpreted as the average value, this result means that $|\langle n \mid a \rangle|^2$ is the probability of finding the state Ψ_a in the state ψ_n. The mean square deviation in a state, as given by (8–34), is in general not zero.

A particularly interesting example is the position operator $\hat{x} = x$. The eigenfunction of this operator is the delta function, because

$$x \, \delta(x - \alpha) = \alpha \, \delta(x - \alpha) \tag{8-40}$$

Thus the function $\delta(x - \alpha)$ gives a state in which the position of a particle is definitely α. For a general state with a wave function of $\Psi_a(x)$ the probability of finding the above state is given by

$$\left| \int \delta(x - \alpha) \Psi_a(x) \, dx \right|^2 = |\Psi_a(\alpha)|^2 \tag{8-41}$$

Therefore $|\Psi_a(\mathbf{r})|^2$ has the physical meaning of the probability of finding a particle at $\mathbf{r}$, and the normalization condition $\int |\Psi_a(r)|^2 \, d\tau = 1$, which we impose on wave functions, has the physical meaning that the particle exists somewhere in space.

Equation (8–37) implies the completeness of eigenfunctions ψ_n. Actually there is no guaranty that eigenfunctions of a Hermitian operator always form a complete set. However, it is known from experience that for all common Hermitian operators which appear in quantum mechanics eigenfunctions do form complete sets. We will, therefore, assume in the following discussions that a set of eigenfunctions is complete, and therefore span a Hilbert space.[5]

---

[5] A Hermitian operator whose eigenfunctions form a complete set, and therefore span a Hilbert space, is called an *observable*.

PROBLEM 8–8. When ψ_n's form a complete set, show that

$$\sum_n \psi_n(\mathbf{r})\psi_n(\mathbf{r'}) = \delta(\mathbf{r} - \mathbf{r'})$$

This relation is called the *closure property*.

HINT. Expand $\delta(\mathbf{r} - \mathbf{r'})$ in the form of (8–37).

PROBLEM 8–9. Show that when the closure property holds, functions ψ_n form a complete set. That is, an arbitrary function $\Psi_a(\mathbf{r})$ can be expanded in the form of (8–37).

PROBLEM 8–10. Assuming that the expansion (8–39) has two terms

$$\Psi_a = \langle 1 \mid a \rangle \psi_1 + \langle 2 \mid a \rangle \psi_2$$

where

$$\hat{f}\psi_1 = f_1\psi_1 \quad \text{and} \quad \hat{f}\psi_2 = f_2\psi_2$$

show that

$$\langle a|(\widehat{\Delta f})^2|a\rangle > 0 \quad \text{if} \quad f_1 \neq f_2$$

HINT.

$$\langle a|(\widehat{\Delta f})^2|a\rangle = (|\langle 1 \mid a \rangle|^2 - |\langle 1 \mid a \rangle|^4)(f_1 - f_2)^2$$

8–4 COMMUTATORS

If $\hat{f}$ and $\hat{g}$ are different operators, $\hat{f}\hat{g}$ is not necessarily equal to $\hat{g}\hat{f}$. The difference between these two product operators is designated as

$$[\hat{f}, \hat{g}] \equiv \hat{f}\hat{g} - \hat{g}\hat{f} \qquad (8\text{–}42)$$

and is called the *commutator* between $\hat{f}$ and $\hat{g}$. If $\hat{f} = x$ and $\hat{g} = \partial/\partial x$, for example, we obtain

$$\left[x, \frac{\partial}{\partial x}\right]\chi = x\frac{\partial}{\partial x}\chi - \frac{\partial}{\partial x}(x\chi) = x\frac{\partial}{\partial x}\chi - \chi - x\frac{\partial}{\partial x}\chi = -\chi \qquad (8\text{–}43)$$

for an arbitrary function χ. Abstracting from the above relation, we obtain the operator relation

$$\left[x, \frac{\partial}{\partial x}\right] = -1 \qquad (8\text{-}44)$$

Thus x and $\partial/\partial x$ do not commute; while obviously x and $\partial/\partial y$ commute.

Quantum mechanics can be formulated from classical mechanics by replacing a classical Poisson bracket relation

$$[f, g]_P = k \qquad (8\text{-}45)$$

by the commutator relation

$$[\hat{f}, \hat{g}] = i\hbar \hat{k} \qquad (8\text{-}46)$$

where

$$\hbar = 1.0545919 \times 10^{-34} \text{ J sec} \qquad (8\text{-}47)$$

Thus, if q_i and p_i are canonically conjugate in classical mechanics, as we saw in (1–67) their Poisson bracket is equal to one; therefore the corresponding quantum mechanical operators $\hat{q}_i$ and $\hat{p}_i$ should have a commutator equal to $i\hbar$. In general, from (1–65), (1–66), and (1–67), we obtain the quantum mechanical relations[6]

$$[\hat{q}_i, \hat{q}_j] = 0 \qquad (8\text{-}48)$$

$$[\hat{p}_i, \hat{p}_j] = 0 \qquad (8\text{-}49)$$

$$[\hat{q}_i, \hat{p}_j] = i\hbar \, \delta_{ij} \qquad (8\text{-}50)$$

A set of Hermitian operators which satisfy these commutator relations are

$$\hat{q}_i = q_i \quad \text{and} \quad \hat{p}_i = -i\hbar \frac{\partial}{\partial q_i} \qquad (8\text{-}51)$$

[6] The relation $[\hat{q}, \hat{p}] = i\hbar$ was first given by M. Born and P. Jordan, *Zeits. f. Physik*, **34**, 858 (1925), which also used the word "quantum mechanics" for the first time. The formulation was subsequently completed by Born, Heisenberg, and Jordan, *Zeits. f. Physik*, **35**, 557 (1926). The correspondence between Poisson brackets and commutators was pointed out by P. A. M. Dirac, *Proc. Roy. Soc. London* (A), **109**, 642 (1925).

Among the many commutation relations which can be obtained from the fundamental relations (8–48), (8–49), and (8–50), those involving the components of angular momentum

$$\hat{\mathbf{l}} = \hat{\mathbf{r}} \times \hat{\mathbf{p}} \tag{8–52}$$

are particularly important. We find from the fundamental commutator relations that

$$[\hat{l}_x, \hat{l}_y] = i\hbar\hat{l}_z \qquad [\hat{l}_y, \hat{l}_z] = i\hbar\hat{l}_x \qquad [\hat{l}_z, \hat{l}_x] = i\hbar\hat{l}_y \tag{8–53}$$

or, in general, if $\mathbf{u}_1$ and $\mathbf{u}_2$ are unit vectors

$$[\hat{\mathbf{l}} \cdot \mathbf{u}_1, \hat{\mathbf{l}} \cdot \mathbf{u}_2] = i\hbar\hat{\mathbf{l}} \cdot (\mathbf{u}_1 \times \mathbf{u}_2) \tag{8–54}$$

From (8–51) we also obtain

$$\hat{l}_x = -i\hbar \left(y \frac{\partial}{\partial z} - z \frac{\partial}{\partial y} \right) \qquad \hat{l}_y = -i\hbar \left(z \frac{\partial}{\partial x} - x \frac{\partial}{\partial z} \right)$$

$$\hat{l}_z = -i\hbar \left(x \frac{\partial}{\partial y} - y \frac{\partial}{\partial x} \right) \tag{8–55}$$

PROBLEM 8–11. Show that

(a) $[\hat{f}, \hat{g}] = -[\hat{g}, \hat{f}]$
(b) $[\hat{h}\hat{f}, \hat{g}] = \hat{h}[\hat{f}, \hat{g}] + [\hat{h}, \hat{g}]\hat{f}$
(c) $[\hat{e}\hat{f}, \hat{g}\hat{h}] = \hat{e}[\hat{f}, \hat{g}]\hat{h} + [\hat{e}, \hat{g}]\hat{f}\hat{h} + \hat{g}\hat{e}[\hat{f}, \hat{h}] + \hat{g}[\hat{e}, \hat{h}]\hat{f}$

The first two of these relations correspond to (1–63) and (1–64), respectively. Note, however, that the last term in (b) should be $[\hat{h}, \hat{g}]\hat{f}$, and not $\hat{f}[\hat{h}, \hat{g}]$.

PROBLEM 8–12. Show that

$$[\hat{h}, [\hat{f}, \hat{g}]] + [\hat{f}, [\hat{g}, \hat{h}]] + [\hat{g}, [\hat{h}, \hat{f}]] = 0$$

This relation corresponds to Jacobi's identity (Problem 1–30), but is much easier to prove.

PROBLEM 8–13. Show that if $\hat{f}$ and $\hat{g}$ are both Hermitian, $[\hat{f}, \hat{g}]$ is anti-Hermitian.

HINT. Problem 8–3.

PROBLEM 8–14. Show that one may also obtain

$$\hat{q}_i = i\hbar \frac{\partial}{\partial p_i} \qquad \hat{p}_i = p_i$$

instead of (8–51). This is called the momentum representation since this expression assumes that the functions on which these operators operate are functions of p_i's, instead of functions of q_i's.

PROBLEM 8–15. Show that

$$\exp\left(\frac{i\boldsymbol{\xi} \cdot \hat{\mathbf{p}}}{\hbar}\right) \chi(\mathbf{r}) = \chi(\mathbf{r} + \boldsymbol{\xi})$$

where $\chi(\mathbf{r})$ is an analytic function of $\mathbf{r}$, and $\boldsymbol{\xi}$ is a constant vector. (See Problem 8–4 for the definition of an exponential operator.)

PROBLEM 8–16. Derive the commutator relations (8–53) and (8–54).

PROBLEM 8–17. Show that

$$[\hat{l}^2, \hat{l}_x] = 0$$

(See Problem 1–33.)

PROBLEM 8–18. Show that

$$\hat{l}_x = i\hbar \left(\sin \varphi \frac{\partial}{\partial \theta} + \cos \varphi \cot \theta \frac{\partial}{\partial \varphi}\right)$$

$$\hat{l}_y = -i\hbar \left(\cos \varphi \frac{\partial}{\partial \theta} - \sin \varphi \cot \theta \frac{\partial}{\partial \varphi}\right)$$

$$\hat{l}_z = -i\hbar \frac{\partial}{\partial \varphi}$$

in spherical coordinates.

PROBLEM 8–19. When the wave function is

$$\Psi = b^{1/2}\pi^{-1/4} \exp\left[-\frac{1}{2}\left(\frac{x}{b}\right)^2\right]$$

where b is a constant, show that the probability of the particle having a momentum between $\hbar k_x$ and $\hbar k_x + \hbar\, dk_x$ is proportional to

$$\rho(k_x)\hbar\, dk_x \propto \exp\left[-(k_x b)^2\right] dk_x$$

(See Fig. 8–1.)

FIGURE 8–1. Probability densities in coordinate space and momentum space (Problem 8–19).

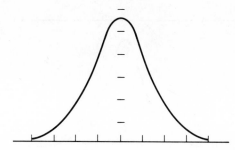

HINT. An eigenfunction of $\hat{p}_x$ $(= -i\hbar\, \partial/\partial x)$ is $\exp(ik_x x)$. In the same way that (8–41) gives the probability of finding a particle at $x = \alpha$, we see that the probability of finding the momentum as $\hbar k_x$ is proportional to

$$\left|\int \Psi^*(x)\,\exp(ik_x x)\, dx\right|^2$$

PROBLEM 8–20. When the wave function is

$$\Psi = \sqrt{\tfrac{1}{2}b}\; e^{-(1/2)b|x|}$$

show that the probability of the particle having a momentum between hk_x and $hk_x + \hbar\, dk_x$ is proportional to

$$\rho(k_x)\hbar\, dk_x \propto \frac{dk_x}{k_x{}^2 + (\tfrac{1}{2}b)^2}$$

(See Fig. 8–2.)

8–5 UNCERTAINTY RELATIONS

When two operators $\hat{f}$ and $\hat{g}$ commute, the eigenfunctions of $\hat{f}$ can simultaneously be the eigenfunctions of $\hat{g}$. In order to prove this theorem, let us assume that $\psi_n, \psi_m, \ldots,$ are the eigenfunctions of $\hat{f}$ with eigenvalues f_n,

FIGURE 8–2. Probability densities in coordinate space and momentum space (Problem 8–20).

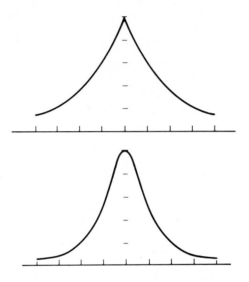

$f_m, \ldots$, respectively. Even if ψ_n was not an eigenfunction of $\hat{g}$, we can still express $\hat{g}\psi_n$ as a linear combination of the eigenfunction ψ_m as

$$\hat{g}\psi_n = \sum_m \psi_m \langle m|\hat{g}|n\rangle \tag{8-56}$$

assuming that the set of the eigenfunctions ψ_m is complete. In this formula $\langle m|\hat{g}|n\rangle$'s are expansion coefficients. Operating the commutator $[\hat{f}, \hat{g}]$ on ψ_n we obtain

$$[\hat{f}, \hat{g}]\psi_n = \hat{f}\hat{g}\psi_n - \hat{g}\hat{f}\psi_n = \sum_m \hat{f}\psi\langle m|\hat{g}|n\rangle - \hat{g}f_n\psi_n$$

$$= \sum_m f_m\psi_m\langle m|\hat{g}|n\rangle - \sum_m f_n\psi_m\langle m|\hat{g}|n\rangle$$

$$= \sum_m \psi_m\langle m|\hat{g}|n\rangle(f_m - f_n) \tag{8-57}$$

If $\hat{f}$ and $\hat{g}$ commute (8–57) should be zero. Consequently, since ψ_m are linearly independent this means that each coefficient must vanish separately, i.e.,

$$\langle m|\hat{g}|n\rangle(f_m - f_n) = 0 \tag{8-58}$$

If there is no degeneracy, that is, if f_n's are all distinct, then (8–58) shows that

$$\langle m|\hat{g}|n\rangle = g_n \, \delta_{m,n} \tag{8–59}$$

Hence ψ_n is also an eigenfunction of $\hat{g}$.

When a k-fold degeneracy exists, that is, when $\psi_1, \ldots, \psi_k$ are the eigenfunctions of $\hat{f}$ with a common eigenvalue f, then (8–58) means that if $\hat{g}$ is operated on one of these functions, the resultant expansion like (8–56) contains, at most, these k functions and no other eigenfunctions. In the matrix formulation, (8–56) becomes

$$\hat{g}(\psi_1\psi_2 \cdots \psi_k) = (\psi_1\psi_2 \cdots \psi_k) \begin{pmatrix} \langle 1|\hat{g}|1\rangle & \langle 1|\hat{g}|2\rangle \cdots \langle 1|\hat{g}|k\rangle \\ \langle 2|\hat{g}|1\rangle & \langle 2|\hat{g}|2\rangle \cdots \langle 2|\hat{g}|k\rangle \\ \cdots\cdots\cdots\cdots\cdots\cdots\cdots\cdots\cdots \\ \langle k|\hat{g}|1\rangle & \langle k|\hat{g}|2\rangle \cdots \langle k|\hat{g}|k\rangle \end{pmatrix}$$
$$\tag{8–60}$$

where $(\psi_1\psi_2 \cdots \psi_k)$ is a row matrix made of the k degenerate eigenfunctions. On the other hand, it is shown in Problem 8–7 that if $\mathbf{U}$ is an arbitrary $k \times k$ unitary matrix, a set of transformed functions

$$(\psi_{1'}\psi_{2'} \cdots \psi_{k'}) = (\psi_1\psi_2 \cdots \psi_k)\mathbf{U} \tag{8–61}$$

are also eigenfunctions of $\hat{f}$ with the same eigenvalue as the untransformed set. For the transformed functions the equation, which corresponds to (8–60), is

$$\hat{g}\psi' = \hat{g}\psi\mathbf{U} = \psi(\langle i|\hat{g}|j\rangle)\mathbf{U}$$
$$= \psi\mathbf{U}\mathbf{U}^{-1}(\langle i|\hat{g}|j\rangle)\mathbf{U} = \psi'\mathbf{U}^{-1}(\langle i|\hat{g}|j\rangle)\mathbf{U} \tag{8–62}$$

where we use the notations

$$\psi \equiv (\psi_1\psi_2 \cdots \psi_k)$$

$$\psi' \equiv (\psi_{1'}\psi_{2'} \cdots \psi_{k'}) \tag{8–63}$$

$$(\langle i|\hat{g}|j\rangle) \equiv \begin{pmatrix} \langle 1|\hat{g}|1\rangle \cdots \langle 1|\hat{g}|k\rangle \\ \cdots\cdots\cdots\cdots\cdots\cdots \\ \langle k|\hat{g}|1\rangle \cdots \langle k|\hat{g}|k\rangle \end{pmatrix}$$

If the unitary transformation matrix **U** is such that

$$\mathbf{U}^{-1}(\langle i|\hat{g}|j\rangle)\mathbf{U} = \begin{pmatrix} g_1 & 0 & 0 & \cdots & 0 \\ 0 & g_2 & 0 & \cdots & 0 \\ 0 & 0 & g_3 & \cdots & 0 \\ \cdots\cdots\cdots\cdots\cdots \\ 0 & 0 & \cdots\cdots & g_k \end{pmatrix} \tag{8–64}$$

or the resultant matrix is diagonal, then the transformed functions, which are still eigenfunctions of $\hat{f}$, are also the eigenfunctions of $\hat{g}$. The required transformation (8–64) can be found in the same way as the transformation (2–90), by solving the secular equation

$$\det (\langle i|\hat{g}|j\rangle - g\, \delta_{ij}) = 0 \tag{8–65}$$

which is analogous to (2–94). Thus we see that the statement given at the beginning of this section is proved even when degeneracies exist. (See section 8–12 for a more complete discussion on the unitary transformation.)

We can give this statement in another form: If $\hat{f}$ and $\hat{g}$ commute we can find a wave function which is a simultaneous eigenfunction of both these operators. The corresponding state is such that the mean square deviation of $\hat{f}$ and $\hat{g}$ are both zero:

$$\langle a|(\widehat{\Delta f})^2|a\rangle = 0 \qquad \langle a|(\widehat{\Delta g})^2|a\rangle = 0 \tag{8–66}$$

where $(\widehat{\Delta f})$ and $(\widehat{\Delta g})$ operators are defined by (8–33).

When $\hat{f}$ and $\hat{g}$ do not commute, and

$$[\hat{f}, \hat{g}] = i\hat{k} \tag{8–67}$$

where $\hat{f}$, $\hat{g}$, and $\hat{k}$ are assumed to be Hermitian (see Problem 8–13), then it follows that

$$[(\widehat{\Delta f}), (\widehat{\Delta g})] = i\hat{k} \tag{8–68}$$

Taking an arbitrary real number α, we obtain

$$\int |(\alpha(\widehat{\Delta f}) - i(\widehat{\Delta g}))\Psi|^2\, dv = \int [(\alpha(\widehat{\Delta f})^* + i(\widehat{\Delta g})^*)\Psi^*](\alpha(\widehat{\Delta f}) - i(\widehat{\Delta g}))\Psi\, dv$$

$$= \int \Psi^*(\alpha(\widehat{\Delta f})^\dagger + i(\widehat{\Delta g})^\dagger)(\alpha(\widehat{\Delta f}) - i(\widehat{\Delta g}))\Psi\, dv$$

$$= \langle(\alpha(\widehat{\Delta f}) + i(\widehat{\Delta g}))(\alpha(\widehat{\Delta f}) - i(\widehat{\Delta g}))\rangle$$

$$= \alpha^2\langle(\widehat{\Delta f})^2\rangle + \alpha\langle\hat{k}\rangle + \langle(\widehat{\Delta g})^2\rangle \tag{8–69}$$

where we used (8–68) and assumed that $\hat{f}$ and $\hat{g}$ are Hermitian. The expectation values are designated by the abbreviated notation $\langle \quad \rangle$. Since the left-hand side of (8–69) cannot be negative, the right-hand side of (8–69) must also be non-negative. Thus

$$0 \leq \alpha^2 \langle (\widehat{\Delta f})^2 \rangle + \alpha \langle \hat{k} \rangle + \langle (\widehat{\Delta g})^2 \rangle = \langle (\widehat{\Delta f})^2 \rangle \left[\alpha + \frac{\langle \hat{k} \rangle}{2\langle (\widehat{\Delta f})^2 \rangle} \right]^2$$

$$+ \frac{1}{4\langle (\widehat{\Delta f})^2 \rangle} [4\langle (\widehat{\Delta f})^2 \rangle \langle (\widehat{\Delta g})^2 \rangle - \langle k \rangle^2] \qquad (8\text{–}70)$$

The first term of the last expression is also non-negative. Therefore

$$\langle (\widehat{\Delta f})^2 \rangle \langle (\widehat{\Delta g})^2 \rangle \geq \tfrac{1}{4} \langle \hat{k} \rangle^2 \qquad (8\text{–}71)$$

Thus unless $\langle \hat{k} \rangle$ is zero, the mean square deviations of $\hat{f}$ and $\hat{g}$ cannot be simultaneously zero. This is called *Heisenberg's uncertainty relation.*[7] Particularly, from (8–50),

$$\langle (\widehat{\Delta x})^2 \rangle \langle (\widehat{\Delta p_x})^2 \rangle \geq \tfrac{1}{4} \hbar^2 \qquad (8\text{–}72)$$

The uncertainty relation clearly shows the difference between the classical and quantum mechanical concepts of coordinates, momenta, and other physical quantities. However, since $\hbar$ is so small, the deviation from classical mechanics is significant (except for a few cases) only in microscopic or atomic and subatomic phenomena.

PROBLEM 8–21. (a) Show that

$$\langle (\widehat{\Delta l_x})^2 \rangle \langle (\widehat{\Delta l_y})^2 \rangle \geq \tfrac{1}{4} \hbar^2 \langle \hat{l}_z^2 \rangle$$

(b) When the wave function does not depend on angles, i.e.,

$$\Psi = \Psi(r)$$

show that this state is an eigenfunction of $\hat{l}_x$, $\hat{l}_y$, and $\hat{l}_z$ simultaneously, with eigenvalues all zero. This appears to be a contradiction of the uncertainty principle because these operators do not commute with each other, as seen in (8–53). Show that the above situation actually does not violate the un-

---

[7] W. Heisenberg, *Zeits. f. Physik*, **43**, 172 (1927).

certainty principle. [This apparent contradiction was argued by E. U. Condon, *Science*, **69**, 573 (1929).]

PROBLEM 8–22. Show that

$$\langle(\widehat{\Delta l_z})^2\rangle\langle(\widehat{\Delta\varphi})^2\rangle \geq \tfrac{1}{4}\hbar^2$$

HINT. See Problem 8–18.

PROBLEM 8–23. Show that when the wave function is

$$\Psi(x) = a^{-1/2}\pi^{-1/4}\exp\left[-\frac{1}{2}\left(\frac{x}{a}\right)^2\right]$$

we obtain

$$\langle(\widehat{\Delta x})^2\rangle\langle(\widehat{\Delta p_x})^2\rangle = \tfrac{1}{4}\hbar^2$$

or the uncertainty is the minimum.

8–6 SCHROEDINGER EQUATION (WAVE MECHANICS)

We have seen in (1–62) that if H is the Hamiltonian function

$$\frac{df}{dt} = [f, H]_P + \frac{\partial f}{\partial t} \tag{1–62}$$

in classical mechanics for any function f. The corresponding quantum mechanical relation may be assumed to be

$$i\hbar\frac{d\hat{f}}{dt} = [\hat{f}, \hat{H}] + i\hbar\frac{\partial\hat{f}}{\partial t} \tag{8–73}$$

where $\hat{H}$ is a Hamiltonian operator. This approach gives the so-called *Heisenberg picture* (section 8–15).

An alternative way of bringing (1–62) into quantum mechanics is to assume that such an equation holds true for the expectation value such as

$$i\hbar\frac{d\langle\hat{f}\rangle}{dt} = \langle[\hat{f}, \hat{H}]\rangle + i\hbar\left\langle\frac{\partial\hat{f}}{\partial t}\right\rangle \tag{8–74}$$

and let the wave function be time dependent. Since the wave function $\Psi(\mathbf{r}, t)$ is a field function, the differentiation of the wave function with respect to time should be done without changing $\mathbf{r}$, because $\mathbf{r}$ is not a dynamical variable but a parameter in a field function. (See the argument in section 3–1.) Thus total differentiation must be replaced by partial differentiation:

$$\frac{d\Psi}{dt} \rightarrow \frac{\partial\Psi}{\partial t} \tag{8–75}$$

Writing the expectation value explicitly as in (8–30) we see

$$i\hbar \frac{d\langle\hat{f}\rangle}{dt} = i\hbar \frac{d}{dt} \int \Psi^*\hat{f}\Psi \, d\tau = i\hbar \int \left(\frac{\partial\Psi^*}{\partial t} \hat{f}\Psi + \Psi^*\hat{f}\frac{\partial\Psi}{\partial t}\right) d\tau$$

$$+ \, i\hbar \int \Psi^* \frac{\partial\hat{f}}{\partial t} \Psi \, d\tau \tag{8–76}$$

where we used (8–75), and in the last term wrote $\partial\hat{f}/\partial t$, instead of $d\hat{f}/dt$, indicating, in the same sense as in (1–59), that only an explicit time dependence of $\hat{f}$ should be considered. For example, $\partial\hat{x}/\partial t = 0$ and $\partial\hat{p}/\partial t = 0$ since they are not explicitly time dependent.

On the other hand, if $\hat{H}$ is Hermitian,

$$\langle[\hat{f}, \hat{H}]\rangle + i\hbar \left\langle\frac{\partial\hat{f}}{\partial t}\right\rangle = \int (-\Psi^*\hat{H}\hat{f}\Psi + \Psi^*\hat{f}\hat{H}\Psi) \, d\tau + i\hbar \int \Psi^* \frac{\partial\hat{f}}{\partial t} \Psi \, d\tau$$

$$= \int (-(\hat{H}\Psi)^*\hat{f}\Psi + \Psi^*\hat{f}\hat{H}\Psi) \, d\tau$$

$$+ \, i\hbar \int \Psi^* \frac{\partial\hat{f}}{\partial t} \Psi \, d\tau \tag{8–77}$$

Comparing (8–76) and (8–77) we see that (8–74) holds true if

$$i\hbar \frac{\partial\Psi}{\partial t} = \hat{H}\Psi \tag{8–78}$$

This is called the Schroedinger wave equation, or simply the Schroedinger equation,[8] and the whole approach in which we assume that the wave function, rather than the operators, carries the essential time dependence, is called the *Schroedinger picture*.

[8] E. Schroedinger, *Ann. d. Physik.* **81**, 109 (1926).

When $\hat{H}$ does not explicitly depend on time, equation (8–78) can be reduced to

$$\mathscr{E}\psi_{\mathscr{E}} = \hat{H}\psi_{\mathscr{E}} \tag{8–79}$$

by assuming that

$$\Psi(\mathbf{r}, t) = \psi_{\mathscr{E}}(\mathbf{r})e^{-i\mathscr{E}t/\hbar} \tag{8–80}$$

Equation (8–79) is called the time-independent Schroedinger equation, and shows that the spatial portion of the total wave function as given by (8–80) is an eigenfunction of $\hat{H}$, and $\mathscr{E}$ is the corresponding eigenvalue of $\hat{H}$, or the energy of that state. A general wave function, which is a solution of the Schroedinger equation (8–78) itself, is a linear combination of (8–80). When a wave function is expressed by a single term like (8–80), it represents a state with definite energy.

Equation (8–74), which may be called a generalized Ehrenfest theorem, means that the average value of a physical quantity follows the classical equations of motion; while we should keep in mind that in general the value of a physical quantity itself fluctuates with a nonzero mean square deviation.

The Hamiltonian operator which appears in the Schroedinger equation can be obtained from its classical expression as follows. For a nonrelativistic particle of mass m, moving under the influence of potential $U(\mathbf{r})$, we see

$$\hat{H} = \frac{\hat{p}^2}{2m} + U(\mathbf{r}) = -\frac{\hbar^2}{2m}\nabla^2 + U(\mathbf{r}) \tag{8–81}$$

When $\hat{H}$ is Hermitian $U(\mathbf{r})$ should be real. In this case the Schroedinger equation, using (8–81), gives

$$
\begin{aligned}
\frac{\partial(\Psi^*\Psi)}{\partial t} &= \Psi\frac{\partial\Psi^*}{\partial t} + \Psi^*\frac{\partial\Psi}{\partial t} = \frac{-1}{i\hbar}(\Psi\hat{H}\Psi^* - \Psi^*\hat{H}\Psi) \\
&= -\frac{i\hbar}{2m}(\Psi\nabla^2\Psi^* - \Psi^*\nabla^2\Psi) \\
&= -\frac{i\hbar}{2m}\nabla\cdot(\Psi\nabla\Psi^* - \Psi^*\nabla\Psi)
\end{aligned}
\tag{8–82}
$$

This equation has the form of the equation of continuity

$$\frac{\partial\rho}{\partial t} + \nabla\cdot\mathbf{j} = 0 \tag{3–35}$$

and can be interpreted as the equation for the conservation of probability, if $\Psi^*\Psi$ is the probability density, as given by (8–41), and if

$$\mathbf{j} = \frac{i\hbar}{2m} (\Psi\nabla\Psi^* - \Psi^*\nabla\Psi) \tag{8–83}$$

is the *probability current density.*

PROBLEM 8–24. Prove the uncertainty relation

$$\langle(\Delta\hat{f})^2\rangle\langle(\Delta\hat{H})^2\rangle \geq \tfrac{1}{4}\hbar^2\langle\dot{f}\rangle^2$$

PROBLEM 8–25. (a) When the wave function is

$$\Psi = \sqrt{\tfrac{1}{2}b}\ e^{-(1/2)b|x|}$$

show that

$$\langle\hat{H}\rangle = \frac{\hbar^2}{8m} b^2 + \langle U\rangle$$

(b) Show that the expectation value of the kinetic energy in the above problem can be estimated from the uncertainty relation (8–72).

PROBLEM 8–26. (a) Show that the probability current density $\mathbf{j}$ is zero if Ψ is real or pure imaginary, and express $\mathbf{j}$ by means of the real and imaginary parts of Ψ.

(b) When the wave function is

$$\Psi = \frac{e^{ikx}}{\sqrt{2\pi}}$$

show that

$$j_x = \frac{\hbar k}{2\pi m} \quad \text{and} \quad \langle p_x\rangle = \hbar k$$

(c) If $\Psi = |\Psi|e^{i\theta}$, show that $\mathbf{j} = (\hbar/m)|\Psi|^2\nabla\theta$.

PROBLEM 8–27. If U has an imaginary part, $U = \mathscr{R}e\, U + i\alpha$, then show that

$$\frac{\partial \rho}{\partial t} + \mathbf{V} \cdot \mathbf{j} = \frac{2}{\hbar} \Psi^* \alpha \Psi$$

Here α gives the rate of creation of the particle.

8–7 CLASSICAL APPROXIMATION. WKB SOLUTION

The Schroedinger equation for one particle moving under the influence of a potential $U(\mathbf{r})$ is

$$i\hbar \frac{\partial \Psi}{\partial t} = \left(-\frac{\hbar^2}{2m} \nabla^2 + U(\mathbf{r}) \right) \Psi \qquad (8\text{--}84)$$

Let us introduce a function S by

$$\Psi(\mathbf{r}, t) = \exp\left[\frac{i}{\hbar} S(\mathbf{r}, t) \right] \qquad (8\text{--}85)$$

then the Schroedinger equation can be expressed as

$$-\frac{\partial S}{\partial t} = \frac{1}{2m} \left[(\nabla S) \cdot (\nabla S) - i\hbar \nabla^2 S \right] + U \qquad (8\text{--}86)$$

or

$$\frac{\partial S}{\partial t} + \frac{1}{2m} (\nabla S) \cdot (\nabla S) + U = \frac{i\hbar}{2m} \nabla^2 S \qquad (8\text{--}87)$$

which reduces to the classical Hamilton–Jacobi equation (1–121) when $\hbar \to 0$. Therefore, classical mechanics can be considered as a limiting case of quantum mechanics.[9]
If we expand

$$S = S_0 + \hbar S_1 + \hbar^2 S_2 + \cdots \qquad (8\text{--}88)$$

[9] Note that this argument is analogous to that of section 5–15 which shows that classical mechanics and geometrical optics are obtained as limiting cases of a wave equation.

the first term S_0 will be classical Hamilton's principal function, and the following terms will give the successive quantum mechanical corrections. Since the time dependence of (8–86) is so simple, we let S_0 contain all the time dependence and assume that S_1, S_2, etc., are time independent. Inserting (8–88) into (8–87) under that assumption, and equating terms of the same power in $\hbar$ to each other separately, we obtain

$$\frac{\partial S_0}{\partial t} + \frac{1}{2m} (\nabla S_0) \cdot (\nabla S_0) + U = 0 \tag{8-89}$$

$$(\nabla S_0) \cdot (\nabla S_1) = i\tfrac{1}{2}\nabla^2 S_0 \tag{8-90}$$

$$(\nabla S_0) \cdot (\nabla S_2) + \tfrac{1}{2}(\nabla S_1) \cdot (\nabla S_1) = i\tfrac{1}{2}\nabla^2 S_1 \tag{8-91}$$

. . . .

To simplify the following discussions let us consider a one-dimensional case as an example. From Problem 1–23, we know that

$$S_0 = -\mathscr{E}t \pm \int \sqrt{2m(\mathscr{E} - U)}\, dx \tag{8-92}$$

when

$$\mathscr{E} > U$$

where $\mathscr{E}$ is the total energy. In classical mechanics we simply neglect regions where $\mathscr{E} < U$, because classically a particle cannot be in regions where the kinetic energy is negative. In quantum mechanics, however, we cannot neglect such regions and, as we will see presently, the solution for the negative kinetic energy region is

$$S_0 = -\mathscr{E}t \pm i \int \sqrt{2m(U - \mathscr{E})}\, dx \tag{8-93}$$

when

$$\mathscr{E} < U$$

Inserting either (8–92) or (8–93) into (8–90) we obtain the same solution

$$S_1 = -i\frac{1}{4}\int \frac{(\partial U/\partial x)}{\mathscr{E} - U}\, dx = i\tfrac{1}{4}\ln (U - \mathscr{E}) \tag{8-94}$$

To this approximation, we thus obtain

$$\Psi \cong \frac{1}{\sqrt{p}} (Ae^{i\int p\,dx/\hbar} + Be^{-i\int p\,dx/\hbar})e^{-i\mathscr{E}t/\hbar} \qquad (8\text{--}95)$$

whenever

$$\mathscr{E} > U \quad \text{and} \quad p = \sqrt{2m(\mathscr{E} - U)} \qquad (8\text{--}96)$$

and

$$\Psi \cong \frac{1}{\sqrt{p'}} (Ce^{\int p'\,dx/\hbar} + De^{-\int p'\,dx/\hbar})e^{-i\mathscr{E}t/\hbar} \qquad (8\text{--}97)$$

whenever

$$\mathscr{E} < U \quad \text{and} \quad p' = \sqrt{2m(U - \mathscr{E})} \qquad (8\text{--}98)$$

For these solutions A, B, C, and D are constants.

Obviously neither of the approximations, (8–95) nor (8–97), are valid near the classical turning point, where $\mathscr{E} = U$. A different approach is necessary to cover this third region. If U is a smooth function of x, and does not change very rapidly in the region near the turning point, we may approximate U over this region as

$$U(x) = \mathscr{E} - \kappa'(x - a) \qquad (8\text{--}99)$$

where $x = a$ is the classical turning point (Fig. 8–3). Assuming that the constant κ' is positive, this corresponds to the case where U is larger than $\mathscr{E}$ when $x < a$. Equation (8–84), using this approximation, becomes

$$\frac{\partial^2 \psi_\kappa}{\partial x^2} + \kappa(x - a)\psi_\kappa = 0 \qquad (8\text{--}100)$$

where

$$\kappa = \frac{2m\kappa'}{\hbar^2} \qquad (8\text{--}101)$$

FIGURE 8–3. The linear potential given by (8–99), and the classical turning point $x = a$.

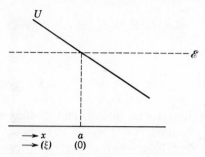

A solution of equation (8–100), which stays finite through the entire region of x is given by the Airy integral Φ as

$$\psi_\kappa(x) = D'\Phi(-\xi) \equiv \frac{D'}{\sqrt{\pi}} \int_0^\infty \cos\left(\tfrac{1}{3}u^3 - \xi u\right) du$$

$$= D'\pi 3^{-7/6} i[J_{1/3}(\tfrac{2}{3}\xi^{3/2}) - J_{-1/3}(\tfrac{2}{3}\xi^{3/2})] \qquad (8\text{–}102)$$

where D' is a constant, J is the Bessel function, and

$$\xi = \kappa^{1/3}(x - a) \qquad (8\text{–}103)$$

This solution has the asymptotic expressions

$$\psi_\kappa(x) \xrightarrow[\xi \to -\infty]{} \tfrac{1}{2}D'(-\xi)^{-1/4} \exp\left[-\tfrac{2}{3}(-\xi)^{3/2}\right] \qquad (8\text{–}104)$$

$$\psi_\kappa(x) \xrightarrow[\xi \to \infty]{} D'\xi^{-1/4} \sin\left(\tfrac{2}{3}\xi^{3/2} + \tfrac{1}{4}\pi\right) \qquad (8\text{–}105)$$

In the case of (8–99), on the other hand, we see that

$$\int_a^x p\,\frac{dx}{\hbar} = \int_a^x \sqrt{2m(\mathscr{E} - U)}\,\frac{dx}{\hbar} = \sqrt{\kappa} \int_a^x \sqrt{x - a}\,dx = \tfrac{2}{3}\xi^{3/2}$$

$$(8\text{–}106)$$

and

$$\sqrt{p} = [2m(\mathscr{E} - U)]^{1/4} = \hbar^{1/2}\kappa^{1/6}\xi^{1/4} \qquad (8\text{–}107)$$

when

$$x > a \qquad \text{or} \qquad \xi > 0$$

while

$$\int_x^a p' \frac{dx}{\hbar} = \tfrac{2}{3}(-\xi)^{3/2} \qquad \sqrt{p'} = \hbar^{1/2} \kappa^{1/6}(-\xi)^{1/4} \qquad (8\text{-}108)$$

when

$$x < a \qquad \text{or} \qquad \xi < 0$$

Therefore, the asymptotic forms (8–104) and (8–105) are exactly the form of the approximate solutions, (8–97) and (8–95), respectively.

Another solution of (8–100), which diverges as $\xi \to -\infty$, has the asymptotic expressions

$$\psi'_\kappa(x) \xrightarrow[\xi \to -\infty]{} \tfrac{1}{2} C'(-\xi)^{-1/4} \exp\left[\tfrac{2}{3}(-\xi)^{3/2}\right] \qquad (8\text{-}109)$$

$$\psi'_\kappa(x) \xrightarrow[\xi \to \infty]{} C'\xi^{-1/4} \cos\left(\tfrac{2}{3}\xi^{3/2} + \tfrac{1}{4}\pi\right) \qquad (8\text{-}110)$$

where C' is a constant.

Consequently, we see that solutions (8–97) and (8–99) should be combined using solutions around the classical turning point to give

$$\psi_{\text{WKB}}(x) = \frac{1}{\sqrt{p'}} \left[C \exp\left(\int_x^a p' \frac{dx}{\hbar} \right) + D \exp\left(-\int_x^a p' \frac{dx}{\hbar} \right) \right] \qquad (8\text{-}111)$$

when

$$x < a \qquad \text{where} \qquad U > \mathscr{E}$$

and

$$\psi_{\text{WKB}}(x) = \frac{2}{\sqrt{p}} \left[C \cos\left(\int_a^x p \frac{dx}{\hbar} + \tfrac{1}{4}\pi \right) + D \sin\left(\int_a^x p \frac{dx}{\hbar} + \tfrac{1}{4}\pi \right) \right]$$

$$(8\text{-}112)$$

when

$$x > a \qquad \text{where} \qquad U < \mathscr{E}$$

These approximate solutions, which are illustrated in Fig. 8–4, are called the WKB solutions, or the WKB approximations.[10]

[10] WKB stands for the authors G. Wentzel, H. A. Kramers, and L. Brillouin, who developed this method.

FIGURE 8–4. Combination of the WKB solutions by means of the Airly integral.

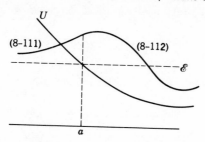

In the same way, when $U > \mathcal{E}$ for $x > a$ and $U < \mathcal{E}$ for $x < a$, we see that (Fig. 8–5)

$$\psi_{\text{WKB}}(x) = \frac{1}{\sqrt{p'}}\left[E \exp\left(\int_a^x p' \frac{dx}{\hbar} \right) + F \exp\left(- \int_a^x p' \frac{dx}{\hbar} \right) \right] \quad (8\text{–}113)$$

when

$$x > a \quad \text{where} \quad U > \mathcal{E}$$

and

$$\psi_{\text{WKB}}(x) = \frac{2}{\sqrt{p}}\left[E \cos\left(\int_x^a p \frac{dx}{\hbar} + \tfrac{1}{4}\pi \right) + F \sin\left(\int_x^a p \frac{dx}{\hbar} + \tfrac{1}{4}\pi \right) \right]$$

$$(8\text{–}114)$$

when

$$x < a \quad \text{where} \quad U < \mathcal{E}$$

In these equations the coefficients C, D, E, and F are determined from specific boundary conditions.

FIGURE 8–5. Combination of the WKB solutions by means of the Airly integral.

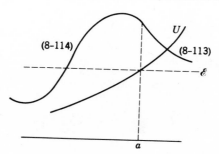

PROBLEM 8–28. When $U < \mathscr{E}$ in the region of $a \le x \le b$ but $U > \mathscr{E}$ everywhere outside the region obtain the *Bohr quantization condition*

$$\int_a^b p \, dx = \pi\hbar(n + \tfrac{1}{2})$$

using the WKB solution. n is a positive integer number (Fig. 8–6).

FIGURE 8–6. A potential dip and the classical turning points a and b.

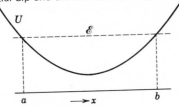

PROBLEM 8–29. Assuming that

$$U = \tfrac{1}{2}m\omega^2 x^2$$

(a) show that the Bohr quantization condition (Problem 8–28) restricts energy to be

$$\mathscr{E}_n = \hbar\omega(n + \tfrac{1}{2})$$

(b) Obtain the WKB solution for this potential.

PROBLEM 8–30. In the momentum representation, where the wave function is assumed to be a function of p, show that the one-dimensional Schroedinger equation for the potential of (8–99) with $a = 0$ is

$$\left(\frac{1}{2m} p^2 + i\kappa'\hbar \frac{\partial}{\partial p}\right) \phi = 0$$

Obtain the eigenfunction $\phi(p)$ for this problem.

HINT. See Problem 8–14.

8–8 DE BROGLIE WAVE

The Hamiltonian of a free particle is

$$\hat{H} = \frac{1}{2m} \hat{p}^2 = -\frac{\hbar^2}{2m} \nabla^2 \qquad (8\text{–}115)$$

Since this operator commutes with the $\hat{\mathbf{p}}$-operator, and since the eigenfunction of the $\hat{\mathbf{p}}$-operator, $e^{i\mathbf{k}\cdot\mathbf{r}}$, are nondegenerate, these eigenfunctions of $\hat{\mathbf{p}}$ are also the eigenfunctions of $\hat{H}$. Note, however, that the converse is not true, because the eigenfunctions of $\hat{H}$ have degeneracies. For example, $\cos(\mathbf{k}\cdot\mathbf{r})$ is an eigenfunction of $\hat{H}$, but is not an eigenfunction of $\hat{\mathbf{p}}$.

A solution of the Schroedinger equation for a free particle is

$$\Psi_k(x, t) = \sqrt{\rho}\, e^{ikx - i\omega t} \tag{8-116}$$

where ρ is the probability density $|\Psi|^2$, and

$$\omega = \frac{\hbar k^2}{2m} = \frac{\mathscr{E}_k}{\hbar} \tag{8-117}$$

where $\mathscr{E}_k$ is the eigenvalue of $\hat{H}$, or the energy of the state given by the wave function (8–116).

The solution (8–116) is a monochromatic wave with wavelength

$$\lambda = \frac{2\pi}{k} = \frac{2\pi\hbar}{p_k} \tag{8-118}$$

where p_k is the eigenvalue of $\hat{p}_x$ for this state. This relation between the wavelength and the momentum is called the de Broglie relation. It was proposed by de Broglie in 1923, and was confirmed by experiment, before quantum mechanics was developed.

From (8–117) and (8–118) we see that the phase velocity of the wave (de Broglie wave) is

$$\frac{\omega}{k} = \frac{\mathscr{E}_k}{p_k} = \frac{p_k}{2m} = \tfrac{1}{2}V \tag{8-119}$$

or one-half of the particle velocity V. However, the group velocity V_g is

$$V_g = \frac{d\omega}{dk} = V \tag{8-120}$$

i.e., is exactly equal to the particle velocity. The phase velocity is not directly related to any observable quantity.

PROBLEM 8–31. Show that (8–116) gives the probability current density

$$j_x = \rho V$$

expected from the particle picture.

PROBLEM 8–32. A wave packet is given by

$$\Psi(x, t) = \frac{1}{\sqrt{2\pi}} \int_{-\infty}^{\infty} \exp\left[-\tfrac{1}{2}a(k - k_0)^2 + i(kx - \omega t)\right] dk$$

Show that when ω is a function of k as

$$\omega = \omega_0 + V_g(k - k_0) + \gamma(k - k_0)^2 + \cdots$$

where V_g is the group velocity, i.e., $V_g = d\omega/dk|_{\omega = \omega_0}$, then

$$|\Psi(x, t)|^2 = \frac{1}{2\sqrt{\tfrac{1}{4}a^2 + (\gamma t)^2}} \exp\left[-\frac{(x - V_g t)^2}{a + ((2\gamma t)^2/a)}\right]$$

This result means that the center of this wave packet moves with velocity V_g. The wave packet, being localized and having the velocity of the corresponding particle as given by (8–120), can be regarded as the best quantum mechanical picture for a classical particle. Note, however, that the spatial dimension of a wave packet increases with time (Fig. 8–7).

FIGURE 8–7. The time development of a wave packet.

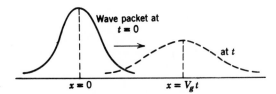

Reflection and Transmission of de Broglie Wave

Let us take a one-dimensional potential (Fig. 8–8) such that

$$\begin{aligned} U &= 0 && \text{if} && x < 0 \\ U &= U_0 && \text{if} && x > 0 \end{aligned} \tag{8–121}$$

If $\mathscr{E} > U_0$, we expect an eigenfunction of $\hat{H}$ in the form of

$$\begin{aligned} \Psi &= \sqrt{\rho}\,(e^{ikx} + Ae^{-ikx})e^{-i\omega t} && \text{if} && x < 0 \\ \Psi &= \sqrt{\rho}\,Be^{ik'x - i\omega t} && \text{if} && x > 0 \end{aligned} \tag{8–122}$$

where

$$k\hbar = \sqrt{2m\mathscr{E}_k} \qquad k'\hbar = \sqrt{2m(\mathscr{E}_k - U_0)} \tag{8–123}$$

FIGURE 8–8. The one-dimensional step potential of (8–121).

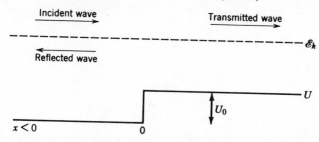

The wave function (8–122) represents a situation where a de Broglie wave comes in from the left with the incident current density ρV. $|A|^2 \rho V$ is reflected back, while $|B|^2 \rho V'$ is transmitted into the $x > 0$ region. Now Ψ and $\partial \Psi / \partial x$ should both be continuous at $x = 0$, since $\partial^2 \Psi / \partial x^2$ is finite. Thus

$$1 + A = B \qquad k(1 - A) = k'B \qquad (8–124)$$

which give

$$A = \frac{k - k'}{k + k'} \qquad B = \frac{2k}{k + k'} \qquad (8–125)$$

In classical mechanics, a particle of $\mathscr{E} > U_0$ is never reflected back, but our result (8–125) shows that in quantum mechanics, there is a nonzero probability that the particle will be reflected back even when it has sufficient energy to escape.

Another effect characteristic of quantum mechanics is the tunnel effect. Take a potential (Fig. 8–9)

$$
\begin{aligned}
U &= 0 && \text{if} && x < 0 \\
U &= U_0 && \text{if} && 0 < x < a \\
U &= 0 && \text{if} && a < x
\end{aligned}
\qquad (8–126)
$$

FIGURE 8–9. The square potential barrier of (8–126).

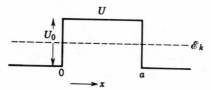

When $\mathcal{E}_k < U_0$, the wave function in the region $0 < x < a$ is

$$\Psi = \sqrt{\rho}\,(Ce^{-\kappa x} + De^{\kappa x})e^{-i\omega t} \qquad 0 < x < a \qquad (8\text{–}127)$$

where

$$\kappa = \frac{\sqrt{2m(U_0 - \mathcal{E}_k)}}{\hbar} \qquad (8\text{–}128)$$

Since this wave function is real, there is no current in the center region; but the probability of finding a particle is not zero in this region, contrary to the classical situation. Assuming the same wave function (8–122) for the other regions, and taking the same boundary conditions as before at $x = 0$ and $x = a$, we obtain the transmission coefficient for $\mathcal{E}_k < U_0$ as

$$|B|^2 = \frac{4(k\kappa)^2}{4(k\kappa)^2 + (k^2 + \kappa^2)^2 \sinh^2(a\kappa)} \qquad (8\text{–}129)$$

When, on the other hand, $\mathcal{E}_k > U_0$, the transmission coefficient is

$$|B|^2 = \frac{4(kk')^2}{4(kk')^2 + (k^2 - k'^2)^2 \sin^2(ak')} \qquad (8\text{–}130)$$

which shows that not all particles with enough energy penetrate (Fig. 8–10).

FIGURE 8–10.   The transmission coefficient for the square potential barrier.

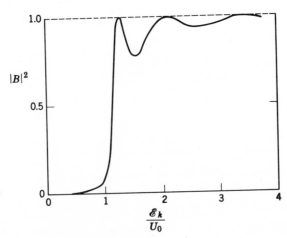

One can take the wave packet as given in Problem 8–34 and discuss its scattering and transmission in the same way as discussed above. A motion picture was made[11] to demonstrate the result of a computer calculation of this problem (Fig. 8–11).

FIGURE 8–11.   Scattering and transmission of a wave packet. Number shown on each picture is the number of that frame in the motion picture.[11]

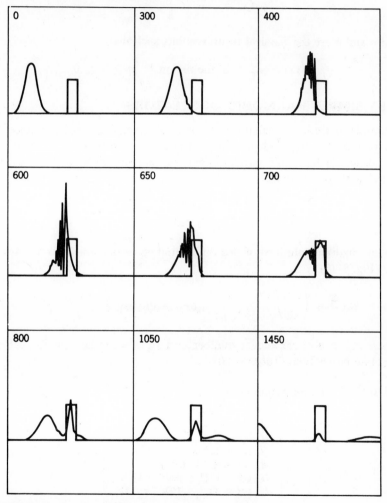

PROBLEM 8–33.   Obtain (8–129) and (8–130).

[11] A. Goldberg, H. M. Schey, and L. Schwartz, *Am. J. Phys.* **35**, 177 (1967).

PROBLEM 8-34. A potential barrier $U(x)$ is given. Using the WKB approximation show that the transmission coefficient is given by

$$\frac{\exp\left(2\int_a^b \kappa(x)\,dx\right)}{\left[\exp\left(2\int_a^b \kappa(x)\,dx\right) + \tfrac{1}{4}\right]^2}$$

where

$$\kappa(x) = \frac{\sqrt{2m(U(x) - \mathscr{E})}}{\hbar}$$

and a and b are the classical turning points such that

$$U(x) > \mathscr{E} \qquad \text{in the region} \qquad a \geq x \geq b$$

8-9 SIMPLE HARMONIC OSCILLATORS

Simple harmonic oscillators in classical mechanics were discussed in Chapter 2. The time-independent Schroedinger equation for a one-dimensional simple harmonic oscillator can be immediately obtained from the classical Hamiltonian as

$$\mathscr{E}\psi_{\mathscr{E}}(x) = \left(-\frac{\hbar^2}{2m}\frac{\partial^2}{\partial x^2} + \tfrac{1}{2}kx^2\right)\psi_{\mathscr{E}}(x) \qquad (8\text{-}131)$$

The normalized solutions of this differential equation, which do not diverge are (Fig. 8-12)

$$\psi_n(x) = \left(\frac{m\omega}{\pi\hbar}\right)^{1/4} \frac{1}{2^{n/2}\sqrt{n!}} H_n(x\sqrt{m\omega/\hbar}) \exp\left(\frac{-\tfrac{1}{2}m\omega x^2}{\hbar}\right) \qquad (8\text{-}132)$$

where n is a positive integer number and H_n is a special function called a Hermite polynomial (Table 8-1):

TABLE 8-1 *Hermite Polynomials*

$$
\begin{aligned}
H_0\,(\xi) &= 1 \\
H_1\,(\xi) &= 2\xi \\
H_2\,(\xi) &= 4\xi^2 - 2 \\
H_3\,(\xi) &= 8\xi^3 - 12\xi \\
H_4\,(\xi) &= 16\xi^4 - 48\xi^2 + 12 \\
H_5\,(\xi) &= 32\xi^5 - 160\xi^3 + 120\xi
\end{aligned}
$$

$$H_n(\xi) = (-1)^n \exp(\xi^2) \frac{d^n}{d\xi^n} \exp(-\xi^2) \qquad (8\text{-}133)$$

The corresponding eigenvalue is (Fig. 8–12)

$$\mathscr{E}_n = \hbar\sqrt{k/m}\,(n + \tfrac{1}{2}) = \hbar\omega(n + \tfrac{1}{2}) \qquad (8\text{–}134)$$

FIGURE 8–12. A simple harmonic potential and its eigenfunctions and eigenvalues.

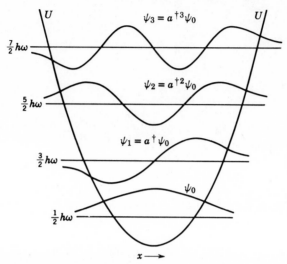

Creation Annihilation Operators

It is instructive to attack this same problem from an operator formalism as follows. The Hamiltonian (8–131) can be rewritten as

$$\hat{H} = \left(\frac{\hat{p}^2}{2m} + \tfrac{1}{4}k\hat{x}^2\right) = \frac{1}{2}\left[\left(\sqrt{\tfrac{1}{2}k}\,\hat{x} + i\,\frac{\hat{p}}{\sqrt{2m}}\right)\left(\sqrt{\tfrac{1}{2}k}\,\hat{x} - i\,\frac{\hat{p}}{\sqrt{2m}}\right)\right.$$

$$\left. + \left(\sqrt{\tfrac{1}{2}k}\,\hat{x} - i\,\frac{\hat{p}}{\sqrt{2m}}\right)\left(\sqrt{\tfrac{1}{2}k}\,\hat{x} + i\,\frac{\hat{p}}{\sqrt{2m}}\right)\right]$$

$$= \tfrac{1}{2}\hbar\omega(\hat{a}\hat{a}^\dagger + \hat{a}^\dagger\hat{a}) \qquad (8\text{–}135)$$

where

$$\hat{a} \equiv \frac{1}{\sqrt{2\hbar}}\left[(mk)^{1/4}\hat{x} + \frac{i\hat{p}}{(mk)^{1/4}}\right] \qquad (8\text{–}136)$$

$$\hat{a}^\dagger \equiv \frac{1}{\sqrt{2\hbar}}\left[(mk)^{1/4}\hat{x} - \frac{i\hat{p}}{(mk)^{1/4}}\right] \qquad (8\text{–}137)$$

and

$$\omega = \sqrt{k/m}$$

Since $\hat{x}$ and $\hat{p}$ are both Hermitian, $\hat{a}$ and $\hat{a}^\dagger$ are Hermitian conjugate to each other.

From (8–136), (8–137), and the basic commutator relation (8–50), we obtain

$$[\hat{a}, \hat{a}^\dagger] = 1 \qquad (8\text{–}138)$$

Therefore

$$[\hat{a}, \hat{H}] = \hbar\omega\hat{a} \qquad (8\text{–}139)$$

and

$$[\hat{a}^\dagger, \hat{H}] = -\hbar\omega\hat{a}^\dagger \qquad (8\text{–}140)$$

Let us assume that ψ_n is an eigenfunction of $\hat{H}$ with an eigenvalue $\mathscr{E}_n$, and that we do not know yet what ψ_n and $\mathscr{E}_n$ are:

$$\hat{H}\psi_n = \mathscr{E}_n\psi_n \qquad (8\text{–}141)$$

Operating $\hat{a}$ from the left on both sides of this equation, and using (8–139), we obtain

$$\mathscr{E}_n\hat{a}\psi_n = \hat{a}\hat{H}\psi_n = (\hat{H}\hat{a} + \hbar\omega\hat{a})\psi_n = (\hat{H} + \hbar\omega)\hat{a}\psi_n \qquad (8\text{–}142)$$

or

$$\hat{H}(\hat{a}\psi_n) = (\mathscr{E}_n - \hbar\omega)(\hat{a}\psi_n) \qquad (8\text{–}143)$$

which means that if ψ_n is an eigenfunction of $\hat{H}$, then $\hat{a}\psi_n$ is another eigenfunction of $\hat{H}$, and that the corresponding eigenvalues differ by $\hbar\omega$. In the same way we also obtain

$$\hat{H}(\hat{a}^\dagger\psi_n) = (\mathscr{E}_n + \hbar\omega)(\hat{a}^\dagger\psi_n) \qquad (8\text{–}144)$$

which shows that $\hat{a}^\dagger\psi_n$ is also an eigenfunction of the Hamiltonian $\hat{H}$.

If the lowest energy state (ground state) has an eigenfunction ψ_0 and an eigenvalue $\mathscr{E}_0$, then (8–143) implies that

$$\hat{a}\psi_0 = 0 \qquad (8\text{–}145)$$

in order to guarantee that there are no other states with lower energy than $\mathscr{E}_0$. In addition (8–144) shows that all other (excited) states are obtained by operating $\hat{a}^\dagger$ a given number of times on ψ_0;

$$\hat{H}(\hat{a}^{\dagger n}\psi_0) = (\mathscr{E}_0 + n\hbar\omega)(\hat{a}^{\dagger n}\psi_0) \qquad (8\text{–}146)$$

To obtain $\mathscr{E}_0$, we express $\hat{H}$ of (8–135) as

$$\hat{H} = \hbar\omega(\hat{a}^\dagger\hat{a} + \tfrac{1}{2}) \qquad (8\text{–}147)$$

using (8–138). Since the ground state is defined by (8–145) we see that

$$\hat{H}\psi_0 = \hbar\omega(\hat{a}^\dagger\hat{a} + \tfrac{1}{2})\psi_0 = \tfrac{1}{2}\hbar\omega\psi_0 \qquad (8\text{–}148)$$

or

$$\mathscr{E}_0 = \tfrac{1}{2}\hbar\omega$$

Thus

$$\mathscr{E}_n = \mathscr{E}_0 + n\hbar\omega = \hbar\omega(n + \tfrac{1}{2}) \qquad (8\text{–}149)$$

The eigenfunctions can be obtained from (8–145). Since $\hat{a}$ is defined by (8–136) we see that

$$\left(\sqrt{mk}\, x + \hbar\frac{\partial}{\partial x}\right)\psi_0 = 0 \qquad (8\text{–}150)$$

is the explicit form of (8–145) in the coordinate representation. Thus

$$\psi_0 = A \exp\left(\frac{-\sqrt{mk}\, x^2}{(2\hbar)}\right) \qquad (8\text{–}151)$$

where $A = (m\omega/\pi\hbar)^{1/4}$, as given in (8–132). Other eigenfunctions are obtained, using (8–137), as

$$\psi_n = B_n\,\hat{a}^{\dagger n}\psi_0 = B'_n\left(\sqrt{mk}\, x - \hbar\frac{\partial}{\partial x}\right)^n \exp\left(\frac{-\sqrt{mk}\, x^2}{(2\hbar)}\right) \qquad (8\text{–}152)$$

where B_n and B'_n are normalization constants.

All these results are obtained from (8–135) and (8–138) only. Operators $\hat{a}$ and $\hat{a}^\dagger$ which satisfy (8–138) are called *step-down* and *step-up* operators, respectively, for the obvious reason, but they are more often called the *annihilation* and *creation* operators, respectively.

Coupled Oscillators

When many particles are coupled to each other by quadratic potentials, one can find the normal coordinates X_i, given by (2–74) and (2–75), such that

$$L = T - U = \frac{1}{2}\sum_i m_i\dot{x}_i^2 - \frac{1}{2}\sum_i k_i x_i^2 + \frac{1}{2}\sum_{i \neq j} k_{ij}x_i x_j$$

$$= \sum_i (\tfrac{1}{2}\dot{X}_i^2 - \tfrac{1}{2}\Omega_i^2 X_i^2) \qquad (8\text{–}153)$$

Therefore

$$P_i = \frac{\partial L}{\partial \dot{X}_i} = \dot{X}_i \qquad (8\text{-}154)$$

and

$$H = \sum_i (\tfrac{1}{2}P_i^2 + \tfrac{1}{2}\Omega_i^2 X_i^2) \qquad (8\text{-}155)$$

In quantum mechanics they are

$$\hat{H} = \sum_i \hat{H}_i = \sum_i (\tfrac{1}{2}\hat{P}_i^2 + \tfrac{1}{2}\Omega_i^2 \hat{X}_i) \qquad (8\text{-}156)$$

and

$$[\hat{P}_i, \hat{X}_j] = -i\hbar\, \delta_{ij} \qquad (8\text{-}157)$$

If

$$\hat{H}_i \psi_{n_i}(X_i) = \mathscr{E}_{n_i} \psi_{n_i}(X_i) \qquad (8\text{-}158)$$

which we just solved, then the eigenfunction of the entire system is obtained as a simple product of all such individual eigenfunctions $\psi_{n_i}(X_i)$; because

$$\hat{H}\psi_{n_1}(X_1) \cdots \psi_{n_i}(X_i) \cdots = \sum_i \hat{H}_i \psi_{n_1}(X_1) \cdots \psi_{n_i}(X_i) \cdots$$

$$= (\sum \mathscr{E}_{n_i})\psi_{n_1}(X_1) \cdots \psi_{n_i}(X_i) \cdots \qquad (8\text{-}159)$$

The energy of this product state is the sum of individual energies.

Since the probability density of this product state is obviously given by the product of individual probability density $|\psi_{n_i}(X_i)|^2$, we see there is no correlation between the motion of different normal coordinates, which is consistent with the classical notion.

PROBLEM 8–35. Obtain (8–139) and (8–140) from (8–135) and (8–138).

PROBLEM 8–36. Show that

$$e^{\xi\hat{a}}\hat{a}^\dagger e^{-\xi\hat{a}} = \hat{a}^\dagger + \xi \qquad e^{\xi\hat{a}^\dagger}\hat{a}e^{-\xi\hat{a}^\dagger} = \hat{a} - \xi$$

where ξ is a constant.

PROBLEM 8–37. Using the result of Problem 8–36, show that

$$e^{\xi\hat{a}}f(\hat{a}, \hat{a}^\dagger)e^{-\xi\hat{a}} = f(\hat{a}, \hat{a}^\dagger + \xi)$$

where $f(\hat{a}, \hat{a}^\dagger)$ is a function given as a power series of $\hat{a}$ and $\hat{a}^\dagger$.

PROBLEM 8–38. Show that

$$e^{\xi \hat{a}^\dagger \hat{a}} \hat{a} e^{-\xi \hat{a}^\dagger \hat{a}} = \hat{a} e^{-\xi}$$

and

$$e^{\xi \hat{a}^\dagger \hat{a}} \hat{a}^\dagger e^{-\xi \hat{a}^\dagger \hat{a}} = \hat{a}^\dagger e^{\xi}$$

where ξ is a constant.

PROBLEM 8–39. Show that

$$\hat{a}^\dagger \hat{a} \psi_n = n \psi_n$$

and

$$\hat{a} \hat{a}^\dagger \psi_n = (n + 1) \psi_n$$

PROBLEM 8–40. Show that if

$$
\mathbf{a} = \begin{pmatrix} 0 & \sqrt{1} & 0 & 0 & \cdots \\ 0 & 0 & \sqrt{2} & 0 & \cdots \\ 0 & 0 & 0 & \sqrt{3} & \cdots \\ \cdots\cdots\cdots\cdots\cdots\cdots \end{pmatrix} \qquad \mathbf{a}^\dagger = \begin{pmatrix} 0 & 0 & 0 & 0 & \cdots \\ \sqrt{1} & 0 & 0 & 0 & \cdots \\ 0 & \sqrt{2} & 0 & 0 & \cdots \\ 0 & 0 & \sqrt{3} & 0 & \cdots \\ \cdots\cdots\cdots\cdots\cdots\cdots \end{pmatrix}
$$

then

$$\mathbf{a}\mathbf{a}^\dagger - \mathbf{a}^\dagger \mathbf{a} = \mathbf{1}$$

which is the same relation as (8–138). The matrix $\mathbf{a}$ is an infinite matrix.

PROBLEM 8–41. (a) Show that if

$$\int |\psi_n(x)|^2 \, dx = 1$$

then

$$\int |\hat{a} \psi_n(x)|^2 \, dx = n$$

and

$$\int |\hat{a}^\dagger \psi_n(x)|^2 \, dx = n + 1$$

(b) Show that

$$\int \psi_{n-1} x \psi_n \, dx = \sqrt{n\hbar/2m\omega}$$

and

$$\int \psi_{n+1} x \psi_n \, dx = \sqrt{[(n+1)\hbar]/2m\omega}$$

HINT. Use the result of (a).

PROBLEM 8–42. Show that Bohr's quantization condition (Problem 8–28), gives an exact result (8–149) in the case of a simple harmonic oscillator.

8–10 TWO THEOREMS CONCERNING EIGENFUNCTIONS OF $\hat{H}$ IN ONE-DIMENSIONAL PROBLEMS

In one-dimensional problems $\hat{H}$ is

$$\hat{H} = -\frac{\hbar^2}{2m}\frac{\partial^2}{\partial x^2} + U(x) \tag{8–160}$$

and $U(x)$ is a real function of x. Let us assume that ψ is an eigenfunction of $\hat{H}$, then:

THEOREM 1. If $\psi = \partial\psi/\partial x = 0$ at some point x, there is no degeneracy in this state.

PROOF: Let ϕ be another eigenfunction of $\hat{H}$ with the same eigenvalue $\mathscr{E}$; then

$$\phi\hat{H}\psi - \psi\hat{H}\phi = -\frac{\hbar^2}{2m}\left(\phi\frac{\partial^2\psi}{\partial x^2} - \psi\frac{\partial^2\phi}{\partial x^2}\right) = -\frac{\hbar^2}{2m}\frac{\partial}{\partial x}\left(\phi\frac{\partial\psi}{\partial x} - \psi\frac{\partial\phi}{\partial x}\right) \tag{8–161}$$

but

$$\phi\hat{H}\psi - \psi\hat{H}\phi = \mathscr{E}\phi\psi - \mathscr{E}\psi\phi = 0 \tag{8–162}$$

Therefore $\phi\,\partial\psi/\partial x - \psi\,\partial\phi/\partial x$ is constant through the entire region of x, and if ψ and $\partial\psi/\partial x$ are simultaneously zero at one point, then this constant is zero, or

$$\frac{1}{\phi}\frac{\partial\phi}{\partial x} = \frac{1}{\psi}\frac{\partial\psi}{\partial x} \tag{8–163}$$

which, when integrated, gives

$$\phi = \psi \times \text{(constant)} \tag{8-164}$$

Q.E.D.

COROLLARY 1. There is no degeneracy for bound states in one-dimensional problems.

   PROOF: For bound states ψ and $\partial\psi/\partial x$ are both zero at infinity. Q.E.D.

As we noticed in Fig. 8-12, eigenfunctions of $\hat{H}$ usually have nodes, where ψ changes sign.

THEOREM 2. If $\hat{H}\psi_1 = \mathscr{E}_1\psi_1$ and $\hat{H}\psi_2 = \mathscr{E}_2\psi_2$, and if $0 > \mathscr{E}_1 > \mathscr{E}_2$, then ψ_1 has more nodes than ψ_2.

   PROOF: Let ψ_2 be zero at $x = a$ and $x = b$, and be either positive or negative definite in the region between a and b. (See Fig. 8-13).

FIGURE 8-13. Theorem 2.

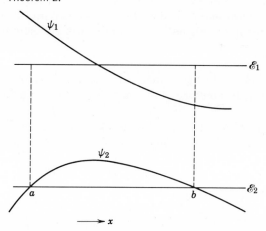

Applying the same calculation of (8-161) to ψ_1 and ψ_2, and then integrating over the region between a and b, we obtain

$$-\frac{\hbar^2}{2m}\psi_1\frac{\partial\psi_2}{\partial x}\bigg|_a^b = -(\mathscr{E}_1 - \mathscr{E}_2)\int_a^b \psi_1\psi_2\, dx \tag{8-165}$$

If ψ_2 is positive definite in this region, then $\partial\psi_2/\partial x$ is positive at a and negative at b. If ψ_1 does not have a node between a and b, then the sign of the right-hand side of (8–165) can never agree with that of the left-hand side, since $\mathscr{E}_1 > \mathscr{E}_2$ is assumed. Thus ψ_1 has to change its sign somewhere in this region, or has to have at least one node. The same conclusion can be obtained when ψ_2 is negative definite in this region.

If both ψ_1 and ψ_2 are bound states, they both should go to zero at $x = +\infty$ and $x = -\infty$; but if ψ_2 does not have a node between b and ∞, the above result shows that ψ_1 should have at least one more node between b and ∞. The same result will be obtained in a region towards $-\infty$. Counting all nodes necessary, we obtain the theorem. Q.E.D.

COROLLARY 2.   For any continuous one-dimensional potential U, the eigenfunction of its lowest bound state (ground state) has no node, the next excited state has one node, and the number of nodes increases by one each time as we go up to higher bound states.

PROOF: This situation is observed in the case of a simple harmonic oscillator. (See Fig. 8–12.) Any given potential can be obtained by continuously changing U starting from that of a simple harmonic oscillator. During that continuous change, both the eigenfunctions and the eigenvalues will be changed continuously; but Theorem 2 still holds throughout such continuous modification. Thus the situation with respect to the number of nodes of a simple harmonic potential problem will also stay the same for all other potentials. Q.E.D.

PROBLEM 8–43.   (a) Show that a bound state eigenfunction Ψ of $\hat{H}$ (assumed to be real) should be of the form of $\psi e^{i\phi}$, where ψ is real and ϕ is a constant.
(b) Show that the current density j is zero for a bound state.
(c) Show that the expectation value of $\hat{p}$ is zero in a bound state.

HINT.   (a) Compare $\hat{H}\Psi = \mathscr{E}\Psi$ with $\hat{H}^*\Psi^* = \mathscr{E}\Psi^*$ and use Corollary 1.
(c) Notice that $\langle\hat{p}\rangle^* = \langle\hat{p}\rangle$ but $\hat{p}^* = -\hat{p}$.

PROBLEM 8–44.   A one-dimensional square well potential of depth U_0 and width l is given (Fig. 8–14). Show that if n is the largest integer such that

$$ n < \frac{l\sqrt{2mU_0}}{\pi\hbar} $$

then there are n bound states.

FIGURE 8–14. The potential well of Problem 8–44.

8–11 MATRIX REPRESENTATIVES

Bra and Ket

We have already used such notations as

$$\langle a|\hat{f}|a\rangle \equiv \int \psi_a^* \hat{f} \Psi_a \, d\tau \qquad (8\text{--}30)$$

and

$$\langle n \mid a\rangle \equiv \int \psi_n^* \Psi_a \, d\tau \qquad (8\text{--}38)$$

as shorthand notations. These notations are very convenient for general discussions. For example, we may want to discuss those integrals in momentum space, where wave functions and operators are given in terms of momenta. Since in the bracket notation we do not have to specify variables we can simply use the same notation and generalize the meaning to include the momentum representation as well as the coordinate representation.

Dirac went further and separated the prefactor $\langle a|$ and the postfactor $|a\rangle$. They are called *bra* and *ket*, respectively[12] (also called bra-vector and ket-vector). Corresponding to (8–15) an *eigenket* of operator $\hat{f}$ is defined by

$$\hat{f}|n\rangle = f_n|n\rangle \qquad (8\text{--}166)$$

When $\hat{f}$ is Hermitian we can define the corresponding *eigenbra* by

$$\langle n|\hat{f}^\dagger = f_n\langle n| \qquad (8\text{--}167)$$

We can represent a state by ket $|a\rangle$ or by bra $\langle a|$ instead of the wave function Ψ_a. Such bra or ket is often called state vector or just state.

When a set of kets $\ldots, |n\rangle, |m\rangle, \ldots$ is complete, or spans a Hilbert space, we can express an arbitrary ket $|a\rangle$ as

$$|a\rangle = \sum_n a_n|n\rangle \qquad (8\text{--}168)$$

[12] P. A. M. Dirac, *Quantum Mechanics*, Oxford University Press, London, 1958.

Then the corresponding expansion of $\langle a|$ is

$$\langle a| = \sum_n a_n^* \langle n| \tag{8-169}$$

The Hilbert space of kets is separated from that of bras, because additions and multiplications by scalars of kets give kets, and those of bras give bras only. But these spaces are *dual* to each other because there exists a one-to-one correspondence given by (8-168) and (8-169), and particularly because the inner product is defined by bra *and* ket:

$$\langle \phi \mid \chi \rangle = \int \phi^* \chi \, d\tau = (\phi, \chi) \tag{8-170}$$

It is convenient to take an orthonormal set of bra and ket such that

$$\langle n \mid m \rangle = \delta_{n,m} \tag{8-171}$$

Then the expansion coefficients in (8-168) and (8-169) are given by

$$a_n = \langle n \mid a \rangle \quad \text{and} \quad a_n^* = \langle a \mid n \rangle \tag{8-172}$$

which are consistent with our previous definition (8-38).

When bra and ket are placed together in the opposite order, for example $|n\rangle\langle m|$, we have an operator. Out of an orthonormalized set of kets the operator $|n\rangle\langle m|$ will select $|m\rangle$ and convert that into $|n\rangle$.

Matrix Representatives

When $\ldots, |n\rangle, |m\rangle, \ldots$ form a complete orthonormal set, we have

$$\hat{g}|n\rangle = \sum_m |m\rangle\langle m|\hat{g}|n\rangle \tag{8-173}$$

for an operator $\hat{g}$. The expansion coefficient is given by

$$\langle m|\hat{g}|n\rangle = (\psi_m, \hat{g}\psi_n) = (\hat{g}^\dagger \psi_m, \psi_n) \tag{8-174}$$

If we arrange all kets in a row matrix

$$(|n\rangle) \equiv (|1\rangle \; |2\rangle \; |3\rangle \cdots) \tag{8-175}$$

then we can express all equations ($n = 1, 2, \ldots$) of (8–173) together in the matrix form

$$\hat{g}(|n\rangle) = (|m\rangle)(\langle m|\hat{g}|n\rangle) \qquad (8\text{–}176)$$

where

$$(\langle m|\hat{g}|n\rangle) = \begin{pmatrix} \langle 1|\hat{g}|1\rangle & \langle 1|\hat{g}|2\rangle & \langle 1|\hat{g}|3\rangle \cdots \\ \langle 2|\hat{g}|1\rangle & \langle 2|\hat{g}|2\rangle & \langle 2|\hat{g}|3\rangle \cdots \\ \langle 3|\hat{g}|1\rangle & \langle 3|\hat{g}|2\rangle & \langle 3|\hat{g}|3\rangle \cdots \\ \cdots\cdots\cdots\cdots\cdots\cdots\cdots\cdots \end{pmatrix} \qquad (8\text{–}177)$$

In this sense $\langle m|\hat{g}|n\rangle$ is called a matrix element of $\hat{g}$.

If $\hat{g}$ is a Hermitian operator, then

$$\langle m|\hat{g}|n\rangle = \langle m|\hat{g}^\dagger|n\rangle = \langle n|\hat{g}|m\rangle^* \qquad (8\text{–}178)$$

or the corresponding matrix is a Hermitian matrix. [See (8–174).]

Taking two operators $\hat{g}$ and $\hat{h}$, we obtain from (8–176) that

$$\hat{h}\hat{g}(|n\rangle) = \hat{h}(|m\rangle)(\langle m|\hat{g}|n\rangle) = (|l\rangle)(\langle l|\hat{h}|m\rangle)(\langle m|\hat{g}|n\rangle) \qquad (8\text{–}179)$$

and

$$\hat{g}\hat{h}(|n\rangle) = (|l\rangle)(\langle l|\hat{g}|m\rangle)(\langle m|\hat{h}|n\rangle) \qquad (8\text{–}180)$$

We call $(\langle m|\hat{g}|n\rangle)$ a matrix representative of $\hat{g}$, because (8–176), (8–179), and (8–180) show that any operator calculations can be done by the corresponding matrix calculations using the matrix representative. (In such calculations, we should remember that matrices do not commute in general.) It should be noted that the explicit expression of the matrix representative depends on the choice of the base vector $(|n\rangle)$.

Regarding $\hat{h}\hat{g}$ as a single operator, (8–176) gives

$$\hat{h}\hat{g}(|n\rangle) = \sum_m (|m\rangle)(\langle m|\hat{h}\hat{g}|n\rangle) \qquad (8\text{–}181)$$

Comparing the *mn*-matrix element of this matrix to that of the product matrices of (8–179), and using the matrix multiplication rule, we obtain

$$\langle m|\hat{h}\hat{g}|n\rangle = \sum_l \langle m|\hat{h}|l\rangle\langle l|\hat{g}|n\rangle \qquad (8\text{–}182)$$

which implies the operator relation

$$\sum_l |l\rangle\langle l| = \hat{1} \qquad (8\text{–}183)$$

This is a generalized closure property. (See Problem 8–8.)

Quite often, the spectrum of eigenvalues of a given operator is continuous, rather than discrete, as we have assumed so far. When it is continuous, we should modify (8–173) to

$$\hat{g}|n\rangle = \int |m\rangle\langle m|\hat{g}|n\rangle \, dm \qquad (8\text{–}173')$$

(8–182) to

$$\langle m|\hat{h}\hat{g}|n\rangle = \int \langle m|\hat{h}|l\rangle\langle l|\hat{g}|n\rangle \, dl \qquad (8\text{–}182')$$

and (8–183) to

$$\int |l\rangle\langle l| \, dl = \hat{1} \qquad (8\text{–}183')$$

PROBLEM 8–45. If we take a column vector $(|n\rangle)^T$, instead of the row vector $(|n\rangle)$, show that

$$\hat{h}\hat{g}(|n\rangle)^T = (\langle l|\hat{g}|m\rangle)^*(\langle m|\hat{h}|n\rangle)^*(|n\rangle)^T$$

assuming that $\hat{h}$ and $\hat{g}$ are both Hermitian.

PROBLEM 8–46. In the case of one particle and the coordinate representation, the right-hand side of (8–182) is

$$\sum_l \int \psi_m^*(\mathbf{r})\hat{h}\psi_l(\mathbf{r}) \, dv \int \psi_l^*(\mathbf{r}')\hat{g}\psi_n(\mathbf{r}') \, dv'$$

Show that the closure property given in Problem 8–8 equates this expression to the left-hand side of (8–182).

PROBLEM 8–47. Show that the generalized closure property (8–183) implies the completeness of $|l\rangle$'s.

8–12 UNITARY TRANSFORMATIONS

Let $\hat{f}$ and $\hat{g}$ be two Hermitian operators, which do not commute with each other, and let $|f_n\rangle$ and $|g_m\rangle$ be their eigenkets, respectively. If each set of eigenkets form a complete set, we obtain

$$|f_n\rangle = \sum_m |g_m\rangle\langle g_m | f_n\rangle \qquad (8\text{–}184)$$

and

$$|g_m\rangle = \sum_n |f_n\rangle\langle f_n | g_m\rangle \qquad (8\text{–}185)$$

In the same way we obtain

$$\langle f_n| = \sum_m \langle f_n | g_m\rangle\langle g_m| \qquad (8\text{–}184')$$

and

$$\langle g_m| = \sum_n \langle g_m | f_n\rangle\langle f_n| \qquad (8\text{–}185')$$

Note that

$$\langle g_m | f_n\rangle = \langle f_n | g_m\rangle^* \qquad (8\text{–}186)$$

It is convenient to combine equations (8–184) for all n's together in matrix form as

$$(|f_n\rangle) = (|g_m\rangle)(\langle g_m | f_n\rangle) \qquad (8\text{–}187)$$

where $(|f_n\rangle)$ and $(|g_m\rangle)$ are row matrices, such as (8–175), while

$$(\langle g_m | f_n\rangle) = \begin{pmatrix} \langle g_1 | f_1\rangle & \langle g_1 | f_2\rangle & \langle g_1 | f_3\rangle \cdots \\ \langle g_2 | f_1\rangle & \langle g_2 | f_2\rangle & \cdots \\ \langle g_3 | f_1\rangle & \langle g_3 | f_2\rangle & \cdots \\ \cdots\cdots\cdots\cdots\cdots\cdots\cdots\cdots\cdots\cdots \end{pmatrix} \qquad (8\text{–}188)$$

Then, from (8–185) and (8–186) we obtain

$$(|g_m\rangle) = (|f_n\rangle)(\langle f_n | g_m\rangle) = (|f_n\rangle)(\langle g_m | f_n\rangle)^\dagger \qquad (8\text{–}189)$$

Inserting (8–189) into (8–187), we obtain

$$(|f_n\rangle) = (|f_n\rangle)(\langle g_m | f_n\rangle)^\dagger(\langle g_m | f_n\rangle) \qquad (8\text{–}190)$$

which shows that

$$(\langle g_m \,|\, f_n \rangle)^\dagger (\langle g_m \,|\, f_n \rangle) = \mathbf{1} \qquad (8\text{-}191)$$

where $\mathbf{1}$ is the unit matrix. A matrix which satisfies the relation (8-191) is called a unitary matrix [Appendix I(b)]. Thus the transformation matrix between two orthonormal complete sets of eigenkets, (8-188), is a unitary matrix. Taking the corresponding determinant relation, we see that (8-191) gives

$$|\det (\langle g_m \,|\, f_n \rangle)|^2 = 1 \qquad (8\text{-}192)$$

Comparing this result to (2-87), we see that a unitary matrix is a generalization of an orthogonal matrix: For an orthogonal matrix, each matrix element is real, while in a unitary matrix, it may be complex. This generalization is necessary simply because the eigenfunctions are in general complex.

From (8-185) and (8-185'), we obtain

$$\begin{aligned}
\langle g_{m'} |\hat{f}| g_m \rangle &= \sum_{n',n} \langle g_{m'} \,|\, f_{n'} \rangle \langle f_{n'} |\hat{f}| f_n \rangle \langle f_n \,|\, g_m \rangle \\
&= \sum_{n',n} \langle g_{m'} \,|\, f_{n'} \rangle f_n \, \delta_{n'n} \langle f_n \,|\, g_m \rangle
\end{aligned} \qquad (8\text{-}193)$$

In matrix notation (8-193) becomes

$$(\langle g_{m'} |\hat{f}| g_m \rangle) = (\langle g_{m'} \,|\, f_{n'} \rangle)(f_n \, \delta_{n'n})(\langle f_n \,|\, g_m \rangle) \qquad (8\text{-}194)$$

where $(f_n \, \delta_{n'n})$ is a diagonal matrix

$$(f_n \, \delta_{n'n}) = \begin{pmatrix} f_1 & 0 & 0 & \cdots \\ 0 & f_2 & 0 & \cdots \\ 0 & 0 & f_3 & \cdots \\ \multicolumn{4}{c}{\cdots\cdots\cdots\cdots\cdots} \end{pmatrix} \qquad (8\text{-}195)$$

and the diagonal elements are the eigenvalues of $\hat{f}$. In the same way that (2-94) was obtained from (2-88), we see that (8-194) shows that the eigenvalues $f_1, f_2, \ldots$ are obtained as solutions of the secular equation

$$\begin{vmatrix} \langle g_1 |\hat{f}| g_1 \rangle - \lambda & \langle g_1 |\hat{f}| g_2 \rangle & \langle g_1 |\hat{f}| g_3 \rangle & \cdots \\ \langle g_2 |\hat{f}| g_1 \rangle & \langle g_2 |\hat{f}| g_2 \rangle - \lambda & \langle g_2 |\hat{f}| g_3 \rangle & \cdots \\ \langle g_3 |\hat{f}| g_1 \rangle & \langle g_3 |\hat{f}| g_2 \rangle & \langle g_3 |\hat{f}| g_3 \rangle - \lambda & \cdots \\ \multicolumn{4}{c}{\cdots\cdots\cdots\cdots\cdots\cdots\cdots\cdots\cdots\cdots\cdots} \end{vmatrix} = 0 \qquad (8\text{-}196)$$

This gives a method for finding the eigenvalues and eigenfunctions of $\hat{f}$, when eigenfunctions of $\hat{g}$ are known.

The matrix relation (8–187) must represent an operator relation

$$(|f_n\rangle) = \hat{U}_{fg}(|g_m\rangle) \qquad (8\text{–}197)$$

where $\hat{U}_{fg}$ is an operator whose matrix representative is the transformation matrix $(\langle g_m | f_n \rangle)$. Since the transformation matrix is a unitary matrix, the operator $\hat{U}_{fg}$ is a unitary operator, which satisfies

$$\hat{U}_{fg}{}^\dagger \hat{U}_{fg} = \hat{1} \qquad (8\text{–}198)$$

Thus we have

$$(|g_m\rangle) = \hat{U}_{fg}{}^\dagger(|f_n\rangle) \qquad (8\text{–}199)$$

The matrix elements of a given operator $\hat{h}$ in two different representations are related by

$$\langle f_{n'}|\hat{h}|f_n\rangle = \langle g_{m'}|\hat{U}_{fg}{}^\dagger\hat{h}\hat{U}_{fg}|g_m\rangle \qquad (8\text{–}200)$$

according to (8–197). Thus, for a transformation between different representations, an operator is transformed by a unitary transformation. Unitary transformations, however, do not change the quantum mechanical canonical relations, the commutator relations (8–46). That is, if

$$\hat{f}' = \hat{U}^\dagger\hat{f}\hat{U}, \qquad \hat{g}' = \hat{U}^\dagger\hat{g}\hat{U}, \qquad \text{and} \qquad \hat{k}' = \hat{U}^\dagger\hat{k}\hat{U} \qquad (8\text{–}201)$$

then

$$[\hat{f}', \hat{g}'] = \hat{U}^\dagger[\hat{f}, \hat{g}]\hat{U} = i\hbar\hat{U}^\dagger\hat{k}\hat{U} = i\hbar\hat{k}' \qquad (8\text{–}202)$$

holds corresponding to (8–46). Thus, if $\hat{q}_i$ and $\hat{p}_j$ satisfy the canonical relation (8–50), then $\hat{q}_i'$ and $\hat{p}_j'$ satisfy the same relation, or the transformed operators are canonically conjugate to each other. We therefore conclude that *unitary transformations play the role corresponding to canonical transformations in classical mechanics.*

It should be noted that a wave function in the coordinate representation is given by

$$\langle \mathbf{r}_0 | \Psi \rangle = \Psi(\mathbf{r}_0) \qquad (8\text{–}203)$$

in Dirac's notation. Here $\langle \mathbf{r}_0|$ is the eigenbra of the position operator $\hat{\mathbf{r}}$ with eigenvalue $\mathbf{r}_0$:

$$\langle \mathbf{r}_0|\hat{\mathbf{r}} = \langle \mathbf{r}_0|\mathbf{r}_0 \qquad (8\text{–}204)$$

This is obvious, because in the coordinate representation,

$$\langle \mathbf{r}_0 \mid \Psi \rangle = \int \delta(\mathbf{r} - \mathbf{r}_0) \Psi(\mathbf{r}) \, dv = \Psi(\mathbf{r}_0) \qquad (8\text{–}205)$$

[See (8–40).] In the same way, the corresponding wave function in the momentum representation is expressed as

$$\langle \mathbf{p}_0 \mid \Psi \rangle = (2\pi)^{-3/2} \int \exp \left(\frac{i \mathbf{p}_0 \cdot \mathbf{r}}{\hbar} \right) \Psi(\mathbf{r}) \, dv = \Phi(\mathbf{p}_0) \qquad (8\text{–}206)$$

where

$$\langle \mathbf{p}_0 | \hat{\mathbf{p}} = \langle \mathbf{p}_0 | \mathbf{p}_0 \qquad (8\text{–}207)$$

PROBLEM 8–48. Show that

$$(\langle g_m \mid f_n \rangle)(\langle g_m \mid f_n \rangle)^\dagger = 1 \qquad \text{or} \qquad \hat{U}_{fg} \hat{U}_{fg}{}^\dagger = \hat{1}$$

Note that this relation is different from (8–191).

PROBLEM 8–49. If $\hat{f}$ is a Hermitian operator, show that $e^{i\hat{f}}$ is a unitary operator.

PROBLEM 8–50. If

$$(|f_n\rangle) = \hat{U}_{fg}(|g_m\rangle) \qquad \text{and} \qquad (|g_m\rangle) = \hat{U}_{gh}(|h_l\rangle)$$

show that $\hat{U}_{fg}\hat{U}_{gh}$ is another unitary operator.

PROBLEM 8–51. Taking a matrix element of the generalized closure property (8–183) in terms of $\langle \mathbf{r} |$ and $| \mathbf{r}' \rangle$, show that the result gives the original closure property, given in Problem 8–8.

PROBLEM 8–52. Calculate $\langle \mathbf{r}_0 \mid \Psi \rangle$ using the momentum representation, and show that the result is still $\Psi(\mathbf{r}_0)$.

8–13 REDUCTION OF MATRIX REPRESENTATIVES USING COMMUTING OPERATORS: GROUP THEORY

Suppose that the eigenfunctions of $\hat{f}$ are known, and there is another operator $\hat{g}$ which commutes with $\hat{f}$. Then it was shown in section 8–5 that the matrix representative of $\hat{g}$ given by taking eigenvectors of $\hat{f}$, $|f_n\rangle$'s, as the base, has the form of

$$(\langle f_{n'}|\hat{g}|f_n\rangle) = \begin{pmatrix} (g_1) & 0 & 0 & \cdots \\ 0 & (g_2) & 0 & \cdots \\ 0 & 0 & (g_3) & \cdots \\ \cdots\cdots\cdots\cdots\cdots\cdots\cdots \end{pmatrix} \qquad (8\text{-}208)$$

where (g_n) is a square submatrix given by those eigenfunctions of $\hat{f}$ which have a common eigenvalue f_n (degeneracy). This matrix is not diagonal in general, and in order to obtain the eigenvalues of $\hat{g}$, we still have to find a proper unitary transformation, as given by the secular equation (8-196) to diagonalize each submatrix (g_n). However, since the matrix is reduced into smaller submatrices, the order of the secular equations is smaller than in the original secular equation. Therefore, the problem of finding eigenvalues and eigenfunctions of a given operator is partially solved, or at least made easier, if we find other operators which commute with the given operator and whose eigenfunctions are known.

This approach can be systematized using group theory. A set of operators, $\hat{f}, \hat{g}, \ldots$, form a group, if this set includes (a) the identity operator $\hat{I}$, such that $\hat{I}\hat{f} = \hat{f}\hat{I} = \hat{f}$, (b) the inverse operator $\hat{f}^{-1}$, such that $\hat{f}^{-1}\hat{f} = \hat{I} = \hat{f}\hat{f}^{-1}$, (c) and the product operator $\hat{f}\hat{g}$ and $\hat{g}\hat{f}$, for every operator $\hat{f}$ or $\hat{g}$ in this set.

If $\hat{f}$ and $\hat{g}$ commute with a given operator, say a Hamiltonian $\hat{H}$, then

$$[\hat{f}\hat{g}, \hat{H}] = \hat{f}[\hat{g}, \hat{H}] + [\hat{f}, \hat{H}]\hat{g} = 0 \qquad (8\text{-}209)$$

or the product operator $\hat{f}\hat{g}$ also commutes with $\hat{H}$. Obviously the identity operator $\hat{I}$ commutes with $\hat{H}$, which also implies that

$$[\hat{I}, \hat{H}] = [\hat{f}^{-1}\hat{f}, \hat{H}] = \hat{f}^{-1}[\hat{f}, \hat{H}] + [\hat{f}^{-1}, \hat{H}]\hat{f} = 0$$

$$(8\text{-}210)$$

i.e., if $\hat{f}$ commutes with $\hat{H}$ so does $\hat{f}^{-1}$. Therefore a set of operators which commute with a given $\hat{H}$ form a group, which we call a *symmetry group* of $\hat{H}$. There can be more than one symmetry group for a given $\hat{H}$.

According to group theory, a given group has a set of irreducible representations, D_1, D_2, etc., and it is guaranteed that one can always find a set of base functions for which the matrix representative of all operators (elements) of this group are simultaneously reduced as

$$
\begin{pmatrix}
D_1(\hat{f}) & 0 & 0 & \cdots \\
0 & D_2(\hat{f}) & 0 & \cdots \\
0 & 0 & D_3(\hat{f}) & \cdots \\
\cdots\cdots\cdots\cdots\cdots\cdots\cdots\cdots\cdots
\end{pmatrix}
\begin{pmatrix}
D_1(\hat{g}) & 0 & 0 & \cdots \\
0 & D_2(\hat{g}) & 0 & \cdots \\
0 & 0 & D_3(\hat{g}) & \cdots \\
\cdots\cdots\cdots\cdots\cdots\cdots\cdots\cdots\cdots
\end{pmatrix}
\text{ etc.} \qquad (8\text{--}211)
$$

where $D_i(\hat{f})$ is a square submatrix of the matrix representative of $\hat{f}$. The base functions which span $D_i(\hat{f})$, $D_i(\hat{g})$, ..., simultaneously are said to belong to the D_i representation.

If $\hat{H}$ commutes with every operator of this group, then its matrix representative, using the same set of base functions as that for which (8–211) are given, will also be of the form

$$
\begin{pmatrix}
D_1(\hat{H}) & 0 & 0 & \cdots \\
0 & D_2(\hat{H}) & 0 & \cdots \\
0 & 0 & D_3(\hat{H}) & \cdots \\
\cdots\cdots\cdots\cdots\cdots\cdots\cdots\cdots\cdots
\end{pmatrix}
\qquad (8\text{--}212)
$$

(This statement will be proved more carefully in section 11–5.) Thus we can classify the eigenfunctions of $\hat{H}$ according to the irreducible representations of the symmetry group. If the group is known, then group theory can tell which irreducible representations exist in this group, so that we see how the eigenfunctions are to be classified. Group theory, however, does not tell us how to obtain the explicit result (8–212). Such a calculation has to be carried out explicitly; only the form of the final result is predicted by group theory.

We will come back to group theory and its application in physics in Chapter 11.

PROBLEM 8–53. The inversion operator $\hat{I}$ is defined by $\hat{I}\Psi(\mathbf{r}) = \Psi(-\mathbf{r})$. Show that $\hat{I}$ and $\hat{1}$ form a group. The group is designated as $\mathbf{C}_i$.

PROBLEM 8–54. If $\hat{H}$ commutes with $\hat{I}$, show that the eigenfunctions of $\hat{H}$ can be classified into parity $+$ and parity $-$ states: If

$$
\hat{H}|\mathscr{E}_n\rangle = \mathscr{E}_n|\mathscr{E}_n\rangle
$$

then either

$$\hat{I}|\mathscr{E}_n\rangle = |\mathscr{E}_n\rangle \qquad \text{parity } +$$

or

$$\hat{I}|\mathscr{E}_n\rangle = -|\mathscr{E}_n\rangle \qquad \text{parity } -$$

PROBLEM 8–55. The exchange operator (12) exchanges the coordinates and momenta of two electrons 1 and 2.
(a) Show that $\hat{I}$ and (12) form a group. The group is designated as $\mathscr{S}_2$.
(b) Show that a Hamiltonian of a two electron system, $\hat{H}(\hat{p}_1, \hat{p}_2, r_1, r_2)$, commutes with (12).
(c) Show that the eigenkets of the above Hamiltonian give either

$$(12)|\mathscr{E}_n\rangle = |\mathscr{E}_n\rangle \qquad \text{or} \qquad (12)|\mathscr{E}_n\rangle = -|\mathscr{E}_n\rangle$$

PROBLEM 8–56. If one particle is in a cylindrically symmetric potential, $(\partial U/\partial\varphi = 0)$, show that the Hamiltonian of this system commutes with $\exp(\alpha\,\partial/\partial\varphi)$, where α is an arbitrary constant. Show also that these exponential operators with different α's commute with each other. The group formed by the exponental operators with all possible α's is designated as $\mathbf{C}_\infty$.

8–14 PERTURBATION METHOD

Group theory enables one to reduce a given Hamiltonian matrix to a sum (direct sum) of submatrices as shown in (8–212), but usually it is not sufficient to completely diagonalize the matrix. We still have to solve secular equations to diagonalize reduced matrices. If a reduced matrix is small enough, modern electronic computers can solve the secular equation, but in many cases reduced matrices are still too large even for such machines. The perturbation method is often useful in finding approximate solutions in such cases.

Let us assume that

$$\hat{H} = \hat{H}_0 + \hat{H}_1 \tag{8–213}$$

and that the eigenvalues and eigenfunctions of $\hat{H}_0$ are known:

$$\hat{H}_0|n^0\rangle = \mathscr{E}_n{}^0|n^0\rangle, \qquad \hat{H}_0|m^0\rangle = \mathscr{E}_m{}^0|m^0\rangle, \ldots \tag{8–214}$$

The problem is to find approximate expressions of $\mathscr{E}_n$ and $|n\rangle$ of $\hat{H}$,

$$\hat{H}|n\rangle = \mathscr{E}_n|n\rangle \tag{8–215}$$

by means of $\mathscr{E}_n{}^0, \mathscr{E}_m{}^0, \ldots$, and $|n^0\rangle, |m^0\rangle, \ldots$.

We assume that no degeneracy exist in (8–214). We will discuss degenerate cases towards the end of this section.

First-Order Perturbation

If we express the Hamiltonian matrix $\hat{H}$ by taking the eigenvectors of $\hat{H}_0$ as the bases, we see that the diagonal matrix elements are of the form of $\mathscr{E}_n{}^0 + \langle n^0|\hat{H}_1|n^0\rangle$ while the nondiagonal matrix elements are of the form of $\langle n^0|\hat{H}_1|m^0\rangle$. The required eigenvalues, $\mathscr{E}_n, \mathscr{E}_m, \ldots$, are obtained as the solutions of the secular equation such as (8–196), which reduces to

$$(\mathscr{E}_n{}^0 + \langle n^0|\hat{H}_1|n^0\rangle - \lambda)(\mathscr{E}_m{}^0 + \langle m^0|\hat{H}_1|m^0\rangle - \lambda)(\cdots - \lambda)$$
$$+ 0(\hat{H}_1{}^2) = 0 \qquad (8\text{–}216)$$

where the last term represents all terms in which nondiagonal elements are involved. These terms are of the order of $\hat{H}_1{}^2$ or higher because the non-diagonal elements always appear in products such as $\langle n^0|\hat{H}_1|m^0\rangle\langle m^0|\hat{H}_1|n^0\rangle$ in the expansion of the determinant. Since all diagonal matrix elements except for the one in which $\mathscr{E}_n{}^0$ appears are of the order of $\hat{H}_1{}^0$, we see that

$$\mathscr{E}_n{}^{(1)} = \mathscr{E}_n{}^0 + \langle n^0|\hat{H}_1|n^0\rangle \qquad (8\text{–}217)$$

is the expression of $\mathscr{E}_n$ correct to the first order of $\hat{H}_1$.

Projection Operator. Green's Operator (Resolvent). Space Contraction Operator

A projection operator $\hat{Q}_n$ is defined by

$$\hat{Q}_n = \hat{1} - |n^0\rangle\langle n^0| = \sum_{m \neq n} |m^0\rangle\langle m^0| \qquad (8\text{–}218)$$

where the closure property (8–183) is used. This operator eliminates $|n^0\rangle$ (or $\langle n^0|$) out of the complete set of eigenkets (or eigenbras) of $\hat{H}_0$.

Another kind of operator useful in discussing perturbation theory is composed of *Green's operators*, more often called *resolvents*.[13] A Green's operator $\hat{G}_\mathscr{E}$ is defined by

$$(\hat{H} - \mathscr{E})\hat{G}_\mathscr{E} = -\hat{1} \qquad (8\text{–}219)$$

or

$$\hat{G}_\mathscr{E} = (\mathscr{E} - \hat{H})^{-1} \qquad (8\text{–}219')$$

---

[13] See T. Kato, *Perturbation Theory for Linear Operators*, Springer-Verlag, New York, 1966, for mathematical discussions of resolvent.

One can see (Problem 8–57) that Green's functions are given as the coordinate representations of the corresponding Green's operators.

We will use Green's operators such as $(\mathscr{E}_n^{(1)} - \hat{H}_0 - \hat{Q}_n\hat{H}_1)^{-1}$ in the following. The following formula is important for a Green's operator made of two operators $\hat{f}$ and $\hat{g}$.

$$(\hat{f} + \hat{g})^{-1} = \hat{f}^{-1}(\hat{f} + \hat{g} - \hat{g})(\hat{f} + \hat{g})^{-1} = \hat{f}^{-1} - \hat{f}^{-1}\hat{g}(\hat{f} + \hat{g})^{-1}$$

$$= \hat{f}^{-1} - \hat{f}^{-1}\hat{g}\hat{f}^{-1} + \hat{f}^{-1}\hat{g}\hat{f}^{-1}\hat{g}\hat{f}^{-1} - \hat{f}^{-1}\hat{g}\hat{f}^{-1}\hat{g}\hat{f}^{-1}\hat{g}\hat{f}^{-1} + \cdots$$

$$(8\text{–}220)$$

Let us define an operator $\hat{S}_n^0$, called a *space contraction operator*, by

$$\hat{S}_n^0 = \hat{I} + (\mathscr{E}_n^0 - \hat{H}_0 - \hat{Q}_n\hat{H}_1)^{-1}\hat{Q}_n\hat{H}_1$$

$$= \hat{I} + (\mathscr{E}_n^0 - \hat{H}_0)^{-1}\hat{Q}_n\hat{H}_1 + (\mathscr{E}_n^0 - \hat{H}_0)^{-1}\hat{Q}_n\hat{H}_1$$

$$\times (\mathscr{E}_n^0 - \hat{H}_0 - \hat{Q}_n\hat{H}_1)^{-1}\hat{Q}_n\hat{H}_1$$

$$= \hat{I} + (\mathscr{E}_n^0 - \hat{H}_0)^{-1}\hat{Q}_n\hat{H}_1\hat{S}_n^0 \qquad (8\text{–}221)$$

where we use the infinite series expansion (8–220) of the Green's operator $(\mathscr{E}_n^0 - \hat{H}_0 - \hat{Q}_n\hat{H}_1)^{-1}$. Note that $\hat{I} - \hat{Q}_n = |n^0\rangle\langle n^0|$.

Since

$$(\hat{H}_0 - \mathscr{E}_n^0)(\mathscr{E}_n^0 - \hat{H}_0 - \hat{Q}_n\hat{H}_1)^{-1}\hat{Q}_n\hat{H}_1$$

$$= (\hat{H}_0 - \mathscr{E}_n^0)\{(\mathscr{E}_n^0 - \hat{H}_0)^{-1} + (\mathscr{E}_n^0 - \hat{H}_0)^{-1}\hat{Q}_n\hat{H}_1$$

$$\times (\mathscr{E}_n^0 - \hat{H}_0 - \hat{Q}_n\hat{H}_1)^{-1}\}\hat{Q}_n\hat{H}_1$$

$$= -\hat{Q}_n\hat{H}_1\hat{S}_n^0 \qquad (8\text{–}222)$$

we can easily see (Problem 8–59) that

$$(\hat{H} - \mathscr{E}_n^0)\hat{S}_n^0 = \hat{H}_0 - \mathscr{E}_n^0 + (\hat{I} - \hat{Q}_n)\hat{H}_1\hat{S}_n^0 \qquad (8\text{–}223)$$

PROBLEM 8–57. Show that (8–219) gives

$$\int \langle\mathbf{r}|(\hat{H} - \mathscr{E})|\mathbf{r}''\rangle\langle\mathbf{r}''|\hat{G}_\mathscr{E}|\mathbf{r}'\rangle \, dv'' = [\hat{H}(\mathbf{r}) - \mathscr{E}]\langle\mathbf{r}|\hat{G}_\mathscr{E}|\mathbf{r}'\rangle = -\delta(\mathbf{r} - \mathbf{r}')$$

This shows that the previously discussed Green's functions $G(\mathbf{r}, \mathbf{r}')$ are given by $\langle\mathbf{r}|\hat{G}_\mathscr{E}|\mathbf{r}'\rangle$ or as a coordinate representation of Green's operator $\hat{G}_\mathscr{E}$.

PROBLEM 8–58.   Show that

$$\hat{G}_{\mathscr{E}} = \sum_n |n\rangle\langle n| \frac{1}{\mathscr{E} - \mathscr{E}_n}$$

where $\hat{H}|n\rangle = \mathscr{E}_n|n\rangle$.

HINT.   Use (8–219') and the closure property (8–183).

PROBLEM 8–59.   Verify (8–223) using (8–222).

PROBLEM 8–60.   Show that

$$\langle m^0|(\hat{H} - \mathscr{E}_n^0)\hat{S}_n^0|\hat{n}^0\rangle = 0 \quad \text{if} \quad m \neq n$$

Third-Order Perturbation

Let us transform the original orthonormal set $|n^0\rangle$, $|m^0\rangle$, ... into $\hat{S}_n^0|n^0\rangle$, $\hat{S}_m^0|m^0\rangle$, ... and see what happens.

The transformed kets and bras are not orthogonal any more, but non-diagonal matrix elements of $\hat{I}$ are of the order of $\hat{H}_1^2$ or higher. This statement can be shown as follows.

Using the rule given in Problem 8–3 and a general assumption that $\hat{H}_0$ and $\hat{H}_1$ are both Hermitian, we see that

$$\hat{S}_m^{0\dagger}\hat{S}_n^0 = \hat{I} + (\mathscr{E}_n^0 - \hat{H}_0)^{-1}\hat{Q}_n\hat{H}_1\hat{S}_n^0 + \hat{S}_m^{0\dagger}\hat{H}_1\hat{Q}_m(\mathscr{E}_m^0 - \hat{H}_0)^{-1}$$
$$+ \hat{S}_m^{0\dagger}\hat{H}_1\hat{Q}_m(\mathscr{E}_m^0 - \hat{H}_0)^{-1}(\mathscr{E}_n^0 - \hat{H}_0)^{-1}\hat{Q}_n\hat{H}_1\hat{S}_n^0 \quad (8\text{–}224)$$

The middle two terms of the right-hand side of this equation are rewritten as

$$(\mathscr{E}_n^0 - \hat{H}_0)^{-1}\hat{Q}_n\hat{H}_1\hat{S}_n^0 + \hat{S}_m^{0\dagger}\hat{H}_1\hat{Q}_m(\mathscr{E}_m^0 - \hat{H}_0)^{-1}$$
$$= (\mathscr{E}_n^0 - \hat{H}_0)^{-1}\hat{Q}_n\hat{H}_1 + (\mathscr{E}_n^0 - \hat{H}_0)^{-1}\hat{Q}_n\hat{H}_1(\mathscr{E}_n^0 - \hat{H}_0)^{-1}\hat{Q}_n\hat{H}_1S_n^0$$
$$+ \hat{H}_1\hat{Q}_m(\mathscr{E}_m^0 - \hat{H}_0)^{-1} + \hat{S}_m^{0\dagger}\hat{H}_1\hat{Q}_m(\mathscr{E}_m^0 - \hat{H}_0)^{-1}\hat{H}_1\hat{Q}_m(\mathscr{E}_m^0 - \hat{H}_0)^{-1}$$
$$= (\mathscr{E}_n^0 - \hat{H}_0)^{-1}\hat{Q}_n\hat{H}_1 + \hat{H}_1\hat{Q}_m(\mathscr{E}_m^0 - \hat{H}_0)^{-1}$$
$$+ \hat{S}_m^{0\dagger}(\mathscr{E}_n^0 - \hat{H}_0)^{-1}\hat{Q}_n\hat{H}_1(\mathscr{E}_n^0 - \hat{H}_0)^{-1}\hat{Q}_n\hat{H}_1\hat{S}_n^0$$
$$+ \hat{S}_m^{0\dagger}\hat{H}_1\hat{Q}_m(\mathscr{E}_m^0 - \hat{H}_0)^{-1}\hat{H}_1\hat{Q}_m(\mathscr{E}_m^0 - \hat{H}_0)^{-1}\hat{S}_n^0$$
$$+ \hat{S}_m^{0\dagger}\hat{H}_1\hat{Q}_m(\mathscr{E}_m^0 - \hat{H}_0)^{-1}(\mathscr{E}_n^0 - \hat{H}_0)^{-1}\hat{Q}_n\hat{H}_1(\mathscr{E}_n^0 - \hat{H}_0)^{-1}\hat{Q}_n\hat{H}_1\hat{S}_n^0$$
$$+ \hat{S}_m^{0\dagger}\hat{H}_1\hat{Q}_m(\mathscr{E}_m^0 - \hat{H}_0)^{-1}\hat{H}_1\hat{Q}_m(\mathscr{E}_m^0 - \hat{H}_0)^{-1}(\mathscr{E}_n^0 - \hat{H}_0)^{-1}\hat{Q}_n\hat{H}_1\hat{S}_n^0$$

$$(8\text{–}225)$$

using (8–221). Putting this result into (8–224) we obtain

$$
\begin{aligned}
\hat{S}_m^{0\dagger}\hat{S}_n^0 = \hat{1} &+ [(\mathscr{E}_n^0 - \hat{H}_0)^{-1}\hat{Q}_n\hat{H}_1 + \hat{H}_1\hat{Q}_m(\mathscr{E}_m^0 - \hat{H}_0)^{-1}] \\
&+ [(\mathscr{E}_n^0 - \hat{H}_0)^{-1}\hat{Q}_n\hat{H}_1 - (\mathscr{E}_n^0 - \mathscr{E}_m^0)^{-1}\hat{H}_1\hat{Q}_m] \\
&\times (\mathscr{E}_n^0 - \hat{H}_0)^{-1}\hat{Q}_n\hat{H}_1 \\
&+ \hat{H}_1\hat{Q}_m(\mathscr{E}_m^0 - \hat{H}_0)^{-1} \\
&\times [\hat{H}_1\hat{Q}_m(\mathscr{E}_m^0 - \hat{H}_0)^{-1} - \hat{Q}_n\hat{H}_1(\mathscr{E}_m^0 - \mathscr{E}_n^0)^{-1}] \\
&+ 0(\hat{H}_1^3) \qquad\qquad\qquad\qquad\qquad\qquad (8\text{–}226)
\end{aligned}
$$

where

$$
\begin{aligned}
0(\hat{H}_1^3) = \hat{H}_1\hat{Q}_m(\mathscr{E}_m^0 - \hat{H}_0)^{-1}&[\hat{H}_1\hat{Q}_m(\mathscr{E}_m^0 - \hat{H}_0)^{-1} + (\mathscr{E}_n^0 - \hat{H}_0)^{-1}\hat{Q}_n\hat{H}_1] \\
&\times (\mathscr{E}_n^0 - \hat{H}_0)^{-1}\hat{Q}_n\hat{H}_1\hat{S}_n^0 + \hat{S}_m^{0\dagger}\hat{H}_1\hat{Q}_m(\mathscr{E}_m^0 - \hat{H}_0)^{-1} \\
&\times [\hat{H}_1\hat{Q}_m(\mathscr{E}_m^0 - \hat{H}_0)^{-1} + (\mathscr{E}_n^0 - \hat{H}_0)^{-1}\hat{Q}_n\hat{H}_1] \\
&\times (\mathscr{E}_n^0 - \hat{H}_0)^{-1}\hat{Q}_n\hat{H}_1 + \hat{S}_m^{0\dagger}\hat{H}_1\hat{Q}_m(\mathscr{E}_m^0 - \hat{H}_0)^{-1}\hat{H}_1\hat{Q}_m \\
&\times (\mathscr{E}_m^0 - \hat{H}_0)^{-1}(\mathscr{E}_n^0 - \hat{H}_0)^{-1}\hat{Q}_n\hat{H}_1 \, (\mathscr{E}_n^0 - \hat{H}_0)^{-1}\hat{Q}_n\hat{H}_1\hat{S}_n^0
\end{aligned}
$$

$$(8\text{–}227)$$

Therefore,

$$
\begin{aligned}
\langle m^0|\hat{S}_m^{0\dagger}\hat{S}_n^0|n^0\rangle = (\mathscr{E}_n^0 - \mathscr{E}_m^0)^{-2}&\{\langle m^0|\hat{H}_1|m^0\rangle + \langle n^0|\hat{H}_1|n^0\rangle\}\langle m^0|\hat{H}_1|n^0\rangle \\
&+ \langle m^0|0(\hat{H}_1^3)|n^0\rangle \quad \text{if} \quad m \neq n \qquad (8\text{–}228)
\end{aligned}
$$

which is of the second order in $\hat{H}_1$.

In a similar way we obtain

$$
\begin{aligned}
\hat{S}_m^{0\dagger}&(\hat{H}_0 + \hat{H}_1)\hat{S}_n^0 \\
= \hat{H}_0 &+ \hat{H}_1 + \hat{H}_0(\mathscr{E}_n^0 - \hat{H}_0)^{-1}\hat{Q}_n\hat{H}_1 + \hat{H}_1\hat{Q}_m(\mathscr{E}_m^0 - \hat{H}_0)^{-1}\hat{H}_0 \\
&+ [(\mathscr{E}_n^0 - \hat{H}_0)^{-1}\hat{H}_1\mathscr{E}_n^0(\mathscr{E}_n^0 - \hat{H}_0)^{-1}\hat{Q}_n\hat{H}_1 \\
&\qquad\quad - (\mathscr{E}_n^0 - \mathscr{E}_m^0)\hat{H}_1\hat{H}_0(\mathscr{E}_n^0 - \hat{H}_0)^{-1}\hat{Q}_n\hat{H}_1 \\
&\qquad\quad + \hat{H}_1\hat{Q}_m\mathscr{E}_m^0(\mathscr{E}_m^0 - \hat{H}_0)^{-1}\hat{H}_1(\mathscr{E}_m^0 - \hat{H}_0)^{-1} \\
&\qquad\quad - \hat{H}_1\hat{Q}_m\hat{H}_0(\mathscr{E}_m^0 - \hat{H}_0)^{-1}\hat{H}_1(\mathscr{E}_m^0 - \mathscr{E}_n^0)^{-1}] \\
&- [(\hat{1} - \hat{Q}_n)\hat{H}_0(\mathscr{E}_n^0 - \hat{H}_0)^{-1}\hat{H}_1(\mathscr{E}_n^0 - \hat{H}_0)^{-1}\hat{Q}_n\hat{H}_1 \\
&\qquad\quad + \hat{H}_1\hat{Q}_m(\mathscr{E}_m^0 - \hat{H}_0)^{-1}\hat{H}_1(\mathscr{E}_m^0 - \hat{H}_0)\hat{H}_0(\hat{1} - \hat{Q}_m)] \\
&+ [\hat{H}_1(\hat{1} - \hat{Q}_m)\hat{H}_0(\mathscr{E}_n^0 - \hat{H}_0)^{-1}\hat{Q}_n\hat{H}_1 \\
&\qquad\quad - \hat{H}_1\hat{Q}_m(\mathscr{E}_m^0 - \hat{H}_0)^{-1}\hat{H}_0(\hat{1} - \hat{Q}_n)\hat{H}_1](\mathscr{E}_n^0 - \mathscr{E}_m^0)^{-1} \\
&+ 0'(\hat{H}_1^3) \qquad\qquad\qquad\qquad\qquad\qquad\qquad\qquad\qquad (8\text{–}229)
\end{aligned}
$$

where

$$
\begin{aligned}
0'(\hat{H}_1{}^3) = & [\hat{H}_1 + \hat{H}_0(\mathscr{E}_n{}^0 - \hat{H}_0)^{-1}\hat{Q}_n\hat{H}_1 + \hat{H}_1\hat{Q}_m(\mathscr{E}_m{}^0 - \hat{H}_0)^{-1}\hat{H}_0] \\
& \times (\mathscr{E}_n{}^0 - \hat{H}_0)^{-1}\hat{Q}_n\hat{H}_1(\mathscr{E}_n{}^0 - \hat{H}_0)^{-1}\hat{Q}_n\hat{H}_1\hat{S}_n{}^0 \\
& + \hat{S}_m{}^0\hat{H}_1\hat{Q}_m(\mathscr{E}_m{}^0 - \hat{H}_0)^{-1}\hat{H}_1\hat{Q}_m(\mathscr{E}_m{}^0 - \hat{H}_0)^{-1} \\
& \times [\hat{H}_1 + \hat{H}_0(\mathscr{E}_n{}^0 - \hat{H}_0)^{-1}\hat{Q}_n\hat{H}_1 + \hat{H}_1\hat{Q}_m(\mathscr{E}_m{}^0 - \hat{H}_0)^{-1}\hat{H}_0] \\
& + \hat{S}_m{}^0\hat{H}_1\hat{Q}_m(\mathscr{E}_m{}^0 - \hat{H}_0)^{-1}\hat{H}_1(\mathscr{E}_n{}^0 - \hat{H}_0)^{-1}\hat{Q}_n\hat{H}_1\hat{S}_n{}^0 \\
& + \hat{S}_m{}^0\hat{H}_1\hat{Q}_m(\mathscr{E}_m{}^0 - \hat{H}_0)^{-1}\hat{H}_1\hat{Q}_m(\mathscr{E}_m{}^0 - \hat{H}_0)^{-1}\hat{H}_0 \\
& \times (\mathscr{E}_n{}^0 - \hat{H}_0)^{-1}\hat{Q}_n\hat{H}_1(\mathscr{E}_n{}^0 - \hat{H}_0)^{-1}\hat{Q}_n\hat{H}_1\hat{S}_n{}^0 \qquad (8\text{--}230)
\end{aligned}
$$

When we calculate the matrix element of (8–229) we see that it reduces to a relatively simple form of

$$
\begin{aligned}
\langle m^0|\hat{S}_m{}^0\hat{H}\hat{S}_n{}^0|n^0\rangle = & \{\mathscr{E}_n{}^0\langle m^0|\hat{H}_1|m^0\rangle + \mathscr{E}_m{}^0\langle n^0|\hat{H}_1|n^0\rangle\}\langle m^0|\hat{H}_1|n^0\rangle \\
& \times (\mathscr{E}_n{}^0 - \mathscr{E}_m{}^0)^{-2} + \langle m^0|0'(\hat{H}_1{}^3)|n^0\rangle \\
& \qquad\qquad\qquad\qquad\qquad \text{if} \quad m \neq n \qquad (8\text{--}231)
\end{aligned}
$$

Our results (8–228) and (8–231) show that all nondiagonal matrix elements of $(\hat{H} - \lambda\hat{1})$ calculated by taking the transformed bases $\hat{S}_n{}^0|n^0\rangle$, $\hat{S}_m{}^0|m^0\rangle$, ... are of the order of $\hat{H}_1{}^2$ or higher. Since those nondiagonal elements always appear as pairs in the expansion of $\det(\mathbf{H} - \lambda\mathbf{1})$, which is the secular equation, we see that approximate eigenvalues of $\hat{H}$ accurate to the order of $\hat{H}_1{}^3$ can be obtained by simply equating the diagonal matrix elements of $(\hat{H} - \lambda\hat{1})$ to zero. [See the argument in relation to (8–216).] Therefore

$$
\mathscr{E}_n{}^{(3)} = \frac{\langle n^0|\hat{S}_n{}^{0\dagger}\hat{H}\hat{S}_n{}^0|n^0\rangle}{\langle n^0|\hat{S}_n{}^{0\dagger}\hat{S}_n{}^0|n^0\rangle} \qquad (8\text{--}232)
$$

gives an expression of an eigenvalue exact to the third order of $\hat{H}_1$.

Fifth-Order Perturbation

We can improve the accuracy of the approximation if we can find a transformation by which the terms of the second order in $\hat{H}_1$ are taken out of (8–228) and (8–231). This turns out to be quite easy. Actually if $\hat{H}_1$ does not have any diagonal matrix elements, which happens often, we see from (8–228) and (8–231) that those nondiagonal matrix elements are already of the order of $\hat{H}_1{}^3$ or higher. Therefore when $\hat{H}_1$ does not have any diagonal matrix elements, $\mathscr{E}_n{}^{(3)}$ given by (8–232) is already exact to the order of $\hat{H}_1{}^5$.

When $\hat{H}_1$ has diagonal matrix elements we can easily find (Problem 8-60) that

$$\hat{S}_n^{(1)} = \hat{1} + (\mathscr{E}_n^{(1)} - \hat{H}_0')^{-1}\hat{Q}_n\hat{H}_1'\hat{S}_n^{(1)} \qquad (8\text{-}233)$$

does the job of removing terms of the order of $\hat{H}_1^2$ from nondiagonal matrix elements (8-228) and (8-231). In (8-233) $\mathscr{E}_n^{(1)}$ is given by (8-217), $\hat{H}_0'$ is $\hat{H}_0$ plus the appropriate diagonal matrix element of $\hat{H}_1$,

$$\hat{H}_0'|m^0\rangle = |m^0\rangle(\mathscr{E}_m^0 + \langle m^0|\hat{H}_1|m^0\rangle) = \mathscr{E}_m^{(1)}|m^0\rangle \qquad (8\text{-}234)$$

and $\hat{H}_1'$ is $\hat{H}_1$ minus all its diagonal matrix elements. We therefore see that

$$\mathscr{E}_n^{(5)} = \frac{\langle n^0|\hat{S}_n^{(1)\dagger}\hat{H}\hat{S}_n^{(1)}|n^0\rangle}{\langle n^0|\hat{S}_n^{(1)\dagger}\hat{S}_n^{(1)}|n^0\rangle} \qquad (8\text{-}235)$$

is correct to the order of $\hat{H}_1^5$.

Degenerate Case

Let us assume that $|n_1^0\rangle$, $|n_2^0\rangle$, ..., $|n_b^0\rangle$ have a common eigenvalue $\mathscr{E}_n^0$ of $\hat{H}_0$. Since the argument in relation to (8-216) still holds true even if such degeneracy exists, the formula (8-217) is correct to the first order of $\hat{H}_1$ without any modification. In order to obtain a third-order perturbation formula we replace $\hat{Q}_n$ by

$$\hat{Q}_N = \hat{1} - |n_1^0\rangle\langle n_1^0| - |n_2^0\rangle\langle n_2^0| - \cdots - |n_b^0\rangle\langle n_b^0| \qquad (8\text{-}236)$$

and replace $\hat{S}_n^0$ by

$$\hat{S}_N^0 = \hat{1} + (\mathscr{E}_n^0 - \hat{H}_0)^{-1}\hat{Q}_N\hat{H}_1\hat{S}_N^0 \qquad (8\text{-}237)$$

We see that the previous results (8-228) and (8-231) still hold true for all nondiagonal matrix elements except for those in the $b \times b$ submatrix which is spanned by $|n_1^0\rangle$, ..., $|n_b^0\rangle$. Since all nondiagonal matrix elements (except for those in the $b \times b$ submatrix) stay in the order of $\hat{H}_1^2$ or higher during the diagonalization of the $b \times b$ submatrix, we see that the solutions of

$$\begin{vmatrix} \langle n_1^0|\hat{S}_N^{0\dagger}\hat{H}\hat{S}_N^0|n_1^0\rangle - \lambda\langle n_1^0|\hat{S}_N^{0\dagger}\hat{S}_N^0|n_1^0\rangle \cdots \\ \qquad\qquad \langle n_b^0|\hat{S}_N^{0\dagger}\hat{H}\hat{S}_N^0|n_1^0\rangle - \lambda\langle n_b^0|\hat{S}_N^{0\dagger}\hat{S}_N^0|n_1^0\rangle \\ \langle n_1^0|\hat{S}_N^{0\dagger}\hat{H}\hat{S}_N^0|n_2^0\rangle - \lambda\langle n_1^0|\hat{S}_N^{0\dagger}\hat{S}_N^0|n_2^0\rangle \cdots \\ \cdots\cdots\cdots\cdots\cdots\cdots\cdots\cdots\cdots\cdots\cdots\cdots\cdots\cdots\cdots\cdots\cdots \\ \langle n_1^0|\hat{S}_N^{0\dagger}\hat{H}\hat{S}_N^0|n_b^0\rangle - \lambda\langle n_1^0|\hat{S}_N^{0\dagger}\hat{S}_N^0|n_b^0\rangle \cdots \\ \qquad\qquad \langle n_b^0|\hat{S}_N^{0\dagger}\hat{H}\hat{S}_N^0|n_b^0\rangle - \lambda\langle n_b^0|\hat{S}_N^{0\dagger}\hat{S}_N^0|n_b^0\rangle \end{vmatrix} = 0$$

$$(8\text{-}238)$$

must be correct to the order of $\hat{H}_1^3$.

In some cases we can find a transformation by which degenerate states are decoupled. If $\hat{H}_1$ does not have any nondiagonal matrix elements (direct or indirect) between degenerate states the problem is reduced to the non-degenerate case, and all previous arguments are applicable.

Perturbation Series

Explicit expressions of the approximate eigenvalues are obtained from (8–221). For example

$$\hat{S}_n^{\ 0}|n^0\rangle = |n^0\rangle + \sum_{m \neq n} \frac{|m^0\rangle(mn)}{m} + \sum_{\substack{m \neq n \\ l \neq n}} \frac{|l^0\rangle(lm)(mn)}{lm}$$

$$+ \sum_{\substack{m \neq n \\ l \neq n \\ k \neq n}} \frac{|k^0\rangle(kl)(lm)(mn)}{klm} + \cdots \qquad (8\text{--}239)$$

where

$$(mn) = \langle m^0|\hat{H}_1|n^0\rangle \qquad (8\text{--}240\text{a})$$

in the numerators, and

$$m = \mathscr{E}_n^{\ 0} - \mathscr{E}_m^{\ 0} \qquad (8\text{--}240\text{b})$$

in the denominators.

We see from (8–223) that

$$\langle n^0|\hat{S}_n^{\ 0\dagger}\hat{H}\hat{S}_n^{\ 0}|n^0\rangle = \mathscr{E}_n^{\ 0}\langle n^0|\hat{S}_n^{\ 0\dagger}\hat{S}_n^{\ 0}|n^0\rangle + \langle n^0|\hat{H}_1\hat{S}_n^{\ 0}|n^0\rangle$$

$$(8\text{--}241)$$

Therefore, $\mathscr{E}_n^{(3)}$ can be immediately obtained from (8–239). We see from (8–233) that

$$(\hat{H} - \mathscr{E}_n^{(1)})\hat{S}_n^{(1)} = \hat{H}_0 - \mathscr{E}_n^{(1)} + (\hat{1} - \hat{Q}_n)\hat{H}_1\hat{S}_n^{(1)} \qquad (8\text{--}242)$$

which corresponds to (8–223). Therefore,

$$\langle n^0|\hat{S}_n^{(1)\dagger}\hat{H}\hat{S}_n^{(1)}|n^0\rangle = \mathscr{E}_n^{(1)}\langle n^0|\hat{S}_n^{(1)\dagger}\hat{S}_n^{(1)}|n^0\rangle$$

$$+ \langle n^0|\hat{H}_1\hat{S}_n^{(1)}|n^0\rangle - \langle n^0|\hat{H}_1|n^0\rangle \qquad (8\text{--}243)$$

which helps our calculation. Results are shown in Table 8–2.

TABLE 8–2 *Perturbation Series*

$$\mathscr{E}_n^{(1)} = \mathscr{E}_n^0 + \langle n^0|\hat{H}_1|n^0\rangle$$

$$\mathscr{E}_n^{(3)} = \frac{\langle n^0|\hat{S}_n^{0\dagger}\hat{H}\hat{S}_n^0|n^0\rangle}{\langle n^0|\hat{S}_n^{0\dagger}\hat{S}_n^0|n^0\rangle}$$

$$= \mathscr{E}_n^0 + \frac{(nn) + \displaystyle\sum_{m \neq n} \frac{(nm)(mn)}{m} + \sum_{\substack{m \neq n \\ l \neq n}} \frac{(nl)(lm)(mn)}{lm} + \cdots}{1 + \displaystyle\sum_{m \neq n} \frac{(nm)(mn)}{m^2} + \sum_{\substack{m \neq n \\ l \neq n}} (nm)(ml)(ln)\left(\frac{1}{m^2 l} + \frac{1}{ml^2}\right) + \cdots}$$

$$\left(\begin{array}{ll} (nm) = \langle n^0|\hat{H}_1|m^0\rangle & \text{in the numerators} \\ m = \mathscr{E}_n^0 - \mathscr{E}_m^0 & \text{in the denominators} \end{array}\right)$$

$$\mathscr{E}_n^{(5)} = \frac{\langle n^0|\hat{S}_n^{(1)\dagger}\hat{H}\hat{S}_n^{(1)}|n^0\rangle}{\langle n^0|\hat{S}_n^{(1)\dagger}\hat{S}_n^{(1)}|n^0\rangle} = \mathscr{E}_n^0 + (nn)$$

$$+ \frac{\displaystyle\sum{}' \frac{(nm)(mn)}{m} + \sum{}' \frac{(nl)(lm)(mn)}{lm} + \sum{}' \frac{(nk)(kl)(lm)(mn)}{klm} + \sum{}' \frac{(nh)(hk)(kl)(lm)(mn)}{hklm} + \cdots}{1 + \displaystyle\sum{}' \frac{(nm)(mn)}{m^2} + \sum{}' (nm)(ml)(ln)\left(\frac{1}{m^2 l} + \frac{1}{ml^2}\right) + \sum{}' (nm)(ml)(lk)(kn)\left(\frac{1}{m^2 lk} + \frac{1}{ml^2 k} + \frac{1}{mlk^2}\right) + \cdots}$$

$$\left(\begin{array}{ll} (nm) = \langle n^0|\hat{H}_1|m^0\rangle & \text{in the numerators} \\ m = \mathscr{E}_n^0 + \langle n^0|\hat{H}_1|n^0\rangle - \mathscr{E}_m^0 - \langle m^0|\hat{H}_1|m^0\rangle & \text{in the denominators} \\ \sum{}' = \text{summation excluding all diagonal elements of } \hat{H}_1 & \end{array}\right)$$

Approximations of a function by means of ratios of power series are called the Padé approximant. The formulas we obtained above are examples of the Padé approximant.[14]

PROBLEM 8–61. Show $\hat{S}_n^{(1)}$ of (8–233) removes terms of the order of $\hat{H}_1^2$ from (8–228) and (8–231).

HINT. Use (8–224) and (8–229); and

$$(\mathscr{E}_m^{(1)} - \hat{H}_0')^{-1}|n^0\rangle = (\mathscr{E}_m^0 - \mathscr{E}_n^0)^{-1}|n^0\rangle - (\mathscr{E}_m^0 - \mathscr{E}_n^0)^{-2}$$
$$\times (\langle m^0|\hat{H}_1|m^0\rangle - \langle n^0|\hat{H}_1|n^0\rangle)|n^0\rangle + \cdots$$

[14] G. A. Baker and J. L. Gammel, eds., *The Padé Approximant in Theoretical Physics* Academic Press, New York, 1970.

PROBLEM 8–62. Verify (8–242) and then (8–243).

PROBLEM 8–63. Show that

$$\hat{S}_n^{(1)'} = \hat{1} + (\mathscr{E}_n^{(1)} - \hat{H}_0)^{-1}\hat{Q}_n\hat{H}_1\hat{S}_n^{(1)'}$$

can do the same job as $\hat{S}_n^{(1)}$ defined by (8–233).

PROBLEM 8–64. Show that

$$|n\rangle = |n^0\rangle + (\mathscr{E}_n^0 - \hat{H}_0 - \mathscr{H})^{-1}\mathscr{H}|n^0\rangle$$

holds exactly if

$$\mathscr{H} = \hat{H}_1 - \mathscr{E}_n + \mathscr{E}_n^0$$

HINT. $(\mathscr{E}_n - \hat{H})|n^0\rangle = -\mathscr{H}|n^0\rangle$, so that

$$(\mathscr{E}_n - \hat{H})(|n\rangle - |n^0\rangle) = \mathscr{H}|n^0\rangle$$

PROBLEM 8–65. Show that

(a) $\displaystyle |n\rangle = \sum_m |m^0\rangle\langle m^0|\hat{H}_1|n\rangle(\mathscr{E}_n - \mathscr{E}_m^0)^{-1}$

$\displaystyle \qquad = |n^0\rangle\langle n^0|\hat{H}_1|n\rangle(\mathscr{E}_n - \mathscr{E}_n^0)^{-1}$

$\displaystyle \qquad + \sum_{m \neq n} |m^0\rangle\langle m^0|\hat{H}_1|n\rangle(\mathscr{E}_n - \mathscr{E}_m^0)^{-1}$

(b) $\displaystyle |n\rangle = C\left[|n^0\rangle + \sum_{m \neq n} |m^0\rangle \frac{(mn)}{m'} + \sum_{\substack{m \neq n \\ l \neq n}} |l^0\rangle \frac{(lm)(mn)}{l'm'} \right.$

$$\left. + \sum_{\substack{m \neq n \\ l \neq n \\ k \neq n}} |k^0\rangle \frac{(kl)(lm)(mn)}{k'l'm'} + \cdots \right]$$

where

$$C = \langle n^0|\hat{H}_1|n\rangle(\mathscr{E}_n - \mathscr{E}_n^0)^{-1}$$

$$(mn) = \langle m^0|\hat{H}_1|n^0\rangle$$

in the numerators

$$m' = \mathscr{E}_n - \mathscr{E}_m^0$$

in the denominators.

(c) $\quad \mathscr{E}_n = \mathscr{E}_n^{\,0} + (nn) + \displaystyle\sum_{m \neq n} \frac{(nm)(mn)}{m'} + \sum_{\substack{m \neq n \\ l \neq n}} \frac{(nl)(lm)(mn)}{l'm'}$

$\qquad + \displaystyle\sum_{\substack{m \neq n \\ l \neq n \\ k \neq n}} \frac{(nk)(kl)(lm)(mn)}{k'l'm'} + \cdots$

The last expression of $\mathscr{E}_n$ is called the Wigner–Brillouin perturbation series. It looks very nice, but the trouble is that $\mathscr{E}_n$ appears on the right-hand side of the equation also.

HINT. Use the closure property (8–183) for (a). Calculate $\langle n^0 | \hat{H}_1 | n \rangle$ for (c).

PROBLEM 8–66. In the one-dimensional simple harmonic oscillator problem, where $\hat{H}_0 = (\hat{p}_x^{\,2}/2m) + \frac{1}{2}kx^2$, we know that $\mathscr{E}_n^{\,0} = \hbar\omega(n + \frac{1}{2})$ and $\langle n|x|n - 1 \rangle = \sqrt{n\hbar/2m\omega}$, $\langle n|x|n \rangle = 0, \ldots$ (Problem 8–41.)
(a) Show that if $\hat{H}_1 = \alpha x$, then

$$\mathscr{E}_n^{\,(3)} = \hbar\omega(n + \tfrac{1}{2}) - \cfrac{\beta + (2n + 1)\beta^2 + \frac{1}{2}(5n^2 + 5n + 3)\beta^3 \\ \qquad + \frac{1}{18}(46n^3 + 69n^2 + 110n + 43.5)\beta^4 + \cdots}{1 + (2n + 1)\beta + \frac{5}{2}(n^2 + n + 1)\beta^2 \\ \qquad + \frac{1}{18}(46n^3 + 69n^2 + 227n + 102)\beta^3 + \cdots}$$

where $\beta = \frac{1}{2}\alpha^2/k$.
(b) Show that the exact solution is

$$\mathscr{E}_n = \hbar\omega(n + \tfrac{1}{2}) - \beta$$

in this case, and that the result of the perturbation calculation agrees with the exact solution up to the order of α^5.

PROBLEM 8–67. In the same problem as the last one, we also know that

$$\langle n|x^2|n \rangle = (2n + 1)\frac{\hbar}{2m\omega} \qquad \langle n|x^2|n + 2 \rangle = \sqrt{(n + 1)(n + 2)}\,\frac{\hbar}{2m\omega} \cdots$$

(a) Show that if $\hat{H}_1 = \frac{1}{2}\alpha x^2$ then

$$\mathscr{E}_n^{\,(5)} = \hbar\omega(n + \tfrac{1}{2}) + (2n + 1)\beta$$
$$- \cfrac{[(2n + 1)/(\hbar\omega + 2\beta)]\beta^2 \\ \qquad + [((2n + 1)(n^2 + n + 3))/(2(\hbar\omega + 2\beta)^3)]\beta^4 + \cdots}{1 + [(n^2 + n + 1)/(2(\hbar\omega + 2\beta)^2)]\beta^2 + \cdots}$$

where $\beta = \alpha\hbar\omega/4k$.

(b) Show that the exact solution is

$$\mathcal{E}_n = \hbar\omega(n + \tfrac{1}{2})\sqrt{(k + \alpha)/k}$$

$$= (n + \tfrac{1}{2})\hbar\omega \left[1 + \frac{\alpha}{2k} - \frac{1}{8}\left(\frac{\alpha}{k}\right)^2 + \frac{1}{16}\left(\frac{\alpha}{k}\right)^3 \right.$$

$$\left. - \frac{5}{128}\left(\frac{\alpha}{k}\right)^4 + \frac{7}{256}\left(\frac{\alpha}{k}\right)^5 - \cdots \right]$$

and that $\mathcal{E}_n^{(5)}$ agrees with this exact solution up to the terms of the order of α^5.

PROBLEM 8–68. A secular equation

$$\begin{vmatrix} -a - \mathcal{E} & b \\ b & a - \mathcal{E} \end{vmatrix} = 0$$

is given. We know the solutions

$$\mathcal{E}_\pm = \pm\sqrt{a^2 + b^2} = \pm\left(a + \frac{b^2}{2a} - \frac{b^4}{8a^3} + \frac{b^6}{16a^5} + \cdots\right)$$

Show that

$$\mathcal{E}_n^{(5)} = \pm a \pm \frac{b^2/2a}{1 + (b/2a)^2} = \pm\left(a + \frac{b^2}{2a} - \frac{b^4}{8a^3} + \frac{b^6}{32a^5} + \cdots\right)$$

PROBLEM 8–69. Show that one of the eigenvalues of

$$\begin{pmatrix} a & b & d \\ c & c & e \\ d & e & 0 \end{pmatrix}$$

is approximately given by

$$\frac{\left[-\dfrac{d^2}{a} - \dfrac{e^2}{b} + \dfrac{2cde}{ab} - \dfrac{(cd)^2}{a^2b} - \dfrac{(ce)^2}{ab^2} + \dfrac{2c^3de}{(ab)^2}\right]}{\left[1 + \dfrac{d^2}{a^2} + \dfrac{e^2}{b^2} - 2cde\left(\dfrac{1}{a^2b} + \dfrac{1}{ab^2}\right)\right]}$$

PROBLEM 8-70. A two-dimensional simple harmonic oscillator

$$\hat{H}_0 = (2m)^{-1}(\hat{p}_x^2 + \hat{p}_y^2) + \tfrac{1}{2}k(x^2 + y^2)$$

is given. We know that each states can be designated by two quantum numbers n_x and n_y, and its energy is $(n_x + n_y + 1)\hbar\omega$ where $\omega = \sqrt{k/m}$. Calculate the perturbation energy due to αxy on the $(n_x = 1, n_y = 0)$ and $(n_x = 0, n_y = 1)$ states.

ANSWER. The exact solution in classical theory is given in section 2-6, according to which ω splits into two values

$$\hbar\omega_{1/2} = \hbar\omega\sqrt{(k \pm \alpha)/k}$$

$$= \hbar\omega \pm \beta - \frac{\beta^2}{2\hbar\omega} \pm \frac{\beta^3}{2(\hbar\omega)^2} - \frac{5\beta^4}{8(\hbar\omega)^3} \pm \frac{7\beta^5}{8(\hbar\omega)^4} - \cdots$$

where $\beta = \alpha\hbar\omega/2k$.
The two states we are considering become (10) and (01) states where designations are $(n_1 n_2)$. Their exact energies $\mathcal{E}(n_1 n_2)$ are

$$\mathcal{E}(10) = \tfrac{3}{2}\hbar\omega_1 + \tfrac{1}{2}\hbar\omega_2$$

$$= 2\hbar\omega + \beta - \frac{\beta^2}{\hbar\omega} + \frac{\beta^3}{2(\hbar\omega)^2} - \frac{5\beta^4}{4(\hbar\omega)^3} + \frac{7\beta^5}{8(\hbar\omega)^4} - \cdots$$

$$\mathcal{E}(01) = \tfrac{1}{2}\hbar\omega_1 + \tfrac{3}{2}\hbar\omega_2$$

$$= 2\hbar\omega - \beta - \frac{\beta^2}{\hbar\omega} - \frac{\beta^3}{2(\hbar\omega)^2} - \frac{5\beta^4}{4(\hbar\omega)^3} - \frac{7\beta^5}{8(\hbar\omega)^4} - \cdots$$

Second-Order Perturbation. Taking αxy for $\hat{H}_1$, we obtain

$$\langle n_x n_y | \hat{S}_N^{0\dagger}(\hat{H} - \lambda\hat{1})\hat{S}_N^0 | n_x' n_y' \rangle$$

for our doubly degenerate space as

$$\begin{pmatrix} 2\hbar\omega\{1 - \tfrac{43}{16}\gamma^4 + \cdots\} - \{1 + \tfrac{1}{2}\gamma^2 - \tfrac{17}{16}\gamma^4 + \cdots\}\lambda \\ \beta\{1 - 2\gamma^2 + \cdots\} + \{\gamma^3 + \cdots\}\lambda \end{pmatrix}$$
$$\begin{pmatrix} \beta\{1 - 2\gamma^2 + \cdots\} + \{\gamma^3 + \cdots\}\lambda \\ 2\hbar\omega\{1 - \tfrac{43}{16}\gamma^4 + \cdots\} - \{1 + \tfrac{1}{2}\gamma^2 - \tfrac{17}{16}\gamma^4 + \cdots\}\lambda \end{pmatrix}$$

where

$$\gamma = \frac{\beta}{\hbar\omega}$$

The solutions of the corresponding secular equation are

$$\mathscr{E}_{\pm}^{(3)} = \frac{\hbar\omega[2 \mp \gamma \pm 2\gamma^3 - (43/8)\gamma^4 + \cdots]}{1 + \tfrac{1}{2}\gamma^2 \pm \gamma^3 - (17/16)\gamma^4 + \cdots}$$

$$= 2\hbar\omega \pm \beta - \frac{\beta^2}{\hbar\omega} \pm \frac{\beta^3}{2(\hbar\omega)^2} - \frac{7\beta^4}{4(\hbar\omega)^3} + \cdots$$

which are correct to the third order of β. Fig. 8–15 gives the necessary matrix elements for this calculation.

FIGURE 8–15. A diagram for calculating the second-order perturbation of Problem 8–70. Each state is designated by $(n_x n_y)$. Nondiagonal matrix elements between states are shown explicitly.

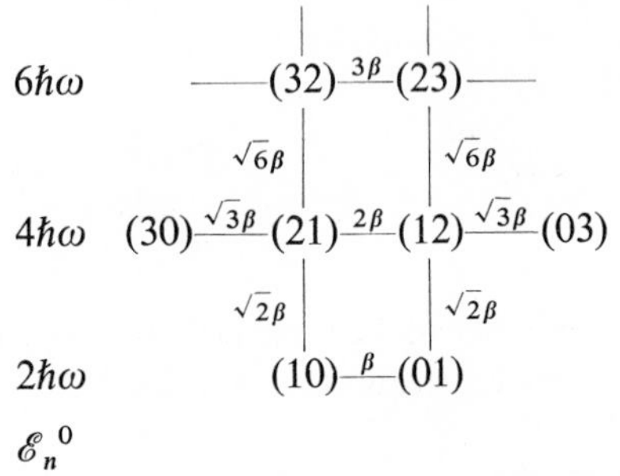

Fifth-Order Perturbation. We can reduce the problem into a nondegenerate one by taking

$$(n_x n_y)_{\pm} = \sqrt{\tfrac{1}{2}}[(n_x n_y) \pm (n_y n_x)]$$

for the bases, because, as shown in Fig. 8–16, αxy does not couple the $+$ and $-$ spaces. Applying $\hat{S}_n^{(1)}$ on each space we obtain

$$\mathscr{E}_{\pm}^{(5)} = 2\hbar\omega \pm \beta - \frac{[2\beta^2/(2\hbar\omega \pm \beta)] + [12\beta^4/(2\hbar\omega \pm \beta)^3] + \cdots}{1 + [2\beta^2/(2\hbar\omega \pm \beta)^2] + \cdots}$$

which indeed agree with the exact answers up to the β^5 terms.

FIGURE 8–16. Diagrams for calculating the fifth-order perturbation of Problem 8–70 (Notice that $(+)$- and $(-)$-spaces are completely separated. $(10)_{\pm} = \sqrt{\tfrac{1}{2}}[(10) \pm (01)]$), etc.

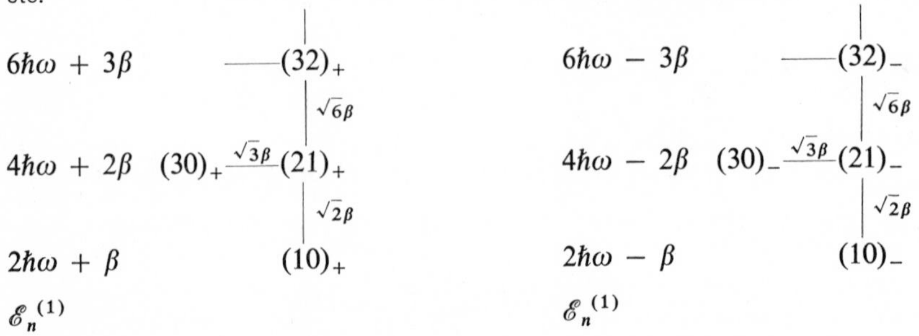

8-15 TIME-DEVELOPMENT OPERATOR. SCHROEDINGER, HEISENBERG, AND INTERACTION PICTURES

Schroedinger's equation (8–78) can be written as

$$i\hbar \frac{\partial}{\partial t} |\Psi, t\rangle = \hat{H}|\Psi, t\rangle \qquad (8\text{–}244)$$

in a general representation. When $\hat{H}$ does not depend on time explicitly, as is often the case, this equation can be formally solved as

$$|\Psi, t\rangle = \exp\left(\frac{-i\hat{H}t}{\hbar}\right) |\Psi, 0\rangle \qquad (8\text{–}245)$$

where $|\Psi, 0\rangle$ is an initial state, while $|\Psi, t\rangle$ is a final state at t, developed out of the initial state. The exponential operator in (8–245) is the time-development operator of a system given by $\hat{H}$. The time-development operator is a unitary operator, when $\hat{H}$ is Hermitian (Problem 8–49). We see clearly, in expression (8–245), that "the motion of particles conforms to the law of probability, but the probability itself is propagated in accordance with the law of causality."[15]

A matrix element of an operator $\hat{f}$ between states at t is

$$\langle\Psi', t|\hat{f}|\Psi, t\rangle = \langle\Psi', 0| \exp\left(\frac{i\hat{H}t}{\hbar}\right) \hat{f} \exp\left(\frac{-i\hat{H}t}{\hbar}\right) |\Psi, 0\rangle \qquad (8\text{–}246)$$

when $\hat{H}$ is Hermitian. This result implies that Schroedinger's picture, which assumes that wave functions are time dependent, but that operators are time independent, is equivalent to another picture, which assumes that wave functions are time independent, but operators are given by

$$\hat{f}_H = \exp\left(\frac{i\hat{H}t}{\hbar}\right) \hat{f} \exp\left(\frac{-i\hat{H}t}{\hbar}\right) \qquad (8\text{–}247)$$

We can easily see that this operator $\hat{f}_H$ satisfies the previous equation (8–73), so that this picture is nothing but the *Heisenberg picture*.

Schroedinger's and Heisenberg's pictures represent the two extremums (Table 8.3), and there can be many other pictures between them, where

[15] M. Born, *Zeits. f. Physik*, **38**, 804 (1926). See B. S. DeWitt's article, "Quantum Mechanics and Reality," *Physics Today*, September 1970, p. 30, for contemporary discussions on this subject.

374 PHYSICS OF THE MICROSCOPIC WORLD

TABLE 8-3 *Schroedinger, Heisenberg, and Interaction Pictures*

	Schroedinger	Interaction	Heisenberg
State	$\|\Psi, t\rangle$ $i\hbar \dfrac{\partial}{\partial t}\|\Psi, t\rangle = \hat{H}\|\Psi, t\rangle$	$\|\Psi_I, t\rangle = e^{i\hat{H}_0 t/\hbar}\|\Psi, t\rangle$ $i\hbar \dfrac{\partial}{\partial t}\|\Psi_I, t\rangle = \hat{H}_I\|\Psi_I, t\rangle$	$\|\Psi_H\rangle = \|\Psi, 0\rangle$ $\dfrac{\partial}{\partial t}\|\Psi_H\rangle = 0$
Time-development operator	$\exp(-i\hat{H}t/\hbar)$ $\hat{U}(t,0) = \hat{1} + \dfrac{1}{i\hbar}\displaystyle\int^t \hat{H}(t)\,dt + \cdots$	$\hat{U}_I(t,0) = \hat{1} + \dfrac{1}{i\hbar}\displaystyle\int^t \hat{H}_I(t)\,dt + \cdots$	$\hat{1}$
Hamiltonian Operator	$\hat{H} = \hat{H}_0 + \hat{H}_1$ $\hat{f}$ $\dfrac{d\hat{f}}{dt} = \dot{\hat{f}}$	$\hat{H}_I(t) = e^{i\hat{H}_0 t/\hbar}\hat{H}_1 e^{-i\hat{H}_0 t/\hbar}$ $\hat{f}_I(t) = e^{i\hat{H}_0 t/\hbar}\hat{f}e^{-i\hat{H}_0 t/\hbar}$ $i\hbar \dfrac{d\hat{f}_I}{dt} = [\hat{f}_I, \hat{H}_I] + i\hbar\dot{\hat{f}}_I$	$\hat{H}$ $\hat{f}_H(t) = e^{i\hat{H}t/\hbar}\hat{f}e^{-i\hat{H}t/\hbar}$ $i\hbar \dfrac{d\hat{f}_H}{dt} = [\hat{f}_H, \hat{H}] + i\hbar\dot{\hat{f}}_H$
Matrix element	$\langle\Psi', t\|\hat{f}\|\Psi, t\rangle = \langle\Psi_I', t\|\hat{f}_I(t)\|\Psi_I, t\rangle = \langle\Psi_H'\|\hat{f}_H(t)\|\Psi_H\rangle$		

both wave functions and operators depend on time. Thus if

$$\hat{H} = \hat{H}_0 + \hat{H}_1 \tag{8-248}$$

we can obtain the same matrix element as (8–246):

$$\langle \Psi', t | \hat{f} | \Psi, t \rangle = \langle \Psi_I', t | \hat{f}_I(t) | \Psi_I, t \rangle \tag{8-249}$$

by taking

$$|\Psi_I, t\rangle = \exp\left(\frac{i\hat{H}_0 t}{\hbar}\right) |\Psi, t\rangle = \exp\left(\frac{i\hat{H}_0 t}{\hbar}\right) \exp\left(\frac{-i\hat{H}t}{\hbar}\right) |\Psi, 0\rangle \tag{8-250}$$

and

$$\hat{f}_I(t) = \exp\left(\frac{i\hat{H}_0 t}{\hbar}\right) \hat{f} \exp\left(\frac{-i\hat{H}_0 t}{\hbar}\right) \tag{8-251}$$

where $|\Psi, t\rangle$ is the wave function in Schroedinger's picture. In this *interaction picture*, or *intermediate picture* as it is often called,[16] we obtain

$$i\hbar \frac{\partial}{\partial t} |\Psi_I, t\rangle = \exp\left(\frac{i\hat{H}_0 t}{\hbar}\right) (\hat{H} - \hat{H}_0) \exp\left(\frac{-i\hat{H}t}{\hbar}\right) |\Psi, 0\rangle$$

$$= \hat{H}_I |\Psi_I, t\rangle \tag{8-252}$$

where

$$\hat{H}_I = \exp\left(\frac{i\hat{H}_0 t}{\hbar}\right) \hat{H}_1 \exp\left(\frac{-i\hat{H}_0 t}{\hbar}\right) \tag{8-253}$$

PROBLEM 8–71. Show that $\hat{f}_H$ of (8–247) satisfies (8–73).

PROBLEM 8–72. Show that

$$e^{\hat{f}} \hat{g} e^{-\hat{f}} = \hat{g} + [\hat{f}, \hat{g}] + \tfrac{1}{2}[\hat{f}, [\hat{f}, \hat{g}]] + \tfrac{1}{6}[\hat{f}, [\hat{f}, [\hat{f}, \hat{g}]]] + \cdots$$

PROBLEM 8–73. When $\hat{H}_0$ does not commute with $\hat{H}_1$, show that

$$[\hat{H}_I(t_1), \hat{H}_I(t_2)] = [[\hat{H}_0, \hat{H}_1], \hat{H}_1](t_1 - t_2)\left(\frac{i}{\hbar}\right) + \cdots$$

assuming that $\hat{H}_0$ and $\hat{H}_1$ do not depend on time explicitly.

[16] S. Tomonaga, *Progr. Theor. Phys.* **1**, 27 (1946).

PROBLEM 8-74. Show that you *cannot* have

$$|\Psi_I, t\rangle = \exp\left(\frac{-i\hat{H}_I t}{\hbar}\right) |\Psi_I, 0\rangle$$

if $\hat{H}_0$ and $\hat{H}_1$ do not commute with each other.

HINT.

$$\exp\left(\frac{-i\hat{H}_I t}{\hbar}\right) = \sum_n \frac{t^n}{n!\,(i\hbar)^n} e^{i\hat{H}_0 t/\hbar} \hat{H}_1^{\,n} e^{-i\hat{H}_0 t/\hbar}$$

which shows that

$$i\hbar \frac{\partial}{\partial t} \exp\left(\frac{-i\hat{H}_I t}{\hbar}\right) \neq \hat{H}_I \exp\left(\frac{-i\hat{H}_I t}{\hbar}\right)$$

PROBLEM 8-75. In a simple harmonic oscillator, where $\hat{H}_0 = \frac{1}{2}\hbar\omega(\hat{a}\hat{a}^\dagger + \hat{a}^\dagger\hat{a})$, show that if

$$\hat{H}_1 = k(\hat{a} + \hat{a}^\dagger)$$

then

$$\hat{H}_I = k(\hat{a}e^{-i\omega t} + \hat{a}^\dagger e^{i\omega t})$$

HINT. See Problem 8-38.

8-16. TRANSITION PROBABILITY

From the probabilistic meaning of wave function, we see that

$$S_{a\to b}(t) = |\langle b| \exp\left(\frac{-i\hat{H}t}{\hbar}\right) |a\rangle|^2 \tag{8-254}$$

gives the probability that a state which was $|a\rangle$ at an initial time finds itself in $|b\rangle$ at time t later. This quantity, therefore, is the transition probability from a to b during the time interval t.

One may be tempted to assume the reversibility $S_{a \to b}(t) = S_{b \to a}(t)$, but that is not quite true. If $\hat{H}$ is Hermitian, we see

$$S_{b \to a}(t) = |\langle a| \exp \left(\frac{-i\hat{H}t}{\hbar} \right) |b\rangle|^2$$

$$= |\langle a| \exp \left(\frac{-i\hat{H}^{\dagger}t}{\hbar} \right) |b\rangle|^2$$

$$= |\langle b| \exp \left(\frac{i\hat{H}t}{\hbar} \right) |a\rangle|^2$$

$$= |\langle b| \exp \left(\frac{-i\hat{H}^{*}t}{\hbar} \right) |a\rangle^*|^2 \tag{8-255}$$

Therefore, the closest statement to the reversibility we can make is that if

$$\hat{H} = \hat{H}^* \tag{8-256}$$

or, if $\hat{H}$ is real, then

$$S_{b \to a}(t) = S_{a^* \to b^*}(t) \tag{8-257}$$

Notice that $|a\rangle^*$, which is a complex conjugate of $|a\rangle$, is not the same state as $|a\rangle$, unless the wave function is real. Thus, if $|a\rangle$ is a plane wave propagating in the direction $\mathbf{k}$;

$$\langle \mathbf{r} | a \rangle = e^{i\mathbf{k} \cdot \mathbf{r}} \tag{8-258}$$

then

$$\langle \mathbf{r} | a \rangle^* = e^{-i\mathbf{k} \cdot \mathbf{r}} \tag{8-259}$$

or the $|a\rangle^*$ is a plane wave propagating in the direction of $-\mathbf{k}$. Fig. 8–17 clearly shows that (8–257) may be true in many cases, but $S_{a \to b}(t) = S_{b \to a}(t)$ is not true in general.

If the initial state $|a\rangle$ happens to be an eigenstate of $\hat{H}$, then

$$\exp \left(\frac{-i\hat{H}t}{\hbar} \right) |a\rangle = \exp \left(\frac{-i\mathscr{E}_a t}{\hbar} \right) |a\rangle \tag{8-260}$$

which shows that the final state is still the same state $|a\rangle$, except for an unimportant phase factor: no transition to other states. In this sense, a state whose wave function is an eigenfunction of $\hat{H}$ is called a *stationary state*.

FIGURE 8–17. $S_{b \to a}(t) = S_{a^* \to b^*}$. This is true in many cases, but $S_{b \to a}(t) = S_{a \to b}(t)$ is not true in general.

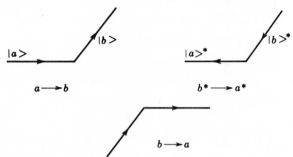

An interesting case occurs when

$$\hat{H} = \hat{H}_0 + \hat{H}_1 \tag{8–261}$$

and the initial and final states are eigenstates of $\hat{H}_0$:

$$\hat{H}_0|a\rangle = \mathscr{E}_a|a\rangle \qquad \hat{H}_0|b\rangle = \mathscr{E}_b|b\rangle \qquad \text{and} \qquad \langle a \,|\, b \rangle = 0 \tag{8–262}$$

If $\hat{H}_1$ is small enough, we can use the expansion formula (Problem 8–78)

$$
\begin{aligned}
\exp\left(\frac{-i\hat{H}t}{\hbar}\right) &= \exp\left(\frac{-i(\hat{H}_0 + \hat{H}_1)t}{\hbar}\right) \\
&= \exp\left(\frac{-i\hat{H}_0 t}{\hbar}\right) - \left(\frac{it}{\hbar}\right)\int_0^1 \exp\left[(1 - \xi)\left(\frac{-i\hat{H}_0 t}{\hbar}\right)\right]\hat{H}_1 \\
&\quad \times \exp\left(\frac{-i\xi\hat{H}_0 t}{\hbar}\right) d\xi + \cdots
\end{aligned} \tag{8–263}
$$

where the neglected terms are of higher order in $\hat{H}_1$. Note that $\hat{H}_0$ and $\hat{H}_1$ may not commute in general. To this approximation we obtain

$$
\begin{aligned}
S_{a \to b}(t) &\cong \left(\frac{t}{\hbar}\right)^2 \left|\int_0^1 \exp\left[(1 - \xi)\left(\frac{-i\mathscr{E}_b t}{\hbar}\right)\right]\langle b|\hat{H}_1|a\rangle \exp\left(\frac{-i\xi\mathscr{E}_a t}{\hbar}\right) d\xi\right|^2 \\
&= \frac{4}{\hbar^2}|\langle b|\hat{H}_1|a\rangle|^2 \frac{\sin^2\left(\tfrac{1}{2}\omega_{ab} t\right)}{\omega_{ab}^{\,2}}
\end{aligned} \tag{8–264}
$$

where

$$\omega_{ab} = \frac{(\mathscr{E}_a - \mathscr{E}_b)}{\hbar} \tag{8–265}$$

We see from (8–264) that the transition probability oscillates with the period of $2\pi/\omega_{ab}$ as shown in Fig. 8–18. This is another characteristic quantum effect.

FIGURE 8–18. $S_{a\to b}(t)$ of (8–248).

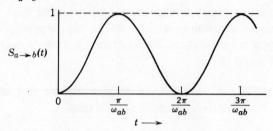

As ω_{ab} approaches zero, the period of oscillation becomes longer, and we see that $S(t)$ grows slowly but steadily, at least when $t < \pi/\omega_{ab}$, and approaches infinity in the limit of $\omega_{ab} \to 0$ (Fig. 8–19). Since

$$\int_{-\infty}^{\infty} \frac{\sin^2 x}{x^2} \, dx = \pi$$

we obtain

$$P_{a\to b}^{(1)} = \lim_{t\to\infty} \frac{S_{a\to b}(t)}{t} = \frac{2\pi}{\hbar^2} |\langle b|\hat{H}_1|a\rangle|^2 \, \delta(\omega_{ab}) \qquad (8-266)$$

which shows that in the limit of very small ω_{ab}, the transition probability per unit time, $P_{a\to b}^{(1)}$, can be nonzero. This formula is often called the *golden rule*. (See Problem 8–97c also.)

FIGURE 8–19 $S_{a\to b}(t)$ in a finite region of t for $\omega_{ab} = \pi$, $\frac{1}{2}\pi$, $\frac{1}{4}\pi$, $\frac{1}{8}\pi$, and $\frac{1}{16}\pi$.

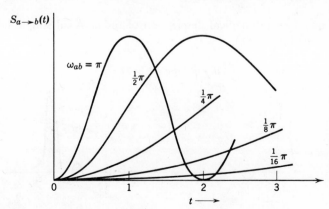

Formula (8–264) shows that if an observation is made at a small finite time Δt, a nonzero transition probability is found between states such that

$$\Delta\mathscr{E}\ \Delta t < \hbar \qquad (8\text{–}267)$$

or the energy conservation law is violated up to about $\hbar/\Delta t$. This relation looks similar to the uncertainty relation (8–72), but the present relation is not a result of any canonical relation; therefore there is a big conceptual difference between (8–267) and (8–72).

PROBLEM 8–76. A Hamiltonian $\hat{H}$ may not be real as assumed in (8–256). If we find a unitary operator $\hat{O}$ such that

$$\hat{O}\hat{H}^*\hat{O}^\dagger = \hat{H}$$

show that

$$S_{b\to a}(t) = S_{a'\to b'}(t)$$

where

$$|a'\rangle = \hat{O}|a\rangle^* \qquad |b'\rangle = \hat{O}|b\rangle^*$$

PROBLEM 8–77. Let $|n\rangle$ and $|n + 1\rangle$ be the eigenkets of $\hat{H}$ of a simple harmonic oscillator, with eigenvalues $n\hbar\omega$ and $(n + 1)\hbar\omega$, respectively. When

$$|a\rangle = \sqrt{\tfrac{1}{2}}\{|n\rangle + |n + 1\rangle\}$$
$$|b\rangle = \sqrt{\tfrac{1}{2}}\{|n\rangle - |n + 1\rangle\}$$

show that

$$S_{a\to b}(t) = \sin^2\left(\tfrac{1}{2}\omega t\right)$$

This is a quantum mechanical description of the classical simple harmonic oscillator. (See Fig. 8–20.)

FIGURE 8–20. States $|a\rangle$ and $|b\rangle$ of Problem 8–77.

PROBLEM 8–78. Show that

$$e^{\hat{f}+\hat{g}} = e^{\hat{f}} + \int_0^1 e^{(1-\xi)\hat{f}}\hat{g}e^{\xi\hat{f}}\,d\xi + \cdots$$

to the first order in $\hat{g}$.

HINT.

$$(\hat{f} + \hat{g})^l = \hat{f}^l + \sum_{m=0}^{l-1} \hat{f}^m\hat{g}\hat{f}^{l-m-1} + \cdots$$

to the first order in $\hat{g}$. Also

$$\int_0^1 (1 - \xi)^m\xi^n\,d\xi = \frac{m!\,n!}{(m + n + 1)!}$$

Dyson Series

In our discussions so far, we assumed that $\hat{H}$ does not depend on time explicitly, which is acceptable when the system is completely quantum mechanical. Sometimes, however, it is more convenient to treat a part of the system classically, and in such cases $\hat{H}$ may depend on time explicitly. For example, an atom may be in a magnetic field produced by an electromagnet, which we can control as we wish. The magnetic field can be treated quantum mechanically, of course, but since it is a macroscopic effect, and since a complete quantum mechanical calculation is rather complicated, it is better to treat the magnetic field classically and let the related part of the Hamiltonian be time dependent.

If $\hat{U}(t, 0)$ is the time development operator for a general case, or if

$$|\Psi, t\rangle = \hat{U}(t, 0)|\Psi, 0\rangle \tag{8–268}$$

then Schroedinger's equation gives

$$i\hbar \frac{\partial}{\partial t} \hat{U}(t, 0)|\Psi, 0\rangle = \hat{H}(t)\hat{U}(t, 0)|\Psi, 0\rangle \tag{8–269}$$

or

$$i\hbar \frac{\partial}{\partial t} \hat{U}(t, 0) = \hat{H}(t)\hat{U}(t, 0) \tag{8–270}$$

A formal solution of this operator equation is

$$\hat{U}(t, 0) = \hat{1} + \frac{1}{i\hbar} \int_0^t \hat{H}(t_1)\hat{U}(t_1, 0) \, dt_1 \qquad (8\text{--}271)$$

where the integration constant is fixed to satisfy the initial condition

$$\hat{U}(0, 0) = \hat{1} \qquad (8\text{--}272)$$

Iterating (8–271) we obtain an expansion formula

$$\hat{U}(t, 0) = \hat{1} + \frac{1}{i\hbar} \int_0^t \hat{H}(t_1) \, dt_1 + \frac{1}{(i\hbar)^2} \int_0^t dt_1 \int_0^{t_1} dt_2 \hat{H}(t_1)\hat{H}(t_2)$$

$$+ \frac{1}{(i\hbar)^3} \int_0^t dt_1 \int_0^{t_1} dt_2 \int_0^{t_2} dt_3 \hat{H}(t_1)\hat{H}(t_2)\hat{H}(t_3) + \cdots$$

$$(8\text{--}273)$$

which is called the *Dyson series*. It is important to keep the chronological order of the Hamiltonians in each term, since $\hat{H}$ at one time does not, in general, commute with $\hat{H}$ at another time.

PROBLEM 8–79. When $\hat{H}$ does not depend on time explicitly, show that the Dyson series reduces to the previous time-development operator $\exp(-i\hat{H}t/\hbar)$.

PROBLEM 8–80. When $\hat{H}$ depends on time explicitly, one may be tempted to express the time-development operator as $\exp(-i \int^t \hat{H} \, dt/\hbar)$. Show that this expression is not correct in general.

PROBLEM 8–81. When

$$\hat{H} = \begin{cases} \hat{H}' & \text{for} & -\infty < t < 0 \\ \hat{H}'' & \text{for} & 0 < t < \Delta \\ \hat{H}''' & \text{for} & \Delta < t < \infty \end{cases}$$

where $\hat{H}'$, $\hat{H}''$, and $\hat{H}'''$ do not depend on time explicitly, and

$$\hat{H}'|n'\rangle = \mathscr{E}_n'|n'\rangle \qquad \hat{H}''|m''\rangle = \mathscr{E}_m''|m''\rangle \qquad \hat{H}'''|l'''\rangle = \mathscr{E}_l'''|l'''\rangle$$

are given (Fig. 8–21),

FIGURE 8–21. The time-dependent potential of Problem 8–81.

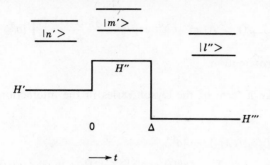

(a) show that

$$S_{n' \to l'''}(t > \Delta) = \left| \sum_m \langle l''' \mid m'' \rangle \langle m'' \mid n' \rangle \exp \left(\frac{-i \mathscr{E}_m'' \Delta}{\hbar} \right) \right|^2$$

(b) show that in the limit of $\Delta \to 0$

$$S_{n' \to l'''}(t > \Delta) \to |\langle l''' \mid n' \rangle|^2$$

This limiting expression is called the *sudden approximation*.

PROBLEM 8–82. A Hamiltonian changes with time such that

$$\hat{H} = \begin{cases} \hat{H}_0 & \text{for} & -\infty < t < 0 \\ \hat{H}_0 + \hat{H}_1(t) & \text{for} & 0 < t < \Delta \\ \hat{H}_0 & \text{for} & \Delta < t < \infty \end{cases}$$

where $\hat{H}_0$ does not depend on time explicitly, but $\hat{H}_1(t)$ may (Fig. 8–22). If

$$\Delta \ll \frac{\hbar}{|\mathscr{E}_{n^0} - \mathscr{E}_{m^0}|}$$

FIGURE 8–22. The time-dependent potential of Problem 8–82.

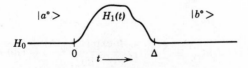

for any pair of the eigenstates of $\hat{H}_0$, and if

$$[\hat{H}_1(t), \hat{H}_1(t')] = 0$$

for any pair of times t and t', as is often the case, then show that

$$S_{a^0 \to b^0}(t > \Delta) = \left| \langle b^0 | \exp \left(\frac{-i \int_0^\Delta \hat{H}_1(\tau) \, d\tau}{\hbar} \right) |a^0\rangle \right|^2$$

is a good approximation.[17]

HINT. Take a term of the Dyson series in the interaction picture, and show that

$$\langle b^0 | \hat{H}_I(t_1) \hat{H}_I(t_2) \hat{H}_I(t_3) \cdots |a^0\rangle$$

$$= \sum_{n,m,\ldots} \langle b^0 | \hat{H}_I(t_1) | n^0 \rangle \langle n^0 | \hat{H}_I(t_2) | m^0 \rangle \langle m^0 | \hat{H}_I(t_3) \cdots |a^0\rangle$$

$$= \sum_{n,m,\ldots} \langle b^0 | \hat{H}_1(t_1) | n^0 \rangle \langle n^0 | \hat{H}_1(t_2) | m^0 \rangle \langle m^0 | \hat{H}_1(t_3) \cdots |a^0\rangle$$

from the first assumption.

PROBLEM 8–83. If $\hat{H}$ varies from $\hat{H}_0$ to $\hat{H}_0 + \hat{h} \, \Delta$ linearly from $t = 0$ to $t = \Delta$, or $\hat{H}(t) = \hat{H}_0 + \hat{H}_1(t) = \hat{H}_0 + \hat{h}t$, and if $\Delta \gg \langle m^0 | \hat{h} | n^0 \rangle / \omega_{mn}$, show, to the first approximation, that the time-developed state in the interaction picture,

$$\hat{U}_I (\Delta, 0) | n_I^0 \rangle \left(\cong \left(1 + \frac{1}{i\hbar} \int_0^\Delta \hat{H}_I(t) \, dt \right) | n_I^0 \rangle \right)$$

is the perturbed state $\hat{S}_n{}^0 | n^0 \rangle$ at the final time, when $\hat{H}_1 = \hat{h} \, \Delta$ (*Adiabatic theorem*).

HINT.

$$\hat{H}_I(t) | n_I^0 \rangle = \sum_m |m_I^0\rangle \langle m_I^0 | \hat{H}_I(t) | n_I^0 \rangle$$

$$= \sum_m |m_I^0\rangle \exp \left[\frac{i(\mathcal{E}_m{}^0 - \mathcal{E}_n{}^0)t}{\hbar} \right] \langle m_I^0 | \hat{h}t | n_I^0 \rangle$$

therefore

$$\int_0^\Delta \hat{H}_I(t) \, dt | n_I^0 \rangle \cong \sum_m |m_I^0\rangle \frac{\hbar \exp (i\omega_{mn} \Delta)}{i\omega_{mn}} \langle m_I^0 | \hat{h} \, \Delta | n_I^0 \rangle$$

$$= \sum_m |m_I^0\rangle \frac{\hbar}{i\omega_{mn}} \langle m_I^0 | \hat{H}_I(\Delta) | n_I^0 \rangle$$

[17] M. Mizushima, *Progr. Theor. Phys. Suppl.* **40**, 207 (1967).

ADDITIONAL PROBLEMS

8-84. Show that
(a)

$$[\hat{p}^2, [\hat{p}^2, \hat{x}^2]] = -8\hbar^2\hat{p}_x^2 \qquad [\hat{x}^2, [\hat{p}^2, \hat{x}^2]] = 8\hbar^2\hat{x}^2$$

(b)

$$[\hat{p}, \hat{l}^2] = i\hbar(\hat{l} \times \hat{p} - \hat{p} \times \hat{l}) \qquad [\hat{r}, \hat{l}^2] = i\hbar(\hat{l} \times \hat{r} - \hat{r} \times \hat{l})$$

8-85. Show that
(a)

$$[\hat{l} \cdot \mathbf{u}_1, \hat{r} \cdot \mathbf{u}_2] + [\hat{r} \cdot \mathbf{u}_1, \hat{l} \cdot \mathbf{u}_2] = 2i\hbar(\mathbf{u}_1 \times \mathbf{u}_2) \cdot \hat{r}$$

$$[\hat{l} \cdot \mathbf{u}_1, \hat{p} \cdot \mathbf{u}_2] + [\hat{p} \cdot \mathbf{u}_1, \hat{l} \cdot \mathbf{u}_2] = 2i\hbar(\mathbf{u}_1 \times \mathbf{u}_2) \cdot \hat{p}$$

(b)

$$\hat{l} \times \hat{r} + \hat{r} \times \hat{l} = 2i\hbar\hat{r} \qquad \hat{l} \times \hat{p} + \hat{p} \times \hat{l} = 2i\hbar\hat{p}$$

8-86.* If

$$[\hat{f}, [\hat{f}, \hat{g}]] = [\hat{g}, [\hat{f}, \hat{g}]] = 0$$

show that

$$e^{\hat{f}}e^{\hat{g}} = e^{\hat{f}+\hat{g}+1/2[\hat{f},\hat{g}]}$$

This is a special case of the Baker–Campbell–Hausdorff formula:[18]

$$e^{\hat{f}}e^{\hat{g}} = exp\,(\hat{f} + \hat{g} + \tfrac{1}{2}[\hat{f}, \hat{g}] + \tfrac{1}{12}[\hat{f}, [\hat{f}, \hat{g}]] + \tfrac{1}{12}[[\hat{f}, \hat{g}], \hat{g}] + \cdots)$$

8-87. Show that

$$[e^{\xi\hat{p}_x}, e^{\eta\hat{x}}] = -2ie^{\xi\hat{p}_x+\eta\hat{x}} \sin{(\tfrac{1}{2}\xi\eta\hbar)}$$

HINT.   Use the result of Problem 8–86.

[18] See R. D. Richtmyer and S. Greenspan, *Commun. Pure Appl. Math.* **18**, 107 (1963).

8-88. Show that[19]

$$[\hat{f}, e^{\alpha\hat{g}}] = \int_0^\alpha e^{(\alpha-\xi)\hat{g}}[\hat{f}, \hat{g}]e^{\xi\hat{g}} \, d\xi$$

8-89. Show that[20]

$$e^{\hat{f}}e^{-\hat{g}} = 1 + \int_0^1 e^{\xi\hat{f}}(\hat{f} - \hat{g})e^{-\xi\hat{g}} \, d\xi$$

8-90. In a simple harmonic oscillator
(a) show that

$$e^{i\alpha\hat{p}/\hbar} = e^{\beta\hat{a}}e^{-\beta\hat{a}^\dagger}e^{\frac{1}{2}\beta^2} = e^{-\beta\hat{a}^\dagger}e^{\beta\hat{a}^\dagger}e^{-\frac{1}{2}\beta^2}$$

where

$$\beta = \alpha \frac{(mk)^{1/4}}{(2\hbar)^{1/2}}$$

(b) show that

$$\langle 0|e^{i\alpha\hat{p}/\hbar}|0\rangle = e^{-\frac{1}{2}\beta^2}$$

where $|0\rangle$ is the ground state. This can be shown using the result of (a).

HINT. (a) can be shown using the result of Problem 8–86 and equations (8–136) and (8–137). Use (8–145) for (b). (b) can also be obtained using the result of Problem 8–15.

8-91. When $\hat{a}^\dagger$ and $\hat{a}$ are the creation and annihilation operators, respectively, show that

$$e^{\alpha\hat{a}^\dagger\hat{a}} = \sum_n \frac{1}{n!} (e^\alpha - 1)^n(\hat{a}^\dagger)^n\hat{a}^n$$

HINT. Operate each side of this equation on $|m\rangle$, which is an eigenfunction of $\hat{a}^\dagger\hat{a}$.

[19] R. Kubo, *Lecture Theor. Phys.* (Univ. Colo.), Vol. **1**, p. 139, Interscience, New York, 1958.
[20] R. P. Feynman, *Phys. Rev.* **76**, 749 (1949).

8-92. When $\hat{T}$ and $\hat{U}$ are the kinetic and potential energy operators, respectively, show that

$$2\langle \hat{T} \rangle = \langle \hat{\mathbf{r}} \cdot \nabla U \rangle$$

or, if U is proportional to r^n, then

$$2\langle \hat{T} \rangle = n\langle U \rangle$$

where $\langle \quad \rangle$ designates the expectation value. This is the quantum virial theorem.

HINT.

$$\frac{d}{dt} \langle \hat{\mathbf{r}} \cdot \hat{\mathbf{p}} \rangle = 0$$

8-93. Solving (8–91) for a one-dimensional space,
(a) show that

$$\frac{dS_2}{dx} = -\frac{1}{4p^2} \frac{d^2p}{dx^2} + \frac{3}{8p^3} \left(\frac{dp}{dx}\right)^2$$

(b) introducing the force $F = p(dp/dx)/m$, obtain

$$S_2 = -\frac{1}{4} \left(\frac{mF}{p^3}\right) - \tfrac{1}{8}m^2 \int \left(\frac{F^2}{p^5}\right) dx$$

8-94. (a) Show that

$$\langle \mathbf{r} \mid \mathbf{p} \rangle = (2\pi)^{-3/2} e^{i\mathbf{p}\cdot\mathbf{r}/\hbar}$$

using the coordinate representation, and then using the momentum representation.
(b) Show that

$$\langle \mathbf{r} \mid \Psi \rangle = \int \langle \mathbf{r} \mid \mathbf{p} \rangle \langle \mathbf{p} \mid \Psi \rangle \, d\mathbf{p}$$

$$\langle \mathbf{p} \mid \Psi \rangle = \int \langle \mathbf{p} \mid \mathbf{r} \rangle \langle \mathbf{r} \mid \Psi \rangle \, dv$$

and compare these expressions with (8–205) and (8–206).

8–95. Show that

$$\sum_m |f_n\rangle\langle g_m|$$

is a unitary operator. $|f_n\rangle$ and $|g_m\rangle$ are eigenkets of $\hat{f}$ and $\hat{g}$, respectively.

8–96. (a) In a harmonic oscillator, show that

$$\langle x|e^{-i\xi\hat{p}/\hbar}|0\rangle$$

is the wave packet, whose center is at $x = \xi$.

(b) Show that the time development of the above wave packet under the simple harmonic potential is given by

$$e^{-i\omega(\hat{a}^\dagger\hat{a}+1/2)t}e^{-i\xi\hat{p}/\hbar}|0\rangle = \exp\,(\eta\hat{a}^\dagger e^{-i\omega t})|0\rangle \exp\,(-\tfrac{1}{2}\eta^2)$$

where

$$\eta = \frac{\xi\sqrt{2}}{\sqrt{m\hbar\omega}}$$

(c) If ξ and $-\xi$ designate the wave packets whose center is at $x = \xi$ and $x = -\xi$, respectively (Fig. 8–23), show that the transition probability from one wave packet to the other is given by

$$S_{\xi\rightarrow-\xi}(t) = \exp\,\left[-4\eta^2\cos^2\,(\tfrac{1}{2}\omega t)\right]$$

FIGURE 8–23. The wave packets under a simple harmonic potential.

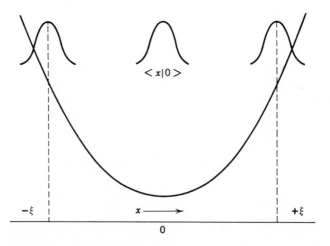

Note that this transition probability oscillates between $\exp(-4\eta^2)$ and one with the period of ω. It is a quantum mechanical description of the classical harmonic oscillator. (See Problem 8–77 also.)

HINT.   For (a), see (8–148). For (b) and (c), see (8–132), (8–133), (8–134), Problems 8–37, 8–83a, and 8–84. $e^{\eta\hat{a}}|0\rangle = |0\rangle$.

8-97.   (a) Show that

$$e^{\hat{f}+\hat{g}} = e^{\hat{f}} + \int_0^1 d\xi e^{(1-\xi)\hat{f}}\hat{g}e^{\xi\hat{f}} + \int_0^1 d\xi \int_0^{1-\xi} d\eta e^{(1-\xi-\eta)\hat{f}}\hat{g}e^{\xi\hat{f}}\hat{g}e^{\eta\hat{f}} + \cdots$$

that is, calculate the next term in the expansion of Problem 8–78.
(b) When $\langle a|\hat{H}_1|b\rangle = 0$, show that

$$S_{a\to b}(t) \cong \frac{4}{\hbar^4\omega_{ab}^2}\left|\sum_c \left[\frac{\sin\left(\frac{1}{2}\omega_{ac}t\right)}{\omega_{ac}}e^{i(1/2)\omega_{ac}t} - \frac{\sin\left(\frac{1}{2}\omega_{bc}t\right)}{\omega_{bc}}e^{i(1/2)\omega_{bc}t}\right]\right.$$
$$\left.\times \langle b|\hat{H}_1|c\rangle\langle c|\hat{H}_1|a\rangle\right|^2$$

(c) If $\omega_{ab} \neq 0$, but states $|c\rangle$ and $|d\rangle$ exist in such a way that $\omega_{ac} \cong 0$ and $\omega_{bd} \cong 0$, then show that

$$P_{a\to b}^{(2)} = \frac{|\langle b|\hat{H}_1|c\rangle|^2 P_{a\to c}^{(1)} + |\langle a|\hat{H}_1|d\rangle|^2 P_{d\to b}^{(1)}}{(\hbar\omega_{ab})^2}$$

where $P^{(1)}$ is defined by (8–266).

8-98.*   If $[\hat{H}_I(t_1), \hat{H}_I(t_2)]$ commutes with $\hat{H}_I$ of any other time, show that

$$\hat{U}_I(t, 0) = \exp\left\{-\frac{i}{\hbar}\int_0^t \hat{H}_I(t_1)\,dt_1 - \frac{1}{2\hbar^2}\int_0^t dt_1 \int_0^{t_1} dt_2[\hat{H}_I(t_1), \hat{H}_I(t_2)]\right\}$$

This is a special case of the Magnus formula:

$$\hat{U}(t, 0) = \exp\left\{(i\hbar)^{-1}\int_0^t \hat{H}_1\,dt_1 + \frac{1}{2}(i\hbar)^{-2}\int_0^t dt_1 \int_0^{t_1} dt_2[\hat{H}_1, \hat{H}_2]\right.$$

$$+ \frac{1}{6(i\hbar)^3}\int_0^t dt_1 \int_0^{t_1} dt_2 \int_0^{t_2} dt_3\{[[\hat{H}_1, \hat{H}_2], \hat{H}_3]$$

$$+ [[\hat{H}_3, \hat{H}_2], \hat{H}_1]\}$$

$$+ \frac{1}{12(i\hbar)^4}\int_0^t dt_1 \int_0^{t_1} dt_2 \int_0^{t_2} dt_3 \int_0^{t_3} dt_4$$

$$\times \{[[[\hat{H}_3, \hat{H}_2], \hat{H}_4], \hat{H}_1] + [[[\hat{H}_3, \hat{H}_4], \hat{H}_2], \hat{H}_1]$$

$$\left.+ [[[\hat{H}_1, \hat{H}_2], \hat{H}_3], \hat{H}_4] + [[[\hat{H}_4, \hat{H}_1], \hat{H}_3], \hat{H}_2]\} + \cdots\right\}$$

$$\text{where}\quad \hat{H}_i \equiv \hat{H}(t_i)$$

8-99. (a) If $\hat{H}_0 = \hbar\omega(\hat{a}^\dagger\hat{a} + \tfrac{1}{2})$ and $\hat{H}_1 = k(\hat{a}^\dagger + \hat{a})$, where k is a constant, show that

$$[\hat{H}_I(t_1), \hat{H}_I(t_2)] = i2k^2 \sin\left[\omega(t_2 - t_1)\right]$$

(b) Show that

$$\hat{U}_I(t, 0) = \exp\left\{\frac{k}{\hbar\omega}\left[(\hat{a}^\dagger - \hat{a})(1 - \cos\omega t) - i(\hat{a}^\dagger + \hat{a})\sin\omega t\right]\right.$$

$$\left. - \frac{ik^2}{\hbar^2}\left(\frac{\sin t}{\omega^2} - \frac{t}{\omega}\right)\right\}$$

HINT. See Problems 8-38, 8-75, and 8-98, and equation (8-138).

Chapter 9   One Particle in a Potential

The structure of atoms is a purely quantum mechanical problem. The stability and the energy levels of the hydrogen atom, the simplest of all atoms, were first explained by Bohr in his quantum theory, and then by Schroedinger as a crucial and successful test of his equation. Scattering theory is based on the same equation, except that the energy is positive. This is in contrast to the case of bound states, where it is negative. However, it is a relatively recent discovery that the positive and negative energy states can be understood simultaneously from analytic properties of a single radial function in the complex k plane. Dispersion relations result from such mathematical investigation. The scattering of electromagnetic waves can be reinvestigated in a quantum mechanical formulation, and the dispersion relation for this case, which is called the Kramers–Kronig relation, is a result.

9-1 HYDROGEN ATOM AND HYDROGEN-LIKE IONS. I

The hydrogen atom is made of one electron and one proton interacting with each other through the Coulomb potential. The Hamiltonian of this system in the nonrelativistic approximation is

$$\hat{H} = \frac{\hat{\mathbf{p}}_p^{\,2}}{2m_p} + \frac{\hat{\mathbf{p}}_e^{\,2}}{2m_e} - \frac{e^2}{4\pi\varepsilon_0 r} \tag{9-1}$$

where the subscripts p and e refer to the proton and the electron, respectively, and r is the distance between them. If the proton is replaced by a heavier nucleus of mass m_n and charge Ze, we have a hydrogen-like ion, whose Hamiltonian is

$$\hat{H} = \frac{\hat{\mathbf{p}}_n^{\,2}}{2m_n} + \frac{\hat{\mathbf{p}}_e^{\,2}}{2m_e} - \frac{Ze^2}{4\pi\varepsilon_0 r} \tag{9-2}$$

Since (9–1) and (9–2) are mathematically equivalent, we will treat the problem using (9–2) in the following discussion.

We have seen in section 1–2 that a two-body problem can be reduced into two one-body problems in Newton's mechanics. The same situation exists in quantum mechanics: Thus if

$$\mathbf{r} = \mathbf{r}_e - \mathbf{r}_n \tag{9-3}$$

and

$$\mathbf{R} = \frac{m_e \mathbf{r}_e + m_n \mathbf{r}_n}{m_e + m_n} \tag{9-4}$$

we can reduce (9–2) to

$$H = H_R + H_r \tag{9-5}$$

where

$$\hat{H}_R = \frac{\hat{\mathbf{p}}_R^{\,2}}{2M} = -\frac{\hbar^2}{2M} \nabla_R^{\,2} \tag{9-6}$$

and

$$\hat{H}_r = \frac{\hat{\mathbf{p}}_r^{\,2}}{2m} - \frac{Ze^2}{4\pi\varepsilon_0 r} = -\frac{\hbar^2}{2m} \nabla_r^{\,2} - \frac{Ze^2}{4\pi\varepsilon_0 r} \tag{9-7}$$

taking the coordinate representation. Here M and m are the total mass and the reduced mass, respectively, of this system. The Hamiltonians (9–6) and

(9-7) are for the external (or center of mass) motion and for the internal motion, respectively. Since the nuclear mass is much larger than the electron mass, the center of mass is very near to that of the nucleus. That is, (9-6) approximately gives the motion of the nucleus in free space, while (9-7) approximately gives the motion of the electron in the field of the nucleus. (See Problem 1-1.)

Since the Hamiltonian is completely separated into two parts, its eigenfunction can be written as a product

$$\hat{H}\psi_R(\mathbf{R})\psi_r(\mathbf{r}) = (\mathscr{E}_R + \mathscr{E}_r)\psi_R(\mathbf{R})\psi_r(\mathbf{r}) \tag{9-8}$$

where

$$\hat{H}_R\psi_R(\mathbf{R}) = \mathscr{E}_R\psi_R(\mathbf{R}) \tag{9-9}$$

and

$$\hat{H}_r\psi_r(\mathbf{r}) = \mathscr{E}_r\psi_r(\mathbf{r}) \tag{9-10}$$

Equation (9-9), for the external motion, is an equation for a free particle. If we express the Laplacian in the Cartesian coordinate system as (5-28), we obtain a solution of (9-9)

$$\psi_k(\mathbf{R}) = (2\pi)^{-3/2}e^{i\mathbf{k}\cdot\mathbf{R}} \tag{9-11}$$

where

$$\mathscr{E}_R = \frac{(\mathbf{k}\hbar)^2}{2M} \tag{9-12}$$

instead of (5-31). We can also express the Laplacian in the spherical coordinate system as (5-30), and obtain (5-41) and (5-49), but

$$\left[\frac{1}{R^2}\frac{d}{dR}\left(R^2\frac{d}{dR}\right) - \frac{l(l+1)}{R^2} + \frac{2M}{\hbar^2}\mathscr{E}\right]f(R) = 0 \tag{9-13}$$

in place of (5-50). Under the assumption of the uniqueness of the eigenfunction, (5-51), l is an integer number. The spherical wave solution is, therefore

$$\psi_{\mathscr{E}lm}(\mathbf{R}) = f_{\mathscr{E}l}(R)Y_{lm}(\theta, \varphi) \tag{9-14}$$

where $Y_{lm}(\theta, \varphi)$ is the spherical harmonic and

$$f_{\mathscr{E}l}(R) = Aj_l(\xi) + Bn_l(\xi) \tag{9-15}$$

in terms of the spherical Bessel function j_l and the spherical Neuman function n_l. In (9–15)

$$\xi = \sqrt{2M\mathscr{E}} \, \frac{R}{\hbar} \tag{9-16}$$

and A and B are constants. Since the Neuman functions diverge at the origin, solutions which stay finite in the entire space are obtained by taking $B = 0$ in (9–15).

PROBLEM 9-1. Obtain (9–5) from (9–2).

PROBLEM 9-2. Obtain $\psi_R(\mathbf{R})$ in the cylindrical coordinate system.

9-2 HYDROGEN ATOM AND HYDROGEN-LIKE IONS. II

Since $\hat{H}_r$, the Hamiltonian for the internal motion, has the Coulomb potential term, a logical choice for the coordinate system is the spherical coordinate system. We see, expressing the Laplacian in the spherical coordinates, that an eigenfunction of $\hat{H}_r$ can be written as

$$\psi_{\mathscr{E}lm}(r) = f_{\mathscr{E}l}(r) Y_{lm}(\theta, \varphi) \tag{9-17}$$

where $f_{\mathscr{E}l}(r)$ is a function of r, and is a solution of

$$\left[\frac{1}{r^2} \frac{d}{dr} \left(r^2 \frac{d}{dr} \right) - \frac{l(l+1)}{r^2} + \frac{2Z}{ar} + \frac{2m}{\hbar^2} \mathscr{E} \right] f_{\mathscr{E}l} = 0 \tag{9-18}$$

where

$$a = \frac{4\pi\varepsilon_0 \hbar^2}{me^2} \cong a_0 = \frac{4\pi\varepsilon_0 \hbar^2}{m_e e^2} = 5.2917715 \times 10^{-11} \text{ m} \tag{9-19}$$

This length a_0 is called the *Bohr radius*, and is a convenient unit of length in which atomic dimensions are expressed.

The solution of the differential equation (9–18) for $\mathscr{E} < 0$, which does not diverge at the origin, is

$$f_{\mathscr{E}l}(r) = N_{\mathscr{E}l} \left(\frac{2r\sqrt{-2m\mathscr{E}}}{\hbar} \right)^l \exp \left(-\frac{r\sqrt{-2m\mathscr{E}}}{\hbar} \right)$$

$$\times F \left(\frac{-Z\hbar}{a\sqrt{-2m\mathscr{E}}} + l + 1, 2l + 2; \frac{2r\sqrt{-2m\mathscr{E}}}{\hbar} \right) \tag{9-20}$$

where F is the confluent hypergeometric function defined by

$$F(\xi, \eta; z) = 1 + \frac{\xi z}{\eta 1!} + \frac{\xi(\xi + 1)z^2}{\eta(\eta + 1)2!} + \cdots \qquad (9\text{-}21)$$

[See Appendix I(c).]

When the real part of z becomes very large, the confluent hypergeometric function approaches

$$F(\xi, \eta; z) \xrightarrow[\mathscr{R}ez \to \infty]{} \frac{\Gamma(\eta)}{\Gamma(\xi)} z^{\xi - \eta} e^z \qquad (9\text{-}22)$$

unless the power series (9-21) terminates at a finite term. We see from (9-20) that, the radial function f diverges as r approaches infinity unless the power series of the confluent hypergeometric function terminates at a finite term. Comparing (9-21) with (9-20), we thus see that in order for the radial function f to stay finite in the entire space we should have

$$-\frac{Zh}{a\sqrt{-2m\mathscr{E}}} + l + 1 = -n' \qquad (n' = 0, 1, 2, \ldots) \qquad (9\text{-}23)$$

Thus

$$\mathscr{E}_n = -\frac{Z^2 \hbar^2}{2a^2 mn^2} = -\frac{Z^2}{2n^2} \left(\frac{m}{m_e}\right) \frac{e^4 m_e}{(4\pi\varepsilon_0 \hbar)^2} \qquad (9\text{-}24)$$

where

$$n = n' + l + 1 = \text{integer} \geq l + 1 \qquad (9\text{-}25)$$

The resultant energy levels are shown in Fig. 9-1 for the hydrogen atom. The last factor in (9-24),

$$\frac{e^4 m_e}{(4\pi\varepsilon_0 \hbar)^2} = \frac{e^2}{4\pi\varepsilon_0 a_0} \qquad (9\text{-}26)$$

is a natural unit of energy in atoms. This energy unit, the Bohr radius, and other units convenient in discussing atomic structure, are often called atomic units. The atomic units are tabulated in Table 9-1.

FIGURE 9–1.   Energy levels of the hydrogen atom (bound states).

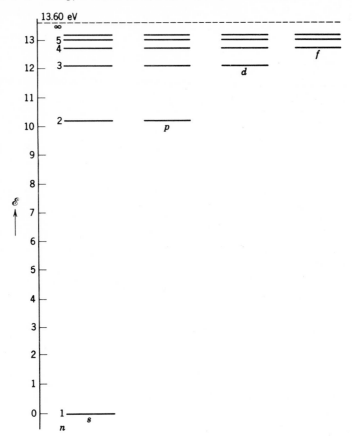

TABLE 9–1   *Atomic Units*[a]

Quantity	Symbol (Name)	Value in mks Units
Mass	m_e (electron mass)	9.109558×10^{-31} kg
Length	a_0 (Bohr radius)	$5.2917715 \times 10^{-11}$ m
Energy	$e^2/(4\pi\varepsilon_0 a_0)$	4.359828×10^{-18} J
Time	$4\pi\varepsilon_0 a_0 \hbar/e^2$	2.418884×10^{-17} sec
Angular momentum	$\hbar$	$1.0545919 \times 10^{-34}$ J sec
Linear momentum	$\hbar/a_0$	$1.9928901 \times 10^{-24}$ kg m/sec
Charge	e	$1.6021917 \times 10^{-19}$ C
Scalar potential	$e/(4\pi\varepsilon_0 a_0)$	27.211652 V

[a] *The following are 1 in a.u.*

Using the above solutions to the angular and radial portions of $\hat{H}_r$, we obtain the eigenfunction (9–17) as

$$\psi_{nlm}(\mathbf{r}) \; (\equiv \langle \mathbf{r} \mid nlm \rangle)$$

$$= N_{nl} \left(\frac{2Z\rho}{n} \right)^l e^{-Z\rho/n} F\left(-(n - l - 1), \, 2l + 2; \frac{2Z\rho}{n} \right) Y_{lm}(\theta, \varphi) \qquad (9\text{–}27)$$

where

$$\rho = \frac{r}{a} \qquad (9\text{–}28)$$

and the normalization constant is

$$N_{nl} = \frac{1}{(2l + 1)!} \sqrt{\frac{(n + l)!}{2n(n - l - 1)!}} \left(\frac{2Z}{na} \right)^{3/2} \qquad (9\text{–}29)$$

Formula (9–27) shows that each eigenfunction is specified by three integer numbers, n, l, and m, which are called the principal, the orbital angular momentum, and the magnetic quantum numbers, respectively. The radial part of the eigenfunctions $f_{nl}(\rho)$ are shown in Table 9–2 and Fig. 9–2 for some states. In Table 9–2 and Fig. 9–1, the conventional spectroscopic notations, $s, p, d, f, h, \ldots$, are used for $l = 0, 1, 2, 3, 4, \ldots$, respectively.

TABLE 9–2   *Some Radial Functions*

nl	$f_{nl}(\rho)(a/Z)^{3/2}$
$1s$	$2e^{-Z\rho}$
$2s$	$\sqrt{\tfrac{1}{2}}(1 - \tfrac{1}{2}Z\rho)e^{-Z\rho/2}$
$2p$	$\dfrac{1}{2\sqrt{6}} Z\rho e^{-Z\rho/2}$
$3s$	$\dfrac{2}{3\sqrt{3}}(1 - \tfrac{2}{3}Z\rho + \tfrac{2}{27}(Z\rho)^2)e^{-Z\rho/3}$
$3p$	$\dfrac{8}{27\sqrt{6}}(1 - \tfrac{1}{6}Z\rho)Z\rho e^{-Z\rho/3}$
$3d$	$\dfrac{4}{81\sqrt{30}}(Z\rho)^2 e^{-Z\rho/3}$

FIGURE 9–2. Hydrogenic radial functions for some lower bound states.

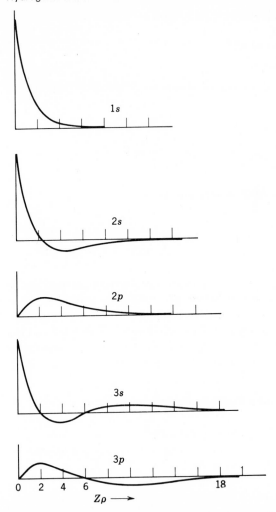

Formula (9–26) shows that the energy of a bound state depends on the principal quantum number only. Since l can take any integer number between 0 and $n - 1$, and m can take any integer number between l and $-l$, we see the number of independent states which have the same n, or the same energy, is n^2. Therefore, each level has n^2-fold degeneracy. The m degeneracy, which is of $(2l + 1)$-fold, is due to the spherical symmetry of the Hamiltonian, and will be discussed in section 11–8. The l degeneracy, i.e., the fact that energy does not depend on the l-quantum number, is sometimes called the accidental degeneracy, since it is found only for the $(1/r)$-potential. (See section 11–17).

Using a general formula, one can calculate the expectation value of a power of ρ. The results are shown in Table 9–3.

TABLE 9–3 *Some Expectation Values*

$$\langle \rho^2 \rangle = \frac{n^2[5n^2 + 1 - 3l(l + 1)]}{2Z^2}$$

$$\langle \rho \rangle = \frac{3n^2 - l(l + 1)}{2Z}$$

$$\left\langle \frac{1}{\rho} \right\rangle = \frac{Z}{n^2}$$

$$\left\langle \frac{1}{\rho^2} \right\rangle = \frac{Z^2}{n^3(l + \frac{1}{2})}$$

$$\left\langle \frac{1}{\rho^3} \right\rangle = \frac{Z^3}{n^3(l + 1)(l + \frac{1}{2})l}$$

PROBLEM 9–3. Show that (9–18) can be written as

$$\left[\frac{d^2}{dr^2} - \frac{l(l + 1)}{r^2} + \frac{2Z}{ar} + \frac{2m}{\hbar^2} \mathscr{E} \right] (rf) = 0$$

PROBLEM 9–4. Using the above result, observe that Theorem 2 of section 8–9 is applicable to the radial functions shown in Table 9–2.

PROBLEM 9–5. Applying the WKB method to find rf of Problem 9–3, show that when $l = 0$ the exact solution (9–24) is obtained if the phase factor $\frac{1}{4}\pi$ is neglected both at $r = 0$ and at the classical turning point.

PROBLEM 9–6. Using Table 9–3, show that the virial theorem (Problem 8–92) holds for a hydrogen-like ion.

PROBLEM 9–7. Using Table 9–3, calculate the mean square deviation $\langle nlm|(\Delta\rho)^2|nlm \rangle$.

PROBLEM 9–8. Show that $\langle (\Delta p_x)^2 \rangle \langle (\Delta x)^2 \rangle = (5n^2 + 1)\hbar^2/18$ in the ns state of a hydrogen-like ion.

HINT. $\langle \hat{p}^2 \rangle = 2m(-\langle U \rangle + \mathscr{E}_n) = -2m\mathscr{E}_n$, using the virial theorem.

PROBLEM 9–9. Find the energy of bound states of a particle under the potential

$$U = \frac{\alpha}{r^2} - \frac{\beta}{r}$$

where α and β are constants.

ANSWER.

$$\mathscr{E}_{vl} = -\frac{2m\beta^2}{[\hbar(2v + 1) + \sqrt{(2l + 1)^2\hbar^2 + 8m\alpha}]^2}$$

where

$$v = 0, 1, 2, \ldots$$

9–3* COULOMB SCATTERING

Let us return to (9–10), which is

$$\left(-\frac{\hbar^2}{2m}\nabla^2 - \frac{Ze^2}{4\pi\varepsilon_0 r} - \mathscr{E}\right)\psi = 0 \qquad (9\text{–}30)$$

and this time, assuming that $\mathscr{E} > 0$, let us find a solution which is as close to a plane wave as possible. For that purpose we introduce k and γ by

$$\mathscr{E} = \frac{(\hbar k)^2}{2m} \qquad (9\text{–}31)$$

$$\frac{Ze^2}{4\pi\varepsilon_0} = \frac{\hbar^2}{2m}(2\gamma k) \qquad (9\text{–}32)$$

and write (9–30) as

$$\left(\nabla^2 + k^2 + \frac{2\gamma k}{r}\right)\psi = 0 \qquad (9\text{–}33)$$

If we assume the solution of this equation in the form of

$$\psi = e^{ikz}F(\zeta) \qquad (9\text{–}34)$$

where

$$\zeta = r - z = 2r \sin^2 \tfrac{1}{2}\theta \qquad (9\text{–}35)$$

a straightforward calculation shows that (9–33) reduces to

$$\left[\zeta \frac{d^2}{d\zeta^2} + (1 - ik\zeta) \frac{d}{d\zeta} + \gamma k\right] F = 0 \qquad (9\text{–}36)$$

a regular solution of this equation is the confluent hypergeometric function $F(i\gamma, 1; ik\zeta)$. Thus we find a solution of (9–33)

$$\psi = e^{ikz} \frac{F(i\gamma, 1; ik\zeta)}{\mathscr{A}} \qquad (9\text{–}37)$$

where $\mathscr{A}$ is an adjustable constant.

The asymptotic expansion of the confluent hypergeometric function at large $k\zeta$ is

$$F(i\gamma, 1; ik\zeta) \rightarrow \frac{(-ik\zeta)^{-i\gamma}}{\Gamma(1 - i\gamma)} + \frac{e^{ik\zeta}}{\Gamma(i\gamma)} (ik\zeta)^{i\gamma - 1}$$

$$= \frac{e^{-\frac{1}{2}\pi\gamma}}{\Gamma(1 - i\gamma)} e^{-i\gamma \ln(k\zeta)} - \frac{e^{-\frac{1}{2}\pi\gamma}}{\Gamma(1 + i\gamma)k\zeta} e^{ik\zeta + i\gamma \ln(k\zeta)}$$

$$(9\text{–}38)$$

Choosing

$$\mathscr{A} = \frac{e^{-\frac{1}{2}\pi\gamma}}{\Gamma(1 - i\gamma)} \qquad (9\text{–}39)$$

we see, from (9–37), that

$$\psi \rightarrow e^{ikz - i\gamma \ln(k\zeta)} - A(\theta) \frac{e^{ikr + i\gamma \ln(2kr)}}{r} \qquad (9\text{–}40)$$

where

$$A(\theta) = \frac{\gamma \Gamma(1 - i\gamma)}{2k\Gamma(1 + i\gamma) \sin^2 \frac{1}{2}\theta} e^{2i\gamma \ln(\sin \frac{1}{2}\theta)} \qquad (9\text{–}41)$$

The first term of (9–40) is essentially a plane wave propagating in the z direction. Actually when the probability current density $\mathbf{j}$ is calculated according to (8–83), we see that j_x and j_y approach zero as $1/r$, while j_z approaches $\hbar k/m$ as r goes to infinity. In the same sense the second term of (9–40) is essentially the spherical wave going out from the center as

e^{ikr}/r (Fig. 9–3). The probability current density of this outgoing spherical wave in a solid angle $d\Omega$ is

$$j_r r^2 \, d\Omega = \left(\frac{\hbar k}{m}\right) |A(\theta)|^2 \, d\Omega \qquad (9\text{–}42)$$

FIGURE 9–3. Scattering. Incident plane wave and scattered spherical wave.

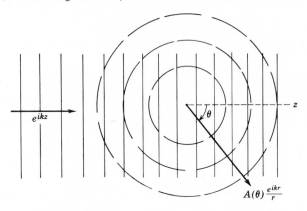

Therefore, the differential scattering cross section, which is the above probability current density divided by that of the incident plane wave, is

$$\frac{d\sigma}{d\Omega} = |A(\theta)|^2 = \left|\frac{\Gamma(1 - i\gamma)}{\Gamma(1 + i\gamma)}\right|^2 \frac{\gamma^2}{4k^2 \sin^4 \tfrac{1}{2}\theta} = \left(\frac{Ze^2}{16\pi\varepsilon_0 \mathscr{E}}\right)^2 \frac{1}{\sin^4 \tfrac{1}{2}\theta} \qquad (9\text{–}43)^1$$

This formula, which is plotted in Fig. 9–4, was derived from classical mechanics by Rutherford and verified experimentally by him. Quantum mechanics gives the same result as classical mechanics in this case.

PROBLEM 9–10. Obtain the Rutherford formula (9–43) in classical mechanics.

HINT. Problem (1–38c) gives the relation between the deflection angle θ and other constants. When the incident beam is homogeneous, the probability of an incident particle being in $b \sim b + db$ is proportional to

---

[1] $\left|\dfrac{\Gamma(1 - i\gamma)}{\Gamma(1 + i\gamma)}\right|^2 = \dfrac{\Gamma(1 - i\gamma)\Gamma^*(1 - i\gamma)}{\Gamma(1 + i\gamma)\Gamma^*(1 + i\gamma)} = \dfrac{\Gamma(1 - i\gamma)\Gamma(1 + i\gamma)}{\Gamma(1 + i\gamma)\Gamma(1 - i\gamma)} = 1$

FIGURE 9–4. $d\sigma/d\Omega$ versus θ according to the Rutherford formula.

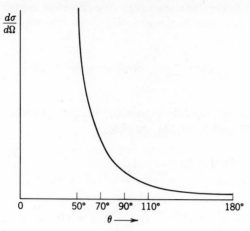

$b\,db$, which, in turn, is proportional to $\cos\left(\tfrac{1}{2}\theta\right)d\theta/\sin^3\left(\tfrac{1}{2}\theta\right)$. On the other hand, $d\Omega = 2\pi\sin\theta\,d\theta = 4\pi\sin\left(\tfrac{1}{2}\theta\right)\cos\left(\tfrac{1}{2}\theta\right)d\theta$. See Fig. 9–5.

FIGURE 9–5. Geometrical interpretation of $d\sigma/d\Omega$ in particle scattering.

$$j_{in}\,d\sigma = j_{sc}r^2\,d\Omega \quad\text{or}\quad d\sigma/d\Omega = r^2(j_{sc}/j_{in})$$

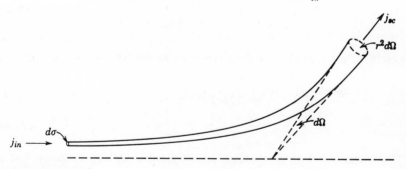

Bound States

Equation (9–33) can also be expressed in spherical coordinates as we did in the previous two sections. Since the only difference between the present case and the case of section 9–2 is that energy $\mathscr{E}$ is now positive, we see that the previous solution (9–17) with (9–20) holds if we generalize the radial function (9–20) by allowing it to be complex. Using our abbreviations (9–31) and (9–32) we obtain

$$f_l(k,\,r) = N(-ikr)^l e^{ikr} F(l + 1 - i\gamma,\, 2l + 2;\, -2ikr) \qquad (9\text{–}44)$$

from (9–20). N is a normalization constant, which is different from the previous one, because we are now considering continuous states.

We have seen in (9–15) that when the particle is free the radial function is the spherical Bessel function $j_l(kr)$. We therefore expect that (9–44) will approach $j_l(kr)$ for large distances r. We know that asymptotically

$$j_l(kr) \xrightarrow[r \to \infty]{} \frac{1}{kr} \cos\left[kr - \tfrac{1}{2}(l + 1)\pi\right] \qquad (9\text{–}45)$$

Using the same asymptotic expansion of the confluent hypergeometric function that we used in (9–38), we obtain

$$f_l(k, r) \xrightarrow[r \to \infty]{} N \, \frac{\Gamma(2l + 2)}{\Gamma(l + 1 - i\gamma)(2i)^l e^{\frac{1}{2}\pi\gamma}} \, \frac{1}{}$$

$$\times \frac{1}{kr} \cos\left[kr - \tfrac{1}{2}(l + 1)\pi + \gamma ln(2ikr) + \tfrac{1}{2}i\pi\gamma\right] \qquad (9\text{–}46)$$

As in (9–40), we see that the Coulomb field has effects even at infinity. However, when γ is small, (9–46) is close to (9–45), if we choose N as

$$N = (2i)^l e^{\frac{1}{2}\pi\gamma} \, \frac{\Gamma(l + 1 - i\gamma)}{\Gamma(2l + 2)} \qquad (9\text{–}47)$$

Therefore the radial function $f_l(k, r)$ is close to the free wave solution $j_l(kr)$ if

$$f_l(k, r) = \frac{\Gamma(l + 1 - i\gamma)}{\Gamma(2l + 2)} \, e^{\frac{1}{2}\pi\gamma}(2kr)^l e^{ikr} F(l + 1 - i\gamma, 2l + 2; -2ikr) \qquad (9\text{–}48)$$

The radial function normalized this way has a striking property. Let us consider the complex k plane. The radial function (9–48) is obtained by taking k on the real axis; while we expect that the radial function for bound states, (9–20), is essentially the same function when k is taken on the positive imaginary axis, i.e.,

$$k = \frac{i\sqrt{-2m\mathscr{E}}}{\hbar} \qquad (9\text{–}49)$$

On the complex k plane (except for $k = 0$), the radial function (9–48) has singular points because of the gamma function $\Gamma(l + 1 - i\gamma)$, while the confluent hypergeometric function and other simple functions are all regular

(Fig. 9–6). The gamma function has simple poles when its argument is zero or a negative integer. Thus $f_l(k, r)$ has simple poles at

$$l + 1 - i\gamma = -n' \ (=0, -1, -2, \ldots) \tag{9-50}$$

FIGURE 9–6. Poles of the hydrogenic radial function $f_l(k, r)$. Poles are at $k = i(\text{constant})/n$ where n is the principal quantum number.

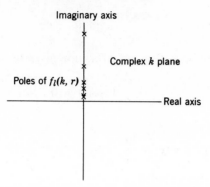

Expressing γ by $\mathscr{E}$, through (9–32) and (9–49), we see that (9–50) gives exactly the previously obtained result (9–25)! The residue of $f_l(k, r)$ at each of these poles gives (9–27), the radial function of a bound state, except for some normalization constant. Therefore, the radial function (9–48), originally constructed for continuous states, also contains all the necessary information for bound states.

PROBLEM 9–11. Obtain (9–46) and (9–44) and the asymptotic expansion formula of the confluent hypergeometric function given in Appendix I(c).

PROBLEM 9–12. Confirm that (9–50) does give (9–25).

9–4 SCATTERING BY A POTENTIAL

Consider one particle under a given potential $U(\mathbf{r})$, which is localized around the origin, tending to zero faster than $1/r$ as $r \to \infty$:

$$\lim_{r \to \infty} rU(\mathbf{r}) = 0 \tag{9-51}$$

The time-independent Schroedinger equation can be written as

$$\left(\nabla^2 + k^2 - \frac{2m}{\hbar^2} U(\mathbf{r})\right) \psi(\mathbf{r}) = 0 \tag{9-52}$$

which reduces to (9–33) for the Coulomb potential.

Equation (9–52) can be converted into an integral form using the Green's function of $\nabla^2 + k^2$, which is found in Problem 5–6:

$$(\nabla^2 + k^2)G_k(\mathbf{r}, \mathbf{r}') = \delta(\mathbf{r} - \mathbf{r}') \qquad (5\text{–}139)$$

$$G_k(\mathbf{r}, \mathbf{r}') = \frac{-1}{4\pi} \frac{e^{ik|\mathbf{r}-\mathbf{r}'|}}{|\mathbf{r} - \mathbf{r}'|} \qquad (5\text{–}141)$$

Thus, if $\phi(\mathbf{r})$ is the free particle wave function given by

$$(\nabla^2 + k^2)\phi(\mathbf{r}) = 0 \qquad (9\text{–}53)$$

then

$$\psi(\mathbf{r}) = \phi(\mathbf{r}) + \frac{2m}{\hbar^2} \int G_k(\mathbf{r}, \mathbf{r}')U(\mathbf{r}')\psi(\mathbf{r}') \, dv' \qquad (9\text{–}54)$$

is equivalent to (9–52).

If we consider the asymptotic region of r such that the potential is zero, we can apply the approximation previously used in (5–146):

$$\exp\left(ik\sqrt{r^2 - 2r \cdot r' + r'^2}\right) \cong e^{ikr}e^{-i\boldsymbol{\kappa} \cdot \mathbf{r}'} \qquad (9\text{–}55)$$

where

$$\boldsymbol{\kappa} = \frac{k\mathbf{r}}{r} \qquad (9\text{–}56)$$

and express (9–54) as

$$\psi(\mathbf{r}) \cong \phi(\mathbf{r}) - \left(\frac{2m}{\hbar^2}\right) \frac{e^{ikr}}{4\pi r} \int e^{-i\boldsymbol{\kappa} \cdot \mathbf{r}'} U(\mathbf{r}')\psi(\mathbf{r}') \, dv' \qquad (9\text{–}57)$$

This expression permits the following simple physical interpretation; the first term is the free incident wave, while the second term, which depends on r as e^{ikr}/r, is the scattered spherical wave, and $\boldsymbol{\kappa}$ is the wave vector of the scattered wave.

The incident wave is usually a plane wave, $(2\pi)^{-3/2}e^{i\mathbf{k}\cdot\mathbf{r}}$. We now express (9–57) in the form

$$\psi(r) = (2\pi)^{-3/2}\left[e^{i\mathbf{k} \cdot \mathbf{r}} - A(k; \theta, \varphi)\frac{e^{ikr}}{r}\right] \qquad (9\text{–}58)$$

where θ and φ are the spherical angles for $\mathbf{r}$ in a frame where $\mathbf{k}$ is along the z axis. Then the probability current density of this state in the radial direction is

$$\mathbf{j} = \frac{i\hbar}{2m}\left(\Psi\mathbf{r}\,\frac{\partial\Psi^*}{\partial r} - \Psi^*\mathbf{r}\,\frac{\partial\Psi}{\partial r}\right) = \mathbf{j}_{inc} + \mathbf{j}_{sc} + \mathbf{j}_{int} \qquad (9\text{-}59)$$

where

$$\mathbf{j}_{inc} = \frac{\hbar}{(2\pi)^3 m}\,\mathbf{k} \qquad (9\text{-}60a)$$

$$\mathbf{j}_{sc} = \frac{\hbar}{(2\pi)^3 m}\,|A(k;\theta,\varphi)|^2\,\frac{\boldsymbol{\kappa}}{r^2} \qquad (9\text{-}60b)$$

and

$$\mathbf{j}_{int} = \frac{-\hbar}{(2\pi)^3 m}\,|A(k;\theta,\varphi)|\,\frac{\cos\left[kr(1-\cos\theta)+\delta\right]}{r}\,(\mathbf{k}+\boldsymbol{\kappa}) \qquad (9\text{-}60c)$$

The quantity δ is the phase of A. When r is very large, j_{int}, which is the current due to the interference between the two terms of (9-58), oscillates rapidly as a function of θ, and its average over a finite angle is negligible. (An exceptional case is when $\theta \cong 0$. See Problem 9-19.) Therefore the current in the direction of $\mathbf{r}$, or $\boldsymbol{\kappa}$ is given by $\mathbf{j}_{sc}$, and the differential cross section of the scattering into a solid angle $d\Omega$ is

$$d\sigma = \frac{|\mathbf{j}_{sc}|r^2\,d\Omega}{|\mathbf{j}_{inc}|} = |A(k;\theta,\varphi)|^2\,\frac{|\boldsymbol{\kappa}|}{|\mathbf{k}|}\,d\Omega = |A(k;\theta,\varphi)|^2\,d\Omega \qquad (9\text{-}61)$$

or

$$\frac{d\sigma}{d\Omega} = |A(k;\theta,\varphi)|^2 \qquad (9\text{-}62)$$

Comparing (9-57) with (9-58), we obtain the expression of the differential scattering cross section

$$\frac{d\sigma}{d\Omega} = \frac{2\pi m^2}{\hbar^4}\left|\int e^{-i\boldsymbol{\kappa}\cdot\mathbf{r}'}U(\mathbf{r}')\psi(\mathbf{r}')\,dv'\right|^2 \qquad (9\text{-}63)$$

Since ψ is not known until the Schroedinger equation (9–52) is solved, the formula (9–63) is, although exact in the limit of large r, only a formal solution of the problem. In order to make it practically useful, we can apply the iteration method. The first approximation can be obtained if we use the zeroth approximation in (9–57); i.e., if the potential is small enough, ψ should not be too much different from the incident free wave ϕ. If the incident wave is a plane wave, (9–63) gives

$$\frac{d\sigma}{d\Omega} \cong \left(\frac{m}{2\pi\hbar^2}\right)^2 \left| \int e^{i(\mathbf{k}-\boldsymbol{\kappa})\cdot\mathbf{r}'} U(\mathbf{r}')\, dv' \right|^2 \tag{9–64}$$

This is called the *Born approximation*, and is the most frequently used formula for practical calculations of cross sections.

PROBLEM 9–13. Obtain (9–60a), (9–60b), and (9–60c).

PROBLEM 9–14. Assuming that

$$U(\mathbf{r}) = U(r) = \begin{cases} U_0 & \text{for} & 0 \le r \le r_0 \\ 0 & \text{for} & r_0 < r \end{cases}$$

calculate the differential cross section in the Born approximation.

ANSWER.

$$\left(\frac{2m}{\hbar^2}\right)^2 \frac{(\sin(Kr_0) - Kr_0 \cos(Kr_0))^2}{K^3}$$

where $K = 2k \sin(\tfrac{1}{2}\theta)$ (Fig. 9–7).

FIGURE 9–7. Geometrical interpretation of K in Problem 9–14.

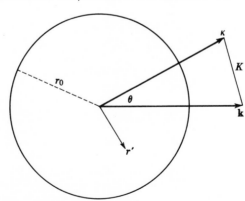

PROBLEM 9–15.   Assuming the Yukawa potential

$$U(r) = U_0 \frac{e^{-\alpha r}}{r}$$

(a) calculate the differential cross section in the Born approximation.
(b) In the limit of $\alpha \to 0$, the Yukawa potential reduces to the Coulomb potential. Show that the above result of (a) gives the Rutherford formula in this limit.

ANSWER.   For (a): $(2U_0 m/\hbar^2)^2/(K^2 + \alpha^2)^2$, where K is given in Problem 9–14.

PROBLEM 9–16.   The Born approximation is the first-order approximation of (9–63). Obtain the second-order approximation.

HINT.   Use (9–54) where $\psi(\mathbf{r}') \cong \phi(\mathbf{r}') = (2\pi)^{-3/2} e^{i\mathbf{k}\cdot\mathbf{r}'}$ in the integral.

PROBLEM 9–17.   Show that (9–54) can be written as

$$|\psi\rangle = |\phi\rangle + \left(\mathscr{E}_\infty - \frac{\hat{p}^2}{2m}\right)^{-1} U |\psi\rangle$$

in the operator form. This form is called the Lippmann–Schwinger equation.

HINT.   Introduce the Green's operator by $(\mathscr{E}_\infty - (\hat{p}^2/2m))\hat{G} = \hat{1}$.

PROBLEM 9–18.   The transition probability per unit time from the incident plane wave to an outgoing plane wave can be calculated from (8–266), using these plane waves for states $|a\rangle$ and $|b\rangle$, and the scattering potential U for $\hat{H}_1$. Obtain the Born approximation (9–64) from the result.

HINT. $\delta(\omega_{k\kappa}) = (m/k\hbar)\,\delta(k - \kappa)$. The number of outgoing waves with κ is $(2\pi)^3 \kappa^2 \, d\kappa \, d\Omega$.

Optical Theorem
   The total current density $\mathbf{j}$ of (9–59) should satisfy the continuity equation, which is $\nabla \cdot \mathbf{j} = 0$ in this case. Let us consider a large sphere around the center of the scattering, and consider the current conservation on its surface. From (9–60a) we see that $\mathbf{j}_{inc}$ coming into the sphere is the same as $\mathbf{j}_{inc}$ going out of the sphere, or $\mathbf{j}_{inc}$ itself satisfies the continuity equation. On the

other hand, $\mathbf{j}_{sc}$, being purely outgoing from this sphere, should be compensated by the incoming $\mathbf{j}_{int}$, in order that the total current satisfies the continuity equation. Thus

$$\oint \mathbf{j}_{total} \cdot d\mathbf{a} = \oint (\mathbf{j}_{sc} + \mathbf{j}_{int}) \cdot d\mathbf{a} = 0 \tag{9-65}$$

where the integration is over the entire surface of the sphere. Obviously

$$\oint \mathbf{j}_{sc} \cdot d\mathbf{a} = \frac{\hbar k}{(2\pi)^3 m} \int |A(k; \theta, \varphi)|^2 \, d\Omega \tag{9-66}$$

In $\mathbf{j}_{int}$, as we noticed before, $\cos [kr(1 - \cos \theta) + \delta]$ oscillates very rapidly, compared to all other factors, with respect to θ. Therefore, the integration can be done partially to give

$$\oint \mathbf{j}_{int} \cdot d\mathbf{a} = \frac{-\hbar}{(2\pi)^3 m} \int |A(k; \theta, \varphi)| \cos [kr(1 - \cos \theta) + \delta] k (\cos \theta + 1) r \, d\Omega$$

$$\cong \frac{-\hbar}{(2\pi)^3 m} \int |A(k; \theta, \varphi)| \, d\varphi \sin [kr(1 - \cos \theta) + \delta]$$

$$\times (\cos \theta + 1)|_{\theta=0}^{\pi}$$

$$= \frac{2\hbar}{(2\pi)^2 m} |A(k; 0, 0)| \sin \delta = \frac{2\hbar}{(2\pi)^2 m} \mathscr{I}mA(k; 0) \tag{9-67}$$

where $\mathscr{I}m \, A(k; 0)$ is the imaginary part of A at $\theta = 0$ (which implies $\varphi = 0$). Therefore, the continuity equation (9-65) gives

$$\int |A(k; \theta, \varphi)|^2 \, d\Omega = -\frac{4\pi}{k} \mathscr{I}mA(k; 0) \tag{9-68}$$

or, from (9-62),

$$\sigma(k) = -\frac{4\pi}{k} \mathscr{I}mA(k; 0) \tag{9-69}$$

where σ is the total scattering cross section. This result is called the *optical theorem*.

PROBLEM 9–19. Show that in the forward direction, i.e., $\theta = \varphi = 0$,

$$\mathbf{j}_{int} r^2 \, d\Omega = \frac{-2\hbar\mathbf{k}}{(2\pi)^3 m} \, \mathscr{R}e A(k; 0) r \, d\Omega$$

where $\mathscr{R}e A(k; 0)$ is the real part of $A(k; 0)$. This means that the interference current j_{int} gives a strong backward current in the forward direction. This backward current produces a shadow of the scattering potential.

9–5 PARTIAL WAVES

In many cases the scattering potential has spherical symmetry, and does not depend on angles:

$$U(\mathbf{r}) = U(r) \qquad (9\text{–}70)$$

When this is the case, the Hamiltonian of a particle is also spherically symmetric, and its eigenfunctions can be expressed as

$$\psi_{klm}(r) = f_l(k, r) Y_{lm}(\theta, \varphi) \qquad (9\text{–}71)$$

where the radial function $f_l(k, r)$ satisfies

$$\left[\frac{1}{r^2} \frac{d}{dr} \left(r^2 \frac{d}{dr} \right) - \frac{l(l+1)}{r^2} - \frac{2m}{\hbar^2} U(r) + k^2 \right] f_l(k, r) = 0 \qquad (9\text{–}72)$$

which is a generalization of (9–18).

Equation (9–72) can be "solved" using the Green's function method. Since the solution of this equation without the potential is the spherical Bessel function $j_l(kr)$, we obtain

$$f_l(k, r) = j_l(kr) + \frac{2m}{\hbar^2} \int G_k^{(l)}(r, r') U(r') f_l(k, r') r'^2 \, dr' \qquad (9\text{–}73)$$

if

$$\left[\frac{1}{r^2} \frac{d}{dr} \left(r^2 \frac{d}{dr} \right) - \frac{l(l+1)}{r^2} + k^2 \right] G_k^{(l)}(r, r') = \frac{\delta(r - r')}{r^2} \qquad (9\text{–}74)$$

We can apply the Bessel transformation, which corresponds to the Fourier transformation in the sense that

$$G_k^{(l)}(r, r') = \int_0^\infty j_l(qr) G_k^{(l)}(q; r') q^2 \, dq \qquad (9\text{–}75)$$

where the spherical Bessel function appears in the integral, instead of the trigonometric functions in the Fourier transformation formula. For the delta function, we have the so-called Bessel–Fourier formula

$$\int_0^\infty j_l(qr)j_l(qr')q^2 \, dq = \tfrac{1}{2}\pi \frac{\delta(r - r')}{r^2} \tag{9-76}$$

Putting (9–75) in (9–73), and using (9–76), we obtain

$$G_k^{(l)}(q; r') = \frac{2}{\pi} \frac{j_l(qr')}{k^2 - q^2} \tag{9-77}$$

Therefore

$$G_k^{(l)}(r, r') = \frac{2}{\pi} \int_0^\infty \frac{j_l(qr)j_l(qr')}{k^2 - q^2} q^2 \, dq = \frac{1}{\pi} \int_{-\infty}^\infty \frac{j_l(qr)j_l(qr')}{k^2 - q^2} q^2 \, dq \tag{9-78}$$

The integral (9–78) can be evaluated using contour integration. If $r > r'$, for example, we first express the integral of (9–78) as

$$G_k^{(l)}(r, r') = \frac{1}{2\pi} \int_{-\infty}^\infty \frac{j_l(qr')h_l^{(1)}(qr)}{k^2 - q^2} q^2 \, dq$$

$$+ \frac{1}{2\pi} \int_{-\infty}^\infty \frac{j_l(qr')h_l^{(2)}(qr)}{k^2 - q^2} q^2 \, dq \tag{9-79}$$

where $h_l^{(1)}$ and $h_l^{(2)}$ are spherical Hankel functions of the first and second kind, respectively. When

$$z = \rho \cos \varphi + i\rho \sin \varphi \tag{9-80}$$

these spherical Hankel functions have the following asymptotic forms:

$$h_l^{(1)}(z) \xrightarrow[\rho \to \infty]{} \frac{(-i)^{l+1} e^{i\rho \cos \varphi} e^{-\rho \sin \varphi}}{z} \tag{9-81a}$$

$$h_l^{(2)}(z) \xrightarrow[\rho \to \infty]{} \frac{i^{l+1} e^{-i\rho \cos \varphi} e^{\rho \sin \varphi}}{z} \tag{9-81b}$$

On the other hand,

$$j_l(z') \xrightarrow[\rho \to \infty]{} \frac{1}{2} \frac{\left[i^{l+1} e^{-i\rho' \cos \varphi} e^{\rho' \sin \varphi} + (-i)^{l+1} e^{i\rho' \cos \varphi} e^{-\rho' \sin \varphi} \right]}{z} \tag{9-81c}$$

Therefore, the first term of (9–79) can be expressed as a contour integral on the real axis closed by the semicircle of infinite radius in the upper half-plane, (sin $\varphi > 0$), while the second term of (9–79) can be expressed as a contour integral closed by the semicircle in the lower half-plane, (sin $\varphi < 0$), as shown in Fig. 9–8.

FIGURE 9–8.   Integral contour for evaluating (9–78). Take the upper semicircle for the first term of (9–79), and the lower semicircle for the second term of (9–79).

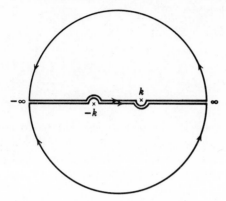

Since the integrand in (9–78) has two poles on the real axis,[2] we obtain different results depending on how the contour goes around these poles. There are four ways to go around these two poles, as shown in Fig. 6–4, which is reproduced in Fig. 9–9.

FIGURE 9–9.   Part of the contours on the real axis.

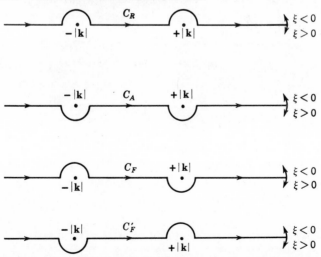

[2] Each of the spherical Hankel functions has an essential singularity at the origin ($q = 0$), but since j_l ($= \frac{1}{2}(h_l^{(1)} + h_l^{(2)})$) is regular, the contributions from the singularity at the origin cancel each other when the sum (9–79) is calculated. Therefore we can ignore the essential singularity at $q = 0$.

The choice among these four ways of going around the poles is determined by the physics of the problem, and in our case the contour C_F turns out to be suitable. Thus the first term of (9–79), which is given by the contour C_F closed by the semicircle in the upper half-plane, is

$$\frac{1}{2\pi} \int_{-\infty}^{\infty} \frac{j_l(qr')h_l^{(1)}(qr)}{k^2 - q^2} q^2 \, dq$$

$$= \frac{-1}{4\pi} \oint_{C_F} \left(\frac{1}{q-k} + \frac{1}{q+k} \right) q j_l(qr')h_l^{(1)}(qr) \, dq$$

$$= -\tfrac{1}{2}ikj_l(kr')h_l^{(1)}(kr) \tag{9–82}$$

since the contour encircles the pole at $+k$ counterclockwise. On the other hand, the contour for the second term of (9–79), which is given by C_F on the real axis and is closed by the semicircle in the lower half-plane, encircles the pole at $-k$ in the clockwise direction: Therefore

$$\frac{1}{2\pi} \int_{-\infty}^{\infty} \frac{j_l(qr')h_l^{(2)}(qr)}{k^2 - q^2} q^2 \, dq$$

$$= \frac{-1}{4\pi} \oint_{C_F} \left(\frac{1}{q-k} + \frac{1}{q+k} \right) q j_l(qr')h_l^{(2)}(qr) \, dq$$

$$= \tfrac{1}{2}i(-k)j_l(-kr')h_l^{(2)}(-kr) = -\tfrac{1}{2}ikj_l(kr')h_l^{(1)}(kr) \tag{9–83}$$

where we used the relations

$$h_l^{(2)}(-x) = (-1)^l h_l^{(1)}(x) \quad \text{and} \quad j_l(-x) = (-1)^l j_l(x) \tag{9–84}$$

Finally

$$G_k^{(l)}(r, r') = -ikj_l(kr')h_l^{(1)}(kr) \quad (r > r') \tag{9–85}$$

is obtained with this choice of the contour.

When this Green's function is used in (9–73), assuming that r is large enough so that U is zero at r and that (9–81a) can be used for the Hankel function, we obtain

$$f_l(k, r) = j_l(kr) - ik\frac{2m}{\hbar^2} \int j_l(kr_<)h_l^{(1)}(kr_>)U(r')f_l(k, r')r'^2 \, dr'$$

$$\xrightarrow[r \to \infty]{} j_l(kr) - (-i)^l \frac{e^{ikr}}{r} \left(\frac{2m}{\hbar^2} \right) \int j_l(kr')U(r')f_l(k, r')r'^2 \, dr' \tag{9–86}$$

where the second term gives an outgoing spherical wave. This justifies the choice of the contour used in evaluating the Green's function. In (9–86) $r_<$ and $r_>$ are the smaller and larger, respectively of r and r'.

PROBLEM 9–20. Show that (9–85) can be generalized as

$$G_k^{(l)}(r, r') = -ikj_l(kr_<)h_l^{(1)}(kr_>)$$

where $r_<$ and $r_>$ are the smaller and larger, respectively, of r and r'.

PROBLEM 9–21. Show that if the contour $C_{F'}$ is used on the real axis, we obtain the Green's function of

$$G_k^{(l)}(r, r') = ikj_l(kr_<)h_l^{(2)}(kr_>)$$

which gives an incoming spherical wave.

9–6 BOUND STATES

So far we have been discussing scattering, or positive energy states. For negative energy states, or bound states in the same potential U, however, the equation is still the same as (9–72), except that the energy k^2 is negative, i.e., k is imaginary. We can, therefore, express the radial function of negative energy states in the same way as (9–86), except that k should be replaced by an imaginary quantity

$$k = \frac{i\sqrt{-2m\mathscr{E}}}{\hbar} = i\alpha \tag{9–87}$$

as we had in (9–49). Thus, a solution of the radial equation for a negative energy state is

$$f_{\mathscr{E}l}(r) = j_l(i\alpha r) + \alpha \frac{2m}{\hbar^2} \int j_l(i\alpha r_<)h_l^{(1)}(i\alpha r_>)U(r')f_{\mathscr{E}l}(r')r'^2 \, dr' \tag{9–88}$$

We notice, however, that

$$j_l(i\alpha r) \xrightarrow[r \to \infty]{} \frac{i^l e^{\alpha r}}{\alpha r} \tag{9–89}$$

and

$$\int j_l(i\alpha r_<)h_l^{(1)}(i\alpha r_>)U(r')f_{\mathscr{E}l}(r')r'^2 \, dr'$$

$$\xrightarrow[r\to\infty]{} (-i)^l \frac{e^{-\alpha r}}{\alpha r} \int j_l(i\alpha r')U(r')f_{\mathscr{E}l}(r')r'^2 \, dr' \qquad (9\text{-}89')$$

or the first term of (9-88) diverges, while the second term of (9-88) converges. The radial function of a bound state should be square integrable, and should not diverge as (9-89). While (9-88) is the radial function for an arbitrary negative energy $\mathscr{E}$, the bound states should be such that at their particular energy $\mathscr{E}_n$ the first term $j_l(i\alpha r)$ does not contribute. Since $j_l(i\alpha r)$ is a regular function this can happen when $f_{\mathscr{E}l}(r)$ has poles at the energies of bound states; namely, of the form

$$f_{\mathscr{E}l} = \sum_n \frac{f_{nl}}{\mathscr{E} - \mathscr{E}_n} \qquad (9\text{-}90)$$

where $\mathscr{E}_n$'s are the energies of bound states. The radial function of the nlth bound state is given as the residue of $f_{\mathscr{E}l}$ at this pole:

$$f_{nl}(r) = (\mathscr{E} - \mathscr{E}_n)f_{\mathscr{E}l}(r)|_{\mathscr{E}=\mathscr{E}_{nl}}$$

$$= N_{nl} \int j_l(i\alpha_{nl}r_<)h_l^{(1)}(i\alpha_{nl}r_>)U(r')f_{nl}(r')r'^2 \, dr' \qquad (9\text{-}91)$$

where N_{nl} is a constant and α_{nl} is α of (9-87) when $\mathscr{E}$ is $\mathscr{E}_{nl}$. We have already seen this relation between the radial functions of continuous states and bound states in section 9-3, for the special case of the Coulomb potential.

The radial function (9-86) may have poles other than those on the positive imaginary axis in the complex k plane. We will see in section 9-9 that these other poles correspond to resonance scattering.

PROBLEM 9-22. Show that

$$f_{nl}(r) \xrightarrow[r\to\infty]{} (\text{constant}) \times \frac{\exp{(-r\sqrt{-2m\mathscr{E}_{nl}}/\hbar})}{r}$$

and

$$f_{nl}(r) \xrightarrow[r\to 0]{} (\text{constant}) \times r^l$$

for any short-range potential.

PROBLEM 9–23.   In a potential well (Fig. 9–10)

$$U(r) = \begin{cases} -U_0 & \text{if} \quad 0 \le r \le r_0 \\ 0 & \text{if} \quad r_0 < r \end{cases}$$

we know that $f_{\mathscr{E}l}(r) = C_{\mathscr{E}l}j_l(k'r)$ for $0 \le r \le r_0$. Here $C_{\mathscr{E}l}$ is a constant and $k' = \sqrt{2m(\mathscr{E} - U_0)}/\hbar$. Taking the case of $l = 0$, find an equation which determines bound state energies from the poles of $C_{\mathscr{E}0}$.

FIGURE 9–10.   Square well potential.

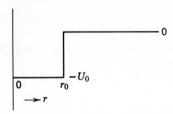

HINT.   At $r = 0$, we have $f_{\mathscr{E}0}(0) = C_{\mathscr{E}0}$. Therefore (9–88) is

$$C_{\mathscr{E}0} = 1 + C_{\mathscr{E}0} \frac{2m}{\hbar^2 k'} U_0 \int_0^{r_0} e^{-\alpha r} \sin(k'r)\, dr$$

The answer is

$$\alpha \sin(k'a) + k' \cos(k'a) = 0$$

PROBLEM 9–24.   We know that (9–72) can be written as

$$\left[\frac{d^2}{dr^2} - \frac{2m}{\hbar^2} U(r) + k^2 \right] rf = 0$$

when $l = 0$. (See Problem 9–3.) Obtain the answer of Problem 9–23, using the method we used in section 8–8 for one-dimensional Schroedinger equations.

PROBLEM 9–25.   For the same potential well as in Problem 9–23, show that the bound state energies should satisfy

$$-\alpha \frac{2m}{\hbar^2} U_0 \int_0^{r_0} j_l(i\alpha r) j_l(k'r) r^2\, dr = \frac{j_l(k'r_0)}{h_l^{(1)}(i\alpha r_0)}$$

where k' is defined in the previous problem.

HINT.   Find the pole of $C_{\mathscr{E}l}$ from equation (9–88) at $r = r_0$.

PROBLEM 9–26.   When the potential is

$$U(r) = -\lambda \, \delta(r - r_0)$$

where λ is a constant, show that there is one and only one bound state with $l = 0$, whose energy is found from

$$\lambda = \frac{2\alpha a}{(1 - e^{-2\alpha a})}$$

where α is given by (9–87).[3]

9–7   PHASE SHIFTS

A general solution of the original Schroedinger equation (9–52) can be expressed as a linear combination of the partial wave solution (9–71), when the potential is spherical as (9–70). As in atomic spectra, partial waves with $l = 0, 1, 2, \ldots$ are called s-, p-, d-, $\ldots$ waves, respectively. The scattering problems, in which an incident plane wave is scattered into a spherical wave, can be treated nicely by means of the partial waves.

The plane wave $\exp(i\mathbf{k} \cdot \mathbf{r})$ can be expanded as

$$e^{i\mathbf{k} \cdot \mathbf{r}} = \sum_{l=0}^{\infty} (2l + 1)i^l j_l(kr)P_l\,(\cos\theta) \qquad (9\text{–}92)$$

where P_l is the Legendre polynomial and θ is the angle between $\mathbf{k}$ and $\mathbf{r}$. This formula can be easily verified by observing that $j_l(kr)P_l\,(\cos\theta)$, being eigenfunctions of the free wave Hamiltonian, form a complete set, and that when $\theta = 0$ both sides are eigenfunctions of $\partial/\partial r$ with the same eigenvalue ik. A solution of the original Schroedinger equation (9–52) is, therefore, obtained from (9–86) as

$$\psi(k, r) = \sum_{l=0}^{\infty} (2l + 1)i^l P_l\,(\cos\theta)f_l(k, r)$$

$$= e^{i\mathbf{k} \cdot \mathbf{r}} - \sum_{l=0}^{\infty} i^{l+1} k \frac{2m}{\hbar^2} (2l + 1)P_l\,(\cos\theta)$$

$$\times \int j_l(kr_<)h_l^{(1)}(kr_>)U(r')f_l(k, r')r'^2 \, dr'$$

$$\xrightarrow[r \to \infty]{} e^{i\mathbf{k} \cdot \mathbf{r}} - \left[\sum_{l=0}^{\infty} \frac{2m}{\hbar^2} (2l + 1)P_l\,(\cos\theta) \right.$$

$$\left. \times \int j_l(kr')U(r')f_l(k, r')r'^2 \, dr' \right] \frac{e^{ikr}}{r} \qquad (9\text{–}93)$$

---

[3] K. Gottfried, *Quantum Mechanics*, W. A. Benjamin, New York, 1966, pp. 131–135.

We see that (9–93) is in the desired form of (9–58), and the scattering amplitude is given by

$$A(k; \theta) = \sum_{l=0}^{\infty} A_l(k; \theta) \qquad (9\text{–}94)$$

where the partial wave scattering amplitude is

$$A_l(k; \theta) = \frac{2m}{\hbar^2}(2l + 1)P_l(\cos \theta) \int j_l(kr')U(r')f_l(k, r')r'^2 \, dr' \qquad (9\text{–}95)$$

Notice that each scattering amplitude is independent of φ, because the potential is angle independent. Since the Legendre polynomials are orthogonal to each other, we obtain, from (9–62),

$$\sigma(k) = \int |A(k; \theta)|^2 \, d\Omega = \sum_{l=0}^{\infty} \sigma_l(k) \qquad (9\text{–}96)$$

where

$$\sigma_l(k) = \int |A_l(k; \theta)|^2 \, d\Omega = \left(\frac{4m}{\hbar^2}\right)^2 \pi(2l + 1) \left| \int j_l(kr')U(r')f_l(k, r')r'^2 \, dr' \right|^2$$

$$(9\text{–}97)$$

As we have seen in Problems 9–3 and 9–24, equation (9–72) can be written as

$$\left[\frac{d^2}{dr^2} - \frac{l(l + 1)}{r^2} - \frac{2m}{\hbar^2} U + k^2 \right] (rf_l) = 0 \qquad (9\text{–}98)$$

When the potential U is real, which we assume here, this equation means

$$(rf_l) \frac{\partial}{\partial r}(rf_l)^* - (rf_l)^* \frac{\partial}{\partial r}(rf_l) = 0 \qquad (9\text{–}99)$$

On the other hand, (9–86) gives

$$rf_l(k, r) \xrightarrow[r \to \infty]{} \frac{1}{2k} \{ e^{-i[kr - (1/2)\pi(l+1)]} + e^{i[kr - (1/2)\pi(l+1)]} \}$$

$$- e^{i(kr - (1/2)\pi l)} \frac{2m}{\hbar^2} \int j_l(kr')U(r')f_l(k, r')r'^2 \, dr' \qquad (9\text{–}100)$$

In order that (9-100) satisfy (9-99) it is required that

$$1 - 2ik \frac{2m}{\hbar^2} \int j_l(kr')U(r')f_l(k, r')r'^2 \, dr' = \exp(2i\delta_l) \qquad (9\text{-}101)$$

where δ_l is a real function of k defined by this equation and called the phase shift. In terms of the phase shift, (9-100) is

$$rf_l(k, r) \xrightarrow[r \to \infty]{} \frac{1}{2k} \{e^{-i[kr - (1/2)\pi(l+1)]} + e^{i[kr - (1/2)\pi(l+1) + 2\delta_l]}\} \qquad (9\text{-}102)$$

which means that the effect of the scattering potential is to make the outgoing spherical partial wave have a phase shift $2\delta_l$ with respect to the incoming partial wave.

Inserting (9-101) into (9-95), we see that the partial wave scattering amplitude can be expressed as

$$A_l(k; \theta) = -k^{-1}(2l + 1)P_l(\cos \theta)e^{i\delta_l} \sin \delta_l \qquad (9\text{-}103)$$

which gives

$$\sigma_l = \int |A_l(\theta)|^2 \, d\Omega = \frac{4\pi}{k^2} (2l + 1) \sin^2 \delta_l \qquad (9\text{-}104)$$

The partial wave formalism of the scattering theory is particularly advantageous when the energy is low. Taking the classical particle picture of incident momentum $\hbar k$, we see that the maximum angular momentum $\hbar l_{\max}$ which can be significantly effected by the scattering potential is $\hbar ka$, where a is the range of the potential. Thus any partial waves whose l is larger than ka will not have an appreciable phase shift, and can be neglected in calculating the cross section. Quite often only s waves, or s and p waves are needed in calculation of the cross section.

PROBLEM 9-27. Show that (9-92) holds.

HINT.

$$(2l + 1) \frac{d}{dz} j_l(z) = l j_{l-1}(z) - (l + 1)j_{l+1}(z)$$

PROBLEM 9-28. Find the θ dependence of the differential cross section $d\sigma/d\Omega$, when only s and p waves have appreciable phase shifts.

PROBLEM 9–29. Prove the optical theorem (9–69) from (9–103) and (9–104).

PROBLEM 9–30. When the scattering potential U is small, the first approximation of A_l can be obtained by taking j_l for f_l in (9–95). If, in addition, U is of a very short range, show that A_l is proportional to k^{2l}, and that $\sigma_l(k)$ is proportional to k^{4l}.

HINT

$$j_l(kr) \cong (kr)^l \times (\text{constant}) \quad \text{if} \quad kr \ll 1$$

PROBLEM 9–31. Show that, if

$$U = \frac{(\hbar^2/2m)\xi}{r^2}$$

where ξ is a constant, then

$$\delta_l = \tfrac{1}{2}\pi(l - l')$$

where

$$l'(l' + 1) = l(l + 1) + \xi$$

PROBLEM 9–32. When the potential is

$$U(r) = \begin{cases} U_0 & \text{if} \quad 0 \le r \le r_0 \\ 0 & \text{if} \quad r_0 < r \end{cases}$$

and the energy is less than U_0; show that

$$\delta_0 = \tan^{-1}(kQ) - kr_0$$

where

$$Q = \frac{\tanh\{r_0\sqrt{(2m/\hbar^2)U_0 - k^2}\}}{\sqrt{(2m/\hbar^2)U_0 - k^2}}$$

PROBLEM 9–33. In the same square well potential as in Problem 9–32, show that $\sigma_l(k) = 0$ when

$$\frac{k j_{l+1}(kr_0)}{j_l(kr_0)} = \frac{\kappa j_{l+1}(\kappa r_0)}{j_l(\kappa r_0)}$$

is satisfied, where

$$\kappa^2 = k^2 - \frac{2m}{\hbar^2} U_0$$

HINT. $f_l(k, r) = C j_l(\kappa r)$ when $r \leq r_0$. C is a constant.

$$\sin \delta_l = -\frac{2mk}{\hbar^2} U_0 \int_0^{r_0} j_l(kr) j_l(\kappa r) r^2 \, dr |C|$$

$$= k r_0^2 [k j_{l+1}(k r_0) j_l(\kappa r_0) - \kappa j_l(k r_0) j_{l+1}(\kappa r_0)] |C|$$

PROBLEM 9–34. When $U_0 = +\infty$ in the square well potential
(a) Show that $\delta_l = -k r_0$.
(b) Show that $\sigma_l \to 4\pi r_0^2$ in the limit of $k \to 0$. Notice that this is four times the classical cross section.

9–8* RESONANCE SCATTERING. THE DISPERSION RELATION

In section 9–6, we saw that if a potential has bound states, they are exhibited as poles of the radial function f_l on the positive imaginary axis in the complex k plane. The radial function f itself, however, may have poles in other parts of the complex k plane, and these other poles do not correspond to bound states.

Let us assume that $f_l(k, r)$ has a simple pole at

$$k = \beta + i\alpha \tag{9–105}$$

This means that in the vicinity of this pole

$$f_l(k, r) \cong \frac{1}{k - \beta - i\alpha} f_l(k, r) \tag{9–106}$$

where f_l is an analytic function. Since, from (9–86) and (9–95),

$$(2l + 1) P_l(\cos \theta) f_l(k, r) \xrightarrow[r \to \infty]{} (2l + 1) P_l(\cos \theta) j_l(kr) - (-i)^l A_l(k; \theta) \frac{e^{ikr}}{r}$$

$$\tag{9–107}$$

and since $j_l(kr)$ and e^{ikr} do not have any singularities in the entire complex k plane, (9–106) means that the scattering amplitude has a pole at the same point of (9–105), or is of the form of

$$A_l(k; \theta) \cong \frac{1}{k - \beta - i\alpha} \mathscr{A}_l(k; \theta) \qquad (9\text{–}108)$$

Thus, from (9–97), we obtain

$$\sigma_l(k) = \int |A_l(k; \theta)|^2 \, d\Omega = \frac{I(k)}{(k - \beta)^2 + \alpha^2} \qquad (9\text{–}109)$$

when the real k is close to β. If $I(k)$ is a slow varying function of k, the cross section σ_l shows a peak at $k = \beta$ with the width of about α. This is called the *resonance scattering*. (Fig. 9–11.)

FIGURE 9–11.   Resonance scattering. Formula (9–109).

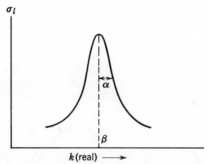

Now we shall show that if $\beta \neq 0$, then α has to be negative; i.e., $f_l(k, r)$ does not have poles in the upper half-plane of the complex k plane except on the imaginary axis.

Since $j_l(kr)$ is analytic in the entire complex k plane, (9–107) and (9–108) show that

$$(2l + 1)P_l(\cos \theta)f_l(k, r) \xrightarrow[r \to \infty]{} \frac{-(-i)^l}{k - \beta - i\alpha} \mathscr{A}_l(k; \theta) \frac{e^{i\beta r}e^{-\alpha r}}{r} \qquad (9\text{–}110)$$

in the vicinity of the pole $\beta + i\alpha$. When α is positive, the last factor $\exp(-\alpha r)$ shows that the state given by f_l is a localized state, or the corresponding probability density $|\psi|^2$ is essentially confined in a region $0 \leq r < \alpha^{-1}$ On the other hand, since the energy of this state, in the region outside of the potential, is $(\hbar k)^2/2m$, the time dependence of the wave function is given by the factor

$$e^{-i\mathscr{E}t/\hbar} = \exp\left[-i\frac{\hbar}{2m}(\beta^2 - \alpha^2)t\right] e^{(\hbar/m)\alpha\beta t} \qquad (9\text{–}111)$$

Thus, if $\alpha\beta \neq 0$, the probability density $|\psi(t)|^2$ has to change with time. This result contradicts the conservation of probability.

The above contradiction is removed if $\alpha < 0$: Because if $\alpha < 0$, then (9–110) shows that the probability $|\psi|^2$ diverges as $r \to \infty$, and the wave function is not square integrable. The time variation of the probability density in a finite region due to the factor of (9–111) can be compensated by the indefinite probability density at infinity. Thus poles of $f_l(k, r)$ may exist in the lower half-plane, but not in the upper half-plane except on the imaginary axis ($\beta = 0$).

The dispersion relation can be obtained using the analyticity of $f_l(k, r)$ or $A_l(k; \theta)$ we just discussed. In the complex k plane take a contour following the real axis in the positive direction and close it with a semicircle in the upper half-plane (Fig. 9–12), and let us evaluate

$$\oint \frac{A_l(k; \theta)}{k - k_0}\, dk \qquad (9\text{–}112)$$

FIGURE 9–12. Poles of $f_l(k, r)$ in the complex k plane, and the contour for the dispersion relation (9–116). Poles on the positive imaginary axis are due to bound states, while those in the lower half-plane give resonance scatterings.

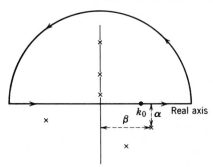

on this contour, assuming that k_0 is a real number. Since the contour goes right over a pole at k_0, and encircles all poles of $A_l(k; \theta)$ on the positive imaginary axis, which correspond to bound states, this contour integral is given by

$$\oint \frac{A_l(k; \theta)}{k - k_0}\, dk = \pi i A_l(k_0; \theta) + 2\pi i \sum_n \mathscr{A}_l(i\alpha_n; \theta)(i\alpha_n - k_0)^{-1} \qquad (9\text{–}113)$$

where $i\alpha_n$ are the values of k at bound states, and

$$\mathscr{A}_l(i\alpha_n;\theta) = \lim_{k \to i\alpha_n} (k - i\alpha_n)A_l(k;\theta) \qquad (9\text{-}114)$$

is the residue of A_l at a bound state. Since $A_l(k;\theta)$ decreases as $1/k$ or faster, as (9-103) shows, we expect that when the radius of the semicircle is infinite, the above contour integral is equal to the integral on the real axis. Taking Cauchy's principal value for this real integral, because we have a pole at k_0, we obtain

$$\oint \frac{A_l(k;\theta)}{k - k_0}\,dk = \mathscr{P}\int_{-\infty}^{\infty} \frac{A_l(k;\theta)}{k - k_0}\,dk \qquad (9\text{-}115)$$

where, on the right-hand side, k is real and $\mathscr{P}$ stands for Cauchy's principal value. Putting (9-113) and (9-115) together, and taking the imaginary part, we obtain

$$\tfrac{1}{2}\mathscr{P}\int_{-\infty}^{\infty} \frac{\mathscr{I}mA_l(k;\theta)}{k - k_0}\,dk = \mathscr{P}\int_0^{\infty} \frac{k\mathscr{I}mA_l(k;\theta)}{k^2 - k_0^{\,2}}\,dk$$

$$= \tfrac{1}{2}\pi\mathscr{R}eA_l(k_0;\theta) + \pi\sum_n \mathscr{R}e\mathscr{A}_l(i\alpha_n;\theta)(i\alpha_n - k_0)^{-1} \qquad (9\text{-}116)$$

using the fact that $\mathscr{I}mA_l(k;\theta)$ is an odd function of k. This is called the *dispersion relation*. A particularly useful dispersion relation is obtained for the forward scattering, where, using the optical theorem

$$\sigma_l(k) = -\frac{4\pi}{k}\,\mathscr{I}mA_l(k;0) \qquad (9\text{-}69)$$

we obtain

$$-\mathscr{P}\int_0^{\infty} \frac{k^2\sigma_l(k)}{k^2 - k_0^{\,2}}\,dk = 2\pi^2\mathscr{R}eA_l(k_0;0)$$

$$+ 4\pi^2\mathscr{R}e\sum_n \mathscr{A}_l(i\alpha_n;0)(i\alpha_n - k_0)^{-1} \qquad (9\text{-}117)$$

It may happen that A_l does not approach zero fast enough as $|k| \to \infty$, that the contribution from the semicircle to the contour integral may not be negligible. When that is the case, one can still employ a similar argument, using

$$\oint \frac{A_l(k;\theta)}{k(k - k_0)}\,dk \qquad (9\text{-}118)$$

instead of (9-112).

PROBLEM 9–35. In the square well potential

$$U = \begin{cases} -U_0 & \text{for} \quad 0 \le r \le r_0 \\ 0 & \text{for} \quad r_0 < r \end{cases}$$

(a) show that

$$u_0 \equiv \frac{2m}{k\hbar^2} U_0 \int_0^{kr_0} j_l(\rho) j_l \left(\frac{\kappa\rho}{k} \right) \rho^2 \, d\rho \xrightarrow[|k| \to \infty]{} \frac{mU_0 r_0}{\hbar^2}$$

where

$$\kappa^2 = k^2 + \frac{2m}{\hbar^2} U_0$$

(b) then show that

$$A_l(k; \theta) \xrightarrow[|k| \to \infty]{} - (2l + 1) P_l(\cos \theta) \frac{u_0}{k^2}$$

and

$$\delta_l \to \frac{u_0}{k}$$

HINT. Use (9–95) and (9–101).

PROBLEM 9–36. In the same square well potential, show that if there is one bound state, there must be at least one resonance scattering state.

HINT. The result of Problem 9–35 implies that the contour integral of $A_l(k; \theta)$ over the complete circle of infinite radius is zero.

PROBLEM 9–37. Show that

$$\mathscr{I}m A_l(k; \theta) = -\mathscr{I}m A_l(-k; \theta)$$

HINT. In (9–93), compare $\psi(k, r)$ to $\psi^*(-k, r)$, using (9–95).

PROBLEM 9–38. Obtain a dispersion relation, similar to (9–117), from (9–118).

9–9 SCATTERING OF ELECTROMAGNETIC WAVES. THE COMPLEX POLARIZABILITY

The scattering of electromagnetic waves was already discussed in Chapter 6, particularly section 6–12, but it is instructive to reformulate the theory in terms of the present formulation of quantum mechanical scattering theory.

Let us concentrate our attention to the electric field. The general solution (6-88) shows that we can write

$$E(r, t) = E_{inc}(r, t) + E_{sc}(r, t) \qquad (9\text{-}119)$$

where E_{inc} is the incident wave which we take as a plane wave propagating in the z direction

$$E_{inc}(r, t) = E_0 e^{ik(z - ct)} \qquad (9\text{-}120)$$

The scattered wave E_{sc} is given by (6-130). In the nonrelativistic limit

$$E_{sc}(r, t) = \frac{q\mu_0}{4\pi} \frac{n \times (n \times \ddot{r}_q)}{R_q} \bigg|_{t_{ret}} \qquad (9\text{-}121)$$

where q and r_q are the charge and position of the scatterer,

$$R_q(t_{ret}) = r - r_q(t_{ret}) \qquad (6\text{-}116')$$

and

$$t_{ret} = t - \frac{R_q(t_{ret})}{c} \qquad (6\text{-}119)$$

The complex polarizability, defined by

$$q r_q(t) = \alpha(\omega) E_{inc}(r_q, t) = \alpha(\omega) E_0 \exp\left[ik(z_q - ct)\right] \qquad (9\text{-}122)$$

relates the displacement r_q and the incident field E_{inc}, including the possible phase difference between them as shown in Problem 6-40. We assume that α is scalar, or the scatterer is spherically symmetric, to simplify the discussion. Notice that E_{sc} does not appear explicitly in (9-122), but is implicitly taken into account in the phase difference mentioned above. We obtain from (9-122) and (6-119) that

$$q \ddot{r}_q(t_{ret}) = -\omega^2 \alpha(\omega) E_0 \exp(ikz_q) \exp\left[ik(R_q - ct)\right] \qquad (9\text{-}123)$$

Assuming, for simplicity, that the scatterer is bound to the origin and its displacement from the origin is negligible compared to r, we can take

$$z_q = 0 \quad \text{and} \quad R_q = r \qquad (9\text{-}124)$$

in (9–123), and express (9–119) as

$$\mathbf{E}(\mathbf{r}) = |\mathbf{E}_0| \left[\mathbf{e}_0 e^{ikz} - \mathbf{A}\, \frac{e^{ikr}}{r} \right] \tag{9–125}$$

where $\mathbf{e}_0$ is a unit vector in the direction of $\mathbf{E}_0$, and

$$\mathbf{A} = \frac{\mu_0 \omega^2}{4\pi} \left[\mathbf{n} \times (\mathbf{n} \times \mathbf{e}_0) \right] \alpha(\omega) \tag{9–126}$$

The result (9–125) is in the same form as (9–58), except that the present one is slightly more complicated because the electromagnetic field is a vector field. The analogy, however, is strong enough so that we can immediately obtain the scattering cross section in the same form as (9–62) as

$$\frac{d\sigma}{d\Omega} = |\mathbf{A}|^2 = \left(\frac{\mu_0 \omega^2}{4\pi} \right)^2 |\alpha(\omega)|^2 [1 - (\mathbf{n} \cdot \mathbf{e}_0)^2] \tag{9–127}$$

which is the same result obtained in Problem 6–40(b). We see that the scattering amplitude is essentially given by the complex polarizability α in this case.

Let us consider what the real and imaginary parts of the complex polarizability mean. For that purpose we take a continuous body made of N scatterers per unit volume, assuming that N is small enough so that the mutual interaction between the scatterers is negligible compared to the interaction between the field and individual scatterer.

Since (5–98) states that the polarization of the medium $\mathbf{P}$ is given by the density of electric dipole moments, we obtain

$$\mathbf{P}(\omega) = Nq\mathbf{r}_q = N\alpha(\omega)\mathbf{E}(\omega) \tag{9–128}$$

and the complex capacitivity of the medium

$$\varepsilon(\omega) = \varepsilon_0 + \varepsilon_0 \chi_e(\omega) = \varepsilon_0 + N\alpha(\omega) = (\varepsilon_0 + N\mathscr{R}e\alpha) + iN\mathscr{I}m\alpha \tag{9–129}$$

from (5–102) and (5–103). In most cases the medium is diamagnetic and nonconducting, so that

$$\mu = \mu_0 \qquad \text{and} \qquad \sigma = 0 \tag{9–130}$$

where σ is the conductivity. The wave equation (5–118) in such medium gives

$$-k^2 + \varepsilon\mu_0\omega^2 = 0 \tag{9-131}$$

when $\mathbf{E}$ is in the form of (9–120). Inserting (9–129) in (9–131), and assuming that N is small enough, we obtain

$$k \cong \frac{\omega}{c}(1 + \tfrac{1}{2}\varepsilon_0^{-1}N\mathscr{R}e\alpha + \tfrac{1}{2}i\varepsilon_0^{-1}N\mathscr{I}m\alpha) \tag{9-132}$$

When k of a plane wave (9–120) is complex, the real part is related to the phase velocity, while the imaginary part is related to the decay, or extinction. The index of refraction, introduced by (5–123) is

$$n = 1 + \tfrac{1}{2}\varepsilon_0^{-1}N\mathscr{R}e\alpha \tag{9-133}$$

while the extinction coefficient e, defined as the rate of decrease of field energy per unit length, is

$$e \equiv -\frac{1}{|\mathbf{E}_0|^2}\frac{d\,|\mathbf{E}_0|^2}{dz} = \frac{\omega N}{c\varepsilon_0}\mathscr{I}m\alpha \tag{9-134}$$

Therefore, the complex polarizability is given by

$$N\alpha(\omega) = 2\varepsilon_0[n(\omega) - 1] + i\frac{c\varepsilon_0}{\omega}e(\omega) \tag{9-135}$$

PROBLEM 9–39. Obtain (9–127) from (9–126), and then show that (9–127) agrees with the result of Problem 6–40(b).

PROBLEM 9–40. Show that (9–131) is obtained from (5–118), and then show that (9–132) is obtained from (9–129) and (9–131).

PROBLEM 9–41. Obtain (9–133) and (9–134) from (9–132).

PROBLEM 9–42. Show that (9–134) is consistent with the result of Problem 6–40(c).

PROBLEM 9–43. Express n as a function of ω explicitly, using $\alpha(\omega)$ given in Problem 6–40(a) (Fig. 9–13). Such frequency dependence of the index of refraction is called the *dispersion*.

FIGURE 9–13. n and ϵ as a function of ω.

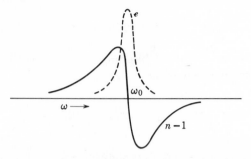

9–10 CAUSALITY AND KRAMERS–KRONIG RELATION

We have seen in the last section that the scattering amplitude of the scattering of electromagnetic waves is essentially given by the complex polarizability, and that the real and imaginary parts of the complex polarizability are essentially given by the index of refraction and the extinction coefficient, respectively. On the other hand, we saw in section 9–8 that if we know the analytic property of a scattering amplitude, we can obtain the dispersion relation, such as (9–117), which relates the real and imaginary parts of the scattering amplitude to each other. Thus, if we know the analytic property of the complex polarizability, we should be able to obtain the dispersion relation between the index of refraction and the extinction coefficient. Kramers and Kronig[4] showed that, in this case, the causality principle is enough to determine the analytic property of the complex polarizability.

Consider a sharp pulse of light propagating in the z direction as shown in Fig. 9–14. A component (either electric or magnetic) of such pulse is expressed as $\delta(z - ct)$, but can also be expressed as a superposition of plane waves as

$$\delta(z - ct) = \frac{1}{2\pi} \int_{-\infty}^{\infty} e^{ik(z - ct)} \, dk \qquad (9\text{--}136)$$

Let us assume that a scatterer is placed at $z = 0$, so that the pulse hits the scatterer at $t = 0$. A scattered wave is created at this instance ($t = 0$) and will propagate through space during the time $t > 0$. Thus, for a positive time, the component of the radiation field should be

$$\delta(z - ct) + F_{sc}(r, t) = \frac{1}{2\pi} \int_{-\infty}^{\infty} \left[e^{ik(z - ct)} + A(k)r^{-1}e^{ik(r - ct)} \right] dk$$

$$t > 0 \qquad (9\text{--}137)$$

[4] H. A. Kramers, *Collected Scientific Papers*, North-Holland, Amsterdam, 1956, p.333. R. Kronig, *J. Opt. Soc. Am.* **12**, 547 (1926).

FIGURE 9–14. Causality in scattering.

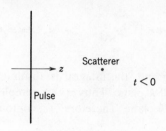

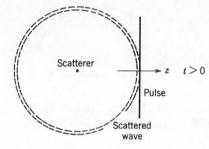

where $A(k)$ is an appropriate component of $\mathbf{A}$, which is given by (9–126). Since the polarizability α is a function of ω, and $\omega = kc$, $A(k)$ is a function of k. In order to simplify the following arguments, let us take a special case of forward scattering, in which $r = z$, and

$$A(k) = -\frac{\mu_0 \omega^2}{4\pi} \alpha(k) \tag{9–138}$$

from (9–126) using the complex polarizability $\alpha(k)$, so that

$$\delta(z - ct) + F_{sc}(z, t) = \frac{1}{2\pi} \int_{-\infty}^{\infty} e^{ik(z-ct)} \left(1 - \frac{\mu_0 \omega^2}{4\pi z} \alpha\right) dk \tag{9–139}$$

We can apply the inverse Fourier transformation to obtain α by means of F_{sc}. In doing so, it is important to note that causality implies that F_{sc} is zero for $t < 0$. Therefore,

$$\alpha(k) = \frac{4\pi z c}{\mu_0 \omega^2} \int_{-\infty}^{\infty} F_{sc}(z, t) e^{-ik(z-ct)} dt$$

$$= \frac{4\pi z c}{\mu_0 \omega^2} \int_{0}^{\infty} F_{sc}(z, t) e^{-ik(z-ct)} dt \tag{9–140}$$

In the complex k plane, where $k = \mathscr{R}e\,k + i\mathscr{I}m\,k$, we obtain

$$\alpha(k) = \frac{4\pi z}{\mu_0 \omega^2} e^{-ikz} \int_0^\infty F_{sc}(z, t) e^{i\mathscr{R}ekct} e^{-\mathscr{I}mkct}\, dt \qquad (9\text{–}141)$$

Since t is positive in the integrand, the integral is absolutely convergent if $\mathscr{I}m\,k > 0$, or in the upper-half k plane. Thus $\alpha(k)$ is completely analytic, or has no poles in the upper-half k plane. Therefore, if a contour is made of the

FIGURE 9–15. Integral contour for the Kramers–Kronig relation. One pole at k_0.

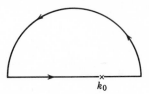

real axis and the semicircle of infinite radius in the upper-half k plane (Fig. 9–15), the integration on this contour gives

$$\oint \frac{\alpha(k)}{k - k_0}\, dk = \mathscr{P} \int_{-\infty}^\infty \frac{\alpha(k)}{k - k_0}\, dk = \pi i \alpha(k_0) \qquad (9\text{–}142)$$

in the same way as before [equations (9–113) and (9–115)], except that there is no residue term other than that at k_0. Using the fact that $\mathscr{I}m\,\alpha(k) = -\mathscr{I}m\,\alpha(-k)$ (Problem 9–37), we obtain

$$\int_0^\infty \frac{k\,\mathscr{I}m\,\alpha(k)}{k^2 - k_0^2}\, dk = \tfrac{1}{2}\pi\mathscr{R}e\,\alpha(k_0) \qquad (9\text{–}143)$$

or, using (9–135),

$$\int_0^\infty \frac{\epsilon(\omega)}{\omega^2 - \omega_0^2}\, d\omega = \frac{\pi}{c}\big[n(\omega_0) - 1\big] \qquad (9\text{–}144)$$

where

$$\omega = kc \qquad \text{and} \qquad \omega_0 = k_0 c$$

In the same way, since $\mathscr{R}e\,\alpha(k) = \mathscr{R}e\,\alpha(-k)$, we obtain

$$\int_0^\infty \frac{\omega_0[n(\omega) - 1]}{\omega^2 - \omega_0^2}\, d\omega = \frac{\pi c}{4\omega_0}\,\epsilon(\omega_0) \qquad (9\text{–}145)$$

Relations (9–144) and (9–145) are called the *Kramers–Kronig relations.*

PROBLEM 9-44. Fill up the steps between (9-143) and the Kramers–Kronig relations.

PROBLEM 9-45. Show that α given in Problem 6-40(a) actually satisfies (9-143).

ADDITIONAL PROBLEMS

9-46. A three-dimensional simple harmonic oscillator is given by

$$\left(-\frac{\hbar^2}{2m}\nabla^2 + \tfrac{1}{2}m\omega^2 r^2\right)\psi = \mathscr{E}\psi$$

(a) Regarding the problem as that of three independent one-dimensional simple harmonic oscillators, obtain all energy levels up to $4\hbar\omega$, and find the degeneracy of each level.

(b) Express ψ as $f_{\mathscr{E}l}(r)Y_{lm}(\theta, \varphi)$, and find a differential equation for $f_{\mathscr{E}l}$.

(c) Knowing that a solution $f_{\mathscr{E}l}(r)$, which does not diverge at the origin, is

$$\exp(-\tfrac{1}{2}\xi^2)\xi^{l+1}F\left(-\frac{1}{2}\left(\frac{\mathscr{E}}{\hbar\omega} - l - \frac{3}{4}\right), l + \frac{3}{2}; \xi^2\right)$$

where $\xi^2 = (m\omega/\hbar)r^2$, find the eigenvalues $\mathscr{E}_n$, and compare the result with that of (a).

9-47. Show that Schroedinger's equation for a hydrogen-like ion gives

$$\left(\frac{1}{2m}\mathbf{p}^2 - \mathscr{E}\right)\psi(\mathbf{p}) = \frac{Ze^2}{(2\pi)^3\varepsilon_0}\int \frac{\psi(\mathbf{p}')}{|\mathbf{p} - \mathbf{p}'|^2}\,d\mathbf{p}'$$

when transformed into the momentum representation.

9-48. Kramers found the recursion relation between the expectation values of ρ^s in a nl state of the hydrogen atom as

$$\frac{s+1}{n^2}\langle\rho^s\rangle - (2s+1)\langle\rho^{s-1}\rangle + \tfrac{1}{4}s[(2l+1)^2 - s^2]\langle\rho^{s-2}\rangle = 0$$

(a) Show that results given in Table 9–3 are consistent with this relation.

(b) Derive this recursion relation.

9–49. Show that the first Born approximation, by which (9–64) is obtained, does not satisfy the optical theorem.

9–50. When a potential is essentially zero outside a range r_0, show that

$$\delta_l \xrightarrow[k \to 0]{} \text{constant} \times (kr_0)^{2l+1}$$

9–51. Using the WKB approximation (section 8–7), and assuming that the scattering potential U is much smaller than $(\hbar k)^2/2m$, obtain

$$\delta_0 = - \int_{r_0}^{\infty} \frac{mU(r)\,dr}{\hbar^2 k}$$

where r_0 is the classical turning point, and δ_0 is the phase shift of the s wave.

9–52. Taking the square well potential of Fig. 9–10, show that

$$A_l(k; 0) \xrightarrow[k \to 0]{} - \frac{r_0}{[1 \cdot 3 \cdots (2l - 1)]^2} \frac{j_{l+1}(\kappa r_0)}{j_{l-1}(\kappa r_0)} (kr_0)^{2l}$$

where

$$\kappa^2 = k^2 + \frac{2m}{\hbar^2} U_0$$

HINT.   Use the result of Problem 9–33.

9–53. Show that there is no resonance scattering in the Coulomb scattering.

HINT.   Use (9–48) and the asymptotic expression of the confluent hypergeometric function given in Appendix I(c).

9–54. Obtain another Kramers–Kronig relation from

$$\oint \frac{\alpha(k)}{k(k - k_0)}\,dk$$

assuming that $\lim_{\omega \to 0} \epsilon/\omega = \epsilon'$ is a finite constant.

Chapter 10 Relativistic Quantum Mechanics

In the same way as the classical relation $\mathscr{E} = p^2/2m$ corresponds to the Schroedinger equation, we should find quantum mechanical equations which correspond to the classical relativistic relation $\mathscr{E}^2 = (cp)^2 + (mc^2)^2$. This relation, which clarified the relation between mass and energy in classical mechanics, turns out to be even more significant in quantum mechanics, for it leads to the concepts of antiparticles and spin angular momentum.

The Klein–Gordon and Dirac theories are discussed in this chapter. The Klein–Gordon theory, in the modern viewpoint, linearizes the energy term only, and gives rise to the concept of an antiparticle, or charge conjugate particle. The Dirac theory, on the other hand, linearizes both energy and momentum terms. The former process gives rise to the concept of antiparticles, as in the Klein–Gordon theory, while the latter process gives rise to the concept of spin angular momentum. Because the most familiar elementary particle, the electron, was known to have spin angular momentum when the theory was proposed in 1928, the Dirac theory was immediately recognized as amazingly successful. Later, when the anti-electron, or the positron, and other antiparticles were found, the theory was completely established. The theory is now known to be applicable to all leptons, including neutrinos as a special case of mass zero. The theory is also applicable to some extent to baryons, including the proton and neutron, but a large deviation is found with respect to their magnetic properties. The Klein–Gordon theory, which predicts no spin, is applicable to an extent to spin zero particles, such as the π meson.

10–1 KLEIN–GORDON EQUATION

The Schroedinger equation for a free particle is given by (8–84) and (8–115) as

$$i\hbar \frac{\partial \Psi}{\partial t} = - \frac{\hbar^2}{2m} \nabla^2 \Psi \tag{10-1}$$

Taking the wave function as

$$\Psi = e^{i(\mathbf{p} \cdot \mathbf{r} - \mathscr{E}t)/\hbar} \tag{10-2}$$

which is essentially the same as (8–116), we obtain from (10–1) the classical relation

$$\mathscr{E} \, (= T) = \frac{\mathbf{p}^2}{2m} \tag{10-3}$$

for the free particle, where T is the kinetic energy. We notice that the wave function (10–2) can be expressed in a covariant form

$$\Psi = \exp \left(\frac{-i p_j x^j}{\hbar} \right) \tag{10-4}$$

where the four vectors p_j and x^j are given by (4–74) and (4–16), respectively. Thus, relativistic quantum mechanics, which gives the classical relativistic relation

$$T^2 = (pc)^2 + (mc^2)^2 \tag{10-5}$$

can be formulated by keeping (10–4) as it is, and simply replacing the Schroedinger equation (10–1) by

$$- \left(\hbar \frac{\partial}{\partial t} \right)^2 \Psi = [-(c\hbar \nabla)^2 + (mc^2)^2] \Psi \tag{10-6}$$

or

$$\hbar^2 \Box \Psi = (mc)^2 \Psi \tag{10-6'}$$

This is called the *Klein–Gordon equation.*[1]

[1] O. Klein, *Zeits. f. Physik* **37**, 895 (1926); W. Gordon, *Zeits. f. Physik* **40**, 117 (1926).

An important difference between the Klein–Gordon equation and the Schroedinger equation is that the time derivative is of the second order in the former, but is of the first order in the latter. This means that while the Schroedinger equation can be transformed into the equation of continuity (8–82), giving the probability density as $\Psi^*\Psi$ and the probability current density as given by (8–83), the same procedure applied to the Klein–Gordon equation gives the same formula for the probability current density as before, but

$$\rho = \frac{i\hbar}{2mc^2}\left(\Psi\,\frac{\partial\Psi^*}{\partial t} - \Psi^*\,\frac{\partial\Psi}{\partial t}\right) \qquad (10\text{--}7)$$

for the probability density. In contrast to the previous expression $|\Psi|^2$, this expression of the probability density is not positive definite, as a probability density should be.

λ-Operator

Since the trouble appeared because the time derivative is of the second order, instead of the first order, in the Klein–Gordon equation, we are motivated to find an equation

$$\left(i\hbar\,\frac{\partial}{\partial t} - \hat{H}_{KG}\right)\Psi = 0 \qquad (10\text{--}8)$$

which is equivalent to (10–6):

$$(\hat{H}_{KG})^2 = -(c\hbar\nabla)^2 + (mc^2)^2 \qquad (10\text{--}9)$$

The linearization can be done if we introduce a $\hat{\lambda}$ operator, whose three components have the properties

$$\hat{\lambda}_1{}^2 = \hat{\lambda}_2{}^2 = \hat{\lambda}_3{}^2 = \hat{1} \qquad (10\text{--}10)$$

$$\hat{\lambda}_1\hat{\lambda}_2 = -\hat{\lambda}_2\hat{\lambda}_1 = i\hat{\lambda}_3 \qquad \hat{\lambda}_2\hat{\lambda}_3 = -\hat{\lambda}_3\hat{\lambda}_2 = i\hat{\lambda}_1$$
$$\hat{\lambda}_3\hat{\lambda}_1 = -\hat{\lambda}_1\hat{\lambda}_3 = i\hat{\lambda}_2 \qquad\qquad (10\text{--}11)$$

$$[\hat{\lambda}_j, \nabla] = 0 \qquad \left[\hat{\lambda}_j, \frac{\partial}{\partial t}\right] = 0 \qquad j = 1, 2, 3 \qquad (10\text{--}12)$$

where the last relations imply that this $\hat{\lambda}$ operator is associated with a new degree of freedom which is not included in those of the ordinary three-dimensional space and time. It is easy to show that

$$\hat{H}_{KG} = -(\hat{\lambda}_3 + i\hat{\lambda}_2) \frac{\hbar^2}{2m} \nabla^2 + \hat{\lambda}_3 mc^2 \qquad (10\text{-}13)$$

indeed satisfies (10–9).

Looking at (10–10) and (10–11), we notice that these are exactly the relations satisfied by the Pauli matrices (4–60). Thus, if we take a two-component spinor

$$\Psi(\mathbf{r}, t) = \begin{pmatrix} \Phi(\mathbf{r}, t) \\ \chi(\mathbf{r}, t) \end{pmatrix} \qquad (10\text{-}14)$$

such that when $\hat{\lambda}$ operates on this spinor, it is transformed as

$$\hat{\lambda}_j \Psi = \sigma_j \Psi \qquad (10\text{-}15)$$

then we see that all relations (10–10) and (10–11) are satisfied.[2] Therefore, the Pauli matrices are the matrix representatives of $\hat{\lambda}$, and the spinor (10–14) belongs to this representation. Taking this spinor for the wave function in (10–8) and using $\hat{H}_{KG}$ of (10–13), we obtain

$$i\hbar \frac{\partial}{\partial t} \Psi + \frac{\hbar^2}{2m} (\sigma_3 + i\sigma_2) \nabla^2 \Psi - mc^2 \sigma_3 \Psi = 0 \qquad (10\text{-}16)$$

which shows that the Klein–Gordon equation is equivalent to the two simultaneous equations:

$$i\hbar \frac{\partial}{\partial t} \Phi + \frac{\hbar^2}{2m} \nabla^2 (\Phi + \chi) - mc^2 \Phi = 0 \qquad (10\text{-}17a)$$

and

$$i\hbar \frac{\partial}{\partial t} \chi - \frac{\hbar^2}{2m} \nabla^2 (\Phi + \chi) + mc^2 \chi = 0 \qquad (10\text{-}17b)$$

Equation of Continuity

Let us introduce another spinor

$$\Psi^\dagger = (\Phi^* \, \chi^*) \qquad (10\text{-}18)$$

[2] Note that $\hat{\lambda}_i \hat{\lambda}_j \Psi = \hat{\lambda}_i (\sigma_j \Psi) = \sigma_i (\sigma_j \Psi) = \sigma_i \sigma_j \Psi$ in the spinor calculus.

which is the Hermitian conjugate of Ψ. Since $(\sigma\Psi)^\dagger = \Psi^\dagger\sigma^\dagger = \Psi^\dagger\sigma$ we see that (10–16) can be rewritten as

$$-i\hbar\,\frac{\partial}{\partial t}\,\Psi^\dagger + \frac{\hbar^2}{2m}\,\nabla^2\Psi^\dagger(\sigma_3 - i\sigma_2) - mc^2\Psi^\dagger\sigma_3 = 0 \qquad (10\text{–}19)$$

Multiplying (10–19) by $\sigma_3\Psi$ from the right, we obtain

$$-i\hbar\left(\frac{\partial}{\partial t}\,\Psi^\dagger\right)\sigma_3\Psi + \frac{\hbar^2}{2m}\,(\nabla^2\Psi^\dagger)(\sigma_3 - i\sigma_2)\sigma_3\Psi - mc^2\Psi^\dagger\Psi = 0 \qquad (10\text{–}20)$$

while multiplying (10–16) by $\Psi^\dagger\sigma_3$ from the left, we obtain

$$i\hbar\Psi^\dagger\sigma_3\,\frac{\partial}{\partial t}\,\Psi + \frac{\hbar^2}{2m}\,\Psi^\dagger\sigma_3(\sigma_3 + i\sigma_2)\nabla^2\Psi - mc^2\Psi^\dagger\Psi = 0 \qquad (10\text{–}21)$$

Subtracting (10–21) from (10–20), we obtain

$$i\hbar\,\frac{\partial}{\partial t}\,(\Psi^\dagger\hat{\lambda}_3\Psi) - \frac{\hbar^2}{2m}\,\nabla\cdot[(\nabla\Psi^\dagger)(1 + \hat{\lambda}_1)\Psi - \Psi^\dagger(1 + \hat{\lambda}_1)\nabla\Psi] = 0$$

$$(10\text{–}22)$$

where we used the $\hat{\lambda}$ operator again. Since this is the desired form of the equation of continuity, we immediately obtain

$$\rho = \Psi^\dagger\hat{\lambda}_3\Psi \qquad (10\text{–}23)$$

and

$$\mathbf{j} = \frac{i\hbar}{2m}\,[(\nabla\Psi^\dagger)(1 + \hat{\lambda}_1)\Psi - \Psi^\dagger(1 + \hat{\lambda}_1)\nabla\Psi] \qquad (10\text{–}24)$$

The probability density ρ is

$$\rho = \Psi^\dagger\hat{\lambda}_3\Psi = \Psi^\dagger\sigma_3\Psi = |\Phi|^2 - |\chi|^2 \qquad (10\text{–}25)$$

which is still not positive definite. However, this form is better than (10–7) in the sense that it allows a new interpretation of ρ, i.e., this is not the particle density, but the charge density. We interpret $|\Phi(\mathbf{r})|^2$ and $|\chi(\mathbf{r})|^2$ as the probability of finding the particle at $\mathbf{r}$ in its positive and negative charge states, respectively. Thus the special spinors

$$\Psi_+ = \begin{pmatrix} \Phi \\ 0 \end{pmatrix} \qquad (10\text{–}26a)$$

and

$$\Psi_- = \begin{pmatrix} 0 \\ \chi \end{pmatrix} \tag{10-26b}$$

represent purely positive and negative charge states, respectively. Since these spinor functions are the eigenfunctions of $e\hat{\lambda}_3$, i.e.,

$$e\hat{\lambda}_3\Psi_+ = e\Psi_+ \tag{10-27a}$$

and

$$e\hat{\lambda}_3\Psi_- = -e\Psi_- \tag{10-27b}$$

we see that $e\hat{\lambda}_3$ is the charge operator. Here e is the elementary, or the electronic charge.

PROBLEM 10–1. Obtain (10–7) and (8–83) from (10–6).

PROBLEM 10–2. Show, from (10–11), that $\hat{\lambda}_j = \hat{\lambda}_j^\dagger$.

PROBLEM 10–3. Show that $\hat{H}_{KG}$ of (10–13) satisfies (10–9).

PROBLEM 10–4. Obtain (10–17a) and (10–17b) from (10–16).

PROBLEM 10–5. Show that

$$(\hat{\lambda}_1 + i\hat{\lambda}_2)\Psi_+ = 0 \quad (\hat{\lambda}_1 + i\hat{\lambda}_2)|\Psi_-| = 2|\Psi_+|$$
$$(\hat{\lambda}_1 - i\hat{\lambda}_2)|\Psi_+| = 2|\Psi_-| \quad (\hat{\lambda}_1 - i\hat{\lambda}_2)\Psi_- = 0$$

HINT. Because of the normalization $|\Phi| = |\chi|$ in (10–26a) and (10–26b).

PROBLEM 10–6. Express **j** by means of two spinor components.

ANSWER.

$$\mathbf{j} = -\frac{\hbar}{m}\mathscr{I}m[(\Phi + \chi)^*\nabla(\Phi - \chi)]$$

PROBLEM 10–7. (a) Show that the Klein–Gordon equation can be obtained from the classical equation (10–5) by the replacements

$$\hat{p}_j = i\hbar\frac{\partial}{\partial x^j}$$

(b) Show that the above result implies relativistic commutation relations

$$[\hat{p}_j, \hat{x}^k] = i\hbar\delta_j{}^k$$

10-2 PLANE WAVE SOLUTIONS OF THE KLEIN–GORDON EQUATIONS

The linearized Klein–Gordon equation (10–16) or (10–17) has a solution of the form

$$\Psi_{k,\omega} = \begin{pmatrix} \xi \\ \eta \end{pmatrix} e^{i(\mathbf{k}\cdot\mathbf{r} - \omega t)} \qquad (10\text{–}28)$$

where $\mathbf{k}$, ω, ξ, and η are constants. From (10–17a) and (10–17b) we see that these constants should satisfy

$$\left(\hbar\omega - mc^2 - \frac{\hbar^2\mathbf{k}^2}{2m}\right)\xi - \frac{\hbar^2\mathbf{k}^2}{2m}\eta = 0 \qquad (10\text{–}29\text{a})$$

$$\frac{\hbar^2\mathbf{k}^2}{2m}\xi + \left(\hbar\omega + mc^2 + \frac{\hbar^2\mathbf{k}^2}{2m}\right)\eta = 0 \qquad (10\text{–}29\text{b})$$

In order to have nontrivial solutions for ξ and η, the secular equation

$$\begin{vmatrix} \hbar\omega - mc^2 - \dfrac{\hbar^2\mathbf{k}^2}{2m} & -\dfrac{\hbar^2\mathbf{k}^2}{2m} \\ \dfrac{\hbar^2\mathbf{k}^2}{2m} & \hbar\omega + mc^2 + \dfrac{\hbar^2\mathbf{k}^2}{2m} \end{vmatrix} = 0 \qquad (10\text{–}30)$$

should be satisfied; so that

$$\hbar\omega = \pm\sqrt{(mc^2)^2 + (c\hbar k)^2} \qquad (10\text{–}31)$$

Solutions (10–31) obviously correspond to the classical relation (10–5), when $\hbar\omega$ and $\hbar k$ are equated to the energy and momentum, respectively. Notice, however, that $\hbar\omega$ can be negative as well as positive, while the concept of kinetic energy does not allow a negative value. We reconcile this conflict by taking

$$\hbar\omega = \pm\mathscr{E} \qquad (10\text{–}32)$$

where $\mathscr{E}$ is a positive quantity, which we call *energy*.

If we take the positive solution ($\hbar\omega = +\mathscr{E}$) of (10–31), put that into (10–29), and normalize the wave function as

$$\Psi^\dagger \hat{\lambda}_3 \Psi = |\xi|^2 - |\eta|^2 = 1 \qquad (10\text{–}33)$$

then we obtain

$$\xi_{+\mathscr{E}} = \frac{\mathscr{E} + mc^2}{2\sqrt{mc^2\mathscr{E}}} \qquad \eta_{+\mathscr{E}} = \frac{\mathscr{E} - mc^2}{2\sqrt{mc^2\mathscr{E}}} \qquad (10\text{–}34)$$

while, if we take the negative solution ($\hbar\omega = -\mathscr{E}$), then we obtain

$$\xi_{-\mathscr{E}} = \frac{mc^2 - \mathscr{E}}{2\sqrt{mc^2\mathscr{E}}} \qquad \eta_{-\mathscr{E}} = \frac{\mathscr{E} + mc^2}{2\sqrt{mc^2\mathscr{E}}} \qquad (10\text{–}35)$$

in the normalization

$$\Psi^\dagger \hat{\lambda}_3 \Psi = -1 \qquad (10\text{–}33')$$

The wave function $\Psi_{k,\omega}$ is finally obtained by putting either set of the above solutions into (10–28).

In the nonrelativistic limit, where $\mathscr{E} \cong mc^2$, we see that

$$\xi_{+\mathscr{E}} = \eta_{-\mathscr{E}} \cong 1 \qquad \eta_{+\mathscr{E}} = \xi_{-\mathscr{E}} \cong 0 \qquad (10\text{–}36)$$

Therefore, in this limit the positive solutions give the pure charge positive states, and the negative solutions give the pure charge negative states. In general, however, a state with definite energy is not a pure charge state.

Antiparticle. Charge Conjugation. (Klein–Gordon Theory)

A particle whose wave function is

$$\Psi_C = \begin{pmatrix} \eta \\ \xi \end{pmatrix} e^{-i(\mathbf{k}\cdot\mathbf{r} - \mathscr{E}t/\hbar)} \qquad (10\text{–}37a)$$

is called the antiparticle of a particle whose wave function is

$$\Psi = \begin{pmatrix} \xi \\ \eta \end{pmatrix} e^{i(\mathbf{k}\cdot\mathbf{r} - \mathscr{E}t/\hbar)} \qquad (10\text{–}37b)$$

If (10–37b) is a solution of the Klein–Gordon equation, (10–37a) is also a solution with the same energy $\mathscr{E}$ (Problem 10–9). These particles propagate to the same direction, along **k**, but their momenta, which are the eigenvalues of $-i\hbar\nabla$, have directions opposite to each other (Fig. 10–1). In addition, their charge densities are opposite to each other:

$$e\Psi_C^\dagger \hat{\lambda}_3 \Psi_C = -e\Psi^\dagger \hat{\lambda}_3 \Psi \tag{10-38}$$

FIGURE 10–1. The charge conjugation $\hat{S}_C'$ in the Klein–Gordon theory. The charge conjugate particles have opposite sign of ε, $\mathbf{p}$, and q.

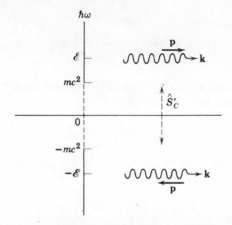

Because of the last relation Ψ_C and Ψ are also called *charge conjugate* to each other. The charge conjugation operator which transforms Ψ into Ψ_C

$$\hat{S}_C'\Psi = \Psi_C \tag{10-39}$$

is easily seen to be

$$\hat{S}_C' = \hat{\lambda}_1 \hat{K} \tag{10-40}$$

where $\hat{K}$ is an operator which transforms Ψ into its complex conjugate Ψ^*.

Since Ψ and Ψ_C are both solutions of the same linearized Klein–Gordon equation, any linear combination of these two wave functions is also a solution. Particularly $\Psi + \Psi_C$ and $\Psi - \Psi_C$ are important. These two solutions represent neutral particles, since the charge density is zero in either case. The difference of these two states appears when the *charge parity* is calculated. If we operate the charge conjugation operator to these wave functions we see that

$$\hat{S}_C'(\Psi \pm \Psi_C) = \pm(\Psi \pm \Psi_C) \tag{10-41}$$

Thus these states are both eigenfunctions of $\hat{S}_C'$ but their eigenvalues are $+1$ or -1. The state $\Psi + \Psi_C$, which is the eigenfunction of $\hat{S}_C'$ with eigenvalue $+1$, is called a state with the charge parity $+$, while the other state $\Psi - \Psi_C$ has the charge parity $-$. Photons and π^0 mesons are typical neutral particles. It is found experimentally that the charge parity of photons is negative, and that of the π^0 mesons is positive.[3]

PROBLEM 10–8. Obtain (10–34) and (10–35) from (10–29).

PROBLEM 10–9. If $\hat{f}$ is an operator $\hat{K}\hat{f}\hat{K}^{-1} = \hat{f}^*$.
(a) Show that

$$\hat{S}'_C \left(i\hbar \frac{\partial}{\partial t} - \hat{H}_{KG} \right) \hat{S}'_C = - \left(i\hbar \frac{\partial}{\partial t} - \hat{H}_{KG} \right)$$

(b) Show that Ψ and Ψ_C of (10–37) and (10–38) are the solutions of the same Klein–Gordon equation with the same $\mathscr{E}$.

HINT. $\hat{\lambda}_1^* = \hat{\lambda}_1$, $\hat{\lambda}_3^* = \hat{\lambda}_3$, but $\hat{\lambda}_2^* = -\hat{\lambda}_2$, since $\sigma_2^* = -\sigma_2$.

PROBLEM 10–10. (a) Show that (10–6) is equivalent to

$$\Psi = \Phi + \chi \quad \text{and} \quad i\hbar \frac{\partial \Psi}{\partial t} = mc^2(\Phi - \chi)$$

where Φ and χ are the spinor components introduced in (10–14).
(b) Show that Ψ^* is the charge conjugate state of Ψ, i.e., Ψ^* is the same state as Ψ_C.

10–3 KLEIN–GORDON EQUATIONS WITH ELECTROMAGNETIC FIELDS

We found in classical physics that the canonical momentum of a particle interacting with an electromagnetic field is

$$\mathbf{p} = m\mathbf{V}\gamma + q\mathbf{A} \tag{6–44}$$

where $\mathbf{A}$ is the vector potential at the position of the particle. The energy of this particle is, therefore,

$$\mathscr{E} = mc^2\gamma + q\phi = c\sqrt{(\mathbf{p} - q\mathbf{A})^2 + (mc)^2} + q\phi \tag{10–42}$$

---

[3] The Klein–Gordon equation describes a spin zero scalar particle, but photons have spin one, and while π mesons have spin zero they are pseudoscalars. The concept of the charge parity, nevertheless is applicable to them.

which can be expressed in the covariant form

$$(p_j - qA_j)(p^j - qA^j) = (mc)^2 \tag{10-43}$$

Replacing the 4-canonical-momentum-vector by the differential operator (Problem 10–7a)

$$\hat{p}_j = i\hbar \frac{\partial}{\partial x^j} \tag{10-44}$$

we obtain the Klein–Gordon equation with the field,

$$\left[\left(i\hbar \frac{\partial}{\partial t} - q\phi \right)^2 - c^2(i\hbar \nabla + q\mathbf{A})^2 - (mc^2)^2 \right] \Psi = 0 \tag{10-45}$$

Hydrogen-Like Ions

Let us calculate the energy levels of hydrogen-like ions from the Klein–Gordon equation. In this case m is the reduced mass of an electron-nucleus system, and

$$q\phi = -\frac{Ze^2}{4\pi\varepsilon_0 r} \qquad A = 0 \tag{10-46}$$

Putting (10–46) into (10–45), and assuming that

$$\Psi = f_{\mathscr{E}l}(r) Y_{lm}(\theta, \varphi) e^{-i\mathscr{E}t/\hbar} \tag{10-47}$$

we obtain

$$[(c\hbar\nabla)^2 + (\mathscr{E} - q\phi)^2 - (mc^2)^2] f_{\mathscr{E}l} Y_{lm} = 0 \tag{10-48}$$

which reduces to

$$\left[\frac{1}{r^2} \frac{d}{dr} \left(r^2 \frac{d}{dr} \right) - \frac{1}{r^2} (l(l+1) - (Z\alpha)^2) + \frac{2Z\alpha\mathscr{E}}{\hbar c r} - \frac{(mc^2)^2 - \mathscr{E}^2}{(\hbar c)^2} \right] f_{\mathscr{E}l} = 0$$

$$\tag{10-49}$$

where

$$\alpha = \frac{e^2}{4\pi\varepsilon_0 \hbar c} = \frac{1}{137.03602} \tag{10-50}$$

is called the fine structure constant.

Comparing equation (10–49) to the previous equation in the Schroedinger theory, (9–18), we see that they are mathematically the same, if we make the replacements,

Schroedinger   Klein–Gordon

$l(l + 1)$ $l(l + 1) - (Z\alpha)^2$

$$\frac{Z}{a} \qquad\qquad \frac{Z\alpha\mathscr{E}}{\hbar c} \qquad\qquad (10\text{–}51)$$

$$\frac{2m\mathscr{E}}{\hbar^2} \qquad\qquad \frac{\mathscr{E}^2 - (mc^2)^2}{(\hbar c)^2}$$

Therefore, if we define σ by

$$\sigma(\sigma + 1) = l(l + 1) - (Z\alpha)^2 \qquad\qquad (10\text{–}52)$$

we see that the hypergeometric function in the solution for the radial function is

$$F\left(\frac{Z\alpha\mathscr{E}}{\hbar c}\ \frac{\hbar c}{\sqrt{(mc^2)^2 - \mathscr{E}^2}} + \sigma + 1, \ldots\right) \qquad\qquad (10\text{–}53)$$

instead of the hypergeometric function in (9–20). The eigenfunctions are obtained as before by requiring that the first argument in the hypergeometric function, as explicitly shown in (10–53), is equal to a negative integer, or zero, $-n'$:

$$\frac{Z\alpha\mathscr{E}}{\sqrt{(mc^2)^2 - \mathscr{E}^2}} + \sigma + 1 = -n' \qquad n' = 0, 1, 2, \ldots \qquad (10\text{–}54)$$

Solving this equation we obtain

$$\mathscr{E} = \frac{mc^2}{\sqrt{1 + (Z\alpha/N)^2}} \qquad\qquad (10\text{–}55)$$

where

$$N = n' + \sigma + 1 \qquad\qquad (10\text{–}56)$$

When $Z\alpha$ is small enough, we obtain

$$N = n' + l + 1 - \frac{(Z\alpha)^2}{2l + 1} - \cdots \qquad\qquad (10\text{–}56')$$

and

$$
\begin{aligned}
\mathscr{E}_{nl} &= mc^2 - \frac{m(cZ\alpha)^2}{2N^2} - \frac{mc^2(Z\alpha)^4}{8N^4} - \cdots \\
&= mc^2 - \frac{m(cZ\alpha)^2}{2n^2} - \frac{mc^2(Z\alpha)^4}{2n^4}\left(\frac{n}{l+\frac{1}{2}} - \frac{3}{4}\right) - \cdots
\end{aligned}
\tag{10-57}
$$

where

$$
n = n' + l + 1 \tag{10-58}
$$

is the principal quantum number introduced in (9–25). Notice that the second term in (10–57) is exactly the previous result (9–24) for the energy in Schroedinger's theory.

The third term in (10–57) shows that the energy depends on l as well as n in this relativistic theory, in contrast to the nonrelativistic theory where the energy was given by a single quantum number n. Thus the l degeneracy of the hydrogen-like ion is removed when relativistic effects are taken into account (Fig. 10–2). For example, energies of the $2s$ and $2p$ levels are exactly the same in Schroedinger's theory, but they are different by $4mc^2(Z\alpha)^4/48$, $2p$ being higher than $2s$, when the relativistic effect is taken into account. This splitting is called the *fine structure*, and the theory and experiment agree as far as the order of magnitude of the splitting is concerned. The nature of each level, however, is rather different from what we find here. We will come back to this point in section 10–12.

FIGURE 10–2. Schematic energy level diagram of a hydrogen-like ion according to the Klein–Gordon theory. For $n = 2$ and 3 levels. The fine structures are very much exaggerated.

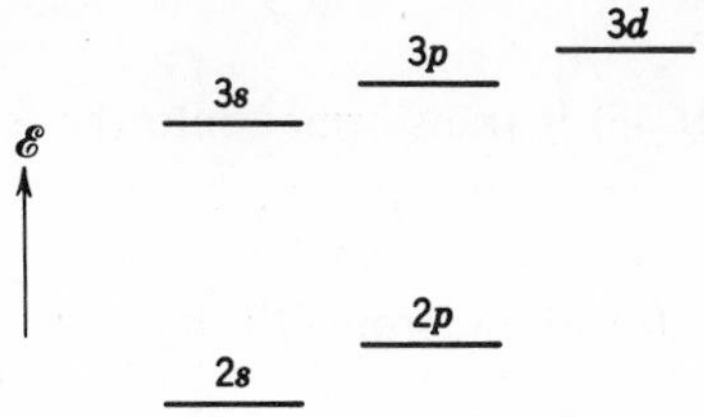

Zeeman Effect

An external field can be given by potentials which satisfy $\phi = 0$ and $\mathbf{V} \cdot \mathbf{A} = 0$, as a special case of the Lorentz gauge. The middle term of (10–45) is then

$$
\begin{aligned}
-c^2(i\hbar\mathbf{V} + q\mathbf{A})^2\Psi &= c^2[(\hbar\mathbf{V})^2 - i\hbar q\mathbf{V}\cdot\mathbf{A} - i\hbar q\mathbf{A}\cdot\mathbf{V} - (qA)^2]\Psi \\
&= [(c\hbar\mathbf{V})^2 + 2c^2 q\mathbf{A}\cdot\hat{\mathbf{p}} - (cq\mathbf{A})^2]\Psi
\end{aligned}
\tag{10-59}
$$

A homogeneous magnetic field **B** is given by the vector potential

$$\mathbf{A} = \tfrac{1}{2}\mathbf{B} \times \mathbf{r} \tag{10-60}$$

Thus (10–59) gives

$$(c\hbar\nabla)^2 + c^2 q(\mathbf{B} \times \mathbf{r}) \cdot \hat{\mathbf{p}} - (\tfrac{1}{2}cq\mathbf{B} \times \mathbf{r})^2$$

$$= (c\hbar\nabla)^2 + c^2 q\mathbf{l} \cdot \mathbf{B} - (\tfrac{1}{2}cq)^2 (r^2 B^2 - (\mathbf{r} \cdot \mathbf{B})^2) \tag{10-61}$$

where **l** is the orbital angular momentum operator. The term which is proportional to B is called the paramagnetic term, while the terms which are proportional to B^2 are called the diamagnetic terms.

Putting (10–61) into (10–45), and remembering that $\mathcal{E}$ is approximately mc^2, we see that the extra energy due to B is

$$\Delta\hat{H} \cong \frac{-q}{2m}\mathbf{l} \cdot \mathbf{B} + \frac{q^2}{8m}(r^2 B^2 - (\mathbf{r} \cdot \mathbf{B})^2) \tag{10-62}$$

in the nonrelativistic limit.

PROBLEM 10–11. Show that

$$\rho = \frac{i\hbar}{2mc^2}\left(\Psi^* \frac{\partial\Psi}{\partial t} - \Psi \frac{\partial\Psi^*}{\partial t}\right) - \frac{q^2\phi_0}{mc^2}\Psi^*\Psi$$

$$\mathbf{j} = i\frac{e\hbar}{2m}[\Psi(\nabla\Psi^*) - \Psi^*(\nabla\Psi)] - \frac{q^2 A_0}{m}\Psi^*\Psi$$

are obtained when (10–45) is transformed into the form of the equation of continuity.

PROBLEM 10–12. (a) Show, from (10–45), that

$$\left[i\hbar\frac{\partial}{\partial t} - q\phi + (\hat{\lambda}_3 + i\hat{\lambda}_2)\frac{1}{2m}(i\hbar\nabla + q\mathbf{A})^2 - \hat{\lambda}_3 mc^2\right]\Psi = 0$$

(b) Show that

$$\left[i\hbar\frac{\partial}{\partial t} + q\phi + (\hat{\lambda}_3 + i\hat{\lambda}_2)\frac{1}{2m}(i\hbar\nabla - q\mathbf{A})^2 - \hat{\lambda}_3 mc^2\right]\Psi_C = 0$$

PROBLEM 10-13. In the gauge transformation (6–5) and (6–6), or

$$A'_j = A_j + \frac{\partial \chi}{\partial x^j}$$

(a) Show that $(p_j - qA_j) = e^{-iq\chi/\hbar}(p_j - qA'_j)e^{iq\chi/\hbar}$.
(b) Show that if Ψ satisfies (10–45), then

$$\Psi' = \Psi e^{iq\chi/\hbar}$$

satisfies the corresponding equation with A'_j, the gauge transformed potentials.

PROBLEM 10-14. Using (10–51), show that

$$f_{\mathscr{E}l}(r) = Ne^{-(1/2)\rho}\rho^{\jmath}F(-\eta + \jmath + 1, 2\jmath + 2, \rho)$$

where

$$\rho = \frac{2r\sqrt{(mc^2)^2 - \mathscr{E}^2}}{\hbar c} \qquad \eta = \frac{Z\alpha\mathscr{E}}{\sqrt{(mc^2)^2 - \mathscr{E}^2}}$$

and N is a normalization constant.

PROBLEM 10-15. Obtain the final result of (10–57) from (1–54).

10-4 SPIN ANGULAR MOMENTUM

The Zeeman effect has been observed in many atoms (electrons), and it was found that in some cases formula (10–62) provided an adequate explanation, but in many other cases it did not. It was found, in the case of the hydrogen atom for example, that a spin angular momentum $\hat{s}$ had to be introduced in such a way that $\hat{s}_z$ could take only two eigenvalues, $\frac{1}{2}\hbar$ and $-\frac{1}{2}\hbar$, and that the paramagnetic term of the perturbation $\Delta\hat{H}$ of (10–62) should be modified as

$$\Delta\hat{H} = \frac{e}{2m}(1 + 2\hat{s}) \cdot \mathbf{B} + \text{(diamagnetic terms)} \tag{10-63}$$

where we took $q = -e$. Since the expectation value of an angular momentum is of the order of $\hbar$, the natural unit of the magnetic moment of an atom (electron) is

$$\mu_B = \frac{e\hbar}{2m_e} = 0.9274096 \times 10^{-23} \text{ A m}^2 \tag{10-64}$$

which is called the *Bohr magneton*. A magnetic moment due to an angular momentum **j** is expressed as

$$\mu = - \frac{g_j \mu_B \mathbf{j}}{\hbar} \tag{10-65}$$

by means of the Bohr magneton and the g factor g_j. According to (10–63),

$$g_l = 1 \qquad g_s = 2 \tag{10-66}$$

The vector operator $\hat{\mathbf{s}}$, being an angular momentum, should satisfy the same commutation relations as **l**, (8–53), namely

$$[\hat{s}_x, \hat{s}_y] = i\hbar\hat{s}_z \qquad [\hat{s}_y, \hat{s}_z] = i\hbar\hat{s}_x \qquad [\hat{s}_z, \hat{s}_x] = i\hbar\hat{s}_y \tag{10-67}$$

The second requirement, that $\hat{s}_z$ has only two eigenvalues, reminds us of the $\hat{\lambda}_3$ operator, which we discussed in section 10–1. Thus, we expect that an eigenfunction of $\hat{s}_z$ is a spinor which, when operated by $\hat{s}_z$, gives the Pauli matrix σ_3 as the matrix representative:

$$\hat{s}_z\Psi = \tfrac{1}{2}\hbar\sigma_3\Psi \tag{10-68a}$$

where $\tfrac{1}{2}\hbar$ appears since the two eigenvalues are $\tfrac{1}{2}\hbar$ and $-\tfrac{1}{2}\hbar$. We see easily that the Pauli matrices σ_1, σ_2, and σ_3 satisfy the commutation relations (10–67), which means that the same spinor as the one in (10–68) gives

$$\hat{s}_x\Psi = \tfrac{1}{2}\hbar\sigma_1\Psi \quad \text{and} \quad \hat{s}_y\Psi = \tfrac{1}{2}\hbar\sigma_2\Psi \tag{10-68b}$$

and

$$\hat{s}^2\Psi = (\hat{s}_x^2 + \hat{s}_y^2 + \hat{s}_z^2)\Psi = (\tfrac{1}{2}\hbar)^2(\sigma_1^2 + \sigma_2^2 + \sigma_3^2)\Psi = \tfrac{3}{4}\hbar^2\Psi \tag{10-69}$$

So far the new degree of freedom $\hat{\mathbf{s}}$ looks similar to the $\hat{\lambda}$ operator, at least mathematically. The expected form of (10–63), however, shows that physically they are entirely different, because (10–63) cannot be obtained from the Klein–Gordon equation. Thus $\hat{\mathbf{s}}$ should be introduced in a different way than $\hat{\lambda}$. Remembering that $\hat{\lambda}$ appeared in reducing $(\partial/\partial t)^2$ to $\partial/\partial t$ in the Klein–Gordon equation, we expect that $\hat{\mathbf{s}}$ may appear in linearizing some other term. Looking at (10–6), one might naturally ask "why not linearize $-(c\hbar\nabla)^2$, or $(c\hat{\mathbf{p}})^2$?"

It can be easily shown from (10–68) that if $\hat{\mathbf{A}}$ and $\hat{\mathbf{B}}$ are vector operators, and if $\hat{\mathbf{A}}$ commutes with $\hat{\mathbf{s}}$, then

$$\left(\frac{2}{\hbar}\right)^2 (\hat{\mathbf{s}} \cdot \hat{\mathbf{A}})(\hat{\mathbf{s}} \cdot \hat{\mathbf{B}}) = \hat{\mathbf{A}} \cdot \hat{\mathbf{B}} + \frac{2i}{\hbar} \hat{\mathbf{s}} \cdot (\hat{\mathbf{A}} \times \hat{\mathbf{B}}) \qquad (10\text{–}70)$$

This relation shows that $\hat{\mathbf{p}}^2$ is equivalent to $(2/\hbar)^2(\hat{\mathbf{s}} \cdot \hat{\mathbf{p}})^2$. When an external field exists, then, why not replace $(\hat{\mathbf{p}} + e\mathbf{A})^2$ by

$$\left(\frac{2}{\hbar}\right)^2 [\hat{\mathbf{s}} \cdot (\hat{\mathbf{p}} + e\mathbf{A})]^2 = (\hat{\mathbf{p}} + e\mathbf{A})^2 + i\frac{2}{\hbar} \hat{\mathbf{s}} \cdot [(\hat{\mathbf{p}} + e\mathbf{A}) \times (\hat{\mathbf{p}} + e\mathbf{A})]$$

$$= (\hat{\mathbf{p}} + e\mathbf{A})^2 + 2e\hat{\mathbf{s}} \cdot (\nabla \times \mathbf{A} + \mathbf{A} \times \nabla)$$

$$= \hat{\mathbf{p}}^2 + e\mathbf{l} \cdot \mathbf{B} + \tfrac{1}{4}e^2(r^2B^2 - (\mathbf{r} \cdot \mathbf{B})^2) + 2e\hat{\mathbf{s}} \cdot \mathbf{B}$$

$$(10\text{–}71)$$

which has exactly the desired form of (10–63)?

PROBLEM 10–16. Show that

$$[\sigma_1, \sigma_2] = 2i\sigma_3 \qquad [\sigma_2, \sigma_3] = 2i\sigma_1 \qquad [\sigma_3, \sigma_1] = 2i\sigma_2$$

PROBLEM 10–17. Show the relation in (10–70).

10–5 DIRAC EQUATION

Using the two operators $\hat{\lambda}$ and $\hat{\mathbf{s}}$, we see that

$$\hat{H}_D = \frac{2c}{\hbar} \hat{\lambda}_1(\hat{\mathbf{s}} \cdot \hat{\mathbf{p}}) + \hat{\lambda}_3 mc^2 \qquad (10\text{–}72)$$

satisfies the general requirement for a relativistic Hamiltonian,

$$\hat{H}_D^2 = (c\hat{\mathbf{p}})^2 + (mc^2)^2 \qquad (10\text{–}73)$$

just as the Klein–Gordon Hamiltonian $\hat{H}_{KG}$. However, our arguments in the last section strongly suggest that $\hat{H}_D$, when generalized, will give the spin angular momentum, which $\hat{H}_{KG}$ does not. The linearized equation

$$\left(i\hbar \frac{\partial}{\partial t} - \hat{H}_D\right) \Psi = 0 \qquad (10\text{–}74)$$

is the Dirac equation, and $\hat{H}_D$ is the Dirac Hamiltonian.[4]

[4] P. A. M. Dirac, *Proc. Roy. Soc.*, **A117**, 601 (1928); **A118**, 351 (1928).

Since $\hat{\lambda}$ and $\hat{s}$ represent different degrees of freedom, the Dirac wave function Ψ should be a *bispinor*: Let formulas (10–14) and (10–15) hold, except that we now regard each component by itself as a spinor, for which (10–68) and (10–69) are satisfied. Then

$$\Phi = \begin{pmatrix} \Phi_1 \\ \Phi_2 \end{pmatrix} \qquad \chi = \begin{pmatrix} \chi_1 \\ \chi_2 \end{pmatrix} \tag{10–75}$$

and

$$\Psi = \begin{pmatrix} \Phi \\ \chi \end{pmatrix} = \begin{pmatrix} \Phi_1 \\ \Phi_2 \\ \chi_1 \\ \chi_2 \end{pmatrix} \tag{10–76}$$

For example

$$\hat{s}_x \Psi = \tfrac{1}{2}\hbar \begin{pmatrix} \sigma_1 \Phi \\ \sigma_1 \chi \end{pmatrix} = \tfrac{1}{2}\hbar \begin{pmatrix} \sigma_1 & (0) \\ (0) & \sigma_1 \end{pmatrix} \begin{pmatrix} \Phi \\ \chi \end{pmatrix} = \tfrac{1}{2}\hbar \begin{pmatrix} 0 & 1 & 0 & 0 \\ 1 & 0 & 0 & 0 \\ 0 & 0 & 0 & 1 \\ 0 & 0 & 1 & 0 \end{pmatrix} \begin{pmatrix} \Phi_1 \\ \Phi_2 \\ \chi_1 \\ \chi_2 \end{pmatrix}$$

$$\tag{10–77}$$

and

$$\hat{\lambda}_1 \Psi = \begin{pmatrix} (0) & (1) \\ (1) & (0) \end{pmatrix} \begin{pmatrix} \Phi \\ \chi \end{pmatrix} = \begin{pmatrix} \chi \\ \Phi \end{pmatrix} = \begin{pmatrix} 0 & 0 & 1 & 0 \\ 0 & 0 & 0 & 1 \\ 1 & 0 & 0 & 0 \\ 0 & 1 & 0 & 0 \end{pmatrix} \begin{pmatrix} \Phi_1 \\ \Phi_2 \\ \chi_1 \\ \chi_2 \end{pmatrix} \tag{10–78}$$

where (0) and (1) are 2×2 zero and unit matrices, respectively.

It is conventional to write

$$\hat{H}_D = c\hat{\boldsymbol{\alpha}} \cdot \hat{\mathbf{p}} + \hat{\beta}mc^2 \tag{10–79}$$

instead of (10–72). Obviously

$$\hat{\boldsymbol{\alpha}} = \frac{2}{\hbar} \hat{\lambda}_1 \hat{s} \qquad \hat{\beta} = \hat{\lambda}_3 \tag{10–80}$$

so that

$$\{\hat{\alpha}_j, \hat{\alpha}_k\} \equiv \hat{\alpha}_j \hat{\alpha}_k + \hat{\alpha}_k \hat{\alpha}_j = 2\delta_{jk} \tag{10–81}$$

$$\{\hat{\alpha}_j, \hat{\beta}\} \equiv \hat{\alpha}_j \hat{\beta} + \hat{\beta} \hat{\alpha}_j = 0 \tag{10–82}$$

and

$$\hat{\beta}^2 = 1 \tag{10–83}$$

where $\{\ ,\ \}$ is called the *anticommutator*. Also if

$$\alpha = \begin{pmatrix} (0) & \sigma \\ \sigma & (0) \end{pmatrix} \quad \text{and} \quad \beta = \begin{pmatrix} (1) & (0) \\ (0) & -(1) \end{pmatrix} = \begin{pmatrix} 1 & 0 & 0 & 0 \\ 0 & 1 & 0 & 0 \\ 0 & 0 & -1 & 0 \\ 0 & 0 & 0 & -1 \end{pmatrix}$$

$$(10\text{–}84)$$

then

$$\hat{\alpha}\Psi = \alpha\Psi \quad \text{and} \quad \hat{\beta}\Psi = \beta\Psi \qquad (10\text{–}85)$$

which show that $\hat{\alpha}$ and $\hat{\beta}$ are Hermitian operators.
Putting (10–79) into (10–74) we obtain

$$\left(i\hbar \frac{\partial}{\partial t} + ic\hbar\hat{\alpha} \cdot \nabla - \hat{\beta}mc^2 \right) \Psi = 0 \qquad (10\text{–}86)$$

If we introduce the Hermitian conjugate bispinor

$$\Psi^\dagger \equiv (\Phi_1{}^* \Phi_2{}^* \chi_1{}^* \chi_2{}^*) \qquad (10\text{–}87)$$

we easily obtain from (10–86), following the familiar procedure, the equation of continuity, in which

$$\rho = \Psi^\dagger\Psi = |\Phi_1|^2 + |\Phi_2|^2 + |\chi_1|^2 + |\chi_2|^2 \qquad (10\text{–}88)$$

$$j = c\Psi^\dagger\hat{\alpha}\Psi = \frac{2c}{\hbar} \Psi^\dagger\hat{\lambda}_1\hat{s}\Psi \qquad (10\text{–}89)$$

PROBLEM 10–18. Obtain (10–81), (10–82), and (10–83) using (10–80).

PROBLEM 10–19. Obtain (10–84) and (10–85) from (10–80).

PROBLEM 10–20. Obtain (10–88) from (10–86).

PROBLEM 10–21. In (10–72) we took $\hat{\lambda}_1$ and $\hat{\lambda}_3$ in the first and second terms, respectively. The choice is actually arbitrary, and we can as well take $\hat{\lambda}_2$ and $\hat{\lambda}_3$, or $\hat{\lambda}_1$ and $\hat{\lambda}_2$, in place of the choice we took. (There are six choices.) Find $\hat{\alpha}$ and $\hat{\beta}$ for the two choices given above.

PROBLEM 10–22. From (10–89), obtain

$$j_x = c(\chi_1\Phi_2{}^* + \chi_2\Phi_1{}^* + \Phi_1\chi_2{}^* + \Phi_2\chi_1{}^*)$$
$$j_y = -ic(\chi_1\Phi_2{}^* - \chi_2\Phi_1{}^* + \Phi_1\chi_2{}^* - \Phi_2\chi_1{}^*)$$
$$j_z = c(\Phi_1\chi_1{}^* - \Phi_2\chi_2{}^* + \chi_1\Phi_1{}^* - \chi_2\Phi_2{}^*)$$

PROBLEM 10–23. (a) Show that a plane wave solution of the Dirac equation which is also an eigenfunction of $\hat{s}_z$ with the eigenvalue $\frac{1}{2}\hbar$ is

$$\Psi = \begin{pmatrix} \xi \\ 0 \\ \eta \\ 0 \end{pmatrix} e^{i(k \cdot r - \omega t)}$$

where ω is given by (10–31).
(b) Show that the ratio $\xi/\eta = (mc^2 + \mathscr{E})/(c\hbar k)$ if $\hbar\omega = \mathscr{E}$, but $\xi/\eta = -c\hbar k/(mc^2 + \mathscr{E})$ if $\hbar\omega = -\mathscr{E}$.

10–6* BILINEAR COVARIANTS. SPACE INVERSION

The Dirac equation (10–86) can be rewritten in the form

$$\left(i\hbar\beta \frac{\partial}{\partial t} + ic\hbar\beta\hat{\alpha} \cdot \nabla - mc^2\right)\Psi = 0 \tag{10–90}$$

which can be recast in the covariant expression

$$\left(i\hbar\hat{\gamma}^j \frac{\partial}{\partial x^j} - mc\right)\Psi = 0 \tag{10–91}$$

or

$$(\hat{\gamma}^j\hat{p}_j - mc)\Psi = 0 \tag{10–92}$$

where

$$\hat{\gamma}^0 = \hat{\beta} \qquad \gamma^1 = \hat{\beta}\hat{\alpha}_x \qquad \hat{\gamma}^2 = \hat{\beta}\hat{\alpha}_y \qquad \hat{\gamma}^3 = \hat{\beta}\hat{\alpha}_z \tag{10–93}$$

and

$$\hat{p}_j \equiv i\hbar \frac{\partial}{\partial x^j} \tag{10–94}$$

The matrix representatives of $\hat{\gamma}^j$ are easily obtained from (10–84).

Since $\hat{\alpha}$ and $\hat{\beta}$ are both Hermitian and anticommute with each other, as (10–82) shows, we see that

$$\hat{\gamma}^{0\dagger} = \hat{\gamma}^0 \qquad \hat{\gamma}^{1\dagger} = -\hat{\gamma}^1 \qquad \hat{\gamma}^{2\dagger} = -\hat{\gamma}^2 \qquad \hat{\gamma}^{3\dagger} = -\hat{\gamma}^3$$

$$(10\text{–}95)$$

In special relativity, where $g_{00} = 1$, $g_{11} = g_{22} = g_{33} = -1$ and all the other components of the metric tensor are zero (section 4–4), the above relations can be written as

$$\hat{\gamma}^{j\dagger} = \hat{\gamma}_j \qquad (10\text{–}96)$$

using the covariant components. From (10–81) and (10–82) we easily obtain

$$\{\hat{\gamma}_j, \hat{\gamma}^k\} = 2\delta_j{}^k \qquad (10\text{–}97)$$

$$\{\hat{\gamma}_j, \hat{\gamma}_k\} = 2g_{jk} \qquad (10\text{–}97')$$

where $\delta_j{}^k$ is the Kronecker tensor defined by (4–56). By expressing the anti-commutation relations in such a form, the theory is presumably applicable to general space–times, including accelerated systems or space with a gravity field.

If we introduce the *adjoint bispinor*

$$\overline{\Psi} = \Psi^\dagger \hat{\gamma}^0 = \Psi^\dagger \hat{\beta} = (\Phi_1{}^* \ \Phi_2{}^* -\chi_1{}^* -\chi_2{}^*) \qquad (10\text{–}98)$$

we see that the eqation of continuity obtained from (10–86) or (10–91) can be expressed as

$$\frac{\partial}{\partial x^j} (\overline{\Psi}\gamma^j\Psi) = 0 \qquad (10\text{–}99)$$

which, in a general space–time, should be generalized to

$$(\overline{\Psi}\gamma^j\Psi)_{;j} = 0 \qquad (10\text{–}99')$$

In any case, these relations show that the $\overline{\Psi}\gamma^j\Psi$ form a 4-vector under relativistic transformations, or the Lorentz transformations if the space–time is inertial.

A scalar is obtained from two 4-vectors as in (4–52). Using relations (10–96) and Table 10–2, we obtain

$$(\overline{\Psi}\hat{\gamma}^j\Psi)(\overline{\Psi}\hat{\gamma}_j\Psi) = (\overline{\Psi}\Psi)^2 \qquad (10\text{–}100)$$

which shows that $\overline{\Psi}\Psi$ is a scalar under proper relativistic (Lorentz) transformations.

By multiplying two vector components, a component of a second rank tensor is obtained, as in (4–53). Thus contravariant components of a second rank tensor satisfy

$$(\overline{\Psi}\hat{\gamma}_i\Psi)\{(\overline{\Psi}\hat{\gamma}^i\hat{\gamma}^j\Psi) + (\overline{\Psi}\hat{\gamma}^i\hat{\gamma}^j\Psi)^*\} = 2\,(\overline{\Psi}\Psi)(\overline{\Psi}\hat{\gamma}_j\Psi) \qquad (10\text{–}101)$$

Since $\overline{\Psi}\Psi$ is a scalar $\overline{\Psi}\gamma^j\gamma^k\Psi$ is a component of a second rank tensor. Note that the second rank tensor obtained in this way is an antisymmetric tensor, since the $\hat{\gamma}$'s anticommute, as (10–97) shows:

$$\overline{\Psi}\hat{\gamma}_j\hat{\gamma}^k\Psi = -\overline{\Psi}\hat{\gamma}^k\hat{\gamma}_j\Psi \qquad (10\text{–}102)$$

In constructing third and fourth rank tensors in this way, one should remember that there are only four $\hat{\gamma}$'s and that the product of two identical $\hat{\gamma}$'s is just $\hat{1}$. Thus, for example, there is only one independent component in the fourth rank tensor:

$$\overline{\Psi}\hat{\gamma}^0\hat{\gamma}^1\hat{\gamma}^2\hat{\gamma}^3\Psi \equiv \overline{\Psi}\hat{\gamma}^5\Psi \qquad (10\text{–}103)$$

where $\hat{\gamma}^5$ is a new operator defined by

$$\hat{\gamma}^5 = \hat{\gamma}^0\hat{\gamma}^1\hat{\gamma}^2\hat{\gamma}^3 \qquad \hat{\gamma}_5 = \hat{\gamma}_3\hat{\gamma}_2\hat{\gamma}_1\hat{\gamma}_0 \qquad (10\text{–}104)$$

Its matrix representative can be easily found from those of other $\hat{\gamma}$'s as

$$\hat{\gamma}^5\Psi = \gamma^5\Psi = \begin{pmatrix} 0 & 0 & -1 & 0 \\ 0 & 0 & 0 & -1 \\ -1 & 0 & 0 & 0 \\ 0 & -1 & 0 & 0 \end{pmatrix}\overline{\Psi} \qquad (10\text{–}105)$$

and its anticommutation relations are

$$\{\hat{\gamma}_j, \hat{\gamma}^5\} = 0 \qquad \{\hat{\gamma}^j, \hat{\gamma}_5\} = 0 \qquad \hat{\gamma}_5\hat{\gamma}^5 = \hat{1} \qquad (10\text{–}106)$$

From the last relation and the obvious relation $\hat{\gamma}^{5\dagger} = \hat{\gamma}_5$, we see that

$$\overline{\Psi}\hat{\gamma}^5\Psi\overline{\Psi}\hat{\gamma}_5\Psi = \overline{\Psi}\Psi\hat{\gamma}_5\overline{\Psi}\hat{\gamma}_5\Psi = (\overline{\Psi}\Psi)^2 \qquad (10\text{–}107)$$

which is a scalar as shown before. Therefore $\overline{\Psi}\hat{\gamma}^5\Psi$ is actually a scalar under proper relativistic transformations.

There are only four independent components in the third rank tensor constructed in this way, a typical one being $\overline{\Psi}\hat{\gamma}^0\hat{\gamma}^1\hat{\gamma}^2\Psi$. Since

$$\hat{\gamma}^5\hat{\gamma}_3 = \hat{\gamma}^0\hat{\gamma}^1\hat{\gamma}^2\hat{\gamma}^3\hat{\gamma}_3 = \hat{\gamma}^0\hat{\gamma}^1\hat{\gamma}^2 \tag{10-108}$$

for example, these four components can be written as $\overline{\Psi}\hat{\gamma}^5\hat{\gamma}_j\Psi$. Now since

$$(\overline{\Psi}\hat{\gamma}^5\hat{\gamma}_j\Psi)(\overline{\Psi}\hat{\gamma}_5\hat{\gamma}^j\Psi) = (\overline{\Psi}\hat{\gamma}_j\Psi)(\overline{\Psi}\hat{\gamma}^j\Psi) = (\overline{\Psi}\Psi)^2 \tag{10-109}$$

as can be shown following a similar procedure, we see that these four components of the third rank tensor actually behave as a 4-vector under proper relativistic transformations.

Space Inversion. Parity Operator

We have found two scalars, $\overline{\Psi}\Psi$ and $\overline{\Psi}\hat{\gamma}^5\Psi$, and two vectors, $\overline{\Psi}\hat{\gamma}^j\Psi$ and $\overline{\Psi}\hat{\gamma}_5\hat{\gamma}^j\Psi$, in addition to one second rank antisymmetric tensor, $\overline{\Psi}\hat{\gamma}^j\hat{\gamma}^k\Psi$. If we consider the space inversion we can classify these five quantities completely.

The space inversion $\hat{I}$ is introduced in Problem 8–53, and has the properties that

$$\hat{I}\nabla\hat{I}^{-1} = -\nabla \quad \text{and} \quad \hat{I}t\hat{I}^{-1} = t \tag{10-110}$$

and does not operate on the $\hat{\gamma}$'s, since $\hat{\gamma}$'s are for other degrees of freedom than **r** or t. If we transform the operator part of equation (10–91) by using $\hat{I}$ we obtain

$$\hat{I}\left(i\hbar\hat{\gamma}^0\frac{\partial}{c\,\partial t} + i\hbar\gamma\cdot\nabla - mc\right)\hat{I}^{-1} = i\hbar\hat{\gamma}^0\frac{\partial}{c\,\partial t} - i\hbar\gamma\cdot\nabla - mc \tag{10-111}$$

On the other hand, we see from (10–97) that

$$\hat{\gamma}_0\left(i\hbar\hat{\gamma}^0\frac{\partial}{c\,\partial t} - i\hbar\gamma\cdot\nabla - mc\right)\gamma^0 = i\hbar\gamma^0\frac{\partial}{c\,\partial t} + i\hbar\gamma\cdot\nabla - mc \tag{10-112}$$

Therefore, the Dirac Hamiltonian is invariant under transformation $\hat{I}\hat{\gamma}_0$, in contrast to the Schroedinger Hamiltonian which is invariant under $\hat{I}$. If the Dirac equation (10–91) holds, then

$$\left(i\hbar\gamma^j\frac{\partial}{\partial x^j} - mc\right)\hat{I}\hat{\gamma}_0\Psi = 0 \tag{10-113}$$

also holds. Therefore, if degeneracy does not exist we conclude (Problem 8–54) that either

$$\hat{I}\hat{\gamma}_0 \Psi(t, \mathbf{r}) = \begin{pmatrix} \Phi(t, -\mathbf{r}) \\ -\chi(t, -\mathbf{r}) \end{pmatrix} = +\Psi(t, \mathbf{r}) \qquad (10\text{--}114\text{a})$$

or

$$\hat{I}\hat{\gamma}_0 \Psi(t, \mathbf{r}) = -\Psi(t, \mathbf{r}) \qquad (10\text{--}114\text{b})$$

The signs on the right-hand side of these equations are called the *parity* of that state, and the operator $\hat{I}\hat{\gamma}_0$ is called the *parity operator*. We may call

$$\hat{S}_P \equiv \hat{\gamma}_0 \qquad (10\text{--}115)$$

the dual inversion operator in this sense.

We can further classify the five quantities we are discussing using the dual inversion operator $\hat{S}_P$ as follows:

$$\hat{S}_P(\overline{\Psi}\Psi)\hat{S}_P^{-1} = \Psi^{\dagger}\hat{\gamma}_0\hat{\gamma}^0\hat{\gamma}^0\Psi = \Psi^{\dagger}\hat{\gamma}^0\Psi = \overline{\Psi}\Psi \qquad (10\text{--}116\text{a})$$

$$\hat{S}_P(\overline{\Psi}\hat{\gamma}^5\Psi)\hat{S}_P^{-1} = \Psi^{\dagger}\hat{\gamma}_0\hat{\gamma}^0\hat{\gamma}^5\hat{\gamma}^0\Psi = \Psi^{\dagger}\hat{\gamma}^5\hat{\gamma}^0\Psi = -\Psi^{\dagger}\hat{\gamma}^0\hat{\gamma}^5\Psi = -\overline{\Psi}\hat{\gamma}^5\Psi \qquad (10\text{--}116\text{b})$$

$$\hat{S}_P(\overline{\Psi}\hat{\gamma}^j\Psi)\hat{S}_P^{-1} = \Psi^{\dagger}\hat{\gamma}_0\hat{\gamma}^0\hat{\gamma}^j\gamma^0\Psi = \Psi^{\dagger}\hat{\gamma}^j\hat{\gamma}^0\Psi = \begin{cases} \overline{\Psi}\hat{\gamma}^0\Psi & \text{if} \quad j = 0 \\ -\overline{\Psi}\gamma^j\Psi & \text{if} \quad j = 1, 2, 3 \end{cases} \qquad (10\text{--}116\text{c})$$

$$\hat{S}_P(\overline{\Psi}\hat{\gamma}_5\hat{\gamma}^j\Psi)\hat{S}_P^{-1} = \Psi^{\dagger}\hat{\gamma}_0\hat{\gamma}^0\hat{\gamma}_5\hat{\gamma}^j\hat{\gamma}^0\Psi = \Psi^{\dagger}\hat{\gamma}_5\hat{\gamma}^j\hat{\gamma}^0\Psi$$

$$= \begin{cases} \Psi^{\dagger}\hat{\gamma}_5\hat{\gamma}^0\hat{\gamma}^0\Psi = -\overline{\Psi}\hat{\gamma}_5\hat{\gamma}^0\Psi & \text{if} \quad j = 0 \\ \Psi^{\dagger}\hat{\gamma}^0\hat{\gamma}_5\hat{\gamma}^j\Psi = \overline{\Psi}\hat{\gamma}_5\hat{\gamma}^j\Psi & \text{if} \quad j = 1, 2, 3 \end{cases} \qquad (10\text{--}116\text{d})$$

In (10–116b), we see that $\overline{\Psi}\hat{\gamma}^5\Psi$ is a scalar, but changes its sign when space-inverted. This quantity is, therefore, called a pseudoscalar. Comparing (10–116d) with (10–116c), we see that $\overline{\Psi}\hat{\gamma}_5\hat{\gamma}^j\Psi$, which behaves as a 4-vector in the proper relativistic transformations, is actually a pseudovector, in contrast to $\overline{\Psi}\hat{\gamma}^j\Psi$, which is a vector. These tensor quantities are called the *bilinear covariants* (see Table 10.1).

TABLE 10–1 *Bilinear Covariants*

	Components	Number of Components
Scalar	$\overline{\Psi}\Psi$	1
Vector	$\overline{\Psi}\hat{\gamma}^j\Psi$	4
Tensor (2nd rank)	$\overline{\Psi}\hat{\gamma}^j\hat{\gamma}^k\Psi$	6
Pseudovector	$\overline{\Psi}\hat{\gamma}_5\hat{\gamma}^j\Psi$	4
Pseudoscalar	$\overline{\Psi}\hat{\gamma}^5\Psi$	1

So far we have introduced λ, $\hat{s}$, $\hat{\alpha}$, $\hat{\beta}$, and $\hat{\gamma}^j$ operators through their commutation or anticommutation relations, and showed their matrix representatives which satisfy the same relations. It should be pointed out, however, that the matrix representatives cannot be uniquely determined from their commutation or anticommutation relations. The matrix representatives we took are the most common ones, called the standard, or the Dirac–Pauli representation, but there can be many other representations. Particularly the fact that the matrix representatives of the second components, $\hat{\lambda}_2$, $\hat{s}_y$, $\hat{\alpha}_y$, and γ^2, are purely imaginary is an accidental result of this particular representation, and does not have any deep significance. We present the standard representation in Table 10–2 for future uses.

TABLE 10–2   *Standard (Dirac–Pauli) Representation*

$$\lambda_1: \begin{pmatrix} 0 & 0 & 1 & 0 \\ 0 & 0 & 0 & 1 \\ 1 & 0 & 0 & 0 \\ 0 & 1 & 0 & 0 \end{pmatrix} \quad \lambda_2: \begin{pmatrix} 0 & 0 & -i & 0 \\ 0 & 0 & 0 & -i \\ i & 0 & 0 & 0 \\ 0 & i & 0 & 0 \end{pmatrix} \quad \lambda_3: \begin{pmatrix} 1 & 0 & 0 & 0 \\ 0 & 1 & 0 & 0 \\ 0 & 0 & -1 & 0 \\ 0 & 0 & 0 & -1 \end{pmatrix}$$

$$\mathbf{s}: \tfrac{1}{2}\hbar \begin{pmatrix} \sigma & (0) \\ (0) & \sigma \end{pmatrix} \qquad s_x: \tfrac{1}{2}\hbar \begin{pmatrix} 0 & 1 & 0 & 0 \\ 1 & 0 & 0 & 0 \\ 0 & 0 & 0 & 1 \\ 0 & 0 & 1 & 0 \end{pmatrix}$$

$$s_y: \tfrac{1}{2}\hbar \begin{pmatrix} 0 & -i & 0 & 0 \\ i & 0 & 0 & 0 \\ 0 & 0 & 0 & -i \\ 0 & 0 & i & 0 \end{pmatrix} \qquad s_z: \tfrac{1}{2}\hbar \begin{pmatrix} 1 & 0 & 0 & 0 \\ 0 & -1 & 0 & 0 \\ 0 & 0 & 1 & 0 \\ 0 & 0 & 0 & -1 \end{pmatrix}$$

$$\beta: \begin{pmatrix} 1 & 0 & 0 & 0 \\ 0 & 1 & 0 & 0 \\ 0 & 0 & -1 & 0 \\ 0 & 0 & 0 & -1 \end{pmatrix} \quad \alpha_x: \begin{pmatrix} 0 & 0 & 0 & 1 \\ 0 & 0 & 1 & 0 \\ 0 & 1 & 0 & 0 \\ 1 & 0 & 0 & 0 \end{pmatrix} \quad \alpha_y: \begin{pmatrix} 0 & 0 & 0 & -i \\ 0 & 0 & i & 0 \\ 0 & -i & 0 & 0 \\ i & 0 & 0 & 0 \end{pmatrix}$$

$$\alpha_z: \begin{pmatrix} 0 & 0 & 1 & 0 \\ 0 & 0 & 0 & -1 \\ 1 & 0 & 0 & 0 \\ 0 & -1 & 0 & 0 \end{pmatrix} \qquad \gamma^0: \begin{pmatrix} 1 & 0 & 0 & 0 \\ 0 & 1 & 0 & 0 \\ 0 & 0 & -1 & 0 \\ 0 & 0 & 0 & -1 \end{pmatrix}$$

$$\gamma^1: \begin{pmatrix} 0 & 0 & 0 & 1 \\ 0 & 0 & 1 & 0 \\ 0 & -1 & 0 & 0 \\ -1 & 0 & 0 & 0 \end{pmatrix} \qquad \gamma^2: \begin{pmatrix} 0 & 0 & 0 & -i \\ 0 & 0 & i & 0 \\ 0 & i & 0 & 0 \\ -i & 0 & 0 & 0 \end{pmatrix}$$

$$\gamma^3: \begin{pmatrix} 0 & 0 & 1 & 0 \\ 0 & 0 & 0 & -1 \\ -1 & 0 & 0 & 0 \\ 0 & 1 & 0 & 0 \end{pmatrix} \qquad \gamma^5: \begin{pmatrix} 0 & 0 & -1 & 0 \\ 0 & 0 & 0 & -1 \\ -1 & 0 & 0 & 0 \\ 0 & -1 & 0 & 0 \end{pmatrix}$$

PROBLEM 10–24. Obtain (10–95).

PROBLEM 10–25. Obtain (10–99) from (10–91).

PROBLEM 10–26. Obtain (10–106).

PROBLEM 10–27. Show that $\hat{\gamma}^5 = (\hat{\gamma}^5)^{-1} = \hat{\gamma}_5$. Show also that $\hat{\gamma}^5 = -\hat{\lambda}_1$.

10–7* CHARGE CONJUGATION. (DIRAC THEORY)

When electromagnetic fields exist, the Dirac equation (10–92) should be generalized as

$$[\hat{\gamma}^j(\hat{p}_j - qA_j) - mc]\Psi = 0 \qquad (10\text{–}117)$$

where $\hat{p}_j$ is still given by (10–94) and A_j is the component of the four potential at the position of the particle. If the particle is the electron, then $q = -e$.

The charge conjugation operator $\hat{C}$ is defined by

$$\hat{C}q\hat{C}^{-1} = -q \qquad (10\text{–}118)$$

or

$$\hat{C}[\hat{\gamma}^j(\hat{p}_j - qA_j) - mc]\hat{C}^{-1} = \hat{\gamma}^j(\hat{p}_j + qA_j) - mc \qquad (10\text{–}119)$$

Let us find a dual charge conjugation operator $\hat{S}_C$ such that $\hat{C}\hat{S}_C$ leaves the Dirac Hamiltonian invariant, or

$$\hat{S}_C[\hat{\gamma}^j(\hat{p}_j + qA_j) - mc]\hat{S}_C^{-1} = \hat{\gamma}^j(\hat{p}_j - qA_j) - mc \qquad (10\text{–}120)$$

In order to find $\hat{S}_C$, first we investigate what the complex conjugation operator $\hat{K}$ can do. This operator was introduced in (10–40). An operator $\hat{f}$ is transformed by $\hat{K}$ as

$$\hat{K}\hat{f}\hat{K}^{-1} = \hat{f}^* \qquad (10\text{–}121)$$

For example

$$\hat{K}\hat{p}_j\hat{K}^{-1} = \hat{K}\left(i\hbar\,\frac{\partial}{\partial x^j}\right)\hat{K}^{-1} = -i\hbar\,\frac{\partial}{\partial x^j} = -\hat{p}_j \qquad (10\text{–}122)$$

The Pauli matrices σ_1 and σ_3 are real, but σ_2 is imaginary, as seen in (4–60). Therefore

$$\hat{K}\sigma_1\hat{K}^{-1} = \sigma_1 \qquad \hat{K}\sigma_2\hat{K}^{-1} = -\sigma_2 \qquad \hat{K}\sigma_3\hat{K}^{-1} = \sigma_3 \qquad (10\text{–}123)$$

The matrix representatives of some operators, such as $\hat{\lambda}_2$, $\hat{\alpha}_y$, and $\hat{\gamma}^2$, are given by σ_2 ,and these operators should have the same transformation properties as σ_2. Thus

$$\hat{K}\hat{\lambda}_2\hat{K}^{-1} = -\hat{\lambda}_2 \qquad \hat{K}\hat{\alpha}_y\hat{K}^{-1} = -\hat{\alpha}_y \qquad \hat{K}\hat{\gamma}^2\hat{K}^{-1} = -\hat{\gamma}^2$$

$$(10\text{-}124)$$

while other components, $\hat{\lambda}_1$, $\hat{\lambda}_3$, $\hat{\alpha}_x$, etc., do not change sign under this transformation. We, therefore, see that

$$\hat{K}[\hat{\gamma}^j(\hat{p}_j + qA_j) - mc]\hat{K}^{-1} = (\hat{K}\hat{\gamma}^j\hat{K}^{-1})[(\hat{K}\hat{p}_j\hat{K}^{-1}) + qA_j] - mc$$

$$= -\hat{\gamma}^j(\hat{p}_j - qA_j) + 2\hat{\gamma}^2(\hat{p}_2 - qA_2) - mc$$

$$(10\text{-}125)$$

The difference between this result and the desired one (10–120) is that the sign of $\hat{\gamma}^j$, except for $\hat{\gamma}^2$, is incorrect. This difference is easily removed by using the anticommutation relation (10–97). Thus

$$\hat{S}_C = \hat{\gamma}^2\hat{K}$$

$$(10\text{-}126a)$$

From (10–80) and (10–93) we see that $\hat{S}_C$ can also be expressed as

$$\hat{S}_C = \beta\hat{\alpha}_y\hat{K} = \frac{2}{\hbar}\hat{\lambda}_3\hat{\lambda}_1\hat{s}_y\hat{K} = i\frac{2}{\hbar}\hat{\lambda}_2\hat{s}_y\hat{K}$$

$$(10\text{-}126b)$$

where the last expression may be compared to the corresponding one in the Klein–Gordon theory, (10–40).

Since $\hat{C}\hat{S}_C$ commutes with the Dirac Hamiltonian, $\hat{C}\hat{S}_C\Psi$ is also a solution of the same equation (10–117) for Ψ. We call $\hat{C}\hat{S}_C\Psi$ the charge conjugate to Ψ, and a particle described by $\hat{C}\hat{S}_C\Psi$ is called the charge conjugate particle or antiparticle of a particle described by Ψ.

Not only the charge, but also the momentum and spin are opposite between a particle and its antiparticle in this definition. From (10–122) and (10–126a) we obtain

$$\hat{C}\hat{S}_C\hat{p}(\hat{C}\hat{S}_C)^{-1} = \hat{K}\hat{p}\hat{K}^{-1} = -\hat{p}$$

$$(10\text{-}127)$$

Therefore, if

$$\hat{p}\Psi = k\hbar\Psi$$

$$(10\text{-}128)$$

then

$$\hat{C}\hat{S}_C\hat{\mathbf{p}}(\hat{C}\hat{S}_C)^{-1}\hat{C}\hat{S}_C\Psi = k\hbar\hat{C}\hat{S}_C\Psi \tag{10-129}$$

or

$$\hat{\mathbf{p}}(\hat{C}\hat{S}_C\Psi) = -k\hbar(\hat{C}\hat{S}_C\Psi) \tag{10-130}$$

In order to prove the corresponding relation for the spin we use (10–126b), from which we obtain

$$\hat{C}\hat{S}_C\hat{s}_x(\hat{C}\hat{S}_C)^{-1} = \left(\frac{2i}{\hbar}\right)^2 \hat{s}_y\hat{s}_x\hat{s}_y = -\hat{s}_x \left(\frac{2i\hat{s}_y}{\hbar}\right)^2 = -\hat{s}_x \tag{10-131a}$$

$$\hat{C}\hat{S}_C\hat{s}_y(\hat{C}\hat{S}_C)^{-1} = \left(\frac{2i}{\hbar}\right)^2 \hat{s}_y\hat{K}\hat{s}_y\hat{K}^{-1}\hat{s}_y = -\hat{s}_y \tag{10-131b}$$

$$\hat{C}\hat{S}_C\hat{s}_z(\hat{C}\hat{S}_C)^{-1} = \left(\frac{2i}{\hbar}\right)^2 \hat{s}_y\hat{s}_z\hat{s}_y = -\hat{s}_z \tag{10-131c}$$

or

$$\hat{C}\hat{S}_C\hat{\mathbf{s}}(\hat{C}\hat{S}_C)^{-1} = -\hat{\mathbf{s}} \tag{10-132}$$

Thus in the same way as (10–128) through (10–130) we can see that the spin of antiparticle $\hat{C}\hat{S}_C\Psi$ is opposite to that of Ψ (Fig. 10–3).

FIGURE 10–3.   The charge conjugation $\hat{C}\hat{S}_C$ in the Dirac theory. The charge conjugate particles have opposite sign of ε, $\hat{\mathbf{p}}$, q, and $\hat{\mathbf{s}}$.

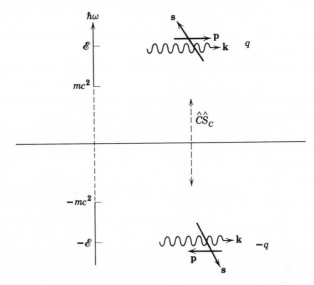

PROBLEM 10–28. Show that $\hat{C}\hat{S}_C\hat{\lambda}_3(\hat{C}\hat{S}_C)^{-1} = -\hat{\lambda}_3$.

PROBLEM 10–29. Show that

$$\hat{C}\hat{S}_C(\hat{I}\hat{S}_P)(\hat{C}\hat{S}_C)^{-1} = -\hat{I}\hat{S}_P$$

What does this imply?

PROBLEM 10–30. Express each component of $\hat{C}\hat{S}_C\Psi$ by those of Ψ.

10–8* TIME REVERSAL

The time reversal operator $\hat{T}$ is defined as

$$\hat{T}t\hat{T}^{-1} = -t \qquad \hat{T}\mathbf{r}\hat{T}^{-1} = \mathbf{r} \tag{10–133}$$

The problem of time reversal in quantum mechanics was discussed in section 8–16, using the time-development operator method. The same discussions are applicable here, if the time-development operator is $\exp(-i\hat{H}_D t/\hbar)$, and $\hat{H}_D$ is the Dirac Hamiltonian. We have seen in section 8–16, particularly in formula (8–257), that if the Hamiltonian which gives the time-development operator is real, time reversal symmetry between complex conjugate states will exist. Since the Schroedinger Hamiltonian of a free particle is real, that argument is widely applicable in nonrelativistic quantum mechanics. The Dirac Hamiltonian, however, is not real even when the particle is free: From (10–79), using (10–122) and (10–124), we see that

$$\hat{H}_D{}^* = \hat{K}(c\hat{\boldsymbol{\alpha}} \cdot \hat{\mathbf{p}} + \hat{\beta}mc^2)\hat{K}^{-1}$$
$$= -c\hat{\boldsymbol{\alpha}} \cdot \hat{\mathbf{p}} + 2c\hat{\alpha}_y\hat{p}_y + \hat{\beta}mc^2 \neq \hat{H}_D \tag{10–134}$$

The problem of the time reversal is, therefore, rather different in Dirac's theory.

The time reversal in (10–133) implies that if there is a current, the direction of the current is reversed by $\hat{T}$, but the charge density is not. The field produced by the current is, therefore, transformed by $\hat{T}$ as

$$\hat{T}\mathbf{B}\hat{T}^{-1} = -\mathbf{B} \qquad \hat{T}\mathbf{E}\hat{T}^{-1} = \mathbf{E} \tag{10–135}$$

where $\mathbf{B}$ and $\mathbf{E}$ are the magnetic flux density and the electric field strength, respectively. This also means that

$$\hat{T}\mathbf{A}\hat{T}^{-1} = -\mathbf{A} \qquad \hat{T}\phi\hat{T}^{-1} = \phi \tag{10–136}$$

because of (6–1) and (6–4).

The time reversed Dirac Hamiltonian is

$$\hat{T}[\hat{\gamma}^j(\hat{p}_j - qA_j) - mc]\hat{T}^{-1} = \hat{\gamma}^j(\hat{p}_j + qA_j) - 2\hat{\gamma}^0(\hat{p}_0 + qA_0) - mc$$

$$(10\text{–}137)$$

This result suggests that $\hat{S}_T$, an operator dual to $\hat{T}$ such that $\hat{T}\hat{S}_T$ commutes with the Dirac Hamiltonian, can be expressed by means of $\hat{S}_C$. The only thing necessary in addition is to change the sign of $\hat{\gamma}^0$, and that can be done by including $\hat{\gamma}^5\hat{\gamma}^0$, because

$$\hat{\gamma}^5\hat{\gamma}^0\hat{\gamma}^0\hat{\gamma}_0\hat{\gamma}_5 = -\hat{\gamma}^5\hat{\gamma}_5\hat{\gamma}^0 = -\hat{\gamma}^0 \qquad (10\text{–}138\text{a})$$

but

$$\hat{\gamma}^5\hat{\gamma}^0\hat{\gamma}^1\hat{\gamma}_0\hat{\gamma}_5 = -\hat{\gamma}^5\hat{\gamma}^0\hat{\gamma}_0\hat{\gamma}^1\hat{\gamma}_5 = \hat{\gamma}^5\hat{\gamma}^0\hat{\gamma}_0\hat{\gamma}_5\hat{\gamma}^1 = \hat{\gamma}^1 \qquad (10\text{–}138\text{b})$$

etc. Therefore, in the Dirac theory, we can express $\hat{S}_T$ as

$$\hat{S}_T = \hat{\gamma}^5\hat{\gamma}^0\hat{C} = \hat{\gamma}^5\hat{\gamma}^0\hat{\gamma}^2\hat{K} \qquad (10\text{–}139)$$

Again, this expression is valid only in the particular representation (standard representation) we took. In other representations $\hat{\gamma}^2$ may be replaced by something else. Expressions of $\hat{S}_T$ by other operators are useful. From (10–80) and (10–93), we see that (10–139) can also be written as

$$\hat{S}_T = \hat{\beta}\hat{\alpha}_x\hat{\beta}\hat{\alpha}_y\hat{\beta}\hat{\alpha}_z\hat{\beta}\hat{\beta}\hat{\beta}\hat{\alpha}_y\hat{K} = \hat{\alpha}_x\hat{\alpha}_y\hat{\alpha}_z\hat{\alpha}_y\hat{K} = -\hat{\alpha}_x\hat{\alpha}_z\hat{K}$$

$$= -\left(\frac{2}{\hbar}\right)^2 \hat{\lambda}_1\hat{s}_x\hat{\lambda}_1\hat{s}_z\hat{K} = i\frac{2}{\hbar}\hat{s}_y\hat{K} \qquad (10\text{–}140)$$

Some of the useful results are summarized in Table 10–3.

TABLE 10–3 *Important Operators and Their Expressions in the Dirac Theory*

Operators	Transformations of					Dual Operators
	t	$\hat{\mathbf{p}}$	ϕ	$\mathbf{A}$	q	
$\hat{I}$ space inversion	t	$-\hat{\mathbf{p}}$	ϕ	$-\mathbf{A}$	q	$\hat{S}_P = \hat{\gamma}_0 = \hat{\lambda}_3$
$\hat{C}$ charge conjugation	t	$\hat{\mathbf{p}}$	ϕ	$\mathbf{A}$	$-q$	$\hat{S}_C = \hat{\gamma}^2\hat{K} = i(2/\hbar)\hat{\lambda}_2\hat{s}_y\hat{K}$
$\hat{T}$ time reversal	$-t$	$\hat{\mathbf{p}}$	ϕ	$-\mathbf{A}$	q	$\hat{S}_T = \hat{\gamma}^5\hat{\gamma}^0\hat{\gamma}^2\hat{K} = i(2/\hbar)\hat{s}_y\hat{K}$
$\hat{K}$ complex conjugation	t	$-\hat{\mathbf{p}}$	ϕ	$\mathbf{A}$	q	

Kramers' Theorem

Kramers showed that each energy level of an electron, or any particle which obeys the Dirac equation, in an electrostatic field is at least doubly degenerate. This double degeneracy, called Kramers' doublet, may be removed if a magnetic field is present, but can never be removed by any electric fields. The theorem can easily be generalized to a system of an odd number of electrons.

Since $\mathbf{A} = 0$ if an external field is electrostatic, the Dirac Hamiltonian in that case is

$$\hat{H}_D = c\hat{\boldsymbol{\alpha}} \cdot \hat{\mathbf{p}} + \hat{\beta}mc^2 + q\phi \qquad (10\text{--}141)$$

where ϕ is the electrostatic potential. Let us assume that ψ is an eigenfunction of this Hamiltonian,

$$\hat{H}_D\psi = \mathscr{E}\psi \qquad (10\text{--}142)$$

where $\mathscr{E}$ is the eigenvalue. The time dependence of ψ can be ignored.

If we operate both sides of (10–142) by $\hat{T}\hat{S}_T$ we obtain

$$\hat{T}\hat{S}_T\hat{H}_D(\hat{T}\hat{S}_T)^{-1}\hat{T}\hat{S}_T\psi = \hat{H}_D\hat{S}_T\psi = \mathscr{E}\hat{S}_T\psi \qquad (10\text{--}143)$$

since $\hat{T}\hat{S}_T$ commutes with the Dirac Hamiltonian and ψ is independent of t.

Since both (10–142) and (10–143) hold with the same eigenvalue $\mathscr{E}$, we see that ψ and $\hat{S}_T\psi$ are either independent, resulting in a double degeneracy, or identical within a phase factor. In the latter case, they are related as

$$\hat{S}_T\psi = e^{i\delta}\psi \qquad (10\text{--}144)$$

where δ is a constant. The situation is, therefore, either (10–144) holds or it does not, and if not, $\hat{S}_T\psi$ is independent from ψ, forming a degenerate pair.

If we operate both sides of (10–144) by $\hat{S}_T$ again, using its expression given in (10–140), we obtain

$$\hat{S}_T^2\psi = \hat{S}_T e^{i\delta}\hat{S}_T^{-1}\hat{S}_T\psi = \left(\frac{2}{\hbar}\right)^2 \hat{s}_y^2(\hat{K}e^{i\delta}\hat{K}^{-1})e^{i\delta}\psi = e^{-i\delta}e^{i\delta}\psi = \psi \qquad (10\text{--}145)$$

but

$$\hat{S}_T^2\psi = \left[i\frac{2}{\hbar}\hat{s}_y\hat{K}\right]^2 \psi = i^2\psi = -\psi \qquad (10\text{--}146)$$

and since they contradict each other, (10–144) cannot hold. Therefore $\hat{S}_T\psi$ and ψ are independent of each other forming a doubly degenerate level.

If a system is made of n electrons, the dual time reversal operator for this system is

$$\hat{S}_T = \left(\frac{2i}{\hbar}\right)^n \hat{s}_y(1)\hat{s}_y(2) \cdots \hat{s}_y(n)\hat{K} \qquad (10\text{–}147)$$

where $s_y(j)$ is the s_y operator for the jth electron. In this case, (10–145) still holds, but (10–146) is generalized to

$$\hat{S}_T^2\psi = i^{2n}\psi \qquad (10\text{–}148)$$

which shows that Kramers' degeneracy appears when n, the number of electrons, is odd, but disappears when n is even.

PROBLEM 10–31. Show that

$$\hat{S}_T\hat{s}\hat{S}_T^{-1} = -\hat{s}$$

(Fig. 10–4.)

FIGURE 10–4. The time reversal operation $\hat{T}\hat{S}_T$ in the Dirac theory. The T-conjugate particles have opposite sign of **k**, $\hat{\mathbf{p}}$, and $\hat{\mathbf{s}}$.

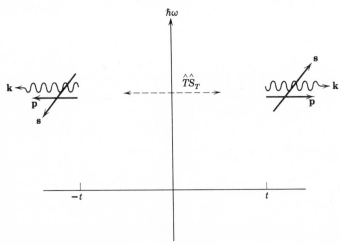

PROBLEM 10–32. Show that

$$\Psi^\dagger \hat{S}_T \Psi = 0 \qquad \text{and} \qquad (\hat{S}_T \Psi)^\dagger \Psi = 0$$

PROBLEM 10–33. Show that

$$\hat{S}_P \hat{S}_C \hat{S}_T = i\hat{\lambda}_1 = -i\hat{\gamma}^5$$

PROBLEM 10–34. Show that

$$\hat{S}_T \hat{S}_C \hat{S}_T^{-1} = -\hat{S}_C$$

10–9* NEUTRINOS

Dirac's equation predicts the existence of spin $\frac{1}{2}$ in addition to the existence of antiparticles, which is also explained by the Klein–Gordon equation. The theory is thus applicable to all leptons (electrons, muons, and neutrinos) and all baryons (nucleons, Λ, Σ, Ξ particles, etc.) as seen in Table 11–20. However, for baryons the magnetic moment associated with the spin is different from that given by the Dirac theory, (10–66), while that of the leptons closely obeys the Dirac theory. Therefore, Dirac's equation can only be used as a starting equation for the treatment of baryons, with serious modifications. On the other hand, leptons, particularly the electron, are found to obey Dirac's theory to a high accuracy as we shall see in the rest of this chapter. We should note, however, that this theory is still not the ultimate theory. Small, but nonzero discrepancies are observed and explained by more elaborate theories (radiative corrections).

Before discussing the electron, let us discuss neutrinos, which are special kinds of leptons, with zero mass and zero charge. The Dirac equation, (10–90) or (10–117), reduces to

$$i\hbar \frac{\partial \Psi}{\partial t} = c\hat{\alpha} \cdot \hat{\mathbf{p}} \Psi = \frac{2c}{\hbar} \hat{\lambda}_1 \hat{\mathbf{s}} \cdot \hat{\mathbf{p}} \Psi = -\frac{2c}{\hbar} \hat{\gamma}^5 \mathbf{s} \cdot \hat{\mathbf{p}} \Psi \qquad (10\text{--}149)$$

when $m = q = 0$. From (10–76) and (10–78), we see that (10–149) can be rewritten as

$$i\hbar \frac{\partial \Phi}{\partial t} = \frac{2c}{\hbar} \hat{\mathbf{s}} \cdot \hat{\mathbf{p}} \chi \qquad (10\text{--}150\text{a})$$

and

$$i\hbar \frac{\partial \chi}{\partial t} = \frac{2c}{\hbar} \hat{\mathbf{s}} \cdot \hat{\mathbf{p}} \Phi \qquad (10\text{--}150\text{b})$$

These two coupled equations can be easily replaced by the following two uncoupled equations:

$$i\hbar \frac{\partial}{\partial t}(\Phi + \chi) = \frac{2c}{\hbar}\, \hat{\mathbf{s}} \cdot \hat{\mathbf{p}}(\Phi + \chi) \qquad (10\text{--}151a)$$

and

$$i\hbar \frac{\partial}{\partial t}(\Phi - \chi) = -\frac{2c}{\hbar}\, \hat{\mathbf{s}} \cdot \hat{\mathbf{p}}(\Phi - \chi) \qquad (10\text{--}151b)$$

Each of these two equations is for a two-component spinor. These equations were obtained by Weyl,[5] but revitalized by Lee and Yang[6] in relation to their celebrated theory of parity nonconservation in weak interactions.

Since Weyl's equations (10–151a) and (10–151b) are independent of each other, we have a choice for the neutrino Hamiltonian between

$$\hat{H}_+ = \frac{2c}{\hbar}\, \hat{\mathbf{s}} \cdot \hat{\mathbf{p}} \qquad (10\text{--}152a)$$

and

$$\hat{H}_- = -\frac{2c}{\hbar}\, \hat{\mathbf{s}} \cdot \hat{\mathbf{p}} \qquad (10\text{--}152b)$$

or between two Weyl's equations

$$i\hbar \frac{\partial}{\partial t}\mathbf{F}_+ = \hat{H}_+\mathbf{F}_+ \qquad (10\text{--}153a)$$

and

$$i\hbar \frac{\partial}{\partial t}\mathbf{F}_- = \hat{H}_-\mathbf{F}_- \qquad (10\text{--}153b)$$

Taking the time dependence of each spinor wave function as

$$\mathbf{F}_\pm = \mathbf{f}_\pm(\mathbf{r})e^{-i\omega t} \qquad (10\text{--}154)$$

[5] H. Weyl, *Zeits. f. Physik*, **56**, 330 (1929).
[6] T. D. Lee and C. N. Yang, *Phys. Rev.* **105**, 167 (1957).

Weyl's equations reduce to the time-independent equations

$$\hbar\omega\mathbf{f}_+ = \frac{2c}{\hbar}\,\hat{\mathbf{s}} \cdot \hat{\mathbf{p}}\mathbf{f}_+ \qquad (10\text{–}155a)$$

and

$$\hbar\omega\mathbf{f}_- = -\frac{2c}{\hbar}\,\hat{\mathbf{s}} \cdot \hat{\mathbf{p}}\mathbf{f}_- \qquad (10\text{–}155b)$$

where the frequency ω can be either positive or negative, and is related to the energy $\mathscr{E}$, which is positive definite, as

$$\hbar\omega = \pm\mathscr{E} \qquad (10\text{–}156)$$

An interesting result of the two-component theory of the neutrino is that $\mathbf{f}_+$ and $\mathbf{f}_-$ are eigenfunctions of $\hat{\mathbf{s}} \cdot \hat{\mathbf{p}}$, as seen in (10–155). If we define the *helicity h* as the eigenvalue of the operator

$$\hat{h} = \frac{2c}{\hbar\mathscr{E}}\,\hat{\mathbf{s}} \cdot \hat{\mathbf{p}} \qquad (10\text{–}157)$$

we see that a stationary state (10–154) has a definite helicity of either 1 or -1. As a matter of fact, the helicity is determined when the sign in (10–156) is given, as shown in Table 10–4. It is reasonable to call a particle, for which

TABLE 10–4   *Helicity of Neutrino and Antineutrino*

		$h\omega$	h		
$\mathbf{f}_+$		$+\mathscr{E}$	$+1$	(neutrino)	nonexistent
		$-\mathscr{E}$	-1	(antineutrino)	
$\mathbf{f}_-$		$+\mathscr{E}$	-1	neutrino	
		$-\mathscr{E}$	$+1$	antineutrino	

$h\omega = +\mathscr{E}$, the neutrino, and the other one, for which $h\omega = -\mathscr{E}$, the antineutrino. The helicity of the neutrino can still be either 1 or -1, depending on which of Weyl's equations, (10–153a) or (10–153b), is chosen by nature.

It so happens[7] that (10–153b) is the equation for the neutrino and anti-neutrino, while (10–153a) is not applicable to nature (Fig. 10–5).

In the two-component neutrino theory, in which we take (10–153b) and disregard (10–153a) completely, it should be noted that the neutrino and antineutrino are not charge conjugate to each other. Using (10–127) and (10–132) we see that

$$\hat{S}_C \hat{\mathbf{s}} \cdot \hat{\mathbf{p}} \hat{S}_C^{-1} = (\hat{S}_C \hat{\mathbf{s}} \hat{S}_C^{-1}) \cdot (\hat{S}_C \hat{\mathbf{p}} \hat{S}_C^{-1}) = (-\hat{\mathbf{s}}) \cdot (-\hat{\mathbf{p}}) = \hat{\mathbf{s}} \cdot \hat{\mathbf{p}} \qquad (10\text{–}158)$$

or

$$\hat{S}_C \hat{H}_+ \hat{S}_C^{-1} = \hat{H}_+ \qquad \hat{S}_C \hat{H}_- \hat{S}_C^{-1} = \hat{H}_- \qquad (10\text{–}159)$$

---

[7] The neutrino and antineutrino are created in the β^- and β^+ decays. Assuming the conservation of the lepton number (1 for a lepton and -1 for an antilepton), we call the one which is created with an electron the antineutrino $\bar{\nu}$, and we call the other one, which is created with a positron (antielectron), the neutrino ν. Thus the β decays of the neutron n and the proton p are

$$n \to p + e^- + \bar{\nu}$$
$$p \to n + e^+ + \nu$$

The second reaction cannot take place when the proton is free, since the rest-mass energy (mc^2) of the proton is less than that of the neutron, but can take place in nuclei, where the nuclear forces make up the energy difference. The neutrino and antineutrino are also created in the $\pi - e$ decay by

$$\pi^- \to e^- + \bar{\nu}$$
$$\pi^+ \to e^+ + \nu$$

In both reactions it is found experimentally [M. Goldhaber, L. Grodzins, and A. W. Sunyan, *Phys. Rev.* **109**, 1015 (1958)] that the helicity h is -1 for the neutrino, and $+1$ for the antineutrino (Fig. 10–5.)

Neutrinos and antineutrinos are created in other reactions where the muon μ^- and antimuon μ^+ appear, instead of the electron and positron. Thus

$$n \to p + \mu^- + \bar{\nu}'$$
$$p \to n + \mu^+ + \nu'$$
$$\pi^- \to \mu^- + \bar{\nu}'$$
$$\pi^+ \to \mu^+ + \nu'$$

It is found experimentally [G. Danby, J-M. Gaillard, K. Goulianos, L. M. Lederman, N. Mistry, M. Schwartz, and J. Steinberger, *Phys. Rev. Letters* **9**, 36 (1962)] that ν', produced by those reactions, reacts with a neutron to produce a muon and a proton,

$$\nu' + n \to \mu^- + p$$

but does not produce an electron and a proton, i.e., the reaction

$$\nu' + n \nrightarrow e^- + p$$

is forbidden. Therefore, ν' is a different neutrino from ν. It is, however, not clear in the present theory just what makes ν' and ν different from each other.

FIGURE 10–5. The neutrino and the antineutrino. The signs of $\hbar\omega$ are given by the definition, but their helicities are determined by experiment.

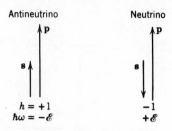

but

$$\hat{S}_C \left(i\hbar \frac{\partial}{\partial t} \right) \hat{S}_C^{-1} = -i\hbar \frac{\partial}{\partial t} \tag{10-160}$$

Therefore $\hat{S}_C \mathbf{F}_-$ is essentially $\mathbf{F}_+$, or the nonexisting particle which obeys the disregarded equation (10–153a). The other way of saying this is that the dual charge conjugation $\hat{S}_C$ changes the sign of $\hbar\omega$ but does not change the helicity, and that transformation brings $\mathbf{f}_-$ to $\mathbf{f}_+$ according to Table 10–4.

On the other hand, the space inversion $\hat{I}$ transforms the helicity operator as

$$\hat{I}\hat{\mathbf{s}} \cdot \hat{\mathbf{p}}\hat{I}^{-1} = (\hat{I}\hat{\mathbf{s}}\hat{I}^{-1}) \cdot (\hat{I}\hat{\mathbf{p}}\hat{I}^{-1}) = \hat{\mathbf{s}} \cdot (-\hat{\mathbf{p}}) = -\hat{\mathbf{s}} \cdot \hat{\mathbf{p}} \tag{10-161}$$

or the helicity changes its sign under the space inversion, but obviously $\hat{I}$ does not change the sign of $\hbar\omega$. Therefore $\hat{I}\mathbf{F}_-$ is again essentially $\mathbf{F}_+$, a nonexisting particle. We notice, however, that in the combined transformation $\hat{I}\hat{S}_C$, the signs of both $\hbar\omega$ and h change, and a neutrino is transformed to an antineutrino, and vice versa, as Fig. 10–6 shows.

PROBLEM 10–35. Obtain the equation of continuity from (10–153) and show that

$$\rho_+ = \mathbf{F}_+^\dagger \mathbf{F}_+ \qquad \mathbf{j}_+ = \frac{2c}{\hbar} \mathbf{F}_+^\dagger \hat{\mathbf{s}} \mathbf{F}_+$$

$$\rho_- = \mathbf{F}_-^\dagger \mathbf{F}_- \qquad \mathbf{j}_- = -\frac{2c}{\hbar} \mathbf{F}_-^\dagger \hat{\mathbf{s}} \mathbf{F}_-$$

PROBLEM 10–36. Taking the plane wave solution $\mathbf{F}_- = \binom{\xi}{\eta} e^{i(kz - \omega t)}$, where k is positive, show directly from (10–155b) that either ξ or η has to be zero depending on the sign of ω.

FIGURE 10–6. The neutrino–antineutrino conjugation is given by $\hat{S}_C\hat{I}$.

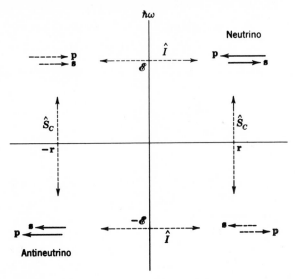

PROBLEM 10–37. Show that

$$(\hat{I} - \hat{\gamma}^5)\Psi = \mathbf{F}_+ \qquad \text{and} \qquad (\hat{I} + \hat{\gamma}^5)\Psi = \mathbf{F}_-$$

($\hat{\gamma}^5$ is sometimes called the chirality in relation to this property.)

10–10 TOTAL ANGULAR MOMENTUM

The Dirac Hamiltonian for an electron in a central potential is

$$\hat{H}_D = c\hat{\boldsymbol{\alpha}} \cdot \hat{\mathbf{p}} + \hat{\beta}mc^2 + U(r) = \frac{2c}{\hbar} \hat{\lambda}_1 \hat{\mathbf{s}} \cdot \hat{\mathbf{p}} + \hat{\lambda}_3 mc^2 + U(r) \qquad (10\text{–}162)$$

where $U(r)$ depends on the radial coordinate only.

Since, as seen in Problem 8–18, the orbital angular momentum operator $\hat{\mathbf{l}}$ contains differentiations with respect to the angular coordinates only, it commutes with $U(r)$, which does not contain any angular coordinates:

$$[\hat{\mathbf{l}}, U(r)] = 0 \qquad (10\text{–}163)$$

The $\hat{\mathbf{l}}$ operator obviously commutes with the second term of $\hat{H}_D$, $\hat{\lambda}_3 mc^2$, since $\hat{\lambda}$ and $\mathbf{r}$ refer to different degrees of freedom.

Let us find if $\hat{\mathbf{l}}$ commutes with the first term of $\hat{H}_D$ or not. Since $\hat{\lambda}$ and $\hat{\mathbf{s}}$ both denote different degrees of freedom than $\mathbf{r}$, they both commute with $\hat{\mathbf{l}}$. Therefore, using the result of Problem 8–85, we see that

$$[\mathbf{u} \cdot \hat{\mathbf{l}}, \hat{\lambda}_1 \hat{\mathbf{s}} \cdot \hat{\mathbf{p}}] = \hat{\lambda}_1 [\mathbf{u} \cdot (\mathbf{r} \times \hat{\mathbf{p}}), \hat{\mathbf{s}} \cdot \hat{\mathbf{p}}] = i\hbar \hat{\lambda}_1 \mathbf{u} \cdot (\hat{\mathbf{s}} \times \hat{\mathbf{p}})$$

$$(10\text{–}164)$$

where $\mathbf{u}$ is a unit vector with an arbitrary direction. In general, therefore, $\mathbf{u} \cdot \hat{\mathbf{l}}$ does not commute with $\hat{H}_D$. However, using this result we can easily see that

$$[\mathbf{u} \cdot (\hat{\mathbf{l}} + \hat{\mathbf{s}}), \hat{\mathbf{s}} \cdot \hat{\mathbf{p}}] = [\mathbf{u} \cdot \hat{\mathbf{l}}, \hat{\mathbf{s}} \cdot \hat{\mathbf{p}}] + [\mathbf{u} \cdot \hat{\mathbf{s}}, \hat{\mathbf{s}} \cdot \hat{\mathbf{p}}]$$

$$= i\hbar \mathbf{u} \cdot (\hat{\mathbf{s}} \times \hat{\mathbf{p}}) + i\hbar \mathbf{u} \cdot (\hat{\mathbf{p}} \times \hat{\mathbf{s}}) = 0 \qquad (10\text{–}165)$$

Since $\hat{\mathbf{s}}$ commutes with other terms of $\hat{H}_D$, we see that

$$[(\hat{\mathbf{l}} + \hat{\mathbf{s}}), \hat{H}_D] = 0 \qquad (10\text{–}166)$$

as long as the potential is central. We call

$$\hat{\mathbf{j}} = \hat{\mathbf{l}} + \hat{\mathbf{s}} \qquad (10\text{–}167)$$

the total angular momentum operator. Equation (10–166) shows that an eigenfunction of $\hat{\mathbf{j}}$ can also be an eigenfunction of $\hat{H}_D$ simultaneously.

Eigenvalues of $\hat{j}^2$ and $\hat{j}_z$

The commutation relation of the $\hat{\mathbf{l}}$ and $\hat{\mathbf{s}}$ operators were obtained before:

$$[\hat{l}_x, \hat{l}_y] = i\hbar \hat{l}_z \qquad [\hat{l}_y, \hat{l}_z] = i\hbar \hat{l}_x \qquad [\hat{l}_z, \hat{l}_x] = i\hbar \hat{l}_y \qquad (8\text{–}53)$$

$$[\hat{s}_x, \hat{s}_y] = i\hbar \hat{s}_z \qquad [\hat{s}_y, \hat{s}_z] = i\hbar \hat{s}_x \qquad [\hat{s}_z, \hat{s}_x] = i\hbar \hat{s}_y \qquad (10\text{–}67)$$

Since these commutation relations for $\hat{\mathbf{l}}$ and $\hat{\mathbf{s}}$ are the same, we obtain

$$[\hat{j}_x, \hat{j}_y] = i\hbar \hat{j}_z \qquad [\hat{j}_y, \hat{j}_z] = i\hbar \hat{j}_x \qquad [\hat{j}_z, \hat{j}_x] = i\hbar \hat{j}_y \qquad (10\text{–}168)$$

These relations show that, in general, it is impossible to find a simultaneous eigenfunction of all three components. (An exceptional case is the s state;

see Problem 8–21.) An eigenfunction of $\hat{j}_z$ is, in general, not an eigenfunction of $\hat{j}_x$ or $\hat{j}_y$. We see, however, that

$$[\hat{j}^2, \hat{j}_z] = [\hat{j}_x{}^2, \hat{j}_z] + [\hat{j}_y{}^2, \hat{j}_z] + [\hat{j}_z{}^2, \hat{j}_z]$$

$$= -i\hbar(\hat{j}_x\hat{j}_y + \hat{j}_y\hat{j}_x) + i\hbar(\hat{j}_y\hat{j}_x + \hat{j}_x\hat{j}_y) + 0 = 0 \qquad (10\text{–}169)$$

This means that an eigenfunction of $\hat{j}_z$ can also be an eigenfunction of $\hat{j}^2$. It is seen directly, using the result of Problem 8–18, that

$$\hat{l}^2 Y_{lm}(\theta, \varphi) = l(l + 1)\hbar^2 Y_{lm}(\theta, \varphi) \qquad (10\text{–}170)$$

and

$$\hat{l}_z Y_{lm}(\theta, \varphi) = m\hbar Y_{lm}(\theta, \varphi) \qquad (10\text{–}171)$$

where $Y_{lm}(\theta, \varphi)$ are the spherical harmonics. We shall show presently that the eigenvalues of $\hat{j}^2$ and $\hat{j}_z$ (or $\hat{l}^2$ and $\hat{l}_z$) are obtained from the commutation relations (10–168) [or (8–53)] directly, without knowing eigenfunctions explicitly.

Dirac's bra and ket notations are convenient in our discussion. Suppose the eigenkets of $\hat{j}^2$ and $\hat{j}_z$ exist and satisfy

$$\hat{j}^2|Am\rangle = A(A + 1)\hbar^2|Am\rangle \qquad (10\text{–}172)$$

$$\hat{j}_z|Am\rangle = m\hbar|Am\rangle \qquad (10\text{–}173)$$

We have taken the same form as (10–170) and (10–171), but here we as yet know nothing about the numbers A and m.

Let us introduce two useful operators

$$\hat{j}_+ = \hat{j}_x + i\hat{j}_y \qquad \text{and} \qquad \hat{j}_- = \hat{j}_x - i\hat{j}_y \qquad (10\text{–}174)$$

Since $\hat{j}_x$ and $\hat{j}_y$ are Hermitian, we have

$$\hat{j}_+{}^\dagger = \hat{j}_- \qquad \hat{j}_-{}^\dagger = \hat{j}_+ \qquad (10\text{–}175)$$

From (10–168) we obtain

$$[\hat{j}_+, \hat{j}_z] = -\hbar\hat{j}_+ \qquad [\hat{j}_-, \hat{j}_z] = \hbar\hat{j}_- \qquad [\hat{j}_+, \hat{j}_-] = 2\hbar\hat{j}_z$$

$$(10\text{–}176)$$

Operating on $|Am\rangle$ by the first relation of (10–176) we obtain

$$\hat{j}_+\hat{j}_z|Am\rangle - \hat{j}_z\hat{j}_+|Am\rangle = -\hbar\hat{j}_+|Am\rangle \qquad (10\text{–}177)$$

which, using (10–173), can be reduced to

$$\hat{j}_z\hat{j}_+|Am\rangle = (m + 1)\hbar\hat{j}_+|Am\rangle \qquad (10\text{–}178)$$

This result shows that $\hat{j}_+|Am\rangle$ is an eigenket of $\hat{j}_z$ with the eigenvalue $(m + 1)\hbar$. Therefore this is $|A\,m+1\rangle$ by definition, except that $|A\,m+1\rangle$ is presumably normalized, but $\hat{j}_+|Am\rangle$ may not be. Thus, if C_m is a normalization constant

$$\hat{j}_+|Am\rangle = C_m|A\,m+1\rangle \qquad (10\text{–}179)$$

which also shows that

$$C_m = \langle A\,m+1|\hat{j}_+|Am\rangle \qquad (10\text{–}180)$$

or

$$\hat{j}_+|Am\rangle = |A\,m+1\rangle\langle A\,m+1|\hat{j}_+|Am\rangle \qquad (10\text{–}181)$$

It is easy to obtain

$$\hat{j}_-|Am\rangle = |A\,m-1\rangle\langle A\,m-1|\hat{j}_-|Am\rangle \qquad (10\text{–}182)$$

in exactly the same way except that in using (10–176) the second relation, instead of the first relation, is taken.

Results (10–181) and (10–182) shows that there is a series of eigenkets of $\hat{j}_z$ each of which is characterized by a quantum number m, which is different by 1 or -1 from that of the adjacent state. They also show that $\hat{j}_+$ and $\hat{j}_-$ are the step-up and step-down operators which connect these eigenstates of $\hat{j}_z$ (Fig. 10–7). This is analogous to the creation and annihilation operators $\hat{a}^\dagger$ and $\hat{a}$ for the simple harmonic oscillator, which we discussed in section 8–9.

FIGURE 10–7. The eigenkets $|jm\rangle$ and the operators $\hat{j}_+$ and $\hat{j}_-$.

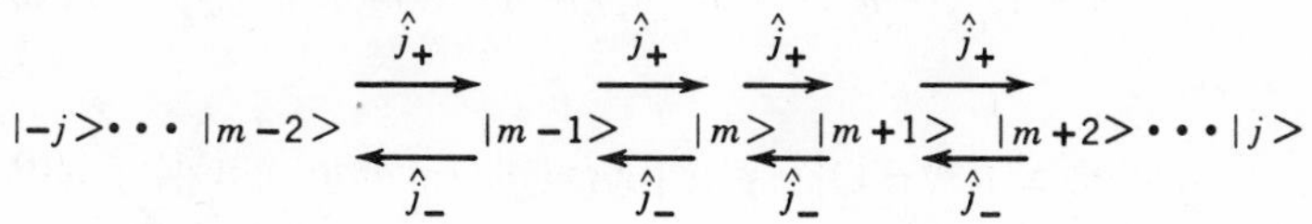

Let us assume that there is a maximum number j that m can achieve. This means that the series of states obtained by a successive operation of $\hat{\jmath}_+$ terminates when the last state is $|Aj\rangle$, or the eigenstate of $\hat{\jmath}_z$ with the eigenvalue $j\hbar$. Thus we require that

$$\hat{\jmath}_+|Aj\rangle = 0 \tag{10-183}$$

Operating on this state $|Aj\rangle$ by the operators in the last relation of (10-176) we obtain

$$\hat{\jmath}_+\hat{\jmath}_-|Aj\rangle = 2\hbar^2 j|Aj\rangle \tag{10-184}$$

which gives

$$\langle Aj|\hat{\jmath}_+|A\,j-1\rangle\langle A\,j-1|\hat{\jmath}_-|Aj\rangle = 2\hbar^2 j \tag{10-185}$$

when (10-181) and (10-182) are used. Since $\hat{\jmath}_-$ is Hermitian conjugate to $\hat{\jmath}_+$, this relation can be rewritten as

$$|\langle Aj|\hat{\jmath}_+|A\,j-1\rangle|^2 = 2\hbar^2 j \tag{10-186}$$

Since, for a general state, we have

$$\begin{aligned}
\hat{\jmath}_+\hat{\jmath}_-|Am\rangle &= \hat{\jmath}_+|A\,m-1\rangle\langle A\,m-1|\hat{\jmath}_-|Am\rangle \\
&= \hat{\jmath}_+|A\,m-1\rangle\langle Am|\hat{\jmath}_+|A\,m-1\rangle^* \\
&= |Am\rangle|\langle Am|\hat{\jmath}_+|A\,m-1\rangle|^2 \tag{10-187a}
\end{aligned}$$

and

$$\hat{\jmath}_-\hat{\jmath}_+|Am\rangle = |Am\rangle|\langle A\,m+1|\hat{\jmath}_+|Am\rangle|^2 \tag{10-187b}$$

the last relation of (10-176), when applied to $|Am\rangle$ gives

$$|\langle Am|\hat{\jmath}_+|A\,m-1\rangle|^2 - |\langle A\,m+1|\hat{\jmath}_+|Am\rangle|^2 = 2\hbar^2 m \tag{10-188}$$

This recursion relation, when combined with the initial value given by (10-186), gives

$$|\langle A\,m+1|\hat{\jmath}_+|Am\rangle|^2 = \hbar^2[j(j+1) - m(m+1)] \tag{10-189a}$$

and

$$|\langle Am|\hat{\jmath}_+|A\,m-1\rangle|^2 = \hbar^2[j(j+1) - m(m-1)] \tag{10-189b}$$

Let us assume, in the same way as (10–183), that there is a minimum number that m can take, or

$$\hat{j}_-|Am_{min}\rangle = 0 \qquad (10\text{–}190)$$

Actually, since $\hat{j}_+$ and $\hat{j}_-$ are Hermitian conjugate to each other and since $m = j$ and $m = -j + 1$ are the only possibilities which make the coefficient (10–189b) zero, if there is a lowest number which the quantum number m can take, $-j$ must be that lowest number, in order that (10–182) is consistent with (10–190). Therefore the possible quantum numbers for m are

$$m = -j, -j + 1, -j + 2, \ldots, j - 2, j - 1, j \qquad (10\text{–}191)$$

The upper and the lower ends of this series, however, can meet only if

$$j = \text{integer} \quad \text{or} \quad \text{half-integer} \qquad (10\text{–}192a)$$

therefore

$$m = \text{integer} \quad \text{or} \quad \text{half-integer} \qquad (10\text{–}192b)$$

The last question is what is A? Since

$$\hat{j}^2 \equiv \hat{j}_x^2 + \hat{j}_y^2 + \hat{j}_z^2 = \tfrac{1}{2}\hat{j}_+\hat{j}_- + \tfrac{1}{2}\hat{j}_-\hat{j}_+ + \hat{j}_z^2 \qquad (10\text{–}193)$$

we obtain

$$
\begin{aligned}
\hat{j}^2|Am\rangle &= \tfrac{1}{2}\hat{j}_+\hat{j}_-|Am\rangle + \tfrac{1}{2}\hat{j}_-\hat{j}_+|Am\rangle + \hat{j}_z^2|Am\rangle \\
&= \{\tfrac{1}{2}[j(j + 1) - m(m - 1)] + \tfrac{1}{2}[j(j + 1) - m(m + 1)] \\
&\qquad\qquad\qquad\qquad\qquad\qquad\qquad + m^2\}\hbar^2|Am\rangle \\
&= j(j + 1)\hbar^2|Am\rangle \qquad (10\text{–}194)
\end{aligned}
$$

using (10–187a), (10–187b), and (10–189a). Comparing this result with (10–172), we see that

$$A = j \qquad (10\text{–}195)$$

It is not necessary but convenient to take as real and positive the coefficients $\langle A\,m+1|\hat{j}_+|A\,m\rangle$ and $\langle A\,m|\hat{j}_-|A\,m+1\rangle$, which are given by

(10–189a) and (10–189b). Thus, in summarizing all the results we have obtained, we write

$$\hat{j}_+ |jm\rangle = \hbar\sqrt{j(j + 1) - m(m + 1)}|j\,m+1\rangle \qquad (10\text{–}196)$$

$$\hat{j}_- |jm\rangle = \hbar\sqrt{j(j + 1) - m(m - 1)}|j\,m-1\rangle \qquad (10\text{–}197)$$

$$\hat{j}_z |jm\rangle = m\hbar|jm\rangle \qquad (10\text{–}198)$$

$$\hat{j}^2 |jm\rangle = j(j + 1)\hbar^2|jm\rangle \qquad (10\text{–}199)$$

It should be noted that these results are obtained from the commutation relations (10–168). Therefore, any operators which satisfy the same commutation relations, including $\hat{l}$ and $\hat{s}$, satisfy the same formulas (10–196) through (10–199). The previous results (10–170) and (10–171) are special cases of (10–199) and (10–198).

PROBLEM 10–38. Confirm (10–165).

PROBLEM 10–39. Obtain (10–176) from (10–168).

PROBLEM 10–40. Obtain (10–187b).

PROBLEM 10–41. Obtain (10–193).

PROBLEM 10–42. (a) Show that

$$\hat{s}_z \begin{pmatrix} 1 \\ 0 \end{pmatrix} = \tfrac{1}{2}\hbar \begin{pmatrix} 1 \\ 0 \end{pmatrix} \quad \text{and} \quad \hat{s}_z \begin{pmatrix} 0 \\ 1 \end{pmatrix} = -\tfrac{1}{2}\hbar \begin{pmatrix} 0 \\ 1 \end{pmatrix}$$

(b) If

$$\hat{s}_+ = \hat{s}_x + i\hat{s}_y \quad \text{and} \quad \hat{s}_- = \hat{s}_x - i\hat{s}_y$$

show that

$$\hat{s}_+ \begin{pmatrix} 0 \\ 1 \end{pmatrix} = \hbar \begin{pmatrix} 1 \\ 0 \end{pmatrix} \quad \text{and} \quad \hat{s}_- \begin{pmatrix} 1 \\ 0 \end{pmatrix} = \hbar \begin{pmatrix} 0 \\ 1 \end{pmatrix}$$

(c) Compare the results of (b) with (10–196) and (10–197).

PROBLEM 10–43. Show that

$$\langle j\,m\pm 1|\hat{j}_x|j\,m\rangle = \tfrac{1}{2}\hbar\sqrt{j(j+1)-m(m\pm 1)}$$

$$\langle j\,m\pm 1|\hat{j}_y|j\,m\rangle = \mp\tfrac{1}{2}i\hbar\sqrt{j(j+1)-m(m\pm 1)}$$

PROBLEM 10–44. Show that

$$|j\,m\rangle = \hbar^{-j+m}\sqrt{\frac{(j+m)!}{(2j)!\,(j-m)!}}\,\hat{j}_-^{\,j-m}|jj\rangle$$

10–11 CLEBSCH–GORDAN COEFFICIENTS

We know that the eigenfunctions of $\hat{l}^2$ and $\hat{l}_z$ are the spherical harmonics, and that the eigenfunctions of $\hat{s}^2$ and $\hat{s}_z$ are the two-component spinors. Our next problem is to express the eigenfunctions of $\hat{j}^2$ and $\hat{j}_z$ explicitly by means of these known eigenfunctions of $\hat{\mathbf{l}}$ and $\hat{\mathbf{s}}$, from which the total angular momentum $\hat{\mathbf{j}}$ is made.

Using the ket notation, we write

$$\hat{l}^2|l\,m_l\rangle = l(l+1)\hbar^2|l\,m_l\rangle \tag{10–200}$$

$$\hat{l}_z|l\,m_l\rangle = m_l\hbar|l\,m_l\rangle \tag{10–201}$$

$$\hat{s}^2|s\,m_s\rangle = s(s+1)\hbar^2|s\,m_s\rangle = (\tfrac{3}{4})\hbar^2|\tfrac{1}{2}\,m_s\rangle \quad \text{(since } s = \tfrac{1}{2}\text{)} \tag{10–202}$$

$$\hat{s}_z|s\,m_s\rangle = m_s\hbar|\tfrac{1}{2}\,m_s\rangle \tag{10–203}$$

In the coordinate representation [see (8–203)],

$$\langle \mathbf{r}|l\,m_l\rangle = Y_{lm_l}(\theta, \varphi) \tag{10–204}$$

and in the spinor representation

$$\langle \xi \mid \tfrac{1}{2}\,\tfrac{1}{2}\rangle = \begin{pmatrix}1\\0\end{pmatrix} \qquad \langle \xi \mid \tfrac{1}{2}\,-\tfrac{1}{2}\rangle = \begin{pmatrix}0\\1\end{pmatrix} \tag{10–205}$$

where ξ is the spin coordinate. It is easy to verify that these spinors are the eigenfunctions of $\hat{s}_z$ with the eigenvalues $\tfrac{1}{2}\hbar$ and $-\tfrac{1}{2}\hbar$, respectively, by using equation (10–68), for instance.

If we define the product ket

$$|l\,m_l\,s\,m_s\rangle \equiv |l\,m_l\rangle|s\,m_s\rangle \qquad (10\text{-}206)$$

then this ket satisfies all four eigenvalue equations, (10–200) through (10–203). Since $\hat{\jmath}$ is made of $\hat{l}$ and $\hat{s}$, and since $|l\,m_l\,s\,m_s\rangle$ form a complete set, we expect that $|jm\rangle$ can be expressed by $|l\,m_l\,s\,m_s\rangle$ as

$$|j\,m\rangle = \sum_{m_l,m_s} |l\,m_l\,s\,m_s\rangle\langle l\,m_l\,s\,m_s|jm\rangle$$

$$= \sum_{m_l} \{|l\,m_l\,\tfrac{1}{2}\,\tfrac{1}{2}\rangle\langle l\,m_l\,\tfrac{1}{2}\,\tfrac{1}{2}|jm\rangle + |l\,m_l\,\tfrac{1}{2}\,-\tfrac{1}{2}\rangle\langle l\,m_l\,\tfrac{1}{2}\,-\tfrac{1}{2}|jm\rangle\}$$

$$(10\text{-}207)$$

where $\langle l\,m_l\,s\,m_s|jm\rangle$ are the expansion coefficients, called the *Clebsch–Gordan coefficients* or the *vector coupling coefficients*. The coefficients which appear in (10–207) are the special ones of more general Clebsch–Gordan coefficients, which are going to be discussed in the next chapter.

Since $\hat{\jmath}_z$ and $\hat{l}_z + \hat{s}_z$ are equivalent to each other, we operate on the left- and right-hand sides of equation (10–207) by $\hat{\jmath}_z$ and $\hat{l}_z + \hat{s}_z$, respectively, and obtain the equation

$$m|jm\rangle = \sum_{m_l} \{(m_l + \tfrac{1}{2})|l\,m_l\,\tfrac{1}{2}\,\tfrac{1}{2}\rangle\langle l\,m_l\,\tfrac{1}{2}\,\tfrac{1}{2}|jm\rangle$$

$$+ (m_l - \tfrac{1}{2})|l\,m_l\,\tfrac{1}{2}\,-\tfrac{1}{2}\rangle\langle l\,m_l\,\tfrac{1}{2}\,-\tfrac{1}{2}|jm\rangle\} \qquad (10\text{-}208)$$

Operating on both sides of this equation by $\langle l\,m_l\,\tfrac{1}{2}\,\tfrac{1}{2}|$, and using the orthonormality of the eigenkets, we obtain

$$m\langle l\,m_l\,\tfrac{1}{2}\,\tfrac{1}{2}|jm\rangle = (m_l + \tfrac{1}{2})\langle l\,m_l\,\tfrac{1}{2}\,\tfrac{1}{2}|jm\rangle \qquad (10\text{-}209)$$

or

$$(m - m_l - \tfrac{1}{2})\langle l\,m_l\,\tfrac{1}{2}\,\tfrac{1}{2}|jm\rangle = 0 \qquad (10\text{-}209')$$

Therefore

$$\langle l\,m_l\,\tfrac{1}{2}\,\tfrac{1}{2}|jm\rangle = 0 \quad \text{unless} \quad m = m_l + \tfrac{1}{2} \qquad (10\text{-}210)$$

In the same way we see that

$$\langle l\,m_l\,\tfrac{1}{2}\,-\tfrac{1}{2}|jm\rangle = 0 \quad \text{unless} \quad m = m_l - \tfrac{1}{2} \qquad (10\text{-}211)$$

Thus the expansion (10–207) reduces to

$$|jm\rangle = |l\,m-\tfrac{1}{2}\ \tfrac{1}{2}\ \tfrac{1}{2}\rangle\langle l\,m-\tfrac{1}{2}\ \tfrac{1}{2}\ \tfrac{1}{2}|jm\rangle$$
$$+\ |l\,m+\tfrac{1}{2}\ \tfrac{1}{2}\ -\tfrac{1}{2}\rangle\langle l\,m+\tfrac{1}{2}\ \tfrac{1}{2}\ -\tfrac{1}{2}|jm\rangle \tag{10–212}$$

where we replaced m_l by $m - \tfrac{1}{2}$ or $m + \tfrac{1}{2}$ in the right-hand side.

In order to find the two Clebsch–Gordan coefficients, which appear in (10–212), let us first take an extreme case of $m - \tfrac{1}{2} = l$, where (10–212) reduces to

$$|j\,l+\tfrac{1}{2}\rangle = |l\,l\ \tfrac{1}{2}\ \tfrac{1}{2}\rangle\langle l\,l\ \tfrac{1}{2}\ \tfrac{1}{2}|j\,l+\tfrac{1}{2}\rangle \tag{10–213}$$

since there is no such thing as $|l\,l+1\ \tfrac{1}{2}\ -\tfrac{1}{2}\rangle$. Equation (10–213) implies that

$$j = l + \tfrac{1}{2} \tag{10–214}$$

and that

$$\langle l\,l\ \tfrac{1}{2}\ \tfrac{1}{2}|l+\tfrac{1}{2}\ l+\tfrac{1}{2}\rangle = 1 \tag{10–215}$$

Equation (10–214) is implied since $l + \tfrac{1}{2}$ is the maximum value that m can achieve in (10–212).

Taking $m - \tfrac{1}{2} = l - 1$ in (10–212), we obtain

$$|j\,l-\tfrac{1}{2}\rangle = |l\,l-1\ \tfrac{1}{2}\ \tfrac{1}{2}\rangle\langle l\,l-1\ \tfrac{1}{2}\ \tfrac{1}{2}|j\,l-\tfrac{1}{2}\rangle$$
$$+\ |l\,l\ \tfrac{1}{2}\ -\tfrac{1}{2}\rangle\langle l\,l\ \tfrac{1}{2}\ -\tfrac{1}{2}|j\,l-\tfrac{1}{2}\rangle \tag{10–216}$$

Since there are two coefficients on the right-hand side of this equation, we can obtain, by adjusting these coefficients, two mutually independent kets. One of them must be $|l+\tfrac{1}{2}\ l-\tfrac{1}{2}\rangle$, or a component of the $j = l + \tfrac{1}{2}$ series, and the other one is a component of a new series of eigenket whose maximum m is $l - \tfrac{1}{2}$. Thus the second eigenket which is included in (10–216) must be a component of a $j = l - \tfrac{1}{2}$ series, which we designate as $|l-\tfrac{1}{2}\ l-\tfrac{1}{2}\rangle$. Since the number of terms does not increase over two for any lower m value, $j = l + \tfrac{1}{2}$ and $l - \tfrac{1}{2}$ are the only possible values of j in this case.

If we operate $|jm\rangle$ by $\hat{\jmath}_+\hat{\jmath}_-$, we obtain

$$\hat{\jmath}_+\hat{\jmath}_-|jm\rangle = \hbar^2[j(j + 1) - m(m - 1)]|jm\rangle$$
$$= |l\,m-\tfrac{1}{2}\ \tfrac{1}{2}\ \tfrac{1}{2}\rangle\hbar^2[j(j + 1) - m(m - 1)]\langle l\,m-\tfrac{1}{2}\ \tfrac{1}{2}\ \tfrac{1}{2}|jm\rangle$$
$$+\ |l\,m+\tfrac{1}{2}\ \tfrac{1}{2}\ -\tfrac{1}{2}\rangle\hbar^2[j(j + 1) - m(m - 1)]$$
$$\times\ \langle l\,m+\tfrac{1}{2}\ \tfrac{1}{2}\ -\tfrac{1}{2}|jm\rangle \tag{10–217}$$

using the expansion (10–212). On the other hand,

$$\hat{\jmath}_+\hat{\jmath}_- = (\hat{l}_+ + \hat{s}_+)(\hat{l}_- + \hat{s}_-) = \hat{l}_+\hat{l}_- + \hat{l}_-\hat{s}_- + \hat{s}_+\hat{l}_- + \hat{s}_+\hat{s}_- \qquad (10\text{–}218)$$

and when this operator operates on the right-hand side of (10–212), the result is

$$|l\, m{-}\tfrac{1}{2}\,\tfrac{1}{2}\,\tfrac{1}{2}\rangle\{[l(l+1) - (m-\tfrac{1}{2})(m-\tfrac{3}{2}) + 1]\langle l\, m{-}\tfrac{1}{2}\,\tfrac{1}{2}\,\tfrac{1}{2}|jm\rangle$$
$$+ \sqrt{l(l+1) - m^2 + \tfrac{1}{4}}\langle l\, m{+}\tfrac{1}{2}\,\tfrac{1}{2}\,{-}\tfrac{1}{2}|jm\rangle\}$$
$$+ |l\, m{+}\tfrac{1}{2}\,\tfrac{1}{2}\,{-}\tfrac{1}{2}\rangle\{[l(l+1) - m^2 + \tfrac{1}{4}]\langle l\, m{+}\tfrac{1}{2}\,\tfrac{1}{2}\,{-}\tfrac{1}{2}|jm\rangle$$
$$+ \sqrt{l(l+1) - m^2 + \tfrac{1}{4}}\,\langle l\, m{-}\tfrac{1}{2}\,\tfrac{1}{2}\,\tfrac{1}{2}|jm\rangle\}$$

$$(10\text{–}219)$$

Since (10–219) should be equal to (10–217) we obtain

$$[j(j+1) - m(m-1)]\langle l\, m{-}\tfrac{1}{2}\,\tfrac{1}{2}\,\tfrac{1}{2}|jm\rangle$$
$$= [l(l+1) - (m-\tfrac{1}{2})(m-\tfrac{3}{2}) + 1]\langle l\, m{-}\tfrac{1}{2}\,\tfrac{1}{2}\,\tfrac{1}{2}|jm\rangle$$
$$+ \sqrt{l(l+1) - m^2 + \tfrac{1}{4}}\langle l\, m{+}\tfrac{1}{2}\,\tfrac{1}{2}\,{-}\tfrac{1}{2}|jm\rangle \qquad (10\text{–}220)$$

by equating the coefficients of $|l\, m{-}\tfrac{1}{2}\,\tfrac{1}{2}\,\tfrac{1}{2}\rangle$. Therefore

$$\frac{\langle l\, m{-}\tfrac{1}{2}\,\tfrac{1}{2}\,\tfrac{1}{2}\mid jm\rangle}{\langle l\, m{+}\tfrac{1}{2}\,\tfrac{1}{2}\,{-}\tfrac{1}{2}\mid jm\rangle} = \frac{\sqrt{l(l+1) - m^2 + \tfrac{1}{4}}}{j(j+1) - l(l+1) - m - \tfrac{1}{4}} \qquad (10\text{–}221)$$

which reduces to

$$\frac{\langle l\, m{-}\tfrac{1}{2}\,\tfrac{1}{2}\,\tfrac{1}{2}\mid jm\rangle}{\langle l\, m{+}\tfrac{1}{2}\,\tfrac{1}{2}\,{-}\tfrac{1}{2}\mid jm\rangle} = \sqrt{\frac{l + \tfrac{1}{2} + m}{l + \tfrac{1}{2} - m}} \qquad \text{if} \quad j = l + \tfrac{1}{2} \qquad (10\text{–}222a)$$

and

$$\frac{\langle l\, m{-}\tfrac{1}{2}\,\tfrac{1}{2}\,\tfrac{1}{2}\mid jm\rangle}{\langle l\, m{+}\tfrac{1}{2}\,\tfrac{1}{2}\,{-}\tfrac{1}{2}\mid jm\rangle} = \sqrt{\frac{l + \tfrac{1}{2} - m}{l + \tfrac{1}{2} + m}} \qquad \text{if} \quad j = l - \tfrac{1}{2} \qquad (10\text{–}222b)$$

TABLE 10–5 $\langle lm_l\tfrac{1}{2}m_s|\, jm\rangle$. $m_l + m_s = m$

$j =$	$m_s = \tfrac{1}{2}$	$m_s = -\tfrac{1}{2}$
$l + \tfrac{1}{2}$	$\sqrt{\dfrac{l + m + \tfrac{1}{2}}{2l + 1}}$	$\sqrt{\dfrac{l - m + \tfrac{1}{2}}{2l + 1}}$
$l - \tfrac{1}{2}$	$-\sqrt{\dfrac{l - m + \tfrac{1}{2}}{2l + 1}}$	$\sqrt{\dfrac{l + m + \tfrac{1}{2}}{2l + 1}}$

The Clebsch–Gordan coefficients are finally obtained from (10–222) and the normalization condition

$$|\langle l\, m-\tfrac{1}{2}\, \tfrac{1}{2}\, \tfrac{1}{2}|jm\rangle|^2 + |\langle l\, m+\tfrac{1}{2}\, \tfrac{1}{2}\, -\tfrac{1}{2}|jm\rangle|^2 = 1$$

$$(10\text{–}223)$$

as shown in Table 10–5. More general discussions of the Clebsch–Gordan coefficients will be given in the next chapter.

Putting the above result into (10–212) we obtain

$$|l+\tfrac{1}{2}\, m\rangle = \frac{1}{\sqrt{2l+1}}\{\sqrt{l+m+\tfrac{1}{2}}|l\, m-\tfrac{1}{2}\, \tfrac{1}{2}\, \tfrac{1}{2}\rangle$$
$$+ \sqrt{l-m+\tfrac{1}{2}}|l\, m+\tfrac{1}{2}\, \tfrac{1}{2}\, -\tfrac{1}{2}\rangle\} \qquad (10\text{–}224\text{a})$$

$$|l-\tfrac{1}{2}\, m\rangle = \frac{1}{\sqrt{2l+1}}\{-\sqrt{l-m+\tfrac{1}{2}}|l\, m-\tfrac{1}{2}\, \tfrac{1}{2}\, \tfrac{1}{2}\rangle$$
$$+ \sqrt{l+m+\tfrac{1}{2}}|l\, m+\tfrac{1}{2}\, \tfrac{1}{2}\, -\tfrac{1}{2}\rangle\} \qquad (10\text{–}224\text{b})$$

and from (10–204) and (10–205)

$$\langle \mathbf{r}\xi \mid l+\tfrac{1}{2}\, m\rangle = \frac{1}{\sqrt{2l+1}}\left\{\sqrt{l+m+\tfrac{1}{2}}\, Y_{lm-(1/2)}(\theta,\varphi)\begin{pmatrix}1\\0\end{pmatrix}\right.$$
$$\left. + \sqrt{l-m+\tfrac{1}{2}}\, Y_{lm+(1/2)}(\theta,\varphi)\begin{pmatrix}0\\1\end{pmatrix}\right\}$$

$$\equiv \mathbf{Y}_{ll+(1/2)m} \qquad (10\text{–}225\text{a})$$

$$\langle \mathbf{r}\xi \mid l-\tfrac{1}{2}\, m\rangle = \frac{1}{\sqrt{2l+1}}\left[-\sqrt{l-m+\tfrac{1}{2}}\, Y_{lm-(1/2)}(\theta,\varphi)\begin{pmatrix}1\\0\end{pmatrix}\right.$$
$$\left. + \sqrt{l+m+\tfrac{1}{2}}\, Y_{lm+(1/2)}(\theta,\varphi)\begin{pmatrix}0\\1\end{pmatrix}\right]$$

$$\equiv \mathbf{Y}_{ll-(1/2)m} \qquad (10\text{–}225\text{b})$$

These spinors are called the *spherical spinors*, designated as $\mathbf{Y}_{ljm}$.

PROBLEM 10–45. Obtain (10–219).

PROBLEM 10–46. Show that the same results as (10–222a) and (10–222b) are obtained by equating the coefficients of $|l\, m+\tfrac{1}{2}\, \tfrac{1}{2}\, -\tfrac{1}{2}\rangle$ of (10–219) and (10–217).

PROBLEM 10–47. Show that

$$|l \; m_l \; s \; m_s\rangle = \sum_j |j \; m\rangle\langle j \; m \mid l \; m_l \; s \; m_s\rangle = \sum_j |j \; m\rangle\langle l \; m_l \; s \; m_s \mid jm\rangle^*$$

PROBLEM 10–48. Show that

(a) $\displaystyle\sum_{j,m} \langle l \; m_l \; s \; m_s \mid jm\rangle^*\langle l \; m_l' \; s \; m_s' \mid j \; m\rangle = \delta_{m_l,m_{l'}}\delta_{m_s,m_s'}$

(b) $\displaystyle\sum_{m_l,m_s} \langle l \; m_l \; s \; m_s \mid j \; m\rangle^*\langle l \; m_l \; s \; m_s \mid j' \; m'\rangle = \delta_{j,j'}\delta_{m,m'}$

10–12 THE HYDROGEN ATOM AND HYDROGEN-LIKE IONS. (DIRAC THEORY)

The Dirac equation for an electron in the Coulomb field of a nucleus of charge Ze is

$$\hat{H}_D\psi_{\mathscr{E}} = \mathscr{E} \, \psi_{\mathscr{E}} \tag{10-226}$$

where

$$\hat{H}_D = c\hat{\boldsymbol{\alpha}} \cdot \hat{\mathbf{p}} + \hat{\beta}mc^2 - \frac{Ze^2}{4\pi\varepsilon_0 r} = \frac{2c}{\hbar} \hat{\lambda}_1\hat{\mathbf{s}} \cdot \hat{\mathbf{p}} + \hat{\lambda}_3 mc^2 - \frac{Ze^2}{4\pi\varepsilon_0 r} \tag{10-227}$$

Since the Coulomb potential is central, this Hamiltonian commutes with the total angular momentum $\hat{\mathbf{j}}$, as was shown in the last section. Therefore, the eigenfunction $\psi_{\mathscr{E}}$ can also be an eigenfunction of $\hat{j}^2$ and $\hat{j}_z$ simultaneously:

$$\hat{j}^2\psi_{\mathscr{E}jm}(\mathbf{r}) = j(j + 1)\hbar^2 \psi_{\mathscr{E}jm}(\mathbf{r}) \tag{10-228}$$

$$\hat{j}_z\psi_{\mathscr{E}jm}(\mathbf{r}) = m\hbar \psi_{\mathscr{E}jm}(\mathbf{r}) \tag{10-229}$$

Another important property of the eigenfunction $\psi_{\mathscr{E}}$ is the parity. We have seen in section 10–6 that $\hat{I}\hat{\gamma}_0 = \hat{I}\hat{\beta}$ is the parity operator for a free Dirac particle. The same argument is applicable to the present case of a Coulomb field, since the Coulomb field is invariant under $\hat{I}$. Therefore each state can be specified as

$$\hat{I}\hat{\beta}\psi_{\mathscr{E}Pjm} = P\psi_{\mathscr{E}Pjm} \tag{10-230}$$

where

$$P = +1 \quad \text{or} \quad -1 \tag{10-231}$$

and is called the *parity* of this state.

We also notice that

$$\hat{I}\beta\hat{\jmath}(\hat{I}\beta)^{-1} = \hat{I}\hat{\lambda}_3(\hat{\mathbf{l}} + \hat{\mathbf{s}})(\hat{I}\hat{\lambda}_3)^{-1} = \hat{I}(\mathbf{r} \times \hat{\mathbf{p}})\hat{I}^{-1} + \hat{\mathbf{s}} = \hat{\mathbf{l}} + \hat{\mathbf{s}} = \hat{\jmath}$$

(10–232)

since $\hat{\mathbf{s}}$ and $\hat{\lambda}_3$ commute to each other, and $\hat{\mathbf{s}}$ does not depend on **r**. Note that the orbital angular momentum $\hat{\mathbf{l}}$ is a pseudovector since $\hat{I}\hat{\mathbf{l}}\hat{I}^{-1} = \hat{\mathbf{l}}$, in contrast to $\hat{I}\mathbf{r}\hat{I}^{-1} = -\mathbf{r}$.

We have seen in section 10–5 that each Dirac wave function is a bispinor

$$\Psi = \begin{pmatrix} \Phi \\ \chi \end{pmatrix} = \begin{pmatrix} \Phi_1 \\ \Phi_2 \\ \chi_1 \\ \chi_2 \end{pmatrix}$$

(10–76)

where each spinor component is a proper base for the spin operator $\hat{\mathbf{s}}$, as seen in (10–77). On the other hand, we saw in the last section that there are two spherical spinors for a given j and m; i.e., those for $l = j - \frac{1}{2}$ and $j + \frac{1}{2}$. It is reasonable to expect, then, that

$$\Phi_{\mathscr{E}Pjm}(\mathbf{r}) = \mathbf{Y}_{j-(1/2)jm}(\theta, \varphi)F_{\mathscr{E}Pj}(r)$$

(10–233a)

and

$$\chi_{\mathscr{E}Pjm}(\mathbf{r}) = \mathbf{Y}_{j+(1/2)jm}(\theta, \varphi)iG_{\mathscr{E}Pj}(r)$$

(10–233b)

or

$$\Phi_{\mathscr{E}P'jm}(\mathbf{r}) = \mathbf{Y}_{j+(1/2)jm}(\theta, \varphi)F_{\mathscr{E}P'j}(r)$$

(10–234a)

and

$$\chi_{\mathscr{E}P'jm}(\mathbf{r}) = \mathbf{Y}_{j-(1/2)jm}(\theta, \varphi)iG_{\mathscr{E}P'j}(r)$$

(10–234b)

where F and G are functions of r only. The complex unit i is in (10–233b) and (10–234b) for convenience. That these expressions have been correctly anticipated can be seen from the parity. Since (Fig. 10–8)

$$\hat{I}Y_{lm}(\theta, \varphi) = Y_{lm}(\theta + \pi, \varphi) = (-1)^l Y_{lm}(\theta, \varphi)$$

(10–235)

by the properties of the spherical harmonics, or the associated Legendre polynomials, given in Appendix I(c), we obtain

$$\hat{I}\mathbf{Y}_{ljm} = (-1)^l \mathbf{Y}_{ljm}$$

(10–236)

FIGURE 10-8. The space inversion $\mathbf{r} \to -\mathbf{r}$ means $\theta \to \theta + \pi$ and $\phi \to \phi$.

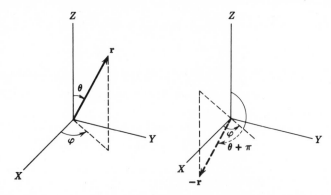

Therefore, when (10–233a) and (10–233b) are used in (10–76), we have

$$\hat{I}\hat{\beta}\Psi_{\mathscr{E}Pjm} = \hat{\beta}\begin{pmatrix}(-1)^{j-(1/2)}\Phi_{\mathscr{E}Pjm}(\mathbf{r}) \\ (-1)^{j+(1/2)}\chi_{\mathscr{E}Pjm}(\mathbf{r})\end{pmatrix} = (-1)^{j-(1/2)}\Psi_{\mathscr{E}Pjm} \qquad (10\text{–}237)$$

which shows that this is an eigenfunction of the parity operator $\hat{I}\hat{\beta}$ with the parity $(-1)^{j-(1/2)}$. The same conclusion will be obtained in the alternative case of (10–234a) and (10–234b), except that the parity will be $(-1)^{j+(1/2)}$. In either case, we notice that the parity P is given by $(-1)^l$ where l is that of the function Φ. Therefore, for a given j, the parity P actually determines the choice between (10–233) and (10–234).

Equations for $F(r)$ and $G(r)$ Functions

When (10–76) is put into (10–226), we see, from (10–78), that the Dirac equation reduces to

$$\frac{2c}{\hbar}\hat{\mathbf{s}} \cdot \hat{\mathbf{p}}\chi_{\mathscr{E}Pjm} + \left(mc^2 - \frac{Ze^2}{4\pi\varepsilon_0 r}\right)\Phi_{\mathscr{E}Pjm} = \mathscr{E}\Phi_{\mathscr{E}Pjm} \qquad (10\text{–}238a)$$

and

$$\frac{2c}{\hbar}\hat{\mathbf{s}} \cdot \hat{\mathbf{p}}\Phi_{\mathscr{E}Pjm} + \left(-mc^2 - \frac{Ze^2}{4\pi\varepsilon_0 r}\right)\chi_{\mathscr{E}Pjm} = \mathscr{E}\chi_{\mathscr{E}Pjm} \qquad (10\text{–}238b)$$

Let us consider the first term $\hat{\mathbf{s}} \cdot \hat{\mathbf{p}}$. Using the identity (10–70) we obtain

$$\begin{aligned}
(\hat{\mathbf{s}} \cdot \mathbf{r})(\hat{\mathbf{s}} \cdot \hat{\mathbf{l}}) &= \tfrac{1}{4}\hbar^2 \mathbf{r} \cdot \hat{\mathbf{l}} + i\tfrac{1}{2}\hbar\hat{\mathbf{s}} \cdot (\mathbf{r} \times \hat{\mathbf{l}}) \\
&= 0 + i\tfrac{1}{2}\hbar\hat{\mathbf{s}} \cdot (\mathbf{r} \times (\mathbf{r} \times \hat{\mathbf{p}})) \\
&= i\tfrac{1}{2}\hbar[(\hat{\mathbf{s}} \cdot \mathbf{r})(\mathbf{r} \cdot \hat{\mathbf{p}}) - r^2(\hat{\mathbf{s}} \cdot \hat{\mathbf{p}})] \qquad (10\text{–}239)
\end{aligned}$$

where we used a vector identity given in Appendix I(a). This result can be rewritten as

$$\hat{\mathbf{s}} \cdot \hat{\mathbf{p}} = r^{-2}(\hat{\mathbf{s}} \cdot \mathbf{r}) \left[\mathbf{r} \cdot \hat{\mathbf{p}} + i \frac{2}{\hbar} \hat{\mathbf{s}} \cdot \hat{\mathbf{l}} \right]$$

$$= \hat{s}_r \left[\hat{p}_r + \frac{2i}{\hbar r} (\hat{\mathbf{s}} \cdot \hat{\mathbf{l}} + \tfrac{1}{2}\hbar^2) \right] \tag{10–240}$$

where

$$\hat{s}_r = \frac{\hat{\mathbf{s}} \cdot \mathbf{r}}{r} \tag{10–241}$$

and

$$\hat{p}_r = r^{-1}(\mathbf{r} \cdot \hat{\mathbf{p}} - i\hbar) = \frac{\hbar}{i} \left(\frac{\partial}{\partial r} + \frac{1}{r} \right) = r^{-1}\hat{\mathbf{p}} \cdot \mathbf{r} \tag{10–242}$$

In (10–240) we observe that

$$(\hat{\mathbf{s}} \cdot \hat{\mathbf{l}} + \tfrac{1}{2}\hbar^2)\mathbf{Y}_{ljm} = [\tfrac{1}{2}(\hat{j}^2 - \hat{l}^2 - \hat{s}^2) + \tfrac{1}{2}\hbar^2]\mathbf{Y}_{ljm}$$

$$\begin{cases} = \tfrac{1}{2}\hbar^2(j + \tfrac{1}{2})\mathbf{Y}_{j-(1/2)jm} & (10\text{–}243\text{a}) \\ = -\tfrac{1}{2}\hbar^2(j + \tfrac{1}{2})\mathbf{Y}_{j+(1/2)jm} & (10\text{–}243\text{b}) \end{cases}$$

The sign is different depending on whether $l = j - \tfrac{1}{2}$ or $j + \tfrac{1}{2}$. The remaining question is what $\hat{s}_r$ does to these spherical spinors. For that purpose we take

$$[(\hat{\mathbf{s}} \cdot \hat{\mathbf{l}}), (\hat{\mathbf{s}} \cdot \mathbf{r})] = -\hbar^2(\hat{\mathbf{s}} \cdot \mathbf{r}) + i\hbar[r^2\hat{\mathbf{s}} \cdot \hat{\mathbf{p}} - (\hat{\mathbf{s}} \cdot \mathbf{r})(\mathbf{r} \cdot \hat{\mathbf{p}})]$$

$$\tag{10–244}$$

which can be shown by a direct calculation, and compare this result with (10–239) to obtain

$$(\hat{\mathbf{s}} \cdot \hat{\mathbf{l}} + \tfrac{1}{2}\hbar^2)\hat{s}_r = -\hat{s}_r(\hat{\mathbf{s}} \cdot \hat{\mathbf{l}} + \tfrac{1}{2}\hbar^2) \tag{10–245}$$

Applying this result to one of the spherical spinors, we obtain

$$(\hat{\mathbf{s}} \cdot \hat{\mathbf{l}} + \tfrac{1}{2}\hbar^2)\hat{s}_r\mathbf{Y}_{j-(1/2)jm} = -\hat{s}_r(\hat{\mathbf{s}} \cdot \hat{\mathbf{l}} + \tfrac{1}{2}\hbar^2)\mathbf{Y}_{j-(1/2)jm}$$

$$= -\tfrac{1}{2}\hbar^2(j + \tfrac{1}{2})\hat{s}_r\mathbf{Y}_{j-(1/2)jm} \tag{10–246}$$

by using (10–243a). When this result is compared with (10–243b) we see that

$$\hat{s}_r Y_{j-(1/2)jm} = A Y_{j+(1/2)jm} \qquad (10\text{–}247a)$$

where A is a constant. In the same way we also obtain

$$\hat{s}_r Y_{j+(1/2)jm} = B Y_{j-(1/2)jm} \qquad (10\text{–}247b)$$

where B is another constant. It is easy to show that

$$[\hat{j}_\pm, \hat{s}_r] = [\hat{l}_\pm, \hat{s}_r] + [\hat{s}_\pm, \hat{s}_r] = 0 \qquad (10\text{–}248)$$

so that the constants A and B do not depend on the m quantum number. In order to evaluate A and B, we, therefore, can take the simplest case of $m = \frac{1}{2}$. In addition, we can take a special case of $\theta = 0$, in this evaluation. Since

$$Y_{l0}(0, \varphi) = \sqrt{\frac{2l + 1}{4\pi}} \qquad Y_{l1}(0, \varphi) = 0 \qquad (10\text{–}249)$$

we have

$$Y_{j-\frac{1}{2}j\frac{1}{2}}(0, \varphi) = \begin{pmatrix} \sqrt{(j + \frac{1}{2})/4\pi} \\ 0 \end{pmatrix} \qquad (10\text{–}250a)$$

and

$$Y_{j+\frac{1}{2}j\frac{1}{2}}(0, \varphi) = \begin{pmatrix} -\sqrt{(j + \frac{1}{2})/4\pi} \\ 0 \end{pmatrix} \qquad (10\text{–}250b)$$

In this special case of $\theta = 0$, $\mathbf{r}$ is in the direction of z, so that $\hat{s}_r$ reduces to $\hat{s}_z$. Since the matrix representative of $\hat{s}_z$ is $\frac{1}{2}\hbar\sigma_3$, we see that

$$A = B = -\tfrac{1}{2}\hbar \qquad (10\text{–}251)$$

or

$$\hat{s}_r Y_{j-\frac{1}{2}jm} = -\tfrac{1}{2}\hbar Y_{j+\frac{1}{2}jm} \qquad (10\text{–}252a)$$

and

$$\hat{s}_r Y_{j+\frac{1}{2}jm} = -\tfrac{1}{2}\hbar Y_{j-\frac{1}{2}jm} \qquad (10\text{–}252b)$$

If we take (10–233a) and (10–233b), where $P = (-1)^{j-(1/2)}$, we see, from (10–240), (10–243), and (10–252), that equations (10–238a) and (10–238b) are reduced to

$$\hbar c \left[-\frac{d}{dr} - \frac{1}{r} - \frac{1}{r}(j + \tfrac{1}{2}) \right] G(r) = \left(\mathscr{E} - mc^2 + \frac{Ze^2}{4\pi\varepsilon_0 r} \right) F(r)$$

$$P = (-1)^{j-(1/2)} \qquad (10\text{–}253\text{a})$$

and

$$\hbar c \left[\frac{d}{dr} + \frac{1}{r} - \frac{1}{r}(j + \tfrac{1}{2}) \right] F(r) = \left(\mathscr{E} + mc^2 + \frac{Ze^2}{4\pi\varepsilon_0 r} \right) G(r)$$

$$P = (-1)^{j-(1/2)} \qquad (10\text{–}253\text{b})$$

If we take the other case of (10–234a) and (10–234b), where $P = (-1)^{j+(1/2)}$, we obtain

$$\hbar c \left[-\frac{d}{dr} - \frac{1}{r} + \frac{1}{r}(j + \tfrac{1}{2}) \right] G(r) = \left(\mathscr{E} - mc^2 + \frac{Ze^2}{4\pi\varepsilon_0 r} \right) F(r)$$

$$P = (-1)^{j+(1/2)} \qquad (10\text{–}254\text{a})$$

and

$$\hbar c \left[\frac{d}{dr} + \frac{1}{r} + \frac{1}{r}(j + \tfrac{1}{2}) \right] F(r) = \left(\mathscr{E} + mc^2 + \frac{Ze^2}{4\pi\varepsilon_0 r} \right) G(r)$$

$$P = (-1)^{j+(1/2)} \qquad (10\text{–}254\text{b})$$

The only difference between (10–253) and (10–254) is the sign of the $(j + \tfrac{1}{2})$-term.

Energy Levels

Let us take the case of (10–253) and obtain $\mathscr{E}$, $F(r)$, and $G(r)$. If we take

$$F(r) = \sqrt{mc^2 + \mathscr{E}}\, F'(r) \qquad \text{and} \qquad G(r) = \sqrt{mc^2 - \mathscr{E}}\, G'(r)$$

$$(10\text{–}255)$$

we see that (10–253a) and (10–253b) can be rewritten:

$$\left[-\frac{d}{dr} - \frac{1}{r} - \frac{1}{r}(j + \tfrac{1}{2}) \right] G' = \left[-\frac{\mathscr{E}'}{\hbar c} + \frac{Z\alpha(mc^2 + \mathscr{E})}{\mathscr{E}'r} \right] F' \qquad (10\text{–}256\text{a})$$

$$\left[\frac{d}{dr} + \frac{1}{r} - \frac{1}{r}(j + \tfrac{1}{2}) \right] F' = \left[\frac{\mathscr{E}'}{\hbar c} + \frac{Z\alpha(mc^2 - \mathscr{E})}{\mathscr{E}'r} \right] G' \qquad (10\text{–}256\text{b})$$

where

$$\mathscr{E}' = \sqrt{(mc^2)^2 - \mathscr{E}^2} \qquad (10\text{-}257)$$

and α is the fine structure constant defined by (10–50). Equations (10–256a) and (10–256b) are equivalent to

$$\left(\frac{d}{dr} + \frac{1}{r} - \frac{\mathscr{E}'}{\hbar c} + \frac{Z\alpha\mathscr{E}}{\mathscr{E}'r}\right)(F' + G') = \frac{1}{r}\left(j + \tfrac{1}{2} - \frac{Z\alpha mc^2}{\mathscr{E}'}\right)(F' - G') \qquad (10\text{-}258a)$$

and

$$\left(\frac{d}{dr} + \frac{1}{r} + \frac{\mathscr{E}'}{\hbar c} - \frac{Z\alpha\mathscr{E}}{\mathscr{E}'r}\right)(F' - G') = \frac{1}{r}\left(j + \tfrac{1}{2} + \frac{Z\alpha mc^2}{\mathscr{E}'}\right)(F' + G') \qquad (10\text{-}258b)$$

Eliminating $F' - G'$ from these two equations, we obtain

$$\left[r^2 \frac{d^2}{dr^2} + 3r\frac{d}{dr} + 1 - \gamma^2 + \frac{2Z\alpha\mathscr{E} - \mathscr{E}'}{\hbar c}r - \left(\frac{\mathscr{E}'}{\hbar c}\right)^2 r^2\right](F' + G') = 0 \qquad (10\text{-}259a)$$

while, eliminating $F' + G'$ we obtain

$$\left[r^2 \frac{d^2}{dr^2} + 3r\frac{d}{dr} + 1 - \gamma^2 + \frac{2Z\alpha\mathscr{E} + \mathscr{E}'}{\hbar c}r - \left(\frac{\mathscr{E}'}{\hbar c}\right)^2 r^2\right](F' - G') = 0 \qquad (10\text{-}259b)$$

The only difference between (10–259a) and (10–259b) is the sign of $\mathscr{E}'$. In these equations we introduced γ by

$$\gamma^2 = (j + \tfrac{1}{2})^2 - (Z\alpha)^2 \qquad (10\text{-}260)$$

Since the only difference between (10–253) and (10–254) is the sign of the $(j + \tfrac{1}{2})$ term, and since that sign disappears in (10–259) as (10–260) shows, we see that the following discussions, including the resulting expression of $\mathscr{E}$, F, and G, are nearly the same for both $P = (-1)^{j-(1/2)}$ and $P = (-1)^{j+(1/2)}$ cases. [An exception appears in (10–264).]

If we introduce $\mathscr{A}$ and $\mathscr{B}$ functions by

$$F' + G' = \frac{\mathscr{A}(r)}{\sqrt{r}} \qquad (10\text{-}261a)$$

and

$$F' - G' = \frac{\mathscr{B}(r)}{\sqrt{r}} \tag{10-261b}$$

we obtain, from (10-259a) and (10-259b), that

$$\left[\frac{d^2}{dr^2} + \frac{2}{r}\frac{d}{dr} - \frac{1}{r^2}(\gamma^2 - \tfrac{1}{4}) + \frac{2Z\alpha\mathscr{E} - \mathscr{E}'}{\hbar cr} - \left(\frac{\mathscr{E}'}{\hbar c}\right)^2\right]\mathscr{A}(r) = 0 \tag{10-262a}$$

$$\left[\frac{d^2}{dr^2} + \frac{2}{r}\frac{d}{dr} - \frac{1}{r^2}(\gamma^2 - \tfrac{1}{4}) + \frac{2Z\alpha\mathscr{E} + \mathscr{E}'}{\hbar cr} - \left(\frac{\mathscr{E}'}{\hbar c}\right)^2\right]\mathscr{B}(r) = 0 \tag{10-262b}$$

These equations have mathematically the same form as (9–18) of the Schroedinger theory and (10–49) of the Klein–Gordon theory, only if we make the substitutions shown in Table 10–6, which includes the substitution (10–51), relating the Schroedinger and the Klein–Gordon theories.

From the first line of Table 10–6, we see that l in the Schroedinger equation is replaced by $\gamma - \tfrac{1}{2}$ in (10–262). Therefore, in analogy to (9–20), or from formulas in Appendix I(c), we immediately obtain the solutions of (10–262a) and (10–262b), which are regular at $r = 0$, as

$$\mathscr{A}(r) = \mathscr{N}_{\mathscr{A}}\left(\frac{2r\mathscr{E}'}{\hbar c}\right)^{\gamma - (1/2)} e^{-r\mathscr{E}'/\hbar c} F\left(\gamma - \frac{Z\alpha\mathscr{E}}{\mathscr{E}'} + 1, 2\gamma + 1; \frac{2r\mathscr{E}'}{\hbar c}\right) \tag{10-263a}$$

$$\mathscr{B}(r) = \mathscr{N}_{\mathscr{B}}\left(\frac{2r\mathscr{E}'}{\hbar c}\right)^{\gamma - (1/2)} e^{-r\mathscr{E}'/\hbar c} F\left(\gamma - \frac{Z\alpha\mathscr{E}}{\mathscr{E}'}, 2\gamma + 1; \frac{2r\mathscr{E}'}{\hbar c}\right) \tag{10-263b}$$

TABLE 10–6 *Correspondence among the Schroedinger, Klein–Gordon, and Dirac Radial Functions of Hydrogen-like Ions*

Schroedinger $f_{\mathscr{E}l}$	Klein–Gordon $f_{\mathscr{E}l}$	Dirac $\mathscr{A}$ function	$\mathscr{B}$ function
$l(l + 1)$	$\mathscr{d}(\mathscr{d} + 1)$ $\equiv l(l + 1)$ $- (Z\alpha)^2$	$\gamma^2 - \tfrac{1}{4} \equiv (j + \tfrac{1}{2})^2 - (Z\alpha)^2$	
Z/a	$Z\alpha\mathscr{E}/\hbar c$	$(Z\alpha\mathscr{E} - \tfrac{1}{2}\mathscr{E}')/\hbar c$	$(Z\alpha\mathscr{E} + \tfrac{1}{2}\mathscr{E}')/\hbar c$
$2m\mathscr{E}/\hbar^2$	$[\mathscr{E}^2 - (mc^2)^2]/(\hbar c)^2$ $\equiv -(\mathscr{E}'/\hbar c)^2$	$-(\mathscr{E}'/\hbar c)^2$	

In the above formulas $\mathcal{N}_{\mathscr{A}}$ and $\mathcal{N}_{\mathscr{B}}$ are constants. Their ratio is determined by substituting (10–261a), (10–261b), (10–263a), and (10–263b) into (10–258a). We obtain, after a short calculation, that

$$\frac{\mathcal{N}_{\mathscr{A}}}{\mathcal{N}_{\mathscr{B}}} = \frac{\gamma\mathscr{E}' - Z\alpha\mathscr{E}}{(j + \tfrac{1}{2})\mathscr{E}' + Z\alpha mc^2} \qquad P = (-1)^{j-(1/2)} \qquad (10\text{–}264a)$$

Note that, since $j + \tfrac{1}{2}$ itself appears in this formula, we obtain a different result if we take (10–254), or the case of $P = (-1)^{j+(1/2)}$, instead of (10–253), in which $P = (-1)^{j-(1/2)}$. It is easy to see that when we start with (10–254), we obtain

$$\frac{\mathcal{N}_{\mathscr{A}}}{\mathcal{N}_{\mathscr{B}}} = \frac{\gamma\mathscr{E}' - Z\alpha\mathscr{E}}{-(j + \tfrac{1}{2})\mathscr{E}' + Z\alpha mc^2} \qquad P = (-1)^{j+(1/2)} \qquad (10\text{–}264b)$$

As discussed in section 9–2, $\mathscr{A}$ and $\mathscr{B}$ functions diverge at $r = \infty$ unless the first parameters in the confluent hypergeometric functions are equal to negative integers or zero. The requirement can be satisfied for both functions simultaneously, if

$$\gamma - \frac{Z\alpha\mathscr{E}}{\mathscr{E}'} = -n' \qquad \text{where} \qquad n' = 0, 1, 2, 3, \cdots \qquad (10\text{–}265)$$

From (10–257) we obtain

$$(Z\alpha\mathscr{E})^2 = (\gamma + n')^2[(mc^2)^2 - \mathscr{E}^2] \qquad (10\text{–}266)$$

and, finally,

$$\mathscr{E} = \frac{mc^2}{\sqrt{1 + (Z\alpha/N)^2}} \qquad (10\text{–}267)$$

where

$$N = n' + \gamma = n' + \sqrt{(j + \tfrac{1}{2})^2 - (Z\alpha)^2} \qquad (10\text{–}268)$$

$$= n' + j + \tfrac{1}{2} - \frac{(Z\alpha)^2}{2j + 1} - \cdots \qquad (10\text{–}269)$$

Since (10–267), (10–268), and (10–269) are the same as (10–55), (10–56), and (10–56'), respectively, except that l is now replaced by j, we obtain

$$\mathscr{E}_{nj} = mc^2 - \frac{m(cZ\alpha)^2}{2n^2} - \frac{mc^2(Z\alpha)^4}{2n^4}\left(\frac{n}{j + \tfrac{1}{2}} - \frac{3}{4}\right) - \cdots \qquad (10\text{–}270)$$

where the principal quantum number n is defined by

$$n = n' + j + \tfrac{1}{2} \qquad (10\text{-}271)$$

instead of (10–58).

When a set of njm quantum numbers is given, there still exist two states, because the parity can be 1 or -1, as shown in (10–233) and (10–234). The energy, however, does not depend on the parity, as can be seen from (10–267) or (10–270). Therefore each nj level is expected to be doubly degenerate with respect to the parity, in addition to the $(2j + 1)$-fold degeneracy with respect to the m quantum number. The parity degeneracy, however, does not exist when $n' = 0$, or $n = j + \tfrac{1}{2}$. In this special case $\mathcal{N}_{\mathcal{A}}$ should be zero, since the confluent hypergeometric function in the $\mathcal{A}$ function is diverging. When $P = (-1)^{j+(1/2)}$ and $n' = 0$, (10–264b) shows, in addition, that $\mathcal{N}_{\mathcal{A}}/\mathcal{N}_{\mathcal{B}} = 0/0$, which means that not only $\mathcal{N}_{\mathcal{A}}$ but also $\mathcal{N}_{\mathcal{B}}$ must be zero. Therefore, in the $P = (-1)^{j+(1/2)}$ case there is no solution of the Dirac equation which stays finite at $r = \infty$ when $n' = 0$.

It is conventional to designate each level by l of the Φ component, rather than by the parity [remember that $P = (-1)^l$] in addition to n and j. Some of the energy levels of a hydrogen-like ion, according to the Dirac theory, are schematically shown in Fig. 10–9. Note that the fine structure of each n state looks about the same as that in the Klein–Gordon theory, as shown in Fig. 10–2, but the physical meaning of each fine structure level is rather different, and the magnitude of the splitting is also slightly different.

FIGURE 10–9. The schematic energy level diagram of $n = 2$ and 3 states of a hydrogen-like ion according to the Dirac theory. The parity degeneracy exists except for the $n' = 0$ states. The fine structures are very much exaggerated.

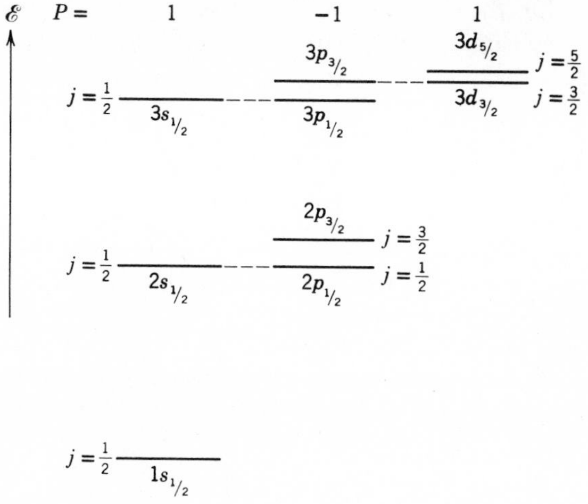

If we introduce N' by

$$N' = N\sqrt{1 + (Z\alpha/N)^2} \tag{10-272}$$

we see, from (10–267), that

$$\frac{\mathscr{E}'}{\hbar c} = \frac{Z}{N'a} \tag{10-273}$$

where a is the Bohr radius defined by (9–19). Using ρ $(= r/a)$, defined by (9–28), instead of r itself, we can express the final form of the $\mathscr{A}$ and $\mathscr{B}$ functions as

$$\mathscr{A}(\rho) = \mathscr{N}n' \left(\frac{2Z\rho}{N'}\right)^{\gamma - (1/2)} e^{-Z\rho/N'} F\left(-n' + 1, 2\gamma + 1; \frac{2Z\rho}{N'}\right) \tag{10-274a}$$

and

$$\mathscr{B}(\rho) = -\mathscr{N}(N' \pm (j + \tfrac{1}{2})) \left(\frac{2Z\rho}{N'}\right)^{\gamma - (1/2)} e^{-Z\rho/N'} F\left(-n', 2\gamma + 1; \frac{2Z\rho}{N'}\right) \tag{10-274b}$$

where the double sign in front of $(j + \tfrac{1}{2})$ means $+$ for the $P = (-1)^{j-(1/2)}$ case while $-$ for the $P = (-1)^{j+(1/2)}$ case. In these formulas $\mathscr{N}$ is a normalization constant. From (10–261a), (10–261b), and (10–255), we finally obtain

$$F(\rho) = \tfrac{1}{2}\mathscr{N}\sqrt{mc^2[1 + (N/N')]} \left(\frac{2Z\rho}{N'}\right)^{\gamma - 1} e^{-Z\rho/N'}$$

$$\times \left\{ (n - j - \tfrac{1}{2}) F\left(-n + j + \tfrac{3}{2}, 2\gamma + 1; \frac{2Z\rho}{N'}\right) \right.$$

$$\left. - [N' \pm (j + \tfrac{1}{2})] F\left(-n + j + \tfrac{1}{2}, 2\gamma + 1; \frac{2Z\rho}{N'}\right) \right\} \tag{10-275a}$$

and

$$G(\rho) = \tfrac{1}{2}\mathcal{N}\sqrt{mc^2[1 - (N/N')]}\left(\frac{2Z\rho}{N'}\right)^{\gamma - 1} e^{-Z\rho/N'}$$

$$\times \left\{ (n - j - \tfrac{1}{2})\, F\!\left(-n + j + \tfrac{3}{2}, 2\gamma + 1; \frac{2Z\rho}{N'}\right) \right.$$

$$\left. + [N' \pm (j + \tfrac{1}{2})]\, F\!\left(-n + j + \tfrac{1}{2}, 2\gamma + 1; \frac{2Z\rho}{N'}\right) \right\}$$

(10–275b)

The normalization constant $\mathcal{N}$ is obtained from

$$\int_0^\infty [|F(r)|^2 + |G(r)|^2]r^2\, dr = 1 \qquad (10\text{–}276)$$

which comes from (10–88). The result is

$$\mathcal{N} = \frac{[\Gamma(2\gamma + n - j + \tfrac{1}{2})]^{1/2}}{\Gamma(2\gamma + 1)\{\Gamma(n - j + \tfrac{1}{2})N'[N' \pm (j + \tfrac{1}{2})]\}^{1/2}}\left(\frac{1}{mc^2}\right)^{1/2}\left(\frac{2Z}{N'a}\right)^{3/2}$$

(10–277)

The Dirac radial functions for the $1s_{1/2}$ state are compared with the Schroedinger radial function for the same state in Fig. 10–10. A large value of Z, 82, is taken to amplify the relativistic effects. When Z is smaller, F approaches f, and G becomes negligible. Table 10–7 gives quantum numbers and the conventional spectroscopic notations for some low energy states.

TABLE 10–7 *Lower Energy States of Hydrogen-like Ions*

Spectroscopic Notation	n	j	γ	n'	N	N'	P
$3d_{5/2}$	3	$\tfrac{5}{2}$	$[9 - (Z\alpha)^2]^{1/2}$	0	$[9 - (Z\alpha)^2]^{1/2}$	3	1
$3d_{3/2}$	3	$\tfrac{3}{2}$	$[4 - (Z\alpha)^2]^{1/2}$	1	$1 + [4 - (Z\alpha)^2]^{1/2}$	$\{5 + 2[4 - (Z\alpha)^2]^{1/2}\}^{1/2}$	1
$3p_{3/2}$	3	$\tfrac{3}{2}$	$[4 - (Z\alpha)^2]^{1/2}$	1	$1 + [4 - (Z\alpha)^2]^{1/2}$	$\{5 + 2[4 - (Z\alpha)^2]^{1/2}\}^{1/2}$	-1
$3p_{1/2}$	3	$\tfrac{1}{2}$	$[1 - (Z\alpha)^2]^{1/2}$	2	$2 + [1 - (Z\alpha)^2]^{1/2}$	$\{5 + 4[1 - (Z\alpha)^2]^{1/2}\}^{1/2}$	-1
$3s_{1/2}$	3	$\tfrac{1}{2}$	$[1 - (Z\alpha)^2]^{1/2}$	2	$2 + [1 - (Z\alpha)^2]^{1/2}$	$\{5 + 4[1 - (Z\alpha)^2]^{1/2}\}^{1/2}$	1
$2p_{3/2}$	2	$\tfrac{3}{2}$	$[4 - (Z\alpha)^2]^{1/2}$	0	$[4 - (Z\alpha)^2]^{1/2}$	2	-1
$2p_{1/2}$	2	$\tfrac{1}{2}$	$[1 - (Z\alpha)^2]^{1/2}$	1	$1 + [1 - (Z\alpha)^2]^{1/2}$	$\{2 + 2[1 - (Z\alpha)^2]^{1/2}\}^{1/2}$	-1
$2s_{1/2}$	2	$\tfrac{1}{2}$	$[1 - (Z\alpha)^2]^{1/2}$	1	$1 + [1 - (Z\alpha)^2]^{1/2}$	$\{2 + 2[1 - (Z\alpha)^2]^{1/2}\}^{1/2}$	1
$1s_{1/2}$	1	$\tfrac{1}{2}$	$[1 - (Z\alpha)^2]^{1/2}$	0	$[1 - (Z\alpha)^2]^{1/2}$	1	1

FIGURE 10–10. The radial functions F and G, multiplied by r, for the $1s_{1/2}$ state of a hydrogen-like ion of $Z = 82$, in comparison to the Schroedinger radial function f, multiplied by r, for the same state and same Z. When Z is smaller G is negligible and F approaches f (Rose).

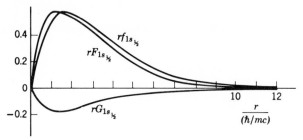

PROBLEM 10–49. Prove (10–244).

PROBLEM 10–50. Prove (10–248).

PROBLEM 10–51. Obtain (10–259a) and (10–259b) from (10–258a) and (10–258b).

PROBLEM 10–52. Obtain (10–263a) and (10–263b) from (10–262a) and (10–262b) in the way suggested in the text.

PROBLEM 10–53. Obtain (10–273).

PROBLEM 10–54. Using (10–270), express the fine structure separation between the $2p_{1/2}$ and $2p_{3/2}$ states by means of fundamental physical constants, and compare the result with the separation between the $2s$ and $2p$ states in the Klein–Gordon theory [formula (10–57)].

PROBLEM 10–55. Show, for the $1s_{1/2}$ $m = \frac{1}{2}$ state, that

$$\Phi = \sqrt{\frac{1 + \gamma_1}{8\pi\Gamma(2\gamma_1 + 1)}} \left(\frac{2Z}{a}\right)^{3/2} (2Z\rho)^{\gamma_1 - 1} e^{-Z\rho} \begin{pmatrix} 1 \\ 0 \end{pmatrix}$$

$$\chi = \sqrt{\frac{1 - \gamma_1}{48\pi\Gamma(2\gamma_1 + 1)}} \left(\frac{2Z}{a}\right)^{3/2} (2Z\rho)^{\gamma_1 - 1} e^{-Z\rho} \begin{pmatrix} \sqrt{2}\cos\theta \\ \sqrt{3}\sin\theta e^{i\varphi} \end{pmatrix}$$

where

$$\gamma_1 = \sqrt{1 - (Z\alpha)^2}$$

HINT. See Appendix I(c) for the spherical harmonics.

PROBLEM 10–56.   Show that when $n = j + \frac{1}{2}$

$$F(\rho) = \mathcal{N}\sqrt{mc^2[1 + (N/N')]}N'\left(\frac{2Z\rho}{N'}\right)^{\gamma - 1} e^{-Z\rho/N'}$$

$$G(\rho) = -\sqrt{[1 - (N/N')]/[1 + (N/N')]}\, F(\rho)$$

and

$$N = \gamma \quad \text{and} \quad N' = j + \tfrac{1}{2}$$

PROBLEM 10–57.   When $Z \ll 137$, show that the ratio $|F|/|G|$ is approximately $2n/(Z\alpha)$.

HINT.   Consider the factor $\sqrt{1 \pm (N/N')}$.

PROBLEM 10–58.   Show from (10–275a) and (10–275b) that in the nonrelativistic limit ($Z\alpha \to 0$), $F(\rho)$ reduces to that part of (9–27), while $G(\rho) \to 0$.

HINT.

$$aF(a + 1, b; z) - aF(a, b; z) = z\frac{a}{b}F(a + 1, b + 1; z)$$

and

$$(1 + a - b)F(a, b; z) - aF(a + 1, b; z) + (b - 1)F(a, b - 1: z) = 0$$

10–13*   ZEEMAN EFFECT OF A HYDROGEN-LIKE ION

If

$$\Psi = \Psi_{\mathscr{E}}(r)e^{-i\mathscr{E}t/\hbar} \tag{10–278}$$

we obtain, from (10–117),

$$\mathscr{E}\psi_{\mathscr{E}} = \hat{H}_D\psi_{\mathscr{E}} \tag{10–226}$$

where

$$\hat{H}_D = c\hat{\boldsymbol{\alpha}} \cdot (\hat{\mathbf{p}} - q\mathbf{A}) + \hat{\beta}mc^2 + q\phi \tag{10–279}$$

For an electron in a hydrogen-like ion under a static magnetic field it reduces to

$$\hat{H}_D = c\hat{\alpha} \cdot \mathbf{p} + \hat{\beta}mc^2 - \frac{Ze^2}{4\pi\varepsilon_0 r} + ce\hat{\alpha} \cdot \mathbf{A}(r) \qquad (10\text{–}280)$$

When the magnetic field is weak, we can treat the additional term $ce\hat{\alpha} \cdot \mathbf{A}(r)$ as a perturbation. From (8–217), the first-order perturbation energy for the (*Pnjm*) state of a hydrogen-like ion can be expressed as

$$\langle Pnjm | ce\hat{\alpha} \cdot \mathbf{A} | Pnjm \rangle = ce \int \Psi_{Pnjm}^\dagger \hat{\lambda}_1 \hat{\mathbf{s}} \cdot \mathbf{A} \Psi_{Pnjm} \, dv$$

$$= ce \left\{ \int \Phi_{Pnjm}^\dagger \hat{\mathbf{s}} \cdot \mathbf{A}\chi_{Pnjm} \, dv + \int \chi_{Pnjm}^\dagger \hat{\mathbf{s}} \cdot \mathbf{A}\Phi_{Pnjm} \, dv \right\}$$

$$(10\text{–}281)$$

When the magnetic field is homogeneous, we can set

$$\mathbf{A} = \tfrac{1}{2}\mathbf{B} \times \mathbf{r} \qquad (10\text{–}60)$$

where $\mathbf{B}$ is the magnetic flux density. In this case

$$\hat{\mathbf{s}} \cdot \mathbf{A} = \tfrac{1}{2}\hat{\mathbf{s}} \cdot (\mathbf{B} \times \mathbf{r}) = \tfrac{1}{2}\mathbf{B} \cdot (\mathbf{r} \times \hat{\mathbf{s}}) \qquad (10\text{–}282)$$

from a vector identity given in Appendix I(a). Since we are free to determine the z axis of the space, it is reasonable to choose the direction of $\mathbf{B}$ as that of the z axis, in which case the perturbation term is essentially $(\mathbf{r} \times \hat{\mathbf{s}})_z$.

It is useful to note that $(\mathbf{r} \times \hat{\mathbf{s}})_z$ commutes with $\hat{j}_z$,

$$[(\mathbf{r} \times \hat{\mathbf{s}})_z, \hat{j}_z] = [(\mathbf{r} \times \hat{\mathbf{s}})_z, \hat{l}_z] + [(\mathbf{r} \times \hat{\mathbf{s}})_z, \hat{s}_z] = 0 \qquad (10\text{–}283)$$

as can be shown directly using the result of Problem 8–85. In general we can show that

$$(\mathbf{r} \times \hat{\mathbf{s}}) \cdot \hat{\mathbf{j}} = \hat{\mathbf{j}} \cdot (\mathbf{r} \times \hat{\mathbf{s}}) \qquad (10\text{–}284)$$

We will see in the next chapter (Problem 11–51) that $\mathbf{r} \times \hat{\mathbf{s}}$ is a vector operator and that as far as the diagonal matrix elements are concerned we can replace $(\mathbf{r} \times \hat{\mathbf{s}})_z$ by

$$(\mathbf{r} \times \hat{\mathbf{s}})_z \rightarrow \hat{j}_z (\mathbf{r} \times \hat{\mathbf{s}}) \cdot \frac{\hat{\mathbf{j}}}{\hat{j}^2} \qquad (10\text{–}285)$$

i.e.,

$$\langle jm|\hat{j}^2(\mathbf{r} \times \hat{\mathbf{s}})_z|jm\rangle \ (= j(j+1)\hbar^2\langle jm|(\mathbf{r} \times \hat{\mathbf{s}})_z|jm\rangle)$$
$$= \langle jm|\hat{j}_z(\mathbf{r} \times \hat{\mathbf{s}}) \cdot \hat{\mathbf{j}}|jm\rangle \ (= m\hbar\langle jm|(\mathbf{r} \times \hat{\mathbf{s}}) \cdot \hat{\mathbf{j}}|jm\rangle) \qquad (10\text{--}286)$$

We can understand (10–286) intuitively in the following way: If a vector operator $\hat{\mathbf{f}}$ is associated with the electron motion, such as $\mathbf{r}$, $\hat{\mathbf{s}}$, or $\mathbf{r} \times \hat{\mathbf{s}}$, the average value, or the expectation value, of $\hat{\mathbf{f}}$ during the electron motion is the same as that of $\hat{\mathbf{j}} \cdot \hat{\mathbf{f}}/|\hat{\mathbf{j}}|$, since $\hat{\mathbf{j}}$ is a conserved vector of the electron motion, and only the component of $\hat{\mathbf{f}}$ along the direction of $\hat{\mathbf{j}}$ contributes to the average value of $\hat{\mathbf{f}}$ (Fig. 10–11). The z component of this effective part of $\hat{\mathbf{f}}$ is, therefore

$$\frac{\hat{j}_z}{|\hat{\mathbf{j}}|} \frac{\hat{\mathbf{j}} \cdot \hat{\mathbf{f}}}{|\hat{\mathbf{j}}|} = \frac{\hat{j}_z \hat{\mathbf{j}} \cdot \hat{\mathbf{f}}}{\hat{j}^2} \qquad (10\text{--}287)$$

If we accept formula (10–286) the only thing we need to know is $(\mathbf{r} \times \hat{\mathbf{s}}) \cdot \hat{\mathbf{j}}$. Now

$$(\mathbf{r} \times \hat{\mathbf{s}}) \cdot \hat{\mathbf{j}} = (\mathbf{r} \times \hat{\mathbf{s}}) \cdot \hat{\mathbf{l}} + (\mathbf{r} \times \hat{\mathbf{s}}) \cdot \hat{\mathbf{s}} = -\hat{\mathbf{s}} \cdot (\mathbf{r} \times \hat{\mathbf{l}}) + (\hat{\mathbf{s}} \times \hat{\mathbf{s}}) \cdot \mathbf{r}$$
$$(10\text{--}288)$$

FIGURE 10–11. In calculating an expectation value of a vector electron operator $\hat{\mathbf{f}}$ (or its z component), you take the effective component $\hat{\mathbf{j}} \cdot \hat{\mathbf{f}}/|\hat{\mathbf{j}}|$ (or its z component $\hat{j}_z\hat{\mathbf{j}} \cdot \hat{\mathbf{f}}/\hat{j}^2$).

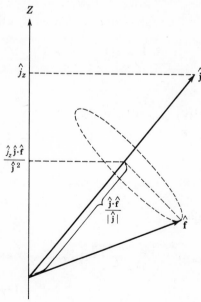

by using a vector identity in Appendix I(a). Notice, however, that unlike an ordinary vector $\hat{\mathbf{s}} \times \hat{\mathbf{s}}$ is not zero, but

$$\hat{\mathbf{s}} \times \hat{\mathbf{s}} = i\hbar\hat{\mathbf{s}} \tag{10-289}$$

from the commutation relations (10–67). On the other hand, we can use (10–239) for the first term on the right-hand side of (10–288): Therefore

$$(\mathbf{r} \times \hat{\mathbf{s}}) \cdot \hat{\mathbf{j}} = \frac{2i}{\hbar}(\hat{\mathbf{s}} \cdot \mathbf{r})(\hat{\mathbf{s}} \cdot \hat{\mathbf{l}}) + i\hbar\hat{\mathbf{s}} \cdot \mathbf{r} = \frac{2i}{\hbar}(\hat{\mathbf{s}} \cdot \mathbf{r})[(\hat{\mathbf{s}} \cdot \hat{\mathbf{l}}) + \hbar^2] \tag{10-290}$$

Since, according to (10–252a) and (10–252b), the operator $\hat{\mathbf{s}} \cdot \mathbf{r}$ converts one spherical spinor to the other, we obtain, if $P = (-1)^{j+(1/2)}$,

$$\int \Phi^{\dagger}_{Pnjm}(\mathbf{r} \times \hat{\mathbf{s}})_z \chi_{Pnjm} \, dv$$

$$= \int F(r)Y^{\dagger}_{j+(1/2)jm} \left[\hat{j}_z(\mathbf{r} \times \hat{\mathbf{s}}) \cdot \frac{\hat{\mathbf{j}}}{\hat{j}^2} \right] iG(r)Y_{j-(1/2)jm} \, dv$$

$$= \frac{m}{j(j+1)\hbar} \int FrGr^2 \, dr \int Y^{\dagger}_{j-(1/2)jm}(\hat{\mathbf{s}} \cdot \hat{\mathbf{l}} + \hbar^2)Y_{j-(1/2)jm} \sin\theta \, d\theta \, d\varphi$$

$$\tag{10-291}$$

The last integral over the angles can be evaluated by using the identity

$$\hat{\mathbf{s}} \cdot \hat{\mathbf{l}} = \tfrac{1}{2}(\hat{j}^2 - \hat{l}^2 - \hat{s}^2) \tag{10-292}$$

as

$$\int Y^{\dagger}_{j-\frac{1}{2}jm}(\hat{\mathbf{s}} \cdot \hat{\mathbf{l}} + \hbar^2)Y_{j-\frac{1}{2}jm} \sin\theta \, d\theta \, d\varphi$$

$$= \langle j - \tfrac{1}{2} j \, m | \hat{\mathbf{s}} \cdot \hat{\mathbf{l}} | j - \tfrac{1}{2} j \, m \rangle + \hbar^2$$

$$= \tfrac{1}{2}[j(j+1) - (j - \tfrac{1}{2})(j + \tfrac{1}{2}) - \tfrac{3}{4}]\hbar^2 + \hbar^2$$

$$= (\tfrac{1}{2}j + \tfrac{3}{4})\hbar^2 \tag{10-293}$$

and we finally obtain

$$\int \Phi^{\dagger}_{Pnjm}(\mathbf{r} \times \hat{\mathbf{s}})_z \chi_{Pnjm} \, dv = \frac{(2j+3)m\hbar}{4j(j+1)} \int F_{Pnj}G_{Pnj}r^3 \, dr$$

$$P = (-1)^{j+(1/2)} \tag{10-294}$$

A similar calculation shows that

$$\int \chi^\dagger_{Pnjm} (\mathbf{r} \times \hat{\mathbf{s}})_z \Phi_{Pnjm} \, dv = \frac{(2j - 1)mh}{4j(j + 1)} \int F_{Pnj} G_{Pnj} r^3 \, dr$$

$$P = (-1)^{j+(1/2)} \qquad (10\text{-}295)$$

for the same parity state. Therefore, from (10–281), (10–282), (10–294), and (10–295), we obtain

$$\langle Pnjm|ce\hat{\boldsymbol{\alpha}} \cdot \mathbf{A}|Pnjm\rangle = \frac{(2j + 1)m}{2j(j + 1)} \int F_{Pnj} G_{Pnj} r^3 \, dr \hbar ceB$$

$$P = (-1)^{j+(1/2)} \qquad (10\text{-}296)$$

Although (10–294) and (10–295) are calculated for one parity state, the final result (10–296) remains nearly the same in the other parity state $(P = (-1)^{j-(1/2)})$, except that $-m$ appears instead of m.

The radial integral in (10–296) can be calculated using (10–275a), (10–275b), (10–277), and a formula in Appendix I(c). The final result is

$$\int_0^\infty F_{Pnj} G_{Pnj} r^3 \, dr = \frac{1}{4m_e c} \frac{1}{N'}$$

$$\times \left\{ \frac{(2\gamma + n - j - \frac{1}{2})(\gamma + n - j - 1)(n - j - \frac{1}{2})}{N' \pm (j + \frac{1}{2})} \right.$$

$$\left. - [N' \pm (j + \frac{1}{2})](\gamma + n - j) \right\} \qquad (10\text{-}297)$$

where the double sign again means $+$ for the $P = (-1)^{j-(1/2)}$ case, and $-$ for the $P = (-1)^{j+(1/2)}$ case. Putting (10–297) into (10–296) and the corresponding result given in Problem 10–61, we obtain the first-order perturbation energy:[8]

$$\langle Pnjm|ce\hat{\boldsymbol{\alpha}} \cdot \mathbf{A}|Pnjm\rangle$$

$$= B\mu_B m \frac{\mp(2j + 1)}{4j(j + 1)N'}$$

$$\times \left\{ \frac{(2\gamma + n - j - \frac{1}{2})(\gamma + n - j - 1)(n - j - \frac{1}{2})}{N' \pm (j + \frac{1}{2})} \right.$$

$$\left. - [N' \pm (j + \frac{1}{2})](\gamma + n - j) \right\} \qquad P = (-1)^{j\mp(1/2)}$$

$$(10\text{-}298)$$

[8] The relativistic formula for the $1s_{1/2}$ state is given by G. Breit, *Nature* **122**, 649 (1928), and those for the $2p_{1/2}$ and $2p_{3/2}$ states are given on page 185 of M. E. Rose, *Relativistic Electron Theory*, Wiley, New York, 1961. The radiative correction gives an extra factor of 1.001159639 for all states.

where μ_B is the Bohr magneton introduced in (10–64). The g factor is defined by expressing the perturbation energy as

$$B\mu_B mg \qquad (10\text{–}299)$$

In the nonrelativistic limit, where $Z\alpha$ is neglected compared to one, we obtain

$$g = \begin{cases} \dfrac{(j + \frac{1}{2})}{j} & \text{for} \quad P = (-1)^{j-(1/2)} \\[2ex] \dfrac{(j + \frac{1}{2})}{(j + 1)} & \text{for} \quad P = (-1)^{j+(1/2)} \end{cases} \quad \text{nonrelativistic}$$

$$(10\text{–}300)$$

PROBLEM 10–59. Prove (10–289), and generalize that to

$$\hat{\mathbf{j}} \times \hat{\mathbf{j}} = i\hbar\hat{\mathbf{j}}$$

PROBLEM 10–60. Obtain (10–295). Note that $\chi^*_{Pnjm} = -iG\mathbf{Y}^*_{Pjm}$.

PROBLEM 10–61. Show that

$$\langle Pnjm|ce\hat{\boldsymbol{\alpha}} \cdot \mathbf{A}|Pnjm \rangle = -\frac{(2j + 1)m}{2j(j + 1)} \int F_{Pnj}G_{Pnj}r^3\, dr$$

when $P = (-1)^{j-(1/2)}$.

PROBLEM 10–62. Obtain the nonrelativistic limit (10–300) from (10–298).

PROBLEM 10–63. When $n = j + \frac{1}{2}$, show that the relativistic g factor is

$$g\,(\text{relativistic}) = \frac{(\gamma + \frac{1}{2})(j + \frac{1}{2})}{j(j + 1)} \qquad \gamma = \sqrt{n^2 - (Z\alpha)^2}$$

For the $1s_{1/2}$ state, it reduces to Breit's formula

$$\tfrac{2}{3}(1 + 2\sqrt{1 - (Z\alpha)^2})$$

10–14* MAGNETIC DIPOLE HYPERFINE STRUCTURE

The proton and neutron are both spin one-half particles like leptons. We may, therefore, assume that they obey Dirac's equation like leptons. Actually, in supporting the applicability of the Dirac theory, there exist as antiparticles the antiproton and antineutron. However, a large deviation from the Dirac theory exists in the nucleons (proton and neutron), as well as in all baryons, with respect to their magnetic moment. As discussed in section 10–4, and to be found in the next section, the Dirac theory predicts that the magnetic moment associated with a spin $\hat{s}$ is $(q/m)\hat{s}$, where q and m are the charge and mass of the particle. Since the charges of the proton and the neutron are $+e$ and zero, respectively, we expect no magnetic moment for the neutron and $(e/M)\hat{I}$ for the proton, where M is the proton mass and $\hat{I}$ is the proton spin. However, it is observed that their magnetic moments are

$$g \frac{e}{2M} \hat{I} \tag{10–301}$$

where

$$g = \begin{cases} 5.58556 & \text{for the proton} \\ -3.8263 & \text{for the neutron} \end{cases} \tag{10–302}$$

Therefore, the Dirac theory is not as applicable for nucleons as for leptons. The quantity

$$\mu_N = \frac{e\hbar}{2M} = 5.050942 \times 10^{-27} \text{ A m}^2 \tag{10–303}$$

is, nevertheless, a convenient unit for nuclear magnetic moments, and is called the *nuclear magneton*.

The vector potential due to a point magnetic moment (10–301) located at the origin is

$$\hat{\mathbf{A}}(\mathbf{r}) = \frac{\mu_0}{4\pi} \frac{g\mu_N}{\hbar} \frac{\hat{I} \times \mathbf{r}}{r^3} \tag{10–304}$$

which can be confirmed by comparing with the result of Problem 5–34. The perturbation term in the Dirac Hamiltonian due to the nuclear magnetic moment is

$$ec\hat{\alpha} \cdot \hat{\mathbf{A}} = \frac{ec\mu_0 g\mu_N}{4\pi\hbar} \hat{\lambda}_1 \hat{s} \cdot (\hat{I} \times \mathbf{r})r^{-3} = \frac{ec\mu_0 g\mu_N}{4\pi\hbar} \hat{\lambda}_1 \hat{I} \cdot (\mathbf{r} \times \hat{s})r^{-3} \tag{10–305}$$

Since we are interested in the expectation values, or the diagonal elements, of this perturbation term, we can replace it by its effective part

$$ec\hat{\alpha} \cdot \hat{A} \rightarrow \frac{ec\mu_0 g\mu_N}{4\pi\hbar} \hat{\lambda}_1(\hat{I} \cdot \hat{j}) \frac{[\hat{j} \cdot (\mathbf{r} \times \hat{s})r^{-3}]}{\hat{j}^2} \qquad (10\text{-}306)$$

in the same way as (10–285). Fig. 10–12 illustrates the present situation.

FIGURE 10–12. An expectation value of $\hat{I} \cdot \hat{f}$, where $\hat{I}$ is the nuclear spin and $\hat{f}$ is an electron operator, is obtained by calculating that of the effective component $(\hat{I} \cdot \hat{j})(\hat{j} \cdot \hat{f})/\hat{j}^2$.

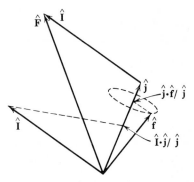

Formula (10–306) shows a coupling between the electron angular momentum $\hat{j}$ and the nuclear angular momentum $\hat{I}$. Because of this coupling neither $\hat{j}$ nor $\hat{I}$ is conserved, and we have to introduce the total angular momentum $\hat{F}$ of the atom:

$$\hat{F} = \hat{j} + \hat{I} = \hat{I} + \hat{s} + \hat{I} \qquad (10\text{-}307)$$

We can follow the same argument we discussed in section 10–10 to define an eigenfunction of $\hat{F}^2$ and $\hat{F}_z$ as

$$\hat{F}^2|Fm_F\rangle = F(F + 1)\hbar^2|Fm_F\rangle \qquad (10\text{-}308a)$$

$$\hat{F}_z|Fm_F\rangle = m_F\hbar|Fm_F\rangle \qquad (10\text{-}308b)$$

and to expand as

$$|Fm_F\rangle = \sum_{mj,mI} |jm_jIm_I\rangle\langle jm_jIm_I \mid Fm_F\rangle \qquad (10\text{-}309)$$

where $|Im_I\rangle$ is an eigenket of $\hat{I}^2$ and $\hat{I}_z$, while $\langle jm_jIm_I|Fm_F\rangle$ are the Clebsch–Gordan coefficients.

Since $\hat{\mathbf{I}} \cdot \hat{\mathbf{J}}$ commutes with $\hat{\mathbf{F}}$, as can be easily shown, the matrix representative of $\hat{\mathbf{I}} \cdot \hat{\mathbf{J}}$ is diagonal in the Fm_F representation. Therefore,

$$\langle Fm_F | ce\hat{\boldsymbol{\alpha}} \cdot \hat{\mathbf{A}} | Fm_F \rangle = \frac{ec\mu_0 g\mu_N}{4\pi\hbar} \langle Fm_F | \hat{\mathbf{I}} \cdot \hat{\mathbf{J}} | Fm_F \rangle$$

$$\times \langle Fm_F | \hat{\lambda}_1 \hat{\mathbf{J}} \cdot (\mathbf{r} \times \hat{\mathbf{s}}) r^- \hat{J}^{3-2} | Fm_F \rangle \qquad (10\text{--}310)$$

From the identity

$$\hat{\mathbf{I}} \cdot \hat{\mathbf{J}} = \tfrac{1}{2}(\hat{F}^2 - \hat{I}^2 - \hat{J}^2) \qquad (10\text{--}311)$$

we obtain

$$\langle Fm_F | \hat{\mathbf{I}} \cdot \hat{\mathbf{J}} | Fm_F \rangle = \tfrac{1}{2}[F(F+1) - I(I+1) - j(j+1)]\hbar^2$$

$$(10\text{--}312)$$

immediately, by using (10–309). As we saw in (10–284), $\hat{\lambda}_1 \hat{\mathbf{J}} \cdot (\mathbf{r} \times \hat{\mathbf{s}}) r^{-3} \hat{J}^{-2}$ commutes with $\hat{\mathbf{J}}$ and certainly with $\hat{\mathbf{I}}$. What this means is that this operator is diagonal in the $jm_j I m_I$ representation, and the diagonal matrix elements are independent of m_j and m_I, as we have already seen in the last section. Using the transformation (10–309), which is obviously a unitary transformation, and the previous result (10–296), we obtain

$$\langle Fm_F | \hat{\lambda}_1 \hat{\mathbf{J}} \cdot (\mathbf{r} \times \hat{\mathbf{s}}) r^{-3} \hat{J}^{-2} | Fm_F \rangle = \mp \frac{2j+1}{2j(j+1)} \int_0^\infty F_{Pnj} G_{Pnj} \, dr$$

$$P = (-1)^{j \mp (1/2)} \qquad (10\text{--}313)$$

The radial integral in (10–313) can be calculated using (10–275a), (10–275b), (10–277), and a formula in Appendix I(c) as

$$\int_0^\infty F_{Pnj} G_{Pnj} \, dr = \frac{Z^3 \alpha}{a^2} \frac{1}{N'^4 \gamma (2\gamma - 1)(2\gamma + 1)}$$

$$\times \left\{ \frac{(2\gamma + n - j - \tfrac{1}{2})(\gamma + n - j - 1)(n - j - \tfrac{1}{2})}{N' \pm (j + \tfrac{1}{2})} \right.$$

$$\left. - (\gamma + n - j)[N' \pm (j + \tfrac{1}{2})] \right\} \qquad (10\text{--}314)$$

where a is the Bohr radius.

Putting (10–312), (10–313), and (10–314) into (10–310), we obtain the perturbation energy:

$$
\langle Fm_F|ec\hat{\alpha}\cdot\hat{A}|Fm_F\rangle
$$

$$
= \frac{\mu_0}{4\pi}\, g\mu_N\mu_B \left(\frac{Z}{a}\right)^3 \frac{[F(F+1) - I(I+1) - j(j+1)]}{2N'^4\gamma(2\gamma-1)(2\gamma+1)} \frac{[\mp(j+\frac{1}{2})]}{j(j+1)}
$$

$$
\times \left\{ \frac{(2\gamma + n - j - \frac{1}{2})(\gamma + n - j - 1)(n - j - \frac{1}{2})}{N' \pm (j + \frac{1}{2})} \right.
$$

$$
\left. - (\gamma + n - j)[N' \pm (j + \frac{1}{2})] \right\} \qquad P = (-1)^{j\mp(1/2)}
$$

$$
\text{(10–315)}
$$

The resulting splitting of a given j level is called the *hyperfine structure*. In the nonrelativistic limit (10–315) reduces to

$$
\frac{\mu_0}{4\pi}\, g\mu_N\mu_B \left(\frac{Z}{a}\right)^3 \frac{F(F+1) - I(I+1) - j(j+1)}{n^3[j(j+1)]^2}
$$

$$
\times \begin{cases} (j+1) & \text{if} \quad P = (-1)^{j-(1/2)} \\ j & \text{if} \quad P = (-1)^{j+(1/2)} \end{cases} \qquad \text{(10–316)}
$$

Since the proton has $I = \frac{1}{2}$, every energy level (nj level) of the hydrogen atom is split into a doublet because of the hyperfine interaction. Some of the doublet splittings are measured, but particularly the one in the $1s_{1/2}$ state has been extremely accurately measured[9] as 1420405751.800 ± 0.028 Hertz. This is one of the most accurate measurements in physics. While both (10–315) and (10–316) gives a fairly good theoretical number of about 1420 MHz, obviously a much more elaborate theory, including that of the proton structure, is necessary to compare with such accurate datum. The relativistic correction for the $1s_{1/2}$ state of a hydrogen-like ion, as obtained from (10–315), is given as the factor

$$
\frac{1}{[2 - 2(Z\alpha)^2 - \sqrt{1 - (Z\alpha)^2}]} \cong 1 + \tfrac{3}{2}(Z\alpha)^2 \qquad \text{(10–317)}
$$

The nonrelativistic formula (10–316) for that state is to be multiplied by this number. This extra factor is about 1.00008 when $Z = 1$, and much smaller

[9] S. B. Crampton, D. Kleppner, and N. F. Ramsey, *Phys. Rev. Letters* **11**, 338 (1963).

than the radiative corrections, which gives the factor 1.00116. The relativistic correction, however, increases as Z increases.[10]

The hyperfine structure is more important for $s_{1/2}$ states than other states with same n. When one tries to find the hyperfine structure in the non-relativistic theory, one finds the same result as (10–316) for states other than s states, but finds an ambiguity for the s states (Problem 10–69). Fermi and Segre[10] have proposed an intuitive method to deal with this ambiguity, but the relativistic theory requires no intuitive argument, and is straightforward for all states including $s_{1/2}$ states.

PROBLEM 10–64. Show that (10–304) gives B of Problem 5–34.

PROBLEM 10–65. Obtain (10–313) in the way suggested in the text.

PROBLEM 10–66. Obtain (10–314).

PROBLEM 10–67. Show that (10–316) is obtained as the nonrelativistic limit of (10–315).

PROBLEM 10–68. When $n = j + \frac{1}{2}$, or $n' = 0$,
(a) show that

$$\langle Fm_F|ce\hat{\alpha} \cdot \hat{A}|Fm_F\rangle = \frac{\mu_0}{4\pi} g\mu_N\mu_B \frac{Z^3}{a}$$

$$\times \frac{F(F+1) - I(I+1) - j(j+1)}{n^2(n^2 - \frac{1}{4})\sqrt{n^2 - (Z\alpha)^2}[\sqrt{n^2 - (Z\alpha)^2} - \frac{1}{2}]}$$

(b) Obtain the relativistic correction factor (10–317) as a special case of the above result.

PROBLEM 10–69. Using the Schroedinger hydrogenic radial function $f_{nl}(r)$, as given in (9–27), we can assume a nonrelativistic wave function with spin as $f_{nl}Y_{ljm}$, which reduces to $\begin{pmatrix} f_{n0} \\ 0 \end{pmatrix}$ and $\begin{pmatrix} 0 \\ f_{n0} \end{pmatrix}$ for the ns states. Taking the magnetic dipole–dipole interaction energy

$$\hat{H}_1 = \frac{\mu_0}{4\pi} 2g\mu_N\mu_B \left[\frac{\hat{\mathbf{I}} \cdot \hat{\mathbf{s}}}{r^3} - \frac{3(\hat{\mathbf{I}} \cdot \mathbf{r})(\hat{\mathbf{s}} \cdot \mathbf{r})}{r^5} \right]$$

[10] For a review of the theories of the hyperfine structure, including that of Fermi and Segre, see Chapter 9 of M. Mizushima, *Quantum Mechanics of Atomic Spectra and Atomic Structure*, W. A. Benjamin, New York, 1970.

as the perturbation, we can calculate the hyperfine structure in the non-relativistic theory. Show that the integral $\int f_{n0} \hat{H}_1 f_{n0}\, dv$, which appears in that theory, is indefinite in the sense that it diverges when the radial integral is done first, while it is zero when the angular integrals are done first. (The integral dv is over the electron coordinates.)

10–15 RELATIVISTIC CORRECTIONS

We have seen in several places, (see, for example, Problems 10–23 and 10–57), that as long as $\hbar\omega$ is positive and close to mc^2, the amplitude of the spinor Φ is much larger than that of the spinor χ. Actually, for a lepton in electromagnetic fields, (10–279) and (10–226) give

$$\frac{2}{\hbar} c\hat{\mathbf{s}} \cdot (\hat{\mathbf{p}} - q\mathbf{A})\chi + (mc^2 - \mathscr{E} + q\phi)\Phi = 0 \qquad (10\text{–}318a)$$

and

$$\frac{2}{\hbar} c\hat{\mathbf{s}} \cdot (\hat{\mathbf{p}} - q\mathbf{A})\Phi + (-mc^2 - \mathscr{E} + q\phi)\chi = 0 \qquad (10\text{–}318b)$$

which show in general that if $\mathscr{E}$ is about mc^2 and if $q\phi$ and $q\mathbf{A}$ are small enough, then

$$\frac{\Phi}{\chi} \cong \frac{mc}{|\hat{\mathbf{p}}|} \cong \frac{c}{V} \qquad (10\text{–}319)$$

where V is the classical particle velocity. Therefore, in the nonrelativistic limit, Φ is the main component, while χ contributes as relativistic corrections.

If we rewrite (10–318b) as

$$\chi = (mc^2 + \mathscr{E} - q\phi)^{-1} \frac{2}{\hbar} c\hat{\mathbf{s}} \cdot (\hat{\mathbf{p}} - q\mathbf{A})\Phi \qquad (10\text{–}320)$$

and put this expression into (10–318a), we obtain

$$\left[\frac{2}{\hbar} c\hat{\mathbf{s}} \cdot (\hat{\mathbf{p}} - q\mathbf{A}) \frac{1}{mc^2 + \mathscr{E} - q\phi} \frac{2}{\hbar} c\hat{\mathbf{s}} \cdot (\hat{\mathbf{p}} - q\mathbf{A}) \right.$$
$$\left. + mc^2 - \mathscr{E} + q\phi \right] \Phi = 0 \qquad (10\text{–}321)$$

This is an equation for the main component, Φ, and therefore, is expected to give a proper expression for the nonrelativistic limit.

Since in the nonrelativistic limit $\mathscr{E}$ is about mc^2 and $q\phi$ is much smaller than mc^2, we can expand

$$\frac{1}{mc^2 + \mathscr{E} - q} = \frac{1}{2mc^2} - \frac{\mathscr{E} - mc^2 - q\phi}{(2mc^2)^2} - \cdots \qquad (10\text{--}322)$$

If we take the first term of this expansion in (10–321), we obtain

$$\left[\frac{2}{\hbar} c\hat{\mathbf{s}} \cdot (\hat{\mathbf{p}} - q\mathbf{A}) \frac{1}{2mc^2} \frac{2}{\hbar} c\hat{\mathbf{s}} \cdot (\hat{\mathbf{p}} - q\mathbf{A}) + mc^2 - \mathscr{E} + q\phi \right] \Phi$$

$$= \left\{ \left(\frac{2}{\hbar}\right)^2 [\hat{\mathbf{s}} \cdot (\hat{\mathbf{p}} - q\mathbf{A})]^2 (2m)^{-1} + mc^2 - \mathscr{E} + q\phi \right\} \Phi$$

$$= \left\{ \frac{\hat{\mathbf{p}}^2}{2m} - \frac{q}{2m} \hat{\mathbf{l}} \cdot \mathbf{B} + \frac{q^2}{8m} (r^2 B^2 - (\mathbf{r} \cdot \mathbf{B})^2) \right.$$

$$\left. - \frac{q}{m} \hat{\mathbf{s}} \cdot \mathbf{B} + q\phi + mc^2 - \mathscr{E} \right\} \Phi = 0 \qquad (10\text{--}323)$$

using the previous result (10–71). The last expression shows that the spin magnetic moment $(q/m)\hat{\mathbf{s}}$ appears in the zeroth-order nonrelativistic limit. The other terms are the ordinary nonrelativistic Hamiltonian terms.

In the second term of (10–322), $(\mathscr{E} - mc^2)/(2mc^2)^2$ is a constant which obviously commutes with $\hat{\mathbf{p}}$. Therefore this term does not change the result (10–323), except that every $1/m$ factor is replaced by

$$\frac{1}{m} \to \frac{1}{m} \left(1 - \frac{\mathscr{E} - mc^2}{2mc^2} \right) \cong \frac{1}{m} \sqrt{\frac{mc^2}{\mathscr{E}}} \qquad (10\text{--}324)$$

which exhibits the relativistic mass defect.

The remaining part of the second term of (10–322), $q\phi/(2mc^2)^2$, being a function of $\mathbf{r}$, does not commute with $\hat{\mathbf{p}}$. Neglecting q^2 terms, this part of the second term contributes

$$\left(\frac{2}{\hbar}\right)^2 c^2 q (2mc^2)^{-2} \hat{\mathbf{s}} \cdot \hat{\mathbf{p}} \phi(\mathbf{r}) \hat{\mathbf{s}} \cdot \hat{\mathbf{p}} = q(2mc)^{-2} \left\{ (\hat{\mathbf{p}}\phi) \cdot \hat{\mathbf{p}} + \frac{2i}{\hbar} \hat{\mathbf{s}} \cdot [(\hat{\mathbf{p}}\phi) \times \hat{\mathbf{p}}] \right\}$$

$$(10\text{--}325)$$

to the nonrelativistic Hamiltonian. We used (10–70) in obtaining the last expression. Now

$$(\hat{\mathbf{p}}\phi) \cdot \hat{\mathbf{p}} = \phi\hat{\mathbf{p}}^2 - i\hbar(\nabla\phi) \cdot \hat{\mathbf{p}} = \phi\hat{\mathbf{p}}^2 + i\hbar\mathbf{E} \cdot \hat{\mathbf{p}} \qquad (10\text{–}326)$$

where $\mathbf{E}$ is the electrostatic field strength. The first term simply gives an extra mass defect correction to the nonrelativistic kinetic energy term, while the second term is a new relativistic effect unique to the Dirac theory. The second term in (10–325), on the other hand, is

$$\frac{2i}{\hbar} \hat{\mathbf{s}} \cdot [(\hat{\mathbf{p}}\phi) \times \hat{\mathbf{p}}] = -2\hat{\mathbf{s}} \cdot (\mathbf{E} \times \hat{\mathbf{p}}) \qquad (10\text{–}327)$$

which is also unique to the Dirac theory.

Accumulating all terms obtained so far, we see that the equation for Φ is approximately

$$\left[\hat{H}_{\text{n.r.}} - \frac{q}{m} \hat{\mathbf{s}} \cdot \mathbf{B} + \frac{iq\hbar}{(2mc)^2} \mathbf{E} \cdot \hat{\mathbf{p}} - \frac{2q}{(2mc)^2} \hat{\mathbf{s}} \cdot (\mathbf{E} \times \hat{\mathbf{p}}) \right] \Phi = \mathscr{E}\Phi$$

$$(10\text{–}328)$$

where

$$\hat{H}_{\text{n.r.}} = \frac{\hat{\mathbf{p}}^2}{2m} + mc^2 + q\phi - \frac{q}{2m} \hat{\mathbf{l}} \cdot \mathbf{B} + \frac{q^2}{8m} (r^2 B^2 - (\mathbf{r} \cdot \mathbf{B})^2) \qquad (10\text{–}329)$$

is the nonrelativistic Hamiltonian. The mass defect effects, (10–324) and the $\phi\hat{\mathbf{p}}^2$ term of (10–326), are neglected.

Among the three relativistic correction terms in (10–328), the first term, which gives the spin magnetic moment, is well known. The second term is called the *Darwin term*, and the last term is called the *spin–orbit interaction term*.

These relativistic terms can be regarded as perturbations to a nonrelativistic problem. If we consider an electron in a free atom we see that

$$\hat{H}_{\text{n.r.}} = mc^2 + \frac{\hat{\mathbf{p}}^2}{2m} + q\phi \qquad (10\text{–}330)$$

If the potential is radial, or $\phi = \phi(r)$, as in many atoms, we obtain Φ in the form of

$$\Phi_{\text{n.r.}}(\mathbf{r}) = f_{\mathscr{E}l}(r)\mathbf{Y}_{ljm}(\theta, \varphi) \qquad (10\text{–}331)$$

where $f_{\mathcal{E}l}(r)$ is the Schroedinger radial function and Y_{ljm} is the spherical spinor. The expectation value of the Darwin term is

$$-e\hbar^2(2mc)^{-2} \int f_{\mathcal{E}l} Y_{ljm}^* \mathbf{E} \cdot \nabla f_{\mathcal{E}l} Y_{ljm} \, dv$$

$$= -\tfrac{1}{2} e\hbar^2(2mc)^{-2} \int \mathbf{E} \cdot \nabla |f_{\mathcal{E}l} Y_{ljm}|^2 \, dv$$

$$= +\tfrac{1}{2} e\hbar^2(2mc)^{-2} \int (\nabla \cdot \mathbf{E}) |f_{\mathcal{E}l} Y_{ljm}|^2 \, dv \qquad (10\text{-}332)$$

where we took $q = -e$. As far as the expectation values are concerned, we, therefore, see that the Darwin term is equivalent to

$$\tfrac{1}{2} e\hbar^2(2mc)^{-2}\nabla \cdot \mathbf{E} = -\tfrac{1}{2} e\hbar^2(2mc)^{-2}\nabla^2\phi = \frac{e\hbar^2}{8\varepsilon_0(mc)^2} \sum_i q_i \, \delta(\mathbf{r} - \mathbf{r}_i)$$

$$(10\text{-}333)$$

where the last expression is obtained from (5–20) and the summation there is over all other electrons and a nucleus in the atom. In the case of a hydrogen-like ion, the Darwin term reduces to

$$\frac{Z(e\hbar)^2}{32\pi\varepsilon_0(mc)^2} \, \delta(r) \qquad (10\text{-}334)$$

the Darwin term in a hydrogen-like ion. When the potential is radial, we have

$$\mathbf{E} = -r^{-1}\frac{d\phi}{dr}\,\mathbf{r} \qquad (10\text{-}335)$$

Therefore, the spin–orbit interaction term is

$$2e(2mc)^{-2}\hat{\mathbf{s}} \cdot (\mathbf{E} \times \hat{\mathbf{p}}) = -2e(2mc)^{-2}r^{-1}\frac{d\phi}{dr}\,\hat{\mathbf{s}} \cdot (\mathbf{r} \times \hat{\mathbf{p}})$$

$$= -2e(2mc)^{-2}r^{-1}\frac{d\phi}{dr}\,\hat{\mathbf{s}} \cdot \hat{\mathbf{l}} \qquad (10\text{-}336)$$

spin–orbit interaction, which gives a coupling between the spin and orbital angular momenta.

In a hydrogen-like ion, we have seen in (9–27) that $f_{\mathcal{E}l}$ at the origin ($r = 0$) is zero if $l \neq 0$, but finite if $l = 0$ (s states). We see that the Darwin term

(10–334) contributes in the s states only, while the spin–orbit interaction term (10–336) contributes in all states other than the s states.

PROBLEM 10–70. Show that the contribution of the Darwin term to the energy of the ns state of a hydrogen-like ion is

$$\frac{mc^2\alpha^4 Z^4}{2n^3}$$

PROBLEM 10–71. Show that (10–324) can be written as

$$\frac{1}{m} \to \frac{\langle\gamma\rangle}{m} \quad \text{where} \quad \langle\gamma\rangle \equiv \frac{1}{\sqrt{1 - \langle(V/c)^2\rangle}}$$

HINT. Use the virial theorem (Problem 9–6).

PROBLEM 10–72. Show that the contribution of the spin–orbit interaction term to the energy of a (nlj) state of a hydrogen-like ion is

$$\frac{mc^2(Z\alpha)^4}{2n^3} \frac{j(j+1) - l(l+1) - \frac{3}{4}}{(l+1)(2l+1)}$$

HINT. Use Table 9–3.

PROBLEM 10–73. The mass defect relativistic correction term, as given from the second term of (10–322), is

$$-\frac{(\mathscr{E} - mc^2 - q\phi)\hat{\mathbf{p}}^2}{(2mc)^2} \cong -\frac{(\mathscr{E} - mc^2 - q\phi)^2}{2mc^2}$$

when $B = 0$. In obtaining this expression we replaced $\hat{\mathbf{p}}^2/2m$ by $\mathscr{E} - mc^2 - q\phi$. Taking the case of $q\phi = -Ze^2/(4\pi\varepsilon_0 r)$, show that the first-order perturbation energy due to this term is

$$-\frac{\alpha^2 Z^4}{2n^3}\left(\frac{1}{l + \frac{1}{2}} - \frac{3}{4n}\right)$$

HINT. Use Table 9–3 for $\langle r^{-1}\rangle$ and $\langle r^{-2}\rangle$, and use the nonrelativistic result, $\mathscr{E} - mc^2 = -m(cZ\alpha)^2/(2n^2)$.

PROBLEM 10–74. Using the results of Problems 10–70 and 10–73, obtain the relativistic correction term in (10–270) for the $ns_{1/2}$ state.

PROBLEM 10–75. Using the result of Problems 10–72 and 10–73, obtain the relativistic correction term in (10–270) for the $np_{1/2}$ state, and observe that the result is the same as that for the $ns_{1/2}$ state.

PROBLEM 10–76. Equation (10–323) shows that to the first order in B,

$$\hat{H} = \hat{H}_0 + \frac{\mu_B}{\hbar} (\hat{\mathbf{l}} + 2\hat{\mathbf{s}}) \cdot \mathbf{B}$$

where $\hat{H}_0$ is the nonrelativistic Hamiltonian of a free hydrogen-like ion. Show that the first-order perturbation energy, proportional to B, is

$$\langle jm| \frac{\mu_B}{\hbar} (\hat{\mathbf{l}} + 2\hat{\mathbf{s}}) \cdot \mathbf{B}|jm \rangle = \mu_B gmB$$

where

$$g = 1 + \frac{j(j + 1) + \frac{3}{4} - l(l + 1)}{2j(j + 1)}$$

HINT. Replace $(\hat{\mathbf{l}} + 2\hat{\mathbf{s}})_z$ by $\hat{j}_z[\hat{\mathbf{j}} \cdot (\hat{\mathbf{l}} + 2\hat{\mathbf{s}})]/\hat{j}^2$.

PROBLEM 10–77. Show that the previous result (10–300) agrees with the above result of g. This expression of g is a special case of the so-called Landé g factor.

HINT. Remember that $P = (-1)^l$, where l is the quantum number for the Φ component.

ADDITIONAL PROBLEMS

10–78. The nonrelativistic Hamiltonian for the hydrogen atom, which is made of a proton and an electron, in an external homogeneous magnetic field $\mathbf{B}$, is

$$\hat{H} = \frac{1}{2m} (\hat{\mathbf{p}}_e + e\mathbf{A}_e)^2 + \frac{1}{2M} (\hat{\mathbf{p}}_p - e\mathbf{A}_p)^2 + U$$

where

$$\mathbf{A}_i = \tfrac{1}{2}\mathbf{B} \times \mathbf{r}_i$$

Reduce this Hamiltonian into

$$\hat{H} = \frac{1}{2m}\,\hat{\mathbf{p}}_e{}^2 + \frac{1}{2M}\,\hat{\mathbf{p}}_p{}^2 + U + \frac{e}{2(m + M)}\left(\frac{M}{m} - \frac{m}{M}\right)\hat{\mathbf{l}}\ \mathbf{B}$$
$$+ \text{(diamagnetic terms)}$$

10–79. (a) Show that if

$$\hat{U} = \frac{[(\hat{\mathscr{E}} + mc^2) + \hat{\lambda}_1(\hat{\mathscr{E}} + mc^2)]}{(4mc^2\hat{\mathscr{E}})^{1/2}}$$

where

$$\hat{\mathscr{E}} = [c^2\hat{p}^2 + (mc^2)^2]^{1/2}$$

then

$$\hat{U}\hat{H}_{KG}\hat{U}^{-1} = \hat{\lambda}_3\hat{\mathscr{E}}$$

where $\hat{H}_{KG}$ is the Klein–Gordon Hamiltonian of a free particle, as given by (10–13).
(b) If $\Psi_{FV} = \hat{U}\Psi$, show that[11]

$$i\hbar\,\frac{\partial}{\partial t}\,\Psi_{FV} = \hat{\lambda}_3\hat{\mathscr{E}}\Psi_{FV}$$

and

$$\int \Psi_{FV}^{\dagger}\hat{\lambda}_3\Psi_{FV}\,dv = \int \Psi^{\dagger}\hat{\lambda}_3\Psi\,dv$$

10–80. (a) Show that Maxwell's equations in free space ($\rho = \mathbf{j} = 0$) can be written as

$$\left(\frac{c}{\hbar}\,\hat{\mathbf{s}}\cdot\hat{\mathbf{p}} - i\hbar\,\frac{\partial}{\partial t}\right)\Psi = 0$$

if

$$\Psi \equiv \begin{pmatrix} B_x - iE_x/c \\ B_y - iE_y/c \\ B_z - iE_z/c \end{pmatrix}$$

[11] H. F. Fechbach and F. Villars, *Revs. Modern Phys.* **30**, 24 (1958).

and

$$\hat{s}_x \Psi = \hbar \begin{pmatrix} 0 & 0 & 0 \\ 0 & 0 & -i \\ 0 & i & 0 \end{pmatrix} \Psi \qquad \hat{s}_y \Psi = \hbar \begin{pmatrix} 0 & 0 & i \\ 0 & 0 & 0 \\ -i & 0 & 0 \end{pmatrix} \Psi$$

$$\hat{s}_z \Psi = \hbar \begin{pmatrix} 0 & -i & 0 \\ i & 0 & 0 \\ 0 & 0 & 0 \end{pmatrix} \Psi$$

(b) Showing that in this case we have

$$\hat{s}^2 \Psi = 2\hbar^2 \Psi$$

interpret Maxwell's equation as Dirac's equation for a particle (photon) of mass zero, spin one, and helicity $+1$. What is an antiphoton?

10–81. In the Heisenberg picture, where

$$\hat{f}_H = \exp\left(\frac{i\hat{H}_D t}{\hbar}\right) \hat{f} \exp\left(\frac{-i\hat{H}_D t}{\hbar}\right)$$

we have

$$i\hbar \frac{d}{dt} \hat{f}_H = [\hat{f}_H, \hat{H}_D]$$

where $\hat{H}_D$ is the Dirac Hamiltonian. Show that

$$\frac{d\mathbf{r}_H}{dt} = c\hat{\boldsymbol{\alpha}}_H \qquad \frac{d}{dt}(\hat{\mathbf{p}}_H - q\mathbf{A}_H) = q(\mathbf{E} + c\hat{\boldsymbol{\alpha}}_H \times \mathbf{B})$$

$$\frac{d\hat{\mathbf{s}}_H}{dt} = -c\hat{\boldsymbol{\alpha}}_H \times \hat{\mathbf{p}}_H \qquad \frac{d\hat{\mathbf{l}}_H}{dt} = c\hat{\boldsymbol{\alpha}}_H \times \hat{\mathbf{p}}_H$$

10–82.* Show, from (10–143) and Problem 8–76, that if $\mathbf{A} = 0$ then

$$S_{a\uparrow \to b\uparrow}(t) = S_{b*\downarrow \to a*\downarrow}(t)$$

where $\uparrow$ and $\downarrow$ designate eigenstates of $\hat{s}_z$ with eigenvalues $\frac{1}{2}\hbar$ and $-\frac{1}{2}\hbar$, respectively.

10–83.* When $\hat{H}_D$ is the Dirac Hamiltonian for a free particle, and

$$\hat{U} = \left[\frac{2\hat{\mathscr{E}}}{(mc^2 + \hat{\mathscr{E}})}\right]^{1/2} \left(\frac{1}{2} + \frac{1}{2}\beta\frac{\hat{H}_D}{\hat{\mathscr{E}}}\right)$$

where

$$\hat{\mathscr{E}} = \sqrt{c^2 \hat{p}^2 + (mc^2)^2}$$

show that

$$\hat{U}\hat{H}_D \hat{U}^{-1} = \hat{\beta}\hat{\mathscr{E}}$$

10–84.* When $\hat{H}_D$ is the Dirac Hamiltonian for a particle in electromagnetic fields, and

$$\hat{U} = \exp\left[\frac{\hat{\beta}\hat{\alpha} \cdot (\hat{\mathbf{p}} - e\mathbf{A})}{2m}\right]$$

show that[12]

$$\hat{U}\hat{H}_D \hat{U}^{-1} = \hat{\beta}mc^2 + e\phi + \frac{1}{2m}\left(\frac{2}{\hbar}\right)^2 \hat{\beta}[\hat{\mathbf{s}} \cdot (\hat{\mathbf{p}} - e\mathbf{A})]^2 + \cdots$$

[11] L. L. Foldy and S. A. Wouthuysen, *Phys. Rev.* **78**, 29 (1950); S. Tani, *Progr. Theor. Phys.* **6**, 267 (1951).

Chapter 11 **Group Theory in Quantum Mechanics**

Prediction, or foretelling the future, is one of the deepest desires of the human mind: One of the most important tasks of physics is to find conservation laws, or to find things or quantities invariant during complicated time-development processes taking place in nature. If a Hamiltonian of a system is given, all operators which commute with the Hamiltonian give conserved quantities of the system. These operators, on the other hand, exhibit symmetries of the Hamiltonian, and form a group called a symmetry group of the Hamiltonian. Investigations of conservation laws, therefore, involve us with the symmetry groups of a Hamiltonian.

Symmetries are seen in the macroscopic world, e.g., homogeneity of time and space; but are more apparent in the microscopic world. Group theory, which enables us to investigate the symmetries of a system, is more useful in relation to quantum mechanics. We shall find that some essential structure of energy levels of a system, particularly the degeneracy, are directly understood through the corresponding symmetry groups.

11–1 SYMMETRIC GROUPS $\mathscr{S}_2$ AND $\mathscr{S}_3$

The *symmetric groups* are typical finite groups. As a matter of fact it is known that every finite group is isomorphic to a subgroup of some symmetric group (Cayley's theorem or Jordan's theorem[1]).

The helium atom is made of two electrons bound to an α particle. The Hamiltonian for these two electrons, which are called electron 1 and electron 2, is a function of their space coordinates, momenta, and spins:

$$\hat{H}_{\text{He}} = \hat{H}_{\text{He}}(\hat{\mathbf{r}}_1, \hat{\mathbf{r}}_2, \hat{\mathbf{p}}_1, \hat{\mathbf{p}}_2, \hat{\mathbf{s}}_1, \hat{\mathbf{s}}_2) \qquad (11\text{--}1)$$

Since electrons are identical to each other, the Hamiltonian must be invariant under the interchange of two electrons:

$$\hat{H}_{\text{He}}(\hat{\mathbf{r}}_1, \hat{\mathbf{r}}_2, \hat{\mathbf{p}}_1, \hat{\mathbf{p}}_2, \hat{\mathbf{s}}_1, \hat{\mathbf{s}}_2) = \hat{H}_{\text{He}}(\hat{\mathbf{r}}_2, \hat{\mathbf{r}}_1, \hat{\mathbf{p}}_2, \hat{\mathbf{p}}_1, \hat{\mathbf{s}}_2, \hat{\mathbf{s}}_1) \qquad (11\text{--}2)$$

Designating the operator which exchanges the electrons 1 and 2 by (12), the above symmetry relation is rewritten as

$$(12)\hat{H}_{\text{He}}(12)^{-1} = \hat{H}_{\text{He}} \qquad (11\text{--}3)$$

or

$$[(12), \hat{H}_{\text{He}}] = 0 \qquad (11\text{--}4)$$

Since if we exchange electrons twice the system is obviously back to its original situation, we obtain

$$(12)(12) = \hat{1} \qquad (11\text{--}5)$$

where $\hat{1}$ is the identity operator. Therefore

$$(12)^{-1} = (12) \qquad (11\text{--}6)$$

and we see that $\hat{1}$ and (12) form a group, which is called the *symmetric group $\mathscr{S}_2$* (Problem 8–55). The number of *elements* in a group is called the *order* of that group, and designated by g. Therefore $g = 2$ for the $\mathscr{S}_2$ group.

A group formed by operators which commute with a given Hamiltonian $\hat{H}$ is called a *symmetry group* of $\hat{H}$. We have seen that the symmetric group

---

[1] See page 16 of M. Hamermesh, *Group Theory*, Addison Wesley, Reading, Massachusetts, 1962.

$\mathscr{S}_2$ is a symmetry group of $\hat{H}_{\text{He}}$, but we will see later that there are other symmetry groups for $\hat{H}_{\text{He}}$. In any case, readers should be careful to distinguish between the symmetric and symmetry groups.

The lithium atom Li has three electrons, which we designate as 1, 2, and 3. The Hamiltonian of this atom $\hat{H}_{\text{Li}}$ must be invariant under all exchanges of these three electrons:

$$[(12), \hat{H}_{\text{Li}}] = [(23), \hat{H}_{\text{Li}}] = [(31), \hat{H}_{\text{Li}}] = 0 \qquad (11\text{--}7)$$

We shall see, however, that $\hat{1}$, (12), (23), and (31) do not form a group, although a relation like (11–5) still holds. In order to see that, let us introduce a set of functions

$$\langle 123 \rangle \quad \langle 132 \rangle \quad \langle 231 \rangle \quad \langle 213 \rangle \quad \langle 312 \rangle \quad \langle 321 \rangle \qquad (11\text{--}8)$$

where 1, 2, and 3 stand for the coordinate, momentum, and spin of each electron, respectively, and these functions are assumed to be different from each other. Then we obtain

$$(12)\langle 123 \rangle = \langle 213 \rangle \qquad (11\text{--}9a)$$

$$(23)\langle 123 \rangle = \langle 132 \rangle \qquad (11\text{--}9b)$$

$$(31)\langle 123 \rangle = \langle 321 \rangle \qquad (11\text{--}9c)$$

For a product operator (12)(23) we obtain

$$\{(12)(23)\}\{(23)(12)\} = (12)(23)(23)(12) = (12)\hat{1}(12) = \hat{1}$$

$$(11\text{--}10)$$

from (11–5). Thus

$$\{(12)(23)\}^{-1} = (23)^{-1}(12)^{-1} = (23)(12) \qquad (11\text{--}11)$$

and from (11–9a) and (11–9b) we obtain

$$(12)(23)\langle 123 \rangle = (12)\langle 132 \rangle = \langle 231 \rangle \qquad (11\text{--}12)$$

Since the resultant function $\langle 231 \rangle$ does not coincide with any of those of (11–9), we see that (12)(23) cannot be expressed by a single exchange operator.

The transformation of (11–12), or

$$\langle 123 \rangle \rightarrow \langle 231 \rangle \tag{11-12'}$$

is a cyclic permutation $1 \leftarrow 2 \leftarrow 3$, and we designate the corresponding operator as (123):

$$(123)\langle 123 \rangle = \langle 231 \rangle \tag{11-13}$$

which implies that

$$(12)(23) = (123) \tag{11-14}$$

We find, on the other hand, that

$$(23)(12)\langle 123 \rangle = (23)\langle 213 \rangle = \langle 312 \rangle \tag{11-15}$$

and that the resulting transformation is another cyclic permutation $1 \leftarrow 3 \leftarrow 2$, which we designate as (132). Therefore

$$(23)(12) = (132) \tag{11-16}$$

At this point we may note that

$$(12) = (21) \qquad (23) = (32) \qquad (31) = (13)$$
$$(123) = (231) = (312) \qquad (132) = (321) = (213) \tag{11-17}$$

in our notation.

From (11–14) and (11–16) we easily see that

$$(123)(132) = \hat{1} = (132)(123) \tag{11-18}$$

or

$$(123)^{-1} = (132) \qquad (132)^{-1} = (123) \tag{11-18'}$$

Following the same procedure as shown above, we can obtain the group table of Table 11–1, which shows that the three exchange operators, the two cyclic permutation operators, together with the identity operator $\hat{1}$, form a group, which we call the symmetric group $\mathscr{S}_3$:

$$\mathscr{S}_3 = \{\hat{1}, (12), (23), (31), (123), (132)\} \qquad g = 6 \tag{11-19}$$

From (8–209) and (11–7) we see that $\mathscr{S}_3$ is a symmetry group of $\hat{H}_{\text{Li}}$.

TABLE 11–1 *Group Table of $\mathscr{S}_3$. (Products $\hat{A}\hat{B}$)*

$\hat{B}\backslash\hat{A}$	$\hat{1}$	(12)	(13)	(23)	(123)	(132)
$\hat{1}$	$\hat{1}$	(12)	(13)	(23)	(123)	(132)
(12)	(12)	$\hat{1}$	(123)	(132)	(13)	(23)
(13)	(13)	(132)	$\hat{1}$	(123)	(23)	(12)
(23)	(23)	(123)	(132)	$\hat{1}$	(12)	(13)
(123)	(123)	(23)	(12)	(13)	(132)	$\hat{1}$
(132)	(132)	(13)	(23)	(12)	$\hat{1}$	(123)

In general, all permutations among n particles, including $\hat{1}$, exchange, and cyclic permutation operators, form the symmetric group $\mathscr{S}_n$, and $\mathscr{S}_n$ is a symmetry group of an n-electron system. Actually the system may be made of any identical particles, such as an n-neutron system or an n-proton system, as well as an n-electron system, to apply this statement.

From the elements of $\mathscr{S}_3$ shown in (11–19) we can select $\hat{1}$ and (12) to form a smaller group $\mathscr{S}_2$. In general, if selected elements of a larger group $\mathbf{G}_1$ form a smaller group $\mathbf{G}_2$, we say that $\mathbf{G}_2$ is a *subgroup* of $\mathbf{G}_1$. Thus $\mathscr{S}_2$ is a subgroup of $\mathscr{S}_3$. On the other hand, if we have two groups

$$\mathbf{G}' = \{\hat{A}'_1, \hat{A}'_2, \cdots, \hat{A}'_{g'}\} \tag{11–20a}$$

and

$$\mathbf{G}'' = \{\hat{A}''_1, \hat{A}''_2, \cdots, \hat{A}''_{g''}\} \tag{11–20b}$$

and if every element of $\mathbf{G}'$ commutes with every element of $\mathbf{G}''$, the *direct product group* $\mathbf{G}' \times \mathbf{G}''$ can be obtained by taking all possible products $\hat{A}'_i \hat{A}''_j (= \hat{A}''_j \hat{A}'_i)$

$$\mathbf{G}' \times \mathbf{G}'' = \{\hat{A}'_1\hat{A}''_1, \hat{A}'_1\hat{A}''_2, \ldots, \hat{A}'_1\hat{A}''_{g''}, \hat{A}'_2\hat{A}''_1, \ldots, \hat{A}'_{g'}\hat{A}''_{g''}\} \tag{11–21}$$

If the orders of $\mathbf{G}'$ and $\mathbf{G}''$ are g' and g'', respectively, the order of $\mathbf{G}' \times \mathbf{G}''$ is $g'g''$. $\mathbf{G}'$ and $\mathbf{G}''$ are subgroups of $\mathbf{G}' \times \mathbf{G}''$.

PROBLEM 11–1. Reproduce Table 11–1.

PROBLEM 11–2. Show that $\mathscr{S}_3$ has at least two subgroups. One of them is $\mathscr{S}_2$ and the other is

$$\mathscr{A}_3 = \{1, (123), (132)\} \qquad g = 3$$

The second group $\mathscr{A}_3$ is called an *alternating group*.

PROBLEM 11–3. Show that

$$\mathscr{S}_2 \times \mathscr{A}_3 = \mathscr{S}_3$$

PROBLEM 11–4. (a) The operator $\hat{C}_3$ rotates a system by 120° around a given axis. Show that $\mathbf{C}_3 = \{\hat{1}, \hat{C}_3, \hat{C}_3{}^2\}$ is an Abelian group. (An Abelian group is a group whose elements all commute with one another.)
(b) The operator $\hat{C}_4$ rotates a system by 90° around a given axis. Show that $\mathbf{C}_4 = \{\hat{1}, \hat{C}_4, \hat{C}_4{}^2, \hat{C}_4{}^3\}$ is an Abelian group.

PROBLEM 11–5. For a four-particle system obtain the following group table (Table 11-2), which shows that these four elements form a group, called the alternating group $\mathscr{A}_4$, of order 4 ($g = 4$).

$$\mathscr{A}_4 = \{\hat{1}, (12)(34), (23)(14), (13)(24)\} \qquad g = 4$$

TABLE 11-2   *Group Table of $\mathscr{A}_4$*

	$\hat{1}$	(12)(34)	(23)(14)	(13)(24)
$\hat{1}$	$\hat{1}$	(12)(34)	(23)(14)	(13)(24)
(12)(34)	(12)(34)	$\hat{1}$	(13)(24)	(23)(14)
(23)(14)	(23)(14)	(13)(24)	$\hat{1}$	(12)(34)
(13)(24)	(13)(24)	(23)(14)	(12)(34)	$\hat{1}$

HINT. $(12)(34)\langle 1234 \rangle = \langle 2143 \rangle$, etc.

11–2   REPRESENTATIONS. (MATRIX GROUPS)

When two groups $\mathbf{G}' = \{\hat{A}'_1, \hat{A}'_2, \ldots, \hat{A}'_g\}$ and $\mathbf{G}'' = \{\hat{A}''_1, \hat{A}''_2, \ldots, \hat{A}''_g\}$ have the same order g, and their elements can be put into one-to-one correspondence, $\hat{A}'_i \leftrightarrow \hat{A}''_i$, when their group tables are compared to each other, we say that $\mathbf{G}'$ and $\mathbf{G}''$ are *isomorphic*:

$$\mathbf{G}' \cong \mathbf{G}'' \tag{11–22}$$

For example, $\mathbf{C}_3 \cong \mathscr{A}_3$, as we shall see in Problem 11–6. If (11–22) holds we can perform any group-theoretic calculations in $\mathbf{G}'$, in terms of those in $\mathbf{G}''$. When the correspondence between the elements of the two groups in their group tables is not one-to-one but one-to-two or more, we say that these two groups are *homomorphic* to each other.

Among several groups which are isomorphic or homomorphic to a given group, there always exist particularly useful groups, called *matrix groups*. Let us consider the symmetric group $\mathscr{S}_3$ as an example. In order to find an isomorphic matrix group, we take a set of functions and obtain matrix representatives of element operators in terms of that set of functions taken as a base. For $\mathscr{S}_3$ we can take the set of six functions which appeared in (11-8). If we form a row matrix (vector) by the six functions and operate on that with (12), we obtain[2]

$$(12)(\langle 123 \rangle \langle 132 \rangle \langle 231 \rangle \langle 213 \rangle \langle 312 \rangle \langle 321 \rangle)$$

$$= (\langle 213 \rangle \langle 231 \rangle \langle 132 \rangle \langle 123 \rangle \langle 321 \rangle \langle 312 \rangle)$$

$$= (\langle 123 \rangle \langle 132 \rangle \langle 231 \rangle \langle 213 \rangle \langle 312 \rangle \langle 321 \rangle)\begin{pmatrix} 0 & 0 & 0 & 1 & 0 & 0 \\ 0 & 0 & 1 & 0 & 0 & 0 \\ 0 & 1 & 0 & 0 & 0 & 0 \\ 1 & 0 & 0 & 0 & 0 & 0 \\ 0 & 0 & 0 & 0 & 0 & 1 \\ 0 & 0 & 0 & 0 & 1 & 0 \end{pmatrix} \quad (11\text{-}23)$$

where the 6×6 matrix is the matrix representative of the exchange operator (12) obtained on the base we chose. Following the same procedure we obtain six matrix representatives as shown in Table 11-3, and one can easily show that these matrices form a matrix group isomorphic to $\mathscr{S}_3$.

In general we designate a matrix representative of an element $\hat{A}_i$ as $D(\hat{A}_i)$, and designate the matrix group isomorphic to $\mathbf{G} = \{\hat{A}_1, \ldots, \hat{A}_g\}$ as Γ, where

$$\Gamma = \{D(\hat{A}_1), D(\hat{A}_2), \ldots, D(\hat{A}_g)\} \quad (11\text{-}24)$$

The matrix group thus obtained is called a *representation* of **G**. The matrix group shown in Table 11-3 is a representation of $\mathscr{S}_3$. Note that different representations are obtained if one uses different bases, so that many representations exist for a given group.

---

[2] In contrast to the spinor calculation we discussed in Chapter 10, it is better here to express the basis vector as a row vector, so that the matrix representatives are ordered in the same way as the operators. Thus if ξ is a row vector we obtain

$$\hat{A}\hat{A}'\xi = \hat{A}\xi D(\hat{A}') = \xi D(\hat{A})D(\hat{A}')$$

Note that in all previous chapters of this book, a row matrix is designated as ξ^T, but we will omit the superscript T in this chapter.

TABLE 11-3 *A Representation (Matrix Group) of* $\mathscr{S}_3$

(12)	(23)	(31)
$\begin{pmatrix} 0 & 0 & 0 & 1 & 0 & 0 \\ 0 & 0 & 1 & 0 & 0 & 0 \\ 0 & 1 & 0 & 0 & 0 & 0 \\ 1 & 0 & 0 & 0 & 0 & 0 \\ 0 & 0 & 0 & 0 & 0 & 1 \\ 0 & 0 & 0 & 0 & 1 & 0 \end{pmatrix}$	$\begin{pmatrix} 0 & 1 & 0 & 0 & 0 & 0 \\ 1 & 0 & 0 & 0 & 0 & 0 \\ 0 & 0 & 0 & 0 & 0 & 1 \\ 0 & 0 & 0 & 0 & 1 & 0 \\ 0 & 0 & 0 & 1 & 0 & 0 \\ 0 & 0 & 1 & 0 & 0 & 0 \end{pmatrix}$	$\begin{pmatrix} 0 & 0 & 0 & 0 & 0 & 1 \\ 0 & 0 & 0 & 0 & 1 & 0 \\ 0 & 0 & 0 & 1 & 0 & 0 \\ 0 & 0 & 1 & 0 & 0 & 0 \\ 0 & 1 & 0 & 0 & 0 & 0 \\ 1 & 0 & 0 & 0 & 0 & 0 \end{pmatrix}$

(123)	(132)
$\begin{pmatrix} 0 & 0 & 0 & 0 & 1 & 0 \\ 0 & 0 & 0 & 0 & 0 & 1 \\ 1 & 0 & 0 & 0 & 0 & 0 \\ 0 & 1 & 0 & 0 & 0 & 0 \\ 0 & 0 & 1 & 0 & 0 & 0 \\ 0 & 0 & 0 & 1 & 0 & 0 \end{pmatrix}$	$\begin{pmatrix} 0 & 0 & 1 & 0 & 0 & 0 \\ 0 & 0 & 0 & 1 & 0 & 0 \\ 0 & 0 & 0 & 0 & 1 & 0 \\ 0 & 0 & 0 & 0 & 0 & 1 \\ 1 & 0 & 0 & 0 & 0 & 0 \\ 0 & 1 & 0 & 0 & 0 & 0 \end{pmatrix}$

The matrix representative of $\hat{1}$ is the 6×6 unit matrix.

The *dimension*[3] of a representation is the dimension of each of the matrix representatives D of that matrix group. Thus the representation shown in Table 11–3 is a six-dimensional representation of $\mathscr{S}_3$.

The representation shown in Table 11–3 is obtained by taking the six functions (11–8) in a particular arrangement as the base. If we arrange these functions in a different order, each matrix of this representation will be changed. If we take another set of six functions, each of which is obtained as a linear combination of the six functions (11–8), but in such a way that the resulting six functions are independent of each other, then each matrix of the representation will again be changed. In these rearrangements or linear transformations of the base set of functions, resultant sets are still equivalent to the original set of six functions. We say, therefore, that those representations which are obtained by equivalent bases are equivalent to each other.

Let us designate a row matrix (vector) given by an arrangement of base functions, such as (11–8), as ξ. Let us designate an equivalent row matrix given by a linear transformation, including simple rearrangements, of the original set of base functions as ξ', and a square transformation matrix, which gives the linear transformation as $\mathbf{X}$, then, by definition

$$\xi' = \xi\mathbf{X} \quad \text{and} \quad \xi = \xi'\mathbf{X}^{-1} \tag{11–25}$$

---

[3] Also called degree.

Let us assume that a matrix representative $D(A)$ of a group element $\hat{A}$ is obtained by taking ξ as the base:

$$\hat{A}\xi = \xi D(\hat{A}) \tag{11-26}$$

Then, from (11-25),

$$\hat{A}\xi' = \hat{A}\xi\mathbf{X} = \xi D(\hat{A})\mathbf{X} = \xi'\mathbf{X}^{-1}D(\hat{A})\mathbf{X} \tag{11-27}$$

That is, $\mathbf{X}^{-1}D(\hat{A})\mathbf{X}$ is the corresponding matrix representative obtained by taking ξ' as the base. Therefore, the representation obtained by taking ξ' as the base is

$$\Gamma = \{\mathbf{X}^{-1}D(\hat{A}_1)\mathbf{X}, \mathbf{X}^{-1}D(\hat{A}_2)\mathbf{X}, \ldots, \mathbf{X}^{-1}D(\hat{A}_g)\mathbf{X}\} \tag{11-28}$$

which is equivalent to the original representation (11-24) according to the above definition.

If the original representation is of dimension d, that is if each D is a $d \times d$ square matrix, then the transformation matrix $\mathbf{X}$ is also a $d \times d$ square matrix, and the equivalent representation (11-28) is still of dimension d. It may happen, however, that there exists a transformation matrix $\mathbf{X}$ such that every transformed matrix representative has the *reduced* form of

$$\mathbf{X}^{-1}D(\hat{A}_i)\mathbf{X} = \begin{pmatrix} D'(\hat{A}_i) & 0 \\ 0 & D''(\hat{A}_i) \end{pmatrix} \tag{11-29}$$

where each of $D'(\hat{A}_i)$ and $D''(\hat{A}_i)$ is a square matrix of dimension smaller than d, and all other nondiagonal matrix elements are zero. If such an $\mathbf{X}$ exists, we say that the original representation Γ is *reducible*.[4]

If the original matrix group Γ is isomorphic or homomorphic to a given group $\mathbf{G}$, then each of

$$\Gamma' = \{D'(\hat{A}_1), D'(\hat{A}_2), \ldots, D'(\hat{A}_g)\} \tag{11-30a}$$

---

[4] Strictly speaking this is *completely* reducible. We say reducible when transformed matrices have a *reduced* form of either

$$\begin{pmatrix} D' & 0 \\ D''' & D'' \end{pmatrix} \quad \text{or} \quad \begin{pmatrix} D' & D''' \\ 0 & D'' \end{pmatrix}$$

One can show, however, that when a matrix group of a finite group or a compact Lie group is reducible it is always completely reducible. See p. 49 of J. S. Lomont, *Applications of Finite Groups*, Academic Press, New York, 1959.

and

$$\Gamma'' = \{D''(\hat{A}_1), D''(\hat{A}_2), \ldots, D''(\hat{A}_g)\} \qquad (11\text{--}30b)$$

is also isomorphic or homomorphic to **G**. Therefore both Γ' and Γ'' are representations of **G**, except that their dimensions are smaller than the original dimension d. We say that Γ is reduced into a *direct sum* of Γ'- and Γ''-representations:

$$\Gamma = \Gamma' + \Gamma'' \qquad (11\text{--}31)$$

When a representation cannot be reduced any more, the representation is an *irreducible* representation.

For the representation of $\mathscr{S}_3$ we obtained in Table 11–3, we can show that if

$$\mathbf{X} = \begin{pmatrix} 1 & 1 & 1 & -1 & 1 & -1 \\ 1 & -1 & 1 & -1 & -1 & 1 \\ 1 & 1 & -1 & -1 & -1 & -1 \\ 1 & -1 & -1 & -1 & 1 & 1 \\ 1 & 1 & 0 & 2 & 0 & 2 \\ 1 & -1 & 0 & 2 & 0 & -2 \end{pmatrix} \qquad (11\text{--}32)$$

then the representation is reduced into a direct sum of four representations of smaller dimensions as shown in Table 11–4. The original representation of Table 11–3 is, therefore, reducible, while at least two of the resultant representations, which are called $\Gamma(3)$ and $\Gamma(1^3)$ and given by

$$\Gamma(3) = \{(1), (1), (1), (1), (1), (1)\} \qquad (11\text{--}33)$$

and

$$\Gamma(1^3) = \{(1), (-1), (-1), (-1), (1), (1)\} \qquad (11\text{--}34)$$

are obviously irreducible, because their dimensions are one. These two representations are homomorphic (not isomorphic) to the $\mathscr{S}_3$ group.

PROBLEM 11–6. (a) Show that $\mathbf{C}_3$ is isomorphic to $\mathscr{A}_3$.
(b) Both $\mathbf{C}_4$ and $\mathscr{A}_4$ have the same order $g = 4$. Are they isomorphic to each other?

PROBLEM 11–7. Obtain the representation given in Table 11–3.

TABLE 11–4 *A Representation of $\mathscr{S}_3$, Equivalent to but of a Reduced Form of That of Table 11–3*

(12)	(23)

$$(12)\quad\begin{pmatrix} 1 & 0 & 0 & 0 & 0 & 0 \\ 0 & -1 & 0 & 0 & 0 & 0 \\ 0 & 0 & -1 & 0 & 0 & 0 \\ 0 & 0 & 0 & 1 & 0 & 0 \\ 0 & 0 & 0 & 0 & 1 & 0 \\ 0 & 0 & 0 & 0 & 0 & -1 \end{pmatrix}\qquad (23)\quad\begin{pmatrix} 1 & 0 & 0 & 0 & 0 & 0 \\ 0 & -1 & 0 & 0 & 0 & 0 \\ 0 & 0 & \tfrac{1}{2} & -\tfrac{3}{2} & 0 & 0 \\ 0 & 0 & -\tfrac{1}{2} & -\tfrac{1}{2} & 0 & 0 \\ 0 & 0 & 0 & 0 & -\tfrac{1}{2} & \tfrac{3}{2} \\ 0 & 0 & 0 & 0 & \tfrac{1}{2} & \tfrac{1}{2} \end{pmatrix}$$

(31)	(123)

$$(31)\quad\begin{pmatrix} 1 & 0 & 0 & 0 & 0 & 0 \\ 0 & -1 & 0 & 0 & 0 & 0 \\ 0 & 0 & \tfrac{1}{2} & \tfrac{3}{2} & 0 & 0 \\ 0 & 0 & \tfrac{1}{2} & -\tfrac{1}{2} & 0 & 0 \\ 0 & 0 & 0 & 0 & -\tfrac{1}{2} & -\tfrac{3}{2} \\ 0 & 0 & 0 & 0 & -\tfrac{1}{2} & \tfrac{1}{2} \end{pmatrix}\qquad (123)\quad\begin{pmatrix} 1 & 0 & 0 & 0 & 0 & 0 \\ 0 & 1 & 0 & 0 & 0 & 0 \\ 0 & 0 & -\tfrac{1}{2} & \tfrac{3}{2} & 0 & 0 \\ 0 & 0 & -\tfrac{1}{2} & -\tfrac{1}{2} & 0 & 0 \\ 0 & 0 & 0 & 0 & -\tfrac{1}{2} & \tfrac{3}{2} \\ 0 & 0 & 0 & 0 & -\tfrac{1}{2} & -\tfrac{1}{2} \end{pmatrix}$$

(132)	Î

$$(132)\quad\begin{pmatrix} 1 & 0 & 0 & 0 & 0 & 0 \\ 0 & 1 & 0 & 0 & 0 & 0 \\ 0 & 0 & -\tfrac{1}{2} & -\tfrac{3}{2} & 0 & 0 \\ 0 & 0 & \tfrac{1}{2} & -\tfrac{1}{2} & 0 & 0 \\ 0 & 0 & 0 & 0 & -\tfrac{1}{2} & -\tfrac{3}{2} \\ 0 & 0 & 0 & 0 & \tfrac{1}{2} & -\tfrac{1}{2} \end{pmatrix}\qquad \hat{I}\quad\begin{pmatrix} 1 & 0 & 0 & 0 & 0 & 0 \\ 0 & 1 & 0 & 0 & 0 & 0 \\ 0 & 0 & 1 & 0 & 0 & 0 \\ 0 & 0 & 0 & 1 & 0 & 0 \\ 0 & 0 & 0 & 0 & 1 & 0 \\ 0 & 0 & 0 & 0 & 0 & 1 \end{pmatrix}$$

PROBLEM 11–8. Obtain the representation given in Table 11–4. Note that

$$X^{-1} = \begin{pmatrix} \tfrac{1}{6} & \tfrac{1}{6} & \tfrac{1}{6} & \tfrac{1}{6} & \tfrac{1}{6} & \tfrac{1}{6} \\ \tfrac{1}{6} & -\tfrac{1}{6} & \tfrac{1}{6} & -\tfrac{1}{6} & \tfrac{1}{6} & -\tfrac{1}{6} \\ \tfrac{1}{4} & \tfrac{1}{4} & -\tfrac{1}{4} & -\tfrac{1}{4} & 0 & 0 \\ -\tfrac{1}{12} & -\tfrac{1}{12} & -\tfrac{1}{12} & -\tfrac{1}{12} & \tfrac{1}{6} & \tfrac{1}{6} \\ \tfrac{1}{4} & -\tfrac{1}{4} & -\tfrac{1}{4} & \tfrac{1}{4} & 0 & 0 \\ -\tfrac{1}{12} & \tfrac{1}{12} & -\tfrac{1}{12} & \tfrac{1}{12} & \tfrac{1}{6} & -\tfrac{1}{6} \end{pmatrix}$$

PROBLEM 11–9. Show that the symmetric group $\mathscr{S}_2$ has two irreducible representations,

$$\Gamma(2) = \{(1), (1)\} \quad\text{and}\quad \Gamma(1^2) = \{(1), (-1)\}$$

PROBLEM 11–10. From (11–25) and (11–32), show that the transformed base functions, in which the reduced representations of Table 11–4 are obtained, are

$$\langle \Gamma(3) \rangle = \langle 123 \rangle + \langle 132 \rangle + \langle 231 \rangle + \langle 213 \rangle + \langle 312 \rangle + \langle 321 \rangle$$

$$\langle \Gamma(1^3) \rangle = \langle 123 \rangle - \langle 132 \rangle + \langle 231 \rangle - \langle 213 \rangle + \langle 312 \rangle - \langle 321 \rangle$$

$$\langle \Gamma'_a \rangle = \langle 123 \rangle + \langle 132 \rangle - \langle 231 \rangle - \langle 213 \rangle$$

$$\langle \Gamma'_b \rangle = -\langle 123 \rangle - \langle 132 \rangle - \langle 231 \rangle - \langle 213 \rangle + 2\langle 312 \rangle + 2\langle 321 \rangle$$

$$\langle \Gamma''_a \rangle = \langle 123 \rangle - \langle 132 \rangle - \langle 231 \rangle + \langle 213 \rangle$$

$$\langle \Gamma''_b \rangle = -\langle 123 \rangle + \langle 132 \rangle - \langle 231 \rangle + \langle 213 \rangle + 2\langle 312 \rangle - 2\langle 321 \rangle$$

and show that these functions are orthogonal to each other if the original six functions of (11–8) are orthonormal.

PROBLEM 11–11. Prove the statement that if Γ of (11–28) is isomorphic or homomorphic to **G** then each of Γ' and Γ'' of (11–30) is also isomorphic or homomorphic to **G**.

HINT. $\mathbf{X}^{-1}D(\hat{A}_i)\mathbf{X}\mathbf{X}^{-1}D(\hat{A}_j)\mathbf{X} = \mathbf{X}^{-1}D(\hat{A}_i)D(\hat{A}_j)\mathbf{X}$.

11–3* SCHUR'S LEMMA

We have seen that a representation is either reducible or irreducible. An immediate question is that given a representation are there any ways to find out if that representation is reducible or not without actually trying all possible similarity transformations. In order to answer this question we start from (11–26) and assume that there exists another set of functions $\xi'_1, \ldots, \xi'_{d'}$, which can also serve as a basis of a representation of the same group,

$$\hat{A}(\xi'_1 \, \xi'_2 \cdots \xi'_{d'}) = (\xi'_1 \cdots \xi'_{d'})D'(\hat{A}) \tag{11–35}$$

but $d' < d$ and is related to the original set as

$$(\xi'_1 \cdots \xi'_{d'}) = (\xi_1 \, \xi_2 \cdots \xi_d) \begin{pmatrix} Y_{11} & \cdots & Y_{d'1} \\ \cdots\cdots\cdots \\ Y_{1d} & \cdots & Y_{d'd} \end{pmatrix} \tag{11–36}$$

or

$$\xi' = \xi \mathbf{Y} \tag{11–36'}$$

Then by definition the original representation Γ is reducible (into Γ' and other representations). If this is the case then

$$\hat{A}\xi' = \hat{A}\xi\mathbf{Y} = \xi D(\hat{A})\mathbf{Y} \qquad (11\text{-}37)$$

from (11–26) and (11–36), so that, comparing this result with (11–35), which is

$$\hat{A}\xi' = \xi' D'(\hat{A}) = \xi\mathbf{Y}D'(\hat{A}) \qquad (11\text{-}38)$$

we obtain

$$D(\hat{A})\mathbf{Y} = \mathbf{Y}D'(\hat{A}) \qquad (11\text{-}39)$$

Conversely if (11–39) holds, that is, if a rectangular matrix $\mathbf{Y}$ by which $D(\hat{A})$ is transformed into a smaller square matrix $D'(\hat{A})$ exists for all elements $\hat{A}$, then

$$\xi D(\hat{A})\mathbf{Y} = \hat{A}\xi\mathbf{Y} = \xi\mathbf{Y}D'(\hat{A}) \qquad (11\text{-}40)$$

that is, $\xi\mathbf{Y}$ ($= \xi'$) can serve as a basis of a representation (of smaller dimension). Therefore (11–39) is a necessary and sufficient condition that Γ is reducible.

THEOREM 1. If Γ and Γ' are two irreducible representations of a group, and if $d > d'$ where d and d' are the dimensions of Γ and Γ', respectively, then a matrix $\mathbf{Y}$ which satisfies (11–39) for all elements $\hat{A}$ is a zero (rectangular) matrix. A zero matrix is a matrix in which all elements are zero.

This is true because otherwise Γ has to be reducible.

THEOREM 2. Every representation is equivalent to a unitary representation.

A unitary representation is a representation in which every matrix representative is a unitary matrix. According to this theorem we can always assume that Γ is a unitary representation without losing generality. In a unitary representation,

$$D(\hat{A}^{-1}) = D(\hat{A})^{-1} = D(\hat{A})^{\dagger} \qquad (11\text{-}41)$$

where $D(\hat{A})^{\dagger}$ is Hermitian conjugate to $D(\hat{A})$. For a proof of this theorem see Wigner's book.[5]

[5] E. P. Wigner, *Group Theory*, Academic Press, New York, 1959, pp. 74 and 75. Also see J. S. Lomont, *Applications of Finite Groups*, Academic Press, New York, 1959, pp. 47 and 48.

THEOREM 3. A matrix which commutes with all matrix representatives of an irreducible representation is a unit matrix multiplied by a constant. In other words, if Γ is irreducible and if

$$D(\hat{A})\mathbf{M} = \mathbf{M}D(\hat{A}) \tag{11-42}$$

then

$$\mathbf{M} = \begin{pmatrix} m & 0 & & & \cdots & 0 \\ 0 & m & 0 & & \cdots & 0 \\ 0 & 0 & m & 0 & \cdots & 0 \\ \vdots & & & \ddots & & \vdots \\ 0 & \cdots & & 0 & m & 0 \\ 0 & \cdots & & 0 & 0 & m \end{pmatrix} = m\mathbf{1} \tag{11-43}$$

where m is a constant and $\mathbf{1}$ is a unit matrix.

Outline of proof. First observe that $\mathbf{M}$ which satisfies (11–42) can always be made diagonal. This is because if we take $\hat{A}^{-1}$ for $\hat{A}$ in (11–42) we obtain

$$D(\hat{A}^{-1})\mathbf{M} = \mathbf{M}D(\hat{A}^{-1}) \tag{11-44}$$

or

$$D(\hat{A})^{\dagger}\mathbf{M} = \mathbf{M}D(\hat{A})^{\dagger} \tag{11-44'}$$

since we can always assume Γ to be a unitary representation according to Theorem 2, and then, taking Hermitian conjugate of (11–44'),

$$\mathbf{M}^{\dagger}D(\hat{A}) = D(\hat{A})\mathbf{M}^{\dagger} \tag{11-44''}$$

Comparing this result with (11–42) we see that $\mathbf{M}$ can always be made Hermitian, and a Hermitian matrix can always be made diagonal when transformed as $\mathbf{V}\mathbf{M}\mathbf{V}^{-1}$. But when (11–42) holds

$$\mathbf{V}D(\hat{A})\mathbf{V}^{-1}(\mathbf{V}\mathbf{M}\mathbf{V}^{-1}) = (\mathbf{V}\mathbf{M}\mathbf{V}^{-1})\mathbf{V}D(\hat{A})\mathbf{V}^{-1} \tag{11-45}$$

also holds, and $\mathbf{V}\Gamma\mathbf{V}^{-1}$ is equivalent to Γ.

If $\mathbf{M}$ is diagonal but of the form of

$$\mathbf{M} = \begin{pmatrix} n & & & & & & \\ & n & & & 0 & & \\ & & \ddots & & & & \\ & & & n & & & \\ & & & & m & & \\ & 0 & & & & m & \\ & & & & & & \ddots \\ & & & & & & & m \end{pmatrix} \qquad n \neq m \tag{11-46}$$

then (11–42) makes $D(\hat{A})$ have a form of

$$D(\hat{A}) = \begin{pmatrix} \rlap{\diagbox{}{}}{} & 0 \\ 0 & \rlap{\diagbox{}{}}{} \end{pmatrix} \tag{11–47}$$

where the shaded areas may contain nonzero matrix elements, but all other matrix elements are zero. Thus Γ must be reducible. Therefore $\mathbf{M}$ must have the form of (11–43) if Γ is irreducible.

THEOREM 4. If two irreducible representations Γ and Γ' of a group have the same dimension, $d = d'$, but are inequivalent to each other, then a matrix $\mathbf{Y}$ which satisfies (11–39) for all elements $\hat{A}$ is a zero (square) matrix.

PROOF. First observe that $det\mathbf{Y} = 0$. This is because if $det\mathbf{Y} \neq 0$ then $\mathbf{Y}^{-1}$ exists, and (11–39) gives $\mathbf{Y}\Gamma'\mathbf{Y}^{-1} = \Gamma$, that is, Γ and Γ' have to be equivalent to each other. Now let us assume that Γ and Γ' are both unitary representations. We can always do so without losing generality according to Theorem 2. Then following the same steps as those from (11–44) to (11–44″) we obtain

$$\mathbf{Y}^\dagger D(\hat{A}) = D'(\hat{A})\mathbf{Y}^\dagger \tag{11–48}$$

from (11–39). Multiplying both sides of this equation by $\mathbf{Y}$ from left and using (11–39) again, we obtain

$$\mathbf{Y}\mathbf{Y}^\dagger D(\hat{A}) = \mathbf{Y}D'(\hat{A})\mathbf{Y}^\dagger = D(\hat{A})\mathbf{Y}\mathbf{Y}^\dagger \tag{11–49}$$

Since this equation holds for all elements $\hat{A}$, $\mathbf{Y}\mathbf{Y}^\dagger$ must be of the same form as $\mathbf{M}$ of (11–43) according to Theorem 3. But we have already seen that $det\mathbf{Y} = 0$. Therefore $\mathbf{Y}$ must be a zero matrix.

Theorems 1, 3, and 4 can be summarized as *Schur's lemma.* If Γ_p and Γ_q are two irreducible representations of a group, and if

$$D_p(\hat{A}_i)\mathbf{Y} = \mathbf{Y}D_q(\hat{A}_i) \tag{11–50}$$

holds for all elements $\hat{A}_i$ $(i = 1, \ldots, g)$, then

$$\begin{cases} \mathbf{Y} \text{ is a zero matrix} & \text{if} \quad p \neq q \\ \mathbf{Y} = m\mathbf{1} & \text{if} \quad p = q \end{cases} \tag{11–51}$$

where m is a constant and $\mathbf{1}$ is a unit matrix.
We claim that

$$\sum_{i=1}^{g} D_p(\hat{A}_i)\mathscr{Y}D_q(\hat{A}_i^{-1}) \tag{11–52}$$

can serve as Y to satisfy (11–50) and therefore must be expressed as (11–51). In (11–52) $\mathscr{U}$ is a rectangular or square matrix of appropriate dimension but of arbitrary matrix elements. We actually see that

$$D_p(\hat{A}_i) \left[\sum_{j=1}^{g} D_p(\hat{A}_j) \mathscr{U} D_q(\hat{A}_j^{-1}) \right] = \sum_{j=1}^{g} D_p(\hat{A}_i \hat{A}_j) \mathscr{U} D_q(\hat{A}_j^{-1}) D_q(\hat{A}_i^{-1}) D_q(\hat{A}_i)$$

$$= \left[\sum_{j=1}^{g} D_p(\hat{A}_i \hat{A}_j) \mathscr{U} D_q(\hat{A}_i \hat{A}_j)^{-1} \right] D_q(\hat{A}_i)$$

$$= \left[\sum_{k=1}^{g} D_p(\hat{A}_k) \mathscr{U} D_q(\hat{A}_k^{-1}) \right] D_q(\hat{A}_i)$$

$$\tag{11-53}$$

We can choose matrix elements of $\mathscr{U}$ to be all zero except for its $\alpha\beta$ element which we choose $\langle \alpha | \mathscr{U} | \beta \rangle = 1$, then Shur's lemma, when applied to (11–52), gives

$$\sum_{i=1}^{g} \langle \varepsilon | D_p(\hat{A}_i) | \alpha \rangle \langle \beta | D_q(\hat{A}_i^{-1}) | \gamma \rangle = \lambda_{\alpha\beta} \delta_{p,q} \delta_{\varepsilon,\gamma} \tag{11-54}$$

where $\langle \varepsilon | D_p(\hat{A}_i) | \alpha \rangle$ is the $\varepsilon\alpha$-matrix element of $D_p(\hat{A}_i)$ and $\lambda_{\alpha\beta}$ is our notation of the constant m of (11–51).

When the representations are unitary as we can always make according to Theorem 2, (11–54) is reduced to

$$\sum_{i=1}^{g} \langle \varepsilon | D_p(\hat{A}_i) | \alpha \rangle \langle \gamma | D_q(\hat{A}_i) | \beta \rangle^* = \lambda_{\alpha\beta} \delta_{p,q} \delta_{\varepsilon,\gamma} \tag{11-55}$$

In order to find the value of $\lambda_{\alpha\beta}$ we let $p = q$ and $\varepsilon = \gamma$ in (11–54) and sum up over ε. Then the left-hand side becomes

$$\sum_{\varepsilon} \sum_{i=1}^{g} \langle \varepsilon | D_p(\hat{A}_i) | \alpha \rangle \langle \beta | D_p(\hat{A}_i^{-1}) | \varepsilon \rangle = \sum_{i=1}^{g} \langle \beta | D_p(A_i^{-1} A_i) | \alpha \rangle$$

$$= g \delta_{\alpha,\beta} \tag{11-56}$$

while the right-hand side becomes $d\lambda_{\alpha\beta}$ where d is the dimension of the Γ_p representation. Therefore

$$\lambda_{\alpha\beta} = \frac{g}{d} \delta_{\alpha,\beta} \tag{11-57}$$

and (11–54) and (11–55) are

$$\sum_{i=1}^{g} \langle \varepsilon | D_p(\hat{A}_i) | \alpha \rangle \langle \beta | D_q(\hat{A}_i^{-1}) | \gamma \rangle = \frac{g}{d} \delta_{p,q} \delta_{\alpha,\beta} \delta_{\varepsilon,\gamma} \qquad (11\text{–}58)$$

and

$$\sum_{i=1}^{g} \langle \varepsilon | D_p(\hat{A}_i) | \alpha \rangle \langle \beta | D_q(\hat{A}_i) | \gamma \rangle^* = \frac{g}{d} \delta_{p,q} \delta_{\alpha,\beta} \delta_{\varepsilon,\gamma} \qquad (11\text{–}59)$$

Note that (11–59) is valid when Γ_q is unitary, but any representation is equivalent to a unitary representation, or can be made unitary by a similarity transformation $V\Gamma V^{-1}$.

PROBLEM 11–12. We see that the representation of $\mathscr{S}_3$ shown in Table 11–4 is not unitary. Show that this representation can be made unitary if transformed by

$$\begin{pmatrix} 1 & 0 & 0 & 0 & 0 & 0 \\ 0 & 1 & 0 & 0 & 0 & 0 \\ 0 & 0 & 2\sqrt{3} & 0 & 0 & 0 \\ 0 & 0 & 0 & 2 & 0 & 0 \\ 0 & 0 & 0 & 0 & 2\sqrt{3} & 0 \\ 0 & 0 & 0 & 0 & 0 & 2 \end{pmatrix}$$

PROBLEM 11–13. Show by direct calculation that (11–58) holds for each of the representations of $\mathscr{S}_3$ given in (11–33), (11–34), and Table 11–4 (for the two remaining two-dimensional ones).

11–4 CHARACTERS. CLASSES

We have seen in section 11–2 that a representation of $\mathscr{S}_3$ given in Table 11–3 can be reduced into the *direct sum* of the four representations,

$$\Gamma = \Gamma(3) + \Gamma(1^3) + \Gamma' + \Gamma'' \qquad (11\text{–}60)$$

where $\Gamma(3)$ and $\Gamma(1^3)$ are given in (11–33) and (11–34), respectively, while

$$\Gamma' = \left\{ \begin{pmatrix} 1 & 0 \\ 0 & 1 \end{pmatrix}, \begin{pmatrix} -1 & 0 \\ 0 & 1 \end{pmatrix}, \begin{pmatrix} \frac{1}{2} & -\frac{3}{2} \\ -\frac{1}{2} & -\frac{1}{2} \end{pmatrix}, \right.$$

$$\left. \begin{pmatrix} \frac{1}{2} & \frac{3}{2} \\ \frac{1}{2} & -\frac{1}{2} \end{pmatrix}, \begin{pmatrix} -\frac{1}{2} & \frac{3}{2} \\ -\frac{1}{2} & -\frac{1}{2} \end{pmatrix}, \begin{pmatrix} -\frac{1}{2} & -\frac{3}{2} \\ \frac{1}{2} & -\frac{1}{2} \end{pmatrix} \right\} \qquad (11\text{–}61)$$

and

$$\Gamma'' = \left\{ \begin{pmatrix} 1 & 0 \\ 0 & 1 \end{pmatrix}, \begin{pmatrix} 1 & 0 \\ 0 & -1 \end{pmatrix}, \begin{pmatrix} -\frac{1}{2} & \frac{3}{2} \\ \frac{1}{2} & \frac{1}{2} \end{pmatrix}, \right.$$
$$\left. \begin{pmatrix} -\frac{1}{2} & -\frac{3}{2} \\ -\frac{1}{2} & \frac{1}{2} \end{pmatrix}, \begin{pmatrix} -\frac{1}{2} & \frac{3}{2} \\ -\frac{1}{2} & -\frac{1}{2} \end{pmatrix}, \begin{pmatrix} -\frac{1}{2} & -\frac{3}{2} \\ \frac{1}{2} & -\frac{1}{2} \end{pmatrix} \right\} \quad (11\text{--}62)$$

are two-dimensional representations. Our question now is, are these two-dimensional representations irreducible or reducible?

As we noticed before, the transformation matrices $\mathbf{X}$ transform one matrix group (representation) into others, which are isomorphic or homomorphic to, but in general different from, the original matrix group (representation). It is known, however, that the trace of a matrix is invariant under similarity transformations:

$$\text{tr} \, (\mathbf{X}^{-1} D \mathbf{X}) = \text{tr} \, D \quad (11\text{--}63)$$

In group theory we call the trace of a matrix representative the *character*, which is designated as

$$\chi_\Gamma(\hat{A}_i) \equiv \text{tr} \, D(\hat{A}_i) \quad (11\text{--}64)$$

All equivalent matrix groups (representations) are characterized by the set of characters, or the character of the representation,

$$\chi_\Gamma = \{\chi_\Gamma(\hat{A}_1), \chi_\Gamma(\hat{A}_2), \dots, \chi_\Gamma(\hat{A}_g)\}$$
$$\equiv \{\text{tr} \, D(\hat{A}_1), \text{tr} \, D(\hat{A}_2), \dots, \text{tr} \, D(\hat{A}_g)\} \quad (11\text{--}65)$$

For example, the character of the six-dimensional representation of $\mathscr{S}_3$ we obtained in Tables 11–3 and 11–4 (transformed) is

$$\chi_\Gamma = \{6, 0, 0, 0, 0, 0\} \quad (11\text{--}66)$$

while those of the reduced representations are

$$\chi_3 = \{1, 1, 1, 1, 1, 1\} \quad (11\text{--}67)$$

$$\chi_{13} = \{1, -1, -1, -1, 1, 1\} \quad (11\text{--}68)$$

$$\chi_{\Gamma'} = \chi_{\Gamma''} = \{2, 0, 0, 0, -1, -1\} \quad (11\text{--}69)$$

Since the trace of a unit matrix is equal to its dimension, the first number of each of χ_Γ, which is $\chi_\Gamma(\hat{1})$, gives the dimension of the representation under consideration.

We can tell if a given representation is reducible or irreducible if we know its character because a *matrix group* $\Gamma = \{D(\hat{A}_1), D(\hat{A}_2), \ldots, D(\hat{A}_g)\}$ *is irreducible if and only if*

$$\sum_{i=1}^{g} |\chi_\Gamma(\hat{A}_i)|^2 = g \tag{11-70}$$

This theorem is immediately obtained from (11–59). Note that the character of a representation is equal to that of its equivalent unitary representation.

In the case of $\mathscr{S}_3$, the order of the group g is 6. For the six-dimensional (reducible) representation, whose character is given in (11–66), we obtain

$$\sum |\chi_\Gamma(\hat{A}_i)|^2 = 6^2 + 0 + 0 + 0 + 0 + 0 = 36 \neq g \tag{11-71}$$

but for the two irreducible representations, whose characters are given in (11–67) and (11–68), we obtain

$$\sum |\chi_3(\hat{A}_i)|^2 = \sum |\chi_{13}(\hat{A}_i)|^2 = 1 + 1 + 1 + 1 + 1 + 1 = 6 \tag{11-72}$$

confirming the theorem (11–70). For the remaining two representations Γ' and Γ'', we obtain from (11–69) that

$$\sum |\chi_{\Gamma'}(\hat{A}_i)|^2 = \sum |\chi_{\Gamma''}(\hat{A}_i)|^2 = 2^2 + 0 + 0 + 0 + 1 + 1 = 6 \tag{11-73}$$

Therefore, according to the theorem, they are both irreducible.

Classes

In (11–66) through (11–69) we notice that the character is common among each of the three sets of elements, $\hat{1}$, $\{(12), (23), (31)\}$, and $\{(123), (132)\}$. In order to understand this result, it is useful to introduce the concept of *classes*.

If elements $\hat{A}_i$ and $\hat{A}_j$ of a group **G** are related through some element $\hat{A}_k$ of **G** by

$$\hat{A}_k \hat{A}_i \hat{A}_k^{-1} = \hat{A}_j \tag{11-74}$$

then we say that $\hat{A}_i$ is *conjugate* to $\hat{A}_j$ and that $\hat{A}_i$ and $\hat{A}_j$ are in a *class*. A class of a group is given by a maximal set of mutually conjugate elements. For example, in $\mathscr{S}_3$, we find from Table 11–1 that

$$(13)(12)(13)^{-1} = (123)(13) = (23) \qquad (11\text{–}75a)$$

$$(23)(12)(23)^{-1} = (132)(23) = (13) \qquad (11\text{–}75b)$$

and so on, which eventually show that this group has three classes

$$1 \qquad \{(12),\ (23),\ (13)\} \qquad \{(123),\ (132)\} : \qquad \mathscr{S}_3 \qquad r = 3$$

$$(11\text{–}76)$$

The number of classes is denoted by r.

Since a representation is isomorphic or homomorphic to a group, relation (11–74) implies

$$D(\hat{A}_k)D(\hat{A}_i)D^{-1}(\hat{A}_k) = D(\hat{A}_j) \qquad (11\text{–}77)$$

if $\hat{A}_i$ and $\hat{A}_j$ are in a class. Therefore

$$\chi_\Gamma(\hat{A}_i)\ (= \text{tr}\ D(\hat{A}_i) = \text{tr}\ D(\hat{A}_j)) = \chi_\Gamma(\hat{A}_j) \qquad (11\text{–}78)$$

as we noticed before in (11–66) through (11–69).

A given finite group has a definite set of inequivalent irreducible representations. The following theorem[6] gives the number of inequivalent irreducible representations.

THEOREM. For a given group, the number of inequivalent irreducible representations is equal to that of the classes r.

Since $r = 3$ for $\mathscr{S}_3$, as we saw in (11–76), there are only three inequivalent irreducible representations in $\mathscr{S}_3$. We have already found all of them, namely, $\Gamma(3)$, $\Gamma(1^3)$, and Γ', or Γ'', or any other equivalent two-dimensional ones, which we designate as $\Gamma(21)$.

Young Diagrams

An explanation of the notations of irreducible representations of symmetric groups may be in order here. These notations are based on a theorem which states that corresponding to each partition of n particles into portions as

$$n = \lambda_1 + \lambda_2 + \cdots + \lambda_\nu \qquad (11\text{–}79)$$

---

[6] This theorem can be proved using (11–87) and (11–88).

where λ's are integer numbers such that

$$\lambda_1 \geq \lambda_2 \geq \cdots \geq \lambda_\nu \geq 0 \qquad (11\text{-}80)$$

there exists one irreducible representation of $\mathscr{S}_n$, which we designate $\Gamma(\lambda_1 \lambda_2 \cdots \lambda_\nu)$. For example $\Gamma(2^2 1)$ ($\equiv \Gamma(221)$) of $\mathscr{S}_5$ corresponds to the partition $5 = 2 + 2 + 1$. Functions which belong to $\Gamma(\lambda_1 \lambda_2 \cdots \lambda_\nu)$ are obtained by first symmetrizing with respect to particles in each portion, and then antisymmetrizing with respect to exchanges among representative particles of these portions. The so called *Young diagrams* are useful to illustrate this procedure. The irreducible representation which corresponds to the partition (11-79) is given by the Young diagram of

1	2	$\cdots$	$\cdots$	$\cdots$	$\cdots$	λ_1
$1'$	$2'$	$\cdots$	$\cdots$	λ_2		
$1''$	$2''$	$\cdots$	λ_3			

$$(11\text{-}81)$$

$1^{(\nu-1)}$	$\cdots$	λ_ν

and its equivalences. A function which belongs to $\Gamma(\lambda_1 \lambda_2 \cdots \lambda_\nu)$ is obtained by first symmetrizing a given function with respect to exchanges of particles in each row, namely, with respect to $\{1, 2, \ldots, \lambda_1\}$, $\{1', 2', \ldots, \lambda_2\}$, and so forth; and then antisymmetrizing the resulting function with respect to $(11')$, $(11'')$, $(1'1'')$, etc., namely, with respect to exchanges among those particles in the first column. For example a function which belongs to the $\Gamma(21)$ representation of $\mathscr{S}_3$ is obtained from a function $\langle 123 \rangle$, by first symmetrizing with respect to particles 1 and 2, to obtain $\langle 123 \rangle + \langle 213 \rangle$, and then antisymmetrizing with respect to particles 1 and 3, to obtain

$$\langle \Gamma_a' \rangle = \langle 123 \rangle - \langle 321 \rangle + \langle 213 \rangle - \langle 231 \rangle \qquad (11\text{-}82)$$

This is one of the functions obtained in Problem 11-10. To obtain the other function $\langle \Gamma_b' \rangle$, which, together with $\langle \Gamma_a' \rangle$, belongs to the same representation, we start with $\langle 312 \rangle$. Symmetrizing with respect to 1 and 2, and then antisymmetrizing with respect to 1 and 3, we obtain

$$\langle 312 \rangle - \langle 132 \rangle + \langle 321 \rangle - \langle 123 \rangle \qquad (11\text{-}83)$$

If the same function $\langle 312 \rangle$ is symmetrized with respect to 1 and 2, but anti-symmetrized with respect to 2 and 3, we obtain

$$\langle 312 \rangle - \langle 213 \rangle + \langle 321 \rangle - \langle 231 \rangle \qquad (11\text{-}84)$$

We see that $\langle \Gamma_b' \rangle$ is obtained when (11-83) and (11-84) are summed up.
The Young diagrams for the above procedures can be given as

$$\langle \Gamma_a' \rangle : \begin{array}{|c|c|} \hline 1 & 2 \\ \hline 3 \\ \cline{1-1} \end{array} \qquad \langle \Gamma_b' \rangle : \begin{array}{|c|c|} \hline 1 & 2 \\ \hline 3 \\ \cline{1-1} \end{array} + \begin{array}{|c|c|} \hline 2 & 1 \\ \hline 3 \\ \cline{1-1} \end{array} \qquad (11\text{-}85)$$

In order to characterize the nature of these functions, or the representation $\Gamma(21)$ itself, we can take the Young diagrams without numerals. Thus some irreducible representations, for example, are expressed by the Young diagrams as

$$\Gamma(3) = \begin{array}{|c|c|c|}\hline \ & \ & \ \\ \hline\end{array} \qquad \Gamma(21) = \begin{array}{|c|c|}\hline \ & \ \\ \hline \ \\ \cline{1-1}\end{array} \qquad \Gamma(1^3) = \begin{array}{|c|}\hline \ \\ \hline \ \\ \hline \ \\ \hline\end{array} \qquad \Gamma(2^2 1) = \begin{array}{|c|c|}\hline \ & \ \\ \hline \ & \ \\ \hline \ \\ \cline{1-1}\end{array}$$

$$(11\text{-}86)$$

All irreducible representations of $\mathscr{S}_2$, $\mathscr{S}_3$, $\mathscr{S}_4$, and $\mathscr{S}_5$ are given in Table 11-5, along with their characters. Since the character is common to all elements in a single class, only representative elements of each class are shown in this table. Notice in the table that each symmetric group has two one-dimensional representations, $\Gamma(n)$ and $\Gamma(1^n)$: These are the totally symmetric and totally antisymmetric representations, respectively. As matrix groups, $\Gamma(n)$ is only homomorphic to $\mathscr{S}_n$ of $n \geq 2$, and $\Gamma(1^n)$ is only homomorphic to $\mathscr{S}_n$ of $n \geq 3$. A theorem states, however, that all other irreducible representations of $\mathscr{S}_n$ have dimensions larger than one, and are isomorphic to $\mathscr{S}_n$. Because of the last reason, those irreducible representations other than $\Gamma(n)$ and $\Gamma(1^n)$ are called *faithful* representations.

Reduction of a Representation Using Character Tables
The most important application of character tables like Table 11-5 in physics is that one can easily reduce a given representation into a direct sum of irreducible representations, if the character of that given representation is known.

TABLE 11–5 *Character Tables of Some Symmetric Groups*

$\mathscr{S}_2$	$g = 2$ $r = 2$		$\mathscr{S}_3$	$g = 6$ $r = 3$	
Class	Î	(12)	Class	Î	(12) (123)
Order	1	1	Order	1	3 2
$\Gamma(2)$	1	1	$\Gamma(3)$	1	1 1
$\Gamma(1^2)$	1	-1	$\Gamma(21)$	2	0 -1
			$\Gamma(1^3)$	1	-1 1

$\mathscr{S}_4$	$g = 24$	$r = 5$		
Class	Î	(12)	(123) (12)(34) (1234)	
Order	1	6	8 3 6	
$\Gamma(4)$	1	1	1 1 1	
$\Gamma(2^2)$	2	0	-1 2 0	
$\Gamma(21^2)$	3	-1	0 -1 1	
$\Gamma(31)$	3	1	0 -1 -1	
$\Gamma(1^4)$	1	-1	1 1 -1	

$\mathscr{S}_5$	$g = 120$	$r = 7$				
Class	Î	(12)	(123) (12)(34) (1234) (12)(345) (12345)			
Order	1	10	20 15 30 20 24			
$\Gamma(5)$	1	1	1 1 1 1 1			
$\Gamma(41)$	4	2	1 0 0 -1 -1			
$\Gamma(32)$	5	1	-1 1 -1 1 0			
$\Gamma(31^2)$	6	0	0 -2 0 0 1			
$\Gamma(2^21)$	5	-1	-1 1 1 -1 0			
$\Gamma(21^3)$	4	-2	1 0 0 1 -1			
$\Gamma(1^5)$	1	-1	1 1 -1 -1 1			

If Γ_p and Γ_q are irreducible representations of a group, then their characters satisfy the following orthonormality relations:

$$\sum_{i=1}^{g} \chi_p^*(\hat{A}_i)\chi_q(\hat{A}_i) = g\delta_{pq} \tag{11–87}$$

$$\sum_{p=1}^{r} \chi_p^*(\hat{A}_i)\chi_p(\hat{A}_j) = \frac{g}{r_i}\delta_{ij} \tag{11–88}$$

where * means complex conjugate. The first relation is an immediate result of (11–59)[7]. In the first relation the summation is over the entire group of elements (g is the order of the group), while in the second relation the summation is over the entire irreducible representations. Remember that r gives the number of classes and the number of (inequivalent) irreducible representations simultaneously. In the second relation r_i is the order of the class to which $\hat{A}_i$ belongs.

When a representation Γ is reduced to a direct sum of irreducible representations Γ_p's as

$$\Gamma = \sum_{p=1}^{r} c_p \Gamma_p \qquad (11\text{–}89)$$

where c_p is the number of times that Γ_p appears, there should be a corresponding relation between their characters as

$$\chi_\Gamma(\hat{A}_i) = \sum_{p=1}^{r} c_p \chi_p(\hat{A}_i) \qquad (11\text{–}90)$$

for every element $\hat{A}_i$. Therefore, from (11–87), we obtain

$$c_p = \frac{1}{g} \sum_{i=1}^{g} \chi_\Gamma^*(\hat{A}_i) \chi_p(\hat{A}_i) \qquad (11\text{–}91)$$

For example, from (11–66) and the character table of $\mathscr{S}_3$ in Table 11–5, we can easily obtain the previous result

$$\Gamma = \Gamma(3) + \Gamma(1^3) + 2\Gamma(21) \qquad (11\text{–}92)$$

when (11–91) is used.

An important aspect in the application of (11–91) is that we do not need to find a representation explicitly in reducing it as (11–89). All we need is the trace, or the diagonal matrix elements of all matrix representatives. Non-diagonal matrix elements have nothing to do with the reduction (11–89).

PROBLEM 11–14. Obtain the result of (11–76) using Table 11–1.

PROBLEM 11–15. Obtain (11–87) from (11–59).

PROBLEM 11–16. Demonstrate that (11–87) and (11–88) hold for the $\mathscr{S}_4$ group.

---

[7] See, for example, p. 110 of M. Hamermesh, *Group Theory*, Addison Wesley, Reading, Massachusetts, 1962, for the derivation of the second relation.

PROBLEM 11-17. Show that χ_Γ of (11-66) can be obtained directly from the fact that none of the six base functions in (11-8) are invariant under any of the exchanges or cyclic permutations of $\mathscr{S}_3$.

PROBLEM 11-18. A representation of $\mathscr{S}_3$ is obtained using the three component vector $(x_1\ x_2\ x_3)$ as the base.
(a) Show that the character of this representation is

$$\chi_\Gamma = \{3, 1, 1, 1, 0, 0\}$$

(b) Show that this representation can be reduced as

$$\Gamma = \Gamma(3) + \Gamma(21)$$

which implies that one cannot obtain a totally antisymmetric function as a linear combination of x_1, x_2, and x_3.

PROBLEM 11-19. A three-dimensional representation Γ of $\mathscr{S}_3$ is obtained by taking the three functions, $x_1 x_2$, $x_2 x_3$, and $x_3 x_1$, as the base. Show that

$$\chi_\Gamma = \{3, 1, 1, 1, 0, 0\}$$

and that

$$\Gamma = \Gamma(3) + \Gamma(21)$$

in this case.

PROBLEM 11-20. A nine-dimensional representation Γ of $\mathscr{S}_3$ is obtained by taking the nine functions, $x_1 y_1$, $x_1 y_2$, $x_1 y_3$, $x_2 y_1$, $x_2 y_2$, $x_2 y_3$, $x_3 y_1$, $x_3 y_2$, and $x_3 y_3$. Show that

$$\chi_\Gamma = \{9, 1, 1, 1, 0, 0\}$$

and that

$$\Gamma = 2\Gamma(3) + \Gamma(1^3) + 3\Gamma(21)$$

PROBLEM 11-21. (a) Prove that

$$\sum_{p=1}^{r} d_p^{\,2} = g$$

where d_p is the dimension of an irreducible representation Γ_p, and the summation is over all inequivalent irreducible representations.
(b) Using Table 11–5, demonstrate that this relation holds in the symmetric groups $\mathscr{S}_3$, $\mathscr{S}_4$, and $\mathscr{S}_5$.

HINT. Let $\hat{A}_i = \hat{A}_j = \hat{1}$ in (11–88).

11–5 SYMMETRY GROUPS OF $\hat{H}$

If a d_p-dimensional representation Γ_p of a group $\mathbf{G} = \{\hat{A}_1, \hat{A}_2, \ldots, \hat{A}_g\}$ is obtained by taking a set of kets (or functions)

$$|f_1^{(p)}\rangle, |f_2^{(p)}\rangle, \ldots, |f_{d_p}^{(p)}\rangle \tag{11-93}$$

as a base, we say that these kets (or functions) belong to the representation Γ_p. We assume that these kets are orthogonal to each other.
If Γ_p and Γ_q are inequivalent irreducible representations, we can prove

$$\langle f_{\alpha'}^{(q)}|f_\alpha^{(p)}\rangle = 0 \quad \text{if} \quad p \neq q \tag{11-94}$$

and

$$\langle f_1^{(p)}|f_1^{(p)}\rangle = \langle f_2^{(p)}|f_2^{(p)}\rangle = \cdots = \langle f_{d_p}^{(p)}|f_{d_p}^{(p)}\rangle \tag{11-95}$$

We shall outline the proof of these relations presently.
If $\langle\beta|D_p(\hat{A}_i)|\alpha\rangle$ is the $\beta\alpha$-matrix element of the matrix representative $D_p(\hat{A}_i)$, then since $|f_\alpha^{(p)}\rangle$ belongs to Γ_p, we obtain

$$\hat{A}_i|f_\alpha^{(p)}\rangle = \sum_{\beta=1}^{d_p} |f_\beta^{(p)}\rangle\langle\beta|D_p(A_i)|\alpha\rangle \tag{11-96}$$

and

$$\hat{A}_i^{-1}|f_{\alpha'}^{(q)}\rangle = \sum_{\beta'=1}^{d_q} |f_{\beta'}^{(q)}\rangle\langle\beta'|D_q(A_i^{-1})|\alpha'\rangle \tag{11-97}$$

which can be rewritten as

$$\{\hat{A}_i^{-1}|f_{\alpha'}^{(q)}\rangle\}^\dagger = \sum_{\beta'=1}^{d_q} \langle\alpha'|D_q(A_i^{-1})|\beta'\rangle\langle f_{\beta'}^{(q)}| \tag{11-98}$$

Therefore,

$$\langle f_{\alpha'}^{(q)}|f_\alpha^{(p)}\rangle = \langle f_{\alpha'}^{(q)}|\hat{A}_i^{-1}\hat{A}_i|f_\alpha^{(p)}\rangle$$
$$= \sum_{\beta,\beta'} \langle\alpha'|D_q(A_i^{-1})|\beta'\rangle\langle f_{\beta'}^{(q)}|f_\beta^{(p)}\rangle\langle\beta|D_p(A_i)|\alpha\rangle \tag{11-99}$$

We have seen in (11–58) that

$$\sum_{i=1}^{g} \langle \alpha' | D_q(A_i^{-1}) | \beta' \rangle \langle \beta | D_p(A_i) | \alpha \rangle = \frac{g}{d_p} \delta_{pq} \delta_{\alpha\alpha'} \delta_{\beta\beta'} \qquad (11\text{–}58')$$

If we sum up both sides of (11–99) over all group elements, and use (11–58'), we obtain

$$g \langle f_{\alpha'}^{(q)} | f_{\alpha}^{(p)} \rangle = \frac{g}{d_p} \sum_{\beta}^{d_p} \langle f_{\beta}^{(q)} | f_{\beta}^{(p)} \rangle \delta_{pq} \delta_{\alpha\alpha'} \qquad (11\text{–}100)$$

When $p \neq q$ this relation reduces to (11–94), while if $p = q$ it gives (11–95). Q.E.D.

When every element of a group **G** commutes with a given Hamiltonian $\hat{H}$, or when

$$\hat{A}_i \hat{H} \hat{A}_i^{-1} = \hat{H} \qquad i = 1, 2, \ldots, g \qquad (11\text{–}101)$$

this group is called a symmetry group of $\hat{H}$ (see section 8–13). If $\hat{H}$ is for a system of n identical particles, such as electrons, the symmetric group $\mathscr{S}_n$ is a symmetry group. However, there usually exist a few symmetry groups other than $\mathscr{S}_n$ for such systems, as we shall see in later sections of this chapter.

When $|\psi_{\alpha}^{(p)}\rangle$ belongs to an irreducible representation Γ_p of a symmetry group of $\hat{H}$, we obtain

$$\hat{A}_i(\hat{H} | \psi_{\alpha}^{(p)} \rangle) = \hat{A}_i \hat{H} \hat{A}_i^{-1} \hat{A}_i | \psi_{\alpha}^{(p)} \rangle = \hat{H} \hat{A}_i | \psi_{\alpha}^{(p)} \rangle$$
$$= \sum_{\beta} (\hat{H} | \psi_{\beta}^{(p)} \rangle) \langle \beta | D_p(A_i) | \alpha \rangle \qquad (11\text{–}102)$$

which shows that $\hat{H} | \psi_{\alpha}^{(p)} \rangle$ also belongs to Γ_p. Therefore, when $p \neq q$, (11–94) shows that

$$\langle \psi_{\alpha'}^{(q)} | \hat{H} | \psi_{\alpha}^{(p)} \rangle = 0 \qquad \text{if} \qquad p \neq q \qquad (11\text{–}103)$$

This proves our previous statement in section 8–13, to the effect that the matrix of $\hat{H}$ can be partially diagonalized if we arrange the base functions according to irreducible representations of a symmetry group, as shown in (8–212).

Since the time-development operator is $\exp(-i\hat{H}t/\hbar)$ as given in (8–245), the result (11–103) implies that if an initial state belongs to an irreducible representation Γ_p of a symmetry group, then the state will never develop out of that representation. An important application of this statement will be seen in the next section.[8]

If $|\psi_\alpha^{(p)}\rangle$ is an eigenfunction of $\hat{H}$, and belongs to an irreducible representation Γ_p of a symmetry group of $\hat{H}$, then (11–95) shows that this state has to have *at least* d_p*-fold degeneracy,* where d_p is the dimension of Γ_p, because

$$\langle\psi_1^{(p)}|\hat{H}|\psi_1^{(p)}\rangle = \langle\psi_2^{(p)}|\hat{H}|\psi_2^{(p)}\rangle = \cdots = \langle\psi_{d_p}^{(p)}|\hat{H}|\psi_{d_p}^{(p)}\rangle$$

(11–104)

Notice, however, that an energy level may have more degeneracy than d_p of one symmetry group, simply because there can be a few other symmetry groups of $\hat{H}$, and some of them may have irreducible representations with higher dimensions than the particular one considered. However, as a guiding principle, physicists assume that whenever degeneracy exists there must be a symmetry group which explains the degeneracy in this way. Thus, as we shall see in section 11–8, the $(2J + 1)$-fold degeneracy of an atomic level of angular momentum J is the result of the fact that the rotation group $\mathbf{R}_3$ is a symmetry group of an atom, while the n^2-fold degeneracy in the non-relativistic hydrogen, which is discussed in section 9–2, is due to the fact that not only $\mathbf{R}_3$ but also $\mathbf{R}_4$ is a symmetry group in the case of this special atom. (This case will be discussed in section 11–17.) A symmetry group which can explain given degeneracies completely is called a complete symmetry group.

PROBLEM 11–22. The space inversion group $\mathbf{C}_i$ (Problem 8–53) is a symmetry group of many Hamiltonians, but no degeneracy is expected from this symmetry. Why?

HINT. The group $\mathbf{C}_i$ has two irreducible representations, but they are both one dimensional, as can be proved using the result of Problem 11–21a.

11–6 THE PAULI PRINCIPLE AND NONRELATIVISTIC ATOMS AND MOLECULES

The concept of identical particles is unique to the microscopic world. In the macroscopic world, where classical physics is applicable, there are particles which are similar or the same kind but none of them are identical

---

[8] Formula (11–103) can be applied in calculating a general matrix element

$$\langle\psi_{\alpha'}^{(q)}|f^{(r)}|\psi_\alpha^{(p)}\rangle$$

where $f^{(r)}$ is an operator which belongs to Γ_r representation. The matrix element is zero unless $f^{(r)}|\psi_\alpha^{(p)}\rangle$ belongs to the Γ_q representation. In this generalized form (11–103) can be applied to find which terms in the perturbation series (Table 8–2) survive, or to find *selection rules* of transitions [formula (8–264)], etc. Note that $f^{(r)}|\psi_\alpha^{(p)}\rangle$ belongs to the direct product representation $\Gamma_r \times \Gamma_p$, which will be discussed in the next section.

to, or indistinguishable from, each other. In the microscopic world, on the other hand, electrons, for example, are identical and indistinguishable from each other. This statement holds for all elementary particles. This concept implies that a Hamiltonian $\hat{H}_n$ of an n-particle (n-electron) system commutes with every operator of the symmetric group $\mathscr{S}_n$. In other words, $\mathscr{S}_n$ is a symmetry group of $\hat{H}_n$.

Since this indistinguishability of identical elementary particles holds, presumably, throughout the entire universe, if an initial state of the whole universe belongs to an irreducible representation of $\mathscr{S}_n$, where n is the number of particles in the whole universe, the universe stays in that irreducible representation forever. The *Pauli principle*, in group theoretic language, states that the electrons, nucleons, and probably all elementary particles with spin $\frac{1}{2}$, are in the totally antisymmetric representation $\Gamma(1^n)$.

The Pauli principle is experimentally verified in atomic, molecular, solid state, and nuclear physics; and to some extent in high-energy physics. The choice of this particular representation by nature is very manifest from its structure. If the representation were the totally symmetric one $\Gamma(n)$, then all electrons would tend to accumulate into aggregates (Bose–Einstein condensation), with the whole universe eventually coagulated into one or two atoms. On the other hand, $\Gamma(n)$ and $\Gamma(1^n)$ are the only stable irreducible representations, in the sense that a function which belongs to one of these representations of $\mathscr{S}_n$, belongs to the same kind of irreducible representation in its subgroups $\mathscr{S}_m$ ($m < n$). Any other irreducible representations of $\mathscr{S}_n$, on the other hand, are reduced into direct sums of more than one irreducible representation in its subgroup $\mathscr{S}_m$. Thus, if the electrons or nucleons belong to any irreducible representations other than $\Gamma(1^n)$ or $\Gamma(n)$, then there has to be more than one kind of smaller atom such as symmetric heliums [each of which belongs to $\Gamma(2)$ of $\mathscr{S}_2$] and antisymmetric heliums [each of which belongs to $\Gamma(1^2)$ of $\mathscr{S}_2$]. This situation could conceivably produce an inhomogeneous universe made of a symmetric world, an antisymmetric world, and so forth.

Nonrelativistic Atoms and Molecules

Let us discuss a system composed of electrons bound to nuclei in either an atom or molecule. As we have seen in the last chapter, each electron has spin as well as space coordinates. The time-independent part of a wave function of an n-electron system is

$$\psi_{\mathscr{E}} = \psi_{\mathscr{E}}(\mathbf{r}_1, \mathbf{r}_2, \ldots, \mathbf{r}_n; \xi_1, \xi_2, \ldots, \xi_n) \qquad (11\text{–}105)$$

where $\mathbf{r}_i$ is the position of ith electron measured from a center (the nucleus if it is an atom), while ξ_i is some spin coordinate of the same electron.

The spin angular momentum $\hat{\mathbf{s}}$ appears only in the relativistic theory of Dirac, while the Hamiltonian of an atom in the nonrelativistic theory of Schroedinger does not depend on $\hat{\mathbf{s}}$ at all, that is, in the Schroedinger theory

$$\hat{H}_n = \hat{H}_n(\hat{\mathbf{p}}_1, \hat{\mathbf{p}}_2, \ldots, \hat{\mathbf{p}}_n, \mathbf{r}_1, \ldots, \mathbf{r}_n) \qquad (11\text{-}106)$$

In the nonrelativistic theory, therefore, the time-independent Schroedinger equation can determine only that part of the wave function (11–105), which depends on space coordinates;

$$\hat{H}_n \phi_{\mathscr{E}}(\mathbf{r}_1, \mathbf{r}_2, \ldots, \mathbf{r}_n) = \mathscr{E} \phi_{\mathscr{E}}(\mathbf{r}_1, \mathbf{r}_2, \ldots, \mathbf{r}_n) \qquad (11\text{-}107)$$

and the wave function itself is expressed as

$$\psi_{\mathscr{E}} = \phi_{\mathscr{E}}(\mathbf{r}_1, \ldots, \mathbf{r}_n)\gamma(\xi_1, \xi_2, \ldots, \xi_n) \qquad (11\text{-}108)$$

where γ is a spin function.

The Pauli principle requires that $\psi_{\mathscr{E}}$ belongs to the totally antisymmetric irreducible representation $\Gamma(1^n)$ of the $\mathscr{S}_n$ group, with respect to the permutations of electrons. In the nonrelativistic case, this implies that the product $\phi_{\mathscr{E}}\gamma$ should belong to $\Gamma(1^n)$, and it is still another problem to which representation each part, $\phi_{\mathscr{E}}$ or γ, belongs.

Direct Product Representations

Since the Schroedinger Hamiltonian (11–107) has the symmetric group $\mathscr{S}_n$ as its symmetry group, we can make the orbital function $\phi_{\mathscr{E}}$ belong to an irreducible representation Γ_p of $\mathscr{S}_n$. In the same way the spin function γ may belong to an irreducible representation Γ_q of $\mathscr{S}_n$. The problem now is to which representation does the product $\phi_{\mathscr{E}}\gamma$ belong?

Let $\phi_1^{(p)}, \phi_2^{(p)}, \ldots, \phi_{d_p}^{(p)}$ be a basis set of functions for a d_p dimensional representation Γ_p of a group $\mathbf{G}$, and let $\gamma_1^{(q)}, \ldots, \gamma_{d_q}^{(q)}$ be a basis set of functions for a d_q dimensional representation Γ_q of the same group. We can form a set of $d_p d_q$ functions of the form $\phi_\alpha^{(p)}\gamma_{\alpha'}^{(q)}$, taking all possible values of α and α' from the above two sets, and obtain a representation of $\mathbf{G}$ in the standard way. The resultant representation, which we designate as $\Gamma_p \times \Gamma_q$, has the dimension of $d_p d_q$ and is called the *direct product representation* of Γ_p and Γ_q.

Since, if $\hat{A}_i$ is an element of **G**,

$$\hat{A}_i \phi_\alpha^{(p)} \gamma_{\alpha'}^{(q)} = \left\{ \sum_\beta \phi_\beta^{(p)} \langle \beta | D_p(\hat{A}_i) | \alpha \rangle \right\} \left\{ \sum_{\beta'} \gamma_{\beta'}^{(q)} \langle \beta' | D_q(\hat{A}_i) | \alpha' \rangle \right\} \qquad (11\text{--}109)$$

we obtain

$$\langle \phi_\alpha^{(p)} \gamma_{\alpha'}^{(q)} | \hat{A}_i | \phi_\alpha^{(p)} \gamma_{\alpha'}^{(q)} \rangle = \langle \alpha | D_p(\hat{A}_i) | \alpha \rangle \langle \alpha' | D_q(\hat{A}_i) | \alpha' \rangle \qquad (11\text{--}110)$$

by assuming that the basis functions are orthonormal. Since the left-hand side of this equation is a diagonal element of the matrix representative of $\Gamma_p \times \Gamma_q$, we obtain

$$\chi_{\Gamma_p \times \Gamma_q}(\hat{A}_i) = \chi_p(\hat{A}_i)\chi_q(\hat{A}_i) \qquad (11\text{--}111)$$

for the character of the direct product representation. Therefore, from (11–91), we see that if

$$c_s = \frac{1}{g} \sum_{i=1}^{g} \chi_p^*(\hat{A}_i)\chi_q^*(\hat{A}_i)\chi_s(\hat{A}_i) \qquad (11\text{--}112)$$

then the direct product representation can be reduced to

$$\Gamma_p \times \Gamma_q = \sum_{s=1}^{r} c_s \Gamma_s \qquad (11\text{--}113)$$

This relation is often called as *Clebsch–Gordan series*.

Using the character tables of the symmetric groups (Table 11–5), we can find the Clebsch–Gordan series of some symmetric groups as shown in Table 11–6.

A pair of representations whose direct product contains the totally anti-symmetric representation $\Gamma(1^n)$ is called a dual pair. The dual pairs are picked up from Table 11–6 and shown in Table 11–7 and Fig. 11–1. It is observed in this table, and can be shown in general, that one irreducible representation cannot be in more than one dual pair. That is, if one representation is given then the partner representation which forms a dual pair is uniquely fixed.

The Pauli principle requires that the orbital function $\phi_{\mathscr{E}}$ and the spin function γ, in a wave function (11–108), belong to irreducible representations of $\mathscr{S}_n$ which form a dual pair.

TABLE 11–6   *The Clebsch–Gordan Series of Some Symmetric Groups* $(A \times B)$

$\mathscr{S}_2$			$\mathscr{S}_3$			
B \ A	$\Gamma(2)$	$\Gamma(1^2)$	B \ A	$\Gamma(3)$	$\Gamma(21)$	$\Gamma(1^3)$
$\Gamma(1^2)$	$\Gamma(1^2)$	$\Gamma(2)$	$\Gamma(1^3)$	$\Gamma(1^3)$	$\Gamma(21)$	$\Gamma(3)$
$\Gamma(2)$	$\Gamma(2)$	$\Gamma(1^2)$			$\Gamma(3) + \Gamma(1^3)$	
			$\Gamma(21)$	$\Gamma(21)$	$+ \Gamma(21)$ symmetric	
			$\Gamma(3)$	$\Gamma(3)$		

$\mathscr{S}_4$					
B \ A	$\Gamma(4)$	$\Gamma(2^2)$	$\Gamma(21^2)$	$\Gamma(31)$	$\Gamma(1^4)$
$\Gamma(1^4)$	$\Gamma(1^4)$	$\Gamma(2^2)$	$\Gamma(31)$	$\Gamma(21^2)$	$\Gamma(4)$
$\Gamma(31)$	$\Gamma(31)$	$\Gamma(21^2) + \Gamma(31)$	$\Gamma(31) + \Gamma(21^2)$	$\Gamma(4) + \Gamma(2^2) + \Gamma(21^2)$	
			$+ \Gamma(2^2) + \Gamma(1^4)$	$+ \Gamma(31)$	
$\Gamma(21^2)$	$\Gamma(21^2)$	$\Gamma(21^2) + \Gamma(31)$	$\Gamma(4) + \Gamma(2^2)$		
			$+ \Gamma(21^2) + \Gamma(31)$		
$\Gamma(2^2)$	$\Gamma(2^2)$	$\Gamma(4) + \Gamma(2^2)$		symmetric	
		$+ \Gamma(1^4)$			
$\Gamma(4)$	$\Gamma(4)$				

TABLE 11–7   *Dual Pairs*

$\mathscr{S}_2$	$\mathscr{S}_3$	$\mathscr{S}_4$
$\{\Gamma(2), \Gamma(1^2)\}$	$\{\Gamma(3), \Gamma(1^3)\}, \{\Gamma(21), \Gamma(21)\}$	$\{\Gamma(4), \Gamma(1^4)\}, \{\Gamma(2^2), \Gamma(2^2)\}$
		$\{\Gamma(31), \Gamma(21^2)\}$

PROBLEM 11–23. Using formula (11–112) and Table 11–5, obtain the Clebsch–Gordan series of $\mathscr{S}_3$ and $\mathscr{S}_4$, as shown in Table 11–6.

PROBLEM 11–24. Two pairs of functions, each of which belongs to the $\Gamma(21)$ representation of $\mathscr{S}_3$, are obtained in Problem 11–10. Using that result, show that $\langle\Gamma'_a\rangle\langle\Gamma''_a\rangle + \frac{1}{3}\langle\Gamma'_b\rangle\langle\Gamma''_b\rangle$ is the function which belongs to the $\Gamma(1^3)$ representation, while $\langle\Gamma'_a\rangle\langle\Gamma''_b\rangle + \langle\Gamma'_b\rangle\langle\Gamma''_a\rangle$ is the function which belongs to the $\Gamma(3)$ representation. (The other two combinations of the two pairs of functions will yield a pair of functions which belongs to the $\Gamma(21)$ representation, according to the Clebsch–Gordan series $\Gamma(21) \times \Gamma(21) = \Gamma(1^3) + \Gamma(21) + \Gamma(3)$, which is given in Table 11–6.)

FIGURE 11–1. The Young diagrams for those pairs of irreducible representations of symmetric groups which form dual pairs.

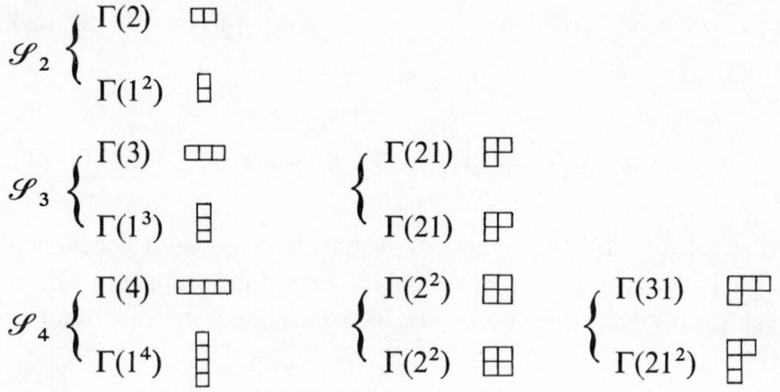

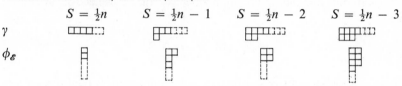

FIGURE 11–2. The Young diagrams for pairs of γ, spin functions, and ϕ_ε, orbital functions, which satisfy the Pauli principle.

11-7 SPIN FUNCTIONS IN SYMMETRIC GROUPS

The spin function γ of (11–108) is made of spin eigenfunctions of individual electrons. The eigenfunctions of the single particle spin operators $\hat{s}^2$ and $\hat{s}_z$ are the spinors $\begin{pmatrix} 1 \\ 0 \end{pmatrix}$ and $\begin{pmatrix} 0 \\ 1 \end{pmatrix}$, as we saw in (10–205), but, in the following discussions, it is more convenient to use the conventional notations

$$\alpha \equiv \begin{pmatrix} 1 \\ 0 \end{pmatrix} \qquad \beta \equiv \begin{pmatrix} 0 \\ 1 \end{pmatrix} \tag{11–114}$$

for these eigenfunctions. Thus if $\hat{s}_i$ is the spin operator for the ith electron, we define $\alpha(i)$ and $\beta(i)$ as

$$\hat{s}_i^2 \alpha(i) = \frac{3\hbar^2}{4} \alpha(i) \qquad \hat{s}_{iz}\alpha(i) = \tfrac{1}{2}\hbar\alpha(i) \tag{11–115a}$$

$$\hat{s}_i^2 \beta(i) = \frac{3\hbar^2}{4} \beta(i) \qquad \hat{s}_{iz}\beta(i) = -\tfrac{1}{2}\hbar\beta(i) \tag{11–115b}$$

From (10–196) and (10–197) it follows that

$$\hat{s}_{i+}\alpha(i) = 0 \qquad \hat{s}_{i-}\alpha(i) = \hbar\beta(i) \qquad (11\text{–}116a)$$

and

$$\hat{s}_{i+}\beta(i) = \hbar\alpha(i) \qquad \hat{s}_{i-}\beta(i) = 0 \qquad (11\text{–}116b)$$

The spin function γ of an n electron system is given by a product of these functions, such as $\alpha(1)\beta(2) \cdots \alpha(n)$, or its linear combination. In the two electron system, for example, there are four independent spin functions

$$\alpha(1)\alpha(2) \qquad \alpha(1)\beta(2) \qquad \beta(1)\alpha(2) \qquad \beta(1)\beta(2) \qquad (11\text{–}117)$$

The total spin angular momentum operator of an n electron system is defined by

$$\hat{\mathbf{S}} = \hat{\mathbf{s}}_1 + \hat{\mathbf{s}}_2 + \cdots + \hat{\mathbf{s}}_n \qquad (11\text{–}118)$$

Since the components of $\hat{\mathbf{S}}$ satisfy the same commutation relations as any angular momenta,

$$[\hat{S}_x, \hat{S}_y] = i\hbar\hat{S}_z \qquad [\hat{S}_y, \hat{S}_z] = i\hbar\hat{S}_x \qquad [\hat{S}_z, \hat{S}_x] = i\hbar\hat{S}_y$$
$$(11\text{–}119)$$

we can follow the same argument as the one we had in section 10–10 to conclude that the eigenfunctions (eigenkets) of $\hat{S}^2$ and $\hat{S}_z$ can be found in such a way that

$$\hat{S}^2|S\,M\rangle = S(S + 1)\hbar^2|S\,M\rangle \qquad (11\text{–}120a)$$

$$\hat{S}_z|S\,M\rangle = M\hbar|S\,M\rangle \qquad M = -S, -S + 1, \ldots, S - 1, S$$
$$(11\text{–}120b)$$

$$\hat{S}_+|S\,M\rangle \equiv (\hat{S}_x + i\hat{S}_y)|S\,M\rangle = \hbar\sqrt{S(S + 1) - M(M + 1)}|S\,M+1\rangle$$
$$(11\text{–}120c)$$

$$\hat{S}_-|S\,M\rangle = \hbar\sqrt{S(S + 1) - M(M - 1)}|S\,M-1\rangle \qquad (11\text{–}120d)$$

It is possible to construct the eigenfunctions $\langle \xi | S\,M \rangle$ from $\alpha(i)$'s and $\beta(i)$'s, using the same procedure which gave $|j\,m\rangle$ in terms of $|l\,m_l\rangle$ and $|s\,m_s\rangle$ in section 10–11. In the case of two electrons, we obviously have

$$\langle \xi \mid 1\ 1 \rangle = \alpha(1)\alpha(2) \qquad \langle \xi \mid 1 -1 \rangle = \beta(1)\beta(2) \qquad (11\text{–}121)$$

where $S = 1$ is determined from $\hat{S}_{+}\alpha(1)\alpha(2) = 0$, as can be seen from (11–116a). The remaining $M = 0$ component of the $S = 1$ spin function is obtained by operating on $\alpha(1)\alpha(2)$ with $\hat{S}_{-}$. We find

$$\langle \xi \mid 1\ 0 \rangle = \frac{1}{\sqrt{2}\,\hbar}\,\hat{S}_{-}\langle \xi \mid 1\ 1 \rangle = \frac{1}{\sqrt{2}\,\hbar}\,(\hat{s}_{1-} + \hat{s}_{2-})\alpha(1)\alpha(2)$$

$$= \sqrt{\tfrac{1}{2}}[\alpha(1)\beta(2) + \beta(1)\alpha(2)] \qquad (11\text{–}122)$$

by using (11–116a) and (11–116b). The other spin function which is orthogonal to the $\langle \xi \mid 1\ 0 \rangle$ spin function must be $\langle \xi \mid 0\ 0 \rangle$:

$$\langle \xi \mid 0\ 0 \rangle = \sqrt{\tfrac{1}{2}}[\alpha(1)\beta(2) - \beta(1)\alpha(2)] \qquad (11\text{–}123)$$

From the viewpoint of the symmetric group $\mathscr{S}_2$, we see that the spin functions (11–121) are symmetric, or belong to the $\Gamma(2)$ irreducible representation, and that $\langle \xi \mid 1\ 0 \rangle$, being obtained by a symmetric operator $\hat{S}_{-}$ from $\langle \xi \mid 1\ 1 \rangle$, has the same symmetry properties as $\langle \xi \mid 1\ 1 \rangle$, and therefore also belongs to $\Gamma(2)$. The only way to obtain $\langle \xi \mid 0\ 0 \rangle$ as a spin function which is orthogonal to $\langle \xi \mid 1\ 0 \rangle$ is to make it an antisymmetric function. Therefore $\langle \xi \mid 0\ 0 \rangle$ has to belong to $\Gamma(1^2)$, and that expectation is confirmed directly in (11–123).

In a general n spin system, the highest value S or M, can take is $\tfrac{1}{2}n$, and taking the spin function α for all electrons, we obtain

$$\left\langle \xi \,\middle|\, \frac{n}{2}\,\frac{n}{2} \right\rangle = \alpha(1)\alpha(2)\cdots\alpha(n) \qquad (11\text{–}124)$$

This spin function is obviously totally symmetric, and consequently belongs to the $\Gamma(n)$ irreducible representation of the symmetric group $\mathscr{S}_n$. Since the $\hat{S}_{-}$ operator, by which all other M-component spin functions are obtained from $\left\langle \xi \,\middle|\, \dfrac{n}{2}\,\dfrac{n}{2} \right\rangle$, is a symmetric operator, we see that all spin functions with $S = \tfrac{1}{2}n$, with any M, belong to $\Gamma(n)$.

Replacing $\alpha(i)$ by $\beta(i)$ in (11–124), we obtain a spin function with $M = \frac{1}{2}n - 1$; and there are n independent spin functions with this M. The symmetric combination of these n functions is the spin function with $S = \frac{1}{2}n$ and $M = \frac{1}{2}n - 1$, and the other $n - 1$ spin functions must be $S = \frac{1}{2}n - 1$ spin functions. Since the only way to make a spin function orthogonal to the totally symmetric function ($S = \frac{1}{2}n$) is to make it antisymmetric with respect to the one spin which is in the β state, those spin functions with $S = \frac{1}{2}n - 1$ have to belong to the $\Gamma(n-1\ 1)$ irreducible representation. Spin functions with $M = \frac{1}{2}n - 2$ are obtained by replacing two α functions by the β functions in (11–124), and there are $n![2!(n - 2)!]$ independent functions available for this M value. Among them one is a component of $S = \frac{1}{2}n$, and $n - 1$ of them are the components of $S = \frac{1}{2}n - 1$ spin functions.

The remaining spin functions, which are orthogonal to these n spin functions, must have $S = \frac{1}{2}n - 2$. Since the two spins in the β state cannot be antisymmetric, the only way to make orthogonal spin functions is to let them belong to the $\Gamma(n-2\ 2)$ irreducible representation. This is the most antisymmetric the functions can be. By going to lower M, finding spin functions with lower S in this orthogonalizing process, we find the results shown in Table 11–8 and Fig. 11–2.

TABLE 11–8   *Relation between S and Irreducible Representations of the Symmetric Group*

$S =$	$\frac{1}{2}n$	$\frac{1}{2}n - 1$	$\frac{1}{2}n - 2$	$\frac{1}{2}n - 3$	$\cdots$	$\frac{1}{2}n - m$
γ	$\Gamma(n)$	$\Gamma(n-1\ 1)$	$\Gamma(n-2\ 2)$	$\Gamma(n-3\ 3)$	$\cdots$	$\Gamma(n-m\ m)$

An important conclusion is that spin functions for a given S belong to one irreducible representation of the symmetric group $\mathscr{S}_n$, and the irreducible representations are different when S is different. This result is due to the fact that individual spin can take only two m_s values, and therefore the spin functions can be either symmetric or antisymmetric, but cannot have any other more complicated symmetry properties.

TABLE 11–9   *Symmetry of Orbital Functions $\phi_\mathscr{E}$ and Associated Total Spin S*

$\phi_\mathscr{E}$	$\Gamma(1^n)$	$\Gamma(21^{n-2})$	$\Gamma(2^21^{n-4})$	$\Gamma(2^31^{n-6})$	$\cdots$	$\Gamma(2^m1^{n-2m})$
S	$\frac{1}{2}n$	$\frac{1}{2}n - 1$	$\frac{1}{2}n - 2$	$\frac{1}{2}n - 3$	$\cdots$	$\frac{1}{2}n - m$

The total wave function $\psi_\mathscr{E}$ is given as a product of an orbital function $\phi_\mathscr{E}$ and a spin function γ, in the nonrelativistic approximation, as given in (11–108). In addition, $\phi_\mathscr{E}$ and γ should have dual symmetry properties, because of the Pauli principle. Therefore we obtain the important conclusion that the orbital functions $\phi_\mathscr{E}$ associated with spin functions γ of different total spin S belong to different irreducible representations of the symmetric group $\mathscr{S}_n$. The explicit relation between S and irreducible representations can be seen in Table 11–9 and Fig. 11–2.

In the nonrelativistic theory of electronic structure of atoms or molecules, each stationary state is obtained as an eigenstate $\phi_\mathscr{E}$ of the nonrelativistic Hamiltonian (11–106). Since the symmetric group $\mathscr{S}_n$ is a symmetry group of this Hamiltonian, its eigenfunction $\phi_\mathscr{E}$ belongs to an irreducible representation of $\mathscr{S}_n$. When the irreducible representation does not appear in Table 11–9, the state is forbidden in nature because of the Pauli principle. When it is in Table 11–9, it is allowed, and the associated total spin S uniquely indicates the symmetry properties of $\phi_\mathscr{E}$.

It is worthwhile to emphasize here that the total spin S appears simply as an emblem of the symmetry properties of the orbital function $\phi_\mathscr{E}$, and not as a result of any spin-dependent interactions, such as spin–spin interaction.

Since there are $2S + 1$ component spin functions for a given S, electronic states with $S = 0, \frac{1}{2}, 1, \frac{3}{2}, \ldots$, are called singlet, doublet, triplet, quartet, … states, respectively. The relativistic effects, such as the spin–orbit interaction (10–336), actually split each level into the $2S + 1$ component levels. The relativistic effects also cause a mixing of states with different S.

PROBLEM 11–25. Show directly that if $\hat{\mathbf{S}} = \hat{\mathbf{s}}_1 + \hat{\mathbf{s}}_2$, then

(a) $$\hat{S}_+\alpha(1)\alpha(2) = 0 = \hat{S}_-\beta(1)\beta(2)$$

(b) $$\hat{S}^2[\alpha(1)\beta(2) + \beta(1)\alpha(2)] = 2\hbar^2[\alpha(1)\beta(2) + \beta(1)\alpha(2)]$$

PROBLEM 11–26. (a) Operating by $\hat{S}_+$, show that

$$\beta(1)\alpha(2)\alpha(3) + \beta(1)\alpha(3)\alpha(2) - \beta(2)\alpha(3)\alpha(1) - \beta(2)\alpha(1)\alpha(3)$$

and

$$-\beta(1)\alpha(2)\alpha(3) - \beta(1)\alpha(3)\alpha(2) - \beta(2)\alpha(3)\alpha(1) - \beta(2)\alpha(1)\alpha(3)$$
$$+ 2\beta(3)\alpha(1)\alpha(2) + 2\beta(3)\alpha(2)\alpha(1)$$

are both spin functions with $S = \frac{1}{2}$.
(b) Show that these two spin functions together belong to the $\Gamma(21)$ representation of $\mathscr{S}_3$.

HINT. See (10–183) for (a) and Problem 11–10 for (b).

PROBLEM 11–27. In a four-spin system, we easily see that independent spin functions for

$$M = 2 \quad \text{is} \quad \alpha(1)\alpha(2)\alpha(3)\alpha(4)$$

$$M = 1 \quad \text{are} \quad \alpha(1)\alpha(2)\alpha(3)\beta(4)$$

and three others,

$$M = 0 \quad \text{are} \quad \alpha(1)\alpha(2)\beta(3)\beta(4)$$

and five others.

(a) A representation $\Gamma(M)$ of $\mathscr{S}_4$ is obtained by taking the set of spin functions for a given M as the base. Calculating the characters, and using Table 11–5, show that

$$\Gamma(M = 2) = \Gamma(4)$$

$$\Gamma(M = 1) = \Gamma(4) + \Gamma(31)$$

$$\Gamma(M = 0) = \Gamma(4) + \Gamma(31) + \Gamma(2^2)$$

(b) From this result, conclude that $\Gamma(S = 2) = \Gamma(4)$, $\Gamma(S = 1) = \Gamma(21^2)$, and $\Gamma(S = 0) = \Gamma(2^2)$, in agreement with Table 11–8.

(c) Show that the sets of spin functions

$$S = 2 \quad \alpha\alpha\alpha\alpha$$

$$S = 1 \begin{cases} \frac{1}{2}(\alpha\alpha\alpha\beta - \alpha\alpha\beta\alpha + \alpha\beta\alpha\alpha - \beta\alpha\alpha\alpha) \\ \frac{1}{2}(\alpha\alpha\alpha\beta - \alpha\alpha\beta\alpha - \alpha\beta\alpha\alpha + \beta\alpha\alpha\alpha) \\ \frac{1}{2}(\alpha\alpha\alpha\beta + \alpha\alpha\beta\alpha - \alpha\beta\alpha\alpha - \beta\alpha\alpha\alpha) \end{cases}$$

$$S = 0 \begin{cases} \frac{1}{2}(\alpha\beta\alpha\beta - \alpha\beta\beta\alpha + \beta\alpha\beta\alpha - \beta\alpha\alpha\beta) \\ \dfrac{1}{\sqrt{12}}(\alpha\beta\alpha\beta + \alpha\beta\beta\alpha + \beta\alpha\beta\alpha + \beta\alpha\alpha\beta - 2\alpha\alpha\beta\beta - 2\beta\beta\alpha\alpha) \end{cases}$$

belong to the irreducible representations $\Gamma(4)$, $\Gamma(21^2)$, and $\Gamma(2^2)$, respectively. This result and the result of part (b) show that these sets are the spin functions for $M_S = S$, with $S = 2, 1$, and 0, respectively.

11–8   ROTATION GROUP R_3

Let us introduce the rotation operator $\hat{R}_\mathbf{u}(\gamma)$, which rotates a physical system by angle γ around a unit vector $\mathbf{u}$. A physical system is described by a wave function $\psi(\mathbf{r})$. When the system, as a whole, is rotated in the way mentioned above, the corresponding wave function for the rotated system, which we designate as $\hat{R}_\mathbf{u}(\gamma)\psi(\mathbf{r})$, has the same functional form as $\psi(\mathbf{r})$, as long as we substitute $\hat{R}_\mathbf{u}(-\gamma)\mathbf{r}$ for $\mathbf{r}$. The situation is illustrated in Fig. 11–3. Thus

$$\hat{R}_\mathbf{u}(\gamma)\psi(\mathbf{r}) = \psi(\hat{R}_\mathbf{u}(-\gamma)\mathbf{r}) \tag{11–125}$$

If we take the rotation axis $\mathbf{u}$ along the Z axis, and express $\mathbf{r}$ in the cylindrical coordinates, (11–125) gives

$$\hat{R}_Z(\gamma)\psi(r) = \hat{R}_Z(\gamma)\psi(z, \rho, \varphi) = \psi(z, \rho, \varphi - \gamma)$$

$$= \sum_n \frac{1}{n!} (-\gamma)^n \left(\frac{\partial}{\partial\varphi}\right)^n \psi(z, \rho, \varphi)$$

$$= \sum_n \frac{1}{n!} (-\gamma)^n \left(\frac{i}{\hbar}\hat{l}_z\right)^n \psi(z, \rho, \varphi)$$

$$= \exp\frac{-i\gamma\hat{l}_z}{\hbar} \psi(r) \tag{11–126}$$

FIGURE 11–3.   Illustration of relation (11–125).

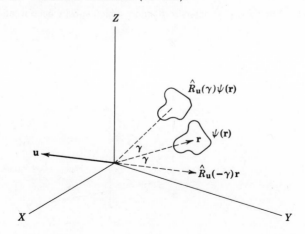

where the z component of the orbital angular momentum operator has been used to replace the differential operator. (See Problem 8–18.) Generalizing this result we obtain

$$\hat{R}_{\mathbf{u}}(\gamma) = \exp \frac{-i\gamma \hat{\mathbf{l}} \cdot \mathbf{u}}{\hbar} \qquad (11\text{–}127)$$

as an expression of the rotation operator.

Each rotation operator $\hat{R}_{\mathbf{u}}(\gamma)$ is given by three parameters: two spherical angles α and β to specify the direction of the rotation axis $\mathbf{u}$, and the rotation angle γ. (See Fig. 11–4.) Since each of these three parameters can be changed continuously, there are infinitely many rotation operators. Note that since

$$\hat{R}_{\mathbf{u}}(\gamma) = \hat{R}_{\mathbf{u}}(\gamma + 2\pi) = \hat{R}_{-\mathbf{u}}(2\pi - \gamma) \qquad (11\text{–}128)$$

the values of the parameters are bounded by

$$0 \leq \alpha < 2\pi$$
$$0 \leq \beta < \tfrac{1}{2}\pi \qquad (11\text{–}129)$$
$$0 \leq \gamma < 2\pi$$

to specify each rotation operator uniquely.

We shall see that these rotation operators form a group, called the rotation group $\mathbf{R}_3$, where the subscript three refers to the three-dimensional space. The $\mathbf{R}_3$ group is a three-parameter continuous group (Lie group), but since the values of the parameters are bounded in finite regions, it is a *compact* group.

FIGURE 11–4.   The three parameters α, β, and γ which specify each rotation operator $\hat{R}_{\mathbf{u}}(\gamma)$.

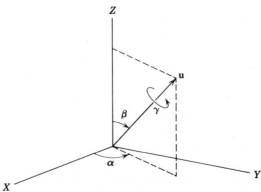

Since $\hat{\mathbf{l}}$ is a Hermitian operator, the rotation operator $\hat{R}_u(\gamma)$ given by (11–127) is a unitary operator (Problem 8–49), such that $\hat{R}_u(\gamma)\hat{R}_u^{\dagger}(\gamma) = \hat{\mathbf{l}}$. A matrix representative $D(\hat{R}_u(\gamma))$, therefore, is a unitary matrix. Also, since $\hat{\mathbf{l}}$ $(= \mathbf{r} \times \hat{\mathbf{p}} = -i\hbar\mathbf{r} \times \nabla)$ is imaginary, $\hat{R}_u(\gamma)$ is a real operator, so that, if we take a real base vector, such as $(x\,y\,z)$, the matrix representative is a real matrix. Thus

$$D(\hat{R}_u(\gamma))D^{\dagger}(\hat{R}_u(\gamma)) = D(\hat{R}_u(\gamma))D^T(\hat{R}_u(\gamma)) = \mathbf{1} \qquad (11\text{–}130)$$

where D^T is the transposed matrix of D, and $\mathbf{1}$ is a unit matrix. This result shows that $D(\hat{R})$ is an orthogonal matrix.

Let us consider an orthogonal transformation

$$(x'\,y'\,z') = (x\,y\,z)\mathbf{X} \qquad (11\text{–}131)$$

where $\mathbf{X}$ is a 3×3 real matrix. The orthogonality condition

$$x'^2 + y'^2 + z'^2 = x^2 + y^2 + z^2 \qquad (11\text{–}132)$$

or

$$(x'\,y'\,z')\begin{pmatrix} x' \\ y' \\ z' \end{pmatrix} = (x\,y\,z)\begin{pmatrix} x \\ y \\ z \end{pmatrix} \qquad (11\text{–}132')$$

implies

$$\mathbf{X}\mathbf{X}^T = \mathbf{1} \qquad (11\text{–}133)$$

which defines the orthogonal matrices. If $\mathbf{X}_1$ and $\mathbf{X}_2$ are two orthogonal matrices, then it is easy to see that $\mathbf{X}_1\mathbf{X}_2$ and $\mathbf{X}_2\mathbf{X}_1$ are also orthogonal matrices, since these compound transformations do not violate (11–132). Therefore, the set of all orthogonal transformation matrices form a group, which we designate as $\mathbf{O}_3$, where the subscript 3 refers to the three-dimensional space. This is a continuous compact group. Each matrix $\mathbf{X}$ has nine elements, but the orthogonality condition (11–133) imposes six conditions between these elements, allowing only three independent parameters for our disposal to specify each transformation. Since each rotation operator is given by three parameters, and since its matrix representative is an orthogonal matrix, we are strongly tempted to say that all rotation operators form a group isomorphic to $\mathbf{O}_3$. There is, however, a difference between rotations and general orthogonal transformations. Taking the determinant equation which corresponds to the matrix equation (11–133), we obtain

$$\det \mathbf{X} = \pm 1 \qquad (11\text{--}134)$$

Thus orthogonal transformations are divided into two kinds, those with $\det \mathbf{X} = +1$ and others with $\det \mathbf{X} = -1$. Taking the first kind of orthogonal transformation, we obtain a subgroup of $\mathbf{O}_3$, called $\mathbf{O}_3{}^+$. Since $D(\hat{R}_u(0))$ is a unit matrix, whose determinant is $+1$, and since $D(\hat{R}_u(\gamma))$ is obtained from $D(\hat{R}_u(0))$ by changing the parameter γ *continuously* from zero to its final value, the determinant of matrix representatives of rotations cannot be -1, which requires a discontinuity. Therefore, the set of all rotation operators form a group $\mathbf{R}_3$ which is isomorphic to $\mathbf{O}_3{}^+$. The isomorphism can be seen by taking a vector $\mathbf{r}$ whose components are x, y, and z, and observing that an orthogonal transformation into $\mathbf{r}'$ whose components are x', y', and z', can be done by one and only one rotation.

PROBLEM 11–28. Show that all rotation operators with a fixed rotation axis $\mathbf{u}$ form an Abelian continuous compact group. This group, which is called $\mathbf{R}_2$, is a subgroup of $\mathbf{R}_3$.

PROBLEM 11–29. Show that the orthogonal transformation matrices with $\det \mathbf{X} = -1$ cannot form a group by themselves. (There is no such group as $\mathbf{O}_3{}^-$.) Show also that

$$\mathbf{O}_3 = \mathbf{C}_i \times \mathbf{O}_3{}^+$$

where

$$\mathbf{C}_i = \{\hat{I}, \hat{1}\}$$

PROBLEM 11–30. In general, rotation operators do not commute with each other.
(a) Illustrate the compound rotations

$$\hat{R}_X(\tfrac{1}{2}\pi)\hat{R}_Y(\tfrac{1}{2}\pi) \qquad \text{and} \qquad \hat{R}_Y(\tfrac{1}{2}\pi)\hat{R}_X(\tfrac{1}{2}\pi)$$

and compare the results.
(b) Express each of these compound rotations by a single rotation.

HINT. See Fig. 11–5. Compound rotations are equivalent to $\hat{C}_3$'s.

FIGURE 11–5. Compound rotations $\hat{R}_Y(\frac{1}{2}\pi)\hat{R}_X(\frac{1}{2}\pi)$ and $\hat{R}_X(\frac{1}{2}\pi)\hat{R}_Y(\frac{1}{2}\pi)$ are not equivalent, but each of them is equivalent to a $\hat{C}_3$ rotation.

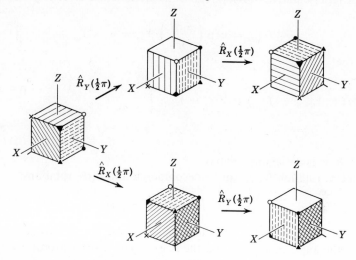

PROBLEM 11–31. Show that

$$\hat{R}_z(\gamma)(x\,y\,z) = (x\,y\,z) \begin{pmatrix} \cos\gamma & -\sin\gamma & 0 \\ \sin\gamma & \cos\gamma & 0 \\ 0 & 0 & 1 \end{pmatrix}$$

which gives a matrix representative of $\hat{R}_Z(\gamma)$.

11–9 IRREDUCIBLE REPRESENTATIONS AND CHARACTERS OF R_3

Since $\hat{R}_u(\gamma)$ of (11–127) is given by the $\hat{\mathbf{l}}$ operator, and since components of $\hat{\mathbf{l}}$, $\hat{l}_z$, $\hat{l}_+$, and $\hat{l}_-$, when operated on Y_{lm}, a spherical harmonic, change m in general, but do not change l-quantum number at all, we see that a matrix representative of $\hat{R}_u(\gamma)$ can be obtained if we take as the basis the $2l + 1$ spherical harmonics Y_{lm}, with a given value of l, namely, Y_{ll}, Y_{ll-1}, ..., Y_{l-l}. The set of the spherical harmonics thus gives a $(2l + 1)$-dimensional irreducible representation of the R_3 group. Therefore, this group has irreducible representations of dimensions $1, 3, 5, \ldots, \infty$, which are designated as D_0, D_1, D_2, etc., respectively.

It is not so easy to obtain an irreducible representation $D_l(\alpha, \beta, \gamma)$ explicitly as a function of the three parameters α, β, and γ, where α and β are the spherical angles for the rotation axis **u**. However, a special case of $D_l(0, 0, \gamma)$, a matrix representative of $\hat{R}_Z(\gamma)$, can be easily obtained as follows. Since

$$\hat{R}_Z(\gamma) Y_{lm} = \exp\left(\frac{-i\gamma\hat{l}_z}{\hbar}\right) Y_{lm} = e^{-i\gamma m} Y_{lm} \qquad (11\text{--}135)$$

we see that

$$\hat{R}_Z(\gamma)(Y_{ll} \, Y_{ll-1} \cdots Y_{l-l}) = (Y_{ll} \cdots Y_{l-l}) \begin{pmatrix} e^{-i\gamma l} & & & \\ & e^{-i\gamma(l-1)} & & 0 \\ & 0 & \ddots & \\ & & & e^{i\gamma l} \end{pmatrix}$$

$$(11\text{--}136)$$

or $D_l(0, 0, \gamma)$ is a diagonal matrix.

A general rotation $\hat{R}_u(\gamma)$ can be decomposed into three rotations,

$$\hat{R}_u(\gamma) = \hat{R}_v(-\beta)\hat{R}_Z(\gamma)\hat{R}_v(\beta) \qquad (11\text{--}137)$$

where **v** is the unit vector perpendicular to **u** and the Z axis (Fig. 11–6), and β is the angle between **u** and the Z axis. The first rotation $\hat{R}_v(\beta)$ brings matter on the **u** axis onto the Z axis, while the last rotation $\hat{R}_v(-\beta)$ brings it back to the **u** axis. The identity (11–137) can be seen intuitively from Fig. 11–7. Since $\hat{R}_u(-\beta)$ is $\hat{R}_u^{-1}(\beta)$, this relation shows that $\hat{R}_u(\gamma)$ and $\hat{R}_Z(\gamma)$ are in a class (see (11–74)). Actually we can find similar relations for any rotation axis **u**, and see that all rotation operators with a given rotation angle γ are in the same class.

Although it is difficult to find $D_l(\alpha, \beta, \gamma)$ itself, it is easy to find the corresponding character χ_l. From (11–136) and (11–137) we see that

$$\chi_l(\gamma) = \sum_{m=l}^{-l} e^{-i\gamma m} = \frac{\sin\{\tfrac{1}{2}\gamma(2l+1)\}}{\sin(\tfrac{1}{2}\gamma)} \qquad (11\text{--}138)$$

which is a *class function*, or is a function of γ only in this case, as we have seen in section 11–4. The orthogonality relation (11–87) is modified as

$$\int_0^{2\pi} \chi_l(\gamma)\chi_{l'}(\gamma) \sin^2(\tfrac{1}{2}\gamma) \, d\gamma = 2\pi\delta_{ll'} \qquad (11\text{--}139)$$

which can be shown directly from (11–138).

If a given representation of $\mathbf{R}_3$ is going to be reduced as[9]

$$D = \sum_l c_l D_l \qquad (11\text{--}140)$$

---

[9] For finite groups we denote Γ for a matrix group and D for each matrix, but for rotation groups it is conventional to use the notation D for both.

FIGURE 11–6. The unit vector **v**, which is perpendicular to **u** and the Z axis.

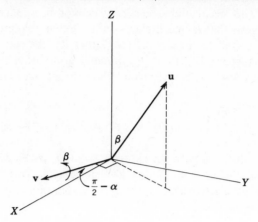

FIGURE 11–7. Illustration of identity (11–137).

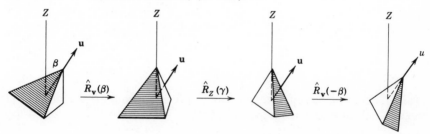

then

$$c_l = \frac{1}{2\pi} \int_0^{2\pi} \chi(\gamma)\chi_l(\gamma) \sin^2 \left(\tfrac{1}{2}\gamma\right) d\gamma \qquad (11\text{–}141)$$

which are the generalizations of (11–89) and (11–91).

Rotation Group and an Atom

Schroedinger's Hamiltonian of the hydrogen atom $\hat{H}_r$, as given by (9–7), commutes with $\hat{\mathbf{l}}$:

$$[\hat{\mathbf{l}}, \hat{H}_r] = 0 \qquad (11\text{–}142)$$

Therefore

$$[\hat{R}_{\mathbf{u}}(\gamma), \hat{H}_r] = 0 \qquad (11\text{–}143)$$

or the rotation group $\mathbf{R}_3$ is a symmetry group of $\hat{H}_r$. The $(2l + 1)$-fold degeneracy of each l level of the hydrogen atom is a direct result of this group-theoretic property of $\hat{H}_r$. We saw, however, in section 9–2 that the hydrogen atom has the additional, so-called accidental degeneracy between ns, np, nd, ... levels. This accidental degeneracy cannot be explained by the $\mathbf{R}_3$ group, and suggests that $\hat{H}_r$ has higher symmetry than that given by (11–142) or (11–143). We shall see in section 11–17 that $\mathbf{R}_4$, the rotation group in a four-dimensional space, is actually a symmetry group of $\hat{H}_r$.

For a complex atom with n electrons, the Schroedinger Hamiltonian is

$$\hat{H}_n = \sum_{i=1}^{n} \left(-\frac{\hbar^2}{2m} \right) \nabla_i^2 + \sum_{i>j} \frac{e^2}{4\pi\varepsilon_0 r_{ij}} - \sum_{i=1}^{n} \frac{Ze^2}{4\pi\varepsilon_0 r_i} \qquad (11\text{–}144)$$

where r_{ij} is the distance between electrons while r_i is the distance between an electron and the nucleus, which has charge Ze. This Hamiltonian is invariant when *all* electrons are rotated by a common angle γ around the nucleus (Fig. 11–8). Therefore

$$[e^{-i\gamma\hat{\mathbf{L}}\cdot\mathbf{u}/\hbar}, \hat{H}_n] = 0 \qquad (11\text{–}145)$$

if

$$\hat{\mathbf{L}} = \sum_{i=1}^{n} \hat{\mathbf{l}}_i \qquad (11\text{–}146)$$

which is the total orbital angular momentum operator of the atom. The generalized rotation operators

$$\hat{R}_\mathbf{u}(\gamma) = e^{-i\gamma\hat{\mathbf{L}}\cdot\mathbf{u}/\hbar} \qquad (11\text{–}147)$$

which replace (11–127), form the rotation group $\mathbf{R}_3$ for the complex atom, and this $\mathbf{R}_3$ is a symmetry group of $\hat{H}_n$. As a direct result we see that each eigenstate of $\hat{H}_n$ has a definite quantum number L as

$$\hat{H}_n\psi_{\mathscr{E}L} = \mathscr{E}\psi_{\mathscr{E}L} \qquad (11\text{–}148a)$$

and

$$\hat{L}^2\psi_{\mathscr{E}L} = L(L + 1)\hbar^2\psi_{\mathscr{E}L} \qquad (11\text{–}148b)$$

and that each L state has the $(2L + 1)$-fold degeneracy. The conventional notations are

$$S \quad P \quad D \quad F \quad G \quad H \quad I\cdots$$

for $\qquad\qquad\qquad\qquad\qquad\qquad\qquad\qquad\qquad (11\text{–}149)$

$$L = 0 \quad 1 \quad 2 \quad 3 \quad 4 \quad 5 \quad 6 \cdots$$

FIGURE 11–8. The rotation operator exp $(-i\gamma\hat{\mathbf{L}} \cdot \mathbf{u}/\hbar)$, which rotates every electron by a common angle γ around an axis **u** which passes through the nucleus.

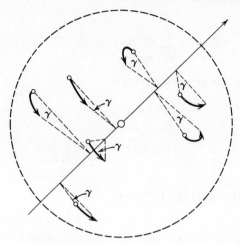

The Hamiltonian $\hat{H}_n$ is also invariant under the space inversion $\hat{I}$. Each eigenfunction of $\hat{H}_n$, therefore, has a definite parity in addition to the L quantum number.

PROBLEM 11–32. Obtain the orthogonality relation of the spherical harmonics

$$\int_0^{2\pi} \int_0^{\pi} Y_{lm}^*(\theta, \varphi) Y_{l'm'}(\theta, \varphi) \sin \theta \, d\theta \, d\varphi = 0 \qquad \text{if} \quad l \neq l'$$

from (11–94).

PROBLEM 11–33. When γ is infinitesimal

$$\hat{R}_{\mathbf{u}}(\gamma) = 1 - i\gamma\hat{I} \cdot \mathbf{u}$$

Assuming that γ is infinitesimal, prove the relation (11–137) analytically.

HINT. Use the result of Problem 8–72 and the commutation relation

$$[\hat{I} \cdot \mathbf{u}_1, \hat{I} \cdot \mathbf{u}_2] = i\hbar\hat{I} \cdot (\mathbf{u}_1 \times \mathbf{u}_2)$$

PROBLEM 11–34. The second orthonormality relation of the characters (11–88) is modified for the rotation group as

$$\sin^2 \left(\tfrac{1}{2}\gamma\right) \sum_{l=0}^{\infty} \chi_l(\gamma)\chi_l(\gamma') = \tfrac{1}{2}\pi\delta(\gamma - \gamma')$$

Prove this relation directly from (11–138).

HINT.

$$\sum_{l=0}^{\infty} \cos (lx) = \pi\delta(x) + \tfrac{1}{2}$$

11–10 ATOMIC SHELL MODEL

It is almost impossible to solve the many-body problem of an n electron atom. The Hamiltonian in the nonrelativistic approximation is known explicitly as (11–144), but no analytical method is known to solve the corresponding differential equation, and even numerical integrations are impractical because of too many variables.

An approximation is obtained by assuming that each electron in a complex atom moves independently under the potential given by the Coulomb field due to the nucleus plus an average Coulomb field due to all other electrons. By the independent motion we mean that

$$\phi_{\mathscr{E}}(\mathbf{r}_1, \mathbf{r}_2, \ldots, \mathbf{r}_n) = \phi'(\mathbf{r}_1)\phi''(\mathbf{r}_2) \cdots \phi^{(n)}(\mathbf{r}_n) \qquad (11\text{–}150)$$

or a wave function is given by a product of individual wave functions. One of them, $\phi'(\mathbf{r}_1)$ for instance, satisfies the eigenvalue equation

$$\hat{H}'(\mathbf{r}_1)\phi'(\mathbf{r}_1) = \mathscr{E}'\phi'(\mathbf{r}_1) \qquad (11\text{–}151)$$

where

$$\hat{H}'(\mathbf{r}_1) = -\frac{\hbar^2}{2m} \nabla_1^2 - \frac{Ze^2}{4\pi\varepsilon_0 r_1} + \frac{e^2}{4\pi\varepsilon_0} \sum_{i=2}^{n} \int |\phi^{(i)}(\mathbf{r}_i)|^2 r_{1i}^{-1} \, dv_i$$
$$(11\text{–}152)$$

where the last terms give the average Coulomb fields due to all other electrons. The set of n equations like (11–151) is called the *Hartree equations*. These equations should be solved self-consistently, because each $\phi^{(i)}(\mathbf{r}_i)$ appears in $n - 1$ of the Hamiltonians like (11–152) and also appears as an eigenfunction of one Hamiltonian. The method is, therefore, called the *self-consistent-field (SCF) method.*

In the effective potential[10]

$$U(\mathbf{r}_i) = \frac{-e^2}{4\pi\varepsilon_0} \left(\frac{Z}{r_i} - \sum_{j \neq i} \int |\phi^{(j)}(\mathbf{r}_j)|^2 r_{ij}^{-1}\, dv_j \right) \qquad (11\text{–}153)$$

the first term is central but other terms, even when summed up, are not central in general. Nevertheless a considerable simplification can be achieved in performing actual calculations in the SCF method if we assume that U is central; or

$$[\hat{\mathbf{l}}_i, U(\mathbf{r}_i)] = 0 \qquad (11\text{–}154)$$

If (11–154) is assumed we immediately obtain

$$\phi^{(i)}(\mathbf{r}_i) = f_i(r_i) Y_{l_i m_i}(\theta_i, \varphi_i) \qquad (11\text{–}155)$$

and the only part left to calculate is the radial function f_i. For practical calculations (11–154), and therefore (11–155), are assumed.

The energy levels given by the Hamiltonian (11–152) cannot be found until the whole procedure of the SCF method is completed. However, a rough prediction of the result can be made from the potential (11–153) if it is assumed to be central. When a spherical charge distribution $\rho(r)$ is given, the electrostatic potential at a distance r from the center is

$$\frac{1}{\varepsilon_0 r} \int_0^r \rho(r) r^2\, dr \qquad (11\text{–}156)$$

Therefore (11–153) implies that

$$-\frac{Ze^2}{4\pi\varepsilon_0 r_i} \xleftarrow[0 \leftarrow r_i]{} U(r_i) \xrightarrow[r_i \to \infty]{} -\frac{(Z - n - 1)e^2}{4\pi\varepsilon_0 r_i} \qquad (11\text{–}157)$$

This potential, as shown in Fig. 11–9, decreases more rapidly than a Coulomb potential as r_i decreases. Because of the deviation from a simple Coulombic form of the potential, the accidental degeneracy, the degeneracy between levels of different l but the same n quantum numbers, which we saw in

[10] Note that the effective potential defined by (11–153) is different for different orbitals. Therefore orbital functions $\phi^{(i)}(\mathbf{r}_i)$ are not orthogonal to each other.

FIGURE 11–9. The effective or average potential given by (11–153). Curves *a* and *b* are the Coulomb potentials $-(Z - n - 1)e^2/(4\pi\varepsilon_0 r)$ and $-Ze^2/(4\pi\varepsilon_0 r)$, respectively.

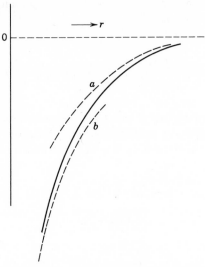

section 9–2, is removed. Since the probability density $|\phi|^2$ is more concentrated near the nucleus for states of smaller l, the removal of the accidental degeneracy is such that the ns level is lower than the np level, and the np level is lower than the nd level, and so on. The energy levels under the potential (11–153), therefore, look like those shown in Fig. 11–10, which may be

FIGURE 11–10. One electron energy level diagram of the atomic shell model.

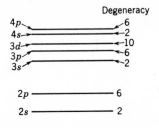

compared with Fig. 9–1 which is the energy level diagram of the hydrogen atom.

In the atomic shell model, which works nicely as a starting point of the SCF calculation, particularly for low-energy states, we determine the electronic configuration by filling those one-electron energy levels obeying the Pauli principle. Since an nl level has the $2(2l + 1)$-fold degeneracy, counting two possible spin orientations, $s, p, d, \ldots$ levels can accommodate at most $2, 6, 10, \ldots$ electrons, respectively. For example, the electron configuration of the lowest energy states of the neutral nitrogen atom,[11] NI, which has seven electrons, is $(1s)^2(2s)^2(2p)^3$, while that of the next lowest energy states is $(1s)^2(2s)^2(2p)^2(3s)$.

Coupling of Orbital Angular Momenta

The assumption of independent motion, which leads to the Hartree equations (11–151), is a crude approximation, and particularly the assignment of the m-quantum number to each electron, as (11–155) implies, is obviously wrong, since $\hat{\mathbf{l}}$ or the individual rotation $\exp{(-i\gamma\hat{\mathbf{l}} \cdot \mathbf{u}/\hbar)}$ does not commute with the correct (nonrelativistic) Hamiltonian $\hat{H}_n$, given by (11–144). As we have seen in the previous section, the total orbital angular momentum $\hat{\mathbf{L}}$ is conserved, but any individual one $\hat{\mathbf{l}}$ is not.

Let us consider the simplest case of two electrons. If we are allowed to take (11–155) for a one-electron wave function, a Hartree wave function (11–150) for the two-electron system is

$$f_1(r_1)f_2(r_2)Y_{lm}(\theta_1, \varphi_1)Y_{l'm'}(\theta_2, \varphi_2) \tag{11–158}$$

Taking all possible values of m and m', we obtain $(2l + 1)(2l' + 1)$ functions of this form. We have seen in section 11–9 that we can obtain the D_l irreducible representation of the $\mathbf{R}_3$ group taking $2l + 1$ of the spherical harmonics Y_{lm} as the base. Therefore, according to the previous discussion in section 11–6, the representation of $\mathbf{R}_3$ obtained by taking the set of $(2l + 1)(2l' + 1)$ functions of the form of (11–158) will be the direct product representation $D_l \times D_{l'}$.

The character χ of this direct product representation can be easily obtained by observing that

$$\exp\left(\frac{-i\gamma\hat{L}_z}{\hbar}\right) Y_{lm}(\theta_1, \varphi_1)Y_{l'm'}(\theta_2, \varphi_2)$$

$$= \exp\left[\frac{-i\gamma(\hat{l}_{1z} + \hat{l}_{2z})}{\hbar}\right] Y_{lm}(\theta_1, \varphi_1)Y_{l'm'}(\theta_2, \varphi_2)$$

$$= \exp\left[-i\gamma(m + m')\right]Y_{lm}(\theta_1, \varphi_1)Y_{l'm'}(\theta_2, \varphi_2) \tag{11–159}$$

---

[11] It is conventional to use the Roman numerals I, II, III, ..., to designate neutral, singly ionized, doubly ionized, ... atoms, respectively.

and remembering the identity (11–137). Thus

$$\chi(D_l \times D_{l'}) = \sum_{m=-l}^{l} \sum_{m'=-l'}^{l'} e^{-i\gamma(m+m')} \tag{11-160}$$

Replacing the above double sum by another equivalent double sum, and comparing the result with (11–138), we obtain (Fig. 11–11)

$$\chi = \sum_{L=|l-l'|}^{l+l'} \sum_{M=-L}^{L} e^{-i\gamma M}$$

$$= \sum_{L=|l-l'|}^{l+l'} \chi_L \tag{11-161}$$

Therefore

$$D_l \times D_{l'} = D_{l+l'} + D_{l+l'-1} + D_{l+l'-2} + \cdots + D_{|l-l'|} \tag{11-162}$$

This is the Clebsch–Gordan series of the $\mathbf{R}_3$ group, which is a special case of (11–113). According to this result, we can arrange the $(2l + 1)(2l' + 1)$ functions of the form of (11–158) to obtain each set of $(2L + 1)$ functions which are eigenfunctions of $\hat{L}^2 \ (= (\hat{\mathbf{l}}_1 + \hat{\mathbf{l}}_2)^2)$ and $\hat{L}_z \ (= \hat{l}_{1z} + \hat{l}_{2z})$ with eigenvalues L and $M \ (= L, L - 1, L - 2, \ldots, -L)$, respectively, and the possible values of L are $l + l', l + l' - 1, l + l' - 2, \ldots, |l - l'|$. The consistency of this conclusion can be tested by observing that

$$[2(l + l') + 1] + [2(l + l' - 1) + 1] + \cdots + \{2|l - l'| + 1\}$$
$$= (2l + 1)(2l' + 1) \tag{11-163}$$

The Clebsch–Gordan series (11–162) can be applied to the coupling of more than two angular momenta. For the lowest energy configuration of the neutral nitrogen atom, which is essentially $(2p)^3$, for example, the Clebsch–Gordan series is

$$D_1 \times D_1 \times D_1 = D_1 \times (D_1 \times D_1) = D_1 \times (D_2 + D_1 + D_0)$$
$$= (D_3 + D_2 + D_1) + (D_2 + D_1 + D_0) + D_1$$
$$= D_3 + 2D_2 + 3D_1 + D_0 \tag{11-164}$$

Therefore, one F, two D, three P, and one S states result from the $(2p)^3$ configuration.

Since the rotation operator $e^{-i\gamma\hat{\mathbf{L}}\cdot\mathbf{u}/\hbar}$ of an n electron system commutes

FIGURE 11–11. Equivalence of the double sums (11–160) and (11–161). Each pair (m, m') is represented by a point of coordinate m and abscissa m'. In (a) horizontal lines indicate summations over m' with fixed m, while in (b) each Γ shape line indicates the summation over M with fixed L, and dotted lines are for points with common M.

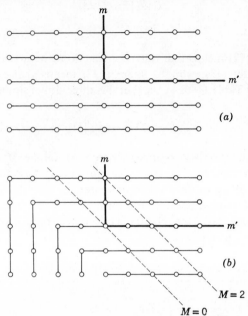

with every operator of the $\mathscr{S}_n$ group, we expect some correspondence between the irreducible representations of the $\mathbf{R}_3$ group and the $\mathscr{S}_n$ group. For the $(2p)^3$ configuration, we see that

$$f(r_1)f(r_2)f(r_3)Y_{11}(\theta_1, \varphi_1)Y_{11}(\theta_2, \varphi_2)Y_{11}(\theta_3, \varphi_3) \qquad (11\text{–}165)$$

is evidently an eigenfunction of $\hat{L}^2$ and $\hat{L}_z$ with eigenvalues $L = 3$ and $M = 3$. Hence this function belongs to the D_3 representation of the $\mathbf{R}_3$ group. We see simultaneously that (11–165) is a totally symmetric function, i.e., it belongs to the $\Gamma(3)$ representation of the $\mathscr{S}_3$ group. Following the same discussion we used with respect to the spin functions in section 11–7, we see that all seven functions of the F state obtained from (11–165) belong to the $\Gamma(3)$ representation.

For $M = 2$ of the $(2p)^3$ configuration, we have three functions

$$Y_{11}Y_{11}Y_{10} \qquad Y_{11}Y_{10}Y_{11} \quad \text{and} \quad Y_{10}Y_{11}Y_{11} \qquad (11\text{–}166)$$

where we neglected the radial parts and arranged the spherical harmonics in the order of electrons 1, 2, and 3. In other words, each one of (11–166) is a function like (11–165) except that one of the m's is changed from 1 to 0. The representation of the $\mathscr{S}_3$ group given by these three functions has the character of

$$
\begin{array}{cccc}
 & \hat{1} & (12) & (123) \\
\chi & 3 & 1 & 0
\end{array}
\qquad (11\text{--}167)
$$

where 1 under (12) means that only one function (the first one) of (11–166) is invariant under (12), while 0 under (123) means that none of them is invariant. Using Table 11–5, we see that this three-dimensional representation is reduced to

$$\Gamma = \Gamma(3) + \Gamma(21) \qquad (11\text{--}168)$$

The $\Gamma(3)$ representation which appears here must be the $M = 2$ component of the F state we found in (11–165). Therefore the two D states we found in (11–164) must belong to the $\Gamma(21)$ representation. (Remember that this is a two-dimensional representation.) Following this procedure we can obtain Table 11–10, which shows the relation between the irreducible representations of the $\mathbf{R}_3$ and $\mathscr{S}_3$ groups.

TABLE 11–10 *Symmetry Properties of Orbital Functions of $(nl)^x$ Configurations*

$l = 1: (np)^x$						
$x = 2$		$x = 3$		$x = 4$		
L		L		L		
2	$\Gamma(2)$	3	$\Gamma(3)$	4	$\Gamma(4)$	
1	$\Gamma(1^2)$	2	$\Gamma(21)$	3	$\Gamma(31)$	
0	$\Gamma(2)$	1	$\Gamma(3) + \Gamma(21)$	2	$\Gamma(4) + \Gamma(31) + \Gamma(2^2)$	
		0	$\Gamma(1^3)$	1	$\Gamma(31) + \Gamma(21^2)$	
				0	$\Gamma(4) + \Gamma(2^2)$	

$l = 2: (nd)^x$						
$x = 2$		$x = 3$		$x = 4$		
L		L		L		
4	$\Gamma(2)$	6	$\Gamma(3)$	8	$\Gamma(4)$	
3	$\Gamma(1^2)$	5	$\Gamma(21)$	7	$\Gamma(31)$	
2	$\Gamma(2)$	4	$\Gamma(3) + \Gamma(21)$	6	$\Gamma(4) + \Gamma(31) + \Gamma(2^2)$	
1	$\Gamma(1^2)$	3	$\Gamma(3) + \Gamma(21) + \Gamma(1^3)$	5	$\Gamma(4) + 2\Gamma(31) + \Gamma(21^2)$	
0	$\Gamma(2)$	2	$\Gamma(3) + 2\Gamma(21)$	4	$2\Gamma(4) + 2\Gamma(31) + 2\Gamma(2^2) + \Gamma(21^2)$	
		1	$\Gamma(21) + \Gamma(1^3)$	3	$3\Gamma(31) + \Gamma(2^2) + 2\Gamma(21^2)$	
		0	$\Gamma(3)$	2	$2\Gamma(4) + 2\Gamma(31) + 2\Gamma(2^2) + \Gamma(21^2) + \Gamma(1^4)$	
				1	$2\Gamma(31) + 2\Gamma(21^2)$	
				0	$\Gamma(4) + 2\Gamma(2^2)$	

In section 11–7, particularly in Table 11–9, we saw that due to the Pauli principle the orbital functions $\phi_{\mathscr{E}}$ which belong to some irreducible representations of the $\mathscr{S}_n$ group are forbidden, while each of those which belongs to one of the other irreducible representations is associated with a definite total spin quantum number S. Comparing Tables 11–10 and 11–9, we find that those states shown in Table 11–11 are permissible for each of the $(nl)^x$ configurations.

TABLE 11–11 *Possible Electronic States due to $(nl)^x$ Configurations*

Configuration			
np, $(np)^5$	^2P		
$(np)^2$, $(np)^4$	$^1S^1D$	^3P	
$(np)^3$	$^2P^2D$		^4S
nd, $(nd)^9$	^2D		
$(nd)^2$, $(nd)^8$	$^1S^1D^1G$	$^3P^3F$	
$(nd)^3$, $(nd)^7$	$^2P^2D^2D^2F^2G^2H$	$^4P^4F$	
$(nd)^4$, $(nd)^6$	$^1S^1S^1D^1D^1F^1G^1G^1I$	$^3P^3P^3D^3D^3F^3F^3G^3H$	^5D
$(nd)^5$	$^2S^2P^2D^2D^2D^2F^2F^2G^2G^2H^2I$	$^4P^4D^4F^4G$	^6S

In Table 11–11, S is given as the multiplicity, and each state is designated by

$$^{2S+1}L \qquad (11\text{–}169)$$

in which the conventional spectroscopic notation (11–149) is used for L. Thus ^2G, for example, means $L = 4$, $S = \frac{1}{2}$.

Hund's Rules

For the neutral nitrogen atom, we have seen that the lowest energy states have the configuration $(1s)^2(2s)^2(2p)^3$. According to Table 11–11 this configuration gives three possible states, ^2P, ^2D, and ^4S. Group theory, however, cannot tell which of these three states is the ground state or how much energy the other states have above a ground state. Elaborate numerical calculations, such as the SCF method, are necessary to find such a quantitative answer, and many have actually been done.[12] Hund, however, found[13] very simple

---

[12] For such calculations see J. C. Slater, *Quantum Theory of Atomic Structure*, McGraw-Hill, New York, 1960, or Chapter 8 of M. Mizushima, *Quantum Mechanics of Atomic Spectra and Atomic Structure*, W. A. Benjamin, New York, 1970.

[13] Hund found these rules in 1925, before quantum mechanics was developed.

empirical rules which tell which of those states given in Table 11–11 have lower energy than the others. His rules are:

(1) *Among possible states of a given configuration, the state or states with the highest S have the lowest energy.*
(2) *Among the states with the same S, the state with the highest L has the the lowest energy.* (11–170)

Therefore, the ground state of the neutral nitrogen atom is ^4S, while the lowest excited state is ^2D, and then comes ^2P, according to these rules. These predictions are in agreement with experiment which is shown in Fig. 11–12.

FIGURE 11–12.   Energy level diagram of the neutral nitrogen atom.

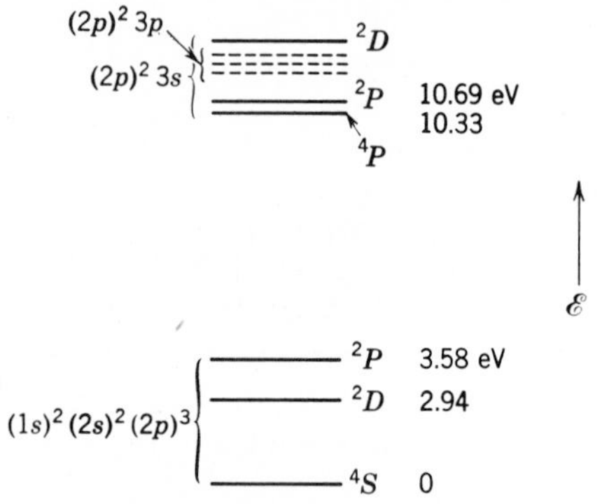

As readers may have noticed already the wave function (11–150) is not properly symmetrized. In order to perform the SCF calculation it is necessary to take wave functions which belong to the proper irreducible representation of the $\mathscr{S}_n$ group as required by the Pauli principle. The symmetrization changes the Hartree equations (11–151) into slightly more complicated equations, which are called the Hartree–Fock equations. The Hartree–Fock equations are solved rather accurately for the ground states of almost all light atoms and some heavy atoms such as *Pr*. Corrections due to the relativistic and correlation effects are also calculated for many atoms.

PROBLEM 11–35.   Show that (11–163) holds.

PROBLEM 11–36.   Confirm Table 11–11 for the $(np)^3$, $(nd)^3$, and $(nd)^4$ configurations.

PROBLEM 11–37.   Configurations like $(np)^6$ and $(nd)^{10}$ are called closed shells. Show that the only possible electronic state due to a closed shell is ^1S.

HINT. Only one wave function, similar to (11–165), is possible for a closed shell, and that has $M_L = 0$ and $M_S = 0$.

PROBLEM 11–38. Find the electronic configurations and three lowest energy states of carbon (6 electrons), phosphorus (15 electrons), and sulfur (16 electrons) atoms.

ANSWER. For the sulfur atom $(1s)^2(2s)^2(2p)^6(3s)^2(3p)^4$, the ground state is ^3P, then ^1D and ^1S in that order.

PROBLEM 11–39. It is pointed out in the last paragraph of this section that the orbital function $\phi_\mathscr{E}$ has to be properly symmetrized. For a two electron system show that the expectation value of $\hat{H}_2$, the nonrelativistic Hamiltonian, is

$$
\mathscr{E} = -\int \phi_1{}^*(r_1) \left(\frac{\hbar^2}{2m} \nabla_1{}^2 + \frac{Ze^2}{4\pi\varepsilon_0 r_1} \right) \phi_1(r_1) \, dv_1
$$

$$
- \int \phi_1{}^*(r^2) \left(\frac{\hbar^2}{2m} \nabla_2{}^2 + \frac{Ze^2}{4\pi\varepsilon_0 r_2} \right) \phi_2(r_2) \, dv_2
$$

$$
+ \iint |\phi_1(r_1)\phi_2(r_2)|^2 \, \frac{e^2}{4\pi\varepsilon_0 r_{12}} \, dv_1 \, dv_2
$$

$$
\pm \iint \phi_1{}^*(r_1)\phi_2{}^*(r_2) \, \frac{e^2}{4\pi\varepsilon_0 r_{12}} \, \phi_2(r_1)\phi_1(r_2) \, dv_1 \, dv_2
$$

where, in the last term, + and − signs are for the singlet ($S = 0$) and triplet ($S = 1$), respectively. Assume that

$$
\int \phi_i{}^*(r)\phi_j(r) \, dv = \delta_{ij}
$$

Notice that the last term of the above equation, which is called the *exchange integral*, gives the energy difference between the singlet and triplet states of the same configuration.

11–11 ROTATIONS INCLUDING SPIN AND THE SU$_2$ GROUP

It is pointed out in section 10–10, particularly in (10–166), that the Dirac Hamiltonian for the hydrogen atom $\hat{H}_D$ does not commute with $\hat{\mathbf{l}}$, but commutes with $\hat{\mathbf{j}}$. Therefore, the rotation group given by the operators $\exp(-i\gamma\hat{\mathbf{l}} \cdot \mathbf{u}/\hbar)$ is not a symmetry group of $\hat{H}_D$, but a group given by exp $(-i\gamma \hat{\jmath} \cdot \mathbf{u}/\hbar)$ is. The group of exp $(-i\gamma \hat{\jmath} \cdot \mathbf{u}/\hbar)$ is actually the same rotation group $\mathbf{R}_3$ as before, since $\hat{l}$ and $\hat{\jmath}$ satisfy exactly the same commutation relations, as seen in (10–168) and (8–53).

In this $\mathbf{R}_3$ group we take the spherical spinors $\mathbf{Y}_{ljm}$ as bases. The spherical spinors are defined by (10–225a) and (10–225b), and are the eigenfunctions of $\hat{l}^2$, $\hat{\jmath}^2$ and $\hat{\jmath}_z$.

As a special case we can take

$$\mathbf{Y}_{0\frac{1}{2}\frac{1}{2}} = \frac{1}{\sqrt{4\pi}} \begin{pmatrix} 1 \\ 0 \end{pmatrix} \quad \text{and} \quad \mathbf{Y}_{0\frac{1}{2}-\frac{1}{2}} = \frac{1}{\sqrt{4\pi}} \begin{pmatrix} 0 \\ 1 \end{pmatrix} \quad (11\text{–}171)$$

as the base functions. In this special case we obtain

$$e^{-i\gamma \hat{\jmath} \cdot \mathbf{u}/\hbar} \mathbf{Y}_{0\frac{1}{2}m} = e^{-i\gamma \hat{s} \cdot \mathbf{u}/\hbar} \mathbf{Y}_{0\frac{1}{2}m} \quad (11\text{–}172)$$

since $\hat{l} \mathbf{Y}_{0\frac{1}{2}m} = 0$. Using the identity (11–137) we obtain

$$e^{-i\gamma \hat{s} \cdot \mathbf{u}/\hbar} = e^{i\beta \hat{s} \cdot \mathbf{v}/\hbar} e^{-i\gamma \hat{s}_z/\hbar} e^{-i\beta \hat{s} \cdot \mathbf{v}/\hbar} \quad (11\text{–}173)$$

where $\mathbf{v}$ is a unit vector perpendicular to both $\mathbf{u}$ and the Z axis, and β is the angle between $\mathbf{u}$ and the Z axis. Since

$$(\hat{s} \cdot \mathbf{v})^2 = (\tfrac{1}{2}\hbar)^2 \quad (11\text{–}174)$$

as (10–68a) and (10–68b) imply, we obtain that

$$
\begin{aligned}
e^{-i\beta \hat{s} \cdot \mathbf{v}/\hbar} &= \sum_{n=0}^{\infty} \frac{1}{n!} \left(\frac{\beta}{i\hbar} \right)^n (\hat{s} \cdot \mathbf{v})^n \\
&= \sum_{n}^{\text{even}} \frac{1}{n!} (i\tfrac{1}{2}\beta)^n + \sum_{n}^{\text{odd}} \frac{1}{n!} (i\tfrac{1}{2}\beta)^{n-1} \left(\frac{\beta \hat{s} \cdot \mathbf{v}}{i\hbar} \right) \\
&= \cos(\tfrac{1}{2}\beta) - i \left(\frac{2\hat{s} \cdot \mathbf{v}}{\hbar} \right) \sin(\tfrac{1}{2}\beta)
\end{aligned}
\quad (11\text{–}175)
$$

If α is the azimuthal angle of $\mathbf{u}$, we see that

$$\hat{s} \cdot \mathbf{v} = \hat{s}_x \sin \alpha - \hat{s}_y \cos \alpha = i\tfrac{1}{2}(\hat{s}_+ e^{-i\alpha} - \hat{s}_- e^{i\alpha}) \quad (11\text{–}176)$$

See Fig. 11–13 for the geometry of the present problem.

FIGURE 11–13. $v_x = \sin \alpha$ and $v_y = -\cos \alpha$. See Fig. 11–6 for the direction of the unit vector **v**.

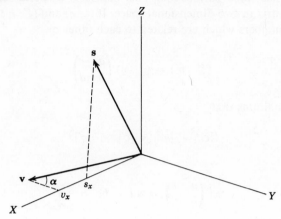

Using (10–196), (10–197), (11–175), and (11–176), we obtain

$$e^{-i\beta \mathbf{s} \cdot \mathbf{v}/\hbar}(\mathbf{Y}_{0\frac{1}{2}\frac{1}{2}} \quad \mathbf{Y}_{0\frac{1}{2}-\frac{1}{2}})$$

$$= (\mathbf{Y}_{0\frac{1}{2}\frac{1}{2}} \quad \mathbf{Y}_{0\frac{1}{2}-\frac{1}{2}}) \begin{pmatrix} \cos(\frac{1}{2}\beta) & \sin(\frac{1}{2}\beta)\, e^{-i\alpha} \\ -\sin(\frac{1}{2}\beta)\, e^{i\alpha} & \cos(\frac{1}{2}\beta) \end{pmatrix} \qquad (11\text{–}177)$$

Therefore, from (11–172), (11–173), and (11–177), we obtain

$$e^{-i\gamma \hat{\mathbf{j}} \cdot \mathbf{u}/\hbar}(\mathbf{Y}_{0\frac{1}{2}\frac{1}{2}} \quad \mathbf{Y}_{0\frac{1}{2}-\frac{1}{2}})$$

$$= (\mathbf{Y}_{0\frac{1}{2}\frac{1}{2}} \quad \mathbf{Y}_{0\frac{1}{2}-\frac{1}{2}}) \begin{pmatrix} \cos(\frac{1}{2}\beta) & -\sin(\frac{1}{2}\beta)\, e^{-i\alpha} \\ \sin(\frac{1}{2}\beta)\, e^{i\alpha} & \cos(\frac{1}{2}\beta) \end{pmatrix} \begin{pmatrix} e^{-i\frac{1}{2}\gamma} & 0 \\ 0 & e^{i\frac{1}{2}\gamma} \end{pmatrix}$$

$$\times \begin{pmatrix} \cos(\frac{1}{2}\beta) & \sin(\frac{1}{2}\beta)\, e^{-i\alpha} \\ -\sin(\frac{1}{2}\beta)\, e^{i\alpha} & \cos(\frac{1}{2}\beta) \end{pmatrix}$$

$$= (\mathbf{Y}_{0\frac{1}{2}\frac{1}{2}} \quad \mathbf{Y}_{0\frac{1}{2}-\frac{1}{2}})$$

$$\times \begin{pmatrix} \cos^2(\frac{1}{2}\beta)\, e^{-i\frac{1}{2}\gamma} + \sin^2(\frac{1}{2}\beta)\, e^{i\frac{1}{2}\gamma} & -i\sin\beta \sin(\frac{1}{2}\gamma)\, e^{-i\alpha} \\ -i\sin\beta \sin(\frac{1}{2}\gamma)\, e^{i\alpha} & \sin^2(\frac{1}{2}\beta)\, e^{-i\frac{1}{2}\gamma} + \cos^2(\frac{1}{2}\beta)\, e^{i\frac{1}{2}\gamma} \end{pmatrix}$$

$$(11\text{–}178)$$

where the last matrix is the two-dimensional representation of the rotation group $\mathbf{R}_3$, called the $D_{1/2}$ irreducible representation.

The matrix we obtained in (11–178) is a unitary matrix, with a determinant of $+1$, and has three parameters: This matrix is a special unitary transformation matrix in two-dimensional space. If (ξ, η) and (ξ', η') are two sets of complex numbers which are related to each other by

$$(\xi' \quad \eta') = (\xi \quad \eta) \begin{pmatrix} a & c \\ b & d \end{pmatrix} \tag{11-179}$$

under the conditions that

$$|\xi|^2 + |\eta|^2 = |\xi'|^2 + |\eta'|^2 \tag{11-180}$$

and

$$\det \begin{pmatrix} a & c \\ b & d \end{pmatrix} = ad - bc = +1 \tag{11-181}$$

then the transformation (11–179) is called the special unitary transformation in two-dimensional space. The transformation matrices with all possible values of parameters form a continuous group called the SU_2 group. Although each matrix has four elements, a, b, c, and d, and each element is complex, giving eight parameters altogether, conditions (11–180) and (11–181) imply the five restrictions

$$c = -b^* \qquad d = a^* \qquad |a|^2 + |b|^2 = 1 \tag{11-182}$$

Therefore each matrix of the SU_2 group is specified by three independent parameters. We thus see that the $D_{1/2}$ matrix representative, obtained in (11–178), is also an element of the SU_2 group. This relation seems to imply the isomorphism between the R_3 and SU_2 group, but that is not true. If we take a special case of $\gamma = 2\pi$ in (11–178) we obtain

$$D_{\frac{1}{2}}(\alpha, \beta, 2\pi) = \begin{pmatrix} -1 & 0 \\ 0 & -1 \end{pmatrix} \tag{11-183}$$

while the rotation operator $e^{-i2\pi \mathbf{j} \cdot \mathbf{u}/\hbar}$ should be equal to the identity operator $\hat{1}$. We see from (11–178) that the unit matrix is obtained not when $\gamma = 2\pi$, but when $\gamma = 4\pi$:

$$D_{\frac{1}{2}}(\alpha, \beta, 4\pi) = \begin{pmatrix} 1 & 0 \\ 0 & 1 \end{pmatrix} \tag{11-184}$$

Therefore the R_3 group is only homomorphic to the SU_2 group, and the correspondence between them is one to two.

Lie Groups

We have seen that the rotation group $\mathbf{R}_3$ can be made of either $e^{-i\gamma \hat{\mathbf{l}} \cdot \mathbf{u}/\hbar}$, or $e^{-i\gamma \hat{\jmath} \cdot \mathbf{u}/\hbar}$, or $e^{-i\gamma \hat{L} \cdot \mathbf{u}/\hbar}$. It does not matter which angular momentum operator is used. The only important things are that every rotation operator is obtained by integrating three independent infinitesimal rotation operators, for example, $1 - i\varepsilon \hat{l}_x/\hbar$, $1 - i\varepsilon \hat{l}_y/\hbar$, and $1 - i\varepsilon \hat{l}_z/\hbar$, and that they are related by the standard commutation relations, for example, $[\hat{l}_x, \hat{l}_y] = i\hbar \hat{l}_z$, etc. A vector space made of three components of an angular momentum, where multiplication rules are given by the standard commutation relations, is called the *Lie algebra* of the $\mathbf{R}_3$ group.

We have also seen the $\mathbf{O}_3$, $\mathbf{O}_3{}^+$, and $\mathbf{SU}_2$ groups. In general we can take a n-dimensional space and find an orthogonal transformation group $\mathbf{O}_n$, a unitary transformation group $\mathbf{U}_n$, and also $\mathbf{SU}_n$. These continuous groups are called *Lie groups*. The structure of each Lie group is given by its Lie algebra.

An atomic shell of orbital angular momentum l can accommodate up to $2(2l + 1)$ electrons. It is possible to apply the $\mathbf{U}_{4l+2}$ group to classify electronic states resulting from the l shell, since the transformations between possible electronic wave functions are unitary transformations, and such applications have been done.[14] We will discuss the $\mathbf{R}_4$ and $\mathbf{SU}_3$ groups later in this chapter.

PROBLEM 11–40.   Confirm the result of (11–175).

PROBLEM 11–41.   Obtain the results of (11–177) and (11–178).

PROBLEM 11–42.   Show that $D_{1/2}(\alpha, \beta, \gamma)$ given by (11–178) is a unitary matrix with determinant $+1$.

PROBLEM 11–43.   Obtain (11–182) from (11–179), (11–180), and (11–181).

PROBLEM 11–44.*   From (11–179) and (11–182) the two-dimensional representation of the $\mathbf{SU}_2$ group is given by

$$\begin{pmatrix} a & -b^* \\ b & a^* \end{pmatrix} \qquad |a|^2 + |b|^2 = 1$$

---

[14] See the article by B. R. Judd in *Group Theory and Its Applications* (edited by E. M. Loebl), Academic Press, New York, 1968. Comprehensive tables for some Lie groups can be found in B. G. Wybourne, *Symmetry Principles and Atomic Spectroscopy*, Wiley-Interscience, New York, 1970, and in Chapter 10 of M. Hamermesh, *Group Theory*, Addison Wesley, Reading, Massachusetts, 1962.

An $(n + 1)$-dimensional representation of the SU_2 group is obtained if we consider the transformation of

$$(\xi^n \ \xi^{n-1}\eta \ \xi^{n-2}\eta^2 \cdots \xi\eta^{n-1} \ \eta^n)$$

Obtain a three-dimensional representation of the SU_2 group in terms of a and b.

11–12 TENSOR OPERATORS. WIGNER SYMBOLS

It has been found so far that the spherical harmonics Y_{lm} belong to the D_l representation while the spherical spinors Y_{ljm} belong to the D_j representation of the R_3 group. Actually these are only two special sets of functions which belong to irreducible representations of this group. In general, we define a rotation operator as

$$e^{-i\gamma\hat{J}\cdot\mathbf{u}/\hbar} \tag{11–185}$$

where $\hat{J}$ is a general angular momentum operator defined solely by the commutation relations

$$[\hat{J}_x, \hat{J}_y] = i\hbar\hat{J}_z \qquad [\hat{J}_y, \hat{J}_z] = i\hbar\hat{J}_x \qquad [\hat{J}_z, \hat{J}_x] = i\hbar\hat{J}_y \tag{11–186}$$

Taking the eigenkets defined by

$$\hat{J}_{\pm}|JM\rangle = \hbar\sqrt{J(J + 1) - M(M \pm 1)}\,|J\,M\pm1\rangle \tag{11–187a}$$

$$\hat{J}_z|JM\rangle = \hbar M|JM\rangle \tag{11–187b}$$

which also imply

$$\hat{J}^2|JM\rangle = \hbar^2 J(J + 1)|JM\rangle \tag{11–187c}$$

we can see that the irreducible representation $D_J(\alpha, \beta, \gamma)$ is obtained as

$$e^{-i\gamma\hat{J}\cdot\mathbf{u}/\hbar}(|JJ\rangle \ |J\,J-1\rangle \cdots |J\,-J\rangle) = (|JJ\rangle \cdots |J\,-J\rangle)D_J \tag{11–188}$$

The fact that D_J which appears in (11–188) is exactly the same representation previously obtained can be seen by observing that the matrix representative of an infinitesimal rotation operator $1 - i\varepsilon\hat{J}\cdot\mathbf{u}/\hbar$ is completely determined by (11–187a) and (11–187b), which are the same relations as those between $\hat{\mathbf{l}}$ and Y_{lm}, and between $\hat{\mathbf{j}}$ and $\mathbf{Y}_{ljm}$. Since any finite rotations are obtained by integrating infinitesimal rotations, the D_J matrix in (11–188) is exactly the same matrix representative we obtained before.

Clebsch–Gordan Coefficients

When $|J_1M_1\rangle$ and $|J_2M_2\rangle$ belong to D_{J_1} and D_{J_2} representations, the product ket $|J_1M_1J_2M_2\rangle$ ($\equiv |J_1M_1\rangle|J_2M_2\rangle$) belongs to the direct product representation $D_{J_1} \times D_{J_2}$. The Clebsch–Gordan series (11–162) guarantees, then, that one eigenket $|JM\rangle$ which belongs to a D_J representation can be obtained as a linear combination of $|J_1M_1J_2M_2\rangle$. Hence the expression

$$|JM\rangle = \sum_{M_1,M_2} |J_1M_1J_2M_2\rangle\langle J_1M_1J_2M_2 \mid JM\rangle \qquad (11\text{–}189)$$

exists, whenever $J_1 + J_2 \geq J \geq |J_1 - J_2|$. The coefficients

$$\langle J_1M_1J_2M_2|JM\rangle$$

in (11–189) are called the *Clebsch–Gordan coefficients*. The previous formula (10–207) is a special case of (11–189), and Table 10–5 gives the Clebsch–Gordan coefficients for that special case.

One can obtain the Clebsch–Gordan coefficients by the same method used in section 10–11. For example one can show that the C–G is zero unless $M = M_1 + M_2$ by operating on both sides of (11–189) with $\hat{J}_z$.

Tensor Operators

A set of tensor operators $\hat{T}_{JM}$ $(M = J, J - 1, \ldots, -J)$ is defined by

$$[\hat{J}_z, \hat{T}_{JM}] = \hbar M\hat{T}_{JM} \qquad (11\text{–}190a)$$

$$[\hat{J}_+, \hat{T}_{JM}] = \hbar\sqrt{J(J + 1) - M(M + 1)}\ \hat{T}_{JM+1} \qquad (11\text{–}190b)$$

and

$$[\hat{J}_-, \hat{T}_{JM}] = \hbar\sqrt{J(J + 1) - M(M - 1)}\ \hat{T}_{JM-1} \qquad (11\text{–}190c)$$

Notice that Y_{JM}, $\mathbf{Y}_{lJM}$, and $|JM\rangle$ satisfy these relations, and therefore, are special tensor operators. However, this definition also shows that operators like

$$\hat{T}_{11} = -\sqrt{\tfrac{1}{2}}\,\hat{J}_+ \qquad \hat{T}_{10} = \hat{J}_z \qquad \hat{T}_{1-1} = \sqrt{\tfrac{1}{2}}\,\hat{J}_- \qquad (11\text{–}191)$$

are tensor operators.

If $|00\rangle$ is a ket which belongs to the D_0 representation, or

$$\hat{\mathbf{J}}|00\rangle = 0 \tag{11–192}$$

we see that

$$\hat{\mathbf{J}}\hat{T}_{JM}|00\rangle = [\hat{\mathbf{J}}, \hat{T}_{JM}]|00\rangle \tag{11–193}$$

From (11–187a), (11–187b), (11–190a), (11–190b), (11–190c), and (11–193) we see that

$$\hat{T}_{JM}|00\rangle = A|JM\rangle \tag{11–194}$$

where A is a normalization constant which is independent of M. Using (11–194) we obtain

$$
\begin{aligned}
e^{-i\gamma\hat{\mathbf{J}}\cdot\mathbf{u}/\hbar}(|JJ\rangle\ |J\,J-1\rangle \cdots |J\,-J\rangle) \\
= A^{-1}e^{-i\gamma\hat{\mathbf{J}}\cdot\mathbf{u}/\hbar}(\hat{T}_{JJ}\ \hat{T}_{JJ-1} \cdots \hat{T}_{J-J})e^{i\gamma\hat{\mathbf{J}}\cdot\mathbf{u}/\hbar}e^{-i\gamma\hat{\mathbf{J}}\cdot\mathbf{u}/\hbar}|00\rangle \\
= A^{-1}e^{-i\gamma\hat{\mathbf{J}}\cdot\mathbf{u}/\hbar}(\hat{T}_{JJ}\ \hat{T}_{JJ-1} \cdots \hat{T}_{J-J})e^{i\gamma\hat{\mathbf{J}}\cdot\mathbf{u}/\hbar}|00\rangle
\end{aligned}
\tag{11–195}
$$

On the other hand,

$$(|JJ\rangle\ |J\,J-1\rangle \cdots |J\,-J\rangle)D_J = A^{-1}(\hat{T}_{JJ}\ \hat{T}_{JJ-1} \cdots \hat{T}_{J-J})D_J|00\rangle \tag{11–196}$$

but (11–195) and (11–196) are equal to each other. Therefore

$$e^{-i\gamma\hat{\mathbf{J}}\cdot\mathbf{u}/\hbar}(\hat{T}_{JJ}\ \hat{T}_{JJ-1} \cdots \hat{T}_{J-J})e^{i\gamma\hat{\mathbf{J}}\cdot\mathbf{u}/\hbar} = (\hat{T}_{JJ} \cdots \hat{T}_{J-J})D_J \tag{11–197}$$

Tensor operators belong to the D_J representation in this sense. It is actually possible to calculate D_J using this relation.

Tensor operators can be coupled to give another tensor operator. If $\hat{T}_{J_1M_1}$ and $\hat{U}_{J_2M_2}$ are two sets of tensor operators operating on different spaces, then

$$\hat{X}_{JM} = \sum_{M_1,M_2} \hat{T}_{J_1M_1}\hat{U}_{J_2M_2}\langle J_1M_1J_2M_2 \mid JM \rangle \tag{11–198}$$

is another tensor operator. Here the coefficients are the Clebsch–Gordan coefficients.

A special case of this coupling is the scalar product of two tensor operators

$$\hat{X}_{00} = \sum_{M=J}^{-J} (-1)^M \hat{T}_{JM} \hat{U}_{J-M} \tag{11-199}$$

For example, the spin–orbit coupling $\hat{s} \cdot \hat{l}$ is the scalar product of two vector operators $(J = 1)$, $\hat{s}$ and $\hat{l}$.

Wigner–Eckart Theorem

It is often necessary to calculate the matrix element of a tensor operator $\hat{T}_{JM}$,

$$\langle \alpha'' J'' M'' | \hat{T}_{JM} | \alpha' J' M' \rangle \tag{11-200}$$

where α' and α'' stand for all other quantum numbers necessary to specify the states.

Formula (11–194) shows that we can find a normalized tensor operator $\hat{t}_{J'M'}$ such that

$$|\alpha' J' M'\rangle = \hat{t}_{J'M'} |\alpha' 00\rangle \tag{11-201}$$

while (11–198) gives

$$\hat{T}_{JM} \hat{t}_{J'M'} = \sum_{J'''} \hat{U}_{J'''M'''} \langle J''' M''' \mid JMJ'M' \rangle \tag{11-202}$$

as its inverse transformation, where $\hat{U}_{J'''M'''}$ is a tensor operator, which we do not have to specify. Putting (11–201) and (11–202) into (11–200) we obtain

$$\begin{aligned}
\langle \alpha'' J'' M'' | \hat{T}_{JM} | \alpha' J' M' \rangle &= \langle \alpha'' J'' M'' | \hat{T}_{JM} \hat{t}_{J'M'} | \alpha' 00 \rangle \\
&= \sum_{J'''} \langle \alpha'' J'' M'' | \hat{U}_{J'''M'''} | \alpha' 00 \rangle \langle J''' M''' \mid JMJ'M' \rangle \\
&= \sum_{J'''} A \langle \alpha'' J'' M'' \mid \alpha' J''' M''' \rangle \langle J''' M''' \mid JMJ'M' \rangle \\
&= A \langle J'' M'' \mid JMJ'M' \rangle \tag{11-203}
\end{aligned}$$

where we have used (11–194) again, and then used formulas (11–94) and (11–95). This result, which is called the *Wigner–Eckart theorem*, shows that the M dependence of the matrix element of a tensor operator is given by the Clebsch–Gordan coefficients.

Wigner Symbols

Wigner introduced the Wigner $3j$ symbols to express formula (11–203) as

$$\langle \alpha''J''M''|\hat{T}_{JM}|\alpha'J'M'\rangle$$
$$= (-1)^{J''-M''} \begin{pmatrix} J'' & J & J' \\ -M'' & M & M' \end{pmatrix} \langle \alpha''J''\|\hat{T}_J\|\alpha'J'\rangle \quad (11\text{–}204)$$

where $\langle \alpha''J''\|\hat{T}_J\|\alpha'J'\rangle$ is an M-independent factor called the double-bar matrix element. The $3j$ symbol is defined, in terms of the Clebsch–Gordan coefficient, as

$$\begin{pmatrix} J'' & J & J' \\ -M'' & M & M' \end{pmatrix} = (-1)^{J-J'+M''} \frac{\langle JMJ'M' \mid J''M''\rangle}{\sqrt{2J''+1}} \quad (11\text{–}205)$$

The Wigner $3j$ symbols, although equivalent to the Clebsch–Gordan coefficients, are more useful because of their high symmetry. It is found that

$$\begin{pmatrix} abc \\ def \end{pmatrix} = \begin{pmatrix} bca \\ efd \end{pmatrix} = \begin{pmatrix} cab \\ fde \end{pmatrix} = (-1)^{a+b+c}\begin{pmatrix} acb \\ dfe \end{pmatrix} = (-1)^{a+b+c}\begin{pmatrix} bac \\ edf \end{pmatrix}$$
$$= (-1)^{a+b+c}\begin{pmatrix} cba \\ fed \end{pmatrix} = (-1)^{a+b+c}\begin{pmatrix} a & b & c \\ -d & -e & -f \end{pmatrix} \quad (11\text{–}206)$$

The $3j$ symbol is zero unless $M + M' - M'' = 0$ and unless J, J', and J'' can form a triangle, as (11–205) clearly shows.

Values of some $3j$ symbols are tabulated in Table 11–12.[15] Some typical double-bar matrix elements are

$$\langle J''\|\hat{\mathbf{J}}\|J'\rangle = \hbar\sqrt{J'(J'+1)(2J'+1)}\ \delta_{J',J''} \quad (11\text{–}207)$$

$$\langle l\|Y_1\|l+1\rangle = -\langle l+1\|Y_1\|l\rangle = -\sqrt{3(l+1)/(4\pi)} \quad (11\text{–}208)$$

$$\langle l''\|Y_k\|l'\rangle = (-1)^{l''}\sqrt{\frac{(2l''+1)(2k+1)(2l'+1)}{4\pi}}\begin{pmatrix} l'' & k & l' \\ 0 & 0 & 0 \end{pmatrix} \quad (11\text{–}209)$$

Quite often a system has two angular momenta $\hat{\mathbf{J}}_1$ and $\hat{\mathbf{J}}_2$ which are coupled to form a total angular momentum $\hat{\mathbf{J}}\ (= \hat{\mathbf{J}}_1 + \hat{\mathbf{J}}_2)$. It is always possible

---

[15] For more values see, for example, A. R. Edmonds, *Angular Momentum in Quantum Mechanics,* Princeton University Press, Princeton, 1959. A table of Clebsch–Gordan coefficients can be found in E. U. Condon and G. H. Shortley, *The Theory of Atomic Spectra,* Cambridge University Press, London, 1937.

to find eigenkets $|J_1 J_2 J M\rangle$ of $\hat{J}_1{}^2$, $\hat{J}_2{}^2$, $\hat{J}^2$, and $\hat{J}_z$, with eigenvalues $J_1(J_1 + 1)\hbar^2$, $J_2(J_2 + 1)\hbar^2$, $J(J + 1)\hbar^2$, and $M\hbar$, respectively. Formula (11–204) is, of course, still applicable to the compound system as

$$\langle \alpha'' J_1'' J_2'' J'' M'' | \hat{T}_{JM} | \alpha' J_1' J_2' J' M' \rangle$$

$$= (-1)^{J''-M''} \begin{pmatrix} J'' & J & J' \\ -M'' & M & M' \end{pmatrix} \langle \alpha'' J_1'' J_2'' J'' \| \hat{T}_J \| \alpha' J_1' J_2' J' \rangle \quad (11\text{–}210)$$

If $\hat{T}_{JM}$ is a tensor operator of $\hat{\mathbf{J}}_1$, but independent of $\hat{\mathbf{J}}_2$, then, using (11–189), we obtain

$$\langle \alpha'' J_1'' J_2'' J'' M'' | \hat{T}_{JM} | \alpha' J_1' J_2' J' M' \rangle$$

$$= \sum_{M_1'', M_2'', M_1', M_2'} \langle J'' M'' \mid J_1'' M_1'' J_2'' M_2'' \rangle \langle J_1' M_1' J_2' M_2' \mid J' M' \rangle$$

$$\times \langle \alpha'' J_1'' M_1'' | \hat{T}_{JM} | \alpha' J_1' M_1' \rangle \langle J_2'' M_2'' \mid J_2' M_2' \rangle$$

$$= \sum_{M_1'', M_2'', M'} (-1)^{2J_2'' - M_1'' - M'' - J_1' - M'} \sqrt{(2J'' + 1)(2J' + 1)}$$

$$\times \begin{pmatrix} J_1'' & J_2'' & J'' \\ M_1'' & M_2'' & -M'' \end{pmatrix} \begin{pmatrix} J_1' & J_2'' & J' \\ M_1' & M_1'' & -M' \end{pmatrix} \begin{pmatrix} J_1'' & J & J_1' \\ -M_1'' & M & M_1'' \end{pmatrix}$$

$$\times \langle \alpha'' J_1'' \| \hat{T}_J \| \alpha' J_1' \rangle \quad (11\text{–}211)$$

where we used (11–204) and (11–205). Since (11–210) and (11–211) should be the same, we can express $\langle \alpha'' J_1'' J_2'' J'' \| \hat{T}_J \| \alpha' J_1' J_2' J' \rangle$ by means of more fundamental double-bar matrix element $\langle \alpha'' J_1'' \| \hat{T}_J \| \alpha' J_1' \rangle$. Such an expression can actually be obtained using the orthogonality relation

$$\sum_{d,e} \begin{pmatrix} abc \\ def \end{pmatrix} \begin{pmatrix} abc' \\ def' \end{pmatrix} = \frac{1}{2c + 1} \delta_{c,c'} \delta_{f,f'} \quad (11\text{–}212)$$

which is just the orthonormality relation of the Clebsch–Gordan coefficients (Problem 11–49). The resulting expression will include products of four $3j$ symbols summed over all M values. The complicated expression was first calculated by Racah, who expressed the final result by his W function. Wigner showed that Racah's formula can be expressed by the Wigner $6j$ symbol, which has higher symmetries than the W function. Racah's formula in terms of the $6j$ symbol is

$$\langle \alpha'' J_1'' J_2'' J'' \| \hat{T}_J \| \alpha' J_1' J_2' J' \rangle$$

$$= (-1)^{J_1'' + J_2'' + J' + J} \sqrt{(2J'' + 1)(2J' + 1)} \begin{Bmatrix} J_1'' & J'' & J_2'' \\ J' & J_1' & J \end{Bmatrix}$$

$$\times \langle \alpha'' J_1'' \| \hat{T}_J \| \alpha' J_1' \rangle \delta_{J_2', J_2''} \quad (11\text{–}213)$$

Values of the $6j$ symbols are tabulated in Table 11–12.

TABLE 11–12 *Wigner Symbols*

3*j* Symbols

Symmetry See (11–206).
Orthogonality

$$\sum_{d,e} \begin{pmatrix} abc \\ def \end{pmatrix} \begin{pmatrix} abc' \\ def' \end{pmatrix} = \frac{1}{2c + 1} \delta_{c,c'} \delta_{f,f'} \tag{11–212}$$

$$\sum_{c=|a-b|}^{a+b} \sum_{f=c}^{c} \begin{pmatrix} abc \\ def \end{pmatrix} \begin{pmatrix} abc \\ d'e'f \end{pmatrix} (2c + 1) = \delta_{d,d'} \delta_{e,e'}$$

Values

$$\begin{pmatrix} abc \\ 000 \end{pmatrix} = \begin{cases} 0 & \text{if} \quad a + b + c \quad \text{is odd} \\ (-1)^p \sqrt{\Delta(abc)}\, \dfrac{p!}{(p - a)!\,(p - b)!\,(p - c)!} & \text{if} \quad a + b + c = 2p \\ & \qquad \text{is even} \end{cases}$$

$$\Delta(abc) = \frac{(a + b - c)!\,(b + c - a)!\,(c + a - b)!}{(a + b + c + 1)!}$$

$$\begin{pmatrix} J & J & 0 \\ M & -M & 0 \end{pmatrix} = (-1)^{J-M} \frac{1}{\sqrt{2J + 1}}$$

$$\begin{pmatrix} J + \frac{1}{2} & J & \frac{1}{2} \\ M & -M - \frac{1}{2} & \frac{1}{2} \end{pmatrix} = (-1)^{J-M-\frac{1}{2}} \sqrt{\frac{J - M + \frac{1}{2}}{(2J + 2)(2J + 1)}}$$

$$\begin{pmatrix} J & J & 1 \\ M & -M & 0 \end{pmatrix} = (-1)^{J-M} \frac{M}{\sqrt{J(J + 1)(2J + 1)}}$$

$$\begin{pmatrix} J & J & 1 \\ M & -M - 1 & 1 \end{pmatrix} = (-1)^{J-M} \sqrt{\frac{(J - M)(J + M + 1)}{2J(J + 1)(2J + 1)}}$$

$$\begin{pmatrix} J + 1 & J & 1 \\ M & -M & 0 \end{pmatrix} = (-1)^{J-M-1} \sqrt{\frac{(J + 1)^2 - M^2}{(J + 1)(2J + 1)(2J + 3)}}$$

$$\begin{pmatrix} J + 1 & J & 1 \\ M & -M - 1 & 1 \end{pmatrix} = (-1)^{J-M-1} \sqrt{\frac{(J - M)(J - M + 1)}{2(J + 1)(2J + 1)(2J + 3)}}$$

Clebsch–Gordan Coefficients

$$\langle J + \tfrac{1}{2}\, M\, J - M - \tfrac{1}{2} \mid \tfrac{1}{2}\tfrac{1}{2} \rangle = (-1)^{J-M-\frac{1}{2}} \sqrt{\frac{J - M + \frac{1}{2}}{(J + 1)(2J + 1)}}$$

$$\langle J\, M\, J - M \mid 1\, 0 \rangle = (-1)^{J+M} \frac{\sqrt{3}\, M}{\sqrt{J(J + 1)(2J + 1)}}$$

$$\langle J\, M\, J - M - 1 \mid 1\, 1 \rangle = (-1)^{J+M} \sqrt{\frac{3(J + M)(J - M + 1)}{2J(J + 1)(2J + 1)}}$$

$$\langle J+1\,M\,J-M \mid 10 \rangle = (-1)^{J+M} \sqrt{\frac{3[(J+1)^2 - M^2]}{(J+1)(2J+1)(2J+3)}}$$

$$\langle J+1\,M\,J-M+1 \mid 11 \rangle = (-1)^{J+M+1} \sqrt{\frac{3(J+M)(J+M+1)}{2(J+1)(2J+1)(2J+3)}}$$

$$\langle J\,M\,J-M \mid 20 \rangle = (-1)^{J+M} \frac{2\sqrt{5}[3M^2 - J(J+1)]}{\sqrt{(2J-1)2J(2J+1)(2J+2)(2J+3)}}$$

$$\langle J\,M\,J-M+1 \mid 21 \rangle = (-1)^{J+M+1}(1-2M) \sqrt{\frac{30(J+M)(J-M+1)}{(2J-1)2J(2J+1)(2J+2)(2J+3)}}$$

$$\langle J\,M\,J-M+2 \mid 22 \rangle = (-1)^{J+M} \sqrt{\frac{30(J+M+1)(J-M)(J-M+1)(J-M+2)}{(2J-1)2J(2J+1)(2J+2)(2J+3)}}$$

$$\langle J+1\,M\,J-M \mid 20 \rangle = (-1)^{J+M}2M \sqrt{\frac{30(J-M+1)(J+M+1)}{2J(2J+1)(2J+2)(2J+3)(2J+4)}}$$

$$\langle J+1\,M\,J-M+1 \mid 21 \rangle$$
$$= (-1)^{J+M}2(J-2M+2) \sqrt{\frac{5(J+M)(J+M+1)}{2J(2J+1)(2J+2)(2J+3)(2J+4)}}$$

$$\langle J+1\,M\,J-M+2 \mid 22 \rangle$$
$$= (-1)^{J+M+1} \sqrt{\frac{20(J+M-1)(J-M)(J+M+)(J+M+2)}{2J(2J+1)(2J+2)(2J+3)(2J+4)}}$$

$6j$ Symbols

Symmetry

$$\begin{Bmatrix} abc \\ def \end{Bmatrix} = \begin{Bmatrix} bac \\ edf \end{Bmatrix} = \begin{Bmatrix} acb \\ dfe \end{Bmatrix} = \begin{Bmatrix} cba \\ fed \end{Bmatrix} = \begin{Bmatrix} bca \\ efd \end{Bmatrix} = \begin{Bmatrix} cab \\ fde \end{Bmatrix}$$

$$\begin{Bmatrix} aef \\ dbc \end{Bmatrix} = \begin{Bmatrix} dbf \\ aec \end{Bmatrix} = \begin{Bmatrix} dec \\ abf \end{Bmatrix}$$

Sum Rules

$$\sum_x (-1)^{2x}(2x+1) \begin{Bmatrix} abx \\ abf \end{Bmatrix} = 1$$

$$\sum_x (-1)^{a+b+x}(2x+1) \begin{Bmatrix} abx \\ baf \end{Bmatrix} = (2a+1)(2b+1)\delta_{f,0}$$

$$\sum_x (2x+1) \begin{Bmatrix} abx \\ cdf \end{Bmatrix} \begin{Bmatrix} abx \\ cdg \end{Bmatrix} = \frac{1}{2f+1}\delta_{f,g}$$

$$\sum_x (-1)^{f+g+x}(2x+1) \begin{Bmatrix} abx \\ cdf \end{Bmatrix} \begin{Bmatrix} bax \\ cdg \end{Bmatrix} = \begin{Bmatrix} adf \\ bcg \end{Bmatrix}$$

TABLE 11–12   *Wigner Symbols* (continued)

Values

$$\begin{Bmatrix} ab0 \\ cde \end{Bmatrix} = (-1)^{a+e+c} \frac{1}{\sqrt{(2a+1)(2c+1)}} \delta_{a,b}\delta_{c,d}$$

$$\begin{Bmatrix} a & a+\frac{1}{2} & \frac{1}{2} \\ b+\frac{1}{2} & b & c \end{Bmatrix} = (-1)^{a+b+c+1} \sqrt{\frac{(a+b-c+1)(a+b+c+2)}{(2a+1)(2a+2)(2b+1)(2b+2)}}$$

$$\begin{Bmatrix} a & a+\frac{1}{2} & \frac{1}{2} \\ b & b+\frac{1}{2} & c+\frac{1}{2} \end{Bmatrix} = (-1)^{a+b+c+1} \sqrt{\frac{(a-b+c+1)(-a+b+c+1)}{(2a+1)(2a+2)(2b+1)(2b+2)}}$$

$$\begin{Bmatrix} aa1 \\ bbc \end{Bmatrix} = (-1)^{a+b+c} \frac{-X}{\sqrt{a(2a+1)(2a+2)b(2b+1)(2b+2)}}$$

$$\begin{Bmatrix} aa2 \\ bbc \end{Bmatrix} = (-1)^{a+b+c}$$

$$\frac{2[3X(X-1) - 4a(a+1)b(b+1)]}{\sqrt{(2a-1)2a(2a+1)(2a+2)(2a+3)(2b-1)2b(2b+1)(2b+2)(2b+3)}}$$

$$X = a(a+1) + b(b+1) - c(c+1)$$

9*j* Symbols

Symmetry

$$\begin{Bmatrix} abc \\ def \\ ghi \end{Bmatrix} = \begin{Bmatrix} adg \\ beh \\ cfi \end{Bmatrix} = \begin{Bmatrix} ifc \\ heb \\ gda \end{Bmatrix}$$

$$(-1)^{a+b+c+d+e+f+g+h+i} \begin{Bmatrix} abc \\ def \\ ghi \end{Bmatrix} = \begin{Bmatrix} def \\ abc \\ ghi \end{Bmatrix} = \begin{Bmatrix} ghi \\ def \\ abc \end{Bmatrix} = \begin{Bmatrix} abc \\ ghi \\ def \end{Bmatrix}$$

$$= \begin{Bmatrix} bac \\ edf \\ hgi \end{Bmatrix} \qquad \begin{Bmatrix} cba \\ fed \\ ihg \end{Bmatrix} = \begin{Bmatrix} acb \\ dfe \\ hih \end{Bmatrix}$$

Orthogonality

$$\sum_{g,h} (2g+1)(2h+1) \begin{Bmatrix} abc \\ def \\ ghi \end{Bmatrix} \begin{Bmatrix} abc' \\ def' \\ ghi \end{Bmatrix} = \frac{1}{(2c+1)(2f+1)} \delta_{c,c'}\delta_{f,f'}$$

Again in a compound system, where $\hat{\mathbf{J}} = \hat{\mathbf{J}}_1 + \hat{\mathbf{J}}_2$, we may want to calculate a matrix element of an operator $\hat{X}_{JM}$ given by two operators $\hat{T}_{J_1M_1}$ and $\hat{U}_{J_2M_2}$ in the form of (11–198), but $\hat{T}_{J_1M_1}$'s are given by $\hat{\mathbf{J}}_1$ only, while $\hat{U}_{J_2M_2}$'s are given by $\hat{\mathbf{J}}_2$ only. The Wigner–Eckart formula (11–204) is still valid and gives

$$\langle \alpha''J_1''J_2''J''M''|\hat{X}_{JM}|\alpha'J_1'J_2'J'M'\rangle$$

$$= (-1)^{J''-M''} \begin{pmatrix} J'' & J & J' \\ -M'' & M & M' \end{pmatrix} \langle \alpha''J_1''J_2''J''\|\hat{X}_J\|\alpha'J'J'J'\rangle \quad (11\text{–}214)$$

and this double-bar matrix element can be expressed, using (11–213) and other formulas, by means of double-bar matrix elements of $\hat{T}_{J_1}$ and $\hat{U}_{J_2}$ and some products of the $6j$ symbols. Wigner introduced the Wigner $9j$ symbol to express the result of this complicated calculation as

$$\langle \alpha''J_1''J_2''J''\|\hat{X}_J\|\alpha'J_1'J_2'J'\rangle$$

$$= \sum_\alpha \langle \alpha''J_1''\|\hat{T}_{J_1}\|\alpha J_1'\rangle\langle \alpha J_2''\|\hat{U}_{J_2}\|\alpha'J_2'\rangle\sqrt{(2J+1)(2J'+1)(2J''+1)}$$

$$\times \begin{Bmatrix} J_1'' & J_1' & J_1 \\ J_2'' & J_2' & J_2 \\ J'' & J' & J \end{Bmatrix} \quad (11\text{–}215)$$

Some values of the $9j$ symbols are tabulated, but they also can be calculated from the $6j$ symbols using the relation

$$\begin{Bmatrix} abc \\ def \\ ghi \end{Bmatrix} = \sum_k (-1)^{2k}(2k+1) \begin{Bmatrix} abc \\ fik \end{Bmatrix} \begin{Bmatrix} def \\ bkh \end{Bmatrix} \begin{Bmatrix} ghi \\ kad \end{Bmatrix} \quad (11\text{–}216)$$

Particularly

$$\begin{Bmatrix} abc \\ def \\ gh0 \end{Bmatrix} = \delta_{c,f}\delta_{g,h} \frac{(-1)^{b+d+c+g}}{\sqrt{(2c+1)(2g+1)}} \begin{Bmatrix} abc \\ edg \end{Bmatrix} \quad (11\text{–}217)$$

Therefore

$$\langle \alpha''J_1''J_2''J''M''|\hat{X}_{00}|\alpha'J_1'J_2'J'M'\rangle$$

$$= \left\langle \alpha''J_1''J_2''J''M'' \left| \sum_M (-1)^M \hat{T}_{JM}\hat{U}_{J-M} \right| \alpha'J_1'J_2'J'M' \right\rangle$$

$$= (-1)^{J_1'+J_2''+J'} \begin{Bmatrix} J' & J_2'' & J_1'' \\ J & J_1' & J_2' \end{Bmatrix} \sum \langle \alpha''J_1''\|\hat{T}_J\|\alpha J_1'\rangle$$

$$\times \langle \alpha J_2''\|\hat{U}_J\|\alpha'J_2'\rangle\delta_{J'',J'}\delta_{M'',M'} \quad (11\text{–}218)$$

PROBLEM 11–45. Show that the three operators shown in (11–191) form a set of tensor operators $\hat{T}_{1m}$.

PROBLEM 11–46. Knowing that $3\hat{J}_z^2 - \hat{J}^2$ is a tensor operator $\hat{T}_{20}$, find the other four components.

ANSWER.

$$\hat{T}_{22} = \sqrt{3/2}\,\hat{J}_+^2 \qquad \hat{T}_{21} = -\sqrt{3/2}\,\hat{J}_+(2\hat{J}_z + 1)$$

$$\hat{T}_{2-1} = \sqrt{3/2}\,\hat{J}_-(2\hat{J}_z - 1) \qquad \hat{T}_{2-2} = \sqrt{3/2}\,\hat{J}_-^2$$

PROBLEM 11–47. As a special case of (11–191), we see that $-\sqrt{\tfrac{1}{2}}\hat{s}_+$, $\hat{s}_z$, and $\sqrt{\tfrac{1}{2}}\hat{s}_-$ form a set of tensor operators with $J = 1$. Using this set in (11–197), obtain the matrix representative D_1 of $e^{i\beta\hat{s}\cdot\mathbf{v}/\hbar}$, which is given by (11–175).

ANSWER.

$$D_1(e^{-i\beta\hat{s}\cdot\mathbf{v}/\hbar}) = \frac{1}{2}\begin{pmatrix} \cos\beta + 1 & \sqrt{2}\,e^{-i\alpha}\sin\beta & -e^{-2i\alpha}(\cos\beta - 1) \\ -\sqrt{2}\,e^{i\alpha}\sin\beta & 2\cos\beta & \sqrt{2}\,e^{-i\alpha}\sin\beta \\ -e^{2i\alpha}(\cos\beta - 1) & -\sqrt{2}\,e^{i\alpha}\sin\beta & \cos\beta + 1 \end{pmatrix}$$

Note that $\hat{s}_+\hat{s}_+ = \hat{s}_-\hat{s}_- = 0$, while $\hat{s}_+\hat{s}_-\hat{s}_+ = \hbar^2\hat{s}_+$ and $\hat{s}_-\hat{s}_+\hat{s}_- = \hbar^2\hat{s}_-$.

PROBLEM 11–48. Using the result of Problem 11–47 obtain $D_1(\alpha, \beta, \gamma)$ of the general rotation $e^{-i\gamma\hat{J}\cdot\mathbf{u}/\hbar}$.

ANSWER. Using (11–137) we obtain

$$\begin{pmatrix} \tfrac{1}{2}(\cos^2\beta + 1)\cos\gamma + \tfrac{1}{2}\sin^2\beta - i\cos\beta\sin\gamma \\ \dfrac{1}{\sqrt{2}}e^{i\alpha}[\sin\beta\cos\beta(\cos\gamma - 1) - i\sin\beta\sin\gamma] \\ -\tfrac{1}{2}e^{2i\alpha}[(\cos^2\beta - 1)\cos\gamma + \sin^2\beta] \\ \dfrac{1}{\sqrt{2}}e^{-i\alpha}[\sin\beta\cos\beta(\cos\gamma - 1) - i\sin\beta\sin\gamma] \\ \sin^2\beta\cos\gamma + \cos^2\beta \\ -\dfrac{1}{\sqrt{2}}e^{i\alpha}[\sin\beta\cos\beta(\cos\gamma - 1) + i\sin\beta\sin\gamma] \\ -\tfrac{1}{2}e^{-2i\alpha}[(\cos^2\beta - 1)\cos\gamma + \sin^2\beta] \\ -\dfrac{1}{\sqrt{2}}e^{-i\alpha}[\sin\beta\cos\beta(\cos\gamma - 1) + i\sin\beta\sin\gamma] \\ \tfrac{1}{2}[(\cos^2\beta + 1)\cos\gamma + \sin^2\beta + 2i\cos\beta\sin\gamma] \end{pmatrix}$$

PROBLEM 11–49. Knowing that (11–189) is a unitary transformation, (a) obtain the orthogonality relation for the Clebsch–Gordan coefficients; (b) obtain the orthogonality relation (11–212) for the 3*j* symbols.

PROBLEM 11–50. Calculate
(a) $\langle SLJM|\hat{S}_z|SLJM \rangle$
(b) $\langle SLJM|(3\hat{S}_z^{\,2} - \hat{S}^2)|SLJM \rangle$
(c) $\langle SLJM|\hat{\mathbf{S}} \cdot \mathbf{L}|SLJM \rangle$
using the method developed in this section.

PROBLEM 11–51. Prove the statement with respect to (10–285), namely if **f** is a vector operator,

$$\langle JM|\hat{f}_z|JM \rangle = \frac{M}{J(J + 1)\hbar} \langle JM|\hat{\mathbf{f}} \cdot \hat{\mathbf{J}}|JM \rangle$$

HINT.   From (11–204)

$$\langle JM|\hat{f}_z|JM \rangle = (-1)^{2(J-M)}\langle J\|\hat{\mathbf{f}}\|J \rangle \frac{M}{\sqrt{J(J + 1)(2J + 1)}}$$

On the other hand,

$$\langle JM|\hat{\mathbf{f}} \cdot \hat{\mathbf{J}}|JM \rangle = \langle JJ|\hat{\mathbf{f}} \cdot \hat{\mathbf{J}}|JJ \rangle = \langle JJ|\hat{f}_z|JJ \rangle\langle JJ|\hat{J}_z|JJ \rangle$$

$$+ \tfrac{1}{2}\langle JJ|\hat{f}_+|JJ - 1 \rangle\langle JJ - 1|\hat{J}_-|JJ \rangle$$

$$= \langle J\|\hat{\mathbf{f}}\|J \rangle\sqrt{J(J + 1)/(2J + 1)}$$

using (10–197), (11–191), and (11–204).

PROBLEM 11–52. Obtain tensor operators $\hat{X}_{2m}$ and $\hat{X}_{1m}$ from $\hat{\mathbf{l}}$ and $\hat{\mathbf{s}}$ in the form of (11–198).

ANSWER.

$$\hat{X}_{22} = \sqrt{6}\,\hat{l}_+\hat{s}_+ \qquad \hat{X}_{21} = -\sqrt{6}\,(\hat{l}_+\hat{s}_z + \hat{l}_z\hat{s}_+)$$

$$\hat{X}_{20} = -\hat{l}_+\hat{s}_- + 4\hat{l}_z\hat{s}_z - \hat{l}_-\hat{s}_+ \qquad \hat{X}_{2-1} = \sqrt{6}\,(\hat{l}_z\hat{s}_- + \hat{l}_-\hat{s}_z)$$

$$\hat{X}_{2-2} = \sqrt{6}\,\hat{l}_-\hat{s}_- \qquad \hat{X}_{11} = \sqrt{2}\,(\hat{l}_+\hat{s}_z - \hat{l}_z\hat{s}_+)$$

$$X_{10} = \hat{l}_+\hat{s}_- - \hat{l}_-\hat{s}_+ \qquad \hat{X}_{1-1} = \sqrt{2}\,(-\hat{l}_z\hat{s}_- + \hat{l}_-\hat{s}_z)$$

11–13 NUCLEAR POTENTIAL IN YUKAWA THEORY, ISOSPIN, AND NUCLEAR SHELL MODEL

Nucleons, i.e., protons and neutrons, are almost as stable as electrons (see Table 11–20). Their masses are about $1840m_e$, where m_e is the electron mass, but their spin is $\frac{1}{2}$ just like the electron. The magnetic moment associated with the spin, however, is quite different from the one predicted by the Dirac theory, as seen in (10–301) and (10–302), which indicates that nucleons may be compound particles. This is also indicated from the fact that nucleons have several excited states, or resonance states.

The nuclear forces which bind nucleons together to form a nucleus are strong but of very short range—about 10^{-15} m. Because the range is so short it is difficult to investigate the nature of nuclear forces as precisely as that of the electromagnetic forces. Many experiments have been done on nucleon–nucleon scattering.[16] (See Fig. 11–14.) As discussed in Chapter 9, it is straightforward to calculate the scattering cross sections once the potential is known. However, it is difficult, if not impossible, to deduce a potential uniquely from given data of scattering cross sections. It is, however, generally agreed that the Yukawa potential $e^{-\kappa r}/r$ represents the r dependence of the nuclear force, when r is larger than $1.5/\kappa$. The constant κ^{-1} gives the range of the Yukawa potential, and is found to be between 10^{-15} m and 1.5×10^{-15} m.

Yukawa Theory

Hideki Yukawa, in his Nobel prize paper[17] of 1935, pointed out that while the Coulomb potential ke/r, where $k = (4\pi\varepsilon_0)^{-1}$, is a solution of

$$\nabla^2 \phi = -4\pi k e\, \delta(\mathbf{r}) \qquad (11\text{–}219)$$

or the Green's function of the Poisson equation (5–20), the Yukawa potential $ge^{-\kappa r}/r$ is a solution of

$$(\nabla^2 - \kappa^2)\Psi = -4\pi g\, \delta(\mathbf{r}) \qquad (11\text{–}220)$$

as seen in Problem 5–61. Since the Poisson equation is the static, or non-relativistic, limit of

$$\square \phi = -4\pi k \rho \qquad (6\text{–}10)$$

---

[16] These data are compiled and reduced to phenomenological phase shifts by MacGregor and his collaborators. For a review, see M. H. MacGregor, *Physics Today*, **22**, 21 (1969). (See Fig. 11–14.)

[17] H. Yukawa, *Proc. Phys.-Math. Soc. Japan*, **17**, 48 (1935). The Nobel prize was awarded to him in 1949.

FIGURE 11–14. Observed phase shifts in nucleon–nucleon scattering.

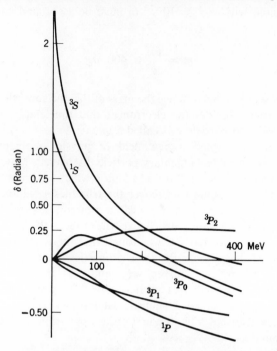

where ρ is the charge density, equation (11–220) is also regarded as a limiting form of

$$(\Box - \kappa^2)\Psi = -4\pi g\rho' \qquad (11\text{–}221)$$

where ρ' is the nucleon particle density. When the fields are free, i.e., $\rho = \rho' = 0$, these two equations reduce to

$$\Box\phi = 0 \qquad (11\text{–}222)$$

and

$$(\Box - \kappa^2)\Psi = 0 \qquad (11\text{–}223)$$

respectively. The former is the equation for the electromagnetic waves, while the latter is the Klein–Gordon equation (10–6′) if

$$\kappa = \frac{mc}{\hbar} \qquad (11\text{–}224)$$

At this point Yukawa concluded that the nuclear force in the form of the Yukawa potential with $\kappa^{-1} \cong 10^{-15}$ m must be associated with a particle of mass

$$m = \frac{\hbar\kappa}{c} \cong 400 \; m_e \qquad (11\text{--}225)$$

where m_e is the electron mass. Since the mass of the nucleons, the only known elementary particles besides the electron at that time, had mass of about $1840 m_e$, the predicted particle was called a *meson*.

Yukawa's theory opened a new field of physics, elementary particle physics. The first unstable elementary particle discovered, however, was not the meson Yukawa predicted, but was a lepton which we know as the muon. It took more than ten years to discover the true meson which gives rise to nuclear force.[18] The meson, which we call a π meson or pion, was found to have a mass m_π of about $273 m_e$, which gives the range of the Yukawa potential $\kappa^{-1} = 1.40 \times 10^{-15}$ m. Later it was found that there are three pions, π^+, π^0, and π^-, which have about the same properties except for their charges, which are $+e$, 0, and $-e$, respectively, and that they are pseudoscalar particles, or spin zero parity -1 particles. Modifying the original Yukawa theory (which may be called a neutral scalar meson theory) according to these experimental discoveries, the first-order approximation to the nuclear potential is found as

$$U_{12}(r) = -\frac{1}{3}\left(\frac{g}{\hbar}\right)^2 \left(\frac{m_\pi}{m_n}\right)^2 \hat{\tau}_1 \cdot \hat{\tau}_2$$

$$\times \left[(\hat{s}_1 \cdot \hat{s}_2) + \left(\frac{1}{(\kappa r)^2} + \frac{1}{\kappa r} + \frac{1}{3}\right)\hat{S}_{12}\right]\frac{e^{-\kappa r}}{r} \qquad (11\text{--}226)$$

where $\hat{\tau}_1$ and $\hat{\tau}_2$ are the *isospin* operators for the nucleons 1 and 2, respectively, and

$$\hat{S}_{12} = 3(\hat{s}_1 \cdot \mathbf{r})(\hat{s}_2 \cdot \mathbf{r})r^{-2} - \hat{s}_1 \cdot \hat{s}_2 \qquad (11\text{--}227)$$

where $\hat{s}_1$ and $\hat{s}_2$ are the familiar spin operators.

Isospin

The isospin operators $\hat{\tau}$ are defined by the same commutation relations as those for the angular momentum operators (10–168), except that their components are not along the XYZ axes, since the isospin space in which

[18] C. M. G. Lattes, H. Muirhead, G. P. S. Occhialini, and C. F. Powell, *Nature*, **159**, 694 (1947).

$\hat{\tau}$'s are defined has nothing to do with ordinary space. If one of the three components of $\hat{\tau}$ is $\hat{\tau}_\zeta$, the commutation relations imply, according to the discussions in section 10–10, that we can find a set of eigenkets such that

$$\hat{\tau}^2|\tau\tau_\zeta\rangle = \tau(\tau + 1)|\tau\tau_\zeta\rangle$$

and

$$\hat{\tau}_\zeta|\tau\tau_\zeta\rangle = \tau_\zeta|\tau\tau_\zeta\rangle \qquad (11\text{–}228)$$

where

$$\tau_\zeta = \tau, \tau - 1, \tau - 2, \ldots, -\tau$$

Since the proton and neutron are very similar in their physical properties, we associate τ of $\frac{1}{2}$ to the nucleon and regard the proton and neutron as the $\tau_\zeta = \frac{1}{2}$ and $-\frac{1}{2}$ substates of the nucleon.[19] The concept of isospin is applicable to all hadrons (mesons and baryons) and their resonance states (section 11–18).

Two-Nucleon System

In addition to the classical degrees of freedom in the three-dimensional space, each nucleon has those of ordinary spin and isospin. Since the nuclear potential energy (11–226) depends on the isospin operators through the scalar product $\hat{\tau}_1 \cdot \hat{\tau}_2$ only, it obviously commutes with the total isospin operator

$$\hat{\mathbf{T}} = \hat{\tau}_1 + \hat{\tau}_2 \qquad (11\text{–}229)$$

If we neglect the Coulomb potential and a small mass difference between the proton and neutron, the Hamiltonian of a two-nucleon system commutes with $\hat{\mathbf{T}}$; or it is invariant under any rotations in the isospin space. If T is the quantum number of the total isospin, such that the eigenvalue of $\hat{T}^2$ is $T(T + 1)$, we see that the eigenstates of this approximate Hamiltonian for the two-nucleon system are classified into isospin triplets ($T = 1$) and isospin singlets ($T = 0$). In this case each eigenfunction is a product of a factor containing the isospin and a factor, which depends on both the ordinary spin and the orbital coordinates. According to the Pauli principle, the product or the eigenfunction itself should be antisymmetric with respect to the exchange of two nucleons. Since the isospin functions have the same mathematical structure as the familiar spin functions discussed in section 11–7,

---

[19] In nuclear physics the proton and neutron are defined as $\tau_\zeta = -\frac{1}{2}$ and $\frac{1}{2}$ components, respectively, while in elementary particle physics or high-energy physics, they are defined as $\tau_\zeta = \frac{1}{2}$ and $-\frac{1}{2}$ components, respectively. We take the latter convention in this book.

the isospin function with $T = 1$ is symmetric and that with $T = 0$ is anti-symmetric with respect to the exchange of nucleons. Therefore, the orbital and ordinary spin part of each eigenfunction must be antisymmetric or symmetric for $T = 1$ or 0, respectively.

The deuteron, which we designate as D, is made of one proton and one neutron, which are bound together with a dissociation energy of 2.226(4) MeV. Since its spin (total angular momentum) is 1 and the magnetic g factor is 1.914708(18), which is very close to the sum of those of proton and neutron [given in (10–302)], the ground state of the deuteron must be almost a ^3S state, or $S = 1$ and $L = 0$. Since ^3S is a symmetric state, it is an isospin singlet state ($T = 0$). No excited state of the deuteron has been observed, but a strong resonance scattering is observed at a very low energy in the nucleon–nucleon scattering when $S = 0$ and $L = 0$, or $T = 1$. The isospin triplet state, which has three components, $pp(T_\zeta = 1)$, $pn(T_\zeta = 0)$, and $nn(T_\zeta = -1)$, is an unstable state in this case. Fig. 11–15 illustrates these levels.

FIGURE 11–15. Energy levels of two nucleon systems. $T_\zeta = 1, 0,$ and -1 represent systems of two proton, neutron and proton (deuteron), and two neutron systems, respectively. The Coulomb potential and the mass difference between proton and neutron are neglected.

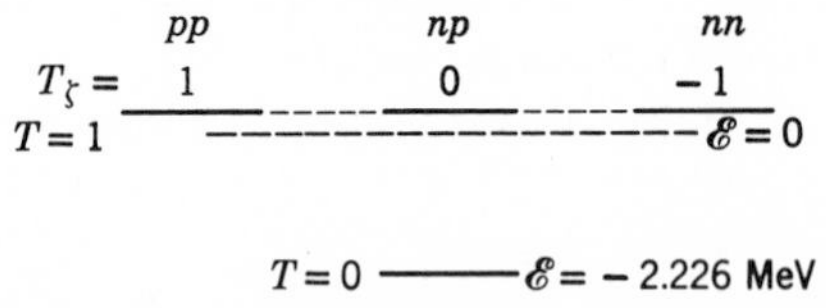

$$
\begin{array}{cccc}
 & pp & np & nn \\
T_\zeta = & 1 & 0 & -1 \\
T = 1 & \text{------------} & \mathscr{E} = 0 \\
\\
T = 0 & \text{------} & \mathscr{E} = -2.226\ \text{MeV}
\end{array}
$$

It is observed that the deuteron has a small but nonzero electric quadruple moment, and that its magnetic g factor is close but slightly different from the sum of those of proton and neutron. These facts show that the ground state of the deuteron is not exactly a ^3S state but a small amount of ^3D is mixed in. This result strongly supports the existence of the tensor force, or the $\hat{S}_{12}$ term, in the nuclear potential (11–226), which, in turn, supports the Yukawa theory with the pseudoscalar pion. The pseudoscalar nature of the pion predicts the existence of the tensor force, and simultaneously predicts the pion to be a particle of spin 0 and intrinsic parity -1: The latter properties of the pion are also confirmed from scattering experiments.

Inter-Nucleon Potentials

The theoretical nuclear potential (11–226) is obtained by the second-order perturbation theory which includes only one pion exchange. The potential is therefore called OPEP (one pion exchange potential). The coupling constant g^2 is observed from pion–nucleon scattering to be 14.7, and m_π

and m_n (nucleon mass) are, of course, known. When one calculates phase shifts of nucleon–nucleon scattering by OPEP one finds appreciable deviations from observed phase shifts. Hamada and Johnston[20] added a hard core (a potential which goes to infinity when $r < r_{core}$) and additional terms to (11–226) which are proportional to $(e^{-\kappa r}/r)^2$ and $(e^{-\kappa r}/r)^3$, respectively, They found that when $r_{core} = 0.343/\kappa$ and suitable coupling constants for the additional terms are chosen, their potential can reproduce observed scattering cross section for a large range of energy and angles.

The correction terms proposed by Hamada and Johnston[20] are purely phenomenological. A more physical potential called OBEP (one boson exchange potential) has been tried by a few other physicists.[21] The idea in OBEP is to take into account the effect of other mesons, namely, η, ρ, and ω on the nuclear potential to the second-order perturbation. The isospin, spin, parity, and mass of each of these mesons are known, and they are also known to interact strongly with nucleons. The coupling constants g^2 were adjusted to fit experimental data. It was found, however, that the existence of one or two scalar mesons of a mass of about $600m_e$ have to be assumed in addition to the known π, η, ρ, and ω mesons, in this scheme. Such scalar mesons have not been observed yet.

Nuclear Shell Model

Goeppert–Mayer and Jensen[22] independently proposed a phenomenological nuclear shell model. They assumed the one-nucleon energy levels as shown in Fig. 11–16, which correspond to Fig. 11–10 of the atomic shell model. These shells, or levels, are filled up according to the Pauli principle by available protons and neutrons in a fashion similar to the method of the atomic shell model, but they assumed that nuclear forces couple each pair of protons or neutrons in a given shell to make the total angular momentum of each pair zero. Thus for even–even nuclei, which have an even number of protons and neutrons, they predict the nuclear spin I, which is actually the total angular momentum of the nucleus, to be zero in their ground states, while for even–odd or odd–even nuclei, which have an even number of protons and an odd number of neutrons or vice-versa, the nuclear spin I is given by j of the odd nucleon in the top unfilled shell. They did not consider odd–odd nuclei.

[20] T. Hamada and I. D. Johnston, *Nucl. Phys.* **34**, 382 (1962).
[21] A. E. S. Green and T. S. Sawada, *Revs. Modern Phys.* **39**, 594 (1967); R. Bryan and B. L. Scott, *Phys. Rev.* **177**, 1435 (1969); G. Saunier and J. M. Pearson, *Phys. Rev. C* **1**, 1353 (1970).
[22] M. Goeppert Mayer, *Phys. Rev.* **75**, 1969 (1949); **78**, 16, 22 (1950). O. Haxel, J. H. D. Jensen, and H. E. Suess, *Phys. Rev.* **75**, 1766 (1949). See A. Bohr and B. R. Mottelson, *Nuclear Structure*, Vol. 1, W. A. Benjamin, New York, 1969.

FIGURE 11–16. One-nucleon levels in the nuclear shell model. Each level is designated by nlj, but n does not correspond to any number of nodes. It is used just to distinguish levels with the same l and j.

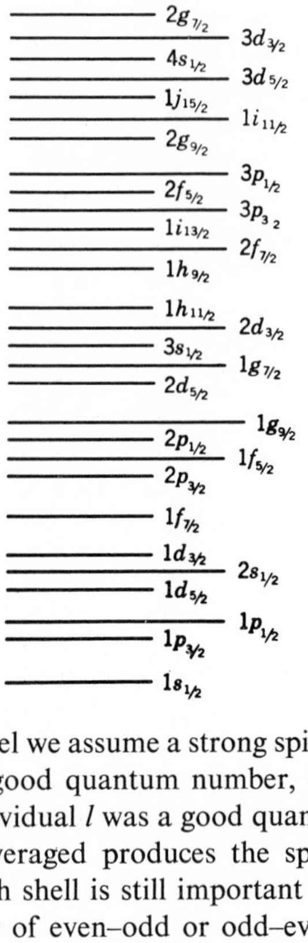

In the nuclear shell model we assume a strong spin–orbit interaction, which makes an individual j a good quantum number, in contrast to the atomic shell model, where an individual l was a good quantum number. Presumably the tensor force when averaged produces the spin–orbit interaction. The l-quantum number of each shell is still important in determining the parity of nuclei. Thus the parity of even–odd or odd–even nuclei in their ground states is given by $(-1)^l$, where l is that of the odd nucleons.

For example, Li^7 is made of three protons and four neutrons. Since the proton configuration in the ground state is $(1s_{1/2})^2(1p_{3/2})$, the parity is -1 and I is $\frac{3}{2}$ in this state, in agreement with experiment.

If we neglect the Coulomb potential and the mass difference between proton and neutron, and assume that the nuclear forces commute with the total isospin operator $\hat{T}$, as (11–226) does, we can reformulate a nuclear shell model more analogous to the atomic shell model.

In the previous example of Li^7, we first neglect the difference between protons and neutrons, and regard the nucleus to be made of seven nucleons. Since the $1s_{1/2}$ state can accommodate four nucleons, counting the possibilities of m_j being $\frac{1}{2}$ and $-\frac{1}{2}$ and those of τ_ζ being $\frac{1}{2}$ and $-\frac{1}{2}$, the nucleon configuration of the ground state is $(1s_{1/2})^4(1p_{3/2})^3$. The angular momenta of the three nucleons in the $1p_{3/2}$ shell are coupled according to the Clebsch–Gordan series

$$D_{3/2} \times D_{3/2} \times D_{3/2} = D_{3/2} \times (D_{3/2} \times D_{3/2})$$
$$= D_{3/2} \times (D_3 + D_2 + D_1 + D_0)$$
$$= D_{9/2} + 2D_{7/2} + 3D_{5/2} + 4D_{3/2} + 2D_{1/2}$$

$$(11–230)$$

The relation between the $\mathbf{R}_3$ group and the $\mathscr{S}_3$ group with respect to these states in (11–230) can be found in the same way as the one discussed in section 11–10, and the results including other typical cases are given in Table 11–13, which corresponds to Table 11–10.

TABLE 11–13 *Symmetry Properties of Spin–Orbital Functions of $(nj)^x$ Configurations*

$j = \frac{1}{2}$ (See Table 11–8.)

$j = \frac{3}{2}$

$x = 2$		$x = 3$		$x = 4$
J		J		J
3	$\Gamma(2)$	$\frac{9}{2}$	$\Gamma(3)$	6 $\Gamma(4)$
2	$\Gamma(1^2)$	$\frac{7}{2}$	$\Gamma(21)$	5 $\Gamma(31)$
1	$\Gamma(2)$	$\frac{5}{2}$	$\Gamma(3) + \Gamma(21)$	4 $\Gamma(4) + \Gamma(31) + \Gamma(2^2)$
0	$\Gamma(1^2)$	$\frac{3}{2}$	$\Gamma(3) + \Gamma(21) + \Gamma(1^3)$	3 $\Gamma(4) + 2\Gamma(31) + \Gamma(21^2)$
		$\frac{1}{2}$	$\Gamma(21)$	2 $\Gamma(4) + \Gamma(31) + 2\Gamma(2^2) + \Gamma(21^2)$
				1 $2\Gamma(31) + \Gamma(21^2)$
				0 $\Gamma(4) + \Gamma(2^2) + \Gamma(1^4)$

The isospin functions of the three-nucleon system take $T = \frac{3}{2}$ and $T = \frac{1}{2}$, and they belong to the $\Gamma(3)$ and $\Gamma(21)$ representations, respectively, according to Table 11–8. Coupling the isospin functions with the spin–orbit functions according to the Pauli principle, we see that for $T = \frac{1}{2}$ one each of the states $I = \frac{7}{2}, \frac{5}{2}, \frac{3}{2}$, and $\frac{1}{2}$ are allowed, while for $T = \frac{3}{2}$ only one state with $I = \frac{3}{2}$ is allowed, for the $(1p_{3/2})^3$ configuration.

Observed energy levels are shown in Fig. 11–17. Notice that $T = \frac{1}{2}$ states have two components; $T_\zeta = \frac{1}{2}$, which corresponds to Be^7 and $T_\zeta = -\frac{1}{2}$, which corresponds to Li^7. When $T = \frac{3}{2}$, each energy level includes that of B$^7(T_\zeta = \frac{3}{2})$ and He$^7(T_\zeta = -\frac{3}{2})$ in addition. According to a calculation, the $T = \frac{3}{2}, I = \frac{3}{2}$ level, predicted by the group theoretic argument, appears

FIGURE 11–17. Calculated and observed energy levels of the $(1s_{1/2})^4(1p_{3/2})^3$ nucleon configuration. The Coulomb potentials are neglected in the calculation, and are corrected in the observed energy levels.

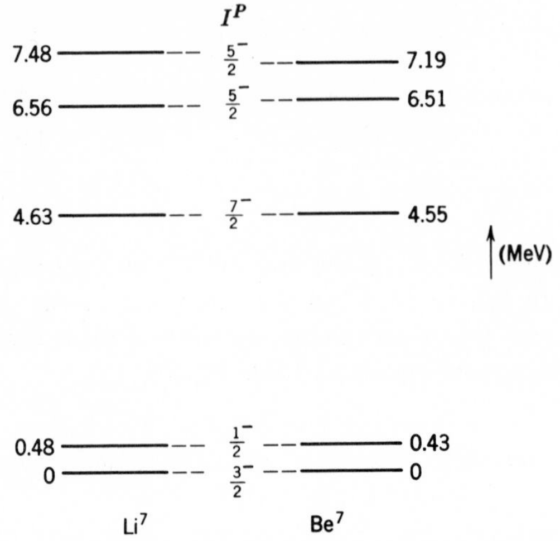

only at a higher energy than those of $T = \frac{1}{2}$. Actually both B^7 and Li^7 are unstable.

The existence of the Coulomb potential and the mass difference between proton and neutron remove the degeneracy of isospin levels. As a result, levels with higher T_ζ have higher energies than those with lower T_ζ with the same T.

The same concept is applicable to heavier nuclei. Some examples are shown in Fig. 11–18.

PROBLEM 11–53. Show that $(\hat{s}_1 \cdot \mathbf{r})(\hat{s}_2 \cdot \mathbf{r})$ term in $\hat{S}_{12}$ of (11–227) does not commute with $\hat{\mathbf{S}} = \hat{s}_1 + \hat{s}_2$, but commutes with $\hat{\mathbf{J}} = \mathbf{r} \times \hat{\mathbf{p}} + \hat{\mathbf{S}}$, where $\hat{\mathbf{p}}$ is the momentum conjugate to $\mathbf{r}$.

PROBLEM 11–54. Show that $\hat{S}_{12}$ of (11–227) can be expressed as a product operator $\hat{X}_{00}$ in the form of (11–199), if we take

$$\hat{T}_{20} = \tfrac{1}{2}(3\hat{s}_{1z}\hat{s}_{2z} - \hat{s}_1 \cdot \hat{s}_2), \quad \hat{U}_{20} = \frac{3z^2 - r^2}{r^2}, \quad \hat{U}_{21} = \frac{-\sqrt{6}\,z(x + iy)}{r^2},$$

and

$$\hat{U}_{22} = \sqrt{\tfrac{3}{2}}(x + iy)^2$$

You can use the result of Problem 11–52 to find the other components of $\hat{T}_{2m}$.

FIGURE 11–18(a). Energy levels of O^{17} and F^{17}. All levels of F^{17} are shifted down by 3.54 MeV to make the comparison easier. The ground state configuration is $(1s_{1/2})^4(1p_{3/2})^8(1p_{1/2})^4(1d_{1/2})$. The first excited state in each case is obtained by exciting a nucleon from $1d_{5/2}$ to $2s_{1/2}$ shell, while most of the other excited states come from $(1p_{1/2})^3(1d_{5/2})^2$ configuration. (A. Bohr and B. R. Mottelson, *Nuclear Structure*, Vol. 1, W. A. Benjamin, New York, 1969.)

FIGURE 11–18(b). Energy levels of K^{39} and Ca^{39}. All levels of Ca^{39} are shifted down by 7.3 MeV to make the comparison easier. The ground state configuration is $(1s_{1/2})^4(1p_{3/2})^8(1p_{1/2})^4(1d_{5/2})^{12}(2s_{1/2})^4(1d_{3/2})^7$. The first excited state in each case is obtained by $(2s_{1/2})^3(1d_{3/2})^8$ configuration. (A. Bohr and B. R. Mottelson, *Nuclear Structure*, Vol. 1, W. A. Benjamin, New York, 1969.)

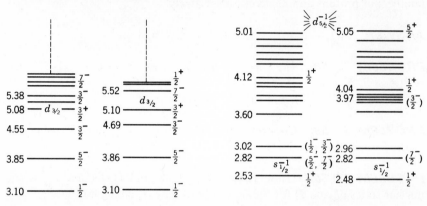

(a)   *(b)*

PROBLEM 11–55. Using the result of Problem 11–54 and (11–218) show that

$$\langle {}^3S_1 M | \hat{S}_{12} | {}^3D_J M \rangle = -15^{-(1/2)} \langle 1 \| \hat{T}_2 \| 1 \rangle \langle 0 \| \hat{U}_2 \| 2 \rangle \delta_{J,1}$$

where $\hat{T}_{2m}$ and $\hat{U}_{2m}$ are the tensor operators defined in Problem 11–54. This result shows that the mixing of the ^3D state in the ^3S state is possible through the $\hat{S}_{12}$ term of (11–227).

PROBLEM 11–56. (a) Show that when $T = 0$ in a two-nucleon system, the possible spin–orbit functions are ^3S, ^3D, ^3G, ..., ^1P, ^1F, ..., etc.
(b) Showing that $\hat{U}$ of (11–226) is invariant under the space inversion $\hat{I}$, show that the matrix elements of $\hat{U}$ between L even states and L odd states are zero.

PROBLEM 11–57. According to the Mayer–Jensen nuclear shell model, obtain the parity and nuclear spin I in the ground states of the following nuclei: $0^{17}(Z = 8, N = 9)$, $A^{39}(Z = 18, N = 21)$, $Cr^{53}(Z = 24, N = 29)$, $Sr^{87}(Z = 38, N = 49)$, and $Mo^{95}(Z = 42, N = 53)$. Z and N are the numbers of protons and neutrons, respectively.

ANSWER. $+1$, -1, -1, $+1$, $+1$; $\frac{5}{2}$, $\frac{7}{2}$, $\frac{3}{2}$, $\frac{9}{2}$, and $\frac{5}{2}$.

PROBLEM 11–58. The same problem as Problem 11–57 for $N^{15}(Z = 7, N = 8)$, $Cl^{35}(Z = 17, N = 18)$, $Co^{57}(Z = 27, N = 30)$, $Ga^{69}(Z = 31, N = 38)$, and $In^{115}(Z = 49, N = 66)$.

PROBLEM 11–59. Confirm Table 11–13 for $j = \frac{3}{2}$, $x = 3$, using the method we used in obtaining Table 11–10.

PROBLEM 11–60. Using Tables 11–7, 11–8, and 11–13, find allowable nuclear states (T and I) for the $(np_{3/2})^3$ and $(np_{3/2})^4$ configurations.

11–14 OCTAHEDRAL POINT GROUP AND OTHER POINT GROUPS

An atom (or ion) in a crystal is under the influence of several neighboring atoms, which destroy the rotational homogeneity of the space at the central atom. However, since the neighboring atoms are located with some regularities, the potential they exert on the central atom maintains some symmetry.

The symmetry of the potential due to neighboring atoms, called crystal fields or ligand fields, depends on each crystal and on each atom considered. As a typical example, let us consider an Na^+ (or Cl^-) ion in the rocksalt, NaCl crystal. This crystal has a simple cubic structure[23] such that if we put the origin of a Cartesian coordinate system at one Na^+ ion and align the crystal properly, we find the six nearest neighboring Cl^- ions on $(0, 0, a)$, $(0, 0, -a)$, $(0, a, 0)$, $(0, -a, 0)$, $(a, 0, 0)$, and $(-a, 0, 0)$, where each parenthesis means (x, y, z) and a is a constant. In the NaCl crystal a is 2.8144×10^{-10} m. See Fig. 11–19 for the illustration.

---

[23] Strictly speaking, the crystal is face center cubic.

FIGURE 11–19. An Na^+ ion and its six nearest neighboring Cl^- ions in the rocksalt crystal.

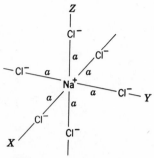

Let us consider the electrons localized in the central Na^+ ion assuming that the neighboring Cl^- ions can simply be regarded as electrostatic potential sources. The potential and the Hamiltonian for those electrons in the Na^+ ion are invariant under several symmetry operations.

The rotation of all electrons around the X axis by 90°, designated as $\hat{C}_4(X)$, leaves the Hamiltonian invariant, or commutes with the Hamiltonian. This operator generates the Abelian group $\mathbf{C}_4(X)$

$$\mathbf{C}_4(X) = \{\hat{I}, \hat{C}_4(X), \hat{C}_4{}^2(X) = \hat{C}_2(X), \hat{C}_4{}^3(X)\} \qquad (11\text{–}231a)$$

(see Problem 11–4), and, as shown in section 8–13, it is a symmetry group of the Hamiltonian of the Na^+ ion. Obviously the same argument is applicable around the Y and Z axes to find that

$$\mathbf{C}_4(Y) = \{\hat{I}, \hat{C}_4(Y), \hat{C}_4{}^2(Y), \hat{C}_4{}^3(Y)\} \qquad (11\text{–}231b)$$

and

$$\mathbf{C}_4(Z) = \{\hat{I}, \hat{C}_4(Z), \hat{C}_4{}^2(Z), \hat{C}_4{}^3(Z)\} \qquad (11\text{–}231c)$$

are also symmetry groups of the Hamiltonian.

Since $\hat{C}_4(X)$ and $\hat{C}_4(Y)$ commute with the Hamiltonian, the compound operators $\hat{C}_4(X)\hat{C}_4(Y)$ and $\hat{C}_4(Y)\hat{C}_4(X)$ also commute with the Hamiltonian. In considering what these compound operators really do, it is convenient to consider the diagram shown in Fig. 11–20, where points 1 to 6 are at the same distance from the origin. From Fig. 11–20 we see that the result of $\hat{C}_4(X)\hat{C}_4(Y)$ is actually obtained by a single operation $\hat{C}_3(111)$ which is the rotation by $2\pi/3$ around the axis which passes through the origin and the point (1, 1, 1). Notice that $\hat{C}_4(Y)\hat{C}_4(X)$ is equivalent to $\hat{C}_3(11-1)$, showing that $\hat{C}_4(X)$ does not commute with $\hat{C}_4(Y)$. Since $\hat{C}_4{}^3 = \hat{C}_4{}^{-1}$, we expect that

FIGURE 11–20. $\hat{C}_4(X)\hat{C}_4(Y) = \hat{C}_3(111)$.

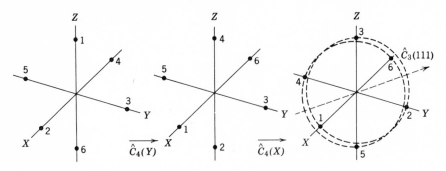

the compound operators like $\hat{C}_4{}^3(X)\hat{C}_4(Y)$, $\hat{C}_4(Y)\hat{C}_4{}^3(X)$, $\hat{C}_4{}^3(X)\hat{C}_4{}^3(Y)$, and $\hat{C}_4{}^3(Y)\hat{C}_4{}^3(X)$ are also equivalent to some $\hat{C}_3$ operators around appropriate axes, respectively. Using the diagrams similar to those in Fig. 11–20, we can actually obtain Table 11–14 which shows that this expectation is true with respect to all 24 compound operators of the type under consideration.

TABLE 11–14 *Equivalence of Twenty-Four Compound Operators with $\hat{C}_3(xyz)$*

$\hat{C}_4(X)\hat{C}_4(Y)$	$\hat{C}_4(Y)\hat{C}_4(X)$	$\hat{C}_4(Y)\hat{C}_4(Z)$
$\hat{C}_3(111)$	$\hat{C}_3(11-1)$	$\hat{C}_3(111)$
$\hat{C}_4(Z)\hat{C}_4(Y)$	$\hat{C}_4(Z)\hat{C}_4(X)$	$\hat{C}_4(X)\hat{C}_4(Z)$
$\hat{C}_3(-111)$	$\hat{C}_3(111)$	$\hat{C}_3(-11-1)$
$\hat{C}_4{}^3(X)\hat{C}_4(Y)$	$\hat{C}_4(Y)\hat{C}_4{}^3(X)$	$\hat{C}_4{}^3(Y)\hat{C}_4(Z)$
$\hat{C}_3(-11-1)$	$\hat{C}_3(-111)$	$\hat{C}_3(-1-11)$
$\hat{C}_4(Z)\hat{C}_4{}^3(Y)$	$\hat{C}_4{}^3(Z)\hat{C}_4(X)$	$\hat{C}_4(X)\hat{C}_4{}^3(Z)$
$\hat{C}_3(1-11)$	$\hat{C}_3(1-1-1)$	$\hat{C}_3(11-1)$
$\hat{C}_4(X)\hat{C}_4{}^3(Y)$	$\hat{C}_4{}^3(Y)\hat{C}_4(X)$	$\hat{C}_4(Y)\hat{C}_4{}^3(Z)$
$\hat{C}_3(1-1-1)$	$\hat{C}_3(1-11)$	$\hat{C}_3(-11-1)$
$\hat{C}_4{}^3(Z)\hat{C}_4(Y)$	$\hat{C}_4(Z)\hat{C}_4{}^3(X)$	$\hat{C}_4{}^3(X)\hat{C}_4(Z)$
$\hat{C}_3(11-1)$	$\hat{C}_3(-1-11)$	$\hat{C}_3(-111)$
$\hat{C}_4{}^3(X)\hat{C}_4{}^3(Y)$	$\hat{C}_4{}^3(Y)\hat{C}_4{}^3(X)$	$\hat{C}_4{}^3(Y)\hat{C}_4{}^3(Z)$
$\hat{C}_3(-1-11)$	$\hat{C}_3(-1-1-1)$	$\hat{C}_3(1-1-1)$
$\hat{C}_4{}^3(Z)\hat{C}_4{}^3(Y)$	$\hat{C}_4{}^3(Z)\hat{C}_4{}^3(X)$	$\hat{C}_4{}^3(X)\hat{C}_4{}^3(Z)$
$\hat{C}_3(-1-1-1)$	$\hat{C}_3(1-11)$	$\hat{C}_3(-1-1-1)$

In Table 11–14 we notice that there are only eight inequivalent $\hat{C}_3$ operators, and only four inequivalent rotation axes, since

$$\hat{C}_3(xyz) = \hat{C}_3{}^2(-x -y -z) = \hat{C}_3{}^{-1}(-x -y -z)$$

$$(11–232)$$

The second type of compound operators, obtained by those operators given in (11–231)'s, are $\hat{C}_4(X)\hat{C}_2(Y)$, $\hat{C}_2(Y)\hat{C}_4(X)$, and so forth. In Fig. 11–21, which corresponds to Fig. 11–20, we see that the result of $\hat{C}_2(Y)\hat{C}_4(X)$ is obtained by a single operation of $\hat{C}_2(01-1)$, which is the rotation by π around the axis which passes through the origin and the point $(01-1)$. There are twelve compound operators of this type, and in this procedure we see that they are equivalent to $\hat{C}_2$ operators with axes which bisect appropriate pairs of coordinate axes. The results are shown in Table 11–15.

FIGURE 11–21. $\hat{C}_2(Y)\hat{C}_4(X) = \hat{C}_2(01 \ -1)$.

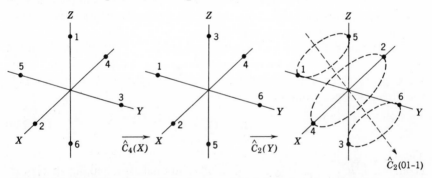

TABLE 11–15 *Equivalence of Twelve Compound Operators with $\hat{C}_2(xyz)$*

$\hat{C}_4(X)\hat{C}_2(Y)$ $\hat{C}_2(011)$	$\hat{C}_2(Y)\hat{C}_4(X)$ $\hat{C}_2(01-1)$	$\hat{C}_2(X)\hat{C}_4(Y)$ $\hat{C}_2(101)$	$\hat{C}_4(Y)\hat{C}_2(X)$ $\hat{C}_2(10-1)$
$\hat{C}_4(Y)\hat{C}_2(Z)$ $\hat{C}_2(101)$	$\hat{C}_2(Z)\hat{C}_4(Y)$ $\hat{C}_2(10-1)$	$\hat{C}_2(Y)\hat{C}_4(Z)$ $\hat{C}_2(110)$	$\hat{C}_4(Z)\hat{C}_2(Y)$ $\hat{C}_2(1-10)$
$\hat{C}_4(Z)\hat{C}_2(X)$ $\hat{C}_2(110)$	$\hat{C}_2(X)\hat{C}_4(Z)$ $\hat{C}_2(1-10)$	$\hat{C}_2(Z)\hat{C}_4(X)$ $\hat{C}_2(011)$	$\hat{C}_4(X)\hat{C}_2(Z)$ $\hat{C}_2(01-1)$

In Table 11–15 we notice that there are only six inequivalent $\hat{C}_2$ operators obtained as the compound operations in this manner. In order to distinguish from the $\hat{C}_2(X)$, $\hat{C}_2(Y)$, and $\hat{C}_2(Z)$ operators, we denote these six $\hat{C}_2$ operators as $\hat{C}_{2'}$.

Finally it is easy to see that the compound operation $\hat{C}_2(X)\hat{C}_2(Y)$ is equivalent to $\hat{C}_2(Z)$. There are six compound operations of this type, but they do not produce any new operators.

Using Tables 11–14 and 11–15 we can calculate any compound operations of higher orders. For example

$$\hat{C}_4(X)\hat{C}_4(Y)\hat{C}_4(Z) = \hat{C}_4(X)\hat{C}_3(111) = \hat{C}_4(X)\hat{C}_4(X)\hat{C}_4(Y)$$

$$= \hat{C}_2(X)\hat{C}_4(Y) = \hat{C}_2(101) \qquad (11\text{–}233)$$

It can be shown, following this procedure, that the eight $\hat{C}_3$ operators and the six $\hat{C}_{2'}$ operators, together with the original $\hat{C}_4$ and $\hat{C}_2$ operators contained in the $\mathbf{C}_4(X)$, $\mathbf{C}_4(Y)$, and $\mathbf{C}_4(Z)$ groups, form a group $\mathbf{O}$, which is called the *octahedron group*. It has been shown that this group is *generated* by the $\hat{C}_4(X)$, $\hat{C}_4(Y)$, and $\hat{C}_4(Z)$ operators.

The $\mathbf{O}$ group has five classes, whose representative elements are $\hat{1}$, $\hat{C}_3$, $\hat{C}_{2'}$, $\hat{C}_2$, and $\hat{C}_4$, respectively, and its order g is 24. As shown in the character table given in Appendix I(d), it has five irreducible representations confirming the theorem presented on page 536.

The crystal field for the Na^+ ion, as shown in Fig. 11–19, has an important symmetry which has not been considered yet; the space inversion $\hat{I}$. From the $\mathbf{O}$ group and the inversion group $\mathbf{C}_i$, which is made of $\hat{1}$ and $\hat{I}$, we obtain the direct product group

$$\mathbf{O}_h = \mathbf{O} \times \mathbf{C}_i \qquad (11\text{–}234)$$

The $\mathbf{O}_h$ group has, in addition to the 24 elements of the $\mathbf{O}$ group, 24 new elements, $\hat{I}$, $\hat{I}\hat{C}_3$, $\hat{I}\hat{C}_{2'}$, $\hat{I}\hat{C}_2$, $\hat{I}\hat{C}_4$, with all rotational axes, making up its order 48. (Note that $\hat{I}\hat{C}_p = \hat{C}_p\hat{I}$.)

The inversion $\hat{I}$ and all compound operators of the form $\hat{I}\hat{C}_p$ are called the second kind operators, in contrast to the pure rotation operators $\hat{C}_p$, which are called the first kind. In Fig. 11–22, in which we use vectors 1 to 6, instead of points used in Figs. 11–20 and 11–21, we see that the compound operator $\hat{I}\hat{C}_2(X)$ is equivalent to a reflection with respect to the YZ plane,

$$\hat{\sigma}(YZ) = \hat{I}\hat{C}_2(X) = \hat{C}_2(X)\hat{I} \qquad (11\text{–}235)$$

where we introduced the notation $\hat{\sigma}$ for the reflection operator. In the same way, we see that $\hat{I}\hat{C}_{2'}$ is also equivalent to a reflection, but in this case with respect to a plane which contains one axis and bisects the other two axes. As an example, $\hat{I}\hat{C}_2(110)$ is shown in Fig. 11–23. This type of reflection is designated as $\hat{\sigma}_d$:

$$\hat{\sigma}_d = \hat{I}\hat{C}_{2'} = \hat{C}_{2'}\hat{I} \qquad (11\text{–}236)$$

FIGURE 11–22. $\hat{I}\hat{C}_2(X) = \hat{C}_2(X)\hat{I} = \hat{\sigma}(YZ).$

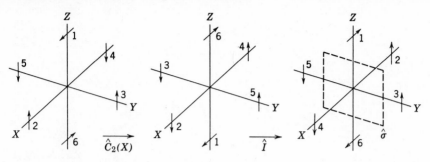

FIGURE 11–23. $\hat{I}\hat{C}_{2'} = \hat{\sigma}_d.$

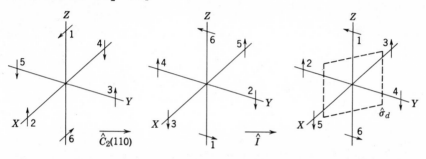

As seen in Fig. 11–24, a compound operator $\hat{C}_3\hat{I}$ $(= \hat{I}\hat{C}_3)$ is equivalent to $\hat{C}_6^{-1}\hat{\sigma}$ $(= \hat{\sigma}\hat{C}_6^{-1})$, where $\hat{C}_6$ is the rotation by $2\pi/6$ around the same axis as the $\hat{C}_3$ rotation and $\hat{\sigma}$ is the reflection with respect to the plane perpendicular to the same rotation axis. It is conventional to denote the latter type of compound operator (second kind) as

$$\hat{S}_p \equiv \hat{C}_p\hat{\sigma} = \hat{\sigma}\hat{C}_p \tag{11–237}$$

Therefore

$$\hat{C}_3\hat{I} = \hat{I}\hat{C}_3 = \hat{S}_6^{-1} = \hat{S}_6^5 \tag{11–238}$$

One may be tempted to say that $\hat{I}\hat{C}_4$ is equivalent to $\hat{S}_8^{-1}$. However, that is not correct, and we can show without too much difficulty that

$$\hat{C}_4\hat{I} = \hat{I}\hat{C}_4 = \hat{S}_4^{-1} \tag{11–239}$$

instead.

FIGURE 11–24. $\hat{I}\hat{C}_3 = \hat{\sigma}\hat{C}_6^{-1}$.

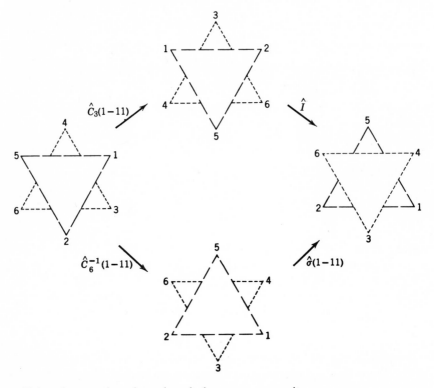

Using the notations introduced above, we can write

$$\mathbf{O}_h = \{\hat{I}, 8\hat{C}_3, 3\hat{C}_{2'}, 6\hat{C}_2, 6\hat{C}_4, \hat{I}, 8\hat{S}_6, 3\hat{\sigma}_d, 6\hat{\sigma}, 6\hat{S}_4\} \qquad g = 48$$

$$(11\text{–}240)$$

The $\mathbf{O}_h$ group has ten classes and ten irreducible representations, which are given in Appendix I(d). Since this group is a symmetry group of the Na^+ ion in NaCl, or any atom (or ion) in a cubic crystal field, we can classify the eigenstates of these atoms according to these ten irreducible representations.

Other Common Point Groups

Combining $\hat{C}_p$, $\hat{I}$, $\hat{S}_p$, and $\hat{\sigma}$ operators we can obtain many other point groups, some of which are applicable to existing crystal fields and molecules.

The simplest point groups are $\mathbf{C}_p$, each of which is Abelian and generated by $\hat{C}_p(Z)$, which is the rotation by $2\pi/p$. Obviously p has to be an integer for

C_p to be a finite group of order p. In the case of crystals it can be shown that only

$$p = 1, 2, 3, 4, \text{ and } 6 \qquad (11\text{-}241)$$

are allowed, but there are molecules for which C_5 is also applicable.[24] The point groups which are applicable to crystals, namely those given by (11-241) and others which are generated by them and additional operators, are called the *crystallographic point groups.*

In addition to $\hat{C}_p(Z)$ we can take a $\hat{C}_2$ operator, which is the rotation by π around an axis perpendicular to the Z axis, and generate a point group called the D_p group. Since the group has to contain p of the $\hat{C}_2$ operator, as seen in Fig. 11-25, the order of the D_p group is $2p$.

The C_{ph} group is generated by $\hat{C}_p(Z)$ and $\hat{\sigma}_h$, the reflection with respect to a horizontal plane which is perpendicular to the Z axis, while the C_{pv} group is generated by $\hat{C}_p(Z)$ and $\hat{\sigma}_v$, the reflection with respect to a vertical plane which contains the Z axis. (See Fig. 11-26.)

To the D_p group we can add a vertical reflection plane in two ways. One way is to let a $\hat{\sigma}_v$ plane contain one of the $\hat{C}_2$ axes as well as the Z axis, while the other way is to let a $\hat{\sigma}$ plane bisect two adjacent $\hat{C}_2$ axes. The first choice generates the D_{ph} group, while the second choice generates the D_{pd} group. The D_{ph} group might be better called the D_{pv} group because of the way it was introduced; but it is more conventional to call it D_{ph}. This does not contradict other conventions since it does contain the $\hat{\sigma}_h$ symmetry (Problem 11-68). The symmetry elements of D_{3h} and D_{2d} (also called V_d) are illustrated in Fig. 11-27.

There are many other point groups, but an important one among them is a tetrahedron group T_d, which is illustrated in Fig. 11-28. This is a symmetry group of CH_4 (methane) and other similar molecules.

FIGURE 11–25. The symmetry elements of D_2 (often called **V**), and D_3.

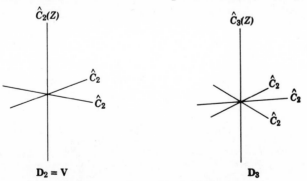

$\hat{C}_2(Z)$ $\hat{C}_3(Z)$

$\hat{C}_2$
$\hat{C}_2$

$\hat{C}_2$
$\hat{C}_2$
$\hat{C}_2$

$D_2 = V$ D_3

[24] Icosahedron which contains many $\hat{C}_5$'s is the closed packing structure of 13 spheres.

FIGURE 11–26. The symmetry elements of $\mathbf{C}_{ph}$ and $\mathbf{C}_{3v}$.

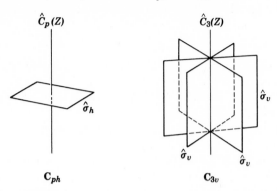

$\mathbf{C}_{ph}$ $\mathbf{C}_{3v}$

FIGURE 11–27.   The symmetry elements of $\mathbf{D}_{3h}$ and $\mathbf{D}_{2d}$.

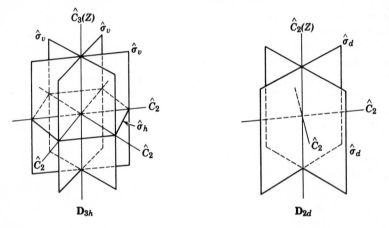

$\mathbf{D}_{3h}$ $\mathbf{D}_{2d}$

FIGURE 11–28.   The symmetry elements of the $\mathbf{T}_d$ group applied to the methane molecule.

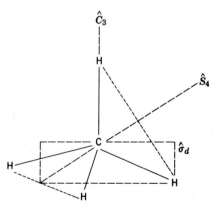

Although there are infinitely many point groups, the number of the crystallographic point groups is, due to (11–242), restricted to 32, all of which are given in Appendix I(d), together with a few other point groups which are applicable to molecules. Some of the results obtained in this section are tabulated in Table 11–16.

TABLE 11–16   *Some Important Point Groups and Their Generators*

Group	Generators	Group	Generators
$\mathbf{C}_p$	$\hat{C}_p(Z)$	$\mathbf{D}_{pd}$	$\hat{C}_p(Z),\ \hat{C}_2,\ \hat{\sigma}_d$
$\mathbf{D}_p$	$\hat{C}_p(Z),\ \hat{C}_2(U_2)$	$\mathbf{O}$	$\hat{C}_4(X),\ \hat{C}_4(Y),\ \hat{C}_4(Z)$
$\mathbf{S}_{2p}$	$\hat{S}_{2p}(Z)$	$\mathbf{O}_h$	$\hat{C}_4(X),\ \hat{C}_4(Y),\ \hat{C}_4(Z),\ \hat{I}$
$\mathbf{C}_{ph}$	$\hat{C}_p(Z),\ \hat{\sigma}_h$	$\mathbf{T}$	$\hat{C}_3,\ \hat{C}_2$
$\mathbf{C}_{pv}$	$\hat{C}_p(Z),\ \hat{\sigma}_v$	$\mathbf{T}_d$	$\hat{C}_3,\ \hat{S}_4,\ \hat{\sigma}_d$
$\mathbf{D}_{ph} = \mathbf{D}_{pv}$	$\hat{C}_p(Z),\ \hat{C}_2,\ \hat{\sigma}_v$	$\mathbf{T}_h$	$\hat{C}_3,\ \hat{S}_4,\ \hat{\sigma}_h$

PROBLEM 11–61. Confirm the equivalence relations given in Table 11–14.

PROBLEM 11–62. Confirm the equivalence relations given in Table 11–15.

PROBLEM 11–63. Show that
(a) $\hat{C}_4^{-1}(Z)\hat{C}_4(Y)\hat{C}_4(Z) = \hat{C}_4(X)$.
(b) $\hat{C}_3^{-1}(111)\hat{C}_4(Y)\hat{C}_3(111) = \hat{C}_4(X)$.
(c) $\hat{C}_2^{-1}(110)\hat{C}_4(Y)\hat{C}_2(110) = \hat{C}_4(X)$.

HINT.   Use Tables 11–14 and 11–15.

PROBLEM 11–64. Show that
(a) $\hat{C}_4^{-1}(Z)\hat{C}_3(111)\hat{C}_4(Z) = \hat{C}_3(-11-1)$.
(b) $\hat{C}_2^{-1}(110)\hat{C}_3(111)\hat{C}_2(110) = \hat{C}_3(11-1)$.

PROBLEM 11–65. Show that
(a) $\hat{C}_4^{-1}(Z)\hat{C}_2(110)\hat{C}_4(Z) = \hat{C}_2(1-10)$.
(b) $\hat{C}_3^{-1}(111)\hat{C}_2(110)\hat{C}_3(111) = \hat{C}_2(101)$.

PROBLEM 11–66. Show that

$$\hat{S}_6{}^2 = \hat{C}_3{}^2 = \hat{C}_3{}^{-1}$$

PROBLEM 11–67. Prove relation (11–239).

PROBLEM 11–68. Show that the $\mathbf{D}_{ph}$ group which is generated by $\hat{C}_p(Z)$ and $\hat{\sigma}_v$ contains $\hat{\sigma}_h$.

11–15 ATOMS IN CRYSTALS

When an atom is free its symmetry group is the rotation group $\mathbf{R}_3$, while if it is under a crystal field the symmetry group is reduced to an appropriate crystallographic point group. Since the point groups of the first kind, $\mathbf{C}_p$, $\mathbf{D}_p$, $\mathbf{T}$, and $\mathbf{O}$, are subgroups of the $\mathbf{R}_3$ group, definite relations between the irreducible representations D_L's of the $\mathbf{R}_3$ group and those of the point groups of the first kind are expected.

Let us consider the D_L representation in the $\mathbf{O}$ group as an example. The character table of the $\mathbf{O}$ group is given in Appendix I(d), but reproduced in Table 11–17 for convenience. The character of $D_L(\hat{R}_u(\gamma))$ is given by

TABLE 11–17   *Characters of the* **O** *Group*

	1	$8\hat{C}_3$	$6\hat{C}_2{}'$	$6\hat{C}_4$	$3\hat{C}_2$	$g = 24$
A_1	1	1	1	1	1	
A_2	1	1	-1	-1	1	
E	2	-1	0	0	2	
F_1	3	0	-1	1	-1	
F_2	3	0	1	-1	-1	

(11–138),

$$\chi_L(\gamma) = \frac{\sin\{\tfrac{1}{2}\gamma(2L + 1)\}}{\sin(\tfrac{1}{2}\gamma)} \tag{11–138}$$

Therefore, in the $\mathbf{O}$ group, we obtain

operators:	$\hat{1}$	$8\hat{C}_3$	$6\hat{C}_{2'}$	$6\hat{C}_4$	$3\hat{C}_2$
γ :	0	$\dfrac{2\pi}{3}$	π	$\tfrac{1}{2}\pi$	π
$\chi_L(\gamma)$:	$2L + 1$	$\left(\dfrac{4}{3}\right)^{1/2}\sin\left(\dfrac{2\pi}{3}L + \dfrac{\pi}{3}\right)$	$(-1)^L$	$\sin(\tfrac{1}{2}\pi L)$ $+\cos(\tfrac{1}{2}\pi L)$	$(-1)^L$

$$\tag{11–242}$$

Putting some values of L into (11–242) we obtain Table 11–18. Thus D_L can be reduced in the **O** group as shown in the same table using the ortho-normality relation (11–87).

TABLE 11–18   *Characters and Reductions of D_L in the* **O** *Group*

Atomic State	L	$\hat{1}$	$8\hat{C}_3$	$6\hat{C}_2{}'$	$6\hat{C}_4$	$3\hat{C}_2$	**O** Group
S	0	1	1	1	1	1	A_1
P	1	3	0	-1	1	-1	F_1
D	2	5	-1	1	-1	1	$E + F_2$
F	3	7	1	-1	-1	-1	$A_2 + F_1 + F_2$
G	4	9	0	1	1	1	$A_1 + E + F_1 + F_2$
H	5	11	-1	-1	1	-1	$E + 2F_1 + F_2$

What we find here is that when an atom is placed in a cubic field the $(2L + 1)$ degeneracy of an L level is partially removed. According to Table 11–18, S and P states maintain their degeneracy, 1 and 3, respectively, but a D state splits into two levels with degeneracies 2 and 3, respectively, and so forth.

Some crystal fields may be approximately cubic, but slightly distorted. For instance if one of the Cl^- ions in Fig. 11–19 is replaced by something else, the crystal field for the Na^+ ion is not cubic but of the symmetry of C_4. The relation between an irreducible representation of the **O** group and those of its subgroups, C_4, D_4, and C_3, helps to predict energy levels when distortions from the cubic field is small enough. Such relations can be easily found by comparing Table 11–17 with the character table of the group of interest given in Appendix I(d). Using the relation (11–87) again, we find that the irreducible representations of the **O** group are reduced in the C_3 group, for instance, as follows:

O Group	$\hat{1}$	$2\hat{C}_3$	C_3 Group	
A_1	1	1	A	
A_2	1	1	A	
E	2	-1	E	(11–243)
F_1	3	0	$A + E$	
F_2	3	0	$A + E$	

In this way we obtain Table 11–19.

TABLE 11–19 *Relations between Irreducible Representations of the* **O** *Group and Those of its Subgroups*

O	D$_4$	D$_3$	D$_4$	D$_2$	C$_4$	D$_3$	C$_3$
A_1	A_1	A_1	A_1	A	A	A_1	A
A_2	B_1	A_2	A_2	B_1	A	A_2	A
E	$A_1 + B_1$	E	B_1	B_1	B	E	E
F_1	$A_2 + E$	$A_2 + E$	B_2	B_3	B		
F_2	$B_2 + E$	$A_1 + E$	E	$B_2 + B_3$	E		

From Table 11–19 we see, for example, that if a crystal field is approximately **O**, but slightly distorted by a **D**$_4$ field, then each F_1 level is slightly split into two levels, A_2 and E (doubly degenerate). If it is again slightly distorted by a **D**$_2$ field, the latter E level is again split into two levels, B_2 and B_3.

When the crystal field has inversion symmetry, the parity of each atomic state is conserved during the process of placing an atom in a crystal. Therefore all parity + atomic states reduce to states with subscript g (gerade) in such a crystal field, while those of parity − reduce to states with subscript u (ungerade). For example, a D parity + atomic state reduces to $E_g + F_{2g}$ when the crystal field has the symmetries of the **O**$_h$ group.

When the relativistic effects, such as spin–orbit or spin–spin interactions, are not negligible, one has to consider the effects of crystal fields on electronic spins. Because of the special property of spin given by (11–183), one has to introduce the so-called *double point groups* in order to discuss the effects of crystal fields on electronic states including spin. Readers may consult other books for such advanced theories.[25]

PROBLEM 11–69. Obtain (11–242) and then Table 11–18.

PROBLEM 11–70. Confirm Table 11–19 using the tables in Appendix I(d).

PROBLEM 11–71. Find all energy levels when an atom with the $(np)^2$ configuration is placed in a crystal field with the symmetries of the **O**$_h$ group. Neglect the fine structure.

[25] See section 10–12 of M. Mizushima, *Quantum Mechanics of Atomic Spectra and Atomic Structure*, W. A. Benjamin, New York, 1970. The double point groups were first discussed by H. A. Bethe, *Ann. Physik*, **3**, 133 (1929).

HINT. Use Tables 11–11 and 11–18. Note that the resulting states are all g states.

PROBLEM 11–72. Find the lowest five levels when an atom with the $(nd)^2$ configuration is placed in a crystal field with the symmetries of the $\mathbf{D}_{4h} (= \mathbf{D}_4 \times \mathbf{C}_i)$ group. Explain what happens if the crystal field is approximately cubic.

HINT. The lowest free atomic state is ^3F, according to Hund's rules.

PROBLEM 11–73. Obtain the following Clebsch–Gordan series for the $\mathbf{O}$ group:

$$A_1 \times A_1 = A_2 \times A_2 = A_1 \qquad A_1 \times A_2 = A_2$$

$$A_1 \times E = A_2 \times E = E \qquad A_1 \times F_1 = A_2 \times F_2 = F_1$$

$$A_1 \times F_2 = A_2 \times F_1 = F_2 \qquad E \times F_1 = E \times F_2 = F_1 + F_2$$

$$E \times E = A_1 + A_2 + E \qquad F_1 \times F_2 = A_2 + E + F_1 + F_2$$

$$F_1 \times F_1 = F_2 \times F_2 = A_1 + E + F_1 + F_2$$

PROBLEM 11–74. (a) Show that a vector $\mathbf{r}$ belongs to the F_1 representation in the $\mathbf{O}$ group.
(b) Using the results of Problems 11–73 and 11–74a, find which of the following matrix elements of $\mathbf{r}$ are definitely zero:

$$\langle A_1|\mathbf{r}|A_2\rangle \qquad \langle E|\mathbf{r}|E\rangle \qquad \langle E|\mathbf{r}|F_2\rangle$$

where $|\Gamma\rangle$ (or $\langle\Gamma|$) is an eigenket (or eigenbra) which belongs to Γ representation of the $\mathbf{O}$ group.

HINT. $\mathbf{r}$ belongs to the D_1 representation of the $\mathbf{R}_3$ group. Use (11–94) for (b). Note that when a matrix element is zero, according to group theory it is definitely zero, but a matrix element can be zero by other reasons than group theory.

PROBLEM 11–75. Reduce the irreducible representations D_0 through D_4 of the $\mathbf{R}_3$ group in the icosahedron group $\mathbf{I}$. See Appendix I(d) for the character table of $\mathbf{I}$.

ANSWER. $D_0 = A$, $D_1 = F_1$, $D_2 = H$, $D_3 = G + F_2$, and $D_4 = H + G$.

PROBLEM 11–76. Obtain the following Clebsch–Gordan series of the **I** group:

$$H \times H = A + F_1 + F_2 + 2G + 2H$$

$$G \times H = F_1 + F_2 + G + 2H$$

$$G \times G = A + F_1 + F_2 + G + H$$

$$F_2 \times G = F_1 + G + H$$

11–16 MOLECULES

Born–Oppenheimer Approximation

A molecule is made of electrons and more than one nuclei. If the coordinate and momentum of an electron are $\mathbf{r}_i$ and $\hat{\mathbf{p}}_i$, respectively, while those of a nucleus are $\mathbf{R}_i$ and $\hat{\mathbf{p}}_i$, respectively, and if the electron mass is m_e and the mass of a nucleus is m_i, then the nonrelativistic Hamiltonian of the molecule can be written as

$$\hat{H} = \sum_{i=1}^{v} \frac{\hat{\mathbf{p}}_i^{\,2}}{2m_i} + \hat{H}_e \qquad (11\text{–}244)$$

where

$$\hat{H}_e = \sum_{i=1}^{n} \frac{\hat{\mathbf{p}}_i^{\,2}}{2m_e} + U(\mathbf{r}_1, \mathbf{r}_2, \ldots, \mathbf{r}_n, \mathbf{R}_1, \ldots, \mathbf{R}_v) \qquad (11\text{–}245)$$

We took the total number of electrons and nuclei as n and v, respectively. In the nonrelativistic limit the potential U is simply the sum of Coulomb potentials between all pairs of particles.

In the center of mass system the nuclear momentum and electron momentum must be of the same order of magnitude, while each nuclear mass m_i is at least 1800 times the electron mass. Therefore, in (11–244), we see that $\hat{H}$ is approximately equal to $\hat{H}_e$. The first approximate solution of the molecular problem is thus obtained by solving

$$\hat{H}_e \phi_n(\mathbf{r}_1, \ldots, \mathbf{r}_n; \mathbf{R}) = \mathcal{U}_n(\mathbf{R}) \phi_n(\mathbf{r}_1, \ldots, \mathbf{r}_n; \mathbf{R}) \qquad (11\text{–}246)$$

where ϕ_n and $\mathcal{U}_n$ are the eigenfunction and eigenvalue of $\hat{H}_e$, and $\mathbf{R}$ represents all nuclear coordinates collectively. Since $\hat{H}_e$ does not contain any differential operators with respect to $\mathbf{R}$, $\mathbf{R}$ can be regarded as a parameter in (11–246).

In order to find the next approximation, taking into account the kinetic energy terms of nuclei, we assume that

$$\psi(\mathbf{r}, \mathbf{R}) = \sum_n \chi_n(\mathbf{R}) \phi_n(\mathbf{r}; \mathbf{R}) \qquad (11\text{–}247)$$

is an eigenfunction of $\hat{H}$. We collectively represented all electron coordinates by $\mathbf{r}$ in (11–247). We thus have

$$
\hat{H}\psi(\mathbf{r}, \mathbf{R}) = \sum_n \left\{ \phi_n \left[\sum_\iota \frac{\hat{\mathbf{p}}_\iota^2}{2m_\iota} + \mathscr{U}_n \right] \chi_n \right.
$$
$$
\left. - \sum_\iota \frac{\hbar^2}{2m_\iota} [2(\mathbf{V}_\iota \phi_n) \cdot (\mathbf{V}_\iota \chi_n) + \chi_n (\mathbf{V}_\iota^2 \phi_n)] \right\}
$$
$$
= \mathscr{E}\psi(\mathbf{r}, \mathbf{R}) \tag{11–248}
$$

Since the ϕ_n's form an orthonormal set with respect to the electron coordinates, we can integrate $\phi_n{}^*(\hat{H} - \mathscr{E})\psi$ over all electron coordinates, to obtain

$$
\left[\sum_\iota \frac{\hat{\mathbf{p}}_\iota^2}{2m_\iota} + \mathscr{U}_n(R) - \mathscr{E} \right] \chi_n(R) = \sum_m \hat{K}_{mn} \chi_m(R) \tag{11–249}
$$

where, from (11–248),

$$
\hat{K}_{mn} = \sum_\iota \frac{\hbar^2}{2m_i} \left[2 \int \phi_n{}^*(\mathbf{V}_\iota \phi_m) \cdot \mathbf{V}_\iota \, dv_1 \cdots dv_n + \int \phi_n{}^* \mathbf{V}_\iota^2 \phi_m \, dv_1 \cdots dv_n \right] \tag{11–250}
$$

The Born–Oppenheimer approximation, or the adiabatic approximation, is obtained by completely neglecting the terms on the right-hand side of (11–249), resulting in

$$
\left[\sum_\iota \frac{\mathbf{p}_\iota^2}{2m_\iota} + \mathscr{U}_n(\mathbf{R}) \right] \chi_{nv}(R) = \mathscr{E}_{nv} \chi_{nv}(\mathbf{R}) \tag{11–251}
$$

This neglect implies that $\mathscr{U}_n(\mathbf{R})$, which is obtained as an eigenvalue of $\hat{H}_e$ in (11–246), can be regarded as the potential in which nuclei move according to the Schroedinger equation. The Born–Oppenheimer approximation is valid when the $\hat{K}_{mn}$ terms can be regarded as small perturbations, i.e., when

$$
\left| \int \chi_{nv}{}^* \hat{K}_{nm} \chi_{mv'} \, dv_R \right| \ll |\mathscr{E}_{nv} - \mathscr{E}_{mv'}| \tag{11–252}
$$

Molecular Shell Model

Since the existence of more than one nuclei destroys the spherical symmetry, the eigenvalue equation (11–246) is much more difficult to solve than the corresponding equation for atoms. Only the simplest cases, such as $H_2{}^+$

(one electron) and H_2 (two electrons), are solved with accuracy comparable to experiment. The Hartree–Fock approximation, in which all correlations between electron motion are neglected, helps to make calculation feasible, and has actually been applied to many molecules.

Some qualitative discussions on the electronic structure of molecules can be done using the *molecular shell model*, in which, as in the atomic and nuclear shell models, we assume that each electron moves in the field of all nuclei with a fixed position and an average field due to all other electrons, and that the latter average field has the highest symmetry possible.

Let us consider a homonuclear diatomic molecule, such as H_2, O_2, or N_2. If we take the axis which passes through two nuclei as the Z axis, the electronic Hamiltonian $\hat{H}_e$ of such a molecule is invariant under the rotation $\hat{R}_Z(\varphi)$ for an arbitrary rotation angle φ. Varying φ continuously from 0 to 2π we obtain infinitely many operators $\hat{R}_Z(\varphi)$, which form a continuous Abelian group called $\mathbf{C}_\infty$, or $\mathbf{R}_2$, the rotation group in two-dimensional space. Since the two nuclei we are considering are identical to each other, the corresponding $\hat{H}_e$ is invariant under $\hat{C}_2$ around an axis perpendicular to the Z axis. Adding such $\hat{C}_2$ to the $\mathbf{C}_\infty$ group we obtain the $\mathbf{D}_\infty$ group. The present $\hat{H}_e$ is also invariant under $\hat{I}$, the inversion with respect to the center of the molecule. We thus obtain the $\mathbf{D}_{\infty h}$ group ($= \mathbf{D}_\infty \times \mathbf{C}_i$) as a symmetry group of $\hat{H}_e$. The characters and notations of some irreducible representations of the $\mathbf{D}_{\infty h}$ and other groups are given in Appendix I(d), and some typical symmetry elements are shown in Fig. 11–29.

FIGURE 11–29. The symmetry elements of the $\mathbf{D}_{\infty h}$ group applied to a homonuclear diatomic molecule.

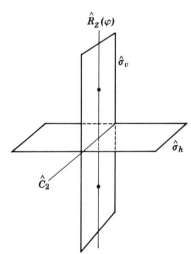

The notations Σ, Π, Δ, ... for the irreducible representations of the C_∞, $D_{\infty h}$, and other groups are taken in direct correspondence to (11–149), since these are the irreducible representations to which eigenfunctions of $\hat{L}_z$ with eigenvalues 0, $\pm\hbar$, $\pm 2\hbar$, ... belong. The subscripts g and u indicate the parity with respect to $\hat{I}$, as before, while the superscripts $+$ and $-$ indicate the symmetry property with respect to $\hat{\sigma}_v$. Thus each electronic state, or each eigenfunction of $\hat{H}_e$, of a homonuclear diatomic molecule belongs to one of Σ_g^+, Σ_g^-, Σ_u^+, Σ_u^-, Π_g, Π_u, Δ_g, ... irreducible representations of the $D_{\infty h}$ group, and the notations indicate its symmetry properties.

In the molecular shell model it is assumed that each electron is in a field of the maximum symmetry, or the symmetry of the $D_{\infty h}$ group in the case of a homonuclear diatomic molecule. Therefore, each one electron wave function also belongs to one of the above-mentioned irreducible representations. Those one-electron wave functions are called the σ_g^+, σ_g^-, σ_u^+, σ_u^-, π_g, π_u, δ_g, ... molecular orbitals, respectively.

When the internuclear distance R is very large, a molecule reduces to two independent atoms for which the atomic shell model is applicable. If $\phi_1^{(a)}(\mathbf{r})$ and $\phi_2^{(a)}(\mathbf{r})$ are atomic one-electron wave functions, which appeared in (11–150) but are now localized to atoms 1 and 2, respectively, a molecular one-electron wave function which belongs to an irreducible representation of $D_{\infty h}$ is obtained as

$$\sqrt{\tfrac{1}{2}}[\phi_1^{(a)}(\mathbf{r}) + \phi_2^{(a)}(\mathbf{r})] \quad \text{or} \quad \sqrt{\tfrac{1}{2}}[\phi_1^{(a)}(\mathbf{r}) - \phi_2^{(a)}(\mathbf{r})]$$

$$(11\text{–}253)$$

These expressions are called LCAO MO, or linear combination of atomic orbitals, molecular orbitals. Note that we have to combine atomic orbitals of the same state [same (a) in the above case] to obtain a MO of proper symmetry properties. In this way, combining two atomic ns orbitals, we obtain $(ns\sigma_g^+)$ and $(ns\sigma_u^+)$ molecular orbitals, respectively. Since the $(ns\sigma_u^+)$ wave function has one extra node, which coincides with the $\hat{\sigma}_h$ plane, compared to the $(ns\sigma_g^+)$ wave function, the former state has higher energy. When two atomic np orbitals are combined the situation is more complicated because of the three-fold degeneracy of each atomic state. In this case Fig. 11–30 helps to show that we obtain four molecular orbitals, $(np\sigma_g^+)$, $(np\sigma_u^+)$, $(np\pi_g)$, and $(np\pi_u)$ (where the last two are doubly degenerate) which are ordered as shown in Fig. 11–31, again according to the number of nodes. (See section 8–10 for the relation between the number of nodes and energy.) Take as an example the nitrogen molecule N_2, which has 14 electrons.

FIGURE 11–30. LCAO MO's obtained by combining two atomic $2p$ orbitals. Vertical dotted lines show the extra nodal planes, which make energy higher.

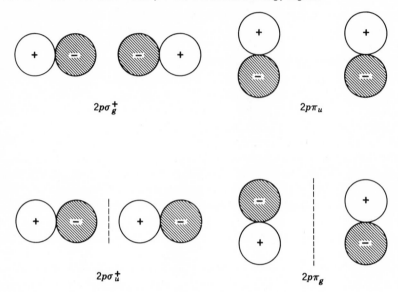

$$2p\sigma_g^+$$

$$2p\pi_u$$

$$2p\sigma_u^+$$

$$2p\pi_g$$

Filling up the molecular shells as shown in Fig. 11–31 from the bottom, according to the Pauli principle, we obtain

$$(1s\sigma_g^+)^2(1s\sigma_u^+)^2(2s\sigma_g^+)^2(2s\sigma_u^+)^2(2p\sigma_g^+)^2(2p\pi_u)^4 : {}^1\Sigma_g^+$$

$$(11\text{--}254)$$

for the electron configuration of the ground state of this molecule. Since in this configuration every shell is filled up completely, the resulting state has to be ${}^1\Sigma_g^+$, where the superscript 1 stands for the singlet, or $S = 0$. Some lower excited states can be obtained by promoting one electron from the $(2p\pi_u)$ shell to the $(2p\pi_g)$ or $(2p\sigma_u^+)$ shell. In doing so we obtain triplets and singlets, but, according to Hund's rules, a triplet is always lower than a corresponding singlet. Fig. 11–32 shows some observed energy levels and their theoretical interpretation.

An even more successful case is the oxygen molecule O_2. With 16 electrons we see in Fig. 11–31 that the last two electrons should be in the $(2p\pi_g)$ shell, which can accommodate four electrons. These two electrons are coupled to give ${}^3\Sigma_g^-$, ${}^1\Sigma_g^+$, and ${}^1\Delta_g$, while other states are prohibited by the Pauli principle. According to Hund's rules the ground state of this molecule is expected to be ${}^3\Sigma_g^-$ in agreement with experiment.

FIGURE 11–31.   Molecular shell structure in homonuclear diatomic molecules.

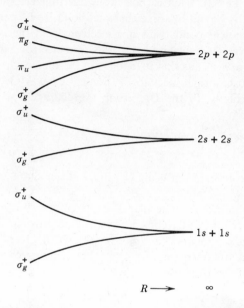

FIGURE 11–32.   Observed electronic levels at the equilibrium nuclear configuration, and their theoretical molecular shell structure (electronic configurations) for the nitrogen molecule N_2. Higher electronic states, which are not shown here, are given in G. Herzberg, *Spectra of Diatomic Molecules*, Van Nostrand, Princeton, 1950.

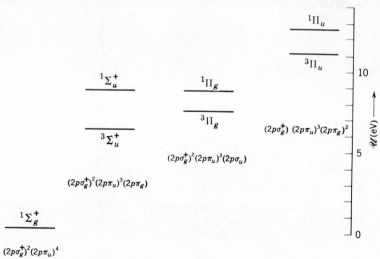

The benzene molecule C_6H_6 is known to have the symmetries of the $\mathbf{D}_{6h}$ group (Fig. 11–33). Each carbon atom has four electrons in its $n = 2$ atomic shell, but is bound by three chemical bonds to its neighbors. Assuming that one chemical bond takes one electron from the $n = 2$ shell, one electron is left free. It is believed that this free electron is in the so-called $2p\pi$ orbital, where π means that its node is in the molecular plane, such that it is antisymmetric with respect to $\hat{\sigma}_h$. Some LCAO MO can be obtained from six $2p\pi$ orbitals. In the $\mathbf{D}_{6h}$ group the character given by these six AO's is

$$\begin{array}{ccccccccccccc}
 & \hat{1} & \hat{C}_6 & \hat{C}_3 & \hat{C}_2 & \hat{C}_2(X) & \hat{C}_2(Y) & \hat{\sigma}_h & \hat{\sigma}_v & \hat{\sigma}_d & \hat{S}_6 & \hat{S}_3 & \hat{I} & (11\text{–}255) \\
\chi & 6 & 0 & 0 & 0 & 0 & -2 & -6 & 0 & 2 & 0 & 0 & 0 &
\end{array}$$

Reducing this representation using (11–91) and the character table given in Appendix I(d), we obtain $A_{2u} + B_{2g} + E_{1g} + E_{2u}$. Counting the number of nodes, we obtain the molecular shell structure shown in Fig. 11–34. With the six electrons under consideration, the electronic configuration for the ground state is obtained as $(a_{2u})^2(e_{1g})^4$, which gives $^1A_{1g}$. The lower excited states are obtained from $(a_{2u})^2(e_{1g})^3(e_{2u})$. Using one of the Clebsch–Gordan series of the $\mathbf{D}_{6h}$ group

$$E_{1g} \times E_{2u} = B_{1u} + B_{2u} + E_{1u} \qquad (11\text{–}256)$$

we see that the above electronic configuration gives triplets and singlets of B_{1u}, B_{2u}, and E_{1u} states, respectively. In Fig. 11–35 these states are shown with observed and calculated energies. The calculation is very elaborate.

FIGURE 11–33. The benzene molecule and some symmetry elements of the $\mathbf{D}_{6h}$ group.

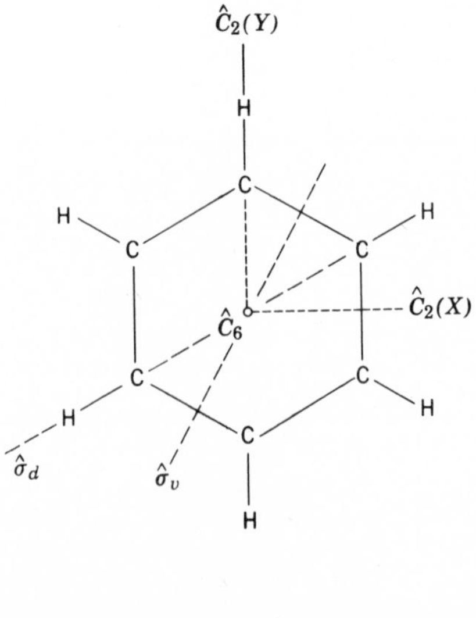

FIGURE 11–34. Molecular shells of the benzene molecule given by the $2p\pi$ atomic orbitals.

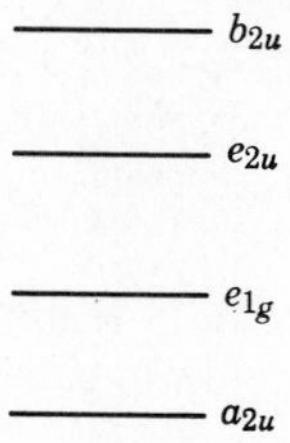

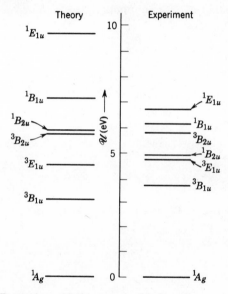

————— b_{2u}

————— e_{2u}

————— e_{1g}

————— a_{2u}

FIGURE 11–35. Observed and calculated electronic energy levels of the benzene molecule at its equilibrium nuclear configuration in the ground state. [D. R. Kearn, *J. Chem. Phys.* **36**, 1608 (1960), and J. W. Moskowitz and M. P. Barnett, *J. Chem. Phys.* **39**, 1557 (1963).]

Vibration and Rotation of Diatomic Molecules

So far we discussed the first approximation (11–246) of the molecular problem given by (11–244). The next approximation in the Born–Oppenheimer approximation is obtained by solving (11–251), for the motion of nuclei under the potential $\mathscr{U}_n(\mathbf{R})$.

For a diatomic molecule, in which there are only two nuclei, equation (11–251) is the Schroedinger equation of a two-body system, and can be reduced to that of a one body as

$$\left[-\frac{\hbar^2}{2m}\frac{\partial^2}{\partial R^2} + \mathscr{U}_n(R) + \frac{\hat{N}^2}{2mR^2} \right] R\chi_{nv}(\mathbf{R}) = \mathscr{E}_{nv}R\chi_{nv}(\mathbf{R}) \quad (11\text{–}257)$$

following the procedure given in section 9–1. In this equation R is the magnitude of the internuclear distance, $\hat{N}$ is the angular momentum due to the nuclei, and m is the reduced mass of the two nuclei. Taking

$$\chi_{nv}(\mathbf{R}) = f_{nvN}(R)Y_{NM}(\theta, \varphi) \tag{11–258}$$

where θ and φ give the orientation of $\mathbf{R}$ in space, (11–257) reduces to

$$\left[-\frac{\hbar^2}{2m}\frac{\partial^2}{\partial R^2} + \mathscr{U}_n(R) + \frac{N(N+1)\hbar^2}{2mR^2} \right] Rf_{nvN}(R) = \mathscr{E}_{nvN}Rf_{nvN}(R) \tag{11–259}$$

When the molecule is stable there must be an internuclear distance R_e at which the potential $\mathscr{U}_n(R)$ is minimum. In the same way as (2–3) we can expand the potential as

$$\mathscr{U}_n(R) = \mathscr{U}_n(R_e) + \tfrac{1}{2}k_n(R - R_e)^2 + \cdots \tag{11–260}$$

where k_n is the force constant. When the higher-order terms in this expansion and the centrifugal potential term in (11–259) are neglected, the problem reduces to that of a simple harmonic oscillator, which has been discussed in section 8–9. In this case we obtain

$$\mathscr{E}_{nv} = \mathscr{U}_n(R_e) + \hbar\omega_n(v + \tfrac{1}{2}) \tag{11–261}$$

where

$$v = 0, 1, 2, \ldots$$

and

$$\omega_n = \sqrt{k_n/m}$$

The effect of the rotational term in (11–259) can be taken into account by means of the first-order perturbation theory (8–217). Since the eigenfunctions of the simple harmonic oscillator are localized around $R = R_e$, we obtain

$$\mathscr{E}_{nvN} = \mathscr{U}_n(R_e) + \hbar\omega_n(v + \tfrac{1}{2}) + B_eN(N + 1) \tag{11–262}$$

where

$$B_e = \frac{\hbar^2}{2mR_e^{\,2}} \tag{11–263}$$

The perturbation theory is applicable for most molecules, where $\hbar\omega_n$ is about 100 times as large as B_e. Fig. 11–36 shows some energy levels of a diatomic molecule schematically.[26]

Vibration of Polyatomic Molecules

When a molecule has v nuclei it has $3v$ degrees of freedom of nuclear motions, including the motion of the center of mass (translational motion) and rotation. Since the degrees of freedom of the translational motion are 3, while those of the rotation are 3 or 2 (diatomic molecules), there are $3v - 6$ or $3v - 5$ vibrational degrees of freedom. When the vibrational amplitudes are small, it was shown in Chapter 2 that we can always find normal coordinates $Q_1, Q_2, \ldots$ such that each of them vibrates in a simple harmonic way independently of each other. Thus a molecular Hamiltonian in the Born–Oppenheimer approximation, in which $\mathcal{U}_n(\mathbf{R})$ appears as the potential, can be approximately separated into the Schroedinger equations of one-dimensional simple harmonic oscillators. Neglecting the translational and

FIGURE 11–36. Molecular energy ε_{nvN}.

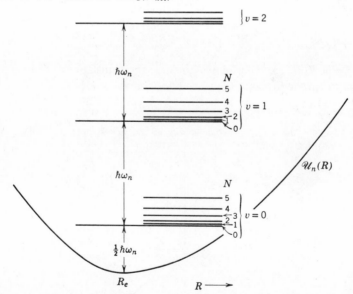

[26] Rotational energy levels given by $B_e N(N + 1)$ are also observed as nuclear levels of some heavier nuclei, such as Gd^{156}, Dy^{160}, Yb^{166}, Hf^{170}, W^{172}, Os^{182}, Pt^{184}, Th^{228}, U^{238}, Pu^{238}, Cm^{248}, and a few others. Vibrational–rotational levels are observed in some cases (Hf^{177} and U^{232}). See section 15.3 of P. Marmier and E. Sheldon, *Physics of Nuclei and Particles*, Vol. II, Academic Press, New York, 1970.

rotational motions, the energy of a polyatomic molecule can be expressed as

$$\mathscr{E}_{n v_1 v_2 \ldots} = \mathscr{U}_n(\mathbf{R}_e) + \hbar\omega_1(v_1 + \tfrac{1}{2}) + \hbar\omega_2(v_2 + \tfrac{1}{2}) + \cdots$$

$$(11\text{–}264)$$

where $v_1, v_2, \ldots$ are the vibrational quantum numbers, each of which can take 0 or any positive integer.

If the molecule has some symmetries in its equilibrium configuration, an appropriate point group can be used to classify the normal vibrations. Take the water molecule H_2O as an example. This molecule has the symmetry of the C_{2v} group. Taking nine infinitesimal displacement vectors as shown in Fig. 11–37, we obtain a 9×9 representation of this point group, whose character can be easily obtained as

$$
\begin{array}{ccccc}
 & \hat{1} & \hat{C}_2 & \hat{\sigma}_v(ZX) & \hat{\sigma}_v(ZY) \\
\chi & 9 & -1 & 1 & 3
\end{array}
\qquad (11\text{–}265)
$$

From the character table of C_{2v} given in Appendix I(d), we see that this representation reduces to

$$3A_1 + A_2 + 2B_1 + 3B_2 \qquad (11\text{–}266)$$

These nine "normal coordinates" include the translational and rotational motions. From the fact that the translational and rotational motions behave as vector and pseudovector, respectively, we can find their characters in the

FIGURE 11–37. The nuclear displacements in the water molecule, which has the symmetries of the C_{2v} group.

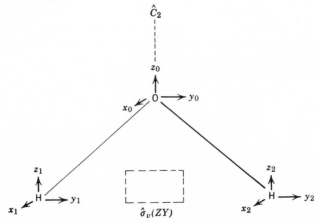

C_{2v} group and see that they belong to $A_1 + B_1 + B_2$ and $A_2 + B_1 + B_2$ representations, respectively. Subtracting these representations from (11–266) we obtain

$$2A_1 + B_2 \tag{11-267}$$

for the normal vibrations of the water molecule. These modes are illustrated in Fig. 11–38.

FIGURE 11–38.   The three normal vibrations of the water molecule.

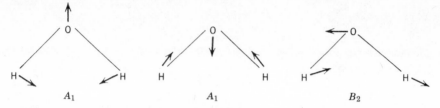

A_1 A_1 B_2

Rotation of Polyatomic Molecules

Assuming that the vibrational amplitudes are small and that the rotational motion can be treated by first-order perturbation theory, the rotational energy of a polyatomic molecule can be obtained by quantizing the Hamiltonian of the classical rigid rotator (3–131). The Hamiltonian of the rotational motion is, therefore,

$$\hat{H}_{\text{rot}} = \frac{1}{2}\left(\frac{\hat{N}_a{}^2}{I_a} + \frac{\hat{N}_b{}^2}{I_b} + \frac{\hat{N}_c{}^2}{I_c}\right) \tag{11-268}$$

where the principal axes and the moments of inertia are defined at the equilibrium configuration. The angular momentum $\hat{N}$ which appears here is due solely to the nuclear motion, and is not the total angular momentum to which electronic motion may contribute.

When I_a, I_b, and I_c are all different from each other, the rotator is called an *asymmetric top* rotator, while if two of them are equal to each other, it is called a *symmetric top* rotator. For example, if

$$I_a = I_b \tag{11-269}$$

(11–268) reduces to

$$\hat{H}_{\text{rot}} = \frac{\hat{N}^2}{2I_a} + \left(\frac{1}{2I_c} - \frac{1}{2I_a}\right)\hat{N}_c{}^2 \tag{11-270}$$

where

$$\hat{N}^2 = \hat{N}_a{}^2 + \hat{N}_b{}^2 + \hat{N}_c{}^2 \qquad (11\text{–}271)$$

is the square of the angular momentum $\hat{N}$ itself.

The principal axes a, b, and c are fixed to and rotate with a molecule. The components of $\hat{N}$, or any angular momenta, with respect to these rotating axes (Fig. 11–39) can be shown to satisfy the commutation relations

$$[\hat{N}_a, \hat{N}_b] = -i\hbar\hat{N}_c \qquad [\hat{N}_b, \hat{N}_c] = -i\hbar\hat{N}_a \qquad [\hat{N}_c, \hat{N}_a] = -i\hbar\hat{N}_b$$
$$(11\text{–}272)$$

in contrast to those for the components with respect to space fixed axes, such as (10–168). However, since (11–272) are reduced to the more familiar ones, such as (10–168), if we take

$$\hat{N}'_a = -\hat{N}_a \qquad \hat{N}'_b = -\hat{N}_b \qquad \hat{N}'_c = -\hat{N}_c \qquad (11\text{–}273)$$

we see that our previous arguments in section 10–10 are applicable to $\hat{N}'$. Thus we are guaranteed to find a set of eigenkets such that

$$\hat{N}^2|NKM\rangle = \hat{N}'^2|NKM\rangle = N(N+1)\hbar^2|NKM\rangle \qquad (11\text{–}274a)$$

$$\hat{N}_c|NKM\rangle = -\hat{N}'_c|NKM\rangle = K\hbar|NKM\rangle \qquad (11\text{–}274b)$$

FIGURE 11–39. The space-fixed coordinate system XYZ and the molecule fixed principal axes a, b, and c.

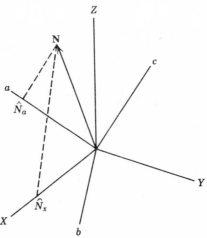

where

$$K = N, N - 1, \ldots, -N$$

We can also prove that

$$[\hat{N}_c, \hat{N}_z] = 0 \qquad (11\text{-}275)$$

which shows that the eigenkets defined above can also be eigenkets of $\hat{N}_z$ simultaneously:

$$\hat{N}_z|NKM\rangle = M\hbar|NKM\rangle \qquad (11\text{-}274c)$$

where

$$M = N, N - 1, \ldots, -N$$

For a symmetric top we can find eigenvalues of $\hat{H}_{rot}$ as

$$\hat{H}_{rot}|NKM\rangle = \left[\frac{1}{2I_a} N(N + 1)\hbar^2 + \left(\frac{1}{2I_c} - \frac{1}{2I_a}\right) K^2\hbar^2\right] |NKM\rangle$$
$$(11\text{-}276)$$

It is considerably more difficult to obtain eigenvalues of $\hat{H}_{rot}$ for a general asymmetric top rotator.

PROBLEM 11-77. Confirm Fig. 11-31.

PROBLEM 11-78. Show that the $(2p\pi_g)^2$ configuration gives ${}^3\Sigma_g^-$, ${}^1\Sigma_g^+$, and ${}^1\Delta_g$.

HINT. First obtain $\Pi_g \times \Pi_g$ using a table in Appendix I(d), then apply the Pauli principle. Note that the π orbital has two components, with $m_z = 1$ and -1.

PROBLEM 11-79. Obtain (11-255), and then show that the six $2p\pi$ AO's in the benzene molecule give $A_{2u} + B_{2g} + E_{1g} + E_{2u}$.

PROBLEM 11-80. If $\phi_1, \phi_2, \ldots, \phi_6$ are the $2p\pi$ AO's of the benzene molecule located at the carbon atoms as shown in Fig. 11-40, show that

$$\sum_{n=1}^{6} \phi_n \exp \frac{inl\pi}{3}$$

FIGURE 11–40. The carbon atoms in the benzene molecule.

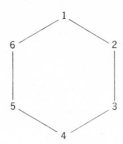

are (unnormalized) MO's such that when $l = 0, 1, 2, 3, 4$, and 5 they belong to the A_{2u}, E_{1g}, E_{2u}, B_{2g}, E_{2u}, E_{1g}, irreducible representations of the D_{6h} group, respectively.

HINT. Find the character.

PROBLEM 11–81. Obtain the following Clebsch–Gordan series of the D_{6h} group:

$$E_{1u} \times E_{1g} = A_{1u} + A_{2u} + E_{2u}$$

$$E_{1u} \times E_{2u} = B_{1g} + B_{2g} + E_{1g}$$

$$B_{2u} \times E_{1u} = E_{2u}$$

PROBLEM 11–82. Find which of the following matrix elements are predicted to be zero in the D_{6h} group:

$$\langle A_{2u}|x|B_{2g}\rangle \qquad \langle B_{2u}|y|E_{1g}\rangle \qquad \langle E_{1g}|x|E_{2u}\rangle \qquad \langle E_{1g}|y|A_{2u}\rangle$$

where $|\Gamma\rangle$ is an eigenket which belongs to Γ representation, and x and y are vectors in the molecular plane.

PROBLEM 11–83. The naphthalene molecule is planar and has the structure shown in Fig. 11–41.
(a) Show that the molecule has the symmetries of the D_{2h} group.
(b) Show that the ten $2p\pi$ electrons of the carbon atoms give

$$2A_{2u} + 3B_{1u} + 2B_{2g} + 3B_{3g}$$

FIGURE 11–41. The naphthalene molecule in its equilibrium nuclear configuration.

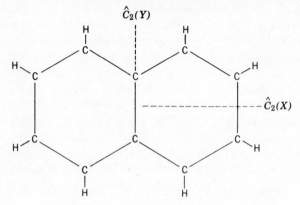

PROBLEM 11–84. Obtain (11–265) and then (11–266).

PROBLEM 11–85. Prove that $[\hat{N}_a, \hat{N}_z] = [\hat{N}_b, \hat{N}_z] = [\hat{N}_c, \hat{N}_z] = 0$.

HINT. Let $\hat{N}_a = \lambda_{aX}\hat{N}_X + \lambda_{aY}\hat{N}_Y + \lambda_{aZ}\hat{N}_Z$, etc., then $-\sqrt{\tfrac{1}{2}}(\lambda_{aX} + i\lambda_{aY})$, λ_{aZ}, $\sqrt{\tfrac{1}{2}}(\lambda_{aX} + i\lambda_{aY})$ are proportional to Y_{11}, Y_{10}, and Y_{1-1}, or form a tensor operator $\hat{T}_{1M}$, for which the commutation relations (11–190) hold with $\hat{N}$.

PROBLEM 11–86.* Prove (11–272).

HINT. Use the hint of Problem 11–85, and then use the relations like $\lambda_{aY}\lambda_{bZ} - \lambda_{aZ}\lambda_{bY} = \lambda_{cX}$, which can be proved by observing that

$$|J| = \lambda_{cX}{}^2 + \lambda_{cY}{}^2 + \lambda_{cZ}{}^2 = 1$$

where

$$dx\, dy\, dz = |J|\, da\, db\, dc$$

[See J. H. Van Vleck, *Revs. Modern Phys.* **23**, 214 (1951).]

11–17* THE R_4 GROUP AS A SYMMETRY GROUP OF THE HYDROGEN ATOM

The nonrelativistic Hamiltonian of the hydrogen atom, given by (9–7), commutes with every rotation operator which is in the R_3 group. As seen from (11–104), this fact allows us to conclude the $(2l + 1)$-fold degeneracy of each nl level of this atom. We have seen, however, in section 9–2, that in this particular case of the hydrogen atom the energy depends on the principal quantum number n only, and that the resulting degeneracy of each n level is n^2. This extra degeneracy, the so-called accidental degeneracy, suggests that the Hamiltonian $\hat{H}_r$ of the hydrogen atom has higher symmetry than that given by the $\mathbf{R}_3$ group. In classical mechanics, in Appendix II(a), it is shown that during the motion of a particle under a potential which is proportional to $1/r$ not only the angular momentum but also the position of the perihelion is conserved. The conservation of angular momentum is a direct result of the fact that the $\mathbf{R}_3$ group is a symmetry group, but the conservation of the position of the perihelion has nothing to do with the $\mathbf{R}_3$ group, and suggests the existence of a larger symmetry group in which the accidental degeneracy can be explained group-theoretically. We will see, in this section, that a four-dimensional rotation group $\mathbf{R}_4$ is such a symmetry group of this problem.

It was shown by Pauli that the so-called Runge (or Lenz) vector

$$\mathbf{A} = (\mathbf{l} \times \mathbf{p}) + mK\frac{\mathbf{r}}{r} \tag{11–277}$$

in classical mechanics is proportional to the position vector of the perihelion, if the potential is K/r. Let us show that the corresponding quantum mechanical operator[27]

$$\hat{\mathbf{A}} = \tfrac{1}{2}[(\hat{\mathbf{l}} \times \hat{\mathbf{p}}) - (\hat{\mathbf{p}} \times \hat{\mathbf{l}})] + mK\frac{\mathbf{r}}{r} \tag{11–278}$$

commutes with

$$\hat{H} = \frac{\hat{\mathbf{p}}^2}{2m} + \frac{K}{r} \tag{11–279}$$

A straightforward calculation shows that

$$[\hat{\mathbf{l}} \times \hat{\mathbf{p}}, \hat{\mathbf{p}}^2] = [\hat{\mathbf{p}} \times \hat{\mathbf{l}}, \hat{\mathbf{p}}^2] = 0 \tag{11–280}$$

while, since $\hat{\mathbf{l}}$ commutes with r, we see that

$$\left[\hat{\mathbf{l}} \times \hat{\mathbf{p}}, \frac{1}{r}\right] = \hat{\mathbf{l}} \times \left[\hat{\mathbf{p}}, \frac{1}{r}\right] = i\hbar\frac{\hat{\mathbf{l}} \times \mathbf{r}}{r^3} \tag{11–281a}$$

[27] Note that in quantum mechanics $\hat{\mathbf{l}} \times \hat{\mathbf{p}} \neq -\hat{\mathbf{p}} \times \hat{\mathbf{l}}$ and $\hat{\mathbf{r}} \times \hat{\mathbf{l}} \neq -\hat{\mathbf{l}} \times \hat{\mathbf{r}}$. See Problem 8–85.

and

$$\left[\hat{\mathbf{p}} \times \hat{\mathbf{l}}, \frac{1}{r}\right] = i\hbar \frac{\mathbf{r} \times \hat{\mathbf{l}}}{r^3} \tag{11–281b}$$

On the other hand, we obtain

$$\left[\frac{\mathbf{r}}{r}, \hat{\mathbf{p}}^2\right] = \frac{1}{2}\left\{\frac{1}{r}\left[\mathbf{r}, \hat{\mathbf{p}}^2\right] + \left[\frac{1}{r}, \hat{\mathbf{p}}^2\right]\mathbf{r} + \left[\mathbf{r}, \hat{\mathbf{p}}^2\right]\frac{1}{r} + \mathbf{r}\left[\frac{1}{r}, \hat{\mathbf{p}}^2\right]\right\}$$

$$= i\hbar\left[\frac{1}{r}\hat{\mathbf{p}} + \tfrac{1}{2}(\hat{\mathbf{p}} \cdot \mathbf{r})\frac{\mathbf{r}}{r^3} + \frac{1}{2}\frac{1}{r^3}(\mathbf{r} \cdot \hat{\mathbf{p}})\mathbf{r}\right.$$

$$\left. + \hat{\mathbf{p}}\frac{1}{r} + \frac{1}{2}\frac{\mathbf{r}}{r^3}(\hat{\mathbf{p}} \cdot \mathbf{r}) + \tfrac{1}{2}\mathbf{r}(\mathbf{r} \cdot \hat{\mathbf{p}})\frac{1}{r^3}\right]$$

$$= \tfrac{1}{2}i\hbar\left[r^{-3}(\hat{\mathbf{l}} \times \mathbf{r}) + (\hat{\mathbf{l}} \times \mathbf{r})r^{-3} - r^{-3}(\mathbf{r} \times \hat{\mathbf{l}}) - (\mathbf{r} \times \hat{\mathbf{l}})r^{-3}\right]$$

$$= i\hbar r^{-3}[(\hat{\mathbf{l}} \times \mathbf{r}) - (\mathbf{r} \times \hat{\mathbf{l}})] \tag{11–282}$$

Therefore

$$[\hat{\mathbf{A}}, \hat{H}] = 0 \tag{11–283}$$

The R₄ Group

The Runge vector $\hat{\mathbf{A}}$ is conserved just like $\hat{\mathbf{l}}$, but $\hat{\mathbf{A}}$ and $\hat{\mathbf{l}}$ are independent from each other. As a matter of fact we can easily show that

$$\hat{\mathbf{A}} \cdot \hat{\mathbf{l}} = \hat{\mathbf{l}} \cdot \hat{\mathbf{A}} = 0 \tag{11–284}$$

by using known commutation relations (Problem 8–85).
We find that

$$[\hat{A}_x, \hat{l}_x] = \tfrac{1}{2}\{[\hat{l}_y\hat{p}_z, \hat{l}_x] - [\hat{l}_z\hat{p}_y, \hat{l}_x] - [\hat{p}_y\hat{l}_z, \hat{l}_x]$$

$$+ [\hat{p}_z\hat{l}_y, \hat{l}_x] + mK\left[\frac{x}{r}, \hat{l}_x\right]\} = 0 \tag{11–285a}$$

$$[\hat{A}_y, \hat{l}_y] = [\hat{A}_z, \hat{l}_z] = 0 \tag{11–285b}$$

and

$$[\hat{A}_x, \hat{l}_y] = [\hat{l}_x, \hat{A}_y] = i\hbar\hat{A}_z \tag{11–286a}$$

$$[\hat{A}_y, \hat{l}_z] = [\hat{l}_y, \hat{A}_z] = i\hbar\hat{A}_x \tag{11–286b}$$

$$[\hat{A}_z, \hat{l}_x] = [\hat{l}_z, \hat{A}_x] = i\hbar\hat{A}_y \tag{11–286c}$$

Using a general relation, as given in Problem 8–11c, we obtain

$$[(\hat{\mathbf{l}} \times \hat{\mathbf{p}})_x, (\hat{\mathbf{l}} \times \hat{\mathbf{p}})_y] = [(\hat{\mathbf{p}} \times \hat{\mathbf{l}})_x, (\hat{\mathbf{p}} \times \hat{\mathbf{l}})_y]$$
$$= -\tfrac{1}{2}\{[(\hat{\mathbf{l}} \times \hat{\mathbf{p}})_x, (\hat{\mathbf{p}} \times \hat{\mathbf{l}})_y] + [(\hat{\mathbf{p}} \times \hat{\mathbf{l}})_x, (\hat{\mathbf{l}} \times \hat{\mathbf{p}})_y]\}$$
$$= -i\hbar\hat{\mathbf{p}}^2\hat{l}_z \qquad (11\text{–}287)$$

and two other similar relations. Also we find that

$$\left[(\hat{\mathbf{l}} \times \hat{\mathbf{p}})_x, \frac{y}{r}\right] + \left[\frac{x}{r}, (\hat{\mathbf{l}} \times \hat{\mathbf{p}})_y\right]$$
$$= -\left[(\hat{\mathbf{p}} \times \hat{\mathbf{l}})_x, \frac{y}{r}\right] - \left[\frac{x}{r}, (\hat{\mathbf{p}} \times \hat{\mathbf{l}})_y\right] = \frac{2i\hbar\hat{l}_z}{r} \qquad (11\text{–}288)$$

From (11–287), (11–288), and other similar relations, we finally find that

$$[\hat{A}_x, \hat{A}_y] = -2i\hbar m\hat{H}\hat{l}_z \qquad (11\text{–}289\text{a})$$

$$[\hat{A}_y, \hat{A}_z] = -2i\hbar m\hat{H}\hat{l}_x \qquad (11\text{–}289\text{b})$$

$$[\hat{A}_z, \hat{A}_x] = -2i\hbar m\hat{H}\hat{l}_y \qquad (11\text{–}289\text{c})$$

where $\hat{H}$ is given by (11–279).

Let us define two vector operators

$$\hat{\mathbf{M}} = \tfrac{1}{2}\left(\hat{\mathbf{l}} + \frac{\hat{\mathbf{A}}}{\sqrt{-2m\hat{H}}}\right) \quad \text{and} \quad \hat{\mathbf{N}} = \tfrac{1}{2}\left(\hat{\mathbf{l}} - \frac{\hat{\mathbf{A}}}{\sqrt{-2m\hat{H}}}\right) \qquad (11\text{–}290)$$

Since $\hat{H}$ commutes with both $\hat{\mathbf{l}}$ and $\hat{\mathbf{A}}$, we easily see that (11–286), (11–289), and the familiar commutation relations for $\hat{\mathbf{l}}$, reduce to

$$[\hat{M}_x, \hat{M}_y] = i\hbar\hat{M}_z \quad [\hat{M}_y, \hat{M}_z] = i\hbar\hat{M}_x \quad [\hat{M}_z, \hat{M}_x] = i\hbar\hat{M}_y$$
$$(11\text{–}291)$$

$$[\hat{N}_x, \hat{N}_y] = i\hbar\hat{N}_z \quad [\hat{N}_y, \hat{N}_z] = i\hbar\hat{N}_x \quad [\hat{N}_z, \hat{N}_x] = i\hbar\hat{N}_y$$
$$(11\text{–}292)$$

and

$$[\hat{\mathbf{M}}, \hat{\mathbf{N}}] = 0 \qquad (11\text{–}293)$$

where the last relation means that every component of $\hat{\mathbf{M}}$ commutes with every component of $\hat{\mathbf{N}}$.

Since each set of (11–291) and (11–292) is a set of commutation relations of three-dimensional angular momentum, we obtain two three-dimensional rotation groups

$$\mathbf{R}_3(\hat{\mathbf{M}}) = \{e^{-i\gamma\hat{\mathbf{M}}\cdot\mathbf{u}/\hbar}\} \tag{11–294}$$

and

$$\mathbf{R}_3(\hat{\mathbf{N}}) = \{e^{-i\gamma'\hat{\mathbf{N}}\cdot\mathbf{u}'/\hbar}\} \tag{11–295}$$

Since both $\hat{\mathbf{M}}$ and $\hat{\mathbf{N}}$ commute with $\hat{H}$, these two rotation groups are both symmetry groups of $\hat{H}$. Besides, since (11–293) shows that every element of $\mathbf{R}_3(\hat{\mathbf{M}})$ commutes with every element of $\mathbf{R}_3(\hat{\mathbf{N}})$, we obtain the direct product group

$$\mathbf{R}_3(\hat{\mathbf{M}}) \times \mathbf{R}_3(\hat{\mathbf{N}}) = \{\exp(-i\gamma\hat{\mathbf{M}}\cdot\mathbf{u}\hbar^{-1} - i\gamma'\hat{\mathbf{N}}\cdot\mathbf{u}'\hbar^{-1})\} \tag{11–296}$$

as a symmetry group of $\hat{H}$. Since we need three parameters to specify each element of a $\mathbf{R}_3$ group, we need six parameters to specify each element of this direct product group. Therefore, the $\mathbf{R}_3 \times \mathbf{R}_3$ group is a six-parameter continuous group (Lie group).

A rotation in a four-dimensional space corresponds to an orthogonal transformation of four real quantities, and such orthogonal transformation is given by a 4 × 4 real matrix

$$\begin{pmatrix} a_{11} & a_{12} & a_{13} & a_{14} \\ a_{21} & a_{22} & a_{23} & a_{24} \\ a_{31} & a_{32} & a_{33} & a_{34} \\ a_{41} & a_{42} & a_{43} & a_{44} \end{pmatrix} \tag{11–297}$$

which has 16 matrix elements. Since there are 10 orthogonality conditions

$$\sum_{j=1}^{4} a_{ij}a_{kj} = \delta_{ik} \tag{11–298}$$

among these matrix elements, there remains $16 - 10 = 6$ quantities at our disposal. Hence each matrix is specified by six parameters. The group formed by all rotations in a four-dimensional space, called $\mathbf{R}_4$, is therefore a six-parameter continuous group. This result strongly suggests that

$$\mathbf{R}_3 \times \mathbf{R}_3 = \mathbf{R}_4 \tag{11–299}$$

One can prove that this is actually true.

Casimir Operators

From the discussions in section 10–10 we are guaranteed to find eigenfunctions of $\hat{M}^2$, $\hat{M}_z$, $\hat{N}^2$, and $\hat{N}_z$, which we designate as $|Mm_M Nm_N\rangle$ such that

$$\hat{M}^2|Mm_M Nm_N\rangle = M(M + 1)\hbar^3|Mm_M Nm_N\rangle$$

$$\hat{M}_z|Mm_M Nm_N\rangle = m_M\hbar|Mm_M Nm_N\rangle$$

$$\hat{N}^2|Mm_M Nm_N\rangle = N(N + 1)\hbar^2|Mm_M Nm_N\rangle \tag{11-300}$$

$$\hat{N}_z|Mm_M Nm_N\rangle = m_N\hbar|Mm_M Nm_N\rangle$$

In the same way as for the $\mathbf{R}_3$ group, we find an irreducible representation D_{MN} of the $\mathbf{R}_4$ group by taking all eigenkets $|Mm_M Nm_N\rangle$ with fixed M and N. Since there are $(2M + 1)(2N + 1)$ such eigenkets, the dimension of this representation is

$$n^2 \equiv (2M + 1)(2N + 1) \tag{11-301}$$

Since $\mathbf{R}_4$ is a symmetry group of $\hat{H}$, an energy level given as an eigenvalue of $\hat{H}$, which belongs to the D_{MN} representation, has the degeneracy n^2, according to (11–104).

The operators $\hat{M}^2$ and $\hat{N}^2$ commute with every element of the $\mathbf{R}_4$ group, given in (11–296), and are called the *Casimir operators* of $\mathbf{R}_4$. Since $\hat{H}$ also commutes with every element of $\mathbf{R}_4$, it must be a function of these Casimir operators:

$$\hat{H} = \hat{H}(\hat{M}^2, \hat{N}^2) \tag{11-302}$$

Since the eigenkets of $\hat{H}$ have to be of the form of $|\alpha Mm_M Nm_N\rangle$, where α is some additional quantum number, or numbers, an eigenvalue of $\hat{H}$ is

$$\hat{H}(\hat{M}^2, \hat{N}^2)|\alpha Mm_M Nm_N\rangle = \mathscr{E}(M(M + 1), N(N + 1))|\alpha Mm_M Nm_N\rangle \tag{11-303}$$

or a function of $M(M + 1)$ and $N(N + 1)$.

In the present problem of the hydrogen atom, we have an additional condition (11–284), which reduces to

$$\hat{M}^2 - \hat{N}^2 = 0 \tag{11-304}$$

which means that only D_{MM} representations are physically meaningful. In this case (11–301) reduces to

$$n^2 = (2M + 1)^2 \tag{11-305}$$

while (11–303) gives an eigenvalue of $\hat{H}$ as

$$\mathscr{E}(M(M + 1), N(N + 1)) \to \mathscr{E}(M^2 + M) = \mathscr{E}(\tfrac{1}{4}(n^2 - 1))$$

$$(11–306)$$

Therefore, the eigenvalue is a function of n^2, the degeneracy of the energy level, in agreement to the previously obtained (section 9–2) result by a direct calculation.

PROBLEM 11–87. Obtain (11–280).

PROBLEM 11–88. Obtain (11–284) directly, without using (11–285).

PROBLEM 11–89. Obtain (11–285).

PROBLEM 11–90. Obtain (11–286a).

PROBLEM 11–91.* Obtain (11–287).

HINT. Straightforward, just takes time.

PROBLEM 11–92. Obtain (11–288).

PROBLEM 11–93. Reduce (11–284) to (11–304).

PROBLEM 11–94. Comparing (11–306) with (9–24), show that we expect

$$\hat{H} = \frac{-mK^2}{2(\hat{l}^2 + \hat{B}^2 + \hbar^2)}$$

where

$$\hat{B} = \frac{\hat{A}}{\sqrt{-2m\hat{H}}}$$

This expectation will be proved to be right in Problem 11–108.

11–18 THE SU₃ GROUP AND ELEMENTARY PARTICLES

Following the discoveries of the muons and π mesons many other elementary particles have been found. (Table 11–20 shows the ground states only. Many other excited or resonance states are also found.) Their masses, spins, parities, magnetic moments, and decay modes are observed to some extent. Attempts have been made to classify these particles from such data.

TABLE 11–20   *The Elementary Particles (Ground States)*[a]

	Name	Charge	Mass (MeV)	Spin Parity	Mean Life (sec)
Photon	γ	0	$< 2 \times 10^{-21}$	1^-	stable
Lepton	ν_e, ν_μ	0	$< 6 \times 10^{-5}$	$\frac{1}{2}$	stable
	e	$-e$	0.5110041(16)$\frac{1}{2}$		stable $> 2 \times 10^{21}$ year
	μ	$-e$	105.659(2)	$\frac{1}{2}$	$2.1983(8) \times 10^{-6}$
Meson	$\pi^\pm$	$\pm e$	139.578(13)	0^-	$2.603(6) \times 10^{-8}$
	π^0	0	134.975(14)	0^-	$0.89(18) \times 10^{-16}$
	$K^\pm$	$\pm e$	493.82(11)	0^-	$1.235(4) \times 10^{-8}$
	K^0	0	497.76(16)	0^-	$\begin{cases} 0.862(6) \times 10^{-10} \\ 5.38(19) \times 10^{-8} \end{cases}$
	η	0	548.8(6)	0^-	
Baryon	p	e	938.256(5)	$\frac{1}{2}^+$	$> 2 \times 10^{28}$ year
	n	0	939.550(5)	$\frac{1}{2}^+$	$0.932(14) \times 10^3$
	Λ	0	1115.60(8)	$\frac{1}{2}^+$	$2.51(3) \times 10^{-10}$
	Σ^+	e	1189.40(19)	$\frac{1}{2}^+$	$0.802(7) \times 10^{-10}$
	Σ^0	0	1192.46(12)	$\frac{1}{2}^+$	$< 1.0 \times 10^{-14}$
	Σ^-	$-e$	1197.32(11)	$\frac{1}{2}^+$	$1.49(3) \times 10^{-10}$
	Ξ^0	0	1314.7(7)	$\frac{1}{2}^+$	$3.03(18) \times 10^{-10}$
	Ξ^-	$-e$	1321.25(18)	$\frac{1}{2}^+$	$1.66(4) \times 10^{-10}$

(Hadron comprises Meson and Baryon groups.)

[a] Taken from A. Barbaro-Galtieri *et al.*, *Rev. Modern Phys.* **42**, 87 (1970). Numbers in parentheses are experimental uncertainties, e.g., 938.256(5) means 938.256 ± 0.005.

From Table 11–20 one notices that the masses of some particles are so close to each other that they can be regarded as in different charge states of a single particle, in the same way as the proton and neutron are regarded as $\tau_\zeta = \frac{1}{2}$ and $-\frac{1}{2}$ states, respectively, of the nucleon (section 11–13). Thus some mesons and baryons are grouped together as isospin substates as shown in Table 11–21.

Gell–Mann[28] and Nishijima[29] found that if a quantum number Y, called hyper-charge, is defined by

$$Y = 2(q - \tau_\zeta) \tag{11–307}$$

[28] M. Gell-Mann, *Phys. Rev.* **92**, 833 (1953).
[29] K. Nishijima, *Progr. Theor. Phys.* **13**, 285 (1955).

where q is the charge in the unit of e, then Y is well conserved in fast decays. The hyper-charge quantum number of some particles are given in Table 11–21.

TABLE 11–21   *Isospins and Hyper-charges of Some Hadrons*

	Mesons								Baryons							
	K^+	K^0	η	π^+	π^0	π^-	$\overline{K^0}$	K^-	p	n	Λ^0	Σ^+	Σ^0	Σ^-	Ξ^0	Ξ^-
τ_ζ	$\frac{1}{2}$	$-\frac{1}{2}$	0	1	0	-1	$\frac{1}{2}$	$-\frac{1}{2}$	$\frac{1}{2}$	$-\frac{1}{2}$	0	1	0	-1	$\frac{1}{2}$	$-\frac{1}{2}$
τ	$\frac{1}{2}$		0	0			$\frac{1}{2}$		$\frac{1}{2}$		0	1			$\frac{1}{2}$	
Y	1		0	0			-1		1		0	0			-1	

The SU_3 Group

Gell–Mann[30] and Ne'eman[31] found that the SU_3 group is useful in understanding relations among mesons and baryons. This group is formed by all special unitary transformations in three-dimensional complex space. In analogy to our previous discussion on the SU_2 group (section 11–11) we see that the transformations are given by 3×3 Hermitian matrices with a determinant of $+1$. A general 3×3 Hermitian matrix is determined by nine parameters, but the last condition on its determinent reduces the number of parameters to eight. Therefore, the SU_3 group is an eight-parameter Lie group.

We have seen that a general 2×2 Hermitian matrix can be expressed by a linear combination of the four Pauli matrices (4–60). In analogy we can see that a general 3×3 Hermitian matrix can be expressed by a linear combination of the following nine matrices:[30]

$$\rho_1 = \begin{pmatrix} 0 & 1 & 0 \\ 1 & 0 & 0 \\ 0 & 0 & 0 \end{pmatrix} \qquad \rho_2 = \begin{pmatrix} 0 & -i & 0 \\ i & 0 & 0 \\ 0 & 0 & 0 \end{pmatrix} \qquad \rho_3 = \begin{pmatrix} 1 & 0 & 0 \\ 0 & -1 & 0 \\ 0 & 0 & 0 \end{pmatrix}$$

$$\rho_4 = \begin{pmatrix} 0 & 0 & 1 \\ 0 & 0 & 0 \\ 1 & 0 & 0 \end{pmatrix} \qquad \rho_5 = \begin{pmatrix} 0 & 0 & -i \\ 0 & 0 & 0 \\ i & 0 & 0 \end{pmatrix} \qquad \rho_6 = \begin{pmatrix} 0 & 0 & 0 \\ 0 & 0 & 1 \\ 0 & 1 & 0 \end{pmatrix}$$

$$\rho_7 = \begin{pmatrix} 0 & 0 & 0 \\ 0 & 0 & -i \\ 0 & i & 0 \end{pmatrix} \qquad \rho_8 = \frac{1}{\sqrt{3}} \begin{pmatrix} 1 & 0 & 0 \\ 0 & 1 & 0 \\ 0 & 0 & -2 \end{pmatrix} \qquad \mathbf{1} = \begin{pmatrix} 1 & 0 & 0 \\ 0 & 1 & 0 \\ 0 & 0 & 1 \end{pmatrix}$$

$$(11\text{–}308)$$

[30] M. Gell-Mann, *Phys. Rev.* **125**, 1067 (1962).
[31] Y. Ne'eman, *Nucl. Phys.* **26**, 222 (1961).

where the last one is the unit matrix. Let us define a three-component complex vector $(q_1 \, q_2 \, q_3)$ and operators $\hat{\rho}_j$ by

$$\hat{\rho}_j(q_1 \, q_2 \, q_3) = (q_1 \, q_2 \, q_3)\rho_j \qquad j = 1, 2, \ldots, 8 \qquad (11\text{--}309)$$

then, from (11–308) one can easily find the commutation relations

$$[\hat{\rho}_j, \hat{\rho}_k] = 2i \sum_{l=1}^{8} f_{jkl}\hat{\rho}_l \qquad (11\text{--}310)$$

and

$$\hat{\rho}_j\hat{\rho}_k + \hat{\rho}_k\hat{\rho}_j = \tfrac{4}{3}\delta_{jk}\hat{1} + 2\sum_{l=1}^{8} d_{jkl}\hat{\rho}_l \qquad (11\text{--}311)$$

where the coefficients f_{jkl} and d_{jkl} are given in Table 11–22.

TABLE 11–22   *Nonzero Elements of f_{jkl} and d_{jkl}*

jkl	f_{jkl}	jkl	d_{jkl}	jkl	d_{jkl}
123	1	118	$1/\sqrt{3}$	366	$-\tfrac{1}{2}$
147	$\tfrac{1}{2}$	146	$\tfrac{1}{2}$	377	$-\tfrac{1}{2}$
156	$-\tfrac{1}{2}$	157	$\tfrac{1}{2}$	448	$-1/(2\sqrt{3})$
246	$\tfrac{1}{2}$	228	$1/\sqrt{3}$	558	$-1/(2\sqrt{3})$
257	$\tfrac{1}{2}$	247	$-\tfrac{1}{2}$	668	$-1/(2\sqrt{3})$
345	$\tfrac{1}{2}$	256	$\tfrac{1}{2}$	778	$-1/(2\sqrt{3})$
367	$-\tfrac{1}{2}$	338	$1/\sqrt{3}$	888	$-1/\sqrt{3}$
458	$\sqrt{3}/2$	344	$\tfrac{1}{2}$		
678	$\sqrt{3}/2$	355	$\tfrac{1}{2}$		

The general expression of the special (unimodular) unitary transformation operator is obtained as

$$\exp\left(i \sum_{j=1}^{8} \gamma_j\hat{\rho}_j\right) \qquad (11\text{--}312)$$

where γ_j's are real parameters. This can be seen since (11–310) and (11–311) guarantee that (11–312) can be expressed as a linear combination of $\hat{1}$ and the eight $\hat{\rho}_j$'s. This expression corresponds to (11–185) of $\mathbf{SU}_2$. Thus the $\hat{\rho}_j$'s are the generators, and (11–310) and (11–311) are the Lie algebra of $\mathbf{SU}_3$.

Irreducible Representations of SU₃

One immediately obtains a three-dimensional irreducible representation by taking $(q_1\ q_2\ q_3)$ as the basis vector. This representation is commonly designated as **3**. Another three-dimensional irreducible representation **3*** is obtained by taking $(q_1{}^*\ q_2{}^*\ q_3{}^*)$, the complex conjugate vector.

When the 9-component vector $(q_1 q_1{}^*\ q_1 q_2{}^*\ q_1 q_3{}^*\ q_2 q_1{}^* \cdots q_3 q_3{}^*)$ is taken as the basis vector, we obtain a reducible representation, because one combination, $q_1 q_1{}^* + q_2 q_2{}^* + q_3 q_3{}^*$, is invariant, and therefore belongs to the one-dimensional irreducible representation **1**. The remaining eight-dimensional one **8** can be shown to be irreducible. Consequently we have a Clebsch–Gordan series

$$\mathbf{3} \times \mathbf{3^*} = \mathbf{1} + \mathbf{8} \tag{11-313}$$

Note that $\mathbf{8^*} = \mathbf{8}$.

Other irreducible representations are obtained by taking λ_1th power of q's and λ_2th power of q^*'s. It is shown[32] that the dimension of the new representation one obtains in this manner is

$$(1 + \lambda_1)(1 + \lambda_2)[1 + \tfrac{1}{2}(\lambda_1 + \lambda_2)] \tag{11-314}$$

Thus SU_3 has irreducible representations **1, 3, 3*, 6, 6*, 8, 10, 10*, 15, 15*, 24, 24*, 27, 28, 28*,** ....

In SU_2 we specified each component of the basis vector which belongs to an irreducible representation D_j by the eigenvalue of $\hat{j}_z$ (i.e., m). In SU_3 we can do that by the eigenvalues of $\hat{\rho}_3$ and $\hat{\rho}_8$, because they are both diagonal as we saw in (11–308). In applying this theory to physics, however, it is more convenient to take

$$\hat{t}_\zeta = \tfrac{1}{2}\hat{\rho}_3 \quad \text{and} \quad \hat{Y}' = \frac{i}{\sqrt{3}}\hat{\rho}_8 \tag{11-315}$$

for the purpose of labeling. From (11–309) we obtain

$$\hat{t}_\zeta(q\ 0\ 0) = \tfrac{1}{2}(q\ 0\ 0) \qquad \hat{Y}'(q\ 0\ 0) = \frac{i}{3}(q\ 0\ 0)$$

$$\hat{t}_\zeta(0\ q\ 0) = -\tfrac{1}{2}(0\ q\ 0) \qquad \hat{Y}'(0\ q\ 0) = \frac{i}{3}(0\ q\ 0) \tag{11-316}$$

$$\hat{t}_\zeta(0\ 0\ q) = 0 \qquad\qquad \hat{Y}'(0\ 0\ q) = -\frac{2i}{3}(0\ 0\ q)$$

[32] J. J. de Swart, *Revs. Modern Phys.* **35**, 916 (1963). Originally given by L'Hopital in 1730.

Therefore, the three components of the basis vector which belongs to **3** are designated by the sets of two numbers (called *weight*), $\frac{1}{2}$, $(\frac{1}{3})$, $(-\frac{1}{2}, \frac{1}{3})$, and $(0, -\frac{2}{3})$, respectively. They are represented by the three points in the *weight diagram*, Fig. 11–42(a). In the same way, the weight diagram (also called eigenvalue diagram) of some other irreducible representations are obtained (Fig. 11–42).

We notice that sometimes one point in a weight diagram represents more than one component (e.g., the center point in **8**). Therefore, we need another number to specify each component completely. For that purpose we take

$$\hat{\tau}^2 \equiv \tfrac{1}{4}(\hat{\rho}_1{}^2 + \hat{\rho}_2{}^2 + \hat{\rho}_3{}^3) \tag{11–317}$$

and express its eigenvalue as $\tau(\tau + 1)$. If the eigenvalues of $\hat{\tau}_\zeta$ and $\hat{Y}'$ are designated as τ_ζ and iY, respectively, then the set of three numbers which we designate as

$$v \equiv (\tau, \tau_\zeta, Y) \tag{11–318}$$

specifies each component.

Clebsch-Gordan Series and Clebsch-Gordan Coefficients

The character of each representation is given by

$$\chi = \sum_j \exp\left(i\gamma_\tau \tau_{\zeta j} + i\gamma_Y Y_j\right) \tag{11–319}$$

where γ_τ and γ_Y are equal to $2\gamma_3$ and $\sqrt{3}\gamma_8$, respectively, of (11–312), and $\tau_{j\zeta}$ and Y_j are the eigenvalues of $\hat{\tau}_\zeta$ and $\hat{Y}'/i$ for the jth component of the basis vector. The summation is over the entire components. Notice that (11–319) is a natural extension of (11–138). Knowing the character (11–319) and the weight diagrams (Fig. 11–42), we can follow our previous discussions on (11–160) and (11–161) and obtain the Clebsch–Gordan series, some of which are given in Table 11–23.

TABLE 11–23   *Some Clebsch–Gordan Series of* **SU**$_3$

$3 \times 3^* = 1 + 8$	$6 \times 6^* = 1 + 27$
$3 \times 3 = 3^* + 6$	$6 \times 6 = 6^* + 15 + 15$
$3 \times 6^* = 1 + 3^* + 6$	$6 \times 8 = 3^* + 6 + 15^* + 24$
$3 \times 6 = 8 + 10$	$8 \times 8 = 1 + 8 + 8 + 10 + 10^* + 27$
$3 \times 8 = 3 + 6^* + 15$	

FIGURE 11–42. Weight diagrams. (a) **3**, (b) **3***, (c) **6**, (d) **6***, (e) **8**.

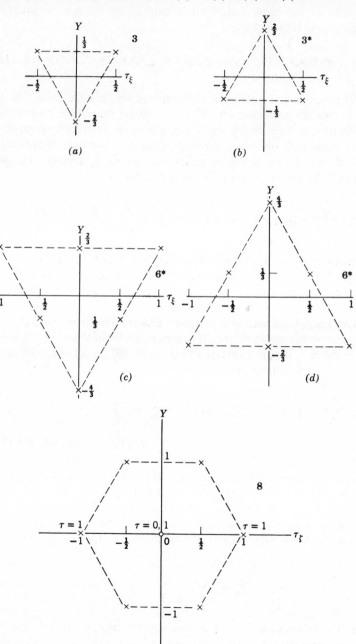

Let us designate an irreducible representation by μ. Then $|\mu v\rangle$ is the vth component of the basis vector which belongs to the μ representation. In analogy to (11–189) we have

$$|\mu_1 v_1 \mu_2 v_2\rangle \ (\equiv |\mu_1 v_1\rangle|\mu_2 v_2\rangle) = \sum_{\mu_\gamma v} |\mu_\gamma v\rangle\langle\mu_\gamma v \mid \mu_1 v_1 \mu_2 v_2\rangle \quad (11\text{–}320)$$

where the expansion coefficients which appear on the right-hand side are the Clebsch–Gordan coefficients of $\mathbf{SU}_3$. Additional number γ is introduced in (11–320) to take care of the cases where a given irreducible representation appears more than once in a direct product. For instance, **8** appears twice in **8** $\times$ **8**. We designate the resultant two **8**'s as $\mathbf{8}_1$ and $\mathbf{8}_2$. De Swart[32] introduced the isoscalar factor as the last factor in

$$\langle\mu_\gamma v \mid \mu_1 v_1 \mu_2 v_2\rangle = \langle\tau\tau_\zeta \mid \tau_1\tau_{\zeta 1}\tau_2\tau_{\zeta 2}\rangle \begin{pmatrix} \mu_1 & \mu_2 & \mu_\gamma \\ \tau_1 Y_1 & \tau_2 Y_2 & \tau Y \end{pmatrix} \quad (11\text{–}321)$$

where the first factor on the right-hand side is the Clebsch–Gordan coefficient of $\mathbf{SU}_2$, which we discussed in section 11–12. Values of the isoscalar factor are obtained by de Swart.[32] A part of his results is given in Table 11–24.

Tensor Operators and the Wigner–Eckart Theorem in SU$_3$

The tensor operators in the $\mathbf{SU}_3$ group are defined in the same way as those in the $\mathbf{SU}_2$ group. Corresponding to (11–197), a set of tensor operators $\hat{T}_{\mu v_1}, \ldots, \hat{T}_{\mu v_\mu}$ is defined by

$$\exp\left(i \sum_{j=1}^{8} \gamma_j \hat{\rho}_j\right) (\hat{T}_{\mu v_1} \cdots \hat{T}_{\mu v_\mu}) \exp\left(-i \sum_{j=1}^{8} \gamma_j \hat{\rho}_j\right)$$
$$= (\hat{T}_{\mu v_1} \cdots \hat{T}_{\mu v_\mu})\mu(\gamma_j) \quad (11\text{–}322)$$

where $\mu(\gamma_j)$ is the matrix representative of the μth irreducible representation which is determined by the eight parameters γ_j's. For example the eight generators $\hat{\rho}_1, \ldots, \hat{\rho}_8$ are tensor operators which belong to **8**.

If $|\alpha'\mu'v'\rangle$ and $|\alpha''\mu''v''\rangle$ belong to the irreducible representations μ' and μ'', respectively, then the Wigner–Eckart theorem (11–203) can be generalized as

$$\langle\alpha''\mu''v''|\hat{T}_{\mu v}|\alpha'\mu'v'\rangle = \sum_\gamma A_\gamma \langle\mu''v'' \mid \mu v \mu'v'\rangle_\gamma \quad (11\text{–}323)$$

where A's are constants which do not depend on v, v', and v'', and γ is the previously introduced numbering index to take care of the cases where a given irreducible representation appears more than once. Thus the theorem

TABLE 11–24 *Isoscalar Factors for* **8** × **8**. $\begin{pmatrix} 8 & 8 & \mu_\gamma \\ \tau_1 Y_1 & \tau_2 Y_2 & \tau Y \end{pmatrix}$

$\tau = 1 \quad Y = 2$

τ_1	Y_1	τ_2	Y_2	27
$\frac{1}{2}$	1	$\frac{1}{2}$	1	1

$\tau = 0 \quad Y = 2$

τ_1	Y_1	τ_2	Y_2	10*
$\frac{1}{2}$	1	$\frac{1}{2}$	1	-1

$\tau = \frac{3}{2} \quad Y = 1$

τ_1	Y_1	τ_2	Y_2	27	10
$\frac{1}{2}$	1	1	0	$1/\sqrt{2}$	$-1/\sqrt{2}$
1	0	$\frac{1}{2}$	1	$1/\sqrt{2}$	$1/\sqrt{2}$

$\tau = \frac{1}{2} \quad Y = 1$

τ_1	Y_1	τ_2	Y_2	27	10	8₁	8₂
$\frac{1}{2}$	1	1	0	$1/\sqrt{20}$	$-\frac{1}{2}$	$3/\sqrt{20}$	$\frac{1}{2}$
1	0	$\frac{1}{2}$	1	$-1/\sqrt{20}$	$-\frac{1}{2}$	$-3/\sqrt{20}$	$\frac{1}{2}$
$\frac{1}{2}$	1	0	0	$3/\sqrt{20}$	$\frac{1}{2}$	$-1/\sqrt{20}$	$\frac{1}{2}$
0	0	$\frac{1}{2}$	1	$3/\sqrt{20}$	$-\frac{1}{2}$	$-1/\sqrt{20}$	$-\frac{1}{2}$

$\tau = 1 \quad Y = 0$

τ_1	Y_1	τ_2	Y_2	27	10	10*	8₁	8₂
$\frac{1}{2}$	1	$\frac{1}{2}$	-1	$1/\sqrt{5}$	$-1/\sqrt{6}$	$1/\sqrt{6}$	$-\sqrt{\frac{3}{10}}$	$1/\sqrt{6}$
$\frac{1}{2}$	-1	$\frac{1}{2}$	1	$1/\sqrt{5}$	$1/\sqrt{6}$	$-1/\sqrt{6}$	$-\sqrt{\frac{3}{10}}$	$-1/\sqrt{6}$
1	0	1	0	0	$1/\sqrt{6}$	$-1/\sqrt{6}$	0	$\sqrt{\frac{2}{3}}$
1	0	0	0	$\sqrt{\frac{3}{10}}$	$\frac{1}{2}$	$\frac{1}{2}$	$1/\sqrt{5}$	0
0	0	1	0	$\sqrt{\frac{3}{10}}$	$-\frac{1}{2}$	$-\frac{1}{2}$	$1/\sqrt{5}$	0

$\tau = 0 \quad Y = 0$

τ_1	Y_1	τ_2	Y_2	27	8₁	8₂	1
$\frac{1}{2}$	1	$\frac{1}{2}$	-1	$\sqrt{\frac{3}{20}}$	$1/\sqrt{10}$	$\sqrt{\frac{1}{2}}$	$\frac{1}{2}$
$\frac{1}{2}$	-1	$\frac{1}{2}$	1	$-\sqrt{\frac{3}{20}}$	$-1/\sqrt{10}$	$\sqrt{\frac{1}{2}}$	$-\frac{1}{2}$
1	0	1	0	$-1/\sqrt{40}$	$-\sqrt{\frac{3}{5}}$	0	$\sqrt{\frac{3}{8}}$
0	0	0	0	$\sqrt{\frac{27}{40}}$	$-1/\sqrt{5}$	0	$-1/\sqrt{8}$

$\tau = \frac{3}{2} \quad Y = -1$

τ_1	Y_1	τ_2	Y_2	27	10*
$\frac{1}{2}$	-1	1	0	$\sqrt{\frac{1}{2}}$	$-\sqrt{\frac{1}{2}}$
1	0	$\frac{1}{2}$	-1	$\sqrt{\frac{1}{2}}$	$\sqrt{\frac{1}{2}}$

$\tau = \frac{1}{2} \quad Y = -1$

τ_1	Y_1	τ_2	Y_2	27	10	8₁	8₂
$\frac{1}{2}$	-1	1	0	$-1/\sqrt{20}$	$\frac{1}{2}$	$-3/\sqrt{20}$	$\frac{1}{2}$
1	0	$\frac{1}{2}$	-1	$1/\sqrt{20}$	$\frac{1}{2}$	$3/\sqrt{20}$	$\frac{1}{2}$
$\frac{1}{2}$	-1	0	0	$3/\sqrt{20}$	$\frac{1}{2}$	$-1/\sqrt{20}$	$-\frac{1}{2}$
0	0	$\frac{1}{2}$	-1	$3/\sqrt{20}$	$-\frac{1}{2}$	$-1/\sqrt{20}$	$\frac{1}{2}$

$\tau = 1 \quad Y = -2$

τ_1	Y_1	τ_2	Y_2	27
$\frac{1}{2}$	-1	$\frac{1}{2}$	-1	1

$\tau = 0 \quad Y = -2$

τ_1	Y_1	τ_2	Y_2	10
$\frac{1}{2}$	-1	$\frac{1}{2}$	-1	1

states that the v dependence of the matrix elements of tensor operators is given by the Clebsch–Gordan coefficients.[32, 33]

Application to the Hadrons

Table 11–20 shows that the masses of the eight mesons are not terribly different from each other, and that those of eight baryons are about the same order of magnitude. It is assumed[30, 31] that each set belongs to the **8** representation of the SU_3 group. Identifying the component numbers τ_ζ and Y of the SU_3 group to the isospin component and hyper-charge given in Table 11–21, we obtain Figs. 11–43a and 11–43b for mesons and baryons, respectively. The basic idea here is that the particles in each set are regarded as in nearly degenerate substates because they have nearly the same mass and other properties.

The degeneracy just mentioned is obviously not well observed since the particles in each set have different masses as seen in Table 11–20. Thus the Hamiltonian, or the mass operator, of these particles is not a scalar in the SU_3 group, but must have terms which belong to irreducible representations other than **1**. Since the hyper-charge Y is well conserved in transitions among these particles, the mass operator, to a good approximation, must be given by the $Y = 0$ components of irreducible representations. The charge is strictly conserved in all known reactions; therefore, the mass operator must have $\tau_\zeta = 0$, from (11–307). In addition, all strong interactions, including the Yukawa potential (11–226), seem to conserve the isospin τ. That is, the mass operator is most likely given by the $\tau = 0$ component. (The electromagnetic mass is given by the $\tau = 1$ component.) We see in Figs. 11–42a through 11–42e that the lowest-order irreducible representations which have the $Y = \tau_\zeta = \tau = 0$ component are **1** and **8**. (The next one is **27**.) Therefore the mass operator $\hat{M}$ is expected to have the form

$$\hat{M} = \hat{T}_{1000} + \hat{T}_{8000} + \cdots \qquad (11\text{–}324)$$

where $\hat{T}_{\mu\nu}$'s are the tensor operators. Using the Wigner–Eckart theorem (11–323) we obtain

$$\langle \alpha 8v | \hat{M} | \alpha 8v \rangle = a \langle 8v \mid 00008v \rangle + b \langle 8v \mid 80008v \rangle_1$$

$$+ c \langle 8v \mid 80008v \rangle_2 + \cdots \qquad (11\text{–}325)$$

---

[33] The proof of this generalized Wigner–Eckart theorem is straightforward. See, for instance, the article by L. O'Raifeartaigh in *Group Theory and Its Applications* (edited by E. M. Loebl), Academic Press, New York, 1968.

FIGURE 11–43.(a) Assignment of eight mesons in **8**. (b) Assignment of eight baryons in **8**.

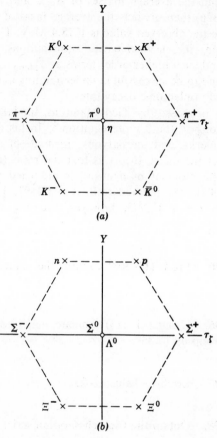

(a)

(b)

Notice that two **8** appear in the matrix element of $\hat{T}_8$ as Table 11–23 shows. Evaluating the Clebsch–Gordan coefficients we obtain

$$\langle \alpha 8v | \hat{M} | \alpha 8v \rangle = a + \tfrac{1}{2}bY + c\left[\tau(\tau + 1) - \tfrac{1}{4}Y^2 - 1\right] + \cdots$$

$$(11\text{–}326)$$

This formula agrees with the empirical formula previously found by Okubo.[34] For example, from the average masses of K and π mesons we obtain $a = 376.30$ MeV, $b = 0$, and $c = -239.00$ MeV. The mass of the η meson predicted by this formula is then 615.3 MeV compared to the observed value

[34] S. Okubo, *Progr. Theor. Phys.* **27**, 949 (1962).

548.8 MeV. For the baryons we obtain $a = 1153.82$, $b = -391.46$, and $c = 38.46$ MeV from the average masses of Λ, Σ, and the nucleons. The mass formula (11–326) then predicts the average mass of the Ξ particles as 1330.32 MeV, while the observed value is 1318.4 MeV. The formula can be refined[32] by including the electromagnetic contributions, which give $\hat{T}_{8100}$, $\hat{T}_{10100}, \ldots$ terms, and such higher-order terms as $\hat{T}_{27000}$.

The theory is found to be successful in understanding many other relations such as leptonic and nonleptonic decay rates.[35]

The quarks are proposed by Gell–Mann as fundamental elementary particles which belong to the **3** representation.[36] In his theory the mesons are made of two quarks, and the baryons are made of three quarks. One remarkable result of the quark theory is that the ratio (magnetic moment of proton)/(magnetic moment of neutron) is predicted to be $-\frac{3}{2}$. As we have seen in (10–302) the observed ratio is -1.4597. The quarks, which must have charges $e/3$ and $2e/3$, however, are not observed yet. (See section 5–1.)

PROBLEM 11–95. From (11–308) confirm the values of f_{jkl} shown in Table 11–22.

PROBLEM 11–96. Using (11–314) tabulate the irreducible representations of the SU_3 group up to **35***.

PROBLEM 11–97. Confirm Table 11–23.

PROBLEM 11–98. Obtain the Clebsch–Gordan series of

$$\mathbf{3} \times \mathbf{10}, \mathbf{3^*} \times \mathbf{10}, \mathbf{6} \times \mathbf{10}, \mathbf{6^*} \times \mathbf{10}, \mathbf{8} \times \mathbf{10}, \mathbf{10} \times \mathbf{10}, \text{ and } \mathbf{10^*} \times \mathbf{10}$$

PROBLEM 11–99. Obtain the weight diagrams of **10**, **10***, and **15**.

ANSWER. For **10** see Fig. 11–44. An easy way to obtain them is to use the Clebsch–Gordan series of **3** × **6** and **3** × **8**.

---

[35] For a review see the article by R. E. Behrends in *Group Theory and Its Applications* (edited by E. M. Loebl).
[36] M. Gell-Mann, *Phys. Letters*, **8**, 214 (1964).

PROBLEM 11–100. Obtain

$$\sum_{\tau_1 Y_1 \tau_2 Y_2} \begin{pmatrix} \mu_1 & \mu_2 & \Big| & \mu_\gamma \\ \tau_1 Y_1 & \tau_2 Y_2 & \Big| & \tau Y \end{pmatrix} \begin{pmatrix} \mu_1 & \mu_2 & \Big| & \mu'_{\gamma'} \\ \tau_1 Y_1 & \tau_2 Y_2 & \Big| & \tau Y' \end{pmatrix} = \delta_{\mu\mu'} \delta_{\gamma\gamma'} \delta_{YY'}$$

and

$$\sum_{\mu\gamma Y} \begin{pmatrix} \mu_1 & \mu_2 & \Big| & \mu_\gamma \\ \tau_1 Y_1 & \tau_2 Y_2 & \Big| & \tau Y \end{pmatrix} \begin{pmatrix} \mu_1 & \mu_2 & \Big| & \mu_\gamma \\ \tau'_1 Y'_1 & \tau'_2 Y'_2 & \Big| & \tau Y \end{pmatrix} = \delta_{\tau_1\tau_1'} \delta_{\tau_2\tau_2'} \delta_{Y_1 Y_1'} \delta_{Y_2 Y_2'}$$

PROBLEM 11–101. Obtain the Okubo-formula (11–326) from (11–325).

PROBLEM 11–102. Ten resonant baryons states, with spin $\frac{3}{2}$ and parity $+$, are observed. They are grouped as (N^{*++}, N^{*+}, N^{*0}, and N^{*-}), (Y_1^{*+}, Y_1^{*0}, and Y_1^{*-}), (Ξ^{*0} and Ξ^{*-}), and Ω^-. Using Fig. 11–44, assign these states to the **10** representation. The mass of these groups of states are observed as 1236, 1382, 1530, and 1676 MeV, respectively. Apply the Okubo-formula (11–326) to them. (The Okubo-formula is for **8**, but turns out to be applicable to **10** also.)

FIGURE 11–44. The weight diagram for **10**.

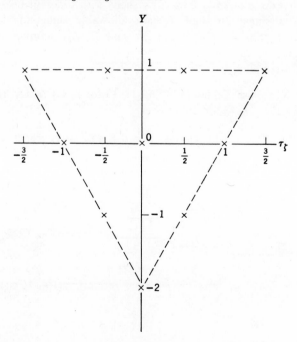

ADDITIONAL PROBLEMS

11–103. The spin–orbit coupling (10–336) gives

$$H' = \sum_{i=1}^{n} \hat{g}_i \hat{\mathbf{s}}_i \cdot \hat{\mathbf{l}}_i$$

for an n-electron system. $\hat{g}_i$ is a scalar operator. Show that

$$\langle \alpha LSJM | \hat{H}' | \alpha LSJM \rangle = g' \langle \alpha LSJM | \hat{\mathbf{L}} \cdot \hat{\mathbf{S}} | \alpha LSJM \rangle$$
$$= \tfrac{1}{2} g' [J(J+1) - L(L+1) - S(S+1)]$$

where g' is independent of J and M.

HINT. From (11–190) show that

$$\langle \alpha LSJM | \hat{g}_i \hat{\mathbf{s}}_i \cdot \hat{\mathbf{l}}_i | \alpha LSJM \rangle = g''' \langle \alpha LSJM | \hat{\mathbf{L}} \cdot \hat{\mathbf{S}} | \alpha LSJM \rangle$$

11–104. The j-j coupling atomic shell model is obtained by assuming that interactions between electrons are negligible compared to spin-orbit coupling, and it is better than the L-S coupling shell model for heavier atoms. In the j-j coupling shell model we take, for example, $2p_{1/2}$ and $2p_{3/2}$ shells with different energy, instead of a single $2p$ shell. Obtain allowable states in the $(2p_{1/2})^3$, $(2p_{1/2})^2(2p_{3/2})$, $(2p_{1/2})(2p_{3/2})^2$, and $(2p_{3/2})^3$ configurations, and compare the result with that of the $(2p)^3$ configuration in the L-S coupling.

HINT. Use Tables 11–8 and 11–13 for $j = \tfrac{1}{2}$ and $\tfrac{3}{2}$, respectively. See Fig. 11–45 for the answer.

FIGURE 11–45. The relation between the L–S and j–j couplings in the $(2p)^3$ configuration.

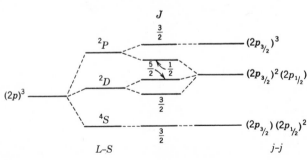

11–105. (a) Show that the Coulomb potential between nucleons 1 and 2 can be expressed as

$$\frac{e^2}{4\pi\varepsilon_0 r_{12}} (\hat{\tau}_{1\zeta} + \tfrac{1}{2})(\hat{\tau}_{2\zeta} + \tfrac{1}{2})$$

by means of the isospin operators.

(b) Show that the magnetic moment operator for a nucleon can be expressed as

$$\frac{e}{2m_p} \{(\hat{\tau}_\zeta + \tfrac{1}{2})\hat{\mathbf{l}} + 2\hat{\tau}_\zeta[(g_p + g_n)\hat{\tau}_\zeta + \tfrac{1}{2}(g_p - g_n)]\hat{\mathbf{s}}\}$$

where m_p is the nucleon mass, $g_p = 5.59$, and $g_n = -3.83$.

(c) Show that the expectation value of the above given magnetic moment operator for the proton state is

$$\frac{e}{2m_p} [(j - \tfrac{1}{2}) + \tfrac{1}{2}g_p] \qquad \text{if} \qquad j = l + \tfrac{1}{2}$$

$$\frac{e}{2m_p} \frac{j}{j + 1} [(j + \tfrac{3}{2}) - \tfrac{1}{2}g_p] \qquad \text{if} \qquad j = l - \tfrac{1}{2}$$

HINT. Use Table 10–5 for (c).

11–106. The ethylene molecule C_2H_4 is planer and has the symmetries of the $\mathbf{D}_{2h}$ group (Fig. 11–46). Find the irreducible representations to which normal vibrations of this molecule belong.

ANSWER.

$$3A_{1g} + A_{1u} + 2B_{1u} + B_{2g} + 2B_{2u} + 2B_{3g} + B_{3u}$$

FIGURE 11–46. The ethylene molecule.

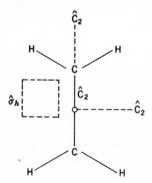

11–107. The ordinary oxygen molecule is made of two 0^{16} nuclei. Since 0^{16} has an even number of nucleons, the exchange of these nuclei should leave the wave function of this molecule invariant. The wave function is

$$\psi = \phi_n(r)\gamma_S f_{nvN} Y_{Nm}\eta$$

in the Born–Oppenheimer approximation. In this expression ϕ_n and γ_S are electronic orbital and spin functions, respectively, f_{nvN} and Y_{Nm} are nuclear vibrational and rotational functions, while η is a function for the structure of nuclei including their spins. Since the nuclear spin of 0^{16} is zero, η is symmetric with respect to the exchange of the nuclei. Show that in the ground electronic state $^3\Sigma_g^-$, the rotational quantum number N has to be odd in order for ψ to be symmetric with respect to the exchange of the nuclei.

HINT. In the Born–Oppenheimer approximation $\phi_n(\mathbf{r})$ is described in terms of a coordinate system fixed to the nuclei, and the exchange of the nuclei is equivalent to $\hat{C}_2$ for ϕ_n.

11–108. (a)* If $\hat{\mathbf{A}}$ is the Runge vector, show that

$$\hat{\mathbf{A}}^2 = (mK)^2 + 2m\hat{H}(\hat{l}^2 + \hbar^2)$$

(b) Using the above result obtain the expression of $\hat{H}$ expected in Problem 11–94.

HINT. (10–242),

$$(\hat{\mathbf{l}} \times \hat{\mathbf{p}}) \cdot (\hat{\mathbf{l}} \times \hat{\mathbf{p}}) = \hat{l}^2 \hat{p}^2 \qquad \hat{\mathbf{p}} \cdot (\hat{\mathbf{l}} \times \hat{\mathbf{p}}) + (\hat{\mathbf{l}} \times \hat{\mathbf{p}}) \cdot \hat{\mathbf{p}} = 2i\hbar \hat{p}^2$$

and other formulas.

Appendix I   Mathematical Formulas

(a) Vector Relations

$$\mathbf{A} \cdot (\mathbf{B} \times \mathbf{C}) = \mathbf{B} \cdot (\mathbf{C} \times \mathbf{A}) = \mathbf{C} \cdot (\mathbf{A} \times \mathbf{B})$$

$$\mathbf{A} \times (\mathbf{B} \times \mathbf{C}) = \mathbf{B}(\mathbf{A} \cdot \mathbf{C}) - \mathbf{C}(\mathbf{A} \cdot \mathbf{B})$$

$$\mathbf{A} \times (\mathbf{B} \times \mathbf{C}) + \mathbf{B} \times (\mathbf{C} \times \mathbf{A}) + \mathbf{C} \times (\mathbf{A} \times \mathbf{B}) = 0$$

$$(\mathbf{A} \times \mathbf{B}) \cdot (\mathbf{C} \times \mathbf{D}) = (\mathbf{A} \cdot \mathbf{C})(\mathbf{B} \cdot \mathbf{D}) - (\mathbf{B} \cdot \mathbf{C})(\mathbf{A} \cdot \mathbf{D})$$

$$(\mathbf{A} \times \mathbf{B}) \times (\mathbf{C} \times \mathbf{D}) = \mathbf{C}[(\mathbf{A} \times \mathbf{B}) \cdot \mathbf{D}] - \mathbf{D}[(\mathbf{A} \times \mathbf{B}) \cdot \mathbf{C}]$$

$$\mathbf{A} \times [\mathbf{B} \times (\mathbf{C} \times \mathbf{D})] = (\mathbf{B} \cdot \mathbf{D})(\mathbf{A} \times \mathbf{C}) - (\mathbf{B} \cdot \mathbf{C})(\mathbf{A} \times \mathbf{D})$$

$$(\mathbf{A} \times \mathbf{B}) \cdot [(\mathbf{B} \times \mathbf{C}) \times (\mathbf{C} \times \mathbf{A})] = [\mathbf{A} \cdot (\mathbf{B} \times \mathbf{C})]^2$$

$$(\boldsymbol{\sigma} \cdot \mathbf{A})(\boldsymbol{\sigma} \cdot \mathbf{B}) = \mathbf{A} \cdot \mathbf{B} + i\boldsymbol{\sigma} \cdot (\mathbf{A} \times \mathbf{B}) \qquad \boldsymbol{\sigma}\text{: Pauli matrices}$$

$$\nabla(\varphi + \psi) = \nabla\varphi + \nabla\psi$$

$$\nabla(\varphi\psi) = \varphi\nabla\psi + \psi\nabla\varphi$$

$$\nabla \cdot (\mathbf{A} + \mathbf{B}) = \nabla \cdot \mathbf{A} + \nabla \cdot \mathbf{B}$$

$$\nabla \times (\mathbf{A} + \mathbf{B}) = \nabla \times \mathbf{A} + \nabla \times \mathbf{B}$$

$$\nabla \times (\varphi\mathbf{A}) = \varphi\nabla \times \mathbf{A} + \nabla\varphi \times \mathbf{A}$$

$$\nabla \times (\mathbf{A} \times \mathbf{B}) = \mathbf{A}(\nabla \cdot \mathbf{B}) - \mathbf{B}(\nabla \cdot \mathbf{A}) + (\mathbf{B} \cdot \nabla)\mathbf{A} - (\mathbf{A} \cdot \nabla)\mathbf{B}$$

$$\nabla(\mathbf{A} \cdot \mathbf{B}) = (\mathbf{A} \cdot \nabla)\mathbf{B} + (\mathbf{B} \cdot \nabla)\mathbf{A} + \mathbf{A} \times (\nabla \times \mathbf{B}) + \mathbf{B} \times (\nabla \times \mathbf{A})$$

$$\nabla \cdot (\varphi \mathbf{A}) = \varphi \nabla \cdot \mathbf{A} + \mathbf{A} \cdot \nabla \varphi$$

$$\nabla \cdot (\mathbf{A} \times \mathbf{B}) = \mathbf{A} \cdot (\nabla \times \mathbf{B}) - \mathbf{B} \cdot (\nabla \times \mathbf{A})$$

$$\nabla \times (\nabla \times \mathbf{A}) = \nabla(\nabla \cdot \mathbf{A}) - \nabla^2 \mathbf{A}$$

$$\nabla \cdot (\nabla \times \mathbf{A}) = 0$$

$$\nabla \times (\nabla \varphi) = 0$$

$$\int_v \nabla \cdot \mathbf{A} \, dv = \oint_S \mathbf{A} \cdot d\mathbf{a} \qquad \text{Gauss' theorem.}$$

$$\int_S (\nabla \times \mathbf{A}) \cdot d\mathbf{a} = \oint_L \mathbf{A} \cdot d\mathbf{l} \qquad \text{Stokes' theorem.}$$

(b) Matrices

Let

$$\mathbf{A} = \begin{pmatrix} A_{11} & A_{12} & A_{13} & \cdots \\ A_{21} & A_{22} & & \cdots \\ A_{31} & A_{32} & & \cdots \\ \cdots\cdots\cdots\cdots\cdots\cdots\cdots \end{pmatrix}$$

then its transposed matrix is

$$\mathbf{A}^T = \begin{pmatrix} A_{11} & A_{21} & A_{31} & \cdots \\ A_{12} & A_{22} & A_{32} & \cdots \\ A_{13} & A_{23} & & \cdots \\ \cdots\cdots\cdots\cdots\cdots\cdots\cdots \end{pmatrix}$$

and Hermitian conjugate matrix is

$$\mathbf{A}^\dagger = \mathbf{A}^{T*} = \begin{pmatrix} A_{11}{}^* & A_{21}{}^* & A_{31}{}^* & \cdots \\ A_{12}{}^* & A_{22}{}^* & A_{32}{}^* & \cdots \\ A_{13}{}^* & A_{23}{}^* & & \cdots \\ \cdots\cdots\cdots\cdots\cdots\cdots\cdots \end{pmatrix}$$

If the unit matrix is

$$\mathbf{1} = \begin{pmatrix} 1 & 0 & 0 & 0 & \cdots \\ 0 & 1 & 0 & 0 & \cdots \\ 0 & 0 & 1 & 0 & \cdots \\ 0 & 0 & 0 & 1 & \cdots \\ \cdots\cdots\cdots\cdots\cdots\cdots \end{pmatrix}$$

and if $\mathbf{AA}^{-1} = \mathbf{1}$, $\mathbf{A}^{-1}$ is called the inverse matrix to $\mathbf{A}$.

Special matrices:

$$\mathbf{A} \text{ is called } \begin{cases} \text{Hermitian if } \mathbf{A}^\dagger = \mathbf{A} \\ \text{orthogonal if } \mathbf{A}^{-1} = \mathbf{A}^T \\ \text{unitary if } \mathbf{A}^{-1} = \mathbf{A}^\dagger \end{cases}$$

Matrix multiplications:

$$\mathbf{AB} = \mathbf{C} \quad \text{means} \quad \sum_k A_{ik}B_{kj} = C_{ij}$$

Then

$$\mathbf{C}^T = \mathbf{B}^T\mathbf{A}^T \qquad \mathbf{C}^\dagger = \mathbf{B}^\dagger\mathbf{A}^\dagger \qquad \mathbf{C}^{-1} = \mathbf{B}^{-1}\mathbf{A}^{-1}$$

Invariants are

$$\text{tr } \mathbf{A} \equiv \sum_i A_{ii}$$

and

$$\det \mathbf{A} \equiv \text{determinant of } \mathbf{A}$$

In general

$$\text{tr } \{\mathbf{AB}\} = \text{tr } \{\mathbf{BA}\}$$

$$\det \{\mathbf{AB}\} = \{\det \mathbf{A}\}\{\det \mathbf{B}\} = \det \{\mathbf{BA}\}$$

Thus

$$\text{tr } \{\mathbf{A}^{-1}\mathbf{BA}\} = \text{tr } \{\mathbf{BAA}^{-1}\} = \text{tr } \mathbf{B}$$

$$\det \{\mathbf{A}^{-1}\mathbf{BA}\} = \det \{\mathbf{BAA}^{-1}\} = \det \mathbf{B}$$

(c) Special Functions

Trigonometric Functions

$$\sin (A \pm B) = \sin A \cos B \pm \cos A \sin B$$

$$\cos (A \pm B) = \cos A \cos B \mp \sin A \sin B$$

$$\sin A \pm \sin B = 2 \sin \left[\tfrac{1}{2}(A \pm B)\right] \cos \left[\tfrac{1}{2}(A \mp B)\right]$$

$$\cos A + \cos B = 2 \cos \left[\tfrac{1}{2}(A + B)\right] \cos \left[\tfrac{1}{2}(A - B)\right]$$

$$\cos A - \cos B = -2 \sin \left[\tfrac{1}{2}(A + B)\right] \sin \left[\tfrac{1}{2}(A - B)\right]$$

$$\sin A \sin B = -\tfrac{1}{2}\left[\cos (A + B) - \cos (A - B)\right]$$

$$\sin A \cos B = \tfrac{1}{2}\left[\sin (A + B) + \sin (A - B)\right]$$

$$\cos A \cos B = \tfrac{1}{2}\left[\cos (A + B) + \cos (A - B)\right]$$

$$\cos A = \tfrac{1}{2}(e^{iA} + e^{-iA}) \qquad \sin A = -\tfrac{1}{2}i(e^{iA} - e^{-iA})$$

Spherical Harmonics

$$Y_{lm}(\theta, \varphi) = \frac{1}{\sqrt{4\pi}} \, \overline{P_l^m}(\cos \theta)e^{im\varphi}$$

$$\begin{cases} = (-1)^m \sqrt{\dfrac{(2l + 1)(l - m)!}{4\pi(l + m)!}} \, P_l^m(\cos \theta)e^{im\varphi} & m \geq 0 \\[3mm] = \sqrt{\dfrac{(2l + 1)(l - |m|)!}{4\pi(l + |m|)!}} \, P_l^{|m|}(\cos \theta)e^{im\varphi} & m < 0 \end{cases}$$

$P_l^m (\cos \theta)$ \qquad associated Legendre function

$$Y_{l-m}(\theta, \varphi) = (-1)^m Y_{lm}^*(\theta, \varphi)$$

$$Y_{l0}(\theta, \varphi) = \sqrt{\frac{4\pi}{2l + 1}} \sum_{m=-l}^{l} Y_{lm}^*(\theta_1, \varphi_1)Y_{lm}(\theta_2, \varphi_2)$$

(See Fig. I–1.)

l	m \quad 0	± 1
0	$1/\sqrt{4\pi}$	
1	$\sqrt{3/4\pi}\,\cos \theta$	$\mp\sqrt{3/8\pi}\,\sin \theta e^{\pm i\varphi}$
2	$\sqrt{5/16\pi}(3 \cos^2 \theta - 1)$	$\mp\sqrt{15/8\pi}\,\sin \theta \cos \theta e^{\pm i\varphi}$
3	$\sqrt{7/16\pi}(5 \cos^3 \theta - 3 \cos \theta)$	$\mp\sqrt{21/64\pi}\,\sin \theta(5 \cos^2 \theta - 1)e^{\pm i\varphi}$

l	± 2	± 3
0		
1		
2	$\sqrt{15/32\pi}\,\sin^2 \theta e^{\pm 2i\varphi}$	
3	$\sqrt{105/32\pi}\,\sin^2 \theta \cos \theta e^{\pm 2i\varphi}$	$\mp\sqrt{35/64\pi}\,\sin^3 \theta e^{\pm 3i\varphi}$

FIGURE I–1. The angles which appear in the sum rule of the spherical harmonics.

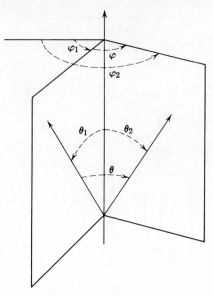

Legendre Polynomials

$P_0(x) = 1$

$P_1(x) = x$

$P_2(x) = \frac{1}{2}(3x^2 - 1)$

$P_3(2) = \frac{1}{2}(5x^3 - 3x)$

$P_4(x) = \frac{1}{8}(35x^4 - 30x^2 + 3)$

$P_1{}^1(x) = \sqrt{1 - x^2}$

$P_2{}^1(x) = 3x\sqrt{1 - x^2}$

$P_2{}^2(x) = 3(1 - x^2)$

$P_3{}^1(x) = \frac{3}{2}(5x^2 - 1)\sqrt{1 - x^2}$

$P_3{}^2(x) = 15x(1 - x^2)$

$P_3{}^3(x) = 15(1 - x^2)^{3/2}$

Special Values

$$P_n(1) = 1 \qquad P_{2n+1}(0) = 0 \qquad P_{2n}(0) = (-1)^n \frac{1 \cdot 3 \cdot 5 \cdots (2n - 1)}{2 \cdot 4 \cdot 6 \cdots 2n}$$

$$P_n(-x) = (-1)^n P_n(x)$$

$$\int_0^{2\pi} \int_0^\pi Y_{lm}{}^*(\theta, \varphi) Y_{l'm'}(\theta, \varphi) \sin \theta \, d\theta d\varphi = \delta_{ll'} \delta_{mm'}$$

Gamma Function

$$\Gamma(z) = \int_0^\infty e^{-t} t^{z-1} \, dt = (z - 1)\Gamma(z - 1)$$

If n is a positive integer

$$\Gamma(n + 1) = n!$$

$$\Gamma(\tfrac{1}{2}) = \sqrt{\pi} \qquad \Gamma(\tfrac{3}{2}) = \tfrac{1}{2}\sqrt{\pi} \qquad \Gamma(1) = 1$$

Stirlin's formula:

$$\Gamma(x + 1) \Rightarrow \sqrt{2\pi}\, x^{x+(1/2)}e^{-x}$$

when $x \to \infty$. (See Fig. I–2.)

FIGURE I–2.   The gamma function.

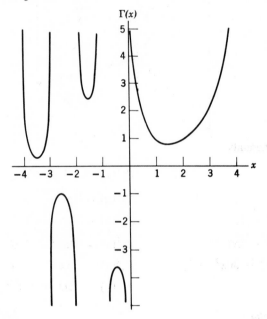

Bessel Functions

$$J_m(z) = (\tfrac{1}{2}z)^m \sum_{k=0}^{\infty} \frac{(-\tfrac{1}{4}z^2)^k}{k!\,\Gamma(m + k + 1)}$$

$$N_m(z) = \frac{1}{\sin(m\pi)}[\cos(m\pi)J_m(z) - J_{-m}(z)] \qquad \text{Neuman functions}$$

$$H_m^{(1)}(z) = J_m(z) + iN_m(z) \qquad \text{First kind Hankel functions}$$

$$H_m^{(2)}(z) = J_m(z) - iN_m(z) \qquad \text{Second kind Hankel functions}$$

(See Fig. I–3.)

FIGURE I-3. $J_0(x)$, $J_1(x)$, $N_0(x)$, and $N_1(x)$.

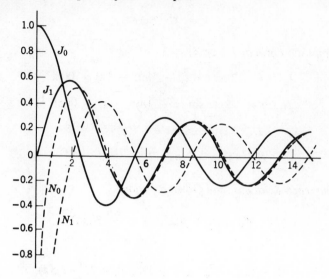

Differential Equations

If

$$\left[z^2 \frac{d^2}{dz^2} + z \frac{d}{dz} + (z^2 - m^2) \right] w = 0$$

then w is $\mathscr{C}_m(z)$, where

$$\mathscr{C}_m(z) = J_m(z),\ N_m(z),\ H_m^{(1)}(z),\ H_m^{(2)}(z)$$

or their linear combinations. If

$$\left(\frac{d^2}{dz^2} - \frac{2m - 1}{z} \frac{d}{dz} + \lambda^2 \right) w = 0$$

then

$$w = z^m \mathscr{C}_m(\lambda z)$$

If

$$\left(\frac{d^2}{dz^2} + \lambda^2 - \frac{m^2 - \frac{1}{4}}{z^2} \right) w = 0$$

then

$$w = z^{1/2}\mathscr{C}_m(\lambda z)$$

Asymptotic Form at Large $|z|$

$$J_m(z) = \sqrt{2/\pi z}\,\cos\,(z - \tfrac{1}{2}m\pi - \tfrac{1}{4}\pi) + 0(|z|^{-1})$$

$$N_m(z) = \sqrt{2/\pi z}\,\sin\,(z - \tfrac{1}{2}m\pi - \tfrac{1}{4}) + 0(|z|^{-1})$$

$$H_m^{(1)}(z) \cong \sqrt{2/\pi z}\,e^{i(z-(1/2)m\pi-(1/4)\pi)}$$

$$H_m^{(2)}(z) \cong \sqrt{2/\pi z}\,e^{-i(z-(1/2)m\pi-(1/4)\pi)}$$

Recurrence Relations, and Other Relations

$$\mathscr{C}_{m-1}(z) + \mathscr{C}_{m+1}(z) = \frac{2m}{z}\,\mathscr{C}_m(z)$$

$$\mathscr{C}_{m-1}(z) - \mathscr{C}_{m+1}(z) = 2\mathscr{C}'_m(z) \equiv 2\,\frac{d}{dz}\,\mathscr{C}_m(z)$$

$$\mathscr{C}_{m-1}(z) + \mathscr{C}_{m+1}(z) = \frac{2m}{z}\,z\mathscr{C}_m(z)$$

$$J_m(ze^{ik\pi}) = e^{ikm\pi}J_m(z)$$

$$e^{(1/2)z(t-1/t)} = \sum_{k=-\infty}^{\infty} t^k J_k(z)$$

$$1 = J_0(z) + 2[J_2(z) + J_4(z) + J_6(z) + \cdots]$$

$$\cos z = J_0(z) - 2J_2(z) + 2J_4(z) - 2J_6(z) + \cdots$$

$$\sin z = 2[J_1(z) - J_3(z) + J_5(z) - \cdots]$$

$$J_m(2z) = \sum_{k=0}^{m} J_k(z)J_{m-k}(z) + 2\sum_{k=1}^{\infty} (-1)^k J_k(z)J_{m+k}(z)$$

$$1 = J_0^2(z) + 2\sum_{k=1}^{\infty} J_k^2(z)$$

$$0 = \sum_{k=0}^{2m} (-1)^k J_k(z)J_{2m-k}(z) + 2\sum_{k=1}^{\infty} J_k(z)J_{2m+k}(z) \qquad m \geq 1$$

Spherical Bessel Functions

$$j_m(z) = \sqrt{\pi/2z}\, J_{m+(1/2)}(z)$$

$$n_m(z) = \sqrt{\pi/2z}\, N_{m+(1/2)}(z) \qquad \text{Spherical Neuman functions}$$

$$h_m^{(1)}(z) = j_m(z) + i n_m(z) \qquad \text{Spherical first kind Hankel functions}$$

$$h_m^{(2)}(z) = j_m(z) - i n_m(z) \qquad \text{Spherical second kind Hankel functions}$$

$$j_m(iz) = i^m \sqrt{\pi/2z}\, I_{m+(1/2)}(z)$$

$$n_m(iz) = i^{-3(m+1)} \sqrt{\pi/2z}\, I_{-m-(1/2)}(z)$$

$I_m(z)$ is a modified Bessel function.

$$j_m(z) = \frac{z^m}{1 \cdot 3 \cdot 5 \cdots (2m+1)}$$

$$\times \left[1 - \frac{\frac{1}{2}z^2}{1!\,(2m+3)} + \frac{(\frac{1}{2}z^2)^2}{2!\,(2m+3)(2m+5)} - \cdots \right]$$

$$n_m(z) = \frac{1 \cdot 3 \cdot 5 \cdots (2m-1)}{z^{m+1}}$$

$$\times \left[1 - \frac{\frac{1}{2}z^2}{1!\,(1-2m)} + \frac{(\frac{1}{2}z^2)^2}{2!\,(1-2m)(3-2m)} - \cdots \right]$$

$$j_0(z) = \frac{\sin z}{z} \quad j_1(z) = \frac{\sin z}{z^2} - \frac{\cos z}{z} \quad j_2(z) = \left(\frac{3}{z^3} - \frac{1}{z} \right) \sin z - \frac{3}{z^2} \cos z$$

$$n_0(z) = -j_{-1}(z) = -\frac{\cos z}{z} \quad n_1(z) = j_{-2}(z) = -\frac{\cos z}{z^2} - \frac{\sin z}{z}$$

(See Figs. I–4 and I–5.)

FIGURE I–4. $j_m(x)$.

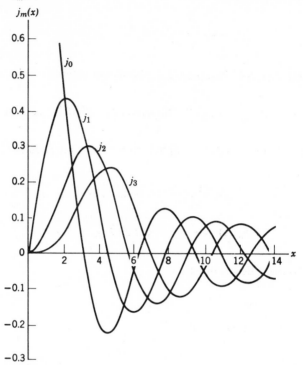

Differential Equations and Other Relations of Spherical Bessel Functions

If

$$f_m(z) = Aj_m(z) + Bn_m(z)$$

where A and B are constants;

$$\left[\frac{1}{z^2} \frac{d}{dz} z^2 \frac{d}{dz} + 1 - \frac{m(m+1)}{z^2} \right] f_m(z) = 0$$

$$(2m+1)f_m(z) = z[f_{m-1}(z) + f_{m+1}(z)]$$

$$(2m+1) \frac{d}{dz} f_m(z) = mf_{m-1}(z) - (m+1)f_{m+1}(z)$$

$$= (2m+1) \left[\frac{m}{z} f_m(z) - f_{m+1}(z) \right]$$

FIGURE I–5. $n_m(x)$.

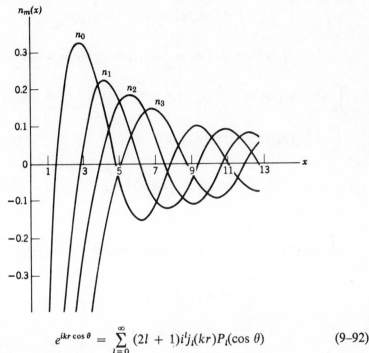

$$e^{ikr \cos \theta} = \sum_{l=0}^{\infty} (2l + 1)i^l j_l(kr) P_l(\cos \theta) \qquad (9\text{–}92)$$

$$1 = \sum_{m=0}^{\infty} (2m + 1)j_m^2(z) \qquad \frac{\sin 2z}{2z} = \sum_{m=0}^{\infty} (-1)^m (2m + 1)j_m^2(z)$$

$$j_m(2z) = -m!\, z^{m+1} \sum_{k=0}^{m} \frac{2m - 2k + 1}{k!\,(2m - k + 1)!} j_{m-k-\frac{1}{2}}(z) n_{m-k+\frac{1}{2}}(z)$$

Asymptotic Form of Spherical Bessel Functions at Large $|z|$

$$j_m(z) \to \frac{1}{z} \cos \left[z - \tfrac{1}{2}\pi(m + 1) \right] \qquad n_m(z) \to \frac{1}{z} \sin \left[z - \tfrac{1}{2}\pi(m + 1) \right]$$

$$h_m^{(1)}(z) \to \frac{1}{z} i^{-m-1} e^{iz} \qquad\qquad h_m^{(2)}(z) \to \frac{1}{z} i^{m+1} e^{-iz} \qquad m\text{: integer.}$$

Integrals

$$\int x^{m+1}\mathscr{C}_m(x)\,dx = x^{m+1}\mathscr{C}_{m+1}(x)$$

$$\int_0^1 x^{m+1}J_m(\lambda x)\,dx = \frac{1}{\lambda}J_{m+1}(\lambda) \qquad \mathscr{R}e\; m > -1$$

$$\int_0^1 x^{m+1}N_m(\lambda x)\,dx = \frac{1}{\lambda}N_{m+1}(\lambda) + 2^{m+1}\frac{\Gamma(m+1)}{\lambda^{m+2}} \qquad \mathscr{R}e\; m > -1$$

$$\int_0^\infty J_{m+1}(\alpha x)J_l(\beta x)x^{l-m}\,dx = 0 \qquad\qquad\quad \text{if}\quad \alpha < \beta$$

$$= \frac{(\alpha^2 - \beta^2)^{m-l}\beta^l}{2^{m-l}\alpha^{m+1}\Gamma(m-l+1)} \qquad \text{if}\quad \alpha \ge \beta$$

$$\int j_l(\alpha r)j_l(\beta r)r^2\,dr = \frac{r^2}{\alpha^2 - \beta^2}\left[\alpha j_{l+1}(\alpha r)j_l(\beta r) - \beta j_l(\alpha r)j_{l+1}(\beta r)\right]$$

$$\int_0^\infty j_l(qr)j_l(qr')q^2\,dq = \tfrac{1}{2}\pi\frac{\delta(r-r')}{r^2} \tag{9-76}$$

$$\int_{-\infty}^\infty j_m(x)j_n(x)\,dx = \frac{\pi}{2n+1}\delta_{n,m}$$

Confluent Hypergeometric Functions (Kummer Functions)

$$F(a, b; z) = 1 + \frac{a}{b}z + \frac{a(a+1)}{b(b+1)}\frac{z^2}{2!} + \frac{a(a+1)(a+2)}{b(b+1)(b+2)}\frac{z^3}{3!} + \cdots$$

Other notations which appear in the literature are

$${}_1F_1(a, b; z) = \Phi(a, b; z) = M(a, b, z) = F(a, b; z)$$

Differential Equations

If

$$\left(z\frac{d^2}{dz^2} + (b-z)\frac{d}{dz} - a\right)w = 0$$

then

$$w = c_1 F(a, b; z) + c_2 z^{1-b}F(a-b+1, 2-b; z)$$

If

$$\left[\frac{d^2}{dz^2} + k^2 - \frac{l(l+1)}{r^2} - \frac{2\gamma k}{r}\right](rf) = 0$$

then

$$f = c_1 e^{-ikr}(kr)^l F(l + 1 - i\gamma, 2l + 2; 2ikr)$$

$$+ c_2 e^{ikr}(kr)^l F(l + 1 + i\gamma, 2l + 2; -2ikr)$$

(Coulomb functions)

are the solutions regular at $r = 0$. c_1 and c_2 are constants.

Asymptotic Form at Large $|z|$

$F(a, b; z)$

$$= \frac{\Gamma(b)}{\Gamma(b - a)}(-z)^{-a}$$

$$\times \left[1 + \frac{a(a - b + 1)}{-z} + \frac{a(a + 1)(a - b + 1)(a - b + 2)}{2!\,(-z)^2} + \cdots\right]$$

$$+ \frac{\Gamma(b)}{\Gamma(a)}e^z z^{a-b}$$

$$\times \left[1 + \frac{(b - a)(1 - a)}{z} + \frac{(b - a)(b - a + 1)(1 - a)(2 - a)}{2!\,z^2} + \cdots\right]$$

Related Formulas

$$F(a, b; 0) = 1$$

$$F(a, b; z) = e^z F(b - a, b; -z) \qquad \text{Kummer transformation}$$

$$aF(a + 1, b + 1; z) = (a - b)F(a, b + 1; z) + bF(a, b; z)$$

$$aF(a + 1, b; z) = (z + 2a - b)F(a, b; z) + (b - a)F(a - 1, b; z)$$

$$zF(a + 1, b + 1; z) = b[F(a + 1, b; z) - F(a, b; z)]$$

$$(b - a)zF(a, b + 1; z) = b(z + b - 1)F(a, b; z)$$
$$+ b(1 - b)F(a, b - 1; z)$$

$$F'(a, b; z) = \frac{a}{b}F(a + 1, b + 1; z) = F(a, b; z) + \left(\frac{a}{b} - 1\right)F(a, b + 1; z)$$

$$zF'(a, b; z) = a[F(a + 1, b; z) - F(a, b; z)]$$
$$= (b - 1)[F(a, b - 1; z) - F(a, b; z)]$$
$$= (b - a)F(a - 1, b; z) - (b - a - z)F(a, b; z)$$
$$= (b - 1)F(a - 1, b - 1; z) - (b - 1 - z)F(a, b; z)$$

$$F' \equiv \frac{d}{dz} F$$

Integrals (Landau and Lifschitz)

$$\int_0^\infty e^{-\lambda z} z^\nu F(a, b; kz) \, dz = \Gamma(\nu + 1)\lambda^{-\nu-1} F\left(a, \nu + 1, b; \frac{k}{\lambda}\right)$$

F: Hypergeometric function

$$\int_0^\infty e^{-kz} z^{\nu-1}[F(-n, b; kz)]^2 \, dz$$

$$= \frac{\Gamma(\nu)\Gamma(n + 1)\Gamma(b)}{k^\nu \Gamma(b + n)}$$

$$\times \left[1 + \frac{n(b - \nu - 1)(b - \nu)}{Bb}\right.$$

$$+ \frac{n(n - 1)(b - \gamma - 2)(b - \gamma - 1)(b - \gamma)(b - \gamma + 1)}{2^2 b(b + 1)} + \cdots$$

$$\left. + \frac{n(n - 1) \cdots 1(b - \nu - n) \cdots (b - \nu + n - 1)}{(n + 1)^2 b(b + 1) \cdots (b + n - 1)}\right]$$

$$\int_0^\infty e^{-\lambda z} z^{b-1} F(a, b; kz) F(a', b; k'z) \, dz$$

$$= \Gamma(b)\lambda^{a+a'-b}(\lambda - k)^{-a}(\lambda - k')^{-a'} F\left(a, a', b; \frac{kk'}{(\lambda - k)(\lambda - k')}\right)$$

Other Elementary Functions Expressed by Confluent Hypergeometric Functions

$$e^z = F(a, a; z) \qquad \sin z = z e^{iz} F(1, 2; -2iz)$$

$$J_n(z) = [\Gamma(n + 1)]^{-1}(\tfrac{1}{2}z)^n e^{-iz} F(n + \tfrac{1}{2}, 2n + 1; 2iz) \qquad \text{Bessel function}$$

$$H_n(z) = 2^n \left[\frac{\Gamma(-\tfrac{1}{2})}{\Gamma(-\tfrac{1}{2}n)} zF(\tfrac{1}{2} - \tfrac{1}{2}n, \tfrac{3}{2}; z^2)\right.$$

$$\left. + \frac{\Gamma(\tfrac{1}{2})}{\Gamma(\tfrac{1}{2} - \tfrac{1}{2}n)} F(-\tfrac{1}{2}n, \tfrac{1}{2}; z^2)\right] \begin{matrix}\text{Hermite}\\\text{polynomial}\end{matrix}$$

$$H_{2n}(x) = (-\tfrac{1}{2})^n \frac{\Gamma(2n + 1)}{\Gamma(n + 1)} F(-n, \tfrac{1}{2}; \tfrac{1}{2}x^2)$$

(d) Character Tables of Some Point Groups

The characters of all 32 crystallographic point groups and a few other point groups ($p = 5$ and ∞) which are applicable to molecules are given. Those irreducible representations to which vectors (x, y, z) and pseudo-vectors (l_x, l_y, l_z) belong are also shown. See section 11–14.

$\mathbf{C_2}$	$\hat{I}$	$\hat{C}_2(Z)$		$\mathbf{C_i = S_2}$	$\hat{I}$	$\hat{I}$		$\mathbf{C_s = C_{1h}}$	$\hat{I}$	$\hat{\sigma}_h$
A	1	1	z, l_z	A_g	1	1	l_x, l_y, l_z	A'	1	1	x, y, l_z
B	1	-1	x, y, l_x, l_y	A_u	1	-1	x, y, z	A''	1	-1	z, l_x, l_y

$\mathbf{C_3}$	$\hat{I}$	$2\hat{C}_3(Z)$		$\mathbf{C_4}$	$\hat{I}$	$\hat{C}_4(Z)$	$\hat{C}_2(Z)$	$\hat{C}_4^3(Z)$	
A	1	1	z, l_z	A	1	1	1	1	z, l_z
E	2	-1	x, y, l_x, l_y	B	1	-1	1	-1	
				E	2	0	-2	0	x, y, l_x, l_y

$\mathbf{C_5}$	$\hat{I}$	$2\hat{C}_5(Z)$	$2\hat{C}_5^2(Z)$		$\mathbf{D_2 = V}$	$\hat{I}$	$\hat{C}_2(Z)$	$\hat{C}_2(Y)$	$\hat{C}_2(X)$	
A	1	1	1	z, l_z	A	1	1	1	1	
E	2	a	$-b$	x, y, l_x, l_y	B_1	1	1	-1	-1	z, l_z
		$a = \frac{1}{2}(\sqrt{5} - 1)$			B_2	1	-1	1	-1	y, l_y
		$b = \frac{1}{2}(\sqrt{5} + 1)$			B_3	1	-1	-1	1	z, l_z

$\mathbf{C_6}$	$\hat{I}$	$2\hat{C}_6(Z)$	$2\hat{C}_3(Z)$	$\hat{C}_2(Z)$		$\mathbf{D_3}$	$\hat{I}$	$2\hat{C}_3(Z)$	$3\hat{C}_2$	
A	1	1	1	1	z, l_z	A_1	1	1	1	
B	1	-1	1	-1		A_2	1	1	-1	z, l_z
E_1	2	1	-1	-2	x, y, l_x, l_y	E	2	-1	0	x, y, l_x, l_y
E_2	2	-1	-1	2						

$\mathbf{D_4}$	$\hat{I}$	$2\hat{C}_4(Z)$	$\hat{C}_2(Z)$	$2\hat{C}_2$	$2\hat{C}_2{}'$	
A_1	1	1	1	1	1	
A_2	1	1	1	-1	-1	z, l_z
B_1	1	-1	1	1	-1	
B_2	1	-1	1	-1	1	
E	2	0	-2	0	0	x, y, l_x, l_y

$\mathbf{D_5}$	$\hat{I}$	$2\hat{C}_5(Z)$	$2\hat{C}_5{}^2(Z)$	$5\hat{C}_2$		
A_1	1	1	1	1	z, l_z	
A_2	1	1	1	-1		$a = \frac{1}{2}(\sqrt{5} - 1)$
E_1	2	a	$-b$	0	x, y, l_x, l_y	$b = \frac{1}{2}(\sqrt{5} + 1)$
E_2	2	$-b$	a	0		

$\mathbf{D_6}$	$\hat{I}$	$2\hat{C}_6(Z)$	$2\hat{C}_3(Z)$	$\hat{C}_2(Z)$	$3\hat{C}_2$	$3\hat{C}_2{}'$	
A_1	1	1	1	1	1	1	
A_2	1	1	1	1	-1	-1	z, l_z
B_1	1	-1	1	-1	1	-1	
B_2	1	-1	1	-1	-1	1	
E_1	2	1	-1	-2	0	0	x, y, l_x, l_y
E_2	2	-1	-1	2	0	0	

$\mathbf{S_4}$	$\hat{I}$	$\hat{S}_4(Z)$	$\hat{C}_2(Z)$	$\hat{S}_4{}^3(Z)$		$\mathbf{S_6}$	$\hat{I}$	$\hat{S}_6(Z)$	$\hat{C}_3(Z)$	$\hat{I}$	
A	1	1	1	1		A_g	1	1	1	1	l_z
B	1	-1	1	-1	z, l_z	B_u	1	-1	1	-1	z
E	2	0	-2	0	x, y, l_x, l_y	E_{1u}	2	1	-1	-2	x, y
						E_{2g}	2	-1	-1	2	l_x, l_y

$$\mathbf{S_6} = \mathbf{C_3} \times \mathbf{C}_i$$

$\mathbf{C_{2h}}$	$\hat{I}$	$\hat{C}_2(Z)$	$\hat{\sigma}_h$	$\hat{I}$		$\mathbf{C_{3h}}$	$\hat{I}$	$\hat{C}_3(Z)$	$\hat{\sigma}_h$	$\hat{S}_3(Z)$	
A_g	1	1	1	1	l_z	A'	1	1	1	1	l_z
A_u	1	1	-1	-1	z	A''	1	1	-1	-1	z
B_g	1	-1	-1	1	l_x, l_y	E'	2	-1	2	-1	x, y
B_u	1	-1	1	-1	x, y	E''	2	-1	-2	1	l_x, l_y

$\mathbf{C}_{4h}$	$\hat{I}$	$\hat{C}_4(Z)$	$\hat{C}_2(Z)$	$\hat{\sigma}_h$	$\hat{S}_4(Z)$	$\hat{I}$	
A_g	1	1	1	1	1	1	l_z
A_u	1	1	1	-1	-1	-1	z
B_g	1	-1	1	1	-1	1	
B_u	1	-1	1	-1	1	-1	
E_g	2	0	-2	-2	0	2	l_x, l_y
E_u	2	0	-2	2	0	-2	x, y

$$\mathbf{C}_{3h} = \mathbf{C}_3 \times \mathbf{C}_s$$
$$\mathbf{C}_{4h} = \mathbf{C}_4 \times \mathbf{C}_i$$
$$\mathbf{C}_{5h} = \mathbf{C}_5 \times \mathbf{C}_s$$
$$\mathbf{C}_{6h} = \mathbf{C}_6 \times \mathbf{C}_i$$

$\mathbf{C}_{5h}$	$\hat{I}$	$2\hat{C}_5(Z)$	$2\hat{C}_5{}^2(Z)$	$2\hat{S}_5{}^3(Z)$	$2\hat{S}_5(Z)$	$\hat{\sigma}_h$	
A'	1	1	1	1	1	1	l_z
A''	1	1	1	-1	-1	-1	z
E'	2	a	$-b$	$-b$	a	2	l_x, l_y
E''	2	a	$-b$	b	$-a$	-2	x, y

$a = \frac{1}{2}(\sqrt{5} - 1)$
$b = \frac{1}{2}(\sqrt{5} + 1)$

$\mathbf{C}_{6h}$	$\hat{I}$	$2\hat{C}_6(Z)$	$2\hat{C}_3(Z)$	$\hat{C}_2(Z)$	$\hat{\sigma}_h$	$2\hat{S}_6(Z)$	$2\hat{S}_3(Z)$	$\hat{I}$	
A_g	1	1	1	1	1	1	1	1	l_z
A_u	1	1	1	1	-1	-1	-1	-1	z
B_g	1	-1	1	-1	-1	1	-1	1	
B_u	1	-1	1	-1	1	-1	1	-1	
E_{1g}	2	1	-1	-2	-2	-1	1	2	l_x, l_y
E_{1u}	2	1	-1	-2	2	1	-1	-2	x, y
E_{2g}	2	-1	-1	2	2	-1	-1	2	
E_{2u}	2	-1	-1	2	-2	1	1	-2	

$\mathbf{C}_{2v}$	$\hat{I}$	$\hat{C}_2(Z)$	$\hat{\sigma}_v(ZX)$	$\hat{\sigma}_v(ZY)$		$\mathbf{C}_{3v}$	$\hat{I}$	$2\hat{C}_3(Z)$	$3\hat{\sigma}_v$	
A_1	1	1	1	1	z	A_1	1	1	1	z
A_2	1	1	-1	-1	l_z	A_2	1	1	-1	l_z
B_1	1	-1	1	-1	x, l_y	E	2	-1	0	x, y, l_x, l_y
B_2	1	-1	-1	1	y, l_x					

$\mathbf{C}_{4v}$	$\hat{I}$	$2\hat{C}_4(Z)$	$\hat{C}_2(Z)$	$2\hat{\sigma}_v$	$2\hat{\sigma}_d$	
A_1	1	1	1	1	1	z
A_2	1	1	1	-1	-1	l_z
B_1	1	-1	1	1	-1	
B_2	1	-1	1	-1	1	
E	2	0	-2	0	0	x, y, l_x, l_y

$\mathbf{C}_{5v}$	$\hat{I}$	$2\hat{C}_5(Z)$	$2\hat{C}_5{}^2(Z)$	$5\hat{\sigma}_v$		
A_1	1	1	1	1	z	
A_2	1	1	1	-1	l_z	
E_1	2	a	$-b$	0	x, y, l_x, l_z	$a = \frac{1}{2}(\sqrt{5} - 1)$
E_2	2	$-b$	a	0		$b = \frac{1}{2}(\sqrt{5} + 1)$

$\mathbf{C}_{6v}$	$\hat{I}$	$2\hat{C}_6(Z)$	$2\hat{C}_3(Z)$	$\hat{C}_2(Z)$	$3\hat{\sigma}_v$	$3\hat{\sigma}_d$	
A_1	1	1	1	1	1	1	z
A_2	1	1	1	1	-1	-1	l_z
B_1	1	-1	1	-1	1	-1	
B_2	1	-1	1	-1	-1	1	
E_1	2	1	-1	-2	0	0	x, y, l_x, l_y
E_2	2	-1	-1	2	0	0	

$\mathbf{D}_{2h} = \mathbf{V}_h$	$\hat{I}$	$\hat{\sigma}(XY)$	$\hat{\sigma}(YZ)$	$\hat{\sigma}(ZX)$	$\hat{I}$	$\hat{C}_2(Z)$	$\hat{C}_2(Y)$	$\hat{C}_2(X)$	
A_g	1	1	1	1	1	1	1	1	
A_u	1	-1	-1	-1	-1	1	1	1	
B_{1g}	1	1	-1	-1	1	1	-1	-1	l_z
B_{1u}	1	-1	1	1	-1	1	-1	-1	z
B_{2g}	1	-1	-1	1	1	-1	1	-1	l_y
B_{2u}	1	1	1	-1	-1	-1	1	-1	y
B_{3g}	1	-1	1	-1	1	-1	-1	1	l_x
B_{3u}	1	1	-1	1	-1	-1	-1	1	x

$\mathbf{D_{3h}}$	$\hat{I}$	$2\hat{C}_3(Z)$	$3\hat{C}_2$	$\hat{\sigma}_h$	$2\hat{S}_3(Z)$	$3\hat{\sigma}_v$	
A_1'	1	1	1	1	1	1	
A_1''	1	1	1	−1	−1	−1	
A_2'	1	1	−1	1	1	−1	l_z
A_2''	1	1	−1	−1	−1	1	z
E'	2	−1	0	2	−1	0	x, y
E''	2	−1	0	−2	1	0	l_x, l_y

$\mathbf{D_{4h}}$	$\hat{I}$	$2\hat{C}_4(Z)$	$\hat{C}_2(Z)$	$2\hat{C}_2$	$2\hat{C}_2'$	$\hat{\sigma}_h$	$2\hat{\sigma}_v$	$2\hat{\sigma}_d$	$2\hat{S}_4(Z)$	$\hat{I}$	
A_{1g}	1	1	1	1	1	1	1	1	1	1	
A_{1u}	1	1	1	1	1	−1	−1	−1	−1	−1	
A_{2g}	1	1	1	−1	−1	1	−1	−1	1	1	l_z
A_{2u}	1	1	1	−1	−1	−1	1	1	−1	−1	z
B_{1g}	1	−1	1	1	−1	1	1	−1	−1	1	
B_{1u}	1	−1	1	1	−1	−1	−1	1	1	−1	
B_{2g}	1	−1	1	−1	1	1	−1	1	−1	1	
B_{2u}	1	−1	1	−1	1	−1	1	−1	1	−1	
E_g	2	0	−2	0	0	−2	0	0	0	2	l_x, l_y
E_u	2	0	−2	0	0	2	0	0	0	−2	x, y

$\mathbf{D_{5h}}$	$\hat{I}$	$2\hat{C}_5(Z)$	$2\hat{C}_5{}^2(Z)$	$\hat{\sigma}_h$	$5\hat{C}_2$	$5\hat{\sigma}_v$	$2\hat{S}_5(Z)$	$2\hat{S}_5{}^3(Z)$	
A_1'	1	1	1	1	1	1	1	1	
A_1''	1	1	1	−1	1	−1	−1	−1	
A_2'	1	1	1	1	−1	−1	1	1	l_z
A_2''	1	1	1	−1	−1	1	−1	−1	z
E_1'	2	a	$-b$	2	0	0	a	$-b$	x, y
E_1''	2	a	$-b$	−2	0	0	$-a$	b	l_x, l_y
E_2'	2	$-b$	a	2	0	0	$-b$	a	
E_2''	2	$-b$	a	−2	0	0	b	$-a$	

$$a = \tfrac{1}{2}(\sqrt{5} - 1) \qquad b = \tfrac{1}{2}(\sqrt{5} + 1)$$

$$\mathbf{D_{2h}} = \mathbf{D_2} \times \mathbf{C_i}$$

$$\mathbf{D_{4h}} = \mathbf{D_4} \times \mathbf{C_i}$$

$$\mathbf{D_{6h}} = \mathbf{D_6} \times \mathbf{C_i}$$

$\mathbf{D}_{6h}$	$\hat{I}$	$2\hat{C}_6(Z)$	$2\hat{C}_3(Z)$	$\hat{C}_2(Z)$	$3\hat{C}_2(Y)$	$3\hat{C}_2(X)$	$\hat{\sigma}_h$	$3\hat{\sigma}_h$	$3\hat{\sigma}_d$	$2\hat{S}_6(Z)$	$2\hat{S}_3(Z)$	$\hat{I}$	
A_{1g}	1	1	1	1	1	1	1	1	1	1	1	1	
A_{1u}	1	1	1	1	1	1	-1	-1	-1	-1	-1	-1	
A_{2g}	1	1	1	1	-1	-1	1	-1	-1	1	1	1	l_z
A_{2u}	1	1	1	1	-1	-1	-1	1	1	-1	-1	-1	z
B_{1g}	1	-1	1	-1	1	-1	-1	-1	1	1	-1	1	
B_{1u}	1	-1	1	-1	1	-1	1	1	-1	-1	1	-1	
B_{2g}	1	-1	1	-1	-1	1	-1	1	-1	1	-1	1	
B_{2u}	1	-1	1	-1	-1	1	1	-1	1	-1	1	-1	
E_{1g}	2	1	-1	-2	0	0	-2	0	0	-1	1	2	l_x, l_y
E_{1u}	2	1	-1	-2	0	0	2	0	0	1	-1	-2	x, y
E_{2g}	2	-1	-1	2	0	0	2	0	0	-1	-1	2	
E_{2u}	2	-1	-1	2	0	0	-2	0	0	1	1	-2	

$\mathbf{D}_{2d} = \mathbf{V}_d$	$\hat{I}$	$2\hat{S}_4(Z)$	$\hat{C}_2(Z)$	$2\hat{C}_2$	$2\hat{\sigma}_d$	
A_1	1	1	1	1	1	
A_2	1	1	1	-1	-1	l_z
B_1	1	-1	1	1	-1	
B_2	1	-1	1	-1	1	z
E	2	0	-2	0	0	x, y, l_x, l_y

$$\mathbf{D}_{3d} = \mathbf{D}_3 \times \mathbf{C}_i$$

$$\mathbf{D}_{5d} = \mathbf{D}_5 \times \mathbf{C}_i$$

$\mathbf{D}_{3d}$	$\hat{I}$	$2\hat{S}_6(Z)$	$2\hat{C}_3(Z)$	$\hat{I}$	$3\hat{C}_2$	$3\hat{\sigma}_d$	
A_{1g}	1	1	1	1	1	1	
A_{1u}	1	-1	1	-1	1	-1	
A_{2g}	1	1	1	1	-1	-1	l_z
A_{2u}	1	-1	1	-1	-1	1	z
E_g	2	-1	-1	2	0	0	l_x, l_y
E_u	2	1	-1	-2	0	0	x, y

D_{4d}	$\hat{I}$	$2\hat{S}_8(Z)$	$2\hat{C}_4(Z)$	$2\hat{S}_8{}^3(Z)$	$\hat{C}_2(Z)$	$4\hat{C}_2$	$4\hat{\sigma}_d$	
A_1	1	1	1	1	1	1	1	
A_2	1	1	1	1	1	-1	-1	l_z
B_1	1	-1	1	-1	1	1	-1	
B_2	1	-1	1	-1	1	-1	1	z
E_1	2	$\sqrt{2}$	0	$-\sqrt{2}$	-2	0	0	x, y
E_2	2	0	-2	0	2	0	0	
E_3	2	$-\sqrt{2}$	0	$\sqrt{2}$	-2	0	0	l_x, l_y

O	$\hat{I}$	$8\hat{C}_3$	$6\hat{C}_2'$	$6\hat{C}_4$	$3\hat{C}_2$	
A_1	1	1	1	1	1	
A_2	1	1	-1	-1	1	
E	2	-1	0	0	2	
F_1	3	0	-1	1	-1	x, y, z, l_x, l_y, l_z
F_2	3	0	1	-1	-1	

O_h	$\hat{I}$	$8\hat{C}_3$	$6\hat{C}_2'$	$6\hat{C}_4$	$3\hat{C}_2$	$\hat{I}$	$6\hat{S}_4$	$8\hat{S}_6$	$3\hat{\sigma}_h$	$6\hat{\sigma}_d$	
A_{1g}	1	1	1	1	1	1	1	1	1	1	
A_{1u}	1	1	1	1	1	-1	-1	-1	-1	-1	
A_{2g}	1	1	-1	-1	1	1	-1	1	1	-1	
A_{2u}	1	1	-1	-1	1	-1	1	-1	-1	1	
E_g	2	-1	0	0	2	2	0	-1	2	0	
E_u	2	-1	0	0	2	-2	0	1	-2	0	
F_{1g}	3	0	-1	1	-1	3	1	0	-1	-1	l_x, l_y, l_z
F_{1u}	3	0	-1	1	-1	-3	-1	0	1	1	x, y, z
F_{2g}	3	0	1	-1	-1	3	-1	0	-1	1	
F_{2u}	3	0	1	-1	-1	-3	1	0	1	-1	

T	$\hat{I}$	$8\hat{C}_3$	$3\hat{C}_2$	
A	1	1	1	
E	2	-1	2	
F	3	0	-1	x, y, z, l_x, l_y, l_z

T_d	$\hat{I}$	$8\hat{C}_3$	$6\hat{\sigma}_d$	$6\hat{S}_4$	$3\hat{C}_2$	
A_1	1	1	1	1	1	
A_2	1	1	-1	-1	1	
E	2	-1	0	0	2	
F_1	3	0	-1	1	-1	l_x, l_y, l_z
F_2	3	0	1	-1	-1	x, y, z

$\mathbf{T}_h$	$\hat{I}$	$8\hat{C}_3$	$3\hat{C}_2$	$3\hat{\sigma}$	$8\hat{S}_6$	$\hat{I}$	
A_g	1	1	1	1	1	1	
A_u	1	1	1	-1	-1	-1	
E_g	2	-1	2	2	1	2	
E_u	2	-1	2	-2	-1	-2	
F_g	3	0	-1	-1	0	3	l_x, l_y, l_z
F_u	3	0	-1	1	0	-3	x, y, z

$$\mathbf{O}_h = \mathbf{O} \times \mathbf{C}_i$$

$$\mathbf{T}_h = \mathbf{T} \times \mathbf{C}_i$$

$\mathbf{I}$	$\hat{I}$	$12\hat{C}_5$	$12\hat{C}_5{}^2$	$15\hat{C}_2$	$20\hat{C}_3$	
A	1	1	1	1	1	
F_1	3	b	$-a$	-1	0	x, y, z, l_x, l_y, l_z
$\dot{F}_2$	3	$-a$	b	-1	0	
G	4	-1	-1	0	1	
H	5	0	0	1	-1	

$$a = \tfrac{1}{2}(\sqrt{5} - 1)$$

$$b = \tfrac{1}{2}(\sqrt{5} + 1)$$

$$\mathbf{I}_h = \mathbf{I} \times \mathbf{C}_i$$

The following semi-infinite point groups are applicable to linear molecules, taking molecular axes as the Z axis.

C_∞	$\hat{R}_z(\varphi)$		$C_{\infty v}$	$\hat{R}_z(\varphi)$	$\hat{\sigma}_v$	
Σ	1	z, l_z	Σ^+	1	1	z
Π	$2\cos\varphi$	x, y, l_x, l_y	Σ^-	1	-1	l_z
Δ	$2\cos 2\varphi$		Π	$2\cos\varphi$	0	x, y, l_x, l_y
Φ	$2\cos 3\varphi$		Δ	$2\cos 2\varphi$	0	
$\cdots\cdots\cdots$			Φ	$2\cos 3\varphi$	0	
			$\cdots\cdots\cdots\cdots$			

$D_{\infty h}$	$\hat{R}_z(\varphi)$	$\hat{\sigma}_h$	$\hat{C}_2$	$\hat{\sigma}_v$	$\hat{S}_z(\varphi)$	$\hat{I}$	
Σ_g^+	1	1	1	1	1	1	z
Σ_u^+	1	-1	-1	1	-1	-1	l_z
Σ_g^-	1	1	-1	-1	1	1	
Σ_u^-	1	-1	1	-1	-1	-1	
Π_g	$2\cos\varphi$	-2	0	0	$-2\cos\varphi$	2	l_x, l_y
Π_u	$2\cos\varphi$	2	0	0	$2\cos\varphi$	-2	x, y
Δ_g	$2\cos 2\varphi$	2	0	0	$2\cos 2\varphi$	2	
Δ_u	$2\cos 2\varphi$	-2	0	0	$-2\cos 2\varphi$	-2	
Φ_g	$2\cos 3\varphi$	-2	0	0	$-2\cos 3\varphi$	2	
Φ_u	$2\cos 3\varphi$	2	0	0	$2\cos 3\varphi$	-2	
....							

Appendix II Kepler's Problem

On May 24, 1543, Nicolaus Copernicus, on the bed in which he was going to die that same day, received the first copy of his book, *On the Revolutions of the Heavenly Spheres*, in which he enunciated his famous idea that the earth, like other planets, rotated around the sun, and not the other way around. The idea was not new, but had been forgotten during the Middle Ages; some Greeks, including Aristarchus of Samos, had already proposed this hypothesis. In any case, modern scientific investigation of the motion of planets, including the earth, around the sun was initiated by Johannes Kepler, who, motivated by the theory of Copernicus, analyzed the vast observational data obtained by his teacher Tycho Brahe, and proposed three laws of the planetary motion. The problem of planetary motion around the sun is, therefore, called the Kepler problem.

(a) Kepler's Laws and Newton's Theory

Kepler, in his book *Astronomia Nova* (*New Aetiological Astronomy or Celestial Physics together with Commentaries on the Movement of the Planet Mars*), printed in 1609, published his discoveries that:

1. Each planet moves in an ellipse, with the sun at one of its foci.
2. The radius vector drawn from the sun to a planet covers equal areas in equal times.

Ten years later, in another book, *The Harmonies of the World*, he added that:
3. The squares of the periods of the different planets are proportional to the cubes of their respective major semiaxes.

The two-body problem of a planet and the sun can be reduced to a one-body problem by the standard procedure (section 1–2). In this case, the mass of the sun is known to be very much larger than that of any planet, so that the center of mass is within the sun, and the reduced mass of the system is essentially equal to that of the planet, which we designate as m.

Since the orbit of each planet is known to be in a plane, the problem is a two-dimensional one, for which the Hamiltonian function is given as

$$H = \frac{1}{2m} p_\rho{}^2 + \frac{1}{2m_\rho{}^2} p_\varphi{}^2 + U \qquad (1\text{–}56)$$

using polar coordinates.

Kepler's second law is nothing but the conservation of p_φ; because the area covered by the radius vector is

$$\frac{1}{2}\int \rho^2 \, d\varphi = \frac{1}{2}\int \rho^2 \dot\varphi \, dt = \frac{1}{2m}\int p_\varphi \, dt \qquad (\text{II}\text{–}1$$

using (1–55), and this is proportional to time only when p_φ is constant. (See Fig. II–1.) From (1–70), this result means that

$$\frac{\partial U}{\partial \varphi} = 0 \qquad (\text{II}\text{–}2)$$

which implies that the gravity potential is central.

FIGURE II–1. $\frac{1}{2}\rho^2 \, d\varphi$ is the infinitesimal area (shaded).

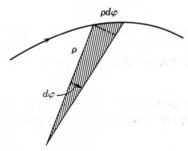

Newton proposed[1] that the gravity potential which attracts a planet to the sun is of the form

$$U = -\frac{k}{r} \tag{II-3}$$

where k is a constant for a given planet. This is a central potential satisfying Kepler's second law just discussed; this proposed form of (II-3) also satisfies Kepler's first and third laws simultaneously as will be shown presently.

Using the Hamilton–Jacobi equation, as given by Problem 1–37(a), the result of Problem 1–37(c) is obtained, which, using (II-3) gives

$$\varphi = \beta + l \int \frac{dr}{r^2 \sqrt{2m(\mathscr{E} - U) - (l/r)^2}}$$

$$= \beta + l \int \frac{dr}{r \sqrt{2m\mathscr{E}r^2 + 2mkr - l^2}}$$

$$= \beta + \sin^{-1}\left(\frac{mkr - l^2}{r\sqrt{(mk)^2 + 2m\mathscr{E}l^2}}\right) \tag{II-4}$$

where $\mathscr{E}$ is the energy,

$$l = p_\varphi \tag{II-5}$$

and we used r instead of ρ, since they are the same. Taking $\beta = -\frac{1}{2}\pi$, we see that (II-4) can be expressed as

$$\frac{C}{r} = 1 + e \cos \varphi \tag{II-6}$$

where

$$e = \sqrt{1 + (2\mathscr{E}\ell^2/mk^2)} \tag{II-7}$$

$$C = \frac{l^2}{mk} \tag{II-8}$$

[1] I. Newton, *Mathematical Principles of Natural Philosophy*, 1686, Book III: *The System of the World.*

Equation (II–6) is the equation for a conic section with one focus at the origin. The constant e is called the eccentricity, classifying the conic section as

$$
\begin{array}{ll}
e = 0 & \text{circle} \\
0 < e < 1 & \text{ellipse} \\
e = 1 & \text{parabola} \\
e > 1 & \text{hyperbola}
\end{array}
\qquad \text{(II–9)}
$$

From (II–7) and (II–9) we see that, for a nonzero l, the orbit of a planet under the potential (II–3) is

$$
\begin{array}{lll}
\text{circle} & \text{if} & \mathscr{E} = -\dfrac{mk^2}{2l^2} \\[2mm]
\text{ellipse} & \text{if} & \mathscr{E} < 0 \text{ in general} \\
\text{parabola} & \text{if} & \mathscr{E} = 0 \\
\text{hyperbola} & \text{if} & \mathscr{E} > 0
\end{array}
\qquad \text{(II–10)}
$$

Since the planets are confined to the solar system, their energies are all negative, so that according to Newton's theory, their orbits are ellipses in general, in agreement with Kepler's first law. (See Fig. II–2.)

In order to explain Kepler's third law, we use (II–1) to express the period of one rotation τ in terms of the area of the ellipse A as

$$
A = \frac{1}{2} \int_0^2 \rho^2 \, d\varphi = \frac{l}{2m} \tau
\qquad \text{(II–11)}
$$

On the other hand, it is known that

$$
A = \pi ab
\qquad \text{(II–12)}
$$

FIGURE II–2. Ellipse.

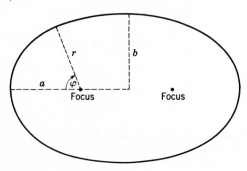

where a and b are the semimajor axis and the semiminor axis of the ellipse, respectively. From (II–6) we obtain

$$a = \frac{C}{2(1 + e)} + \frac{C}{2(1 - e)} = \frac{C}{(1 - e^2)} \qquad \text{(II–13)}$$

and

$$b = a\sqrt{1 - e^2} = \sqrt{aC} \qquad \text{(II–14)}$$

Thus

$$\tau = \frac{2m}{l} A = \frac{2\pi m}{l} ab = \frac{2\pi m\sqrt{C}}{l} a^{3/2} = 2\pi a^{3/2}\sqrt{m/k} \qquad \text{(II–15)}$$

in agreement with Kepler's third law.

(b) Special Relativistic Effects

Staying in the xy plane we obtain the special relativistic Lagrangian function, from (4–85), as

$$
\begin{aligned}
L &= -mc^2\sqrt{1 - \beta^2} - U \\
&= -mc\sqrt{c^2 - \dot{r}^2 - r^2\dot{\varphi}^2} + \frac{k}{r} \qquad \text{(II–16)}
\end{aligned}
$$

where Newton's expression of U, given by (II–3), is taken. Since this Lagrangian function does not depend on φ,

$$p_\varphi = \frac{\partial L}{\partial \dot\varphi} = \frac{mr^2\dot\varphi}{\sqrt{1 - \beta^2}} = l \qquad \text{(II–17)}$$

is conserved, and the total energy

$$\frac{mc^2}{\sqrt{1 - \beta^2}} + U = \mathscr{E} \qquad \text{(II–18)}$$

is also conserved.

From (II–17) we have

$$\left[\left(\frac{mcr}{l}\right)^2 + 1\right] r^2\dot{\varphi}^2 + \dot{r}^2 = c^2 \tag{II–19}$$

while from (II–18) we obtain

$$(\mathscr{E} - U)^2(\dot{r}^2 + r^2\dot{\varphi}^2) = c^2[(\mathscr{E} - U)^2 - m^2c^4] \tag{II–20}$$

From (II–19) and (II–20) we obtain

$$(cl)^2(\dot{r}^2 + r^2\dot{\varphi}^2) = [(\mathscr{E} - U)^2 - m^2c^4]r^4\dot{\varphi}^2 \tag{II–21}$$

which can be written as

$$\frac{1}{r^4}\left(\frac{dr}{d\varphi}\right)^2 = \frac{-(mc)^2}{l^2} + \frac{(\mathscr{E} - U)^2}{(cl)^2} - \frac{1}{r^2} \tag{II–22}$$

When the previous expression of $U = -k/r$ is used, we see that (II–22) can be written as

$$\xi \, d\varphi = \frac{l' \, dr}{r^2\sqrt{2m(\mathscr{E}' - U) - (l'/r)^2}} \tag{II–23}$$

where

$$\xi = \sqrt{1 - [k/(cl)]^2} = 1 - \frac{k^2}{2(cl)^2} + \cdots \tag{11–24}$$

$$l' = l\xi\sqrt{\frac{mc^2}{\mathscr{E}}} \tag{II–25a}$$

and

$$\mathscr{E}' = \frac{1}{2}\left(\mathscr{E} + \frac{m^2c^4}{\mathscr{E}}\right) \tag{II–25b}$$

Comparing (II–23) with the nonrelativistic expression (II–4), we see that the resulting orbit is

$$\frac{C'}{r} = 1 + e' \cos(\xi\varphi) \tag{II–26}$$

where e' and C' are obtained from (II–7) and (II–8), respectively, by replacing l with l' and $\mathscr{E}$ with $\mathscr{E}'$.

An important difference between this result and the nonrelativistic case is that since $\xi \neq 1$, the orbit is not stable: If the planet is at distance r from the sun at a given angle φ, it returns to that distance r at $\xi\varphi + 2\pi$, not at $\varphi + 2\pi$. Thus during each rotation around the sun, the perihelion moves (Fig. II–3) by the angle of

$$\frac{2\pi}{\xi} - 2\pi \cong \pi \left(\frac{k}{cl}\right)^2 \qquad \text{(II–27)}$$

(c) Kepler's Problem in General Relativity

The space–time around the sun is given by the Schwarzschild solution

$$ds^2 = \left(1 - \frac{2GM}{c^2 r}\right) c^2\, dt^2 - \frac{dr^2}{1 - (2GM/c^2 r)} - r^2\, d\theta^2 - r^2 \sin^2 \theta\, d\varphi^2 \qquad \text{(7–137)}$$

and the motion of a planet in this space–time is given by the geodesic equation

$$\frac{d^2 x^i}{ds^2} + \Gamma_{jk}{}^i \frac{dx^j}{ds}\frac{dx^k}{ds} = 0 \qquad \text{(7–18)}$$

of general relativity.

The Christoffel symbols, in this case, are

$$\Gamma_{10}{}^0 = -\Gamma_{11}{}^1 = \frac{GM}{(rc)^2 - 2GMr}$$

$$\Gamma_{22}{}^1 = -r\left(1 - \frac{2GM}{c^2 r}\right)$$

$$\Gamma_{00}{}^1 = \frac{GM}{(cr)^2}\left(1 - \frac{2GM}{c^2 r}\right) \qquad \text{(II–28)}$$

$$\Gamma_{33}{}^1 = -r \sin^2 \theta \left(1 - \frac{2GM}{c^2 r}\right)$$

$$\Gamma_{33}{}^2 = -\sin \theta \cos \theta \qquad \Gamma_{12}{}^2 = \Gamma_{13}{}^3 = \frac{1}{r}$$

$$\Gamma_{23}{}^3 = \cot \theta$$

FIGURE 11–3. Precession of perihelion. (Very much exaggerated.)

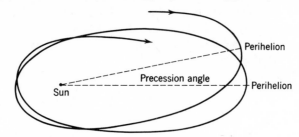

so that the geodesic equations are

$$\frac{d^2t}{ds^2} + \frac{2GM}{(rc)^2 - 2GMr}\frac{dt}{ds}\frac{dr}{ds} = 0 \tag{II–29}$$

$$\frac{d^2r}{ds^2} + \frac{GM}{r^2}\left(1 - \frac{2GM}{c^2r}\right)\left(\frac{dt}{ds}\right)^2 - \frac{GM}{(rc)^2 - 2GMr}\left(\frac{dr}{ds}\right)^2$$

$$-r\left(1 - \frac{2GM}{c^2r}\right)\left(\frac{d\theta}{ds}\right)^2 + \sin^2\theta\left(\frac{d\varphi}{ds}\right)^2 = 0 \tag{II–30}$$

$$\frac{d^2\theta}{ds^2} + \frac{2}{r}\frac{d\theta}{ds}\frac{dr}{ds} - \sin\theta\cos\theta\left(\frac{d\varphi}{ds}\right)^2 = 0 \tag{II–31}$$

$$\frac{d^2\varphi}{ds^2} + 2\cot\theta\frac{d\theta}{ds}\frac{d\theta}{ds} + \frac{2}{r}\frac{dr}{ds}\frac{d\varphi}{ds} = 0 \tag{II–32}$$

Now let $\theta = \frac{1}{2}\pi$, $d\theta/ds = d^2\theta/ds^2 = 0$, as before. Then (II–32) gives

$$\frac{d^2\varphi}{ds^2} + \frac{2}{r}\frac{dr}{ds}\frac{d\varphi}{ds} = 0 \tag{II–33}$$

which shows that

$$\frac{d\varphi}{ds} = \frac{L}{r^2} \tag{II–34}$$

where L is a constant, which is related to l of the previous cases. On the other hand, (II–29) gives

$$\frac{dt}{ds} = \frac{K}{1 - (2GM/c^2r)} \tag{II–35}$$

where K is another constant. The original Schwarzschild solution (7–137) then gives

$$\left(\frac{dr}{ds}\right)^2 = c^2K^2 - \left[1 + \left(\frac{L}{r}\right)^2\right]\left[1 - \frac{2GM}{c^2r}\right] \tag{II–36}$$

Eliminating ds in (II–36), using (II–34), we obtain

$$\frac{1}{r^4}\left(\frac{dr}{d\varphi}\right)^2 = \left(\frac{cK}{L}\right)^2 - \left(\frac{1}{L^2} + \frac{1}{r^2}\right)\left(1 - \frac{2GM}{c^2r}\right) \tag{II–37}$$

which is similar to the special relativistic equation (II–22), except that our equation (II–37) has a term proportional to r^{-3}.

When the orbit is not far from the nonrelativistic one (II–6), and when the eccentricity e is much less than one, we can approximate

$$r^{-3} \cong C^{-3}(1 + e\cos\varphi)^3 \cong C^{-3}(1 + 3e\cos\varphi + 3e^2\cos^2\varphi)$$

$$\cong 3C^{-1}r^{-2} - 3C^{-2}r^{-1} + C^{-3} \tag{II–38}$$

In this approximation, the term proportional to r^{-2} in (II–37) is

$$-\frac{1}{r^2}\left(1 - \frac{6GM}{c^2C}\right) = -\frac{1}{r^2}\left(1 - \frac{6GMmk}{(cl)^2}\right) = -\frac{1}{r^2}\left(1 - \frac{6k^2}{(cl)^2}\right) \tag{II–39}$$

Comparing (II–37), in the approximation (II–39), with (II–4), we see that the resulting orbit is given by

$$\frac{C''}{r} = 1 + e'' \cos(\xi_g \varphi) \tag{II–40}$$

where

$$\xi_g = \sqrt{1 - [6k^2/(cl)^2]} \tag{II–41}$$

Comparing this result with (II–24), we see that the precession of the orbit is 6 times faster than in the case of special relativity.

(d) Simplified Theory of Gravitation

It is instructive to note than in order to obtain the result (II–41) we do not have to start from the Schwarzschild solution of Einstein's equations.

If we simply assume that

$$ds^2 = c^2 \, dt^2 - \left(1 + \frac{2GM}{c^2 r}\right) dr^2 - r^2 \, d\theta^2 - r^2 \sin^2 \theta \, d\varphi^2 \quad \text{(II-42)}$$

we see that the Christoffel symbols are the same as (II–28), those obtained for the Schwarzschild solution, except that

$$\Gamma_{10}{}^0 = -\Gamma_{11}{}^1 = \Gamma_{00}{}^1 = \frac{GM}{(rc)^2} \quad \text{(II-43)}$$

In this case, (II–34) still holds, but (II–35) is replaced by

$$\frac{dt}{ds} = K \exp \left(\frac{GM}{c^2 r}\right) \cong K \left(1 + \frac{GM}{c^2 r}\right) \quad \text{(II-44)}$$

Therefore, rather than (II–36), we obtain

$$\left(\frac{dr}{ds}\right)^2 = c^2 K^2 \left(1 - \frac{2GM}{c^2 r}\right) - \left[1 + \left(\frac{L}{r}\right)^2\right] \left(1 - \frac{2GM}{c^2 r}\right) \quad \text{(II-45)}$$

neglecting terms of the order of G^2. Since (II–34) still holds, we find instead of (II–37) that

$$\frac{1}{r^4} \left(\frac{dr}{d\varphi}\right)^2 = \left(\frac{cK}{L}\right)^2 \left(1 - \frac{2GM}{c^2 r}\right) - \left(\frac{1}{L^2} + \frac{1}{r^2}\right) \left(1 - \frac{2GM}{c^2 r}\right) \quad \text{(II-46)}$$

The only difference between this result and the previous one, (II–37), is in the coefficient of the r^{-1} term, and no charge is seen in the other terms proportional to r^{-2} and r^{-3}, respectively. Since the precession of the perihelion results from the coefficients of the r^{-2} and r^{-3} terms, this means that we obtain the same result (II–41) from (II–46), that is, from (II–42).

Index

MALAYA

ESE AMPHIBIOUS
NGS 10 DEC.

CHINA SEA

TRENGGANU

26 Dec

1 Jan KUANTAN

X *'Prince of Wales' and
'Repulse' sunk by air
action. 10th. Dec. 1941*

JAPANESE AMPHIBIOUS
LANDINGS 26 JAN.

ENDAU

MERSING

MALIM

21 Jan

GEMAS
16 Jan.   *JOHORE*

ALA LUMPUR
I Jan

19
Jan   26 Jan   KHOTA
TINGGI

SEREMBAN   YONG PENG
24 Jan.   X

15 Jan   16 Jan   AYER   31 Jan
HITAM   JOHORE BHARU

MALACCA   MUAR   26   28
BATU   Jan   Jan
TS OF   PAHAT   SINGAPORE
ACCA   JAPANESE   (Surrender 15 FEB.1942),
AMPHIBIOUS   JAPANESE AMPHIBIOUS
LANDINGS   LANDINGS ON SINGAPORE
18 JAN.   ISLAND NIGHT OF 8 FEB.

FORTRESS

FORTRESS

The Story of the Siege and
Fall of Singapore

by

KENNETH ATTIWILL

DOUBLEDAY & COMPANY, INC.

GARDEN CITY · NEW YORK · 1960

To Evadne

Last Flight From Singapore, by A. G. Donahue. Copyright 1943 by The Macmillan Company. Reprinted by permission of the publishers.

ACKNOWLEDGEMENTS

The Author wishes to express thanks to all those people—those quoted in the text and those who for various reasons do not wish their names made public—who gave generously of their time and memories during many interviews which have so greatly helped to make this book.

The Author also wishes to express gratitude to the Authors and Publishers of those many books and official dispatches without whose views and conclusions so many aspects of the Singapore story could not have been known. A list of these sources is given below; but it must be understood that none but the present Author is in any way responsible or liable for any statements or conclusions published in the text.

The Author also wishes to thank Mr. Douglas Warner for his assistance in preparing the manuscript for press.

Among the books and dispatches consulted were:

Imperial War Museum, for *The Affair at Alexandra Hospital*, (Singapore), 14th February 1942. Copy of original manuscript of G. E. Manley held in Imperial War Museum Library.

History of the Argyll and Sutherland Highlanders 2nd Battalion, by Brigadier I. McA Stewart D.S.O., B.E., M.C. Reprinted by permission of Thomas Nelson & Sons, Ltd., Edinburgh.

Behind Bamboo, by Rohan Rivett (Angus & Robertson, Ltd., 1946).

The War Against Japan, Vol. I, "The Loss of Singapore", by Major General S. Woodburn Kirby and others; *ABDA Command*, by General Sir Archibald Wavell. Extracts from Her Britannic Majesty's Stationery Office publications are reproduced by permission of the Controller of H.M.S.O.

The Second World War and *Secret Session Speeches*, by Sir Winston Churchill (Cassell & Co., Ltd.).

Eastern Epic, by Compton Mackenzie, and *The Jungle Is Neutral*, by F. Spencer Chapman (Chatto & Windus, Ltd.).

Malaya and Its History, by Sir Richard Winstedt; *The Civil Defence of Malaya*, by Sir George Maxwell, for the Association of British Malaya, and *Singapore and After*, by Lord Strabolgi (Hutchinson & Co.).

Malayan Climax, by Carline Reid (Mercury Press, Hobart).

White Coolies, by Betty Jeffrey (Angus & Robertson, Ltd.).

Malaya Upside Down, by Chin Kee Onn (Jitts & Co., Singapore).

Director and Staff of the British Imperial War Museum, Lambeth, London.

Curator & Staff of the British Museum Newspaper Library, Colindale, London.

Colonel R. B. Muir, 15th Coy. Madras Sappers & Miners, for permission to reproduce his map of Malaya.

Why Singapore Fell, by H. Gordon Bennett.

Malayan Postscript, by Ian Morrison (Faber & Faber, Ltd.).

Major General J. F. D. Steedman, D.S.O., for portions of his diary.

Who Dies Fighting, by Angus Rose (Jonathan Cape, Ltd.).

The Japanese Thrust (Official Australian War History), by Lionel Wigmore.

In Seventy Days, by E. M. Glover (Frederick Muller, Ltd.).

Of Death But Once, by Roy Bulcock (F. W. Cheshire).

Singapore to Freedom, by O. W. Gilmour (Edward J. Burrow & Co., Ltd.).

Retreat in the East, by O. D. Gallagher (George G. Harrap & Co., Ltd.).

Suez to Singapore, by Cecil Brown (Random House).

Impregnable Fortress, by John C. Sharp.

OFFICIAL DISPATCHES BY:

Air Chief Marshal Sir Robert Brooke-Popham (Suppl. *London Gazette*, 22.1.48).

Vice-Admiral Sir Geoffrey Layton (Suppl. *London Gazette*, 26.2.48).

Lt.-Gen. A. E. Percival (2nd Suppl. *London Gazette*, 26.2.48).

Air Vice Marshal Sir Paul Maltby, (3rd Suppl. *London Gazette*, 26.2.48).

CONTENTS

ILLUSTRATIONS

*Pictures are reproduced by courtesy of Keystone Press
Agency Ltd., and the Imperial War Museum*

MAPS

INTRODUCTION

On the morning of January 31, 1942, under the bright light of the equatorial moon, the remnants of the British Imperial troops who had retreated down the length of the Malay Peninsula, filed across the causeway into Singapore Island and prepared for a siege defence.

A little more than a fortnight later—at 7:50 P.M. on February 15—Lt.-Gen. A. E. Percival, General Officer Commanding Malaya, signed the instrument of British surrender to Lt.-Gen. Tomoyuki Yamashita, commander of the victorious Japanese 25th Army.

Singapore—the "impregnable fortress", the "bastion of Empire", the "Gibraltar of the Far East"—keystone to British naval strategy from the Indian Ocean to Australia—had fallen.

The shock went round the world.

To Germans, Italians, Japanese and all enemies of British colonial rule, the reaction was unrestrained glee and triumph. The pride of Great Britain had been humbled in the dust of Singapore. The weak hand in the velvet glove diplomacy of British civilization had been stripped of its pretensions.

To millions of people in Asia, the ugly side of frightened, demoralized white humanity had been exposed to public view like the flayed skin of an Aztec sacrifice.

To Britons, long persuaded to picture Singapore as a distant Gibraltar or Malta, the news was shattering. The skies could have fallen more easily than the famed and boasted Singapore fortress.

This was disaster. Mr. Churchill, the British Prime Minister, described it as "the worst disaster and largest capitulation in British history". The retreat from Dunkirk in 1940 had been a heavy blow, but the heroes of Dunkirk had not surrendered. Singapore was collapse, cease fire, surrender.

A sense of shock and shame has persisted through seventeen years. It was implicit in Mr. Ian Morrison's indignant refutation of cowardice in his book *Malayan Postscript*, written only three months after the surrender. It is implicit in the stubborn refusal of successive British Parliaments of both parties to conduct an official inquiry into all the circumstances of the collapse—though the American Congress investigated Pearl Harbour immediately after the devastating Japanese air raid in December 1941, and within two months had named the men who failed and the reason why. It is implicit in the urbane smothering of vital facts amid a mass of technical detail in *The War Against Japan*, Vol. 1, published in 1957 as part of the British history of World War II. It is implicit in the refusal of many men and women who were in Singapore to help by telling their part in the true story. They prefer to let the truth lie. "I do not think that any useful purpose would be served," said one man. "After all, what good will it do?"

I believe there are millions of people who want to know the truth. I wanted to know it, for the surrender of Singapore cost me three and a half years' suffering in a Japanese prisoner-of-war camp. During and after the fighting we felt resentful. In the prison camps we swore there would be a big showdown when we came home and told our story. We were not to blame. We blamed those in authority, the British Government, the generals. Some people blamed the Australians, the Chinese, the Malays, the civilians; others blamed the Navy, the Army, the Air Force. Few people at the time of the fall of Singapore were aware of what Mr. Churchill later called "the hideous efficiency of the Japanese war machine". After the war, the official dispatches and

Cabinet secrets were revealed; and we who fought in and around Singapore felt that many of the explanations were glib and glossy and inadequate.

The full and proper answer to all the doubts and suspicions and rumours would have been a Royal Commission on Singapore, while men who knew the truth were still alive. That opportunity was allowed to pass.

After the war, impoverished Britain wanted to borrow money from the U.S.A.; and it may have been decided that a Royal Commission on Singapore would reveal the awkward fact that in December 1941 America knew Japan was going to attack Siam and Malaya, but failed to give assurance of armed support in the event of war until too late for Britain to take forestalling action. The old story of A unwilling to "pull B's chestnuts out of the fire."

Just before the fire exploded, Cecil Brown, reporting for Columbia Broadcasting and *Life*, asked the Japanese Consul in Singapore: "What will stop war breaking out between your country and mine?" The Jap Consul blandly replied: "Draw a line down the middle of the Pacific—you stay one side, we stay the other." It stayed that way until Japan sneaked across the line and bombed Pearl Harbour.

Pearl Harbour and Singapore were strategically linked and interdependent. "United we stand, divided we fall." That is the essential lesson for democracy from Singapore. Have we learned it? Do we now apply it?

In the following pages is an attempt to encompass the whole range of events which culminated in that dismal surrender to the Japanese. I have read as much as I could find, interviewed widely, studied official documents. I have tried to eliminate personal feeling and bias in order to record the dramatic, heroic, tragic story of Singapore as readably and as faithfully as I am able.

London
December 1958.

I

High Explosive Overture

Sunday in Singapore was a time for relaxation. The week's exertions had ended and there was the week-end peace to be enjoyed before work was resumed. There were long, cool drinks to combat the heat; cricket, golf, tennis, yachting, swimming, or bridge for entertainment, according to personal taste and means. There were dinners and dances at luxury hotels, or conversation at club or home over whisky and cocktail and iced beer. It was the customary life of the white man in the tropics.

Life was like that on Sunday, December 7, 1941, in Singapore. The city on its island, suspended like an emerald below the scimitar blade of the Malayan peninsula, was wrapped in its blanket of security. The streets, their shops and bazaars gay with holly and mistletoe and Christmas toys, were thronged with people—Europeans and Chinese, Malays and Indians, civilians and soldiers—especially soldiers. There were British and Australian, Moslem and Sikh, Gurkha and Malay. There were airmen from Great Britain, Australia and New Zealand, and the Netherlands East Indies. There were European and Asiatic members of the Volunteer Forces. At dances and in the crowded cinemas, three-quarters of the audiences were in uniform. They were a great comfort to the civilians.

Above the island droned the aircraft—Buffalo fighters, looking speedy and manoeuvrable to all those people who had never seen Hurricanes or Spitfires; Blenheim bombers, Hudsons, Wirraways, Catalina flying-boats, torpedo-carrying Vildebeestes. In the great docks of the Naval Base lay more comfort still—the ships of the British Eastern Fleet, including the battleship *Prince of Wales* and battle-cruiser *Repulse*, which had arrived amid the thankful cheers of the onlookers only a few days before. The Army, the Navy and the Air Force were ready on Singapore Island for anything that might happen; and for thousands of trusting people this meant that nothing *would* happen. If it did, if the Japanese were foolish enough to try anything, it would soon be over. The enemy wouldn't stand a chance.

Conversation in the clubs and hotels and comfortable tropical lounges centred mainly on the current war scare which flared into prominence a week before when the civilian Volunteers were mobilized. But it was well spiced with business discussion, gossip about forthcoming social events and sporting fixtures, plans for Christmas. In spite of reports of mounting tension in the Pacific as peace talks between America and Japan reached an impasse, despite notices flashed on cinema screens to recall troops to barracks, armchair strategists held forth at length on the reasons why there would be no war in Malaya. The Japanese were bluffing. Japan could not go to war. Economic sanctions imposed against her a few months earlier by the United States and Britain meant that Japan could not afford a war. The vital materials without which war cannot be fought—oil, manganese, bauxite, tin, rubber—had been denied her. Besides, she had her own worries nearer home. She was still deeply committed to the "Incident" in China. Her ancient enemy, Russia, though fighting for her life against the Nazi onslaught in the West, still contrived to maintain her Far Eastern forces on the frontiers of Manchuria—Japan therefore could not afford to weaken her defences in case Russia and China should act to overwhelm the puppet-state. It was true that for several months Japanese forces had been infiltrating south into French Indo-China to fill the power vacuum left by France's defeat in Europe; but the Japanese would not be so foolish as to risk a war against the combined strength of the United States, the British Empire, and the Dutch.

There were some "defeatists" who believed that war with Japan was not only possible but certain, and that the Japanese would make headway. But these pessimists did not disturb Singapore's complacency. If war came, there could be only one answer: the crushing defeat of the aggressors. The optimists pointed to the "well-known" inefficiency of the Japanese, who in four and a half years had failed to defeat the Chinese. There was the north-east monsoon to handicap any attempted Japanese invasion of Malaya; and the jungle was generally considered impenetrable to any large body of troops. As for Singapore itself, the island was said to be impregnable, with its great fortress guns mounted in concrete. Singapore could repel invasion from the sea. Now that the Eastern Fleet had arrived, any cause for uneasiness had been removed.

There was a further picture, latent in the minds of the Europeans in Singapore, which coloured the views of the optimists with a rosy glow—the cartoon of the Jap as a small man with a tombstone grin and large spectacles, who had no inventive genius and could only copy his rivals' wares. Nobody could take that comic fellow seriously. He could not see to shoot straight, he could not sail ships, he could not fly aircraft.

To be fair, the amateur strategists had a certain amount of official backing for their arguments. Only four days earlier, Air Chief Marshal Sir Robert Brooke-Popham, Commander-in-Chief Far East, had told a Press conference: "There are clear signs that Japan does not know which way to turn. Tojo is scratching his head. There are no signs that Japan is going to attack anyone." He had added that his greatest worry had been fighter aircraft, but he was satisfied with the Brewster Buffalo. "We can get on all right with Buffaloes out here," he said, "but they haven't got the speed for England. Let England have the super-Spitfires and hyper-Tornadoes. Buffaloes are quite good enough for Malaya."

So much for European reaction to the explosive situation in the Far East up to December 7, 1941. Complacency ruled the island—solid, more impenetrable than the Malayan jungle. It was a compound of easy living, wishful thinking, misinformation, ignorance, and deceptive appearance.

But the British were only a minority of the population—and a tiny, *élite* minority at that. What did the Chinese think, for instance? The Chinese totalled more than three-quarters of the population of Singapore Island. Their homeland had been under Japanese attack for more than four years. True, they were torn asunder by political differences. There was the Kuomintang faction (the party of Chiang Kai-shek, who headed the official Government of China) and the Communists. There was also a large proportion of people who had weaker ties with China, Straits-born Chinese who had never seen their mother country, whose parents had lived in Singapore for four and five generations. If war came, would not the Chinese fight against the aggressor who had devastated their native homeland? And would not the Singapore Chinese discover a civic pride in the city which had been their home for so long? Most of the Europeans did not know, and made no effort to find out. The Chinese Communist Party in Singapore had been outlawed the year before, and had not been rehabilitated when Russia was attacked in the previous July. The Communists were dangerous, and you could never be sure that the Kuomintang Party was entirely free from them. As everyone knows, you can't tell one Chinese from another—especially if you never look at them. Apart from tenuous business contacts with the richer and more powerful Chinese in commerce, White and

Yellow did not mix. The only thing common to all was the desire to make money.

If they did not understand the feelings of the Chinese, still less did the Europeans understand the Malays and the Indians. The Malays were courteous, good-mannered, easy-going and easily governed. Most of the Indians were Tamils from South India. They were the poorest members of the community, a source of cheap manual labour on rubber estates, docks, railways and in the cities. Few cared about the Indians' viewpoint, if they had one.

There was a wide gulf between officers and other ranks. Officers had their own clubs and drinking places, the men had theirs. The best European hotels were out of bounds to the ranks. There was, with a few honourable exceptions, little fraternization and a good deal of segregation. Officers were welcomed in the homes of the rich Europeans, other ranks found hospitality where they could. It was a division to which even the Australians, much more democratic than the British about officer-men relationships, were not immune. And it was a wider gap than any which existed in Britain after two years of war. In short, it was the British Colonial pattern—a narrow *élite* of Service, commercial and plantation people at the top, living comfortably in huge houses or luxurious flats; and below were the lesser Europeans, the Chinese, the Malays, the Indians, the Eurasians, the Arabs and the Jews, all living their separate lives at varying levels according to their means. Between these disunited sections of the crowded community was a gulf of selfish ignorance and prejudice, which led to grotesque misjudgement of public morale, hastened the crumbling of resistance when things went wrong, and prompted some ludicrous public pronouncements by the military High Command and the Government.

II

These things worried very few people in Singapore on December 7, 1941. The traditional pattern of living was agreeable—why should it change? As daylight vanished under the swift onset of equatorial night, Singapore City blazed with light. Like a confirmed rake bent on self-destruction, the community moved about its complex teeming way of life in streets and kampongs, homes and fields. Only in the rarefied atmosphere of High Command was there concern and the furrowing of brows. Thirty-six hours earlier—at midday on December 6—a lone Hudson bomber had reported Japanese convoys with naval escort steering west and northwest from near Cape Cambodia in Indo-China. Were the movements innocent, or did they presage an attack on Siam or Northern Malaya, or both? Or was it political bluff? The High Command could not make up its mind, and bad weather prevented any new sightings which would resolve the dilemma.

At nine o'clock that night, the full moon rose above Malaya. About the same time, further reports of movements of Japanese vessels in the Gulf of Siam reached the headquarters of Far East Command in Singapore. Shortly after midnight, troops defending the beaches above Kota Bharu, north-eastern tip of Malaya, reported that transport vessels were anchoring off the coast and that the shore defences were being shelled. Two hours later, the main Japanese invasion force landed in Siam, and Japanese fighter and bomber aircraft flocked across from South Indo-China.

At 3:30 A.M. Fighter Control Operations Room reported unidentified aircraft 140 miles from Singapore, approaching from the north-east. Service defences were alerted. No warning could be given to the civil A.R.P. headquarters because they were unmanned. Nobody answered the telephone, which went on ringing, and the city's lights stayed on. No siren sounded for the civilians.

At four o'clock, bombs began to fall on Singapore.

III

An Englishwoman, who lived in a flat above her fashion business in the city centre, was hurled violently out of bed. She heard the roar of heavy gunfire, followed by the terrific crash of an explosion which shuddered the floor under her feet. She ran to the window in time to see Guthrie's import house, just along the road, crumbling in a mass of rubble and dust and twisted steel. She telephoned the police.

"There's an air raid on! Why doesn't somebody put out the lights?"

"Don't be alarmed," said a reassuring voice. "It's only a practice."

"Well, tell them they're overdoing it," she shouted. "They've just hit Raffles Place and knocked Guthrie's for six."

Military defences were ready as the 17 Japanese Navy bombers flew in. Searchlights and A.A. guns were fully manned and three R.A.F. Buffalo fighters were warmed up. "The most perfect night-fighter target I have ever seen," said one of the Buffalo pilots. But he was not allowed to go into action. Air Command kept the Buffaloes on the ground because the defences were not sufficiently practised in night control of aircraft, searchlights and guns. Eager gunners might become confused and shoot down the wrong planes.

The A.A. guns blazed away—ineffectually. None of the Japanese planes was damaged, and the raiders dropped their bombs at will on Tengah and Seletar R.A.F. stations, on the commercial heart of the city, and on a densely-populated area

of Chinatown. They left a trail of wrecked buildings, 63 dead
and 133 injured. Most of the casualties were Sikh watchmen
and Chinese shopkeepers.

Eventually the sirens were sounded, after a direct approach
had been made to the Governor, but the city remained
brilliantly illuminated. The man with the keys to the master
switches could not be found.

When daylight came, people flocked to gape at the bomb
holes. Apologists explained that the absence of a black-out
could not have made much difference anyway because the
moonlight illuminated the targets. On that same day, *The
Times* of London published a two-column descriptive military dispatch from the Singapore correspondent under the
heading "Singapore Prepared". The story brisked with confidence.

Also on that morning appeared in Singapore and elsewhere
an Order of the Day issued jointly by the Far East Commander, Air Chief Marshal Sir Robert Brooke-Popham, and
the Admiral, China Station, Vice-Admiral Sir Geoffrey
Layton. It had been prepared six months earlier to allow time
for translation into the Asiatic languages. Among other
things it said:

"We are ready: we have had plenty of warning and our
preparations are made and tested.

"We do not forget at this moment the years of patience
and forbearance in which we have borne with dignity and
discipline the petty insults and insolences inflicted on us by
the Japanese in the Far East. We know that those things
were only done because Japan thought she could take advantage of our supposed weakness. Now, when Japan herself has
decided to put the matter to a sterner test, she will find out
that she has made a grievous mistake.

"We are confident. Our defences are strong and our
weapons efficient. . . .

"What of the enemy? We see before us a Japan drained for

years by the exhausting claims of her wanton onslaught on China. . . .

". . . Confidence, resolution, enterprise and devotion to the cause must and will inspire every one of us in the fighting services. . . ."

Seldom in military history has a High Command proclamation contained so many falsehoods and inaccurate assumptions.

2

The Four Defeats

"We are ready. . . . Our defences are strong and our weapons efficient. . . ."—Order of the Day, Singapore, December 8, 1941.

Singapore had been bombed, and Britain was at war with Japan, but it made little apparent difference to life on the island except to galvanize the civil defence organizations into activity. Ian Morrison in *Malayan Postscript* recorded: "Two nights later . . . there was hardly a light to be seen. Zealous Chinese volunteers . . . threw themselves into the task of blacking out Singapore as they might have thrown themselves into some game. The slightest chink of light at a window would provoke frenzied shouts from the street below. Drivers of cars were stopped if their lights were thought to be the smallest degree on the bright side. Anyone attempting to light a cigarette on the street would be fiercely pounced upon. . . ." Otherwise it was largely "business as usual". Shops and bazaars carried on "business as usual", commerce continued, housewives shopped or played bridge or went to the cinema or discussed the war. Kota Bharu was a long way off—about four hundred miles—and there were the impenetrable jungle and the prepared British troops with their tested defences and efficient weapons to safeguard Singapore. It is doubtful whether a single unofficial person in Singapore had seriously thought that the Japanese would even obtain

Above, left: Lt.-Gen. A. E. Percival, G.O.C. Malaya.   *Above, right:* Lt.-Gen. Tomoyuki Yamashita, commander of the victorious Japanese forces.   *Below, left:* Sir Shenton Thomas, Governor of the Straits Settlements and High Commissioner for the Malay States. *Center:* Air Vice-Marshal Sir Robert Brooke-Popham, C-in-C. Far East at the time of the invasion of Malaya; General Sir Archibald Wavell, Supreme Commander ABDA Command; and Vice-Admiral Sir Geoffrey Layton, second in command Eastern Fleet. *Below, right:* Maj.-Gen. H. Gordon Bennett, commander of the Australian Imperial Force, Malaya, during the campaign. *(Photos by courtesy of the Imperial War Museum)*

Above: British reinforcements for Malaya disembark from a troop transport at Keppel Harbour, Singapore. *Below*: Australian Army nurses arrive on the Island. It is believed that some of the nurses in this picture were among those massacred by the Japanese on Muntok Beach, Banka Island, when only one, Sister Vivian Bullwinkel, survived. *(Photos by courtesy of the Imperial War Museum)*

a foothold on Malaya. Certainly no one considered that Singapore might be attacked in strength. The first three High Command communiqués aroused no alarm, though they admitted Japanese landings at Kota Bharu ("confused fighting" was taking place) and at Singora and Patani in Siam. Airfields in Northern Malaya had been raided. One communiqué spoke of "mopping up" operations, which was soothing.

The following day, another longish communiqué was issued. It stated that "large numbers of Japanese aeroplanes" had been committed to the fighting, that "twenty-five further transports" were ferrying more Japanese troops to the landing points, and that "severe fighting" was taking place in the Kota Bharu area. But it was also sprinkled with more soothing phrases like "a large measure of control has been achieved by our forces" . . . "reinforcements should reach there during the day. . . ." "The move [to obtain control of northern Malaya] was always foreseen to be a likely one, and the disposition of our troops was designed to meet it."

Ian Morrison remarks: "We did not know down in Singapore (nor, to do them justice, did the High Command) what was really going on in north-east Malaya. . . ."

The people of Singapore never learned the full truth of those early days in Malaya—though eventually they understood that things had gone wrong. In fact, there were three battles for Singapore: the first was in northern Malaya; the second was in Johore, and the third was on the Island itself. The first of these battles was decisively lost in 108 hours. Defeat was administered in four instalments, and it was the initial setback which ensured the ultimate surrender of Singapore. To understand why Singapore fell, it is necessary to examine the cause. The four instalments of defeat were: in strategy, at sea, in the air, and on the land.

II

It has been repeatedly stated that the defenders of Singapore never considered the possibility of an attack from the landward side. That is not true. Maj.-Gen. W. G. S. Dobbie, G.O.C. Malaya in 1937, considered that the chief danger to Singapore was the undefended north. His recommendations to the Chiefs of Staff in London led to a grant of £60,000 for defence works in Johore and on Singapore Island. More important, however, than General Dobbie's far-sighted forecast of the probable intentions of an invader was the fact that, by linking the defence of Singapore to the defence of Malaya, he unwittingly led lesser men to think of the defence of the Naval Base without defending the whole of the island. The strategists reasoned like this:

"The increased striking power of aircraft means that the Naval Base cannot be held unless the airfields of Malaya are also held. In addition, the Island itself is not ideal for defence —despite the natural barrier of water—so let us look further afield for the natural defence line. We shall find it on the borders of Siam, where the neck of Malaya is at its narrowest. An invader must have bases from which to launch his land and air forces; therefore it is probable (as General Dobbie has pointed out) that the invader will launch his initial attacks in the Kra Isthmus on Singora and Patani, in Siam, and at Kota Bharu in northern Malaya. Therefore, defence strategy should be designed to deny these three points to the enemy."

In 1941, accordingly, Brooke-Popham worked out a scheme, which was given the code name of *Matador*, to achieve these ends. When the time was ripe, a British force would strike north from Malaya across the Siamese frontier to occupy Singora and Patani, and a second force, called Krohcol, would advance into Siam to occupy the position known as The Ledge—the best defensive point in the area. From these positions the defender would be in an immeasurably better position to protect Singapore, as well as the tin and rubber treasure-house of Malaya. The Chiefs of Staff in London approved the plan but refused to grant Brooke-Popham the initiative to move into Siam. Britain could not afford to provoke Japan into war, and it was believed that the Japanese were trying to force Britain into acting first in order to brand her as the aggressor and give Japan the excuse for entering Siam to "protect" her. Nevertheless, although the refusal to grant Brooke-Popham the initiative meant an inevitable reduction in the time at his disposal to launch *Matador* successfully, he went ahead and made his dispositions. Brooke-Popham estimated that he would need thirty-six hours' notice of Japan's threat to Siam because it would take the enemy only thirty-three hours to reach Singora from his bases in Indo-China.

On December 1—three days before his "Tojo is scratching his head" announcement—Brooke-Popham mobilized the Volunteers and brought the entire garrison of Malaya Command to "second degree of readiness" (not for the first time—and so it was not taken seriously by some of the officers and men, who thought it was either an exercise or another false alarm). On December 4 Brooke-Popham told the Chiefs of Staff in London that in the circumstances described to him by London and elsewhere, it was unnecessary to try to avoid war with Japan. Japan was going to war anyway. He therefore asked for permission to order *Matador* if, *in his opinion*, the circumstances should arise. On December 5 permission was granted with the reservation: "should the Japanese violate any part of Siam, or if there was good information that a Japanese expedition was advancing with the apparent intention of landing."

On December 6, at 12:30 P.M., the Japanese were spotted.

The man who saw them was Fl. Lt. Jack Ramshaw, captain of an R.A.A.F. Hudson patrol. This with further sightings revealed one battleship, five cruisers, seven destroyers and twenty-five merchant ships. Low monsoon clouds and fierce rain squalls caused the sightings to be lost, but later on the 6th information was received that Japanese convoys had left Camranh Bay, in south Indo-China, and Saigon. It was also said that Siamese frontier guards had begun to erect road blocks on the trunk road to Singora and on the Kroh-Patani Road—the routes the British would take in any attempt to forestall the enemy.

These were the hours of decision. If *Matador* had been ordered that day it might have delayed the Japanese. It was not ordered. Brooke-Popham, despite the logic of facts and the bulwark of his own opinions, felt himself bound by the British Government's desire to avoid a war which he, the Commander-in-Chief, thought unavoidable. He hesitated. He decided that on the evidence of these sightings he could not assume the responsibility for ordering *Matador*. He did, however, order the forces to be brought to first degree of readiness. The troops took up their positions in pouring rain, and waited.

Air reconnaissance was stepped up but bad weather prevented any further sightings until 9 A.M. on the 7th, when Far East Command headquarters were told that three merchant vessels and five warships had been sighted moving in the general direction of Singora, Patani and Kota Bharu.

Earlier that day Brooke-Popham received the latest of a series of telegrams from Sir Josiah Crosby, the British Minister in Bangkok, who loved Siam and was convinced that he had the Siamese in his pocket. His telegram said: "For God's sake do not allow British forces to occupy one inch of Thai territory unless and until Japan has struck the first blow at Thailand."

No doubt Sir Josiah's conviction, coupled with hesitancy in London, helped to bewilder the Commander-in-Chief.

Late on the evening of the 7th, Brooke-Popham and Admiral Sir Tom Phillips, C.-in-C. Eastern Fleet, held a conference to decide what action should be taken. The movements might be a deliberate Japanese attempt to trap the British into attacking Siam, but even if they were bound for invasion, *Matador* could not now forestall them. Yet for the second time, Brooke-Popham was undecided. Instead of either calling it off or deciding to act on his own initiative, he chose to wait for a dawn reconnaissance before coming to a decision. But at 11:20 P.M. Lt.-Gen. Sir L. Health, Commander of III Indian Corps, was warned to be ready to put *Matador* in action at dawn on the 8th.

It was too late. Even then inaction followed indecision. The news of the Japanese landings in Siam at 2:30 A.M. on the 8th should have been the signal for the second offensive plan to go into action with the advance of Krohcol into Siam to capture The Ledge. Nothing was done. Any lingering doubt in Brooke-Popham's mind about the attitude of London should have been dispelled at 8 A.M. when the Chiefs of Staff gave him a free hand to act as he chose. Still nothing was done. News of the Japanese landings at Singora was not received at Far East Command headquarters until 9:45 A.M.—six hours after they occurred; and though it was then formally decided to abandon *Matador* and adopt an alternative plan—a fall back to a defensive position at Jitra to protect Alor Star airfield—nothing was done about Krohcol. It was three o'clock in the afternoon before Krohcol crossed the frontier—and the Japanese thus had a start of twelve hours.

The War Against Japan comments: "It is possible that he [Brooke-Popham] did not fully realize the importance of speed. . . . The need for a quick decision was not apparently realized at headquarters, Malaya Command. . . . The enemy was thus given a start of some ten hours over III Corps; this was to prove disastrous."

The first engagement with the enemy had been lost.

III

Defeat in the Air. Three quotations are all the evidence necessary to explain the loss of air supremacy in the Malayan campaign within twenty-four hours:

> At a meeting of the Chiefs of Staff on April 25, 1941, the Vice-Chief of the Naval Staff advocated the despatch of Hurricane fighters to Malaya. The Vice-Chief of the Air Staff thereupon stated that Buffalo fighters would be more than a match for the Japanese aircraft which were not of the latest type.—*The War Against Japan*, p. 240 footnote.

> Sir Robert [Brooke-Popham] had also said that his greatest worry as C.-in-C., Far East, was fighter aircraft, but he said he was satisfied with the Brewster Buffalo. "We can get on all right with Buffaloes out here, but they haven't got the speed for England. Let England have the super-Spitfires and hyper-Tornadoes. Buffaloes are quite good enough for Malaya."
> If Sir Robert really believed what he told those war correspondents, how lamentable was the Intelligence on which he had to rely for his appreciation of the situation. Within little more than a week of this press conference, Japan had complete mastery in the air over Malaya.—O. D. Gallagher, *Daily Express*, reporting Brooke-Popham's Press Conference of December 3, 1941.

> In May, 1941, a Japanese Zero fighter was shot down in China. Details of the armament and tankage reached Singapore and were passed to the Air Ministry on July 26, 1941, as well as to headquarters Air Command, Far East. Later, the Air Attaché, Chungking, forwarded estimated performance figures which subsequently proved to be reasonably accurate.

On Sept. 29, 1941, the Combined Intelligence Bureau trans-
mitted this data to the same two authorities. Faulty organization at headquarters, Air Command, whose establishment did
not include an intelligence staff, resulted in this valuable
report remaining unsifted from the general mass of intelligence information, and in no action being taken on it.—*The
War Against Japan*, p. 240 footnote.

For "remaining unsifted" read "nobody read it".

The battle figures reveal the result of inefficiency and
ignorance.

When the Japanese began their attack, the total number of
British aircraft based on airfields in northern Malaya was 110.
At the end of the first day only 50 were fit for operations.
The Japanese possessed 530 aircraft—and in less than twenty-
four hours had destroyed the British air effort as a means of
defence or of support for the forward land operations. The
destruction was so swift that when the Commanders-in-Chief
Far East and Eastern Fleet signalled London on December 9
asking urgently for two squadrons of long-range bombers and
two squadrons of night fighters to maintain British air
superiority, British air superiority had already been lost. Air
Command were forced to pull back aircraft from five airfields
so that by the evening of the 9th—about thirty-six hours after
fighting began—the combined British bomber and fighter
force in northern Malaya contained *only ten* serviceable aircraft. Within another twenty-four hours—by December 10—
all the remaining fighters and most of the remaining bombers
had been withdrawn from Malaya into Singapore Island to
protect the Naval Base. At the same time daylight bombing
was cancelled because of the heavy losses it would entail—
and as a result Dutch reinforcements from Java, including 22
Glenn Martin bombers, had to be sent back again. The
Dutch pilots had no experience of night bombing and returned to the Dutch East Indies to complete their training. . . .

IV

Defeat At Sea. On the evening of Invasion Day (December 8) "Force Z" of the Eastern Fleet, comprising the battleship *Prince of Wales*, the battle-cruiser *Repulse* and the four destroyers *Electra, Express, Vampire* and *Tenedos,* sailed from the Naval Base for the battle areas of Kota Bharu and Singora. Admiral Sir Tom Phillips believed that with fighter cover and the element of surprise he could smash the Japanese landings. His commanders and staff officers agreed unanimously that inactivity in Singapore was impossible and that the sudden raid, though hazardous, should be attempted.

Phillips was warned before he sailed that it might not be possible to give his fleet fighter protection, but he sailed. In a signal to his ships' crews he said:

"The enemy has made several landings on the north coast of Malaya and has made local progress. . . . This is an opportunity before the enemy can establish himself. . . . Whatever I meet I want to finish quickly and so get well clear to the eastward before the Japanese can mass too formidable a scale of attack against us. So shoot to sink."

Capt. William Tennant, of *Repulse,* added his word:

"We are off to look for trouble. I expect we shall find it."

On December 9, Rear-Admiral A. F. E. Palliser, Chief of Staff, Eastern Fleet, signalled Phillips from Singapore:

"Fighter protection on Wednesday 10th [the estimated time of the Fleet's attack] will not, repeat not, be possible. . . ."

No fighter protection. One of Phillips's two conditions for successful attack could not be fulfilled. Nevertheless, there remained the possibility of surprise. A grey sky with heavy rain made Force Z feel comfortable and undetected, but about tea-time the rain stopped and the sky cleared. A plane was seen out on the westward horizon. Three more planes were seen later. Capt. Tennant broke the news to his ship's company:

"I have just received a signal from the C.-in-C. He very much regrets to say that he has had to abandon the operation. We were shadowed by three planes in the evening after dodging them all day. The enemy convoy will have largely disappeared when we arrive and we shall find enemy aircraft waiting for us. We are now returning to Singapore."

The wardroom groaned. One man remarked: "This ship will never get into action. It's too lucky."

Force Z did not return direct to Singapore. Instead—and minus one of the destroyers, *Tenedos*, which was short of fuel (one of the minor unexplained mysteries of the war is why she was short of fuel only thirty-six hours after leaving on what would have been at least a four-day mission)—the Fleet changed course to investigate a report that the Japanese were invading Malaya at Kuantan, only 140 miles north of Singapore. Admiral Phillips considered Kuantan to be of military importance and decided to take the risk of going to the help of the defenders. The report was false—bred out of the panic which was even then beginning to grip the British forces. No landings had been made or were ever attempted at Kuantan. When the town was eventually attacked it was by a Japanese force which came overland down the east coast. It is not known who originated the rumour, but it was to have serious consequences.

Also, Admiral Phillips had underestimated the Japanese efficiency. They knew all about the *Prince of Wales* and *Repulse* and had laid their plans. No doubt they would have attacked the Fleet anyway—at the Naval Base, for instance—but they must have rejoiced at the target now offered to them: the only British warships in sight, without Base guns or fighter escort to defend them. When the news reached the

Japanese the 22nd Air Flotilla, specially trained in bombing and torpedo attacks, and manned by crack pilots, was ordered to attack. The flotilla numbered 34 high-level bombers and 51 torpedo bombers.

Meanwhile, at 8 A.M. on the 10th, Force Z arrived off Kuantan. The two heavy ships flew off their aircraft for reconnaissance and the *Express* reconnoitred the harbour. The reports said "complete peace".

Course was then set northward and later eastward because Admiral Phillips wanted to look at a small ship with some junks and barges which had been sighted at extreme range at 7 A.M. Shortly after 10 A.M. *Tenedos* reported from 140 miles southward that she was being bombed. Twenty minutes later a shadowing aircraft was sighted by the *Prince of Wales* and "first degree readiness" was assumed. *Repulse*'s radar picked up enemy aircraft a few minutes afterwards, and at 11 A.M. nine aircraft were seen approaching at a height of 10,000 feet. At 11:30 A.M. all ships within range opened fire.

The enemy aircraft did not change formation. At high level they began bombing *Repulse* with great accuracy. One bomb actually landed amidships but failed to pierce the armoured deck.

Twenty minutes later nine torpedo-bombers attacked. *Repulse* "combed" the torpedo tracks but the *Prince of Wales* was struck twice. Both her port propeller shafts stopped, the steering gear failed, and all her secondary armament except one gun was silenced. From that moment the *Prince of Wales*, listing 13 degrees and at a reduced speed of 15 knots, was never under complete control.

Just before noon *Repulse* avoided a second attack; but a few minutes later, while she was close to the *Prince of Wales*, asking for information and offering help, the final attacks were launched. Six aircraft attacked the *Prince of Wales* and hit her with three torpedoes. Three more attacked *Repulse* and hit her once. Within a few minutes torpedo-bombers appeared "from all directions" and hit *Repulse* four times.

Capt. Tennant, realizing that his ship was doomed, gave the order for everyone to come on deck and cast loose the lifesaving floats. O. D. Gallagher, of the London *Daily Express*, who was on board, wrote:

"Loudspeaker: 'All hands on deck. Prepare to abandon ship. God be with you.' It was 12.25 p.m.

"We all trooped down the ladders, most orderly, except one lad who climbed the rail and was about to jump to a lower deck. . . .

"An officer called 'Now then—come back—we're all going the same way.' He rejoined the line.

"I was tempted to leap to the lower deck, but the calmness was catching. We were able to walk down the side of the ship to the sea. She lay so much over to port that it was a steep but easy walk even upright.

"I had a last glimpse of *Repulse* as a wave swung me round. She was about 100 yards away. Her underwater plates were a bright, light red. Her bows rose high as the air trapped in the underwater forward compartments tried to escape.

"I thought of the officer who had fallen from a ladder previously and was now with his broken ribs in the sick bay. *Repulse* hung, bows high in the air, for a second or two, then slipped easily out of sight.

"I had a tremendous feeling of loneliness. I could see nothing capable of carrying me. . . . I could not swim.

"I kicked, lying on my back. My eyes were burning as oil crept over me: in my mouth, nostrils, ears, hair. Water that I spurted out of my mouth was black. I came across two men hanging on to a round lifebelt. They were black. 'You look like a couple of Al Jolsons,' I told them. 'So do you—that makes three of us.'

"We were picked up by the Australian destroyer *Vampire*. Captain Tennant was also picked up by *Vampire*. Admiral

Phillips went down. It was a grave loss. We needed gallant
little Admiral Tom Phillips."

Capt. Tennant reported:

"When the ship had a 30 degrees list to port I looked over
the starboard side of the bridge and saw the Commander and
two or three hundred men collecting on the starboard side.
I never saw the slightest sign of panic or ill-discipline. I told
them from the bridge how well they had fought the ship and
wished them good luck. The ship hung for at least a minute
and a half to two minutes with a list of 60 degrees or 70
degrees to port and then rolled over at 12.33. . . ."

Just after *Repulse* sank, nine high-level bombers attacked
the *Prince of Wales*, then steaming at eight knots with her
quarter-deck almost awash. They hit her only once, but it was
clear that she was sinking. By 1:10 P.M. she was settling fast.
Ten minutes later she heeled over sharply, capsized and sank.
Twenty officers and 309 men were lost. In *Repulse* the losses
were 27 officers and 486 ratings—a total of 840.

For some reason never explained, Admiral Phillips failed
to notify Singapore of his change of plan, but maintained
radio silence. It was Captain Tennant of *Repulse* who broke
the silence with an emergency signal that the ships were under attack. Eleven Buffaloes standing by on Sembawang,
Singapore, to give protection were ordered to the scene, but
by then—like everything else in the defence of Malaya—it was
too late.

Fl. Lt. T. A. Vigors, O.C. 453 Squadron (R.A.A.F.) whose
aircraft was the first on the scene, paid this tribute to the
men of the Eastern Fleet:

"I witnessed a show of that indomitable spirit for which
the Royal Navy is so famous. I passed over thousands who
had been through an ordeal of greatness of which they alone
can understand. . . .

"It was obvious that the three destroyers were going to
take hours to pick up those hundreds of men clinging to bits

of wreckage, and swimming around in the filthy oily water.
. . . Every man must have realized that. Yet, as I flew
around, every man waved and put his thumb up . . . men in
dire danger waving, cheering and joking as if they were holiday makers at Brighton. I take off my hat to them, for in
them I saw the spirit which wins wars."

The loss of the men was bad. The loss of the ships was
appalling.

Only the night before, in the Cabinet Room at No. 10
Downing Street, the Prime Minister and Admiralty chiefs
had met to review the naval position. Mr. Churchill thought
it would be wise to link up the *Prince of Wales* and *Repulse*
with the remnants of the U. S. Pacific Fleet which had escaped the holocaust at Pearl Harbour. He wrote afterwards:

"It would be a proud gesture at this moment, and would
knit the English-speaking world together. . . . We were all
much attracted by this line of thought. But as the hour was
late we decided to sleep on it, and settle the next morning
what to do with the *Prince of Wales* and the *Repulse*."

In the morning the telephone rang beside the Prime
Minister's bed. It was a call from the Admiralty. The First
Sea Lord spoke:

"Prime Minister, I have to report to you that the *Prince of
Wales* and the *Repulse* have both been sunk by the Japanese
—we think by aircraft. Tom Phillips is drowned."

"Are you sure it's true?"

"There is no doubt at all."

The Prime Minister put down the telephone. He was
thankful to be alone.

"In all the war," he wrote, "I never received a more direct
shock. . . . As I turned over and twisted in bed the full
horror of the news sank in upon me. There were no British or
American capital ships in the Indian Ocean or the Pacific
except the American survivors of Pearl Harbour, who were
hastening back to California. Over all this vast expanse of

waters Japan was supreme, and we everywhere were weak and naked."

In Singapore the news was stunning. Major F. Spencer-Chapman, No. 101 S.T.S. (Guerrillas), wrote in *The Jungle Is Neutral*:

"I shall never forget the sense of utter calamity with which we heard the news. Until this disaster neither I nor any of those with whom I had discussed the war situation had any real doubt that we should be able to hold the Japs, even if it meant giving ground until we had regained the initiative.

"Now our sense of security and faith in the future vanished completely. For the first time I began to consider the possibility of losing Malaya.

"But even then one felt certain that it would be possible to hold the island fortress of Singapore until reinforcements arrived from the Middle East and Europe."

v

Defeat On Land. It is important to glance at the British defence position in the Far East as it was when the Japanese attacked; otherwise it is impossible to understand the appallingly rapid breakdown of morale and the lack of serious resistance to the enemy.

The Air Force totalled 141 operationally serviceable aircraft—425 *below* Malaya Command's estimated requirement and 195 *below* the estimate agreed by the Chiefs of Staff in London. All were obsolete or obsolescent—the Vildebeeste torpedo-bombers, for instance, could not fly at more than 100 m.p.h.

The Army was 17 divisions *below* the minimum strength agreed between the High Command in Malaya and the Chiefs of Staff as essential until the Air Force was brought up to strength.

These numerical shortages were not the fault of the High Command in Malaya. Brooke-Popham and his A.O.C. had sent three urgent signals on June 30, August 18 and August 20, 1941, describing the condition of affairs and asking for more aircraft. "At present," said Brooke-Popham, "not only is our ability to attack shipping deplorably weak, but we have not the staying power to sustain even what we could now do. As our air effort dwindles . . . so will the enemy's chance of landing increase." He ended by once more emphasizing his main preoccupation: "I have no doubt what our first requirement here is. We want to increase our hitting power against ships and our capacity to go on hitting."

His warnings did not fall on deaf ears, but the Chiefs of Staff were at that time powerless to aid him. All available aircraft were committed to other theatres of war. Neither could the army be reinforced up to the strength regarded as essential as long as the air force was below requirement.

These were the numerical shortages. In fighting efficiency the discrepancy was even greater. Lt.-Gen. A. E. Percival, G.O.C. Malaya, reported afterwards: "Most of our troops except those of the permanent garrison were inexperienced and semi-trained. . . . Even the regular units had been so diluted as to lose some at least, and in some cases a great deal, of their pre-war efficiency. . . . The lack of leaders, and even of potential leaders, reached a dangerously low level. . . . The great majority of the troops were young and inexperienced. The Australian units . . . suffered from a lack of leaders with a knowledge of modern warfare. The same applied in some degree to the British units. . . . No units had had any training in bush warfare before reaching Malaya. . . . Successful fighting in jungle country is largely a question of the confidence and self-reliance of the individual. These cannot be acquired without a reasonable period of training in such conditions."

But the training could not be given in the required

amounts—largely because the troops had to build their own defences. The Services had to compete with industry for local labour—and industry offered better wages and conditions than London would allow Malaya Command to offer. There was no compulsion of labour. Naturally the coolies chose to work for industry's higher wages, so the soldiers had to do their own coolie work, and training for war took second place. Percival had hoped for another three months, from December 1941, to improve the training. These were the fighting months.

Even this was not all. Defensive fighting under the best conditions is a poor policy—to be used only as a means of holding a line or delaying an enemy advance until forces can be mounted behind the line for a counter-thrust. In the worst conditions—the jungle is the worst—it is virtually impossible. The jungles, mangrove swamps and thickly-treed areas of cultivation present a peculiar and difficult problem. Visibility is limited, the opportunity for ambush lies almost everywhere. There are no fields of fire, tactical features tend to lose significance and roads and tracks become important. All-round protection is essential. Movement, though possible, is severely restricted—yet static defence spells defeat. Movement is easier in rubber plantations, but the interminable lines of evenly spaced trees and the limited view make it difficult for troops to keep direction. Control in rubber and jungle is difficult, emergencies crop up unexpectedly, and junior commanders' actions assume greater influence. Their errors of tactics, judgement and decision may easily decide an action.

Brooding above all, adding weakness to morale as well as to military efficiency, lies the jungle itself—a terrifying morass of tangled vegetation, steamy heat, nerve-racking noises and the discomfort of insects; mosquitoes by the myriad, moths, beetles, insects of all kinds, biting, buzzing, irritating and debilitating. Rubber, too, with its gloom, dampness and sound-deadening effect breeds a feeling of isolation. The enemy may be anywhere—everywhere—in front or behind— to left or to right. Noise is difficult to pinpoint; men appear and disappear like wraiths. Rumour begins to spread. In the monsoonal season there is the added handicap of torrential rain, hissing down incessantly upon the greenery, dripping dankly on heads and bodies, humid, sweaty, destructive. . . .

It was like this for the young and inexperienced troops who took up their places for the first defensive battle of the Malayan campaign, a battle which was noteworthy for two reasons—it was Britain's first defeat in the jungle; it was the pattern of future defeat in all the attempted defensive actions down the Malayan peninsula.

The seeds of defeat had been sown by indecision at the highest levels over *Matador*. The troops—the 11th Indian Division—had prepared defensive positions north of the village of Jitra; then they had been ordered to offensive positions; finally, after standing by for two days in the monsoonal rains, keyed up for the attack, they had been ordered back to their original line. Because there was no local labour, the troops had erected their own defence positions, in their spare time from training for *Matador*, which had priority. As a result, the defences were not finished when the men moved forward; when they returned, they found their work had been undone by the rains, and waterlogged. They had to buckle down like navvies to the tasks of erecting barbed wire, laying anti-tank mines, and providing signals communications. There was no time to do more than lay the field telegraph wires on the soaking ground, with the result that signals within the division were often defective.

When the battle of Jitra began on the 10th the troops were off-balance, dispirited, and strung out over a seven-mile front. Commanders and staffs had been given insufficient time to learn how to handle their formations in battle. Many of the Indian troops had never seen a tank (the British had no tanks in Malaya because London thought they would be no use in jungle warfare) and were demoralized when the Japanese armour came through. Finally, the divisional commander, Maj.-Gen. D. M. Murray-Lyon, was refused permission by headquarters in Singapore to withdraw 30 miles south to a better defensive position, on the ground that it would be bad for morale. He was also warned that his Division was the only one available for the defence of northern Malaya, and he must not risk heavy losses.

A mere two battalions of Japanese infantry, supported by a company of tanks, defeated the 11th Indian Division and drove it from its initial position within 36 hours. Losses in men, guns, equipment and transport were heavy: the effect on morale was shattering. Jitra was a major disaster.

At the same time, the small force called Krohcol, affected by indecision at headquarters, moved forward cautiously into Siam, met a little unexpected resistance from the Siamese, and needed twenty-four hours to overcome it. On the morning of the 10th, when still six miles short of their objective, they met the leading Japanese troops and were forced to withdraw.

Meanwhile, at Kota Bharu, where one brigade had to hold 30 miles of beaches, the Japanese had gained a foothold despite fierce resistance during which every Dogra manning the beach pillboxes was killed. Confused fighting took place throughout December 8. About four in the afternoon someone unknown—apparently influenced by the whizzing of stray bullets—spread a rumour that the Japanese had reached the perimeter of the airfield and another unknown (and unauthorized) individual gave the order to begin the denial scheme. The airfield buildings were set on fire and the staff began the evacuation. By the time the error was discovered it was too late. The R.A.F. station and squadron maintenance staffs had left to catch a train. They did not even carry out their scorched earth policy properly, for they had left untouched the stocks of bombs and petrol and had failed to make the runways unfit for enemy use. Partly as a result of this early panic, the British began to withdraw. By half-past six the next morning Kota Bharu was in Japanese hands.

The rapid spread of panic, far beyond the fighting line, is illustrated by the evacuation of Kuantan, 200 miles south of Kota Bharu, on the second day of the war. Because of the disastrous losses of aircraft in the first day, the Kuantan airfield was ordered to be cleared of all but two squadrons. At midday Japanese bombers attacked the airfield—which had no A.A. gun defences—and destroyed seven aircraft. P.O. Roy Bulcock, son of Queensland poetess Emily Bulcock, afterwards described the evacuation:

"Number —— Squadron had rushed to their aircraft, dumped all the bombs and spare equipment on the aerodrome, loaded up with 16 or 17 men, and taken off for Singapore, leaving some slightly-damaged machines behind.

"The balance of the ground staff, seeing their officers depart, commandeered the station transport and drove madly for the railway station at Jerantut, 102 miles away. Without waiting for kit or money or personal possessions, the whole crowd had vanished."

Over at the Army mess, Bulcock heard a British captain of the 22nd Indian Brigade say: "I saw a bloody flight-lieutenant soon after the raid started. He ran down to the side road with two friends, blew the door lock off a private car with his pistol, and drove the car on to the main road so fast that he skidded into a deep drain and couldn't get out. Then he held up the Post Office bus with his gun, made all the passengers and driver get out, and drove off at full speed."

"Yes, I saw him hold up the bus," said a major. "I was too far away to do anything. I wondered why he wouldn't take the passengers." He added to the four remaining Air Force officers: "You'll find one of your tractors missing. A chap was driving it off at about five miles an hour, with five other blokes sitting on the tanker. They'll have sore bottoms before they get there."

Bulcock added: "For the first and last time in my life I felt ashamed of being an Australian."

Air Vice-Marshal Sir Paul Maltby, assistant A.O.C., Far East Command, omitted the sordid details in his dispatch. He wrote: "The withdrawal of the ground party from Kuantan might have been better controlled."

VI

Defeat at Jitra was only the first of a series of disastrous engagements for the army of Malaya, and it laid the pattern for the remainder. Jitra was lost on December 12. Three days later the position at Gurun was given up, on the 17th the line at the Muda River—the last defensive position north of Penang Island—had gone. Failure to scuttle or remove all the boats in Penang Harbour allowed the Japanese to gain the use of 24 self-propelled boats and many large junks and barges, in which they shipped troops southwards down the coast to attack the defenders in the rear. Krian River was lost by the 20th, Kuala Kangsar was abandoned on the 22nd, and Kampar was gone by January 2. In all these withdrawals, valuable British material and equipment were lost. On January 7 there occurred the major disaster of Slim River, and on the 11th both Kuala Lumpur and Port Swettenham were in the hands of the enemy, who also picked up large stores of valuable fighting material which the British were unable to destroy or remove. Within a little more than a month the Japanese had advanced from Siam all the way down the peninsula into northern Johore—and the British were back to their last line of defence outside Singapore Island.

If there were reasons why the speed tactics and fire power of the enemy should have bewildered and surprised the British commanders, their tactics should now have been plain. Time and again the enemy had surprised the defenders by his ability to move more rapidly over all types of terrain; time and again he had used enveloping tactics rather than the direct onslaught. Yet in Johore the defending commanders chose to try to defend one strongpoint, only to be destroyed piecemeal. Elsewhere, a weak brigade was asked to hold the most likely road for the Japanese advance, while three brigades were allocated to the area they did not assault directly. The result was defeat again and again in Johore, until it became apparent by the last week of January 1942, that a withdrawal into Singapore Island was inevitable. In its long and detailed account of the campaign in Malaya, *The War Against Japan* tells the same story over and over again—though it is necessary to prise the revealing facts from a mass of tactical and operational detail which obscures the essentials:

"There was considerable confusion" . . . "many of the British and Indian troops were in no condition to withstand an attack in strength" . . . "owing to orders going astray" . . . "troops who had been fighting or on the move for a week . . . had once again to set about this arduous work [erecting defences]" . . . "the Japanese advanced more quickly than had been expected" . . . "the Japanese tanks came as a great surprise" . . . "Penang broadcasting station was not destroyed and they failed to remove or scuttle all ships and craft in the harbour" . . . "in the confusion of withdrawal stocks of high octane petrol were left intact" . . . "at Perak River General Heath decided he could not afford to dispute the crossing at Kuala Kangsar and the Japanese, to their astonishment, were able without opposition to cross the biggest obstacle they had yet met" . . . "The troops were very tired . . . Officers and men moved like automata and often could

not grasp the simplest order" . . . "owing to a misunder-
standing of orders" . . . "their unopposed bombing and
machine-gunning affected the morale" . . . "the battalion
was dead tired . . . the spirit of the men was low and the
battalion had lost 50 per cent of its fighting efficiency" . . .
"the jungle gave the men a blind feeling" . . . "the action at
Slim River was a major disaster" . . . "Gen. Gordon Bennett
evidently had not . . . appreciated the Japanese strategy or
grasped the full extent of the disaster at Muar" . . . "this
decision to withdraw the 45th Brigade was taken twenty-
four hours too late" . . . "having practically annihilated the
British troops" . . . "a scene of indescribable confusion
ensued . . . and the column as such ceased to exist. . . ."

As the Great Run neared Singapore the situation grew
worse. Squadrons experienced in torpedo-bombing were sent
to attack ships with bombs. Half the aircraft were lost, many
of the others badly damaged, both C.O.s were killed and
many of their crews wounded. The aircraft employed were
99 m.p.h. Vildebeestes, known to the R.A.F. as "flying cof-
fins".

There was the error which split the 22nd Indian Infantry
Brigade from the rest of the troops moving south to evacuate
across the Causeway, and resulted in the complete destruction of the Brigade as a fighting force. For four days and
nights the remnants struggled through trackless forest until
all but a handful gave up and surrendered.

That was the pattern—errors by commanders; insufficient
or inadequately trained troops; continuous under-estimation
of a savage, speedy, highly skilled and highly mobile
enemy; spreading confusion and panic. As the Great Run
lengthened, eyes grew red with sleeplessness, limbs became weary with incessant effort, minds sluggish with exhaustion, nerves frayed, morale and discipline and efficiency
constantly diminishing.

To the troops who did not know the truth about Singapore, the island loomed in their minds like an oasis. In Singapore there would be proper accommodation, proper food; above all, a place to lie down in peace for the one night's sleep they all craved.

3

The Great Myth

"I began to consider the possibility of losing Malaya. But even then one felt certain that it would be possible to hold the island fortress of Singapore. . . ."—Major F. Spencer-Chapman, No. 101 S.T.S. (Guerrillas) after the sinking of the Prince of Wales and Repulse.

The more diligently the inquirer delves into the story of Singapore, the more evident it becomes that the chief architect of defeat was the Great Myth, and the confusion it caused. Now that the Myth lies in ruins it is difficult to understand how it ever bamboozled anyone. It is like a conjurer's trick which baffles until it is explained. Once you know how it is done the puzzle is to understand how you were ever baffled. Mr. E. M. Glover, managing editor *Malayan Tribune* group of newspapers, expressed it succinctly in 1947:

"How we came to be deluded into a false sense of security in the Far East will never be fully understood. The facts to lead any ordinarily intelligent person to take a rational view of the situation were well known, and apparently appreciated in the right quarters, but not acted upon."

The trouble with Myths is, not that they deceive people who have no means of finding out the truth, but that they fool those who should know better. Few more shocking revelations have ever been made about the Second World War than Mr. Churchill's admission that he did not know

Above: The last of a group of 131 men and one Chinese girl leave their bomb-shattered converted Yangtze River steamer in a lifeboat during their evacuation from Singapore. *Below:* A Malay child lies dead after an air raid on Singapore, while the mother and a relative weep. *(Photos by Keystone)*

Left: Admiral Sir Thomas Phillips, C.-in-C. Eastern Fleet, who went down when the battleship *Prince of Wales* and the battle-cruiser *Repulse* were sunk off Kuantan by Japanese aircraft on December 10, 1941. With him on the left is his Chief of Staff, Admiral A. F. E. Palliser. *Below:* The *Prince of Wales* is sinking. Sailors scramble over her side—a few on to the deck of a destroyer, the majority into the oily sea. *(Photos by courtesy of the Imperial War Museum)*

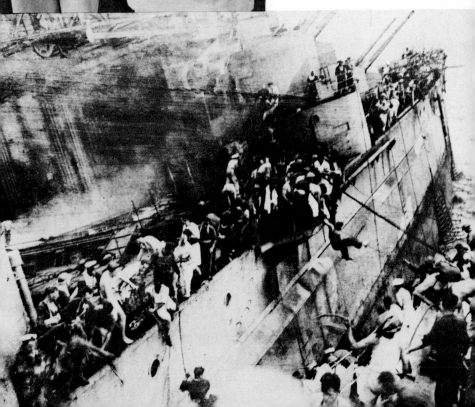

until January 16, 1942—one month before Singapore surrendered—that the Island Fortress was in fact not a fortress.

Here is part of the blistering letter he wrote to General Ismay, for the Chiefs of Staff Committee (he has not made public the contents of paragraph 2):

> I must confess to being staggered by Wavell's telegram of the 16th and other telegrams on the same subject. It never occurred to me for a moment, nor to Sir John Dill, with whom I discussed the matter on the outward voyage, that the gorge of the fortress of Singapore, with its splendid moat half a mile to a mile wide, was not entirely fortified against an attack from the northward. What is the use of having an island for a fortress if it is not to be made into a citadel? To construct a line of detached works with searchlights and cross-fire combined with immense wiring and obstruction of the swamp areas; and to provide the proper ammunition to enable the fortress guns to dominate enemy batteries planted in Johore, was an elementary peace-time provision which it is incredible did not exist in a fortress which has been twenty years building. If this was so, how much more should the necessary field works have been constructed during the two and a half years of the present war? How is it that not one of you poined this out to me at any time when these matters have been under discussion? More especially should this have been done because in my various minutes extending over the last two years I have repeatedly shown that I relied upon this defence of Singapore Island against a formal siege, and have never relied upon the Kra Isthmus plan. In England at the present time we have found it necessary to protect the gorges of all our forts against a landing raid from the rear, and the Portsdown Hill forts at Portsmouth show the principles which have long prevailed. . . . 3. Seaward batteries and a naval base do not constitute a fortress, which is a *completely encircled* strong place. Merely to have seaward batteries and no forts or fixed defences to protect their rear is not to be excused on any ground. By such neglect the whole security of the fortress has been at the mercy of 10,000 men breaking across the straits in small boats. I warn you this will be one of the greatest scandals that could possibly be exposed.

Mr. Churchill shouldered his part of the blame. "I ought to have known," he said. "My advisers ought to have known and I ought to have been told, and I ought to have asked."

His advisers—or some of them—did know. Brooke-Popham, in his post-war dispatch, wrote: "One of the steps taken to discourage the Japanese from starting war was to emphasize the growing strength of our defences in Malaya. The Chiefs of Staff in May 1941 stated that they saw no objection to this policy and we were aided by directions from the Ministry of Information in London to their representative in Singapore."

As a result, Sir Earle Page, Australian delegate in London, could say even in November 1941: "Singapore, heart of the British Empire, and Malaya, and other parts of the Far East, are being defended by troops from the United Kingdom, India, Canada and Australia. Enormous advances have been made which have made Singapore not merely impregnable but able if necessary to be the spearhead of a great offensive."

One phrase in Mr. Churchill's letter, taken in conjunction with the statements made in General Percival's dispatch to Parliament dated February 26, 1948, reveals the true extent of the confusion which existed between the Prime Minister and the generals in Malaya. It is this:

". . . over the last two years I have repeatedly shown that I relied upon this defence of Singapore against a formal siege, and have never relied upon the Kra Isthmus plan. . . ."

General Percival stated in his dispatch:

Paragraph 18. "It cannot be too strongly stressed that the object of the defence was the protection of the Naval Base. . . ."

Paragraph 24. "In August 1940 . . . the necessity for holding the whole of Malaya . . . was now recognized. . . ."

Paragraph 254. "Our object remained as before, i.e., the Defence of the Naval Base."

Paragraph 292. ". . . (a) *Intention.* The intention was to continue to ensure the security of the Singapore Naval Base."

Paragraph 336. "Our object remained as before, i.e., the Defence of the Naval Base."

Paragraph 390. ". . . I therefore decided to authorize a withdrawal to Singapore Island, even though this meant failure to achieve our object of protecting the Naval Base. . . ."

Paragraph 439. "Singapore was not a Fortress. . . . The coasts facing the Straits of Johore, were, when war broke out, completely undefended. . . . It is certain that the troops returning from the mainland, many of whom had never seen Singapore before, were disappointed not to find the immensely strong fortress which they had pictured."

Consider one final quote from this revealing document. Paragraph 451, under the heading "Appreciation of the Situation, 31st January 1942" reads: "Object: . . . we had clearly failed to achieve our object of protecting that [Naval] Base. *From now onward our object was to hold Singapore.*" (Author's italics.)

General Percival wrote those words (and many thousands more) for publication six years after the fall of Singapore. Nowhere in his lengthy dispatch did he mention the signal sent by Mr. Churchill on December 15—only seven days after the invasion—from H.M.S. *Duke of York*, when the Premier was on his way to visit President Roosevelt. It said:

"Beware lest troops required for ultimate defence of Singapore Island and fortress are not used up or cut off in Malay Peninsula. Nothing compares in importance with the fortress."

Percival knew that Singapore was not a fortress (though he referred to it as a fortress four times in a proclamation issued the day after the siege began); Churchill—and the troops who thought they were fighting to defend it—believed it was. Percival was fighting a war to protect the Naval Base; Churchill was asking for the defence of Singapore Island. Out of this division and confusion came disaster.

II

"Singapore was not a Fortress." Exactly how little of a fortress was it? That depends what you mean by "Singapore" —and some of the trouble undoubtedly arose out of loose terminology. To the geographers, there are two Singapores: first, the island, which has roughly the shape and position of the Isle of Wight in relation to England. It is 27 miles long at its widest east-west point and 13 miles wide at its broadest north-south point. Including the islands nearby, it has a coastline 72 miles long. The water barrier between the north coast and the mainland, the Strait of Johore, varies in width from 600 to 5,000 yards, and the Causeway which links the island to the peninsula is 1,115 yards long and carries road, rail and water pipes. There are a few hills, otherwise the island is covered with rubber and other plantations, extensive mangrove swamps on the north and west coasts—and the built-up area of the city of Singapore, which is the geographers' second definition. Singapore city is modern, thriving and important. Its normal resident population in 1941 was about 550,000, but by the end of January 1942 it had been almost doubled by refugees. Three-quarters of the inhabitants were Chinese.

In the military sense, Singapore meant "Singapore Fortress," and "The Singapore Fortress Command" included siege guns and reserve troops for a close defence of the Island and south-east Johore. The "Fortress" area was 35 miles long and 15 wide—centered on the £63,000,000 Naval Base which had been built on the north-eastern shore of the island—the coast closest to an invader from Malaya.

The fixed defences of Singapore Fortress Command were sited to repel a sea attack. Only a few guns could be traversed to fire northward—the 15 inch and 9.2 inch—but the 15 inch guns had no high-explosive ammunition, only armour-piercing shells for use against enemy warships, and the 9.2s had only 30 rounds per gun to withstand a siege which, it had been estimated, might last 180 days before the Fleet could effect relief. On that basis, each 9.2 could afford to fire one H.E. shell every six days.

The beach defences covered the *south* coast—which appeared suitable for enemy landings from troop transports invading from the sea. Defences on the west of the island—known as the Jurong Line—had been reconnoitred and a layout prepared but no works had been built or trenches dug.

There were no defences at all on the north coast—the position most vulnerable to an attack from the land—because it had never really been considered attackable! The Naval Base had to be protected, and as we have seen, the strategists considered that the best line of defence was on the frontiers of Siam.

There were no deep air-raid shelters or concrete surface shelters because, says *The War Against Japan*, "the construction of deep shelters on anything approaching an adequate scale would be exceptionally difficult, for the water level of the island was so near the surface . . ." and "Sufficient open spaces in which to construct above ground the large numbers of shelters needed did not exist."

Hutted accommodation outside the town was therefore provided—though it was not planned to enforce the evacuation of the population.

Finally, there was no conscription of labour. The idea had been rejected as unworkable because of "administrative difficulties".

Such was the nakedness of Singapore on December 8, 1941. Who were the men who would lead its defence? There were two: Sir Shenton Thomas, Governor of the Straits Settlements and High Commissioner for the Malay States, and Lt.-Gen. A. E. Percival, G.O.C. Malaya, who was responsible for operations under a succession of Commanders-in-Chief: Sir Robert Brooke-Popham (until December 25 when he was relieved, at his own request made before the invasion), Sir Henry Pownall (until early January) and finally Sir Archibald Wavell, Supreme Commander ABDA (Australian, British, Dutch, American) Area.

Ian Morrison, who met Sir Shenton and General Percival on the same day during the war, has recorded his impressions of them as they appeared at the height of hostilities.

"The Governor was very affable, as always. He was a short, rather stout man, with a red face, of a most sanguine temperament. I think he prided himself on being a good mixer who could get on with all sorts of people. . . . He was one of those solid, imperturbable, unimaginative Englishmen who would face up to anything. . . . The Governor was a good solid official who had spent much of his life in administering our African colonies. He took his responsibilities to the native population seriously and the Chinese leaders knew he would not let them down. He had that stolid British doggedness. But he was not a realistic thinker or even a very clear-headed one. He was sanguine to the verge of complacency. He had risen to his position, not by virtue of outstanding ability, but by dint of long years of steady conscientious work. There was no colour or forcefulness about him, nor much decisiveness. . . . The Governor was the last man to rally people in a crisis and inspire them to suffering, sacrifice and heroism."

Percival he describes as "a tall, thin person, whose most conspicuous characteristics were two protruding rabbit teeth.

"Percival was a man of considerable personal charm, if one met him socially. He was an able staff officer with a penetrating mind, although a mind that saw the difficulties to any scheme before it saw the possibilities. But he was a completely negative person, with no vigour, no colour, and no conviction. His personality was not strong, and as a leader he did not appeal either to the troops (to whom he was unknown except by name) or to the general public."

III

Singapore was in the war and the *Prince of Wales* and *Repulse* had been sunk—but life did not change much for the vast majority of the population. There were to be no more air raids on Singapore until late in December, and this immunity gave the islanders a false feeling of security. News of fighting on the mainland was continually disappointing, but the Japs were still too distant to seem a menace, and most people imagined the tide would soon be turned. Meanwhile, cinemas and dance-halls stayed open and it was easy to eat out lavishly. Officially, hotels and restaurants were supposed to observe two meatless days a week, but as game and poultry were not counted as meat, nobody noticed the stricture. Sugar and milk were not rationed and the meat and butter allowances were three or four times higher than the ration in England. There was plenty to drink, too.

Some of the efforts of the military to impose war conditions were met with remarks which indicated the refusal in some quarters to understand that there was a war on. When the Secretary of the Golf Club was told that the Club was to be turned into a strongpoint, he replied that it would require a special committee meeting. Maj. Angus Rose, of the Argylls, who tells this story in *Who Dies Fighting*, adds that when he wanted to cut down a row of banana trees to improve his field of fire, he was told it would require written permission from the competent authority.

The unreality of the life was noticeable even to those people who had no thorough inside knowledge of the full state of affairs; and to the refugees who reached Singapore later in December after fleeing from their homes on the peninsula, it seemed cruel and heartless. Mr. Eric Pretty, acting Federal Secretary for the Federated Malay States, who—with others—lost everything he possessed in the world at Kuala Lumpur—furniture, books, silver, glass, cutlery, linen, all his wedding presents—was horrified at the lackadaisical attitude among officials in Singapore. In Government offices people passed each other minute papers on humdrum subjects as if nothing at all unusual was happening.

Red-tape still strangled initiative in certain sections of the Army. For instance, Maj. Rose was doing a staff job at Combined Operations Headquarters when the Japanese started their blitz. He chafed in the office job, which he considered any competent clerk could do, and thought up a scheme to go Jap-hunting behind the enemy lines.

"Tell the General about it," said his wife.

Maj. Rose was a Regular soldier. He knew all about "proper channels". He also knew "if I do this through the proper channels, the Japs will be at the Johore Straits before the General has even heard of it".

He went direct to Percival.

" 'Sit down, Rose,' said the General. 'Now—what's all this about?'

"I outlined my plan. The General meditated for a bit. . . .

" 'All right,' said the General. 'You seem to have the right offensive spirit, Rose, so you can run the show yourself. Tell the B.G.S. tomorrow. . . .'

"Next morning . . . the G.S.O.1 . . . asked me since when I thought my middle name was Wavell and . . . he'd speak to the Brigadier about it. . . .

"I sat down in a chair and watched the minutes slowly ticking past. The B.G.S. [Brig. K. A. Torrance] sucked his pipe in silent meditation for what seemed hours. When the silence was broken . . . he was not amused by my strategy.

He would speak to the General on his return, not only about my strategy but also about my unbecoming conduct. . . .

"One thing which I had learned in my twelve years' service was to accept a 'whortleberry' from my superiors in a gentlemanly fashion. . . ."

In spite of this obstruction, Maj. Rose's plan went forward. Twenty Australians volunteered. Cdr. Victor Clarke, R.N., sailed them up the west coast. They ambushed a Jap column moving south, created havoc, and brought back valuable information. News of the raid raised army morale all round.

It is difficult to blame the ordinary inhabitants of Singapore for refusing to take the war seriously when authority seemed to take it lightly, too.

Yet even in the first days of the war in Malaya, there were people who knew what was coming. Maj. Rose "knew that Singapore was going to be knocked for six" before December 16—and booked his wife on the next available ship out of Singapore Island. General Percival (according to his dispatch to Parliament in 1948) thought the threat was there. "Not long after the commencement of the Malayan campaign it became apparent that we might be driven back to Johore or even to Singapore Island." Was this after Churchill's "beware" signal of December 15? If so, why did Percival wait a whole week, until December 23, before ordering the Commander, Singapore Fortress, "to arrange for reconnaissance" of the north shore of Singapore Island to select positions for the defence of possible landing places? Nothing more was done that month.

Another move, on December 10, which came to nothing was the appointment of Mr. Duff Cooper as Resident Minister for Far Eastern Affairs with Cabinet rank. It did not last long—he left Singapore early in January—and could scarcely have been a great success since his terms of reference prevented him from superseding the existing authority of the

Commanders-in-Chief or His Majesty's civil officials. His authority was largely confined to giving decisions which would otherwise be referred to Whitehall. He did, however, write to Churchill on December 18, saying that he was dissatisfied with the civil defence preparations and that "some senior officials" did not appear able to adjust their mentality to war conditions. On December 31 he expressed this dissatisfaction at the War Council. He said that he foresaw a breakdown of civil defence which would nullify the efforts of the military to defend the city. Some of the population were uneasy about the civil defence preparations, and he urged that one man should be appointed who would have control, unhampered by petty restrictions and applications to committees, to force through at top speed the priority work which needed to be done. He proposed that Brig. I. Simson, Chief Engineer, Malaya Command, should be appointed Director-General of Civil Defence, and the War Council unanimously agreed to give Simson plenary powers. It was also decided that the Governor should be told when he got back from Kuala Lumpur that afternoon and that another discussion should take place next day on the Director-General's proposed terms of reference.

The War Against Japan reports, noncommittally:

> In anticipation of the Council's approval Mr. Duff Cooper informed Brigadier Simson that he was appointed Director-General and handed him terms of reference which gave him plenary powers for Singapore Island and Johore and informed him that all executive departments of the Government would be under his control in matters affecting civil defence. He sent a copy of this to the Governor.
>
> The Minutes of the War Council meeting on the following day record "The Governor will issue a statement which briefly is to the effect that Brigadier Simson would be responsible to the Governor (who would appoint him), who would report in turn to the War Council." The communiqué as issued made no mention of plenary powers nor of any authority in the State

of Johore; it merely substituted Brigadier Simson for the Colonial Secretary as the head of the existing civil defence organization. Simson therefore had no special powers to enable him to compel Government departments and civilian organizations to take such action as he considered necessary, and, further, his activities were confined to Singapore Island only. Mr. Duff Cooper's plan to appoint one man who would have unhampered control did not materialize.

The War Against Japan does not explain why it did not materialize, but we may guess, especially in the light of an earlier obstruction by the civil arm of authority. In October 1940, a group of authorities in Malaya, comprising the ex-Service Association, the Chamber of Mines and the United Planters' Association, criticized the Government's failure to provide an adequate civil defence organization in Malaya and asked for "an officer of the fighting Services" to be placed in control. *The War Against Japan* reports: "The Acting Governor [Sir Shenton Thomas was on leave in England] passed their views to the Colonial Office, expressing resentment at the criticism. He said that he was strongly opposed to the military taking over civil defence." The Colonial Secretary did not share the Acting Governor's disapproval and resentment—but the organization was not changed. There can be little doubt that even at the beginning of 1942, when the Japanese were everywhere pursuing the British forces down the peninsula of Malaya, civil and military rivalries and jealousies were undermining the slim chances of holding the Island.

By the middle of December, General Percival was basing his tactics in Malaya on the estimated time of arrival of the reinforcements he had asked for and been promised: that is why he asked his tiring troops to defend the airfields of northern and central Malaya, to deny their use by the enemy and so prevent Japanese bombers from obtaining fighter support for attacks on incoming British reinforcement convoys.

By Christmas Day, 7 Blenheim bombers had arrived. Altogether, 18 had been sent from the Middle East—the other 11 had either crashed or made forced landings on the way. On Boxing Day Mr. Bowden, Commonwealth Commissioner in Singapore, painted a gloomy picture of the situation to the Australian Government. Singapore, he said, could only be saved by powerful reinforcements of latest fighter aircraft, with ample operationally-trained crews, and military reinforcements in divisions, not brigades. And on New Year's Day the Chiefs of Staff in London reinstated Singapore as Number Two priority for defence, after the United Kingdom —a position which it had formerly occupied and later lost to the Middle East and Russia.

On that day—New Year's Day, 1942—the *Straits Times* published the following editorial:

> Goodbye to the year 1941—and good riddance! We enter into a new year in local conditions that are simply fantastic. There is no better word to describe the situation. Even a month ago we were preparing for the usual New Year land and sea sports in Singapore. All accommodation at the hill stations was booked, prospects for the Penang race meeting were being discussed, and hotels were announcing that very few tables were still available for the New Year's festivities. . . .
>
> Terrible changes have taken place with a rapidity that still leaves us a little bewildered, but as we recover from the initial shock so do we become better able to see things in their proper perspective. We are not overwhelmed: we shall not be overwhelmed. . . .
>
> We shall be rejoicing before 1943 comes round.

The views of the Editor of the *Straits Times*, who could write in this fashion two and a quarter years after the war had broken out in Europe, might be compared with the comments by Cecil Brown, *Life* and Columbia Broadcasting

Corporation correspondent in Singapore, in his Christmas broadcast to the United States:

"In some respects Singapore is a confused man walking about as in a dream, knowing something terrible might happen and not quite sure how to meet it. . . . This is a grim Christmas in Singapore because the British out here are getting *ready* for war with the war going on. . . ."

Time was running out swiftly for the confused man, but it was true: something *was* being done, as the calendar of events in January shows.

4

Cloud-Cuckoo Island

"Let a plan be made at once. . . . The entire male population should be employed on defence works. The most rigorous compulsion is to be used. . . . The defences of Singapore Island [must] be maintained by every means."—From a message from Mr. Churchill to the Chiefs of Staff, January 19, 1942.

If Singapore Island was to be capable of withstanding a siege —and each day of January brought the possibility nearer certainty—a few fundamental tasks had to be undertaken with the minimum delay. Defences needed to be built on the northern shore—which meant workmen to build them; troops were needed to man them—which meant that reinforcements were required and that every life possible must be spared in the Malayan campaign; women and children ("the useless mouths") had to be evacuated—which meant decision on who should go and who should stay. Above all, the inhabitants of all races, creeds and classes had to be welded into a solid mass of war-alert hands, eyes and brains, aware of the problems, ready to meet them, and under a forceful, purposeful, unrelenting mind that knew how to lead, instruct and inspire. These were the minimum needs, for a siege of Singapore meant resistance against all-round attack: it meant the horrors of indisciminate bombing, of eventual shortages of food and water, the hardship of isolation from a world which could not help—perhaps the ultimate savagery of hand-to-hand fighting from house to house.

Instead, the story of Singapore in January 1942 is a hideous Mad Hatter's Castle, a month of cakes and corpses, death and dance-bands, blood and "bull". A few of the many advertisements in the *Straits Times* illustrate the cloud-cuckoo atmosphere which clung to the island. They are naïvely revealing—in more senses than one.

"Raffles Hotel. Tonight. Dinner 7.30 to 9 p.m. Dance 8 p.m. to midnight." "Adelphi Hotel. Grill Room, Dining Room, Table d'hôte lunch, dinner. Orchestral selections by the Reller Band." (Sea View Hotel same as Adelphi.) "Goodwood Park Hotel. Charmingly situated. Ideal for visitors and tourists." "Today—2, 4, and 6 p.m. Alhambra. Greta Garbo and Ramon Navarro in *Mata Hari*."

One optimist advertised: "For Sale. European Guest House in select non-military area. Now absolutely full, good proposition, would also consider lease for fixed period. Reply with bank references to Box No. 309."

A bold-type front-page government notice preached: "Workers! Every hour counts in the battle for Singapore! Don't let the sirens stop your work!"—but on the next page the Cathay Cinema advertised a "rowdy racy riotous comedy" inviting Singapore to "fling your troubles away and have a fling at love and laughter!"

Military censorship was so strict at Press correspondent level that news agency reports cabled to London had to be cabled back to Singapore for publication. They cloaked the news—but Signalman J. C. Sharp, Royal Corps of Signals, and a good many other people followed the progress of the victorious Japanese down the Malayan peninsula by watching the daily advertising of the Hong Kong & Shanghai Banking Corporation with its list of "the under-mentioned branches will be closed until further notice".

For several days at the end of January the *Straits Times*

carried a quarter-page Government public notice, authorized by Straits Colonial Secretary Hugh Fraser, ordering all civilians to evacuate a mile belt along the northern coastline "purely for precautionary measures".

The watchful pen at Number 10 Downing Street was provoked:

"I am concerned about the fullness of the information given in the Singapore newspapers. For instance, why is it necessary to state that a mile has been evacuated for defensive purposes on the north side of the island? . . . They seem to give everything away about themselves in the blandest manner. After all, they are defending a fortress and not conducting a Buchmanite revival."

An increasing number of personal advertisments told the heartbreak story of separated families:

"Will anyone knowing the whereabouts of Mrs. Spencer Lock and children please communicate with Spencer Lock, c/o Mr. A. J. Schooling, 16 Dunearn Road off Bukit Timah."

"Will Mrs. Stanley-Cary contact the 1st Malaya General Hospital for news of her husband."

"Mrs. J. Norman Milne of Lower Perak and her two children are staying at 27 Newton Road, Singapore, and anxious for any news of her husband, Sgt. J. N. Milne, a/c Coy Perak Battalion."

"Will Miss Jennie Lim please write to her sister Lamoon as soon as possible as she is now without a home."

"Mrs. W. T. Rowe and Mrs. Gilliland are safe and well at 10 Lewis Road, Singapore. Phone 6416. Anyone seeing their husbands please pass on this information."

"News is wanted of Mr. Lim Chong Chee of Kuala Lumpur, and Mr. Khoo Leong Sim of Ipoh. . . ."

"Dick. Will you please communicate Stella quickly. If no reply will leave for Australia shortly. . . ."

Page after page of advertisements merely reflected the true picture of confusion and "business as usual".

The unselfish queued at the Medical College to offer their blood to Dr. G. V. Allen's Transfusion Service, for which the doctor had sacrificed his own blood not once but six times: others queued at Mr. E. G. Staunton's office for February petrol coupons.

Voluntary canteen services like that of the Rotary Club appealed for funds and cars and helpers to comfort soldiers and civil defence squads on all-night duties; and Mr. S. J. Cooper waited hopefully in Flat 3, Meyer Mansions, to collect debts due to the refugee Penang Harbour Board.

While Mr. Churchill was asking Whitehall why his December instruction to evacuate the Singapore "useless mouths" had not been obeyed, the evacuated Cameron Highlands School had opened at Nassim Lodge (by kind permission of the owner) for "the benefit of anyone wanting to further the education of their children".

Wrecked sanitation in some areas stank to high heaven, and there was not enough labour available to repair it; but municipal employees—Tamils—scrupulously cleaned the city streets each dawn, and scythed the verges of rural roads.

Bodies which quickly putrefied were laid "for a few hours at most" at identification centres, to be claimed for burial "if relatives and friends so wish"; while a few streets away spotless servants were laying clean tables for dinner and dancing at the exclusive Raffles Hotel; and the New World was advertising "cabaret, non-stop dancing, and the usual tiffin dance on Sunday 10.20 a.m. to 2.30 p.m."

While in Malaya the onslaught of murder, rape and loot flowed rapidly south, in Singapore people still tried to carry on "business as usual"—selling, insuring, banking and accounting, sandwiched between Local Defence and Civil Defence duties. This was approved. The Governor broadcast: "During the raids, shut up your shops and take cover, and as soon as the raiders have passed overhead, open your shops and carry on business as usual."

Everything was chaotic when the bombers came over. Most of the work and traffic stopped. Banks and business closed the till. People ran for basements, leaped into slit trenches or big open storm drains, or fatalistically waited in their frail shanty homes and crowded tenement cubicles hoping the bombs would fall somewhere else.

Then for an hour or two came the fire-fighting and the clearing away of debris, the rescuing and bandaging, the laying out for identification, the death-trucks taking gruesome loads for mass burial in the various cemeteries where pits, 40 feet square, had been dug. The people who performed these grim services were volunteers, and they showed tremendous courage.

There was a willing spirit in Singapore, but no one man was there to guide it. Nobody knew to whom to look for leadership—to General Wavell in Java, to Lt.-Gen. Percival in Flagstaff House, to Sir Shenton Thomas in Government House, or to someone in London. The result was separate direction, confusion of purpose, argument and indecision.

Nowhere was this confusion greater than in the reception and handling of the unexpected and none-too-welcome influx from up-country of a quarter of a million refugees streaming over the Causeway into Isle-of-Wight-size Singapore.

There was a classic muddle over Perak. Civilians had been ordered south by the military, and a large crowd went by train to Kuala Lumpur. There they waited at the station for instruction or advice. A billeting officer arrived with a telegram from Sir Shenton Thomas to the effect: "The evacuation of Perak is entirely unofficial and unnecessary, and liable to cause a panic among the Asiatics. Singapore is already crowded, and no one may come here." The bewildered refugees returned by train to Perak. They got as far as Tanjong Malim on the frontier between Selangor and Perak

States, and were stopped by the military who ordered them back to Kuala Lumpur.

There was a row over that at a meeting of the Far East War Council. Duff Cooper, having heard the facts, was reported to have said: "After all, General Heath *is* going to fight the battle and he must decide when evacuation is necessary."

Then came Penang. The island had not even one anti-aircraft gun for its defence. After four days' severe air bombardment the place was a shambles and panic reigned. European women and children were hurriedly ferried to the mainland and railed to Singapore, bombed at intervals all the way. Finally, the European Volunteers were ordered out, and all the remaining Europeans were "advised" to go. They took it as an order, and all but a few crowded on board the ferries with the Volunteers and steamed down the coast to Singapore.

When they came to their senses, many of the European women from Penang were bitterly ashamed, and said so on arrival at Singapore railway station. They had been passed from town to town in hot, overcrowded railway carriages. They had come away in such a hurry that some of the smaller children were still in night-clothes. They were not allowed to stay in Singapore. They were shunted straight on board a cargo boat under filthy conditions and shipped to Batavia. The whole affair was a shambles of divided command, unpreparation and mismanagement.

Duff Cooper worsened it, adding insult to Asiatic injury, by broadcasting that the population of Penang had been evacuated. He meant the European population. Sir Shenton Thomas had to rub out the error.

"We stand by the ship, gentlemen," he told the mixed Legislative Assembly of the Straits Settlements next day. "In any withdrawal or movement of population, there will be no distinction of race. No European male or female will be ordered by the civil government to withdraw. We stand by the people of this country, with whom we live and work, in this ordeal."

The ghost of Sir Stamford Raffles, the Englishman who founded the Singapore Settlement in 1819 and framed its equitable laws, would have applauded. But in which boat home would go the Asiatic woman and her children? To an England at war? To India, where she could not speak the language? To colour-bar-ridden South Africa? To White Australia? The Commonwealth was willing to accept 1,500 Chinese. Where would the other thousands go?

The question of deciding who was "essential" and who was a "useless mouth" raised issues which only those on the spot could answer, and which London seemed not to understand or care about. In the military sense, children were "useless mouths"; and, where possible, parents had sent them away. But hundreds remained because their mothers could not afford to buy a cheap dress, let alone pay the fare required to ship a child to India or Australia. Many refugee European women with small children were in a deplorable way for housing, clothing and money. The old and infirm were in a tragic plight.

Singapore was not without charity, and Mrs. Elizabeth Curran Sharp wrote publicly to thank "all those kind people at Raffles College who gave us good food, a bath, a bed, and lots of little comforts on our arrival from Kuala Lumpur. I am sure all others are with me in my grateful thanks." But others found rent racketeers and black marketeers, demanding high prices and phoney receipts. Many British women with children were obliged to live in billets, with one suitcase and the contents of a purse. In many cases, loyal Chinese servants accompanied their employers, sharing their sudden poverty and caring for the children while the mothers went out to war work.

And in the fashionable Sea View Hotel out on Meyer

Road, a select society of British women and their off-duty "honey boy" escorts sang "There'll Always Be An England" when the band played popular selections on Sunday mornings.

On the civil official front, a web of red tape hampered every effort. Events moved too fast for pedestrian minds to keep pace. Late in January, the Governor sent a circular letter to the Malayan Civil Service, and copies to the Press. The letter stated:

> The day of minute papers has gone. There must be no more passing of files from one department to another. It is the duty of every officer to act, and if he feels the decision is beyond him he must go and get it. Similarly, the day of letters and reports is over. All written matter should be in the form of short notes, in which only the important points are mentioned. Every officer must accept his responsibility to the full in the taking of decisions. In the great majority of cases, a decision can be taken or obtained after a brief conversation, by telephone or direct. The essential thing is speed in action. Nothing matters which is not directly concerned with defence, and no one should be troubled with it. Officers who show that they cannot take responsibility should be replaced by those who can. Seniority is of no account.

When this was published Singapore, alone of the Straits Settlements, remained in British hands, and the Japanese were beginning to mop up Johore. "The announcement is about two and a quarter years too late," said the *Straits Times*.

A second exhortation appeared in a public notice advertisement in the *Straits Times* after the market gardens of Johore had been lost and when shipping was uncertain. Sir Shenton Thomas's administration had set up a Food Production Office to write dig-for-victory propaganda and issue seeds and fertilizers. . . . "Don't put it off till tomorrow, start growing vegetables today!" said the official advertisement.

II

The progress of the defence work revealed the lackadaisical attitude which prevailed in official quarters. True, the northern coast had been reconnoitred towards the end of December—but when Wavell arrived in Singapore on January 7 he found that no defence works had been built along the northern coastline or even planned in detail. He ordered work to be put in hand immediately; and on January 9, the 1st and 2nd Malay Regiments were ordered to reconnoitre specific sections of the north coast.

Ten days later Mr. Churchill sent the following message to the Chiefs of Staff about the defence of Singapore Island:

> Let a plan be made at once to do the best possible while the battle in Johore is going forward. The plan should comprise:
>
> (a) An attempt to use the fortress guns on the northern front by firing reduced charges and by running in a certain quantity of high explosive if none exists.
>
> (b) By mining and obstructing the landing-places where any considerable force could gather.
>
> (c) By wiring and laying booby-traps in mangrove swamps and other places.
>
> (d) By constructing field works and strong points, with field artillery and machine-gun cross-fire.
>
> (e) By collecting and taking under our control every conceivable small boat that is found in the Johore Straits or anywhere else within reach.
>
> (f) By planting field batteries at each end of the straits, carefully masked and with searchlights, so as to destroy any enemy boat that may seek to enter the straits.
>
> (g) By forming the nuclei of three or four mobile counterattack reserve columns upon which the troops when driven out of Johore can be formed.
>
> (h) The entire male population should be employed upon construction defence works. The most rigorous compul-

sion is to be used, up to the limit where picks and shovels are available.

(*i*) Not only must the defence of Singapore Island be maintained by every means, but the whole island must be fought for until every single unit and every single strong point has been separately destroyed.

(*j*) Finally, the city of Singapore must be converted into a citadel and defended to the death. No surrender can be contemplated.

The Chiefs of Staff sent it on to Wavell—after deleting the dramatic final three paragraphs—and Wavell signalled back that steps were being taken.

At the same time Mr. Churchill also sent a telegram to Wavell. It concluded: "I want to make it absolutely clear that I expect every inch of ground to be defended, every scrap of material or defences to be blown to pieces to prevent capture by the enemy, and no question of surrender to be entertained until after protracted fighting among the ruins of Singapore City."

On January 20, Wavell was in Singapore again—and found that little had been done about the Island defences. However, on the 23rd an *outline plan* was issued, and on January 28 the detailed plan came out. But by this time it was too late. There was no civil labour to do the work. The air raids had frightened them all away and neither the stick of compulsion nor the carrot of improved wages and conditions had been used to prevent the desertions.

The best method of recruiting civil labour had been frequently discussed by the War Council, without agreement. The Navy and Air Force wanted compulsion; the Director-General of Civil Defence and the Army thought that the offer of good wages, meals while at work, protection on the job and compensation for injury would work better. The War Council finally decided on a measure of compulsion plus more attractive conditions—but rejected the payment of a special rate of wages for work in dangerous areas on the ground that it would lead to inflation! It was not until January 21 that agreement was reached in principle; not until the 26th that full agreement was reached by all concerned; not until the 29th that compulsory labour regulations came into force and not until the 31st that London gave Singapore a free hand to fix wages and conditions. That day the siege of Singapore began.

The meagre effect achieved by Churchill's telegram with orders for defence in galvanizing the Island officials into action can be illustrated by two incidents.

On Sunday, January 25—four days after the signals were received—I was at my post with Right Troop of 15/6 H.A.A. Regiment R.A. at our ack-ack gunsite on the Keppel Golf Links, under orders to protect the Empire Dock.

Our four 3·7-inch mobile guns engaged when Jap planes came over, but they weren't much use. Our height-finder stands had inadvertently gone on to Basra, and the Singapore Ordnance had not yet dug up a replacement issue.

Between raids, life on the pretty green golf links was intensely boring. Nobody came from surrounding houses to offer tea or conversation. The troops were browned off because they couldn't hit any Jap planes, and all leave passes to Singapore Town had been cancelled. Slit trenches had been dug. The regimental picks and shovels lay idle in the "Q" tent. Nobody came to borrow them.

I sat in a soggy tent near the guns and wrote a personal diary: "A lousy day. Rained most of the day. Meant to play golf this evening—but it was too wet and the course was sodden, so really did nothing."

The second incident is more poignant and terrible. On that same Sunday Lt.-Col. J. F. D. Steedman, C.R.E. 11th Indian Division, 3rd Corps, who had supervised his sappers blowing road and railway bridges all the way down from Siam, crossed to Singapore under orders to reconnoitre ma-

Above: An Australian anti-tank unit covers a road block of felled trees, and picks off the Japanese tanks before they can get across. *Below:* Wrecked Japanese tanks burn in the jungle road. *(Photos by Keystone)*

Above: Japanese troops, fully equipped, wade shoulder-high across a stream as their advance draws closer to Singapore Island. *Below:* Japanese army lorries crossing an improvised bridge after the famous Causeway linking Singapore Island to Malaya had been breached by the retreating British Imperial troops. *(Photos by Keystone)*

terial requirements for 3rd Corps sector in the coming battle to hold the Island.

"I was dumbfounded when I sent to the Chief Engineer's office and learned that there was not a trench and not a bit of barbed wire on the whole of the north-east and north-west side of the island," he said. "We thought that during our time up-country all these defences would have been installed."

So Lt.-Col. Steedman went straight to the Base Ordnance Depot at Alexandra to get barbed wire and start digging and fencing NOW. Another shock greeted him.

In his diary for that day, January 25, 1942, he wrote the damning words: "Half holiday for the B.O.D."

While Sikhs, Australians and Loyals tackled the enemy with fierce bayonet charges through the breathless jungle around Kluang and Ayer Hitam, in bloody attempts to save the Naval Base at Singapore, the soldier staff at the Base Ordnance Depot had taken the Sunday afternoon off to go to the pictures.

In Johore the Australians were worn out after fighting non-stop for a fortnight. How much wearier were the men of the 11th Indian Division—men from Surrey and Leicester; Jats, Punjabis and Gurkhas—who had slogged at the beginning and slogged all the way down until they officially "ceased to exist as an effective fighting formation", and yet had to go on slogging to the bitter end. There were no Army reserves in Malaya, so there was no relief; and reinforcements late in January were not enough to replace losses.

To all these men, who walked fatigued as if they were doped, Singapore beckoned as an oasis where they could get the one night's rest they craved. No desert wanderer was ever cursed by a more mocking mirage.

"What's wrong with your bods?" Colonel Steedman was asked at Malaya Command headquarters. "Retreat complex? Even the Australians have caught it."

"Retreat complex be damned! Our men are healthy enough. All they want is sleep, and they'll be on top of their form. I must say I thought you fellows down here would have got cracking and made a start with dannert wire and a few trenches."

III

By the middle of January it seemed clear to Mr. Churchill that Singapore would almost certainly fall. The only question was how long it could hold out under siege, and the Prime Minister was faced with an agonizing decision. If Singapore was already doomed, was it necessary or desirable to reinforce the island? The 18th British Division was on the way. But the reinforcements could be given only at the expense of Burma, now also threatened by the Japanese. Burma was the gateway to India and the supply route to Chiang Kai-shek's Chinese. It was vitally important to hold it. If nothing could be done to save Singapore, why send further reinforcements there to fall into prison camps? If Singapore fell, India would be shocked. Reinforcements to Burma would strengthen the resistance there, and so ease the shock to India.

Mr. Churchill told his Chiefs of Staff that the question of a switch should be put bluntly to General Wavell, who must then decide at his discretion. "We may, by muddling things and hesitating to take an ugly decision, lose *both* Singapore and the Burma Road," he warned.

The Chiefs of Staff hesitated. "We all suffered extremely at this time," Mr. Churchill wrote in retrospect. It *was* an ugly decision to take; and before it was taken, Australia got wind of the idea. Australia had sent some of her best men to Singapore. If Singapore fell, Australia would be threatened.

Australia had a Labour Prime Minister, Mr. John Curtin, who did not believe in beating about the bush. He signalled Mr. Churchill: "After all the assurances we have been given, the evacuation of Singapore would be regarded here and elsewhere as an inexcusable betrayal."

Mr. Churchill knew that "betrayal" was not in the picture, but such a statement from one of the family hit hard—as Mr. Curtin intended it should; and the British Prime Minister's advisers in London were reluctant to be practical. Not for the first time Mr. Churchill had to mark time while faint-hearts caught up.

"I was conscious . . . of a hardening of opinion against the abandonment of this renowned key point in the Far East. The effect that would be produced all over the world, especially in the United States, of a British scuttle while the Americans fought on so stubbornly at Corregidor, was terrible to imagine. There is no doubt what a purely military decision should have been.

"By general agreement or acquiescence, however, all efforts were made to reinforce Singapore and to sustain its defences. The 18th Division, part of which had already landed, went forward on its way."

The majority of the reinforcements arrived between January 22 and January 29 against a background of impending disaster. Earlier reinforcements—land and air—had already been put into action and had suffered heavy losses. Behind the scenes urgent signals were flying between London and Singapore. Churchill cabled angrily: "We have already committed exactly the error which I feared when I sent my 'beware' telegram [of December 15]. Forces which might have made a solid front in Johore, or at any rate along the Singapore waterfront, have been broken up piecemeal."

By January 23 the last of the airfields on the mainland had been lost. Only the four on Singapore Island remained in British hands. Because these airfields were vulnerable to air attack, the evacuation of the remnants of the air force to Sumatra had begun and by the end of January every bomber had gone.

On January 26 Percival signalled Wavell: "Consider general situation becoming grave. . . . We are fighting all the way but may be driven back into the Island within a week."

He had already drawn up a plan for the retreat into Singapore and ordered it to be shown to only a few selected high-ranking officers.

On January 27 Percival wrote a personal message to Wavell. He said: "Very critical situation has developed. The enemy has cut off and overrun majority of forces on west coast . . . it looks as if we should not be able to hold Johore for more than another three or four days. We are going to be a bit thin on the Island unless we can get all remaining troops back. Our total fighter strength is now reduced to nine."

Three or four days. . . .

The plan for evacuation, which had been originally agreed for the night of January 31–February 1, was brought forward by twenty-four hours. On Friday, January 30, the ships which brought the 18th British Division discharged their troops—and prepared to sail.

Friday, January 30, was the curtain-raiser to siege.

5

Last Boat Home

"The failure to distinguish between 'effectives' and 'non-effectives' of the women of all nationalities was the main cause of the confusion, congestion and misery of the European 'non-effectives'."—The Civil Defence of Malaya, p. 127.

That day the docks of Singapore were ablaze. From ten o'clock in the morning until four in the afternoon formations of Japanese aircraft had been pattern-bombing the three-mile-long commercial waterfront with high explosive and incendiaries.

The Jap twin-engined bombers, based on the captured British airfields in Malaya, came in waves of 27 in perfect V formation at a height of 20,000 feet. Waspish Navy Zero fighters circled and weaved in escort; occasionally one looped, and flashed a reflection of the bright tropical sunshine.

There was little to stop them. On the bomb-blasted airfields of Singapore, fighter opposition had dwindled to 20 serviceable Hurricanes and six Brewster Buffaloes; and each time they rose to attack one or more was shot down. Puffs of anti-aircraft shellburst followed each formation of raiders, but the Japs seldom broke their tidy pack.

Government notices in the newspapers and pep talks over the radio, urging workers to carry on in spite of air raids, could not compete with a Japanese voice on the air from Penang calling:

"Hello Singapore! How do you like our bombing? You saw what happened yesterday? That is a trifle to what is in store for you. Singapore will soon be a heap of ruins with not a living thing in sight. Today we are coming to bomb the docks."

The previous day four big troopships had arrived in convoy bringing the main body of the 18th British Division to reinforce the crumbling army of Malaya.

They had run the gauntlet of Bomb Alley—the hazardous stretch of reef-strewn water between Singapore and Banka— but Allied fighter planes, now operating from bases in Sumatra, had given convoy protection and beaten off the Japanese raiders.

The four ships had nosed their way alongside Main Wharf on the seaward side of Empire Dock where they were now berthed—the *Empress of Japan, West Point, Wakefield* and *Duchess of Bedford*.

Throughout the night troops had sweated, and were still sweating, to unload military stores and equipment. A procession of revving army lorries came empty and went away full by Main Entrance Road, along Anson Road and up through Singapore Town. Their task was doubly urgent for the ships were due to sail within hours, this time as mercy ships carrying away to safety more than 5,000 women and children— some of the "useless mouths". It was a race to get them on board before it was too late.

The ships had hardly tied up before the Japs resumed their attacks. They tried again and again throughout the night. With a full white moon to outline their target, they plastered the three-mile Harbour Board area, despite the glare of search-lights pin-pointing their flight for ack-ack gunners.

The night's bombing started fires in godowns and warehouses, wrecked small craft and damaged harbour installations, but the four big troopships survived. Some of the undershots left death and chaos along busy Anson Road, around the railway station where wounded troops were arriving by train from Johore, and around the mean streets of Chinatown off Tanjong Pagar Road. Overshots fell in the sea.

All the dockside coolies employed by Singapore Harbour Board had fled the wharves in terror. They had no confidence in the above-ground air-raid shelters which the Harbour Board had provided.

Only philosophic Chinese fishermen working at their normal job found life easier. As they rowed out before dawn they found harvests of fish, stunned by high explosive, lying on the surface. Fresh fish had been scarce in Singapore since the departure of all the Japanese fishermen—along with the Japanese photographers, hairdressers, brothel-keepers, harlots and clerks who had either disappeared from their usual haunts or been sent to Ceylon for internment. The Chinese fishermen had turned their harvest of fish into a harvest of dollars by the time the Japanese resumed their attacks later in the morning.

The early warning system by observers up-country in the Malay peninsula had been swept away by the tide of the enemy advance. The only warning now came from radar on the Island. Bombs fell within minutes. Often the siren's eerie warble, echoing about the Island, mingled with the muffled roar of massed engines flying high.

All the enemy air raids were carried out in the form of pattern-bombing. The leading plane in each formation carried an accurate bomb-aim whose operator signalled the following planes when to release their bombs. The entire load crashed simultaneously with a devastating roar which shook the Island. The raids came every hour, interspersed with sneak raids by single planes which gunned the waterfront haphazardly.

It was a humid morning, characteristic of the worst time

of the year in the wet tropics. The sky was a pale steamy blue, tufted with dazzling white cloud. Hardly a breath of air stirred. The north-east monsoon was exceptionally dry at this period. A week ago it had poured floods of rain but today the sun pressed down like a hot iron, paralysing the land and flattening the sea to a vast glittering sheet of corrugated glass. The wharves were baking underfoot, and the air reeked of timber and tar, rope and machinery, ships' bilge and slops, and the stench of burning.

Offshore, a doomed fuel tanker which had received a direct hit poured out a heavy column of smoke. Up by the Blue Funnel wharves, the freighter *Talthybius* was on fire and sinking. Nosing out through the minefields went the little *Wu Sueh*, dressed in white with a green stripe and a red Geneva cross. She was an ex-Yangtze River boat of 3,400 tons and shallow draught, not suitable for ocean-going but the best Singapore could manage in the way of a hospital ship. She had narrowly avoided the last load of bombs, and was now ferrying wounded to Java.

Here, until recently, had passed the busy traffic of the fifth greatest port in the world, where east met west and traded cargoes—food and manufactures from Europe, the Americas and the Antipodes in exchange for rubber, tin, oil, silks, spices, gems and all the colourful wares of Asia.

Now it was just another military port in a world at war. Where lines of ships would normally be tied up, stem to stern, with gangs of coolies at work amid all the noise and clatter of a great international waterfront, most of the docks stood empty and the wharves deserted. The blaze of newly-started fires and smouldering wreckage from previous bombing emphasized the desolation.

"Makes you bloody sweat just to stand and look at it," remarked Private "Tich" Markwell, of the 4/54th Suffolks. "Give me Saxmundham of a frosty morning any day."

Camped nearby were Lt. Keith Campbell, R.A., and a

party of ack-ack gunners. They belonged to 15 Battery of the 6th H.A.A. Regiment, whose personnel had been divided. One battery had gone to Johore, the rest to Palembang in Sumatra to join the transferred members of the operational R.A.F. from Singapore. Campbell and his men were waiting for a boat to take the guns to Palembang, to protect the airfields P.1 and P.2. Boats were scarce and crews were even scarcer, and the gunners had to queue at the docks with the R.A.F. All available shipping was being pressed into service to move the remaining aircraft of the Air Force across to Sumatra.

Gunner Arnold watched a group of the 18th British Division taking small arms ammunition boxes from a sling and loading them into a three-ton lorry. He'd spent most of the night sipping stolen beer and helping to put out fires. He was browned off and wanted a bath. He went over to the group.

"What did you want to come 'ere for, mate?" he asked, genuinely curious. "Nobody knows what to do with all the booze and ammo, so we're burnin' the bloody place down."

The 18th British Division were equally browned off after being cooped up for three months on a world tour in crowded troopships. They did not like the look of the local surroundings, and they concentrated their dislike on Gunner Arnold's dirty uniform.

"I've heard Singapore's famous for Chinks, drinks and stinks," said a Sherwood Forester. "Now I've met you I believe it."

Being a London Cockney, Arnold enjoyed a bit of give-and-take. "Wait till you smell the local Soho, mate," he grinned. "Wait till you go down Lavender Street. Wait till you get a dekko at all the brass hats and redcaps prowlin' about. Makes your flippin' arm ache. You got a dirty big shock comin' to you lot, mate. Not 'alf."

The Gunner walked away in search of anything to make life tolerable before the next air raid. A bottle of beer would do. He turned several crates, kicked open one with his hob-nailed army boot. It was vinegar. Nobody came to arrest him. He rejoined Campbell's group, sat on one of the gun platforms and stared at the sea. "Blimey," was all he said.

Beyond the wharves and godowns, beyond the large rectangular basin of the Empire Dock and up along the city front, occasional palm trees stood bright green and motionless. Close-packed buildings—old-fashioned oriental squatting among modern steel-and-concrete—appeared to radiate sizzling heat from their bright, glaring walls.

The 18th British Division had no chance to go down Lavender Street or study brass hats and redcaps. They were trucked straight through Singapore Town and out along the Bukit Timah Road. Being fresh, they were to meet the brunt of any land assault on the island. For this task, they were without their heavy equipment which, they were told, was coming in a later convoy.

They had never been in action. They had been trained for desert warfare in the Middle East, and had no knowledge of jungle conditions. They knew nothing about Jap tactics. In the official jargon, they were "fit but soft".

They wondered what had become of their 53rd Brigade Group which had arrived two weeks ahead of the main body. Maj.-Gen. H. Beckwith-Smith, popular commander of the 18th British Division, soon knew. His Norfolks and Cambridgeshires under Brig. Duke had been sent straight from troopship to jungle, and had been cut up around Batu Pahat. Those who survived had struggled to the coast where they were now stranded on the wrong side of the Japanese. The Navy were sending small craft around to the west coast each night to rescue them. Some of the wounded had already been brought out.

They were better off than Col. Anderson's 2/19th and 2/29th Australians, hemmed in on all sides at Parit Sulong under a murderous barrage from snipers, mortars, field artillery and tanks, with wounded refusing to surrender.

They were better placed than the Sikhs and Garhwalies of the 22nd Indian Brigade, lost near Layang Layang, marching their wounded on stretchers, fighting through ambush, covering a few desperate miles a day through swamp and trackless forest in search of Singapore.

Maj.-Gen. Beckwith-Smith heard the latest situation report from Malaya Command, and wondered why he had been sent here. By tomorrow there would be half a dozen generals in Singapore. They would be falling over each other's command.

By tomorrow the Japs would be in Johore Bharu, at the other end of the heavy stone Causeway. The Causeway would be breached, but the Japanese had shown remarkable efficiency in the speed at which they could repair broken bridges within a few hours after major demolitions.

The Jap Army had tanks—Singapore had none. The Japs had a strong well-balanced Navy—Singapore had a few small craft. The Japs had at least 350 fast modern bomber and fighter aircraft in Malaya—Singapore had two dozen fighters, most of which would be gone by tomorrow, and one or two Tiger Moths for reconnaissance.

The Japs could either invest and by-pass the Island, or try the well-proven up-country tactics of beach landing and infiltration, supported by artillery firing from Johore, tanks if they could get them across, aircraft overhead and warships all round. Numerically, the opposing forces were about equal. In weapon power, the odds were about a hundred to one against Singapore.

II

For many of the Europeans and their wives there was a more
important matter to think about—more important than air
raids or lack of northern defences or the imminent danger of
Japanese troops on the other side of the Causeway. They
were thinking about the last boat home.

On January 27 Mr. Churchill had asked Parliament for a
vote of confidence in his leadership. In a masterly survey of
the war, he had warned Great Britain to expect "more bad
news" from the Far East. It was broadcast by the B.B.C.,
relayed by Delhi and heard in Singapore.

"That means us," said the Islanders. "Churchill's given us
up."

Today the ships were in harbour and many eyes turned
that way. Word had gone round that Army, Navy and Air
Force families were being sent away. Husbands who had left
a complex problem for wives to decide, wives who had put
up a brave show, people in the "don't know" class and people
who caught fear from the uncertainty of others, were gal-
vanized. Afraid and confused, many people allowed everything else to go by the board. Northern defences could wait.
Sentiment was given priority. Sympathetic commanders
handed out compassionate leave so that brigadier and private, and many men on the key civilian front, could see a
wife and child off to safety. The flight of "useless mouths"
was on—more than a month late.

It was incredible. Over the telephone, and wherever they
met, women asked each other the burning question "Are you
going?" as if they were discussing some forthcoming social
event. Many women realized this, and hated their position.
Lack of any official order either to go or not to go forced the
issue as a personal one. It was a hard choice, and those in

doubt naturally wanted to know what others were doing. Some said it was defeatist even to think of running away. So much depended on the work of civilians maintaining civil defence and keeping esssential services going. All but a fraction of the population were Asiatics, who were quick to detect any sign of weakness among Europeans. The presence of European women had a steadying effect, whereas the effect would be disastrous if the Europeans went at such a critical time. British prestige was at stake. Against these arguments loomed the spectre of the Japanese sword and bayonet, now fewer than 30 miles from Singapore and rapidly approaching.

"They say it's the last boat home."

The news was passed from mouth to mouth. For security, and to avoid alarm, there was no public announcement. Everything had to be done through commanding officers, civil administrators and business principals, and trickled along the grapevine.

Many Service families had already gone, and those who remained were now told more or less firmly that they must go. Other wives had to be cajoled or persuaded by anxious husbands, and some refused to go. There was anger against the Governor for not giving a definite order. Nobody liked to admit being a "useless mouth", though of course there *were* useless mouths. Everyone agreed that elderly women and those with young children were more hindrance than help, and ought to be evacuated; but the great majority of women were torn between conflicting tides of love and fear. There was heart-searching and bewilderment. There were also the stories of Japanese atrocities.

Stories of the horrors of Japanese conquest had been told by escapers from Hong Kong after the surrender there on Christmas Day. "Asia for the Asiatics!"—"Burn the white devils!"—"Drive the suckers of blood from the East!" screamed Jap propaganda, and Jap soldiers translated the words into action. Finding large stocks of whisky when they entered Hong Kong, the Japs drank to excess, adding alcohol to inflame the intoxication of victory, and ran amok.

Fifty officers of the British Army were roped hand and foot and bayoneted to death. Women—Asiatic and European —were raped and then butchered. Army nurses were dragged out of hospitals and savaged, and one entire Chinese district was declared a soldiers' brothel regardless of who lived there.

Then stories trickled down from nearer home. Malaya was receiving the same brutal treatment as Hong Kong.

As Japanese shock troops arrived to occupy town or village, their method was to spray the tommy-gun around to establish law and order. Village women fled to the long grass and jungle. Finding no women, Japanese went out with bayonets to round them up for brothel purposes. Asiatic men were forced to watch their wives and children outraged in their own homes. Any objection was punished by bullet or bayonet.

Captured soldiers were used for Samurai sword practice. In one instance, 110 Australian and 35 Indian prisoners were massacred at Parit Sulong.

Main streets in Ipoh and Kuala Lumpur dripped with blood from Chinese heads staked on poles to impress the need of obedience to the will of Nippon. Photographs of these gruesome heads hoisted in the Federal capital showed that some of the victims were children. The Japs were wreaking a savage vengeance for China's homeland resistance.

As the Japs advanced, they compelled all and sundry to assist in the work of repairing broken bridges, damaged airfields and other tasks. Unco-operative Malays were thrashed and tied to trees for ants and sun and thirst to destroy them. Holy places and holy books were desecrated.

Everywhere the Japs were infiltrating, dressed as Chinese coolies or wearing Malay sarongs which concealed tommy-guns or grenades. British, Indian and Australian soldiers could not tell spies from loyal Chinese and Malays.

These were the stories reaching Singapore. Who could say how many Japs or pro-Japs were on the Island? Who could tell what danger might lurk next door? Yet the majority of British residents went about assuring each other that the fortress was impregnable. The Governor had said so. "Singapore must not and shall not fall," he broadcast. If the Governor believed this, why should Mrs. Edith Rattray of the Green Cow Tavern at Cameron Highlands presume to argue? Mrs. Rattray had been evacuated against her will from "the highest hotel in Malaya". She refused to be pushed any further. "I've been turned out of my home, but I'll not be turned out of my country," she said. "Whatever the Japs may do, I'll stay put."

But, as the days wore on, more and more women decided to go.

The Government had centralized all passenger bookings in the P. & O. Company. To reduce the danger of bombing casualties among the waiting crowds, the Company's office was moved to Cluny, five miles out of town. Cluny was inconvenient but safer than the high exposed office in the city overlooking Collyer Quay and The Roads. The Cluny house was on a hill, reached by a steep private drive off the road to Johore. Cars had to be parked along the main road and you walked the rest of the way. It was hot sticky going uphill on foot. The sun was broiling. Slow-going women and children kept to the sides under shady trees, moving up one side and down the other.

Inside the house Mr. Frank Hammond, with his assistants Charles Jenkins and Philip Barnes and a number of Asiatic clerks, sat at tables taking particulars and handing out embarkation slips as fast as they could write. They waived questions of payment and there were no favours. It was the same for the Eurasian girl who had married a South African miner as it was for the Scottish planter's lady from Perthshire, the Straits-born Chinese woman, and the wife of the Bombay jeweller: they were equally British. The bright-eyed Chinese clerk smiled as he bowed the way in and out, treating all with equal courtesy.

Those P. & O. men did a twenty-four-hour job, but managed to look cool and composed, answering the silliest questions with patient politeness. They were efficient and selfless. Nothing was too much trouble. They were a credit to their Company.

For those in the queue it was like some grotesque reunion. There were government men, fellow-members of "The Spotted Dog", men who had played rugger together, heads of big business, well-known figures from all over Malaya. Everyone looked strained and tired, particularly women with children in arms or clutching at mothers' skirts. Smiles were bright and brittle, sometimes verging on tears.

There were groans from people who drew embarkation slips marked *Empress of Japan*. "Oh, no! Fancy sending us away in a ship with *that* name!"

"They're going to change it to *Scotland* in honour of the Argylls as soon as we get to Colombo."

"Colombo? Evelyn said we were going to England."

"What does it matter where we go as long as we go there?"

"She says we're not allowed to take out any jewellery."

"Oh, she's government and biased. What next will they think of? What am I supposed to do with my diamond bracelet, throw it away? I'm jolly well going to push all I can in a pot of face cream and to hell with them."

It went on all day and all night. At one time, Japanese planes dive-bombed and machine-gunned the cars parked along Orchard Road. The only shelter was to crouch in the concrete canal at the roadside. Fortunately there were few, if any, casualties, though some of the cars were wrecked. The unlucky owners had to beg the help of others.

From Cluny a procession went back to homes and billets to collect their effects. Every woman was allowed to take on board what she could carry. The lucky ones had husbands or friends to help. Bombs had wrecked the eastern approach to the Empire Dock. Cars had to be parked way back. Women unaccompanied by menfolk gave their cars to friends or abandoned them. To reach the ships they had to be ferried across the open western end of Empire Dock between air raids.

"When they arrived there," says *The Civil Defence of Malaya*, "they found that all Asiatic labour had completely deserted the dockside area. European civilians who had come down to see them off and say good-bye, and naval ratings, helped them to carry such luggage as they had for the journey. Later, at the shipside, there was the wildest confusion and, regardless of embarkation slips, passports and inventories, the women and children poured on board."

"The last boat home" had a magic appeal to all those who had regarded Malaya as a pleasant means of earning enough capital or pension to buy a house at Henley or Hove and live in comfortable retirement.

To wives like Mrs. R. Mullaly, whose lovely home now stood looted in Seremban in the Federated Malay State of Selangor, it meant only heartache when she left Malaya. She and her husband had lived and worked on rubber plantations for twenty years. Their home in Seremban was bought out of their own earned capital, Mr. Mullaly having risen to become an independent agent. Their loss was a personal catastrophe, but they put aside their despair and got on with the job. Mr. Mullaly was a member of the Selangor State Council, and came down to Singapore with the Volunteers. Mrs. Mullaly had worked in first-aid, in clothing refugees, driving lost ones by car to Singapore, and was now washing dishes in the Salvation Army canteen.

When told by her husband that he had booked her passage to England, she at first refused it.

"You're cramping my style, Rube," he complained. "You've lost your home, why don't you go?"

"Dickie, I can't," she said, "not while Mollie Sheehan and all my friends are still here."

"Why don't they go? This is no place for women and children. You should all have been sent away in December."

The same scene, with variations, was talked out between many husbands and wives behind the glaring walls which made Private "Tich" Markwell sweat this hot burning Friday, January 30, 1942.

Mr. E. M. Glover, managing director of the *Malaya Tribune* newspaper group, began to wonder if he was perhaps selfish to want his wife to stay. She did not want to go. They had built themselves an island paradise on high ground out west at Holland Park. He believed the Island was impregnable.

Mr. Edwin Fozard of John Little & Company, import and export merchants who ran one of the big European stores in Raffles Place, had a Yorkshire birth, 1914–18 war experience, and twenty years in Singapore to back his judgement.

"For goodness' sake pack your bag, Peg, and don't unpack it like you did last week. I worry about you every time there's an air raid."

Mrs. Fozard was from Yorkshire, too. "I can't go, Ted, not while there are younger women with small children in front of me. I'll go when my turn comes." She thought it was ridiculous for women with children to stay. She felt lonely in the steel-meshed Censorship Bureau without her friend Mollie Popple, who had answered an advertisement for young ladies to work at ABDA Headquarters in Java. Mollie was now one of "General Wavell's Young Ladies". Peg would have gone, but she could not bear parting from Ted.

Mr. J. Bennett came straight to the point, which was one of the reasons why he was manager of the big Borneo Company. "It's a man's job from now on," he told his wife. She had stayed after they had sent the four children to Australia. "You've done all you can for Singapore. Your duty now is to the children."

Mr. A. Corbet, managing director of Borneo Motors, one of the Borneo Company subsidiaries, had booked a seat for his wife on the K.N.I.L.M. plane for Batavia.

The Dutch had re-routed their Java–Singapore service through Sumatra. The K.N.I.L.M. planes were unarmed. The Corbets were about to leave for Kallang civil airport when a message came to say the plane had been shot down in the Malacca Straits soon after leaving Medan. Future K.N.I.L.M. flights to Singapore were uncertain.

"Perhaps it's an omen. Perhaps I'm not meant to go," she suggested.

Mr. Corbet wasted no time on the occult. He took her passport to the P. & O. house, not far from his own place. The thump and thunder of bombs down at Keppel Harbour made him more determined than ever to get his wife away from the terror.

Mrs. Andre, wife of the seconded commanding officer of the 1st Battalion The Malay Regiment, realized that the time had come to take the children home to Lincolnshire. She hated leaving Toby, but she knew he wanted her to go.

Mrs. Jean Graham had gone in the previous convoy, taking her small son and a pram to India. She had worked part-time at the Naval Base until her husband Ian had been sent to Java to join General Wavell's H.Q. Staff.

Maj. Graham had given her a revolver but she didn't care for it. She expressed the view of a soldier's wife in Singapore. "It isn't like being at home. There's a feeling of insecurity I can't explain."

The Evacuation Committee were at first unwilling to give her exit priority. "There are other ladies waiting with more than one child," they said primly.

"Yes, well, I'm going to have another any minute now,"

she said, "and I want to have it in peace, not with bombs banging all round."

They hastily signed her permit, and she left the same day.

Mrs. A. Chamier, wife of the Commandant, Intelligence Corps, Malaya, was a permanent resident of Singapore. Her son was at school in Melbourne, her young daughter at home. Mrs. Chamier worked as a clerk in R.A.S.C. headquarters, Fort Canning. She wanted to stay and see it through. "If I go, a soldier will have to take my place. As long as I stay in Singapore, I'm worth one bayonet," she said.

Mrs. Dorothy Thatcher had gone to Australia after her job at the Naval Base had melted from under her feet.

"You ought to pack it in, miss," the young R.N. office ratings had joked. "No good staying here. This place is expendable."

She went to work the day the Naval Base was expended. She watched the Royal Navy carry out their suitcases and trunks and deed-boxes. "There was no panic," she said. "They told me they were going to Colombo, and invited me to go along. It was all quite calm. They might have been going to a jolly regatta."

Thinking back, Mrs. Thatcher had no illusions about the future after she packed a brief suitcase and said good-bye to her home at Taiping in Perak. If the Japs could overrun Malaya, they could do the same at Singapore. She saw no point in staying to be interned. She went to the Evacuation Committee for a permit to leave. When they asked why, she told them. She went to Mr. Middlebrook for a passport visa, and to the P. & O. for a ticket. Then she went to a fashion shop to buy a warm coat for the voyage.

"You're not running away?" exclaimed the unmarried saleslady.

"I'm going to Australia to look after my child."

"But why? We're impregnable. We can stand a siege for years."

"Can you? I can't." Mrs. Thatcher bought a white steamer coat and went to Australia in the *Narkunda*.

Col. John Dalley, F.M.S. Police Director of Intelligence Bureau, who was given Army rank and *carte blanche* at his discretion to expand his Chinese guerrilla band—known officially as *Dallforce*, unofficially as "Dalley's Desperadoes"—sent Mrs. Dalley and the child packing. "There's going to be a party at Singapore, and you're not invited," he said, kindly but firmly. Mrs. Dalley packed and went.

The same forceful decision in all quarters could have given Singapore some resemblance to a determined fortress rather than a tragic picnic.

"There's a boat leaving tonight, Ruby, you must be on board. It's the last boat home."

"Are *you* going?"

"No, *I* can take it."

Mrs. Mullaly thought of the charm of Seremban, the view from her bedroom window; that lovely hour when she did the flowers, planned the meals, gave Cookie his shopping list; Dickie coming home from work, tea on the cool veranda, tennis at the Club. Like so many others, she had clung to the hope that the Japanese would be halted and pushed back, and she could return and put her house in order again.

Mr. Mullaly was more realistic. Seremban was a loss. He had written it off. "I'm tired of hearing people say 'Oh yes! When we get down to proper fighting we'll soon see the Japs off.' Go to Colombo, Rube, and I'll come on the next convoy."

Practically everyone on Singapore Island this hot burning Friday reduced the situation to a personal point of view. They were not considering Mr. Churchill's anxiety to get rid of the "useless mouths". They were in a world of their own, remote from General Wavell's Burma–Australia strategy. "Fancy making his headquarters in Java," Mrs. Fozard had said when ABDA Command was formed. "He can't think much of our chances in Singapore." People could see the big upheaval affecting their immediate plans: few comprehended or could even perceive that this was a main crossing in human progress—the end of one march, and a new beginning.

Among the few treasures Mrs. Mullaly had managed to bring away from Seremban was her husband's brand new set of golf clubs. Now she was taking them with her to the ship. "You're mad, Rube," he said, "what do you want to lug those around for?"

Mrs. Mullaly had once graced the English stage in plays by Barrie and Lonsdale. Few people outside the theatre world would understand the heart of a woman who has toured the provincial stage and lived in digs. "You saved up for them," she said now, "and all the bombs in Japan won't stop me. I'll keep them till you can use them again."

Some people tried to preserve the tight chest of drawers into which British Malayan society divided itself. One young Singapore woman, widowed in 1940, was crying because she'd left a hundred dollars' worth of Elizabeth Arden face preparations, and here she was "on this filthy ship and no cabin."

Most of the women were quiet and sensible, too overwrought to bother about anything but finding a place below decks to rest and pray. All the men who came on board—except a few invalids—said good-bye and went ashore, leaving saris, sarongs, coats and skirts standing side by side sharing a common heartache.

Actress Marie Ney had been persuaded to leave London during the 1940 blitz. At a farewell dinner given by the David Lows, H. G. Wells had said to her, "Malaya? My dear, you're running right into the war." Now she was running away from it. She was haunted by the eyes of the pretty little Ceylonese lift attendant in the Cathay Building. Once she had comforted her when the girl was frightened. Now she felt like the blackest deserter, and hated all the circumstances which had brought about this shameful position.

Her husband brought her to the ship. Mr. "Tommy" Menzies had given the best years of his life to Malaya, helping to pioneer and build the rubber industry. He couldn't understand why anyone should want to leave.

He stood at the foot of the gangway while she went on board. She yearned for him to say something, some word of forgiveness because she was going away. He said nothing. He was too bewildered to say anything. He looked up at the ship, turned and walked away. . . .

The long day closed with a blood-red sunset reflected over sea and sky. It was breathlessly hot and calm—almost suffocating below decks. The red world turned to pink and yellow and then was suddenly night. From the ship, Singapore looked dark and lonely except in patches where fires still blazed. It was a quarter to seven, and the alert Chinese A.R.P. wardens allowed no nonsense about brown-out. Any light that showed was promptly fired at.

Some people said the Japanese must have known that women and children were being evacuated and did not press home the attack, otherwise how came it that there was not a holocaust that day at the Empire Dock? Some said it was a miracle, as if the hand of God had deflected the bombs.

The *West Point* and *Wakefield* sailed on the Friday. The *Duchess of Bedford* and *Empress of Japan* steamed away in the full glow of noon next day. Indian soldiers were unloading ammunition from the *Duchess of Bedford* until the moment she sailed.

All four ships, each carrying up to 1,500 European and Asiatic women and children, were attacked by Jap high-level bombers until they were out of range, but ultimately reached their destinations.

And as the first ships steamed out from Singapore, the exhausted Army of Malaya began the march in from Johore.

6

Across the Causeway

"We are going to be a bit thin on the Island unless we can get all remaining troops back."—Lt.-Gen. Percival's message to General Wavell, January 27, 1942.

Apart from the war, the last few days of January 1942 were beautiful. The north-east monsoon had abated, leaving the sun to shine on rich tropical jungle ablaze with flowers. "Kingfishers flashed like jewels along the river bank," said one who was there. "The water and the sky were blue and the scene altogether lovely." At night the velvet heaven was spangled with brilliance. Fireflies glowed among the sweating troops. Presently the full moon would rise above the dark hills of Johore and bathe with light the weary fighting men and the white strip of stone which was their only path into Singapore.

The survivors of battles at Jitra and Kota Bharu via Krian River, Perak River, Slim River, Muar River, Kangsar, Telok Anson, Trolak, Kampar, Kuantan and Mersing talked about today and yesterday. They talked not because they were excited, but only to keep each other awake lest they go to sleep on the march.

Most of the Gunners were already back on the Island, having withdrawn from Johore in organized stages during the last two or three nights. Their artillery now covered the bottleneck of Johore Bharu and the vital Causeway.

Field hospitals and headquarters were already withdrawn, together with all possible supplies, stores and equipment. The rest—as far as time and battle allowed—had been destroyed.

Now the infantry had to thin out of battle and the rearguard had to break contact with the enemy without too much indication of withdrawal.

The plan was to withdraw in three columns of co-ordinated movement under cover of night, with rear troops in motor transport. For rearguard, each column had to provide one battalion to form an outer bridgehead in an arc about four miles outside Johore Bharu, the Sultan of Johore's State capital, which stood at the northern end of the Causeway. When all three columns had passed, the outer bridgehead would then withdraw through an inner bridgehead of about one mile radius.

It fell to the 22nd Australian Brigade and the 2nd Gordon Highlanders to hold the outer bridgehead. The honour of holding the last position, the inner bridgehead, was given to the 2nd Argyll and Sutherland Highlanders, now only 250 strong. Their C.O., Col. Ian Stewart, was in command of both bridgeheads.

In the last few cloudless days of January, the Japanese put on the pressure day and night in an all-round effort to trap the British forces on the mainland. Fresh coastal landings east and west aimed at cutting off any further British retreat, so that the Japs could cross the Johore Strait and move on to Singapore Island without any serious resistance.

The outnumbered defenders were equally determined to fight back to the Island. It was a tense, anxious time. In some of the fiercest encounters of the whole campaign fighting was hand-to-hand, making it difficult to break contact and keep to plan.

Taken at random from the chronicle of heroic battle against odds during those last days in Johore is the story of the force at Bakri surrounded by the enemy. In this trap were the 2/19th and 2/29th Australians and the remnants of the lately-arrived 45th Indian Brigade. The first report said that the whole force had been annihilated. Then it was learned that they were cut off, with tanks and artillery and strong road blocks established behind them. The force was under the command of Lt.-Col. C. G. W. Anderson, A.I.F. Under his cool, inspiring leadership the Australians and Indians fought at heavy cost through seven miles of road blocks. At one point they had to stand under fire hacking a felled tree apart before they could push a Bren-carrier forward and destroy a house full of Japanese with machine-guns. After all this effort, they were finally halted by Japanese who were holding a bridge position. They were then hemmed in on all sides.

This bridge had been abandoned by two British platoons who had no rations for two days and were out of communication with their headquarters.

The 2/19th Australians attacked again and again to regain control of the bridge but the enemy was too strongly entrenched. In the rear, the 2/29th Australians had to beat off Japanese attacking with tanks. Ten tanks were destroyed by the Australians. Casualties mounted.

Colonel Anderson sent an ambulance filled with seriously wounded to the enemy post at the top of the Parit Sulong Bridge. This was a Chinese-style bridge with a steep rise to the centre. The Japs allowed the ambulance to approach to the top of the bridge. They agreed to let the wounded through on condition that the rest of the force surrendered. The wounded refused. The Japs then refused to allow the ambulance to return. It stayed on the bridge until dark. The driver, an officer who was himself wounded, released the brake and the ambulance ran quietly backward off the bridge, and kept going until it reached the Australian lines.

Efforts were made to extricate the force. A unit of the 11th

Indian Division was ordered to counter-attack the Japs, but this did not eventuate. The trapped force, under constant air attack, fought on until Col. Anderson received a wireless signal which told him: "You may at your discretion leave wounded with volunteers, destroy heavy equipment, and escape. Sorry unable help after your heroic effort. Good luck." About 350 Australian and 50 Indian troops found their way out of the trap to the British lines. More than 200 were captured in the attempt. Volunteers stayed with the wounded under the Red Cross: all were massacred. This was when the Japanese took 110 Australians and 35 Indians to the edge of a river, and beheaded them one by one.

Lance-Havildar Benedict of the 13th Field Company, Q.V.O. Madras Sappers and Miners, managed to break free and dived into the river. He avoided a hail of bullets by swimming under-water to the opposite bank. He hid in the undergrowth and watched the massacre. When darkness fell, he heard a cry for help. He swam across and found two Indian sappers, one with his head half severed, the other badly gashed. He got both men across the river, and nursed them for fifteen days. They lived.

Lt.-Col. Anderson, the 45-year-old commander of the 2/19th Australians from New South Wales, was awarded the Victoria Cross for "brave leadership, determination and outstanding courage. He not only showed fighting qualities of a very high order, but, throughout, exposed himself to danger without any regard for his own personal safety."

Then there was the story of Lt.-Col. Parkin commanding the 5/11th Sikhs. He was a diminutive man with the drive and heart of a giant. He just would not let his desperately tired men give in. They were part of the lost 22nd Indian Brigade, bearing their wounded on stretchers and pursued by the enemy.

At one point, the greater part of C Company went over to the enemy. Col. Parkin rallied the remainder and told them to stand and fire. They did. When they ran out of ammunition they fixed their bayonets. By that time they were only 100 strong. A little while later they were down to 30, the others having stayed on the ground after the last rest. A man was sent back to rouse the layabouts, but he failed to return. Col. Parkin was heartbroken, and took two of his Indians aside to ask how this could happen to the magnificent 5/11th Sikhs.

It was due to a piece of dirty white cloth. Someone had tied it to the branch of a tree and given it to an orderly to carry, to keep the discipline of march by marking the position of brigade H.Q. They were all so physically tired and mentally numbed, they didn't think what this might mean. To the men following, the bit of white cloth was taken as a sign of surrender, and the men who saw it gave up the fight.

With his handful of Sikhs, the indomitable Col. Parkin struggled on. Those few eventually reached the south coast opposite the Singapore Naval Base, and were taken off by a naval patrol boat.

High praise was given to the men of the Royal Navy, who succeeded in rescuing a large number of troops from the beaches along Batu Pahat, on the west coast of Johore about 60 miles north of Singapore. These were the survivors of the 6th Norfolks and the 2nd Cambridgeshires, the British Battalion of East Surreys and Leicesters, two companies of The Malay Regiment, gunners and administrative units who had made up what was known as the Batu Pahat force. The Jap landing at Batu Pahat had forced the British to withdraw, but another Jap coastal infiltration to the southward had cut the road to Singapore. Attacks to break through had failed, with heavy loss in men and equipment, and the survivors were ordered to make for the coast.

Information of their plight reached Singapore, and the gunboats *Dragonfly* and *Scorpion* set out with a fleet of small boats in tow to effect a rescue. While the larger vessels stood off the shallow west coast, the little boats sneaked in among the estuaries and collected all the troops they could find. In all, more than 2,000 men were rescued in these operations, which continued for five successive nights without the Japanese finding out what was going on.

One of the heroes of this remarkable feat was Lt.-Comd. Victor Clarke, R.N., a survivor from H.M.S. *Repulse*, who was in command of the *Dragonfly*.

Capt. Mike Bardwell, of the 2nd Argylls, had an exciting story to tell. He was among those cut off at Slim River. He joined up with a number of other Argylls, and they made their way to the west coast. They bought a Chinese junk and sailed across to Sumatra. In due course, Capt. Bardwell turned up again in Singapore, almost in rags. Apart from re-joining his Regiment in time to help hold the bridgehead at the Causeway, there was another good reason why he pre-ferred to be in Singapore. Her name was Kate Londun, and they were engaged to be married.

Maj.-Gen. H. Gordon Bennett, commanding the A.I.F. Malaya, had made great friends with the Sultan of Johore. He called on him to say good-bye, as the Sultan was remain-ing with his people.

"He realizes that he will be a poor man in Malaya," wrote Gordon Bennett, "and considers it possible that the Japs might treat him roughly. He said: 'I suppose I can live on rice and fish, like the rest of my people.' He is very annoyed that the British heads of his many State Departments have left him without even saying good-bye. Only two have de-cided to stay with him. He is extremely annoyed that the medical staff at his hospital should have left."

The A.I.F. Commander took a last look at Johore. He recorded in his diary: "Derelict cars and destroyed houses and bomb holes everywhere. There was a deathly silence. There was not the usual crowd of chattering Malays and busy Chinese. The streets were deserted. It was a funeral march.

I have never felt so sad and upset. Words fail me. This defeat should not have been. The whole thing is fantastic. There seems no justification for it. I have always thought we would hold Johore. Its loss was never contemplated."

II

There had been the question of postponing withdrawal across the Causeway for some hours in the hope that the lost 22nd Indian Brigade would appear, but that hope had faded. Now it was the night of January 30, the time fixed for the final withdrawal. What remained of a total force of three divisions formed up and moved in along the road from Mersing on the east coast, from Tebrau in the north down the main trunk road and railway, and from the west coast area on the road through Skudai. All roads led to the bottleneck of Johore Bharu, and on to the Causeway.

The greatest anxiety was whether the Japanese would send their aircraft to pattern-bomb and dive-bomb the only way out, thereby pinning the retreat. Alternatively, they might infiltrate in force and cut the retreat on one or all of the three main routes to Johore Bharu, causing delays which would upset the time schedule and throw the withdrawal into confusion. Johore was a fabulously rich State with an excellent road system which favoured the Jap advance as much as the British retreat. All three routes were vulnerable. The final route across the Causeway was wide open. If the worst happened, and the Causeway was blocked, the Navy had assembled a fleet of boats, the largest of which were two ex-Yangtse pleasure steamers, to ferry the troops across. It seemed too much to hope that the Japanese, with complete air supremacy, would allow the withdrawal to go forward unmolested.

All the retreating traffic was blacked out except for a

shaded pin-point of tail light on each vehicle. The brilliant moonlight helped driving, but hindered secrecy.

"Look at it," said Signalman Fred Mutton of 75 Section attached to 28th Indian Brigade, reproachfully. "Can't even rely on the ruddy moon now. Any other night it would be raining cats and dogs. Now it's worse than ruddy floodlights at Aldershot Tattoo. Roll on when we're past the Skudai cross-roads." . . . He added later:

"Our orders were to form up in convoy and follow the tail light in front. We knew we were going to Singapore. We carried on all through the night. A very rough long drive. It was nose to tail, pitch dark when we started off, everything blacked out but tail lights, which had to be shielded.

"Watching that tail light in front, it was like a cigarette, just a glimmer. It sort of hypnotized you. It was all you had to go by, and you dared not take your eyes off it. Up to then we'd been out every night, laying out signal lines, reeling in lines, so we'd had hardly any rest since Gawd knows when.

"It was a nightmare journey. I was driving the Signal Section truck. I had an Anglo-Indian Warrant Officer in front with me. He had to keep nudging me to keep me awake, talking like mad. It was a terribly slow drive. All normal convoy discipline went overboard. Only a yard separated vehicles. Sometimes we stopped by bumping into the lorry in front. Soon as I stopped, it was a fight to keep my eyes open. If I let them close, I was asleep.

"D.R.s and Movement Control were up and down the column all the time, shaking drivers awake whenever they stopped. All right as long as you kept moving, and someone was there with you to keep talking. We crept nose to tail to the Causeway, and then I got a funny feeling up and down the spine. It was a ghostly effect. The water looked dark, and the Causeway looked like a white strip stretching right away across.

"This Warrant Officer came from Secunderabad, and kept telling me about it from his childhood on. When he ran out of stories about himself, he talked about his comrades, where they came from, what they did. I reckon that W.O. told me the history of India that night, but I don't remember a thing. It kept me awake, though, hearing his voice.

"We were in with the Gurkhas, because the 28th Brigade was all Gurkhas. They're the boys! We used to get on well with all the Indian troops, but everybody had a special soft spot for the Gurkhas. They were a cheerful crowd, always laughing. I always felt safe when they were around. I wouldn't like to be a Jap when the Gurkhas greased their bare shoulders and went out with their *kukri* knives looking for stray sentries. Whoosh—there goes one more Jap. Sliced his head off clean as a whistle. They've got to learn to do that before they pass as Gurkhas. Don't ask me how they learn it, all I know is they know how!

"The big question when we set off was the Skudai crossroads. Our Intelligence Officer—his name was Tiny—he told us: 'Once we get past the Skudai corner, we're okay. If there's going to be trouble, that's where we expect it.' We breathed a sigh of relief when we got clear of Skudai. I was thinking to myself, 'I'd hate to die in Skudai.' Got a lonely sort of name in my remembrance, has Skudai.

"Johore Bharu looked like something out of a nightmare. Big and black and ghostly, with towers silhouetted against the sky, and a sort of unearthly quiet like the dead of Sunday, except for the racket of the convoy. You felt there were thousands of squinty eyes watching you, and all their fingers itching on the trigger. I was glad when we nosed out over the Causeway. The thing that puzzled us was why the Japs weren't bombing hell out of it.

"The Causeway was wide enough, but the way I was going it felt like a tight-rope. I had to keep my eyes glued to the lorry in front. I was so close I couldn't see anything but the outline of this lorry in front. Half the time I couldn't even see the cigarette-glow of his tail light. It was nerve-racking.

"All the time I could hear the snores coming from the back. I'd got the truck loaded with drums of cable and laying gear, and four or five linemen mixed up with it, flat out asleep and snoring like pigs. I wished I could get my head down. I'd got a big tiger-skin I'd found in my travels down country. Can't remember where I got it, I just found it. Came in handy for a blanket. I'd lost all my heavy kit up in Kota Bharu, and it gets chilly at nights in Malaya. It had a big stuffed head, that tiger-skin. The Indians used to admire it.

"I never want another drive like that as long as I live. We eventually crawled over to Singapore Island, and fetched up in Reuter's Field. I've never known such tiredness. I was too dog tired to worry about anything. I just pulled the old tiger-skin round my shoulders and went to sleep in the truck."

Signalman Mutton was among the lucky ones who came out first. A long convoy stretched back into Johore, still grinding along nose to tail through the night.

III

The long night of January 31 lightened in the east. The moon had gone, and now the stars began to dim. The Japanese had not bombed the Causeway. They had not come at all. It was extraordinary that they had not come—they must have known what was happening.

Lt.-Cdr. J. O. C. Hayes, R.N., was Naval Liaison Officer to Col. Stewart, who was in command of the two bridge-heads, during the withdrawal over the Causeway. Cdr. Hayes was in charge of the Royal Navy side of the operation, with parties—mostly survivors from the sunken *Prince of Wales* and *Repulse*—ready to ferry the army across if necessary. They stood by all through the night. Cdr. Hayes recalled:

"All the men were through the outer bridgehead and the outer bridgehead itself was on the move. This was, I felt, the most anxious moment; for if the Australians were in contact with the enemy at the moment they were due to retire, they would have to fight their way back to the Causeway, eventually leaving the Argyll and Sutherland Highlanders—rather too literally—as the Horatius of the bridge. We all breathed easier when we learnt that in no sector were the Australians in contact.

"The rumbling of the mechanized army subsided and died away. The Causeway seemed empty for a little while; and then the infantry began to march across. You couldn't see them very clearly and the tramp of their feet on the asphalt was the only sound in the stillness. It was a tired tramp. It was nearly dawn now, and with the dawn we expected a certain amount of interruption.

"If I keep one single impression, more vivid than them all, of those last days at Singapore, it will be of the dawn of that day. To the east, where the sky was lightening, our two river steamers were the only movement on a sheet of glass. Below, you could see the khaki forms now, plodding across the bridge; and over the town of Johore itself the full moon was setting among the buttresses of the Sultan's palace. The sky was like that of a cheap postcard which you think is impossibly coloured.

"And then, as the Gordons marched out on to the Causeway, a single piper began to play. You could hear every note. It covered the tramping of feet. I didn't leave the terrace until the sound of the pipes had died away. The Gordons had passed. It was daylight now. I signalled to the boat points 'Finish'."

After the Gordons came the two battalions of Australians. The outer bridgehead was safely across, and now Maj. Angus Rose signalled "Finish" to the Argylls. The Battalion, numbering 880 all ranks in December, had returned to Singapore Island after the Slim River battle 90 strong. All the cooks and bottlewashers at the Argylls' base camp at Tyersall Park had rushed to fill the ranks, and they made up to 250 strong for the inner bridgehead role. Tomorrow they would be joined by volunteers from the Royal Marines of the *Prince of Wales* and *Repulse* survivors, and they would be 400 strong . . . and they would be called "The Plymouth Argylls".

Their present task completed, they marched across the Causeway led by their two remaining pipers—Stewart and Maclean—for all the world as if they were a hundred pipers and all. They had played the Australians across with "Jenny's Black E'en" and "Bonnets Over The Border", and now they piped themselves across with "Hielan' Laddie".

The last man to leave the mainland was Col. Stewart of the Argylls. He had just been told that he had won the D.S.O. for his part in the Malayan campaign. His comment was characteristic: "It's for the Regiment."

Col. Stewart wrote of the crossing of the Causeway: "The unexpected had happened. A withdrawal of more than 30,000 men that ought to have cost us dearly had been carried out without the loss of a single casualty."

When they were all across, Lt.-Col. Muir, commanding the 15th Company Q.V.O. Madras Sappers and Miners, passed the signal to breach the Causeway. Plans for this operation had been completed earlier, and the demolition was regarded as the most important in the campaign.

At 8:15 on that morning of January 31, the sound of a great explosion was heard as far away as Singapore City on the south side of the Island. Men who saw the explosion described how steelwork, rails, rocks, water mains and lock gates hurtled to the sky. "That should stop the little bastards," said one man; and authority probably shared his view, for when the smoke cleared, the sea was flowing through a gap of 70 feet.

The same day, Lt.-Gen. Tomoyuki Yamashita, who had been detailed by the Japanese Imperial General Head-quarters to conquer Malaya, reached the south coast of Johore with the Imperial Guards and two other divisions. From his headquarters in Johore Bharu he could look across the sunlit waters of the narrow straits to his next goal, the Island supposed to be impregnable.

The siege of Singapore had begun.

And one of the first things "Tiger of Malaya" Yamashita learned was that the water in the gap of the Causeway was not more than four feet deep at low tide.

7

The Siege Begins

"The battle of Malaya has come to an end and the battle of Singapore has started. . . . Today we stand beleaguered in our island fortress. Our task is to hold this fortress until help can come—as assuredly it will come . . ."—From a public announcement by Lt.-Gen. Percival, G.O.C. Malaya, February 1, 1942.

Singapore awoke on the morning of February 1—and was astonished to find itself besieged. The secret of the British withdrawal into the Island had been kept so well that most of the civilians and many of the garrison troops were unaware of it until they woke up nearly twenty-four hours later and were told that the Causeway had been breached and that the Japanese were across the narrow water in Johore. No wonder the man in the street was bewildered by the speed of events. Even Mr. Glover did not know. As the head of one of the English-language newspapers and in contact with authority, he might reasonably be expected to have heard a whisper of it; but he had received nothing more than mysterious hints from highly-placed friends and suggestions that it would be a good idea if he allowed his wife Julienne to leave the Island. "Singapore is no place for women now, Jimmie," said his friend Mr. Loder-Waters. "If the Government have bungled evacuation I see no reason why I should keep Pauline here to face the music. We are going to have a rotten time. Take my tip and get Julienne out of the way." Mrs. Glover had refused to go.

At midnight on the Saturday Mr. Glover had telephoned the night-editor of the *Sunday Tribune* for the latest news. The night-editor had read a short official communiqué, released through the Press Bureau, which gave Saturday's casualties from air raids on Singapore—"a formidable total amounting to hundreds". It contained nothing about Johore or the Causeway. Mr. Glover had thanked his office and asked to be kept informed of "anything special".

Mr. Glover did not go to bed. He sensed that something important was happening or was about to happen, for he could hear—for the first time—the sound of heavy gunfire from the mainland.

At one o'clock his telephone rang. It was the night-editor. The latest communiqué had just been received, "as brief as it was alarming." It read: "Last night, in accordance with our prearranged plan, our forces which had been in operation in southern Johore were withdrawn on to the Island of Singapore. The enemy made little effort to interfere with this operation. The Johore Causeway was successfully breached, the Royal Navy and the R.A.F. co-operating."

"There was little sleep for us during the rest of the night," wrote Mr. Glover in his book *In Seventy Days*. "Even in the seclusion of our air-conditioned bedroom we could plainly hear the angry noise of heavy guns announcing to the whole community that the war was upon us."

His comment reflected Singapore's peculiar mental isolation from the storm of world upheaval. The war had been upon them since 1939, and more particularly since 1940 when the Japanese had occupied French Indo-China.

"Let's go to the market and do some shopping before the raids start," said Mrs. Glover. They were up and dressed soon after five o'clock. Their house at Holland Village was five miles out of town. It was still dark when they drove along the Holland Road past the Botanical Gardens and into the city.

The Singapore Cold Storage in Orchard Road had not yet opened, so they went to the municipal vegetable and fruit market. Under normal conditions the market would have been a hive of industry, men and women of all ages and nationalities jostling and chatting as they bargained among the heaped stalls. Early morning marketing was the custom in Singapore. It was usually done for the Europeans by their Chinese servants.

This morning the market had a gaunt look. The stalls were practically empty, the place almost deserted. A few blacked-out candle lamps flickered here and there. The Chinese stall-holders were eager to get rid of their produce and get away before the air raids started. They knew what was happening. For all its big significance, Singapore was a very small island. News travelled fast along the grapevine. Long before the Sunday newspapers appeared, those stall-holders knew that the Japanese were in Johore Bharu.

While his wife searched the stalls, Mr. Glover pondered the new situation. A few days earlier he had sought a heart-to-heart chat with Mr. R. H. Scott, head of the Far East Bureau of the British Ministry of Information.

"At the present rate of progress," Mr. Glover had said, "the Japanese will be at Johore Bharu in a fortnight."

Scott's answer had startled him. "They may be here this week-end. We've got to face up to the fact that they are stronger in every way than we are, and the best we can do is to hold them off until help arrives."

"How long have we got?"

"Four weeks, maybe three. I think we can do it. The Japs know as well as we do that our real defences are on this island, and they know they will find them a tough proposition to break down."

Even at that late date Mr. Scott apparently thought—or

allowed the editor to think—that Singapore's real defences were on the Island.

As Mr. Glover waited for his wife to finish her shopping, he watched the coolies laying big concrete-spun pipes along the city street as makeshift air-raid shelters. "A little late in the day," he thought. "Our one-man Government is always late. Please God the reinforcements will not be late."

Mr. Glover had been in Malaya since 1927 and had seen the building of the great Naval Base. "Hills had been removed, a river deflected and huge swamps converted into dry land," he wrote. "The Base was regarded as the greatest naval-equipment station in an area that stretched from the Pacific coast of the United States to the shores of Great Britain, and was certainly the greatest naval project undertaken by the British Empire for a quarter of a century. In this matter at least the British Government had turned a deaf ear to all the pleas of the pacifists. While the actual strength of its fortifications was a closely-guarded secret so far as the British Press was concerned, we believed that the Base was the greatest arsenal of democracy in south-eastern Asia. It was reassuring to contemplate the power of the big guns, and the havoc they would be able to cause should a landing be attempted. The enemy could only hope to starve us into submission, and to make life difficult and hazardous by intensified air attack. We had more than enough food for a six months' siege."

Mr. Scott had said that if Singapore could hold out for three weeks, sufficient reinforcements would arrive to lift the siege. He was head of Information, so he should know. Surely it would not be difficult to hold out for three weeks?

Continuous day and night air raids were telling their tale. The life of the city was slowing down perceptibly. The streets were avoided by all who could stay from them, but the municipal services were functioning as near normally as conditions permitted, while the volunteers in the Nursing

Services, the A.R.P., the Auxiliary Fire Service, the Special Constabulary and other relief organizations were sticking to their posts with grim determination.

Mr. Glover knew that the main water supply was piped across the Causeway from Johore. A hole in the Causeway would obviously put a hole in the water supply. Singapore would have to rely on the supply from the two small reservoirs on the Island itself. Used carefully, this supply could be made to last probably two months. (In fact, by stringent economy, and without wastage, they could supply the island indefinitely.) He made a mental note that his newspaper would have to "plug" the water situation in Monday's editorial. He would get one of his ace reporters, perhaps American Mrs. Freddie Retz, to write a special article on how to save water in the household.

By the time the Cold Storage was opened for business, a crowd waited to buy the better-class food which came from all over the world and was kept fresh in the refrigerators. Here was where the Europeans bought fresh meat. Before the days of air raids, shopping at the Cold Storage was one of life's pleasanter things. You could wander about the attractive glass shelves for an hour or so without getting bored. There was still any amount of canned food in stock, but all the fun had gone out of shopping. One had to be content nowadays with plain square household bread. They had stopped baking the popular French bread and dainty pastries. Milk was still delivered to outlying suburbs by boys on push tricycles. The Cold Storage dairy herd was one of the outstanding domestic features on Singapore. Butter was rationed, but there was plenty on sale today. The sudden exodus of 6,000 butter-eating European women and children made a big difference.

A week ago there had been a near panic in the shop. It was during an air raid, and one of the bombs landed just outside the back wall where the refrigeration plant was housed.

Damage caused ammonia fumes to escape into the basement where a crowd of shoppers had gone for shelter. "Gas!" yelled a frightened woman, and there was a wild scramble to get out.

There was no dawdling over shopping this morning. People came with lists, bought what they needed, and went home as fast as they could.

It was daylight when the Glovers drove home to Holland Village. The streets were busy with military traffic, tearing along at breakneck speed. Guns were positioned in unexpected places with teams of gunners trying to conceal them under camouflage nets and branches hacked from the nearest trees. Vehicles were parked under trees in suburban gardens, and more guns looked absurdly out of place in backyard orchards.

The muffled growl of gunfire in the north had stopped, but soon the bombers would be over in their undisturbed formations of 27 raining death and destruction below. Heavy smoke was rising from the Naval Base, where several of the big oil tanks had been set on fire in last night's air raids.

It seemed incredible that these disturbing things were really happening on Singapore, so fresh and green and peaceful when viewed from the quiet house on the hill at Holland Village this balmy Sunday morning. In one direction lay the valley across to Bukit Timah, where the dark brooding hills of Johore filled the northern background. Eastward stretched the vivid panorama of undulating bright green dotted with white and gaudy-coloured rooftops pitched and domed, with tall white office buildings marking Singapore City. Westward lay the jungle around the River Jurong. Southward lay the flat calm silver sea, whose limit was lost in hazy horizon where sea and sky merged as one. This would be a perfect day to set off in a motor-boat or a yacht for a picnic on St. John's Island, where one could sit on the rocks and gaze down at the clear colourful marine life two fathoms deep.

On the other side of the Island, at Reuter's Field, Signal-man Fred Mutton and his cargo of linemen and equipment had had no time to go shopping. On arrival after the retreat across the Causeway Fred had fallen asleep over the driving wheel. Cramp woke him up. It was still dark and he was still dog tired. He smelt the balmy night air, climbed down and went to the back of his truck. The linemen were still stretched out among laying gear, snoring and grunting. Fred shook his mate. "Come on, mate, we've arrived. It's a perfect night. Let's get our heads down proper. Stretch out on the ground on my tiger rug instead of being humped in with all that there gear."

He said later: "We no sooner get our heads down than bang! smack! Caw love a duck, what's that? Then we wake up. We're in the middle of the artillery. We're thinking we've been brought out of Johore for a rest. We reckon the reinforcements have arrived. Us and the Gurkhas have been brought out to rest and re-form while the new mob take over, then back we go and have another bash.

"So here we are supposed to be having a decent night's kip, and we're in the middle of the artillery. We see pop pop pop in the distance—gun-flashes coming from Johore. Firing at us. That shook us.

"Our artillery opens up with a perishin' roar just behind us . . . then the Japs come across with a creepin' barrage trying to find our guns. I says to my mate, 'Well, that's our perfect night's sleep gone for a Burton.' And he says, 'I was fast asleep in the lorry. What the hell did you wake me up for?'

"Then it got light, and bang! This time it's the Causeway being blown up. We guess the whole army must've been with-drawn and it's a siege. We haven't stopped to think much about anything. No sleep . . . you can't keep going forever. On top of the rest, the darkness of the drive made us more tired.

"Well, before we know where we are, the Japs are out on the Causeway having a look at the damage. The A.I.F. machine-guns open up, and the Japs disappear. Time for breakfast."

Mr. and Mrs. Ted Fozard awoke on the Sunday to see a column of heavy black smoke rising from the north. Although they did not know it, the oil tanks had been fired. They heard the heavy rumble of gunfire from the north and felt the earth-shaking crumps that followed the drone of the bomber engines.

"I don't know what is happening," said Mr. Fozard, "but I don't like the looks of it. I wish to God you had gone on one of those boats, Peg. You ought to be hundreds of miles away from here by now."

"Well, it's too late now," she answered. Not for worlds would she admit that she was terrified every time there was an air raid. She was determined to stick it out as long as the powers that be would let her. Like her husband, she was Yorkshire and proud of it.

When Mr. Corbet, managing director of Borneo Motors, heard the news on Sunday he was extremely glad he had persuaded his wife to go away in the *Duchess of Bedford*. He now contemplated joining the Air Force. It was no use going on pretending "business as usual".

Out walking, he met Brig. Paris, who had been placed in command of the 11th Indian Division when Maj.-Gen. Murray-Lyon was relieved after the Jitra disaster. In turn, Paris had been relieved after the Slim River disaster. Brig. Key, now Major-General, commanded the long-suffering 11th Indian Division, and Brig. Paris had reverted to command of the shattered 12th Brigade.

"Hallo, Archie," greeted Mr. Corbet. "Glad to see you've survived. What are you doing now?"

"I've been put in charge of plans for defence against landings from the mainland," said Brig. Paris.

"I suppose you have dug out all the plans filed over the past fifty years and settled on the best one?"

"There aren't any plans," said Brig. Paris. "Nobody has ever formulated a plan for the defence of the Island against attack from the mainland. It has been talked about in a vague sort of way, but never with any serious idea that attack from that quarter was at all likely. Or if anybody did think it was possible, nothing was done."

The civilian Corbet listened to this revelation with dismay. "What is going to happen?" he asked.

"The Japs will bring their heavy equipment and launch an attack."

"When?"

"I should say in about six or eight days from now. They can't make the necessary preparations in less than a week, and they are not likely to waste any more time than they can help. At the most I should say ten days."

"What about our side?"

Brig. Paris was not hopeful. "It depends entirely on how the men behave. If the old soldiers will fight, we stand a chance. If the new soldiers fight, we stand a chance. If neither of them fight, we've had it."

"Good lord! Well, you have been in the thick of it with them. Do you think they will fight?"

"I know some who will. For the others, sufficient unto the day."

Statements like these—and the items of news which crept through Singapore—deepened bewilderment into uneasiness. Word went round that the Naval Base had been abandoned. Everyone knew that most of the aircraft had gone—there was only the token force of fighters left—and the headlong retreat down the peninsula had done nothing to bolster morale. People—civilians and military—felt confused. Worse, they felt insecure. The pompous pronouncements of the past fifty-five days, the blanket of censorship, the indecision in high quarters over such matters as labour compulsion, evacuation and civil defence, the feeling that no single mind was controlling events; and above all, the air raids—all these things helped to add to the mounting fear.

Authority was aware of the fear, spreading and menacing. Efforts were made to counter it by more exhortations. Lt.-Gen. Percival, who knew that Singapore was not a fortress, issued his pronouncement. It said:

"The Battle of Malaya has come to an end and the Battle of Singapore has started. For nearly two months our troops have fought an enemy on the mainland who has had the advantage of great air superiority and considerable freedom of movement by sea.

"Our task has been both to impose losses on the enemy and to gain time to enable the forces of the Allies to be concentrated for this struggle in the Far East. Today we stand beleaguered in our island fortress. Our task is to hold this fortress until help can come. This we are determined to do.

"In carrying out this task we want the help of every man and woman in the fortress. There is work for all to do. Any enemy who sets foot in our fortress must be dealt with immediately. The enemy within our gates must be ruthlessly weeded out. There must be no more loose talk and rumour-mongering. Our duty is clear. With firm resolve and fixed determination we shall win through."

The Jap radio was broadcasting that the Governor and his wife had fled to India, and the rest of the British were preparing to leave the Asiatics to their fate.

The Governor went to the microphone to state a denial. He broadcast: "Today we begin the battle of Singapore. The first thing we must be determined is that we are going to win it. In the days that are coming, history will be written. It is for all of us—men and women alike, to ensure that it will be a glorious chapter.

"The G.O.C. has told you that our forces are here and will stay here, to carry out the task of defending the Island and to fight until victory is won. For the rest of us, here we are and here we stay, each of us to do our bit. This is total war, in which the whole population is involved, and there can be no question of some standing aside and leaving others to do the work. It will be grim, no doubt, but no more grim than in Britain, Russia and China, and if the people of those countries can stand up to total war, so can we.

"Let not the Asiatic population of the Island imagine that one day they will find themselves abandoned. That will never be. Europeans, Indians, Chinese, Malays—we all stand together side by side, shoulder to shoulder. We will continue to try and send away as many women and children as may wish to leave, whatever their race, and I am glad to say that the Government of Australia has opened the doors of that great country to Chinese and Eurasians, subject to a guarantee of good repute, which it will be my Government's business to decide.

"Do not be alarmed. There must be no panic, no spreading of rumours, no defeatist tongues, no slackers, but steadfastness, redoubled energy and iron determination to win.

"Within the last few days, substantial reinforcements have been received. They are proof, if proof were needed, that Singapore is intended to be held. All we have to do is to hang on grimly and inexorably, and for not very long, and the reward will be freedom, happiness, and peace for every one of us."

Mr. Glover commented later: "Foolish promises and empty words. How foolish and empty we were to discover all too soon."

"We all stand together, shoulder to shoulder." To the gullible—if there were any left on Singapore Island—it gave the impression of men of diverse race and rank united in a common cause. Unfortunately it did not match the facts, of

which many people were now aware. People wondered why
Mr. S. W. Jones, the Colonial Secretary under the Governor,
had departed in the middle of January. Some people said
Jones was a pessimist and a defeatist—but many members of
the Malayan Civil Service signed a letter to the Governor
expressing their confidence in Mr. Jones. The *Malaya
Tribune* came out with a full page protest that S. W. Jones
had been made a scapegoat for the inefficiency of others.

There was disunity between the military and the civilians.
Advice tendered by the military to the civil authority had
been frequently ignored or challenged; on the other hand,
the military hated any interference from the civilians. There
was also the feeling that all was not well among the members
of the High Command. There were some who said that one
of General Percival's difficulties was the fact that General
Heath, commander of 3rd Corps, who had come to Malaya
with a background of successful operations in Ethiopia, had
found himself subordinate to the G.O.C. Others murmured
about the number of generals who were "cluttering up" the
Island. Intelligent men thought it was farcical that Singapore
should possess General Percival's Malaya Command, General Simmonds's Fortress Command, Heath's 3rd Corps,
Beckwith-Smith's 18th British Div., and Gordon Bennett's
A.I.F. Command. Percival, the G.O.C., was a three-star general; so was Heath, serving under him; and there were four
two-star generals—Gordon Bennett, Simmonds, and two
divisional commanders. With the generals went a multitude
of staff officers.

That there were differences is not surprising—it would
have been sensational if there had been none. There were, for
instance, those for Brig. Paris and those against. In Col.
Steedman's opinion, "I had no faith in his command of the
11th Indian Division. When he was relieved of the command
and Billy Key came, it put everybody's pecker up at once.
Paris was in charge of allocating people to the defence. He

had no belief that the Island could be defended, and he let it be known."

Again, Rear-Admiral Spooner—and his wife, who was caustic about all those women who had left Singapore—believed that the Island could be held. "I met the Admiral," said Col. Steedman, who had been given the job of denying the Naval Base to the enemy, "and he was still talking of the place being relieved and the Japs being driven back, and therefore we must not do too much demolition. That was the Singapore outlook. I pointed out that we were in the front line and that if we failed to do preliminary demolitions immediately, we might be caught on the wrong foot and not be able to do very much at the last moment. These things have to be prepared, you know."

All was not well among the troops. There were the disappointed, who had imagined that they were being withdrawn to Singapore in order to rest and refit; but found themselves still in the front line. There were the exhausted, who had believed that Singapore was a strong fortress; but found that they had to go out, weary as they were, and dig slit trenches and put up barbed wire and lay communications—the thousand and one jobs that must be done before a position can be counted properly defended. There were raw troops and untrained troops, and rivalries born out of national differences. There were the Hyderabads, who had mutinied shortly before the invasion of Malaya, and been broken up and spread among other units. There had been trouble, too, with the Bharwalpures. There was a deep-rooted mutual detestation between Gurkha and Sikh. There was antagonism between British troops (especially the Argylls) and the brash younger Australians, which at times deepened almost to the point of feud. It was said that some of the Australians had behaved tactlessly when they first arrived in Singapore, saying to the established fighting men "get out of the way—we'll show you what to do". Their idea of being tough was to go to some café where the waitresses were Chinese girls, get fresh, and then disappear when the Redcaps came along. But it was with the Argylls that these Australians (not the Aussies of the 29th Division who were good troops and excellent fighting men) fell really foul. Even before the invasion they had been at each other's throats; afterwards it got worse. One particular bone of contention was the Argylls' special preserve, the Presbyterian Services' Club in Orchard Road. The Argylls resented the intrusion of any other troops. They would tolerate men of the Manchesters and Loyals, but "foreigners" like the Australians were regarded with clannish suspicion. When the Australians came in—though they had their own club, the Anzac—they were quickly sent into a corner while hostile Argyll eyes followed them, their owners ready to doff bonnets and wade in.

Many of these divisions and differences were human and therefore inevitable. They occur in all aggregations of men; and if the war had been fought successfully, nothing would have been heard of them. They would have disappeared under the surge of events and served only as amusing memories between rivals. But when a war is not fought successfully, when nerves become strained, when mistrust deepens, when the fighting strength of any army is sapped by failure, minor differences and genuine rivalries become magnified into resentment, with consequent disunity.

The differences were there, spreading and poisoning the "shoulder to shoulder" stand which should have been the ideal. Mr. Corbet of Borneo Motors expressed a civilian view when he said: "In your heart you had no confidence in your own troops. There had been this retreat all the way down and nobody had stood up to the Japs. You knew it was pretty well all over. You couldn't see any chance of getting away from the Island."

II

Lack of confidence in the leadership was heightened when troops and civilians learned that the £63,000,000 Naval Base —the *raison d'être* for the campaign in Malaya and the focal point of British strategy in the Far East—had been abandoned, and demolitions begun, even before the troops had been withdrawn from the mainland. It surprised General Gordon Bennett, who wrote afterwards: "I was surprised to hear that the naval dockyard had been denied. We were here to defend the Naval Base rather than the city of Singapore. This demolition of the docks even before we withdrew from the mainland reflects the lack of confidence in our cause. The morale of the men is undoubtedly affected when they find demolitions going on behind them. It is an admission of defeat."

Signalman Fred Mutton was surprised. "The worst part," he said, "was that we heard the Navy had gone and they had started blowing up the Naval Base—at least, the floating dock had been sunk in Johore Strait. We thought they'd fired the oil wells, then we heard a rumour that it was Jap bombs. A dirty big cloud of smoke was pouring up and stretching away down south.

"My mate said, 'What the hell? Time we were getting weaving. We're supposed to protect the Naval Base, now they've blown it up and ratted, so—what are we waiting for?'

"Then our Skipper came up and said for us to take the lorry to the Base and get any spare stores we could find. The Navy's evacuated, he said, so it's a fair scrounge.

"I'll say it was. I'd worn my boots out. Then I'd found some brown shoes in a kampong up-country. Half a size too small but I was that pleased to get them I didn't mind being a bit crippled. It's murder with no boots.

"We got to the Naval Base and there were a few matelots knocking about. They were loading little boats. It was open shop for anybody who liked to go there—us, of course, not anybody. We loaded our trucks with tins of bully. I got three dozen pairs of shorts as well as new boots and socks. They were a blessing—I hadn't changed my shorts since I dunno when. Why three dozen pairs? Well, I got mates and all."

Ian Morrison visited the Base and described it as "my most tragic memory of the whole Malayan campaign".

"There were some Indian sentries at the gate," he wrote, "but they did not bother to ask us what we wanted. The barrack-like buildings just inside the gate which used to house part of the labour corps of 12,000 Asiatics were completely deserted. A little further on were the headquarters of the Naval Base police, who had been disbanded several days previously. An area of one acre outside the office was knee-deep in their uniforms. Shirts, turbans, truncheons, gas-masks, leather belts, wooden lockers, were lying about in a wild mess. . . . We walked through the various machine-shops and storehouses. Here were the great furnaces where huge blocks of iron and steel could be forged and rolled, enormous hydraulic presses, vast troughs into which the molten metal was poured. Here were lathes of many types. . . . Here was the huge boiler-shop, with great boilers still there. . . . Here was the storeroom for aeronautical equipment, spare floats for seaplanes, propellers, struts, pieces of fuselage and wing. Here was the storeroom for wireless equipment, rows and rows of shelves with every conceivable form of wireless device and gadget. Stacked on the floor were hundreds of boxes of those large electric bulbs used in radio work. Towering up into the sky over the storerooms and machine-shops was the great crane which could lift an entire gun-turret out of a battleship. . . . Just off-shore was the floating dock, its upper works protruding above the surface of the water, the great floating dock which could accommodate a

45,000-ton battleship, towed all the way out from England
by Dutch tugs. It had been dynamited. . . . We struck in-
land up another long line of storehouses. One was filled with
huge coils of rope and hawser and wire and cord. . . . The
Administration buildings . . . were completely deserted.
. . . This was the brain-centre and nerve-centre, not only of
the Navy but of the entire British fighting organization in
the Far East. . . . Now it was completely bare. . . ."

Abandoned, empty, dead. . . . It seemed crazy. Morale
fell with a bump—and it was in an effort to restore it that
Percival issued his public statement of February 1. Later, on
the 5th, he called a Press conference and tried to explain
why the Base had been evacuated. His explanation was not a
great success. Had he told the full story of the confusion behind it, morale would have slipped still further.

The Chiefs of Staff in London had signalled the Admiral
on January 21 that he must evacuate as many as possible of
the skilled personnel in time if Singapore were beleaguered.
On January 28, accordingly, Rear-Admiral Spooner transferred the entire European naval and civilian dockyard staff
to Singapore City, and on the 31st he sent the greater part of
them to Ceylon. He kept a few personnel in the city to maintain local defence facilities and to give technical advice to
the Army in the work of demolition. Says *The War Against
Japan:* "No doubt the Admiral wished to seize the opportunity of evacuating valuable technicians while shipping was
still available, but the hurried evacuation of the Base left an
unfortunate impression in the minds of many soldiers who
did not know that the Admiral, although perhaps precipi-
tately, was acting under instructions." Rear-Admiral Spooner
also failed to discuss with General Percival his decision to
hand over scheme "Q", the denial plan, to the 11th Division;
neither Headquarters Malaya Command nor 3rd Corps H.Q.
was told. Lt.-Col. Steedman, commander of the 11th Division sappers, knew. He was met on February 1 by Col.

A. M. L. Harrison, 11th Division G.S.O. 1, who had just come from Malaya Command.

"I've got a present for you," said Col. Harrison, and handed Steedman what looked like two volumes of the London Telephone Directory.

"Great Scott! What the devil is this?" asked Steedman.

"They seem to want you to demolish the Naval Base. Here is the scheme in detail."

Col. Steedman flipped over a few pages incredulously.

"By now," he said, "it was rather a different pair of boots to what had been visualized in those volumes. It was obviously quite beyond us. I had a Company there, and we were allowed to do a certain amount of preliminary demolition. We went round with hammers and broke up the work-shops with the help of some naval ratings. We were not allowed to do any major demolitions."

Certain work had already been done. It had proved impossible to tow the huge floating dock away from the Johore Strait, so it had been scuttled to clear the line of fire across the Straits; all navigable boats and ships and the smaller floating dock had been sailed to Keppel Harbour on the south of the Island, the pumping machinery of the King George V graving dock had been destroyed, and on the 5th the caisson was wrecked.

In a signal to London on February 2, Percival told the Chiefs of Staff that if he carried out complete denial work immediately, its effect would be a collapse of morale so alarming that it would seriously prejudice his chances of holding the Island. He also told the Chiefs of Staff that he had been given two jobs—to hold Singapore to the last, and to ensure that the scorched earth policy would be complete. He could not do both. In the event, he was unable to do either.

8

The Vital Eight Days

". . . substantial reinforcements have been received. They are proof, if proof were needed, that Singapore is intended to be held. . . ."—From the Governor's broadcast, February 1, 1942.

The eight and a half days from the morning of January 31, when it was known that the last of the mainland troops had been safely withdrawn to Singapore Island, to the evening of February 8, when the Japanese launched their attack across the Straits of Johore, were vital for the defence of the Island. It was known that it would take the enemy several days to mount his invasion attempt. Brig. Paris had estimated six to eight days. General Percival expected it to be "at least a week" according to his dispatch of 1948.

For the defenders the issue was now much more clear cut. Instead of weary battling in jungle and rubber down the length of Malaya—opposed to a phantom enemy who lived off the land, struck from flank and rear, and used tanks wherever he could—the foe was, temporarily at least, just another soldier. To gain his objective he now had to breach a siege defence. He had to force his way across a stretch of water, and while he was on the water he would be more vulnerable to attack. If he could be kept off the Island, then life in Singapore, though it would not be pleasant against shelling and bombing, would be no worse than siege conditions anywhere in modern war. To gain his objective, the enemy would have to blast the defenders out of every strong-point, or reduce them by starvation and thirst. Eventually the defenders would be so weakened by continual battering that, in the absence of relief or reinforcements, an invasion would inevitably succeed. But time was the vital factor. It is said that General Wavell wanted five weeks to give him a chance to relieve or reinforce the Island. It is now known that the Japanese had planned to be at Fort Canning, the Headquarters of Malaya Command on the Island, in 100 days. They succeeded in seventy—and the thirty days they gained made up the vital month.

In their explanations of the fall of Singapore after only seven days' fighting, the generals make a great deal of capital out of their difficulties. It is now tacitly assumed that nothing could have prevented the collapse of Singapore in those few days, given the initial weaknesses in the defence which had existed before the invasion began. This assumption overlooks the vital issue, which is not whether Singapore could have been held indefinitely, but whether Singapore could have made a better fight of it. It was the suddenness of the collapse, more than the collapse itself, which shocked the Commonwealth and the United States. The test of this assumption is merely to ask the question: If Singapore had held out without relief or reinforcements until (say) April or May of that year; if it had been conquered only after bitter house-to-house fighting; if it had surrendered only when the defenders were too weak to hold their guns, when disease was rampant, when food, water and ammunition were almost exhausted; if these things had happened, would Singapore still lie dishonourably across British history? Churchill labelled Singapore 1942 as "the worst disaster and largest capitulation in British history"; and in a Secret Session speech to the House of Commons in April of that year he stated, "After five or six days of confused but not very severe fighting the army and fortress [note "fortress"—though he knew it wasn't] surrendered . . . it does not seem that there was very much bloodshed." That resentful description alone cried out for the true answer of a Royal Commission.

Because of the false impression created throughout the world by the Great Myth, that Singapore was an "impregnable fortress", people compared its inglorious sudden fall with sustained stubborn defence of hopeless positions like Malta, Tobruk, Corregidor, Stalingrad; or, looking back, with Ladysmith and Mafeking, and beyond to time immemorial. Even when the besieged have been slaughtered or forced to surrender, their conquerors have applauded their steadfast courage.

The comparison is unfair, because the conditions and most of the factors at Singapore were entirely different. Malta and Corregidor had the natural advantage of being isolated and rugged; Tobruk and Corregidor had the advantage of being military positions unhampered by civilian problems; Stalingrad had the advantage of being held by one race of people with a single fiercely-faithful purpose—fight to victory, or die for Russia.

Nobody wanted to die for Singapore—which was neither fish, fowl, nor good red herring, but a bastard money-making mixture called "the melting pot". Every separate section of the community was itself divided by aim, race, politics and religion. Every section was selfish, unco-operative with and jealous of the others. East was East, and West was West, and the twain did not meet except to exchange dollars or back horses. From the point of view of solidarity in time of war, it was the worst possible place in the world Britain could have chosen as the site for a great operational strategic keypoint. Sydney, the originally proposed alternative site, would have been far safer; but jealousy by Britain's Service Chiefs ruled out Sydney—they feared they might lose too much control to their Australian cousins!

Yet with all its disadvantages, Singapore fell too soon. Why?

First and foremost, because it was the tail-end of a disease which had allowed the manhood of great nations to dodge world responsibility and remain impotent, disarming, appeasing Italy over Albania, Japan over China, and *Deutschland Uber Alles* at Munich.

Second, it is necessary to consider briefly the conditions of siege warfare, examine the factors governing the Singapore defence, and then judge how far short of ideal the Singapore defence fell—how far the failure was due to basic weakness, and how far it was due to failure by the individual defenders.

II

The first important difference between siege and open warfare is that everything and everybody within the beleaguered area becomes a potential asset or liability. In open warfare, where opposing armies push the battlefront this way or that way, civilians are merely an incidental part of the country traversed. Once the armies have moved on, civilian survivors —who had either fled elsewhere or gone to ground—cautiously re-emerge and continue as best they can with peaceful living. On an island besieged, flight is impossible: there is nowhere else for the mass of civilians to go. They can only wait and tremble—as did most of the humble native population of Singapore, *who were not allowed to bear arms even if they had wanted to or knew how to*—or fight as civilians and die like the soldiers.

In the famous false Order of the Day issued in Singapore on December 8, 1941 by Air Chief Marshal Brooke-Popham and Vice-Admiral Layton were these words:

"Let us all remember that we here in the Far East form part of the great campaign for the preservation in the world of truth and justice and freedom; confidence, resolution, enterprise and devotion to the cause must and will inspire every one of us in the fighting services, while from the civilian population—Malay, Chinese, Indian, or Burmese—we expect that patience, endurance and serenity which is the great virtue of the East and which will go far to assist the fighting men to gain final and complete victory."

Patience, endurance and serenity! What the civilian population needed was rifles, bullets and bayonets, and a little bit of instruction in how to use them. The excusists said there weren't enough rifles, bullets and bayonets—Britain could not spare them. It was a lame excuse. Singapore could have manufactured them, if a start had been made in 1939 when the British Empire went to war.

The truth of the matter is that Sir Shenton Thomas and his administration were more interested in maximum production of Malayan rubber and tin than they were in defending Malaya or building defence works round the place they called Fortress. Moreover, the authorities seemed more afraid of arming the Malayan Asiatics than they were of the menacing Japanese. They were afraid that if the Malayan Chinese were armed, the balance of commercial power in Malaya might be taken out of British hands, swamping the Malays whom Britain was bound by treaty to protect. Because of those factors, the mass of civilians was left as helpless as children caught up in the tide of war. By the time the siege began, all but the Singapore Malays were swamped, and it was too late to arm the Singapore Chinese. Most of the terrified labour had fled to the rural areas and could not be regimented. But for a few thousand volunteers of all nationalities who had taken the war seriously before it reached Singapore—who learned how to fight fires, clear débris, rescue wounded, drive ambulances and give first-aid—conditions on the bombarded island would have been even more chaotic. So much for the civilian potential.

For the military, the task was simple—to prevent the enemy from landing his troops on the island. The method was not so simple, because half the island coastline faced the narrow Strait of Johore, and who could predict at what point the wily Japanese would attack? The initiative lay with the enemy: the defenders had to stand guarding all points of the compass, waiting for attack. In the words of the G.O.C. Malaya, "Any enemy who sets foot in our fortress must be dealt with immediately." That was the intention. The next sentence in Lt.-Gen. Percival's Public Announcement that sunny First of February had the effect of planting a large-sized unexploded bomb behind everybody's back. "The enemy within our gates must be ruthlessly weeded out," he stated.

Every fighting soldier on Singapore Island—excepting the two battalions of The Malay Regiment—had a fixed belief that the whole of Malaya from Siam to Singapore was riddled with Japanese Fifth Column. The G.O.C.'s announcement confirmed their belief. It must be remembered that every fighting soldier in Malaya, except The Malay Regiment, was on foreign soil—men from the British Isles, from India, from Australia. They had no great love for Singapore, no devotion to Malaya Command. All they had was pride of race, pride of regiment, loyalty one to another. And some of them did not even have that.

As for the unarmed simple-minded three-quarters of a million native population, most of whom had come from Mother China in search of dollars, every section turned a suspicious eye on every other section. Who was "the enemy within our gates"?

If the G.O.C. had racked his brains for a year in search of a way to undermine the morale of the troops he was leading, he could not have devised a more deadly weapon. Every "foreign" soldier on the island kept one eye on the enemy in front, and one eye over his shoulder.

Since then, they have all had their say why Singapore fell. Each has found this reason or that reason, generally blaming someone else. All have shirked or failed to recognize the truth. None of the chroniclers of Singapore 1942 has yet exposed the truth. Here is the truth, innocently expressed in the words of the law-abiding G.O.C.—the "enemy within our gates" was a compound of Fear and Distrust, which spread like a rotting mildew from top to bottom, and marched with every fighting unit. Fear made confusion of command—distrust suppressed the Asiatic effort—fear robbed Wavell of the vital month he needed to reinforce and mount his counter-attack. Fear made unit commanders order retreat too soon: fear and distrust held troops in reserve instead of rushing forward to fill every gap till the last man and the last bullet were spent. Compared with the requirement to which Governor Sir Shenton Thomas gave lip-service—"Singapore must stand, it shall stand"—there was not one individual life on the island worth preserving. But fear drove thousands to the waterfront in search of a boat to "freedom". When there were no more boats, fear led the rest of us to surrender and be herded into prison camps; and the hell which followed was compensating retribution paid by us on behalf of all who made cowards of us all. It is possible that Singapore had to fall, because of its basic material weakness. But it did not have to be thrown away inside a week.

III

There was virtually no Navy to defend Singapore.

There was on the Island a purely token force of fighter aircraft—Buffaloes and Hurricanes—supported in theory by bombers based on Sumatra.

There was a total garrison strength of about 85,000 men, equal to approximately four and a third divisions—and

Percival estimated the forces ranged against him at three Japanese divisions, with perhaps other divisions in reserve in Malaya and still others in Indo-China. The Japanese had strong armoured forces (if they could be landed on the Island) and might have some airborne troops.

The garrison strength consisted of thirty-eight Infantry battalions (thirteen British, six Australian, seventeen Indian and two Malay); three machine-gun battalions (one Australian and two British); one reconnaissance battalion, three battalions of Straits Settlements Volunteer Forces (S.S.V.F.) and four Indian State Force battalions organized for airfield defence only. There were also four heavy anti-aircraft regiments plus one battery, two light anti-aircraft regiments less one battery; about 150 anti-aircraft guns and one searchlight regiment.

The famous "big guns of Singapore" were: five 15-inch guns, six 9.2-inch guns, sixteen 6-inch guns. They were the Fixed Defences—The Fortress. They were installed as anti-ship guns. In addition to the A-A guns, spread over the island were artillery units whose equipment ranged from 75-mm. field guns to 25-pounders and 4.5 howitzers. There were six anti-tank batteries and one mountain regiment.

If this force appeared formidable on paper, it was far from formidable in the field. The efficiency of the A.A. defences had been impaired by the loss of the warning system on the mainland. Six of the British infantry battalions had only recently landed on the Island after many weeks cooped up at sea. The other seven British battalions were below strength— and one machine-gun unit and the reconnaissance battalion, together with a great deal of ammunition, were not expected to arrive until February 5. Of the Australian battalions only the machine-gun unit was complete. All the A.I.F. Infantry battalions had been brought up to strength with untrained reinforcements. Only one of the Indian battalions was up to strength. Of the other sixteen Indian units, three had only recently arrived and were semi-trained, nine had been hastily reorganized and included a high percentage of young recruits as well as being short of officers, and the other four were being reformed and in consequence were not fit for action. The two Malay battalions, which had never been in action as a whole, were an unknown quantity under fire. The Volunteers (Europeans apart) were considered fit only for static defence. Many units were short of weapons after the long retreat down the peninsula, and their morale, already low, had sunk lower. Finally the total garrison of 85,000 contained a long "tail"—about 15,000 strong—of base, administrative and non-combatant troops.

Too much hope was centred on the reinforcements, most of whom had arrived in the last ten days of January, and the part they would play in the defence of Singapore. Behind the figures of the late arrivals, there were some disappointing truths. They comprised the 44th Indian Infantry Brigade and 7,000 Indian reinforcements, 1,900 Australian reinforcements, one Australian machine-gun battalion and the main body of the 18th British Division. *The War Against Japan* comments: "The 44th Brigade was no better-trained than the ill-fated 45th" (which had been thrown into battle almost immediately after its arrival early in January and had been surrounded and destroyed as a fighting force on January 19) "and . . . was retained on the island for further training. The 7,000 Indian reinforcements were made up largely of young and only partly-trained recruits and there were few potential leaders among them; the great majority had therefore to be sent to reinforcement camps for further intensive training, instead of being drafted directly to the sorely pressed and depleted units in the field. . . . The fighting efficiency of the 1,900 Australian reinforcements was a grievous disappointment. Some of them had sailed within a fortnight of enlistment and had not even learned to handle their weapons, few of them had had time to assimilate even the rudiments of discipline or were in any sense trained. The decision to select these untrained Australian reinforcements for Malaya was unfortunate."

In mid-December—the time when the reinforcements were offered by the Australian Cabinet—there were 16,600 Australians in the Middle East pool. These could have been drawn upon—indeed, many tried to obtain a transfer to Malaya without success. Alternatively, Singapore could have been supplied from the 12,500 former Australian militiamen who had enlisted into the Australian Imperial Forces after the Japanese invasion of Malaya. The A.I.F. commander, Gordon Bennett, wrote that his reinforcements were "so raw and incomplete they cannot be counted on. Untrained men, most of whom have not even fired a rifle, should not be sent to fight well-trained Japanese." Inexperience also explained the poor quality of some of the Indians—their army had been expanded too quickly under the needs of war elsewhere with the result that trained officers were scarce and potential officers scarcer.

In addition, there is evidence that best use was not made of those formations which were intact. Mr. H. E. I. Phillips, a former member of the 18th Division, wrote to *The Times* on November 10, 1947: ". . . even then the division was not permitted to fight as a division. Already one brigade had been detached, and this was followed by the arbitrary creation of artificial 'forces' constituted of units and sub-units drawn from its other formations. Thus battalions were divorced from brigades, and companies from battalions. Unnecessary difficulties of administration were created and the chain of command disrupted. . . ."

Considering all the circumstances, it is not surprising that General Percival was a very worried man. He toured the defences on February 1 with General Gordon Bennett. "How are we going to defend it?" asked Percival. "More men," answered Bennett. Together they agreed that there were not enough troops to establish an effective defence.

If it is not surprising that Percival should have had doubts about his defenders, it *is* surprising that he did not make better use of the ones he had got.

IV

In his dispatch of February 26, 1948, General Percival explains at length his estimate of the Japanese intentions and his plans for defence. This is what he says:

"I anticipated that the enemy . . . would certainly attack from the mainland but he might also simultaneously with this launch a seaborne and/or an airborne attack. I expected the attack to develop from the west, combined perhaps with a seaborne attack via the Straits of Malacca. I thought it probable that another force would come down the Johore River to attack either Tekong Island or Changi . . . a direct seaborne attack . . . would probably be directed against the south coast east of Singapore Town, while the objectives of an airborne attack would probably be the aerodromes. . . .

"*Plan of Defence.*—The following two alternatives were, broadly speaking, open to us:

"(*a*) to endeavour to prevent the enemy landing or, if he succeeded in landing, to stop him near the beaches and destroy him or drive him out by counter-attack, or

"(*b*) to hold the coast-line thinly and retain large reserves with a view to fighting a battle on Singapore Island.

"The disadvantage of (*a*) was that it was not possible with the forces at our disposal, owing to the extent of the coast-line, to build up a really strong coastal defence. On the other hand, as regards (*b*) there was a lack of depth in which to fight a defensive battle on Singapore Island in front of the vital town area. The Naval and Air Bases, depots, dumps and other installations were dispersed all over the Island and some of them would certainly be lost if the enemy was allowed to get a footing on the Island. Further, the close nature of the country and the short visibility would favour the enemy . . . the moral effect of a successful enemy landing would be bad both on the troops and on the civil population.

"For these reasons alternative (*a*) was adopted."

Two important factors emerge from this revealing document. The first is the transformation accredited to the enemy during the eight-week campaign on the mainland. From gross under-estimation of a comic fellow with no ability to fight, he had apparently evolved into the ubiquitous enemy who could fight anywhere and everywhere at the same time. "I expected the attack to develop from the west, combined perhaps with a seaborne attack. . . . I thought it probable that another force would come down . . . a seaborne and/or airborne attack." And so General Percival, though he admitted that he did not possess the troops to do it, decided to guard the entire 72-mile-long coastline. This decision is the second revealing point. General Percival was attempting the ideal defence although he knew he did not have the troops to carry it out.

One of the axioms of war is that you attack your enemy at his weakest point. It should have been clear to Percival and his generals that the weakest point was the north-west coast where the only defences were those being hastily thrown up by the troops who were detailed to defend them. The generals should also have been aware by that time of the enemy tactics of avoiding strongpoints whenever he could and infiltrating behind the defence lines. The north-west coast was ideal for infiltration. It should also have been obvious by that time that the Japanese were under no delusions about the strength of the defences. Their intelligence organization was excellent. As they knew so much about Malaya, it was morally certain that they knew all about Singapore as well.

General Percival would have been taking a risk if he had concentrated his forces in sufficient strength to throw back an assault from the north-west at the expense of his defences elsewhere. But when you don't possess the troops for an ideal defence line, you are forced to take risks, provided that they are calculated. In fact, Percival did precisely the opposite. Although he said he expected an attack from the west, although all the signs during the week before the assault indicated attack from the north-west, the G.O.C. disposed his forces in such a fashion that density of defence both in infantry and artillery was *less* in the north-west than anywhere else, and practically the whole of the garrison was committed to static defence.

Was the G.O.C.'s memory faulty? *The War Against Japan* gives a somewhat different version of Percival's estimate of the direction of the main Japanese thrust. It states: "During his visit to Singapore on January 20, Wavell had taken the opportunity of discussing with Percival the general dispositions for the defence of the Island. He had suggested that 18th Division, as the freshest and strongest formation, should be placed in that part of the Island most likely to be attacked, which he thought would be the north-west coast; 8th Australian Division in the next most dangerous centre, the north-east coast, and the remainder of the two Indian divisions, when re-formed, as a reserve behind them. Percival, however, had expressed the opinion that the main Japanese attack would be made down the Johore River and the north-east of the Island and he therefore proposed to place 18th Division on the north-east and the Australians on the north-west. Finding that Percival, who had studied the problem for so long, seemed convinced of the probable direction of the enemy attack, Wavell allowed him to dispose the troops as he wished. On his return to Java on January 21 Wavell confirmed the outcome of his conversations with Percival by letter. Believing that decisions as to the dispositions should be made by the man on the spot, he wrote to him saying: 'My idea of the best layout of the defences would be for the Australian division and the 18th Division each to take one of the east or western sectors, and the present garrison the southern sector with the Indians forming a central reserve strengthened possibly by the Gordons or some of the other British troops. Please let me have your comments on this layout. . . .' Percival replied on the 24th that he proposed to divide the Island into three sectors: Southern, Western and Northern, placing the Australians in the Western and the 18th in the Northern, in which case the Indian formations would be in Command Reserve. He had, in fact, issued orders on the 23rd for the defence of the Island on these lines. . . ."

If this is the correct version, Wavell must shoulder some of the blame for the rapid collapse by refusing to override Percival's opinion.

That General Gordon Bennett, who was given command of the vital area with the 8th Australian Division and 44th Indian Brigade—one weakened by 1,900 untrained reinforcements and the other only partly trained—was not happy, is revealed in his diary. "To hold the Australian front," he wrote, "I have only four reliable infantry battalions and two companies of the well-trained machine-gun battalion. This area is the most uninhabited part of the Island.

"I held a conference of brigade commanders and my senior staff and discussed the problem of defence of the Island. All agreed that we were undermanned. As they left I realized the unfairness of asking them and their men to fight with such meagre resources. . . .

"I visited the 2/18th and 2/19th Battalions. They are in position on the north-western corner of the Island. This part of the Island is thickly covered with timber, mostly rubber, with thick mangrove growing right down to the water's edge. The posts, which are many hundreds of yards apart, have a field of fire of only 200 yards. The gaps are patrolled regularly. I am beginning to worry about the extreme weakness."

The task of patrolling the gaps had been given to the men of *Dallforce*—locally-formed guerrillas. *Dallforce* is worth examination not only for the rôle it played in the defence of Singapore, but for the rôle it might have played even earlier, and the light it sheds on the mentality which continued to exist in Singapore even during these vital eight days before the Japanese struck.

v

"Arise, arise, those who do not want to be slaves. Build a new Great Wall with your flesh and blood."

So ran the words of a song specially written and chanted by the Chinese in Singapore as they rallied under the inspiring leadership of Col. John Dalley, director of Intelligence Bureau, Malayan Security Police, to form a force of irregulars for the battle of Singapore. They came from all walks of life—teachers, students, clerks, labourers, ricksha boys and sellers of fancy goldfish. They came from rich homes and shanties, men of middle-age and boys in their teens. They buried their political differences and put aside Secret Society feuds to join common cause against the Japs.

Officially they were called *Dallforce* after Col. Dalley. Unofficially they were called "Dalley's Desperadoes" . . . or "John Dalley's band of Chinese cut-throats".

Some members of *Dallforce* had fought with the guerrillas in China, and were sent from Chungking to instruct Singapore. Some had fought with the 19th Route Army in Shanghai. Some were Communists, released from prison to enlist; and Chinese girl Communists, fired by stories of women in action in Russia, volunteered for front-line fighting. They were bitterly disappointed when they were only allowed to do Red Cross work.

Europeans in non-combatant jobs crowded to Col. Dalley's headquarters to offer their services.

"You're *doing* something," said Gerald Hawkins of the Malayan Civil Service. "For God's sake count me in. I've got an office, and there's nothing to do there. I hear you're going to fight. Look at me—I've just been outside shovelling coal with the coolies, just to feel I was *doing* something."

There was a dash of cavalier adventure about the guerrillas which appealed to men of action. The absence of "proper channels" choked with red tape was an added attraction.

In 1940 Colonel Dalley had written a report called *Jungle Ambush Patrols* which suggested the formation of a network of guerrilla parties throughout Malaya ready to counter Japanese infiltration through the jungle. He had no doubt the Japs were coming that way. Jap agents were all over the country, mapping and photographing.

The report was sent to Malaya Command in December 1940. It stayed there undisturbed until the Japanese were halfway down the peninsula in December 1941, when somebody found it, sent it to General Heath at 3rd Corps H.Q. and Col. Dalley was asked: "Lay it on!"

"It's a bit late," said Dalley. "Six months ago, even three months, I could have networked the country. We shan't be properly organized. We haven't selected our men. We haven't got our food and munitions cached. They should all be in position. Still, I'll do what I can."

The framework of *Dallforce* was formed in Kuala Lumpur just before Christmas. By that time, Col. Dalley had made contact with Col. Alan Warren of the Marines, Col. Jim Gavin, R.E., one of the originators of the Lochailort School of guerrilla warfare—and Maj. Freddie Spencer Chapman. They had been in Singapore for months trying to persuade Sir Shenton Thomas to let them organize parties of military,

police, planters and miners—European officers leading mixed
Asiatics—to train as guerrillas and be ready to operate in
country they knew. Each party would speak the local language.

"No," said Sir Shenton Thomas. He told them he had
conferred with Lt.-Gen. Percival and they agreed (*a*) it
would not be practical; (*b*) it was not desirable; (*c*) the
Chinese must not be armed because some were members of
the illegal Malayan Communist Party; (*d*) in any event, to
admit the possibility of enemy penetration into Malaya
would be disastrous for Asiatic morale.

The grand plan was to establish a guerrilla network up
through Malaya, Siam, Indo-China, to Hong Kong, linking
with China, Burma and India.

"No," said Sir Shenton Thomas.

"As a concession we were told that there would be no
objection to local formation commanders making use of us
as part of the military defence of Malaya—as if we were a
sort of glorified Home Guard!" wrote Spencer Chapman in
The Jungle Is Neutral. "We came to the conclusion that
neither the Governor nor the G.O.C. Malaya was in the least
interested in No. 101 S.T.S." Yet these men were trained
guerrillas. Since early 1940 they had gone through long tough
courses so that they could go out as instructors. Someone at
the War Office had sent them to Singapore for a specific
purpose. For six months they were hamstrung by petty local
authority.

Then came the Jap blitz, and Sir Shenton Thomas hastily
changed his mind. Permission was given for Gavin and
Spencer Chapman to go ahead with the plan to organize and
train, as submitted and refused four months earlier.

"By this time it was far too late for the plan to be effective," wrote Spencer Chapman. "We had to improvise at the
last moment what we hoped to organize beforehand in a
careful and orderly manner."

Nevertheless, the improvisers achieved a good deal of local success—quite enough to show what a formidable obstacle they would have been to the Japanese if they had been allowed to train and organize.

By the time the siege of Singapore began, *Dallforce* was between 3,000 and 4,000 strong—a figure which could have been ten times higher if the equipment and training facilities had been available. They were not. But the Chinese formed an Overseas Volunteer Corps, a Women Workers' Anti-Japanese Association, and a Singapore Chinese Anti-Aggression Council. They met and unanimously passed anti-Jap resolutions, and paraded the town with flags of the Allied Nations. All they wanted were guns to fire at the Japanese.

Col. Dalley made his headquarters in a Chinese school in Holland Road. He used to stand on a table and address the crowds who gathered round. Many were Communists of the illegal M.C.P. Some he had pursued and gaoled. Now he was their hero—"Dalley Sin Sang".

Col. Dalley said: "When I was ordered back to Singapore from the mainland with what force I had left, I had to do some quick recruiting. My job was to cover the flanks of other units all round the Island. Above all, I needed more British officers to train and lead. The only way to get them was to call for volunteers from fixed formations—not a very popular move.

"General Heath agreed. I knew people who wanted to join *Dallforce*—I'd heard from quite a number. The men I wanted were chaps from the police, civil service, and various departments of government. They spoke the language and had some form of military training—especially the police, which was a highly-trained military force. I knew that many of these chaps weren't doing anything frightfully important. They told me they were just kicking their heels and doing jobs they felt weren't worth doing.

"Heath told me to see Percival. I saw Percival and he told me to see the local commanders.

"I went out to see the Volunteers, told them what I wanted, and about 50 per cent stepped up and begged to join. I was tremendously pleased.

"I went to see their commander, Brigadier Moir. He flatly refused to allow any officer or man in the Volunteers to join *Dallforce*. Brigadier Newbiggin of Malaya Command was there at the time, and he supported Moir.

"I knew a large number of police wanted to join. I made out a list and asked the Inspector-General of Police, Dickinson, to release them. He flatly refused.

"I was up against it. Time pressed. I went to Percival and reported that I couldn't get anybody. Percival said he would speak to Sir Shenton Thomas, who was head of civil matters, and as far as the police was concerned this was a civil matter.

"All this was before the siege, but on the day after I got back to Singapore, and the siege was on, a truck loaded with Malayan Volunteer officers arrived at my headquarters.

" 'What's this?' I said. 'Has Brigadier Moir relented?'

" 'No sir,' they said. 'We've deserted the Volunteers. We're now officers in *Dallforce!*' They told me they had nothing to do but sit around.

"Brigadier Moir was furious. He sent Captain Ackhurst to *Dallforce* H.Q. with orders to arrest my new officers for desertion. However, beer was produced and Captain Ackhurst said: 'I'll go back and tell him I couldn't find you.' "

Of the spirit of his Chinese irregulars, Col. Dalley said:

"One of my companies was based in Kuomintang Headquarters. I met George Yeh, and he was upset because I had recruited so many Communists. I told him they'd stepped up quicker, and he promised to find as many anti-Communists as we had Communists. He did, too.

"To keep everybody happy we got some red cloth and made a red triangle on the shoulder for the Communist half.

Quite a bit of my time was taken up in diplomacy. One chap thought he ought to be a general. I made him a sergeant, so that was all right.

"Our arms were shotguns, sporting rifles—anything we could lay hands on. The Army had none to spare. They'd lost so much equipment there weren't enough rifles for all the regulars. We were hoping to be fully equipped when the *Empress of Asia* arrived. She was bringing in masses of rifles and automatics for the 18th British Division."

Unfortunately, the *Empress of Asia* did not arrive.

9

Waiting for the Blow

"There must be no panic, no spreading of rumours, no defeatist tongues, no slackers, but steadfastness, re-doubled energy and iron determination to win."—The Governor's broadcast, February 1.

While General Percival planned his defences and the other generals passed on the orders, the people of Singapore waited. The Japanese opened harassing fire from the mainland on February 1, and on February 5 the full weight of their artillery was brought to bear on the three northern airfields, the Naval Base and key points on the Island's main roads. The British artillery did its best to reply, though handicapped by an allowance of shells which averaged twenty rounds per gun per day. Says *The War Against Japan*: "Ammunition stocks, especially for the 25-pounder guns, were low and Percival was planning for three months' siege."

There was no total compulsion of manpower for defence. True, men were not supposed to leave the Island (unless they could obtain special permission or could plead sickness or advanced age) and *The Straits Times* urged people to save water and grow vegetables, but there was no compulsion. Food rationing was not tightened—stocks were sufficient for a long siege—and there were no attempts to mobilize civilians into a siege mentality. Mr. W. C. S. Corry commented: "I think all non-combatants should have been told to go out with a rifle in our hands and fight. We weren't. Obviously many staff officers became redundant, and with nothing to do morale began to go down. And then there was this extraordinary idea that you must hang on. Civilians had it. They thought there was something rather grand in staying to the bitter end, whereas they were only a nuisance."

Not only were useless men—or those who perhaps could not have been employed as military because of shortage of weapons or ammunition—not sent away, but men who had been on leave—even men whose jobs lay in Northern Malaya —were allowed to return. Maj. J. E. A. Clark, second in command of the 3rd Battalion S.S.V.F. in Penang, complained that he and other senior officers had little to do. No vacancies could be found for them, though a few had been absorbed into the 18th British Division, and they spent the week training recruits—mostly Chinese—at Changi.

Shops continued to do business. The brickworks continued to make bricks. Borneo Motors kept open for emergency repairs and petrol, servicing, and so on. At the end they had nothing to sell except liquor and outboard motors, some of which were bought for cash by people who intended to attempt a getaway. Robinson & Company, like the other European stores, carried on. Their store in Raffles Place had been among the first to be hit by bombs on December 8. In their new shop they bore a charmed life and weren't hit again.

Under the manager, Mr. L. C. Hutchings, there were eight European departmental managers and buyers, and the Asiatic staff. People still wanted clothing and footwear, blackout curtains and shaving gear, and all the other domestic needs. Robinson's were there to supply them.

The buyer for the ladies' department stood by her costumes and corsets and undies. "I refuse to leave Singapore," she said. "I'm not afraid of the Japs."

Mansfield & Company, shipping managers and agents,

played a big part in the affairs of Singapore. There was hardly a committee remotely connected with shipping which did not have a member of Mansfields.

They managed the Straits Steamship Company's vessels which were requisitioned by the Navy. They also handled the troopships.

Mr. Toms, manager for the Straits Steamship Company, worked like a Trojan combing the East Indies for ships which might have been overlooked and could be useful in Singapore.

Mr. S. G. R. Chandra wrote to the editor of *The Straits Times:*

"There are easily 15,000 to 20,000 Sikhs (civilians resident or refugee in Singapore) excluding the troops from Patiala—enough to provide four or five battalions of able-bodied troops. If the necessary arms and equipment were made available, this would be a formidable addition to the defence strength." A good idea, but too late.

Miss Carline Reid was a Tasmanian girl on a working holiday in Malaya when the Japanese blitz arrived. She promptly attached herself to the Selangor Defence Corps as a sort of managing secretary. She horrified them the way she requisitioned things with cheerful abandon.

When they got to Singapore, she housed them by requisitioning a Chinese millionaire's palace big enough for a regiment.

When the Defence Corps dwindled through men joining active service formations, Carline joined the Army and went to Combined Operations H.Q. at Sime Road. She stayed on when others left.

"I had my blood up," she wrote, "and definitely wanted to see the thing through. It seemed frightful that we had to be driven down the Peninsula like a flock of sheep, penned up meekly in Singapore Island, while the entire army of Nipponese dogs barked at us across the Straits."

Fl. Lt. A. G. Donahue of Minnesota, the first American to go into air combat in Britain with the R.A.F., was one of 48 pilots who flew Hurricanes off H.M.S. aircraft-carrier *Indomitable* from a point west of Java on January 26. They flew straight on up to Sumatra, their guns still protected with anti-corrosion grease against damage during sea voyage.

Cleaning these guns was a slow job without proper facilities, and it was some days before they were fit for action.

Most of the 48 Hurricanes went to P.2 airfield, Sumatra. Fifteen flew straight up via P.1 to Singapore. Piloting one of these was Arthur Donahue.

They arrived in time to see the huge black cloud of smoke from some of the oil storage tanks near the Naval Base set on fire by Jap bombs.

"A Chinese business man in Singapore had a standing offer of a bottle of champagne for every Jap plane destroyed," wrote Donahue, "so that evening Red and Denny [fellow-Americans] accompanied by some of the others, drove into town to collect the two bottles they had earned by their victories."

Red's plane was shot down next day, like most of the 51 Hurricanes which had arrived in mid-January. . . .

One of the jobs of Mr. H. P. Bryson, first assistant secretary to the Colonial Secretary's office, and member of the Local Defence Corps, the Singapore equivalent of the Home Guard, was to lock up someone who was supposed to be waiting for a court-martial. Nothing happened, and he never knew what it was all about.

Mrs. Bryson had refused a passage on the four big troopships which had sailed on Saturday. She was working on blood transfusion at the College of Medicine. Early on the 1st she went to work and found the place empty except for two other women. She rang her husband. "What am I to do?" she asked.

"You're going by the first available boat," said Mr. Bryson. There was a small Dutch cattle boat leaving for Batavia in a few hours. Arrangements were quickly made. Mrs. Bryson and half a dozen other European women were given camp-beds on the deck, and off they went. Mr. Bryson then closed up the home and went for full-time duty with the L.D.C. The H.Q. was in an English boys' school. Bryson's job was to detail men for guard duties. The L.D.C. guarded V.P.s like the Cable and Wireless Office, and Post Office. Another section guarded the oil stores on Pulau Bukum Island.

The Japanese elevated an observation balloon over Johore Bharu. "Look at it," said Ted Fozard, disgusted. "Sitting up there like a blooming great sausage on a stick. It makes me so mad to see that blasted thing there, and we obviously haven't got anything to shoot it down with. I wish to God you were on that boat miles away from here, Peg. What chance have we got when they're up there looking in our back window? They'll be able to pick us off one by one."

The Japs started shelling the northern airfields. Donahue's squadron were having lunch in the mess at Tengah Airfield when whoomph! bang! the shells arrived. "We all knew what it was without being told," wrote Donahue. "The time had come for what was to be my first of five evacuations under fire in two weeks."

The squadron of fighters were only there at the urgent request of Command to buck up morale. There were four main airfields on Singapore—Sembawang, Tengah, Seletar, Kallang—and after February 5 only Kallang could be used. Some emergency landing strips had been half-made, and were now being hastily unmade to prevent the enemy landing on them.

R.A.F. officers were billeted in the ritzy Seaview Hotel. Donahue described his bedroom suite as fit for a Hollywood film star: ". . . I had no trouble finding room for my wardrobe of shirts and shorts when I climbed into that huge luxurious bed."

Most of the hotel guests were friendly, and kept a score of the Japanese planes the pilots shot down. Others didn't care so much for a crowd of boisterous young men in sweaty shorts and shirts who came back after a day of taut nerves to relax and roar with laughter. Donahue wrote:

"Some of the boys went for a dip in the hotel swimming pool on this hot afternoon. . . . One grumpy old codger staying at the hotel came snooping around and tried to chase them out, saying they couldn't swim there because they hadn't been introduced.

"Among these boys was Brownie, and he rose nobly to the situation. 'Well, my name's Browne. I guess that introduces me!' The others followed suit, bowed, and dived back into the pool.

"The poor fellow retired perplexed—which, whether he knew it or not, was a good thing for him. The boys were trying to have enough fun to loosen themselves from the strain they were under, and would have thrown him in if he'd bothered them any more."

One afternoon, Donahue went with three of his pals to the Alhambra Theatre, and escaped from reality by watching the film *Ziegfeld Girl*. "The shock was quite rude for us when it was over; completely lost in the lovely atmosphere of American girls and song and gaiety and peace, we stepped outside into the teeming oriental traffic and the sweltering tropical sun, to be reminded that we were half-way round the world from America, with our enemies only a few miles away."

Mr. Ian Blelloch, the Legal Adviser in Kedah and now a refugee in Singapore, lost count of the number of times he told people that there was no truth, as far as he knew, in the rumour that an officer had been arrested for traitorous conduct up north.

"It was a vague story," he explained, "that an officer in Kedah—sometimes said to be R.A.F., British Army, Irish, A.I.F.—had been summarily shot, or court-martialled with varying consequences, or taken under escort to Singapore, or sent out of the country for trial or to serve a sentence.

"Who he was, and what exactly he was supposed to have done, was never very clear. Whether there was any substance behind these rumours, and if so what it was, I never discovered. It was a fantastic time."

On the military front, troops on the north-west sector dug slit trenches and threw up barbed wire—mostly at night because their positions could be seen by the Japanese during the day—while others, like Signalman Fred Mutton, laid and maintained the communications. "We thought we were coming into a fortress—well, it spells its own name, don't it?—and thought we'd be sitting it out like they did in Tobruk. That's what we thought. Instead, there we were laying out perishing lines and mending breaks, same as we did up-country.

"The road which our section had to maintain ran right along the coast; and way in the distance, somewhere in Johore, there was a building—it might have been a temple or something like that—that our people reckoned was a Japanese observation post.

"We were out four times one day trying to find the breaks in the line. As fast as we mended them those perishers shelled the road and blew out another bit and we'd got to go out again and mend that bit. My mate went sick with malaria and I had to go out with one of the Indians. I was driving and he was on the running board looking for the break. 'Whoa!' he shouted. 'There it is!' Eyes like a bloody hawk he had. How he saw it I dunno—the perishing cable was lying on the side of the road half hidden in tall grass. You never saw nothing mended so quick. Going along that road

I put my foot down, I can tell you. If it wasn't shells that were whizzing across, they were lobbing mortars—ruddy dangerous those mortars."

The Japanese also dropped collections of leaflets in attempts to turn the non-European population on their side. They bore pictures in crude and lurid colours with captions in semi-literate English, Chinese, Malay and Tamil. "We are waging war against the White devils" was the principal message.

One leaflet showed a fat European planter at rest under a tree surrounded by beer bottles, watching his Tamil workers toiling in the blazing sun. Another showed a red-faced Tommy grilling a beef steak under the envious eyes of emaciated Indian soldiers. A third warned that any Asiatic servant who stayed with a European employer would be put to death. One leaflet showed two pictures. In the upper cartoon could be seen an audience of British officers enjoying a concert with a buxom woman singer on the platform. The lower half showed a gruesome heap of charred bodies amid smoking ruins. Yet another claimed that a riot had broken out in Singapore and that British and Australian soldiers were being evacuated secretly from the Island, leaving the Asiatics to their fate. It urged Malayan and Indian soldiers "Pack up your troubles in your old kit-bag and co-operate with the Nippon Army". In another set, the pictures threatened outrages to European women. One was a highly-coloured drawing, Japanese style, showing a fat and naked European woman being dragged behind a cart in front of a row of Japanese soldiers.

One was addressed:

To the British Army
    The conduct which you the British soldiers behaved badly to the Japanese is never forgiven by both the God and the humankind. That is, you imprisoned the Japanese and put

them in the leper-house, into the oil-tanks and moreover slaughtered the Japanese non-combatants.

Impress on your mind that you expect to be revenged fifty one hundred times as many as you behaved once.

The Nippon Army

Crude they undoubtedly were—but they were not funny when the authors were only half a mile across the Strait of Johore, when troops on the north coast could hear them chopping and hammering, when they commanded so complete a mastery of the sky that their radio broadcasts could announce with impunity the Japanese air targets. Day by day it warned the Asiatic population: "We're going to bomb the Government offices now." "We're going to bomb the Customs godown tomorrow." "We're going to bomb the station," or the airfield, or whatever it might be, followed by a warning to the Malays and Tamils to keep out of the way.

Not only could the Japanese be heard—they could also be seen. General Gordon Bennett, out on a visit to the 2/26th Battalion, A.I.F., at Kranji, "gazed across the straits towards Johore Bharu with a sad heart and a full memory of my happy sojourn there. The place was deserted. I saw a Japanese staff car driving along the water-front. The occupant alighted and from concealment in some shrubs had a long gaze at the Island."

Gordon Bennett also "visited the 2/10th Australian General Hospital and was shocked to see the overcrowding. The hospital was in two private homes. Every room was packed with beds, practically all of which were occupied. A few malaria cases were arriving and I expected more. . . . The men, especially the wounded, were very cheerful. I decided that serious cases must be evacuated from Singapore by hospital ship to make room for fresh cases, and asked Malaya Command if ships could be provided." One day the Japanese command announced that they would shell the hospital and later dropped fifteen shells on it. Gordon Bennett noted: "The nurses were cool and courageous throughout the shelling, neglecting their own safety to protect their patients. Those nurses are the nearest things to angels I can imagine. They devote themselves wholeheartedly to their heavy task, frequently working continuously for over twenty-four hours to deal with a rush of casualties. They never complain and always have a smile and a kindly word for our wounded and sick men."

Towards the end of that first week Gordon Bennett and Percival toured 44th Indian Brigade Headquarters and then went on to the south coast to a hill from which they could see the land for miles, laid out like a carpet. To their left they could see the positions occupied by two companies of a Punjab regiment with a mangrove swamp in between. The gap between the two companies was at least a mile and a half wide.

They could also see the *Empress of Asia,* the ship which was to have brought valuable rifles, automatics and other war material, burning. Her loss—she was the only vessel out of all the reinforcement convoys into Singapore which did not arrive—was a severe blow.

II

She was an old ship, coal-fired and manned by a scratch crew, most of whom were known as "bomb-dodgers". They were the wartime scrapings of Liverpool. There was a lot of trouble with the firemen during the voyage and the captain said he would arrange to discharge them at Cape Town, but replacements could not be found and the authorities in South Africa would not allow any of the crew to be discharged. They had to continue the voyage, and trouble reached a climax when they found they were going to Singapore. They said that conditions in the stokehold were bad, and that the fire-fighting equipment was in poor shape.

The *Empress of Asia* was one of a convoy of four ships that came through the Sunda Strait and up past Banka, approaching Singapore on Thursday, February 5.

The story went round that the "bomb-dodgers" refused to maintain steam in the *Empress* so that she fell astern. It was early morning. The Japanese singled the lone ship out for attack. Nine dive-bombers set her on fire from end to end.

The Captain decided to keep going and ran her aground on the Sultan Shoal near the Horsburgh Light. When his distress signal reached Singapore Harbour there were no crews for the available tugs—the Malays had bolted—and volunteers were called for.

Mr. J. P. Wormald, ship and engineer surveyor to Lloyd's Register of Shipping, tells how he and three others manned the tug *Varuna* and went to the rescue:

"There were two of us below and two on deck. We had to raise steam and were slow moving, so we were one of the last rescue ships to reach the *Empress of Asia*. The Navy took off most of the troops and crew.

"We picked up a number of survivors, none of whom were hurt or burned, but none had much clothing, and we returned with them to the dockyard.

"As we were pulling away from the *Empress*, the Master and the Pilot left her and got on board the pilot cutter. Just at that moment the burning mainmast crumpled and fell with a crash. Fortunately it did not injure anyone. It was a sad sight to see her burning, with so much precious equipment on board.

"Some time later I was telephoned at my house and told that an R.N.V.R. patrol boat had reported that the fire had burned itself out. The authorities were anxious to know if the equipment was undamaged, and whether anything could be done about it.

"Having had previous experience of ship and bunker fires, I was very doubtful, and said I thought the chances of salvage were slender. However, I agreed to go out to the *Empress* if the military would make the necessary arrangements.

"When I arrived at the rendezvous, it was deserted. Eventually an Army officer arrived, and he didn't know what it was all about. After a considerable amount of running backward and forward, a small party of us set out in a sort of Air Force Rescue craft.

"As I had anticipated, we could not get within 100 yards of the *Empress* for the heat. There was no means of getting on board.

"Schemes were suggested for trying to tow the ship off and into Singapore Harbour, but they were hopeless.

"The loss of life was small, fortunately, but most of the men who came by this convoy arrived just in time to walk into prison without firing a shot."

10

The Storm Breaks

*"All ranks must be imbued with the spirit of attack. . . .
The endeavour of every soldier must be to locate the enemy
and . . . to close with him."—From an instruction issued by
General Percival to all formation commanders, February 3,
1942.*

In the middle of the "quiet" week the Japanese increased
their air and artillery bombardment. The relentless formations of bombers swept down from Johore and hour after
hour pattern-bombed docks, Kallang, depots and barracks,
vital store dumps and outer headquarters and residential
areas, especially the densely-populated Asiatic quarters.
When the formation bombers were absent the dive-bombers came in to shower anti-personnel "grass-cutters"—small
bombs which burst above ground and scattered arcs of sharp-
edged shrapnel—and the fighters cruised up and down the
roads machine-gunning anything they could see from Army
transport to rickshas and bicycles. Even the milk roundsmen
from the Cold Storage were riddled as they pedalled their
tricycle vans around the suburbs.

On February 4 and 5 the Japanese began shelling the Island. They positioned a long-range gun on high ground in
Johore and brought the whole of Singapore within shellfire.
Their attacks were of pinpoint accuracy and though the guns
of Singapore—mobile 25-pounders and 4.5 howitzers, and
the fixed fortress guns—retaliated, there was little they could

do except make a terrifying noise. Mrs. Dick Mullaly thought the noise of the big guns "shocking" and found them more frightening than the enemy bombs. Ted and Peg Fozard, lying awake in the bedroom of their bungalow listening to the whine and crash of shellfire and the bang of answering guns, thought it was like a thunderstorm that just went on and on. They, too, found the noise of the defence guns worst. They deafened the ears and made the temples ache. Peg Fozard, scared, asked: "Do you think Singapore is going to fall?" Her husband said: "I can't see what's to stop it. We obviously haven't the strength. No planes, so we're bombed to hell. No ships, so the Naval Base is on fire. I'm going to get you away, Peg."

The defenders worked on as best they could. The token force of fighters—one squadron of Hurricanes now—were airborne almost constantly during the daylight hours. But the absence of long-range warning and the better manoeuvrability of the Japanese fighters at low levels—plus the shortage of defending planes—meant that heroism could do nothing effective. Bomber raids from British aircraft in Sumatra became fewer as the Japanese attack on the Dutch East Indies gained in intensity.

On a 4.5 howitzer gunsite which commanded Tengah airfield and a section of the Johore Strait, Gunner-Signaller Marshall helped to dig ammunition deep into ground pits ready for the hour when it would be required for action, and mended broken signal communications.

Fred Mutton lost all count of time. "It was just one long day," he said. "We had no set home. We took our kit with us and our home was where we stopped. We slept in the trucks, when we slept at all. Once we had orders to go and break up some big telegraph place. We went with sledgehammers and wrecked everything that might be useful to the Japs—permanent lines and switchboards mostly. You just kept going automatically, never mind thinking what you were doing or what might happen next."

The generals held conferences—depressing affairs, said Gordon Bennett afterwards. "The civil control came in for some severe criticism, especially in regard to the labour question. The unloading of the ships" (the convoy which included the ill-fated *Empress of Asia*) "was slow, the men working only one shift per day. That leaves the ship idle for two-thirds of the day in a constantly-bombed area." And the training of the troops went on. One particular method, records *The War Against Japan*, "was fraught with danger even for trained troops, but, with one of the battalions concerned containing a high percentage of untrained men, the danger was even greater." This method was teaching the men of the 22nd Australian Brigade that if surrounded and in danger of being overwhelmed while holding a forward defensive position, they should fight their way back to their company headquarters and there form defensive perimeters. If these, in turn, were overrun, the companies were to fall back to previously selected areas and establish battalion perimeters which, loosely linked together, would form a final defensive zone.

News of the Japanese preparations for the assault was meagre, because nobody had considered before the withdrawal from the mainland the possibility of leaving small parties behind in Johore to provide information. Instead, patrols were sent across the Straits. On February 5 and 6, enemy artillery fire was concentrated mainly on the northeastern coast and on the area around the southern end of the Causeway, with intermittent harassing fire on the more vulnerable north-western area. It might have been detected as a "blind", for though no patrols had been sent from the Australian front, patrols sent east of the Causeway reported no enemy concentrations there. Indeed, by February 6 Malaya Command Intelligence staff had decided that the attack would come from the north-west. "Accordingly," says *The War Against Japan*, "Western Area" (the Australian front) "was ordered to send over patrols that night to search for enemy concentrations and to ascertain the cause of the sound of chopping and hammering which had been heard in the area.

"During the night of the 7th/8th a boat, carrying some 30 Japanese, approaching the island just west of the Causeway, was destroyed and its occupants killed by machine-gun fire; a small party of enemy troops landed on Palau Ubin and occupied that island, patrols from 4th Norfolk withdrawing to Changi. The two patrols from 22nd Australian Brigade, which had made a reconnaissance of the coast between the Sungei Malayu and the Sungei Pendas to a depth of about one and a half miles, returned and reported that large enemy forces were concentrating in the rubber plantations and around the river estuaries and that there was much movement of motor transport on the roads in that area. They also brought back information of the location of some Japanese units and their headquarters. They had not seen any landing craft, for they had not penetrated as far as the Sungei Skudai or to the upper reaches of the Sungei Malayu, the only areas where these craft could be launched. These reports clearly confirmed the views of the Intelligence staff that the assault was likely to be launched against the Western Area, but gave no definite indication of the date of attack. They were not, however, passed to Malaya Command until about 3.30 p.m. on the afternoon of the 8th. Bennett then asked for a reconnaissance aircraft to check their accuracy and to observe for his artillery. No aircraft was available so, late that afternoon, Western Area artillery put down unobserved harassing fire on the areas indicated by the patrols' reports. It should now have been clear to all concerned that the stage had been set and that the curtain was about to rise."

II

On the morning of February 8 Ted Fozard at last persuaded Peg to leave. Three ships were due to sail that day: the *Devonshire*, the *Plancius*, and the Free French transport *Felix Roussel* which had arrived with the *Empress of Asia* convoy. Slowly Mrs. Fozard packed, ticking off the items they told her she would need—blanket, cup, plate, knife, fork and spoon, towel, toilet things. She added what clothing she could squeeze in. She had no heart in leaving. She was scared by the bombing but she wanted to stay beside Ted. She felt she was being forced to desert.

Ted Fozard said: "When it came time for Peg to go, I got twenty-four hours' leave from the Local Defence Corps. Things were very hectic by now.

"We went to the P. & O. at Cluny, queued for the ticket, and that night we all congregated at Clifford Pier in a big open place. Then the long trek started to the docks. We travelled at the rate of inches at a time. It was bumper to bumper all the way to the docks, with an Army dispatch rider leading the column. It was a frightening drive, all these cars packed so close. By now it was night . . . no lights on the cars.

"When we got to the docks it was a solid mass of cars parked all over the place. Our friends stayed in the car to look after it, and we walked the rest of the way. Peg was allowed only as much luggage as she could carry on board, two small suitcases. We got to the ship at last, and a sergeant of Marines stood at the foot of the gangway. 'Only passengers allowed on board, no visitors,' he said.

"Peg showed her embarkation paper, issued and signed by the P. & O. I was carrying her suitcase. 'Your wife, sir?' the sergeant asked me.

" 'Yes.'

" 'That her luggage?'

" 'Yes, I'm just going to take it to the top of the gangway.'

"He said: 'Right. Take it up . . . and take my advice and stay up there. You won't be the only one.'

"I was in my khaki uniform, equipment on and rifle slung over my shoulder. I said: 'I don't think I can do that.'

" 'It's up to you,' he said. 'Go on board, anyway. Please yourself whether you come off. Take my advice and stay with your wife. We've had it.'

"On board it was just plain hell. We found that the continental ships have the reverse lettering to the decks. Somebody directed us to 'A' Deck. It was in the hold, right down in the depths of the ship.

"What a greeting! 'A' Deck, and you find yourself in the middle of a large empty space—no table, no chair, no furniture, nothing but hammocks. That was Peg's cabin! Lit by hurricane lamps. This ship was allegedly Free French. She'd brought troops from India. You could hardly walk the decks for dirt and grease. The stink in the hold was awful. Still, it was a ship, and Peg was on it.

"The Asiatics got there first and grabbed the cabins. Quite a lot of Eurasians as well as Indians. The British were all down in the holds.

"It was the last straw for Peg. She was worn out with fright and worry, weeping helplessly. She clung to me and begged me to stay. 'The sergeant said you wouldn't be the only one,' she pleaded. 'What good can you do if you go back?' She hardly knew what she was saying, she was so upset.

"There was no tussle in my mind. 'I can't, Peg. If I hadn't got leave, we wouldn't have got the pass and got you on to the ship. I gave my promise I'd be back. You don't want me to rat, do you?' "

Another man who was saying good-bye to his wife was Mr. E. M. Glover. He said: "As I turned on my heel the voice of an officer I knew by sight but could not name said: 'Get out now while the going's good—Singapore won't last a week.' I ignored the advice but I remembered it later."

On board the *Felix Roussel*, Mrs. Fozard told her side of their story. . . .

"I can't tell you what I felt when Ted tore himself away and said good-bye and went back. Something inside me seemed to break in two, and I just sat down on the floor and cried my eyes out. I was past caring if the ship was bombed or what happened. There didn't seem to be anything left. I knew Ted had sent his money to London for me, but that didn't mean a thing at the time.

"My little Australian friend from the Cable Censorship was so kind. I think we all somehow managed to help each other. We scrounged a mattress between us and slept together on the deck. She was little and I was rather large, so she slept on the mattress and made room for my head and shoulders. The rest of me just slept on the deck.

"That was after a day or two. The night we sailed we were so exhausted with all the grief and excitement, we sank down on the bare deck in our clothes and fell asleep. A soldier came and shook us and said: 'I think you ladies ought to get below. We've got some dangerous characters at large on this deck.'

"I said: 'Oh, leave us alone. We're too tired. We don't care.' I found out afterwards that the dangerous characters were those Liverpool stokers who had mutinied on the *Empress of Asia*. They were locked up in our ship, being sent out to India.

"We were supposed to sail at dawn, but it was midnight. I think they gave out a false time in case of fifth column. It didn't really matter, because Jap planes followed us out and bombed us, but we managed to dodge and all the bombs fell in the sea.

"The alert went so often, and each time we were herded

below. We got sick of it and stayed below and prayed—not for ourselves but for those we were leaving in Singapore."

Ted picked up the story after he went ashore. . . .

"The husbands felt pretty grim and miserable. The authorities had given out that the *Felix Roussel* was to sail about six or seven next morning. No point standing on the quayside for hours. We went back to the car and drove to a friend's bungalow half a mile away.

"He was alone, too—his wife and child had left on a previous ship. He had the fridge well stocked, so we had a meal and a bottle of Scotch. We'd been up since dawn and were dead tired. We decided to snatch a few hours' sleep. What a hope!

"The Japs really started bombarding from Johore. It was pandemonium. They must've had hundreds of guns there. No use trying to sleep. We got up and dressed, and phoned L.D.C. headquarters.

" 'Do you want us to report back now?' we asked.

" 'No, stay where you are. Nothing much you can do up here tonight. Report back in the morning.'

"Daybreak came. Very low cloud. Heavy gunfire up by the Johore Straits. Shells popping about everywhere. Then we heard planes, very low, seemed to be buzzing around the bungalow. We went outside, but couldn't see them for cloud.

"We heard a rush of bombs. We dived flat on our faces. The bombs dropped quite close on a small coconut palm plantation. Lumps of trees hit the bungalow, stones and earth as well.

"That passed over, and it was time to return to duty. On the way we had to pass the *Felix Roussel*. No sign of the ship. The godown nearby was now a vacant lot—two or three godowns got it in that last raid. At first we thought they'd got the ship. Then we decided she must have sailed earlier.

It was three and a half years before we found out she sailed at midnight."

It was not until an hour or two later that Ted Fozard and his friends, and all the people of Singapore, learned that the Japanese had landed on the Island.

III

Perhaps it should have been clear to Malaya Command that the balloon was about to go up. The estimates offered by Percival and Paris about the likely date of a Japanese attack were almost up, and the Japanese gunfire and bombing had been intensified. From dawn the enemy air activity increased with bombing and machine-gunning attacks on the forward defences in the Australian sector. The artillery bombardment began a little later and the shells came down in a barrage which by half past one in the afternoon had become a murderous drumfire. Headquarters and communications received special attention and by the time there was a lull in the barrage—at sunset—all the telephone lines had been cut. After sunset the bombardment reached even greater intensity. Says *The War Against Japan:* "Neither Malaya Command nor Western Area Headquarters was however seriously perturbed by this, each apparently thinking that it was the first of a number of days of softening up, or that the enemy would switch the bombardment back next day to the Causeway and north-eastern shores of the Island. As a result no orders were given during the evening for artillery fire to be brought down on the probable enemy forming-up places."

It was half past ten that night when the first Japanese landing-craft were seen. They attacked the entire front between Tanjong Buloh and Tanjong Murai under cover from artillery and mortar fire.

The approach of the enemy should have been the signal

for a blaze of searchlights to light them up for the defenders and a savage counter-barrage to blow the enemy out of the water. The searchlights stayed dark and the guns remained silent. Strict instructions had been given to the searchlight crews that they were not to expose beach lights except on the specific instructions of the unit commanders—it was thought that, once exposed, they would inevitably be destroyed—but by the time the assault was launched all the communications had been cut and those lights which had survived the barrage were never exposed. Because calls for the planned defensive artillery fire never reached the batteries the guns did not open up and it was not until the S.O.S. light signals sent up by the infantry were seen that any of the guns went into action. In any event, there wasn't enough artillery to cover the whole brigade front; and in some areas—particularly on the front of the 2/18th and 2/19th A.I.F. Battalions—the infantry was left without artillery support. Some gun positions did not see the warning Very light and stayed silent.

The Australians fought in darkness broken only by gun flashes and the blaze from burning barges, and without the comfort of a counter-barrage from their own artillery. Too widely spaced, and with untrained reinforcements in their ranks, it is not surprising that the Australians could not keep the Japanese off the Island. The enemy was soon ashore—though in places only at the second or third attempt—and they quickly infiltrated the gaps between the Australian positions.

The jungle became a nightmare. The Japanese, using the tactics which had swept them to victory on the mainland, seemed to be here, there and everywhere. They were guided through the night by leaders with compasses strapped to their wrists.

By one o'clock on the morning of the 9th—only three and a half hours after the invasion had been first sighted—the Australians in the forward areas were withdrawing, under

Above: Enemy infantry creep forward behind a tank on the outskirts of Singapore City. Even the troops' boots are adapted for jungle fighting—cloven rubber-soled "cow-boots" to facilitate climbing trees and marching through swamp. *Below:* Watchful Japanese rest on a hillside outside the city, while smoke from the oil fires darkens the sky. *(Photos by Keystone)*

Above: A launch leaves Singapore Island during the evacuation on Black Friday, February 13, 1942. Keppel Harbour and the southern coastline burn in the background. *(Photo by courtesy of the Imperial War Museum)* *Below:* A volcano-like pillar of heavy black smoke rises over Singapore City from blazing oil tanks at the Naval Base fired by the British. This was one of the last pictures to leave Singapore before the surrender. *(Photo by Keystone)*

orders, to their battalion perimeters. The danger of this training method was now revealed. The Australians did not possess men with compasses strapped to their wrists, and for many of them the jungle was a terrifying maze even without enemy troops lurking inside it. Groups of men became separated from their comrades in the bewildering darkness. Others lost their way. Many died. Some straggled back as far as Bukit Timah. Others even reached Singapore City; and long before they could be picked up, re-organized and sent back, the disorganization was complete. The effect of the withdrawals was to dislocate the whole Brigade area, and by ten o'clock on the morning of the 9th—less than twelve hours after the assault had been sighted—the 22nd Australian Brigade, on whose fighting power had rested the defence of the north-western area of the Island, was no longer a cohesive fighting force.

IV

In the light of day it became possible to piece together some of the story. Col. Dalley, for instance, had posted an augmented company of *Dallforce* by the mouth of the River Kranji, where the Japanese had first landed, under the command of a tin-miner named Harte-Barry. Dalley saw them during the evening before the assault began and gave them orders not to retreat. During the visit he noticed that the nearby Australian machine-gun battalion was looking unhappy under the bombardment. Australian war reporter Ian Morrison, who was with Dalley, confessed: "I was windier that day than I had ever been during the Malayan campaign. And if I was windy, who had been bombed and dive-bombed pretty frequently up at the front and broken in to noise, what must it be like for the young officer who was going up to the front for the first time? We came across some

Australians in a rubber plantation. They had dug shallow
trenches and were sitting in them. They were jittery, no mistake about it. We came across a party with a young Australian soldier from one of the 25-pounder batteries who had his
leg blown off. His mates did not think he would live. They
were carrying him back to the aid post. Thunder and lightning began. The uproar of bombardment was enough as it
was without the heavens also taking part. Only six hours
later the Japs landed in that very sector."

In the morning Dalley went back to his company—and
found it, or what was left of it. One British officer and five
other ranks had fallen back; the remainder of the 200—
mostly Chinese—had stood and fought it out against a Japanese machine-gun battalion. "It was a frightful sight," he
said. "They'd been shot to pieces. They'd used up all their
ammunition. There were no wounded to bring back. They'd
stood their ground. They'd had orders to stay and they
stayed. And they all died.

"The Australian machine-gun battalion did not stand.
They moved back. They were frightened and make no mistake. When I walked back to the road where I'd left my car
I went up to Col. Asherton, A.I.F.—a fine chap—and said: 'I
don't think your chaps are very happy. They aren't liking
this very much.' I walked round with him for half an hour—
he could hold those men when he was around. Their spirits
went up at once. I heard afterwards that he was killed very
early in the piece. I felt that battalion wouldn't be so good
without that particular C.O., and that's what happened.
They were all newcomers—hardly trained. It was very sad."

Gunner-Signaller Marshall had a close-up view. His gun
battery was placed to cover Tengah Airfield. They did not see
the warning Very light—if one was ever fired—and although
they had four guns with a range of four miles and 4,000
rounds of high explosive ammunition they did not fire a shot
while the Japs were rushing the Island. The first thing they

knew was that the enemy had landed and were threatening Tengah Airfield, two miles from their position. Less than an hour later the Australians came down the road from the front line. "It is the most vivid memory I have of the Singapore campaign," he said. "They came moving at a half-trot, panic stricken. I've never seen anything like it. It was pouring with rain and most of them were clad only in shorts. Few were wearing boots, and some of the men's feet were cut to ribbons —they'd come across rivers, through mangrove swamps, through the bush, then out along the Jurong Road. They'd scrapped everything that could hold them back. They'd thrown aside their rifles and ammunition. We watched them in amazement. We were sorry for them. We realized they must have come through a very rough time and something had got hold of them.

"Some of our chaps gave them boots and we gave them what food we had. We asked what had happened and where they were going. 'We're off to the docks,' they said. It seemed to be an entire battalion or what was left of a battalion. Among them was one Aussie soldier fully equipped: rifle, ammunition, cape, shirt, shorts and boots. He came across the road to three of us who were standing watching. 'What's happened?' we asked him. He looked at his fellow-Australians. 'They're finished,' he said. He was quite calm. He was a boy of about nineteen, a private. He seemed sorry for his mates. 'Can I join up with you blokes?' he asked.

"He came along with us. I don't know what happened to him later but I remember that one chap among that horde. He had control of himself. The others had lost control. They were panting, incoherent, a rabble.

"Behind the Aussies came the Indians. The Australians had come through their lines and the Indians had caught the panic. The Indians were mostly young boys. They were without any officers. They didn't stop. They just went hurrying down the Jurong Road heading for Singapore Town. Naturally we were alarmed. We asked the Aussies what had happened. 'The Japs are a quarter of a mile up the road,' they said.

"Meantime a Bren-carrier manned by Argylls had been sent up the Jurong Road on reconnaissance, to see if in fact the Japs were on the Aussies' heels. Presently we heard the roar of its engine approaching on the return journey. The Aussies saw it, thought it was the Japs and just seemed to dive head first off the road. 'Look out,' they yelled. 'It's the Japs!' They dived into storm ditches, into the bush, anywhere out of sight. In seconds the road was as clear as if the rabble had never been there. The Argylls stopped by our position. 'We've been half a mile up the road and never a bloody Jap in sight, so you're all right,' they said."

In Singapore City the shock of the news was stunning. Ian Morrison wrote: "Early the following morning [February 9] I went down to the press room in the Cathay Building and began to hammer out a story on the bombardment of the previous night. Hardly had I begun when Ian Fitchett (official A.I.F. observer) came bursting in with the breathless news, which he himself had only just heard, that the Japs had landed in the north-west of the Island.

"Shortly afterwards the Singapore radio broadcast a short announcement that the enemy had landed and fighting was in progress. An official communiqué issued later by Malaya Command said that offensive action was being taken to 'mop up' the enemy.

"Again that unfortunate term . . . suggested that the Jap attack was merely a slight infiltration, not an all-out large-scale assault."

Working feverishly to try and piece together what had happened, Morrison scrapped the bombardment story and began again:

"When dawn broke, some three or four thousand Jap shock troops had secured a firm foothold on the Island. One column was pressing south from Kranji, the other was pressing east from Pasir Laba, in an attempt to cut off all our troops in the north-western sector. The danger of encirclement became so great that these troops were withdrawn and a rough line established between Kranji and Pasir Laba. The line was bulging inwards all the time. Fighting was in progress round Tengah airfield.

"A counter-attack was scheduled for eleven o'clock. It was our last chance of driving the Japs off the Island. If that counter-attack failed, nothing on earth could save Singapore."

The counter-attack never took place. In his diary Gordon Bennett recorded: "On inquiring of Brigadier Taylor [22nd Australian Brigade] how his attack was progressing, he replied that just as the advance was to commence, the enemy attacked in strength and that his line had fallen back behind the [Tengah] aerodrome. This meant that the aerodrome was in enemy hands.

"For some time communication with the brigade was cut. Liaison officers had a bad time passing between my H.Q. and brigade H.Q. through the heavy and constant artillery and air bombardment on the roads. During the afternoon, General Percival called and seemed very worried.

"While Percival was there, word came through that the 22nd Brigade had been driven back to a line between River Kranji and River Jurong. This was a terrible shock. I rang Taylor, who said the enemy had got in behind his right flank. He seemed confused."

That same day—only two hours after the first broadcast of the Japanese landing—Rohan Rivett, a news-reader of the Singapore Radio, went to see Gordon Bennett at his H.Q. Rivett wrote in his book *Behind Bamboo:* "As the house was being sought by Japanese bombers, and as it had been shelled on the previous day, we had to leave the car half a mile away and go up along the road, under the shelter of the bushes, in single file. Finally we were all gathered in the presence of the red-haired Australian commander, about whom so much controversy has raged. As usual, Gordon Bennett was blunt and perfectly frank and anxious to disabuse the world of the false ideas of Singapore's security still promulgated through blind censorship and propaganda. He indicated that there was no possibility of preventing the Japs from reinforcing their bridgehead or from extending it each night along the northern shore of the Island. After all the lies and absurdities handed out to correspondents through official channels, the Australian general told us the plain unvarnished facts and left us to draw our own conclusions. We were so sick of being given statements which everyone in Singapore and, above all, the Japanese enemy, knew to be transparently false, that the interview was a refreshing draught of sanity in the midst of a flood of phantasmagoria from Alice's Wonderland.

"All that afternoon we, the Censorship, Malaya Command and Gordon Bennett, fought a four-cornered fight as to how much of his statement could be published. The General stood by all his guns and said that it was far too late for anything but the truth. The wiser of the military and censorship stood with him, but masses of red-tape and prejudice had to be trampled on before a limited account could be released.

"On return from Australian Field H.Q. I was glad to give an account of the General's views to the daily confidential conference of departmental heads in the Cathay Building. Sir George Sansom, Chief of the Ministry of Economic Warfare, Mr. R. H. Scott, Chief of Information, Mr. Eric Davis, Chairman of the Malayan Broadcasting Company, and other executives received the blunt tidings with a mixture of resignation and dogged determination, then we all went on with our respective jobs."

II

The Dying Island

"Please inform my wife . . . that I have left this land of the living, and the dying."—From a dispatch by Yates Mc-Daniel, Associated Press of America correspondent, Singapore, February 12.

When allowance is made for all the original blunders in London and elsewhere which left Malaya and the East Indies weakly exposed to Japanese invasion, the blame for collapse of his forces against the final attack on Singapore must be awarded to Lt.-Gen. Percival, who was the G.O.C. By miscalculating or failing to appreciate that the chief danger of attack was on the north-west coast, he left that area too lightly defended, and gave the Australian 22nd Brigade an impossible task. The troops fought as best they could against greatly superior fire power: those who broke down and ran cannot be altogether blamed. The story of their collapse is as much a commentary against the High Command as against themselves.

From the events of the desperate days of Monday the 9th, Tuesday the 10th, Wednesday the 11th, and Thursday the 12th, there emerges a painful catalogue of error, folly, indecision and indescribable confusion. When the evidence is sifted, it no longer remains a source of pained astonishment that Singapore held out for only seven days. The miracle is that it held out for so long. Indeed, there was a moment on

the 11th when the Japanese had split the defence so successfully that their tanks could have driven down the Bukit
Timah road into Singapore City with virtually no opposition.
They did not do so. They stopped at Bukit Timah to consolidate and had to wait four more days for surrender. No
doubt, despite the evidence of tactical weakness they had
accumulated during the two months since December 8, even
the Japanese could not believe that the British defence would
be so easily overcome.

The counter-attack to force the Japanese off the Island
was not made, because Western Area reacted too slowly to
the news. Instead of hurrying every available man and gun to
the area at once, there was a delay of several hours before two
battalions, fit only for static defence, reached Tengah airfield
between six o'clock and a quarter to eight that morning.
Malaya Command did not help, either. Thinking the assault
on the north-west might be a feint designed to lure the defenders away from the north-east, headquarters waited until
the routine morning situation report told them that all was
quiet on the north-east front. Only then did Malaya Command release 12th Brigade, the only command reserve, for
the counter-attack. By that time it was too late. Says *The
War Against Japan:* "The fleeting opportunity to launch a
counter-attack . . . and confine the enemy to the north-west
corner of the Island was lost."

Once it had become clear that nothing could stop the
enemy from pouring in troops and armour through the
bridgehead he had established and consolidated by the capture of Tengah airfield, the defence of the Island rested almost entirely on the holding of the easily defended, partly
reconnoitred and prepared Jurong Line, and the launching
from it of a deliberate counter-offensive with every possible
man and weapon. But the ability of the defence to hold the
Line depended in turn on the security of the defences around
the Causeway area. If the Causeway sector were lost, the

Jurong Line could be turned. *The War Against Japan* explains: "Since by that time [the afternoon of the 9th] it was abundantly clear that the enemy's attack on the north-west coast was in great strength and the increased artillery fire in the Kranji-Woodlands [Causeway] area indicated that he might shortly make an attempt to widen his bridgehead, there was on the evening of the 9th an immediate need to strengthen the defences near the Causeway and to create a powerful mobile reserve in the Bukit Timah area for the all-important counter-offensive. The only way such a reserve could have been found was to leave the north-eastern coast-line defended by a skeleton garrison with a high proportion of machine-guns, and to withdraw 18th Division, concentrating it on the Bukit Timah Road by the morning of the 10th ready for any eventuality. The risk was not great; the threat of a landing on the north-eastern sector had receded and a seaborne landing on the south-east coast could by that time have been almost entirely eliminated."

Percival did not do this. He committed the error of jumping his fences before he came to them, while ignoring the one which had to be leaped immediately. Fearing that the Japanese would hit him in the north-east as soon as his troops were in the west he decided to base his defence on holding the Jurong Line and the Causeway sector and sent his last reserve (15th Brigade) to reinforce Western Area—a move which, says *The War Against Japan*, was of little value unless at the same time he took steps to create a fresh general reserve. The rankling fear of that north-east coast attack meant that he immobilized his troops in that area and the G.O.C. lost his only chance of gaining the initiative. What Percival did do late on that day—the 9th—was to prepare a plan for the defence of Singapore City in the event of an enemy breakthrough and advance down the Bukit Timah Road. He decided that in this case the troops should fall back on a perimeter round the town which would include

Kallang airfield and the MacRitchie and Pierce Reservoirs. During the evening Percival outlined the scheme verbally to Heath and Keith Simmons and soon after midnight issued it in writing as a secret and personal instruction for the information of his three senior commanders and senior members of his own staff.

While this was happening at Command H.Q., Percival was being poorly served by his commanders in the field. Brig. Maxwell, who was in command of the 27th Australian Brigade charged with the task of defending the Causeway area, learned of the Japanese landings early on the morning of the 9th, and immediately began to worry about an attack across the Sungei Peng Siang to take his brigade in the rear. He asked Bennett for permission to withdraw from the Causeway area and readjust his line to face west—a request which Bennett refused. During the day the Japanese stepped up their artillery fire on the sector held by the 27th Australian Brigade and about half past eight that evening a battalion of the Japanese 4th Guards Regiment launched an assault. By midnight the three forward companies of the defending 2/26th Battalion, though suffering heavy casualties, were concentrated in a strong position across the neck of the Kranji peninsula about 500 yards from the shore—and there they stuck. All the Japanese efforts to dislodge them failed so hopelessly that, as we now know, the general in command of the assault troops asked Yamashita for permission to withdraw and go in elsewhere the following morning. Before he gave permission Yamashita called for a new report on the Causeway front—and was told that soon after four o'clock in the morning all resistance had melted away. The withdrawal took place for reasons which are still obscure—Maxwell says he was eventually given permission to retire to the new positions he had planned; Bennett says he was not—and indeed Bennett could scarcely have done so in the light of instructions for fighting the battle which he had been given by

Percival. Whatever happened, the fact was that by daybreak on the 10th the Causeway had been abandoned, the trunk road south to Singapore City uncovered, the enemy allowed to consolidate his landings without opposition, and a gap of 4,000 yards left undefended between the 27th Australian Brigade and the flank of the 11th Division to the east. The pivotal position at Kranji had been thrown away just at the moment when the enemy had been brought to the edge of withdrawal. *The War Against Japan* comments that Maxwell's fear of an attack across the Sungei Peng Siang was magnified since "it was scarcely likely that the Japanese, with two good roads leading straight to the heart of the Island, would trouble to cross this mangrove-fringed river in any strength unless they were unable to make progress by the more direct route. . . . It is quite clear from the evidence available that neither he (Bennett) nor his staff knew anything of the events at the Causeway during the night of the 9th/10th and were as surprised as 11th Division was when, in the early hours of the morning, they learnt that 27th Brigade had withdrawn during the night. . . ."

There was another error during that night which had serious and far-reaching effects on the battle. Someone in Western Area Headquarters issued General Percival's secret and personal perimeter defence plan to the subordinate commanders as an order to be obeyed in certain circumstances—an order for retreat. The Commanders in the field received the instructions between 9 A.M. and 10:30 A.M. on the vital 10th, and acted upon them. They also acted on their own initiative all that day—with the result that by the evening the Jurong Line had been lost, too. *The War Against Japan* sums up the confused and confusing accounts of the day's battle details: "The result of the various errors of judgement committed on the 9th was that by dawn on the critical 10th Feb. the pivot position at Kranji had been abandoned; the two battalions of 27th Australian Brigade, out of touch with their Brigadier, had taken up positions which gave protection to neither the right flank of the Jurong Line nor the left flank of 3rd Corps. Bennett had committed his last reserve formation to the static defence of part of the Jurong Line and Percival had no other reserve immediately to hand. But this was not all. Western Area had issued to the commanders of its subordinate formation an order to be acted on in certain circumstances, based on Percival's secret and personal instructions to his senior commanders regarding the occupation of a perimeter position around Singapore Town. To these weary and distracted officers, sorely in need of reinforcements and encouragement to fight on, despite their difficulties, the receipt of such an order was tantamount to an admission that the higher command regarded the situation as hopeless. The psychological effect of this order undoubtedly had a considerable bearing on their actions during the 10th.

"A study of the events on that day shows clearly that throughout it Western Area failed to co-ordinate the actions of its subordinate formations. Paris, finding that 27th Australian Brigade had left the Causeway sector, withdrew his Brigade on his own initiative from the Jurong Line to the Bukit Panjang village area to guard against any enemy attempt to strike through to the Bukit Timah area. The other Brigade commanders, influenced by either what they knew of the situation or by their interpretation of the orders relating to the occupation of a perimeter position around Singapore, all acted independently. Thus, despite the fact that the enemy during the day had done nothing more than probe the defences with his advance troops, the Jurong Line was abandoned. . . ."

General Wavell did not help matters. He paid his last visit to Singapore on the 10th and, hearing that the Jurong Line had been abandoned, ordered an immediate counter-attack to recapture it. He considered its retention vital. The counter-attack was mounted, was impossible to carry out, and proved disastrous. It might have been successful if fresh formations had been available and it could have been launched from stable positions. But no fresh reserve had been created, the front in the Western Area was utterly disorganized and no steps had been taken to ensure that the Causeway pivot was firmly held. "The attempt to launch it resulted only in further confusion and in the destruction in detail of the forces involved," says *The War against Japan*. The Japanese forced their way through to Bukit Timah, where they captured large quantities of stores and war material. It was at this point on the 11th that they could have driven straight down into Singapore Ctiy.

Wavell's intervention on the 10th was his second error of judgement—and for a precisely opposite reason from his first. When Percival outlined his plans for the defence of the Island, Wavell allowed him to go ahead, though he, the Supreme Commander, thought that the north-west coast should be defended. Wavell's reason for giving Percival his head was that the local commander ought to know best. On the 10th Wavell intervened in disregard of Percival at a time when he could not have known the full position, and so made matters worse.

The dwindling force of Hurricanes continued in action throughout the hours of daylight on the 9th. Four finished the day by responding to an urgent call for air support and, under cover of smoke from the burning oil tanks, succeeded in driving off bombers which were harassing the troops. But the air position was now hopeless. Only Kallang airfield was usable and with Percival's approval Air Vice-Marshal Pulford withdrew the remnants of the squadron to Sumatra. It was intended to use Kallang only as an advanced landing base; in fact, from that day onward no British aircraft was seen in the sky above Singapore.

Fl. Lt. Arthur Donahue was ordered to fly one of the last

three operational Hurricanes out to Palembang. The guests at the Seaview Hotel were dismayed. "No matter how bad the news," Donahue wrote, "as long as they could see the R.A.F. still flying they felt there was hope.

"And still it was cool, quiet and peaceful where we sat on the veranda of that hotel that morning, only a few miles from the fighting. The artillery had quietened down with the coming of daylight.

"Denny and I were enthralled for a while watching an exotic, dark-haired English girl clad in shorts and a light sweater, exercising her two greyhounds among the palm trees out on the lawn.

"She was swinging a cloth about for them to leap at. Her movements and theirs were so graceful that I thought she must be a dancer, but someone said she was a nurse.

"It seemed that either she or the approaching enemy and the terrible fighting must be unreal. It just didn't make sense—but neither did a lot of things, in the last days of Singapore.

"My final memory of Singapore, as it appeared to me looking back for the last time, is of a bright green little country, resting on the edge of the bluest sea I'd ever seen, lovely in the morning sunlight except where the dark tragic mantle of smoke ran across its middle and beyond, covering and darkening the city on the seashore.

"The city itself, with huge leaping red fires in its north and south parts, appeared to rest on the floor of a vast cavern formed by the sinister curtains of black smoke which rose from beyond and towered over it, prophetically, like a great over-hanging cloak of doom."

Donahue, who was officially presumed killed in 1942, won the D.F.C. for his part in the Malayan campaign. The citation read: "This officer carried out low level reconnaissance sorties and successfully attacked enemy shipping and ground objectives. On one occasion while attacking enemy troops who were attempting a landing in the Singapore area he silenced the enemy fire and enabled the rest of the Squadron to press home attacks with impunity. He has destroyed several enemy planes."

II

The utter confusion of those days is revealed by everyone who was actually in the battle. Col. Dalley remembers little except that he was constantly dashing from one position to another, north, south, east and west, in efforts to keep in touch with his units—he had been unable to establish any other method of communication. Gunner Marshall, whose battery had not fired a shot while the Japanese were landing and capturing Tengah airfield, tells the story of his battery's movements from the morning of the 9th:

"We wondered what was going to happen next. If the Japs had taken Tengah airfield"—they never knew officially and therefore dared not open fire in case the rumour was untrue and they shelled British troops—"they must be coming on. We might be infiltrated or overrun at any moment. We were out of touch. I had my wireless truck but the Japanese had been jamming us for days. All I got on my wireless set was a high-pitched shriek.

"Captain Peach pushed off to Battery H.Q. to report the position and get instructions. Our gun position officer was worried because he had four guns and 4,000 rounds of ammo —and no target. It was now late morning. The situation was urgent. We had all this ammo on our hands. We thought we were in a static position, so we had stocked our ammo in various sites, well dug in.

"Our Troop Sergeant-Major returned from B.H.Q. with orders to move. Get the guns out, the R.A.S.C. would send trucks to collect the ammo. The guns were dug out, limbered up, and off they went in the direction of B.H.Q. at Bukit Timah cross-roads. The G.P.O., his truck-driver—a cheerful character called Eddie Hyde—and I were ordered to guard the ammo till the R.A.S.C. came.

"We waited there till it was dusk. The R.A.S.C. never arrived. We started to destroy the ammo by taking out the cordite charges, but it was a hopeless proposition. We heard firing just on our right. The G.P.O. decided we'd done our best, and we shoved off down the Jurong Road.

"We caught up with our guns three miles down, waiting to get through. It was chaos. The road was packed with vehicles and troops and guns. Our orders were to go to Changi—diagonally opposite to where we were, the other corner of the island! By now it was dark. The rain had stopped.

"We had to go through Bukit Timah cross-roads. It was dark, and there was a hell of a confusion. Vehicles going in all directions, or trying to. I noticed a large staff car with what appeared to be four generals or brigadiers inside, poring over a map. They looked anything but happy. I thought: 'Good heavens, doesn't look as if those blokes know what they're going to do next.'

"We got to Changi about 2.30 a.m. on the 10th. It was chaos all the way. I remember we stopped the O.P. truck at a cross-roads. We were leading the guns. There was an M.P. directing traffic.

"Captain Peach leaned out and asked: 'Which is the quickest way to Changi?'

"The M.P. said: 'My God, you're not going to Changi, are you? The Japs are knocking hell out of it.'

"Captain Peach rapped him over the tin hat with his cane. 'I'm not asking for your windy description, I'm asking the quickest way to Changi.'

"We got to Changi and put the guns in position. The same rôle as at Jurong, to cover the beaches. We were told the

Japs were attempting a north-east landing from the mainland, using Pulau Ubin as a screen.

"We were immediately forward of the Johore Battery, the static 15-inch gun taken from the old battleship *Queen Elizabeth*. This was the only static gun that could cover the mainland.

"This gun was firing armour-piercing ammo. They had no H.E. They hit the Johore Railway and made a considerable mess of it. I got that from a Jap later, who said there were very heavy Jap casualties among troop concentrations at the Railway.

"They also hit the headquarters at the base of the observation balloon. It was about 1,500 feet up, with a telephone cable to the ground. I got this from a Jap.

"We put up quite a barrage against Pulau Ubin. Then we suddenly got orders to withdraw from that position. As we pulled out of Changi, the Johore Battery spiked their guns— the 15-inch gun, the 9-inch and 6-inch guns, all spiked. One up and one down—bang! Finish! It was sad for gunners to see those barrels fall open like peeled bananas.

"At this point, men began to grouse and grumble, mainly through not knowing what was going on. We had no idea why we were ordered back. Men were saying 'Looks as if we've had it.'

"We moved back to a position in the residential quarters along Beach Road in support of the Manchesters and Loyals.

"We heard that General Wavell had flown in from Java and ordered Percival and Gordon Bennett to counter-attack. It didn't seem to have much effect. The Japs still infiltrated and we still moved back.

"We heard there'd been some pretty rough fighting around Bukit Timah, and that most of our big store dumps had fallen intact to the Japs. All this was rumour, we never got anything official.

"This move to Beach Road must have been about the 12th. It was hard to keep track of time. By now there seemed no overall command. If you were interested, you did what you could to help your infantry or whoever else was in the vicinity.

"We ran into a position in the gardens of some residential houses, and went into action, firing on a map reference given us by the Manchesters. As soon as we opened up we came under heavy mortar fire.

"That was the first time we really felt we were getting our teeth into it. Morale went sky high and personal safety was forgotten. The mortar fire stopped and we reckoned we'd knocked it out. That gave us a big lift up.

"We'd left our bell tents at Jurong. Sleep was a thing of the past. We took it in turns having a catnap on the truck or wherever we could find a quiet spot.

"We had a Gurkha battalion on our left. They were dealing with Jap infiltrators, and warned us that the Japs were riding bicycles dressed as Malays with grenades and light automatics hidden under their sarongs.

"The Japs never came out in the open and made a fight of it. Where we would form a front line and steam-roll a way through, if the Jap met any resistance he went round and shot up from behind.

"The Loyals moved back towards Singapore Town and we moved back to protect Kallang airfield. Our G.P.O. truck had gone by this time—a bomb or a shell got it. We found a Post Office van abandoned, so we used that. By now we were well armed with rifles and grenades we found abandoned in one of the Beach Road houses.

"We shot off to Beach Road with the guns. Beach Road was now under machine-gun fire from Japs who came down through Changi. We crossed the bridge by Kallang Road, went round by the Gas Works and into Beach Road. The guns were ahead of us. We told them to go ahead, and we'd follow on foot, and then we found this van.

"Suddenly we realized we were being machine-gunned by a Jap fighter chasing us along Beach Road. We swerved into a side road. The Jap circled and we made sure he was after us. Bob stopped the truck, and I piled out.

"I heard the plane zip down. It was a Navy Zero. I heard the rattle of his guns and the bullets smacking the road. I looked for Bob. He was still in the truck. He'd pulled up by a lamp-post and couldn't get out of his seat in time, so just sat there and hoped for the best. The van was hit but the bullets missed Bob.

"I had dived through a hole in the wall into a basement. I found out it was the basement of Raffles Hotel. That plane was still circling outside, so Bob followed me into the basement.

"We found a group of soldiers there. They were Sherwood Foresters from the 18th Division. They were finished. They had their rifles with them. 'What's up, mate?' we asked. 'What are you doing here?'

"'We've chucked it,' they said. They'd been cut off from their units, and didn't know where to go or what was happening, so they decided to sit it out under cover. Once they got into that frame of mind it was hopeless, they were finished.

"Our next position was in the yard of the Volunteer Barracks. We fired over open sights across the Kallang airfield in support of the Loyals. We fired at the Japs who were now in the residential area we'd left half an hour ago. By this time we were on our own, acting independently as a troop. There was no front line in Singapore. The front was where you happened to be and that was all over the place. There was plenty of fighting spirit among all the troops but it wasn't properly used. We weren't given our opportunity to stand and fight. Too many withdrawals."

Signalman Mutton said: "We had no radio, we never saw

a newspaper, we were completely out of touch with everything except our job. When we weren't out laying lines we used to get our heads down whenever we could for a catnap. We were given a job to do—'lay a line here', 'lay a line there' —that's all we knew. We had no idea how the fighting was going. Whenever we saw anyone we used to exchange rumours of what was happening."

Mutton remembers one vivid incident, during shelling in the northern sector. "We were up there and things were getting hotter and hotter. I don't remember just when but there was a direct hit on one of the attap houses. Col. Toosey went in and after a bit he came out with a little dead Malay girl in his arms. She was all limp and he stood there holding her. I'll never forget the look on his face. He stared towards Johore, where the shells were coming from, and he sort of muttered: 'You bastards, you'll pay for this!'

"Then he ordered every gun in the sector to open up, never mind saving the ammo. He gave them Japs hell for killing that little Malay girl."

By this time, units were all mixed up—signals, artillery, infantry, Indians and all. It was no longer any use trying to mend the breaks in the lines, which were being cut to ribbons by mortars.

"Once there was an order sent around," said Mutton. "It was the only official piece of information I got. One of our operators got hold of it in the signal truck and he showed it to me. It was from General Wavell and it said something like: 'When you see the enemy, fix your bayonets and charge and they will melt away like butter.'

"That gave me a laugh. I hadn't even seen a Jap."

III

In view of the alarming military situation on January 9, when it was known that the Japanese had landed on the Island, the Governor ordered the civil denial scheme to be put into effect. Teams drawn from the Public Works Department, and the Royal Engineers, Excise and Customs Officers, volunteers from the Observer Corps, Chinese and Indians, worked night and day. They were handicapped in their task of smashing plant and machinery by people representing vested interests who did their utmost to prevent the destruction of private property. They lodged appeals with the local authorities. Some companies with head offices in Britain, Australia or India petitioned their home Governments. A few actively intervened to prevent the denial teams getting to work. It was more important to these people—unnamed—that their property should be available for them at some future time, if and when the invader was defeated, than that war potential should be denied to the enemy.

Denial was further hampered by the Governor's orders on the 10th to withdraw all the European Supervisory Staff from the Singapore Harbour Board installations; and again by the unauthorized departure of some officials and key civilians who ran away from Singapore on February 9, 10 and 13. "Nevertheless," says *The War Against Japan*, "with certain exceptions the work was completed by the day of surrender. The exceptions were the installations captured by the enemy before they could be destroyed and those exempted by the Government on the plea of the effect of the denial operations on morale." They were still worrying about morale! The result was that the installations of about forty Chinese firms and the workshops and vehicles of two large motor dealers and their subsidiaries remained intact, and the

Japanese received "a welcome present of new vehicles and well-equipped workshops".

Two of the tasks which were accomplished successfully were the destruction of the State money and private stocks of liquor. Mr. Eric Pretty, acting Federal Secretary, burned 5,000,000 Straits dollars in the Treasury—the equivalent of about £600,000. "I never imagined I'd have so much money to burn," he said. Memories of Japanese atrocities in Hong Kong, when their troops got drunk on looted liquor and ran amok, prompted Sir Shenton Thomas's ban on liquor. It came into force at noon on Friday the 13th, but long before then work had begun on the colossal task of destroying every drop of strong drink on the Island.

The Customs staff smashed thousands of bottles in bonded warehouses. The big importers of wines and spirits— the Borneo Company, Caldbeck McGregor, John Little & Co., Robinson & Co.—had large stocks in their own warehouses. Private citizens took a last nip or two and poured the rest down the drain.

Mr. L. C. Hutchings, general manager of Robinson & Co., took half a dozen of his staff and went into action. "We hurled bottles against a high wall at the back of the store," he said. "It took a team of men a day and a half to smash all our stocks. The place smelt like sixteen breweries.

"It might have been the fumes or it might have been emotion, but tears rolled down my face when we came to twelve cases of Napoleon brandy. I decided to pop one bottle away to celebrate the completion of the job. I looked forward to that last nip. But when we'd finished all the smashing, and I went for the bottle I'd hidden . . . it was gone."

Mr. Jack Bennett, manager of the Borneo Company, and his friend Charles Martine, of the Kuching Branch, Sarawak, smashed the Borneo Company's stock of liquor. "It took hours," said Mr. Bennett, "and when we'd finished there was fifty thousand pounds' worth down the drain.

"I'll never forget, when we'd finished, we looked a bit scruffy and bloody. We came out into the street, and an old Chinese stopped us and opened a tin of cigarettes and offered us a smoke. I suppose he thought we'd come straight from the battle. In a way we had. We took the cigarette and he was pleased. That was the spirit in Singapore—real warm friendliness."

Altogether one and a half million bottles of spirits and 60,000 gallons of Samsu (Chinese spirit) were destroyed.

Gradually the life of Singapore was slowing but still Reller's Band played selections at the Adelphi Hotel and people queued for the cinemas. Apologists said it was good for morale, as when the people danced in Brussels on the eve of Waterloo. Raffles Hotel and the clubs and canteens were packed with staff officers who did not know what to do except sit and hope for the best. "All the red-tape from all over Malaya was packed into this little Island," said Capt. Charles Corry, Malayan Civil Service Volunteer with 3rd Corps.

"As a mere regimental officer, and proud of it in this disastrous campaign," wrote Maj. G. Wort, Wiltshire Regiment, seconded to The Malay Regiment and Adjutant to the 1st Battalion, "one feels that the gilded staff would have got on a lot better and conducted the war out here far more effectively if they had carried out some of the lessons they tried to lay down for others. We would like to have seen more of them out and about, instead of sitting in the 'Battle-Box' at Fort Canning, that 'seat' of Malaya Command. It was always felt that Fort Canning had more staff officers than any H.Q. ever known and all 'so busy'—yet ask the Regimental Officers from any unit that fought in this country what they thought."

Of Singapore in general, Maj. Wort wrote: "It seemed that the spirit of holding on and, if necessary, dying at your post had gone out of the make-up of things altogether. Perhaps in some cases it was because no definite orders were issued by the local Government as to policy. In many cases, one got the feeling that the one idea was to get to hell out of it, and the country too, if you could. One has only to ask people who were trying to get ships' passages for their womenfolk, what took place in the Steamship Offices. I was told that there were even some fit men trying to get out of the country by saying that they were sick or over sixty."

The streets were daily becoming more dangerous. Fifth Columnists were liable to be anywhere in Singapore now. It was pitch dark by seven o'clock in the evening, and a long night lay ahead with plenty of opportunities for disguised Japanese and pro-Jap quislings to sneak into the city and occupy good sniping positions in the many flats and offices that were now empty. Sniping was a constant danger along the main streets. Ted Fozard and Eric Pretty said that they dodged bullets near the Supreme Court building. There were plenty of windows in multi-storied business houses, tall blocks of flats, hotels and Chinese lodging-houses—though heaven help the pro-Jap if the loyal Chinese got hold of him.

A.R.P. wardens, most of whom were Chinese, challenged anything that moved by night, their fingers itching on the triggers.

Lt.-Col. Caister, one-time secretary of Essex Cricket Club and now Assistant Provost Marshal Malaya Command, kept getting complaints from people who had been sniped at along the main Orchard Road.

About nine o'clock on Thursday night two of Squadron-Leader Pat Atkins' R.A.F. redcaps spotted a light up in Amber Mansions, fronting Orchard Road. This was reported to the A.P.M. A mixed posse of redcaps went along to Amber Mansions with tommy-guns, hand grenades and revolvers drawn. From the outside, Amber Mansions was in total darkness. It looked sinister. The redcaps searched the ground floor. Nothing. They searched the first floor and so on until they reached the top. Then they found it—a glimmer of

THE SUNDAY TIMES

"Singapore Must Stand; It SHALL Stand"—H.E. the Governor

SUNDAY, FEBRUARY 15, 1942.

STRONG JAP PRESSURE

Defence Stubbornly Maintained

VOLUNTEERS IN ACTION

BRITISH, Australian, Indian and Malay troops, and including now men of the Straits Settlements Volunteer Force,, are disputing every attempt, by the Japanese to advance further towards the heart of Singapore town.

THERE was a strange dimunition of activity during last night, with little shelling and no bombing of the city area. Reasons for this were still unapparent when we went to Press at 7 a.m. British guns had been heard again from 6 a.m. and after a while there was another lull.

The official communique issued at 5.30 p.m. yesterday (Saturday) stated:

"During yesterday afternoon. enemy attacks developed in the Paya Lebar area and in the West. Both were in considerable strength.

"To-day the enemy has maintained his pressure, supporting his attacks with a number of high level bombing raids by large formation of aircraft, continual shelling by his artillery, and low dive-bombing attacks. His artillery have also shelled the town intermittently throughout the night and this morning

"Our troops—British, Australian, Indian and Malay—are disputing every attempt to advance further towards the heart of Singapore town.

"In the town itself, the civil defence services are making every effort to deal with the damage and civil casualties caused by hostile shelling and bombing."

It is understood that our artillery engaged some of the enemy forces with considerable success inflicting about 100 casualties.

Lord Moyne's Message To Singapore

THE following telegram has been received by His Excelleney from the Secretary of State for the Colonies to all those who are so gallantly and doggedly helping in the defence of Singapore

"You are going through a great trial but I knew you are doing everything you can and are resolved to continue to do so. I send to all of you my grateful thanks for your devoted assistance

JAPANESE CLAIMS MORE SUBDUED

THE Australian Army Minister. Mr Forde. said at 6 a.m. yesterday that he was expecting a cable from Lieut.-Gen. Percival, G.O.C. in Singapore, but the absence of this cable did not necessarily mean that the news was bad

The Japanese have almost ceased to broadcast claims of striking successes. The Domei Agency could only say, for instance, that the British continued to counter-attack yesterday, with an intense British bombardment from Blakan Mati and elsewhere.

A London view is that the Japanese have met a far more determined resistance than they expected.

Tokyo radio last night admitted that the Japanese forces have to "evade a rain of enemy bullets, and grudingly declared: "It appears to be the British plan to defend the fortress to the last" It also stated that the Singapore causeway has been destroyed a second time."

More Chinese Troops In Burma

FROM Burma. there is news of the arrival of more Chinese reinforcements. These troops are veterans of Gen. Chiang Kai-shek's armies and old campaigners against the Japanese. They are among the best troops China is able to put into the field.

THE first Victoria Cross to be awarded for fighting in Malaya, goes to Lieut.-Col. Wright Anderson of the Australian Imperial Forces, says the B.B.C.

FIRST V'C' WON IN MALAYA

THE first Victoria Cross to be awarded for fighting in Malaya, goes to Lieut.-Col. Wright Anderson of the Australian Imperial Forces, says the B.B.C.

NEWS IN BRIEF

Reuter Radio Service

REUTER news in brief (specially broadcast to Singapore last night from London), excluding items given elsewhere .

Heavy fighting is going on in the Paan area in Burma. The situation is rather obscure. It is not clear that the Japanse have succeeded in crossing the Salween River at Paan.

When he visited the Khyber Pass. Gen. Chiang Kai-shek met Afridi tribesmen, who assured him of support for the democracies

Batavia : A Japanese parachute attack took place at Palembang i n Sumatra. No details are available. There have been enemy reconnaissance flights over several parts of the Outer Provinces, with attacks here and there

Australian units are apparently still holding out in scattered strong points in the island of Ambonia The attacking Japanese force comprised 13 transports and several warships, and landings were made at three points.

FILL YOUR BATHS

Residents of Singapore are advised to conserve water very carefully and to use every receptacle possible—bottles, baths etc.—to keep water to combat fire or for other emergencies.

AMERICAN ARMY IN JAVA

THE New York Times reported yesterday morning that American troops had arrived in Java, and are stationed there alongside British and Australian forces.—By Radio.

LADY • THOMAS

We are happy to be able to state that Lady Thomas. who became ill several days ago, is now much better.

(Printed and published by the Straits Times Press, Ltd., Cecil Street, Singapore)

PASS THIS PAPER ON TO YOUR NEIGHBOURS

Above: Lt.-Gen. Percival (extreme right) and some of his staff officers march under Japanese guard to Lt.-Gen. Yamashita's H.Q. at Bukit Timah on Singapore Island. One British officer carries the Union Jack, another the white flag of surrender. *Below:* Yamashita (seated, centre) thumps the table with his fist to emphasize his terms—unconditional surrender. Lt.-Gen. Percival sits between his officers, his clenched hand to his mouth. *(Photos by courtesy of the Imperial War Museum)*

light showing under one of the doors. They crept along, tommy-guns leading, kicked open the door—and rushed in.

For a few tense moments there was dead silence, then a woman screamed. The room was a bathroom and in the bath sat an elderly Chinese *amah*, naked, holding in her two hands the expensive scent-spray which had belonged to her employer who had evacuated and left the flat in her care. She was giving her body the Coty treatment . . . and half a dozen red-faced redcaps apologized and withdrew.

In the absence of reliable news in the newspapers, rumour was everywhere. There were nearly as many rumours as bullets. Rumour said that when Wavell flew in on the 10th he created considerable hell at Headquarters because the Japanese had been allowed to land on Singapore Island and had not been pushed off again. Rumour said that he tore strips off General Percival. Another rumour said he told Gordon Bennett to get to hell out of it and take his bloody Australians with him.

The last complete editions of the Singapore newspapers were issued on Wednesday the 11th. On that day all the European newspaper managements and staffs went off to Batavia, along with the last of the overseas war correspondents. The Malayan Broadcasting Corporation transmitters were smashed and the European staff joined the scramble for Java. Cable and Wireless smashed their transmitters and departed. The Ministry of Information went and so did the Department of Information and Publicity, with the exception of Mr. W. A. Wilson, who had spent many years in Malaya as a journalist.

The war reporters and Singapore European newspaper staffs were collected by Capt. Henry Steele like a busy sheep-dog herding his flock. "If Singapore must fall," wrote *Tribune* editor E. M. Glover, "we would clearly be of more use to the Allied war effort away from the Island. Our

minds were made up for us by Captain Steele returning and
urging us to hurry."

"If you don't, the Japs will have your ears for breakfast,"
said the irrepressible Henry Steele—who three and a half
years later was the first Englishman to parachute back to
Singapore.

Sir Shenton Thomas took over the *Straits Times* as a Government Printing Office and appointed Mr. Wilson as acting
editor. Under Mr. Wilson a small staff of non-European
sub-editors and reporters, Ceylonese linotype operators and
a few Chinese printers, who all camped in the office,
managed to assemble and print a single-sheet quarto-size
edition of the *Straits Times*. The Tamil newsboys had disappeared from the streets, but distribution of the 7,000 copies
was effected through dispatch riders who delivered bundles
to A.R.P. posts. To look at these news-sheets now is to wonder whether the courage and enterprise of the staff were
worth it. The first issue, on the morning of the 12th, carried
six news items. Two had nothing to do with the war in
Singapore, one acknowledged a contribution of nearly 800
Straits dollars to the War Fund, the fourth remarked that
"congested areas in Singapore City may be bombed at any
time" and urged the civil population to disperse into open
spaces "so far as possible having regard to their occupation".
The fifth was an intriguing story about an anonymous Australian sergeant who had shown unusual initiative in destroying snipers. The sergeant placed a ventriloquist some
distance away and began to stalk the enemy. The Japanese
were so interested in the source of the mysterious voices
thrown by the ventriloquist that the sergeant was able to spot
them up their trees and kill four before he returned to his
platoon.

The sixth—and main—story was an eight-paragraph official
communiqué headed: "Fighting Continues on West Front
—Ack-Ack Gunners Destroy Three Jap Raiders." Its bland,

censor-inspired prose included the phrases: "Elsewhere on the Island there is no change in the situation . . ." "It is hoped to stabilize our position." . . . "continuous enemy pressure . . . slackened during the night."

A more faithful picture was drawn by Mr. Yates Mc-Daniel, Associated Press of America's correspondent in Singapore, and one of the last Pressmen to leave the Island. In his dramatic last message on Thursday, February 12, he cabled:

"The sky over Singapore is black with smoke from a dozen huge fires this morning as I write my last message from this once beautiful, prosperous and peaceful city. The roar and crash of cannonade and bursting bombs which are shaking my typewriter, and my hands which are wet with the perspiration of fright, tell me without need of an official communiqué that the war which started nine weeks ago 400 miles away is now in the outskirts of this shaken bastion of Empire.

"I am sure a bright tropic sun is somewhere overhead, but in my many-windowed room it is too dark to work without electric lights.

"Under the low rise where the battle is raging I see relay after relay of Japanese planes circling, then going into murderous dives on our soldiers, who are fighting back in the hell over which there is no screen of our own fighter planes. But the Japanese are not completely alone in the skies. I just saw two Vildebeestes—obsolete biplanes operating at 100 miles an hour—fly low over the Japanese positions and unload their bombs with a resounding crash. It makes me ashamed, sitting here with my heart beating faster than their old motors, when I think of the chance those lads have of getting back. If ever brave men earned undying glory, those R.A.F. pilots have this tragic morning.

"There are many other brave men in Singapore today. Not far away are A.A. batteries. They are in open spaces because they must have a clear field of fire. (Please pardon the break in continuity, but a packet of bombs just landed so close that I ducked behind the wall, which I hoped would, and did, screen the blast.) But those gun crews keep on fighting and peppering the smoke-limited ceiling every time Japanese planes come near, which is almost constantly.

"The all-clear has just sounded—a grand joke, because from my window I can see three Japanese planes hedge-hopping under a mile away. I heard a few minutes ago a tragic telephone conversation. Eric Davis, director of the Malayan Broadcasting Corporation, urged Sir Shenton Thomas, Governor, for permission to destroy the outlying broadcasting station. Sir Shenton demurred, saying the situation was not too bad. Davis telephoned the outlying station and instructed the staff to keep on the air, but to stand by for an urgent order. We tuned in to that station, and then, in the middle of a broadcast in the Malayan language, urging Singapore people to stand firm, the station went dead.

"Other men who stuck to their jobs until the last moment were F. L. Y. Duckworth, chief Press censor, of Surrey; Captain K. P. Fearon, Press adviser to the censorship; and Philip William Welby from Grimsby, of the Press censorship staff, who have stayed by us through peace and war.

"My colleagues left last night by ship, and the military spokesman gave his daily talk to an audience of three—representatives of two local papers and myself.

"We have less than a fifty per cent chance of getting clear. I am leaving now by car. I swear before I embark to put the car in gear and head it straight for the Straits of Malacca. Do not expect to hear from me for many days, but please inform my wife, Mrs. McDaniel, Hotel Preanger, Bandoeng, Java, that I have left this land of the living, and the dying."

12

The Tightening Noose

"All we have to do is to hang on grimly and inexorably, and not for long, and the reward will be freedom, happiness and peace for every one of us."—From the Governor's broadcast, February 1, 1942.

By the morning of Black Friday the Thirteenth, British withdrawal to the perimeter positions was completed; and except in the south-west and north, where the Japanese continued to make gains, the line was more or less stable. It was a false picture, for Singapore's situation was irretrievable. You would not have guessed it, however, from the second issue of the one-page *Straits Times* which appeared that morning. The title had been garnished with the Governor's famous slogan, printed in heavy type: "Singapore Must Stand; it SHALL Stand." The three columns carried nine news items from which the beleaguered defenders and civilians of Singapore could read that Chiang Kai-shek had seen Mr. Nehru of India, that the United States was planning to raise its Air Force to an ultimate total of two million men, and that Chinese forces had been successful in a small engagement against the Siamese on the Indo-China frontier. It also announced in a half-column story that an air battle had been fought between Japanese and Dutch East Indies forces over the Java Sea. Closer at home, the inhabitants of rural areas were warned that it was an offence to refuse to accept Malayan currency notes; families of Volunteers were told where to find temporary accommodation, potential readers of the sheet were told how to obtain their copies, free; and it was formally announced that from noon it would be an offence to possess intoxicating liquor. The "lead" story, which was headed "Japanese Suffer Huge Casualties In Singapore—R.A. Gunners Stick To Their Posts", contained only five paragraphs—all quotations from British and Australian newspapers. You could not guess that in the streets of Singapore City, black under the pall of oil smoke, the dead lay uncollected and unburied. Administration had broken down. The Army was hard pressed and running short of everything it needed—food, water, petrol, ammunition. Anti-aircraft guns—especially the Bofors on top of buildings —still blazed away at the non-stop procession of enemy planes which dropped their bombs and then returned to Johore to re-load, but they made little impact. Fighters machine-gunned the arterial roads with impunity. Shelling was incessant and no area was spared. Now that the defence line was drawn tight round the City, the whole of Singapore was a legitimate military target. General Gordon Bennett wrote in his diary:

"As I made my way through the now deserted streets of Singapore, streets that previously were a seething mass of industrious humanity, I could smell the blast of aerial bombs in the air. There was devastation everywhere. The shops were shuttered and deserted. There were hundreds of Chinese civilians who refused to leave their homes.

"Bombs were falling in a nearby street. On reaching the spot, I saw that the side of a building had fallen on an airraid shelter, the bomb penetrating deep into the ground and the explosion forcing in the sides of the shelter.

"A group of Chinese, Malays, Europeans and Australian soldiers were already at work shovelling and dragging the debris away.

"Soon there emerged from the shelter a Chinese boy, scratched and bleeding. He said: 'My sister is under there.' The rescuers dug furiously among the fallen masonry, one little wiry old Chinese man doing twice as much as the others, the sweat streaming from his body.

"At last the top of the shelter was uncovered. Beneath was a crushed mass of old men, women young and old, and young children—some still living, the rest dead.

"The little Oriental never stopped with his work, his sallow face showing the strain of his anguish. His wife and four children were there. Gradually he unearthed them—dead.

"He was later seen holding his only surviving daughter, aged ten, by the hand, watching them move away his family and the other unfortunates.

"This was going on hour after hour, day after day, and the same stolidity and steadfastness among the civilians was evident in every quarter of the City."

The City was crowded with armed deserters. An officer patrol rounded up 1,100 of them in the last few days, but there were still too many to be controlled by the Military Police. They skulked through the town, or hid in the basements of concrete buildings, or looted Chinese shops. Col. Chamier saw them going about with their rifles on their shoulders and their shirts stuffed with cigarettes and tinned food. Mr. Dick Mullaly described the municipal building and all the big buildings of Singapore on the seafront as "solid blocks of Australians sitting on the stairs". Some tried to get aboard the boats that were still leaving the stricken Island. There were instances of deserters threatening and even firing their rifles to try to force their way on board. It is impossible to say how many got away. Lionel Wigmore, in *The Japanese Thrust*, a volume of the Australian official war history, reports that Mr. Bowden cabled the Australian Government on February 12 that "a group of Australians and others had boarded a vessel without authority and in it had sailed to the Netherlands East Indies". He also records that on February 14 groups of soldiers, including Australians, were at large in Singapore seeking to escape and that a few, armed with tommy-guns and hand grenades, threatened to open fire on the launch *Osprey* unless they themselves were taken on board. "When the *Osprey* moved out at 11.30 p.m. with thirty-eight on board, rifle shots were fired at it but no one was hit." On Friday the 13th Australian 1st Corps headquarters in Java learned that "a party of 100 reported to be deserters" had arrived from Singapore. On the 14th, one officer and 165 other ranks "presumed to be deserters" were placed under guard in a Dutch prison in Batavia. When questioned, the men said: "Singapore was in a state of confusion, men were unable to contact their units and were ordered by officers whom they did contact to make their way to port and evacuate." In the absence of an official charge or evidence from the master of the vessel that had carried them, or O.C. troops, it was decided that their retention in prison was illegal. They were set free, and fought with Blackforce during the operations in Java.

The War Against Japan describes those troops who left the Island, or tried to, as "most . . . were deserters from administrative units, and men who had recently arrived in Malaya as reinforcements, inadequately trained and disciplined".

By this time, more and more exhausted troops were losing confidence in their leaders and morale was beginning to crack. But the well-trained units still fought stubbornly. Someone asked Dalley: "Are our chaps fighting?" He replied, drily: "Well, I keep seeing a lot of our bodies with bullet holes in them. They must be somewhere near the Japs to get shot."

Lt.-Col. J. R. G. Andre, O.C. 1st Battalion The Malay Regiment—soon to be awarded the D.S.O. for having "commanded his battalion with great gallantry combined with extreme judgment and calmness"—began his diary for Friday the Thirteenth with the words "George Wort's birthday". That day The Malay Regiment fought a fierce battle around the south coast, and before his birthday was out Maj. Wort was in Alexandra Hospital. His left hand up to his wrist-watch was undamaged; his elbow was shattered. His arm was amputated that night.

Private Ashley Warman, of 4th Suffolks, 5th Infantry Brigade, 18th British Division, gave special thanks for the large Chinese obelisk tombstones. "The Japs kept up an in-cessant barrage of mortar fire," he said, "and pinned the Suffolks in the big storm drains in the cemetery. I dodged behind those tombstones. Then I got mine—a nasty one in the groin—but I reckon that if it hadn't been for the tomb-stone I wouldn't be here now."

Into the narrowing City perimeter crowded the hapless civilians. Mr. Eric Pretty had to move from the bungalow he was sharing with friends when the military walked in to req-uisition it. He went to live at the Singapore Club. So, too, did the Governor, who evacuated Government House after several bombs had whizzed in and killed some of his serv-ants, including his chauffeur. So, too, did about a hundred other leading officials and prominent citizens. Lt.-Gen. Percival had moved in from Flagstaff House and put down his bed in Fort Canning. Sir Percy McElwaine, the Chief Justice of the Straits Settlements, and the judges from Penang and the Federated Malay States, took food and bed-ding and moved into the Supreme Court Building. Sir Percy, unruffled, remarked: "In 1688 seven bishops were confined in the Tower of London. In 1942 eight of His Majesty's judges took shelter in the Supreme Court, Singapore, prior to being confined in Changi Gaol."

Just before the ban on liquor came into force at noon that day, Mr. Jack Bennett and two friends went to the Singa-pore Club, had a few rounds of final drinks and signed the

usual bar chits. In 1946 he received his Club bill, including
those chits. "In spite of the battle raging all around, someone
was looking after the books," he said.

At two o'clock in the afternoon Percival held a conference
at Fort Canning. Others present were Lt.-Gen. Heath, Maj.-
Gens. Gordon Bennett, Keith Simmons, Beckwith-Smith,
and Key, and senior staff officers. They took stock of the
situation.

Food was down to seven days' supply of military stocks,
though there were still ample civilian supplies. There was
sufficient ammunition for the 25-pounders, the Bofors A.A.,
and the mortars. Petrol was down to one small dump plus
the amount still in vehicle tanks, plus the Asiatic Petroleum
reserve tanks on Palau Bukum Island. Water was running to
waste because breaks in the mains were gaining on repairs.
Only low pressure was available at street level, and hospitals
were running short. At the docks all civil labour had disappeared, in the town area debris from bombing and shelling
remained untouched and the dead lay unburied.

Percival gave orders for the fuel reserve at Palau Bukum to
be destroyed—a task which was only partially successful—
and discussed the question of a counter-attack. His generals
were unanimous that no counter-attack could be launched
because the men had been fighting day and night and
were approaching complete exhaustion. Gordon Bennett recorded:

"It was unanimously considered that new enemy attacks
would succeed and that sooner or later the enemy would
reach the streets of the City, which were crowded with battle
stragglers. It was also realized that the civilian population of
the City, which could not escape through the enemy cordon,
were the main sufferers. The heavy air blitz and artillery fire
were causing great havoc, and killing and maiming thousands
of innocent victims. It was decided to send a message to

General Wavell urging him to agree to immediate capitulation."

When Wavell's reply arrived shortly afterwards it said uncompromisingly:

"You must continue to inflict maximum damage on enemy for as long as possible by house-to-house fighting if necessary. Your action in tying down enemy and inflicting casualties may have vital influence in other theatres. Fully appreciate your situation but continued action essential."

The conference also discussed the evacuation of selected personnel from the Island in the final withdrawal of boats. Gordon Bennett said: "Owing to the limitation of shipping, accommodation for only 1,800 from the Army could be arranged. The A.I.F. allotment was 100. It was decided that only those whose capabilities would help our ultimate war effort should be evacuated and also that the proportion of officers to other ranks should be as one is to fifteen."

All that day and throughout the night the chosen few—and some who had not been chosen—made their final preparations. When the men who could not be evacuated heard of it, their morale was not improved.

II

One of Mr. Jack Bennett's managers of a subsidiary company to Borneo Motors was among those who considered that the time was ripe for him to leave Singapore. "As a precaution against going short of money," said Mr. Bennett drily, "he took all the cash from the company's safe. As another precaution, to keep the matter strictly legal, he left his cheque for the amount he removed. By a freak coincidence I came into possession of that company's cash box, found the cheque and kept it. When I got back to London in 1946 I was able to present the cheque to the man who had left it in

Singapore on Friday the 13th. That man's face was red."

Quite a number of people decided that the time was ripe to leave, permission or no permission. The Governor had given strict instructions that male civilians could leave only with his approval, but somehow senior officials like Brig. Simson and Mr. A. Bisseker were signing evacuation passes. Someone returned the favour and signed Mr. Bisseker's pass.

Mr. O. W. Gilmour, deputy municipal engineer in charge of debris and rescue, believed he would be of value somewhere else and later wrote a book describing how and why he left.

The Municipal President, Mr. Rayman, handed him a pass and told him not to tell anyone, even the heads of other departments, nor was anyone else to see him going.

"I had to return to the Control Room," wrote Mr. Gilmour, "and there I met Mr. Fyfe, the Municipal Engineer, who had heard that the President had sent for me. He inquired why, and I had to lie to him. Later I met a Chinese clerk who was a particularly good fellow. He had done his job and any other I asked of him all through. Now he wanted me to make certain arrangements for the next morning. It wasn't easy to tell him I would do this and that, knowing that I would not, and I had an almost uncontrollable desire to shake hands and wish him luck—but that would have been contrary to my instructions."

The title of Mr. Gilmour's book was *Singapore To Freedom.* He returned to Singapore after the war but did not stay long.

The game Tasmanian girl, Carline Reid, was given an exit pass at Fort Canning and told to go. Of her journey to the docks she wrote: "I got lost again. There were several troops taking cover in a drain. One of them nobly came out and guided me to a police station. It was empty. Footsteps echoed hollowly, but when I penetrated to the back I found two Hume Pipe shelters crammed with police and soldiers and at last got good directions.

"On one of the empty stretches I came on a straggling group of soldiers hurrying along. They looked as if they were going somewhere, so I pulled up to ask if this was the way to the docks. They turned out to be the defeated, unshaven, difficult kind, asked was I going to catch a ship as they were coming, too, and with that they climbed all over the car. The poor little machine was fragile at the best of times, and just would not go at all with the extra weight.

"I told them that at this rate none of us would get there, so they philosophically got down, saying, 'Good luck to you.' After another deserted stretch I came to a gate guarded by a soldier who approved my pass—and wished me luck!—and I drew up at the gangway."

The diary of Lt. G. Hutchinson, Johore Volunteers, records: "I had put a mattress under a tree at the cross-roads opposite the Goodwood Park Hotel, and told MacArturn that I wanted to sleep. I was pretty tired as I used to go round our posts two or three times every night, and during the day I was on duty, too. Anyway, I had just gone to sleep when MacArturn woke me and told me that a party of British officers were walking along our front, making towards Ardmore Flats. I went to see. It was the Lieut.-Colonel and H.Q. of one of the Norfolk Battalions.

"He asked me the way to the Harbour. I said 'Why?'

"He said: 'It's all over and we have been told to try and get away.'

"I said: 'We have a bunch of your chaps here, and what about the rest of your mob up near the Race Course?'

"He said: 'I have my orders. It's all over and I'm not going to argue with you.'

"Off they went. His Adjutant held back and said to me: 'You don't think we're running away, do you?'

"I said: 'It looks bloody like it to me.'"

Lt.-Col. Steedman, C.R.E. 11th Indian Division, said: "As a matter of fact, we weren't running away. We had been ordered to go. You see, I was in a similar party to the one that Hutchinson challenged. I can understand the feelings of men who saw us pushing off. There must have been a lot of misunderstandings like this."

Captain Charles Corry, Malayan Civil Service, was one of two Volunteer staff officers told to organize the get-away party nominated from the 3rd Indian Corps. Captain Corry said: "By this time *Dallforce* was being disbanded to protect the Chinese volunteers from Jap reprisal. Colonel Dalley wanted to get certain of his people away to Sumatra to join Colonel Warren and the stay-behind guerrilla organization. Dalley got a launch called the *Mary Rose*, which I think he took by force from some private source in Singapore. It was anchored outside the Master Attendant's (harbourmaster's) steps. We had a rendezvous at 7 p.m. Things were pretty sticky at this time. Our launch was the *Osprey* which I had managed to persuade the Master Attendant to hand over. We found there was no water on board, and had to spend some hours watering her. Some of the officers didn't turn up, but Dalley's people all turned up. Those we eventually took away were—Dalley, several of his officers, Bowden, the Australian Commissioner, and two of his assistants.

"I had about four of Heath's staff officers, including some Japanese-speakers. There were Mervyn Wynne of the Special Branch Police, and another man called Sim of the Police, who spoke Japanese and had a great deal to do with breaking up Jap spy rings. Wynne and Sim were ordered away by Shenton Thomas because he felt they would be immediately executed if they were found in Singapore.

"The *Mary Rose* was away at anchor. We were getting all our party on the *Osprey* and then were going to split between the two. We had a certain amount of trouble from deserters who wanted to get on the launch. We were threatened and fired at several times. One or two came along and got pushed into the drink."

Altogether about 3,000 people were sent away officially that night—Army nurses, Service officers and technicians, and civilian specialists. Others—the runaways—went furtively. Women and children were pushed on board wherever room could be found. They sailed from Singapore in a miscellaneous fleet of small boats—the last of the gunboats and patrol motor-launches, coastal steamers, tugs, harbour launches, outboard motor-boats, yachts, dinghies, tongkans and sampans. Some were seaworthy and some were not.

Tokyo Radio said that night: "There will be no Dunkirk at Singapore. The British are not going to be allowed to get away with it this time. All ships leaving will be destroyed."

Many were destroyed. Some never got past the minefields and harbour wreckage. Most were attacked by Japanese light naval craft and aircraft in the Banka Strait. None knew when they sailed that the Japanese Fleet now stood between Singapore Island and the haven of the Dutch East Indies. Singapore could not tell them because signals from Sumatra were in code—and the man with the code-book had taken it away with him to Java. Singapore received the messages but could not read them. . . .

Capt. Corry's launch *Osprey* fouled in some wreckage, and her passengers were transferred to the *Mary Rose*. They reached the Muntok Strait between Banka Island and Sumatra, and answered a signal from a vessel which they took to be a Dutch guard ship outside the Moesi River. Following instructions, the *Mary Rose* proceeded up the river until a second vessel appeared. Dawn came, and Capt. Corry and the passengers on the *Mary Rose* found themselves under the guns of two armed Japanese trawlers.

Lt.-Col. Steedman was lucky—he reached Padang on the west coast of Sumatra, and got a ship to India. Rear-Admiral Spooner and Air Vice-Marshal Pulford were among the ill-fated. Their motor-launch was wrecked on a malarial island; and before the survivors were taken off by the Japanese three months later, both Spooner and Pulford had died. Just before he left Singapore, Pulford said to Percival, "I suppose you and I will be blamed for this, but God knows we have done our best with what we have been given."

Lt. H. Gordon Riches, R.N.V.R. (Malaya) took his 75-foot motor-launch out from Singapore as one of six vessels in tow by a tug. All were sunk at the breakwater except the tug, which took survivors on board and reached Palembang. On the way up the Moesi River they picked up two R.A.F. pilots who were sitting out in trees overhanging the river. They had been shot down.

A group of Army nurses and soldiers from one of the wrecked ships reached Muntock Beach on Banka Island, and surrendered to the Japanese. A party of Jap soldiers marched the men along the beach out of sight; and presently returned, wiping their bayonets. The nurses, who wore the Red Cross emblem, were then ordered to walk into the sea. When they were waist deep in the water, still facing out to sea, the Japanese machine-gunned them. Only one soldier and one nurse survived that outrage.

The *Li Wo*, a former Upper Yangtse river steamer of 707 tons, requisitioned by the Royal Navy for Malayan patrol duty in 1940 and commanded by Lt. T. S. Wilkinson, R.N.R., sailed from Singapore with many passengers in addition to her eight officers and 68 crew. She was attacked repeatedly by aircraft but managed to evade the bombs. On the morning of the 14th she encountered part of the Japanese invasion convoy bound for Sumatra. Although armed with only one old 4-inch gun—for which she had just thirteen shells left—and two machine-guns, Lt. Wilkinson turned the *Li Wo* to attack the Japanese transports and hit one and set it on fire. For an hour and a half she fought against hopeless odds, taking heavy punishment from the escorting Japanese cruiser. When all thirteen shells had been fired, Lt. Wilkinson turned his crippled vessel full steam ahead—and rammed the burning Japanese transport. The cruiser then blew the *Li Wo* out of the water. There were only ten survivors, and the gallant Lt. Wilkinson was not one of them. Years later, when the story at last reached London, Wilkinson was posthumously awarded the Victoria Cross.

Most of the people who left Singapore in the last few days failed to reach freedom. The majority were cast far and wide among the myriad islands south of Singapore and suffered tormenting privations. Some were rescued by fishermen and taken step by step along an escape route which remained open for several weeks. Some were picked up by the enemy and taken back to Singapore, Sumatra or Java. Some were never picked up. Many were never heard of again after they left Singapore.

III

On Singapore Island the noose tightened remorselessly round the tortured body of the City. In the north and in the south-west the Japanese advanced steadily against a desperate defence. That night—the night of the 13th/14th—the Japanese ran amok in Alexandra British military hospital. The story of that black episode has been preserved in a typescript of unknown authorship which has found its way to the British Imperial War Museum. It states:

"During the morning the water supply was cut off. Shelling and air activity was intense. Bursting shells, mortar bombs, with an occasional shot from our own artillery.

"The enemy were approaching the rear of the hospital from the Ayer Rajah Road. The number of incoming patients had lessened considerably and there was little or no traffic in the wards. During the morning, routine work continued. Japanese troops were seen for the first time at about 13.40 hours, attacking towards the Sisters' quarters.

"Japanese fighting troops were about to enter the Hospital from the rear. Lt. Weston went from the reception room to the rear entrance with a white flag to signify the surrender of the Hospital. The Japanese took no notice of the flag and Lt. Weston was bayoneted to death by the first Japanese to enter the Hospital. The troops now entered the Hospital and ran amok. They were excitable and jumpy, and neither the Red Cross brassards nor the shouting of the word 'hospital' had any effect.

"The following events started at about the same time.

"(*a*) One party entered the theatre block. At this time operations were being prepared in the corridors between the Sisters' bunks and the main theatre, this area being the best lighted and the most sheltered part of the block. The Japs climbed in through the corridors and at the same time a shot from the window was fired, wounding Pte. Lewis in the arm. About ten Japs came into the corridor and all the R.A.M.C. personnel put up their hands. Capt. Smiley pointed to the Red Cross brassards but the enemy appeared very excited and took no notice. The Japs then motioned them to move along the corridor, which they did, and then for no apparent reason, set upon them with bayonets. Lt. Rogers was bayoneted through the throat twice and died at once. Capt. Parkinson was bayoneted to death as also was Cpl. McHewan and Pte. Lewis. A patient on the operating table was bayoneted to death. He was later identified as Cpl. Holden of the 2nd Loyals. Capt. Smiley was bayoneted but he struck the blade away and it hit his cigarette case which was in his left breast-pocket. He was again lunged at and was wounded in the left groin, the previous thrust having wounded his left forearm. He then pretended to be killed, and pushed Pte. Sutton to the floor calling to the others to be quiet. The Japanese then left the corridor.

"(*b*) Another party of Japs went into the wards and ordered the Medical Officer and those patients who could walk outside the Hospital. In one ward two patients were bayoneted. The Japs then went upstairs and gave similar instructions. These Japs appeared to be more human and motioned the patients on stretchers to stay behind. Patients and personnel numbering about two hundred were taken outside, their hands tied behind them with a slip knot, one length of cord being used for about 4 or 5 persons. Some of the patients could only hobble, others had only one arm, while some were still in plaster and others were obviously very ill.

"Many of the seriously ill showed signs of great distress, one or two collapsed and had to be revived. The party was marched by a circular route to the old quarters, where they were herded into rooms, 50 to 70 persons per room, the sizes of which varied from 9 by 9 to 10 by 12 feet, where they literally jammed in and it took minutes to raise their hands above their heads. Sitting down was out of the question and they were forced to urinate against each other. During the night many died, and all suffered severely from thirst and the suffocating atmosphere. Water was promised but none came. When dawn arrived Japs could be seen with cases of tinned fruit which they kept for themselves. By the evening the shelling was at its maximum and shells were bursting all round. One struck the roof, blowing off doors and windows and injuring some of the prisoners. When this happened, 8 men tried to escape, some were successful but others were hit by machine-gun bullets. Previous to this Japs were leading small parties out of sight where afterwards we heard yells and screams, then a Jap soldier returning and wiping blood from his bayonet which left little doubt as to their fate.

"Except for a few that escaped none of these parties were ever seen again. Capt. Alderdyce, who could speak a little bit of Japanese, thought that he would be taken as hostage. Cpls. McDonough and Wilkins wanted them to take care of wounded and they were not seen again since that night and for the last time, and it was assumed that Cpl. Wilkins and McDonough suffered the same fate as the doomed 200 in the servants' quarters. The body of Cpl. McDonough was found outside the Hospital and it appeared that he had been killed by shrapnel.

"(c) A party of Japs came into the reception room shouting and threatening the staff and patients who had congregated there. Sgt. Sheriff was killed. It was difficult to understand the reason for this barbaric attack on the Hospital. Investigations were carried out to find a possible explanation —rumours had it that a party of Sappers and Miners were digging a trench at the rear of the Hospital and on the approach of the Japs made a run for it. When the Japs passed through the Hospital there were about 40 or 50 herded into the corridor and a guard placed on them. Later the guard went away, and Capt. Barlett went out but could find no trace of the Japs."

13

The Fortress Falls

"It will be disgraceful if we cannot hold our boasted fortress of Singapore to inferior enemy forces. There must be no thought of sparing the troops or civilian population, and no mercy must be shown in any shape or form to any weakness. Commanders and senior officers must lead the troops and, if necessary, die with them. There must be no thought of surrender. Every unit must fight it out to the end and in close contact with the enemy."—From General Wavell's signal to General Percival, February 10, 1942.

Early on the 14th, General Percival was told that a complete failure of the water supply was imminent. Breaks in the water mains caused by bombing and shelling meant that more than half the supply from the reservoirs was being lost. The supply might last forty-eight hours, perhaps for only twenty-four hours. Sir Shenton Thomas warned the G.O.C. of the danger of an epidemic if a large proportion of the population was suddenly deprived of water; but Percival replied that though the shortage was serious it did not make the defence of the city impossible. He intended to go on fighting. Sir Shenton signalled the Colonial Office in London, pointing out that one million people were concentrated within a radius of three miles, with water for only twenty-four hours longer. Percival signalled Wavell in Java, asking, "Would you consider giving me wider discretionary powers?" Wavell replied, "In all places where sufficiency of water exists for troops they must go on fighting. Your gallant stand is serving purpose and must be continued to limit of endurance."

South-west of the city there was further heavy fighting, but the enemy attacks were beaten off by the 1st Malaya Brigade after bitter hand-to-hand fighting. The Australians were subjected to heavy shelling; and were mortified to observe the Japanese moving west, without being able to shell them. Ammunition was being rationed.

Singapore City was again heavily bombed and shelled. Streets were blocked by the wreckage of smashed buildings, broken telegraph poles and tangled wires. The civil hospitals were crowded with wounded, and the ground floors of many large hotels and buildings were being used to house casualties. The grounds of the Singapore General Hospital were a shambles. Its once smooth, bright green lawns and rich flower-beds were gashed with huge pits dug for the burial of the dead. Bodies were carted out of the hospital and added to the stench, and lime was thrown on top. It was a gruesome business. Inside it was even worse. There were blood and guts and death everywhere. The stench was foul despite the disinfectant. Yet the Matron, Miss Kathleen Stewart, the Sisters and nurses worked on amid the human butchery. Patients, some alive, some dead, were in thousands. They were on the beds, under the beds, between the beds, overflowing out of the wards and filling the corridors. They were even lying on the stairs. Doctors, helped by nurses and orderlies, moved about finding the dead and hauling them out for burial in the loathsome pits outside. Army and civilian doctors had joined forces and worked as a team. The operating theatres were like butchers' yards. Surgeons amputated shattered limbs as fast as they could operate. There was no time for the delicate treatment which might have saved many limbs. The nursing girls stood by and helped. "Those girls were bathed in blood the last few days," said Sqdn. Ldr. Pat Atkins. "Sister Mollie Hill was a very brave girl. She was the head nursing Sister and she worked the clock round. A pretty woman in her late thirties. South African. They were all brave girls. They had no thought for themselves. Completely selfless." Water was short and instruments were sterilized in water that had been used again and again. Towards the end nurses were washing their hands in bottled mineral water.

During the night of the 14th/15th there was little enemy activity and the general line still held at daybreak on the 15th, despite some gains by the enemy in the west. This was the only crumb of comfort for the defenders as the sun rose on the burning city. Even the ocean seemed to be on fire. The blazing soap factory on Singapore River threw grotesque patterns on the nearby buildings. Up north the blazing oil tanks at the Naval Base—now in Japanese hands—continued to pour a vast mantle of filthy black smoke overhead. Out west at Normanton more oil tanks were burning. Alexandra was ablaze. Timber sheds at Kallang shot up huge tongues of flame—one observer said they reached a height of 600 feet. Godowns and warehouses along the waterfront were afire. Paint and oil from anchored lighters spilled and spread on the water, still blazing. Stocks of rubber smouldered and stank. Across the sea, orbs of fire shone from the islands of Palau Bakum and Sambau, where more oil reserves had been fired to deny them to the enemy. It looked as if minor setting suns had come to rest on the horizon and burst into flame, to ooze liquid fire over the becalmed water.

The Governor watched it from the windows of the Singapore Club. Beside stood Mr. Pretty. "My chief thought was tremendous relief that my wife and children had gone early on and were safe," said Mr. Pretty. "From where we stood, overlooking the Inner Basin, the whole place seemed to be a blazing inferno. The Governor was very upset. He kept saying 'Isn't this terrible?' or words to that effect."

General Gordon Bennett wrote of Black Sunday the 15th:

"Today opened with a hopeless dawn of despair. There is
no hope or help on the horizon. The tropical sun is sending
its steamy heat on to the dying city which is writhing in its
agony.

"The flanks of the Army continue to fall back. The enemy
has advanced past Pasir Panjang towards the City. Enemy
troop movement along Bukit Timah Road has been shelled
by our artillery, as also have enemy troops opposite our own
A.I.F. front, but the momentum of the enemy advance goes
unchecked."

Early that morning Wavell signalled Percival:

"So long as you are in a position to inflict losses and
damage to enemy and your troops are physically capable of
doing so, you must fight on. Time gained and damage to
enemy are of vital importance at this juncture. When you
are fully satisfied that this is no longer possible I give you
discretion to cease resistance. Inform me of intentions.
Whatever happens I thank you and all your troops for gallant efforts of last few days."

It was the wider discretion which Percival had requested
two days earlier. The decision was now in his hands.

On that same day Wavell sent a long telegram to the Prime
Minister in London. Its terms were very different. According to one staff officer, "It was mainly about personalities
and it did not pull any punches."

In England, Percival's daughter Margery was celebrating
her twelfth birthday.

Lt.-Gen. Percival went to early morning church service at
Fort Canning and then called a conference of all senior commanders and key civic officials. The Governor was not
present. The G.O.C. told them that the Supreme Commander had granted him the power to capitulate and then
they discussed the situation. In Percival's opinion there were
two alternatives: to counter-attack in an effort to regain the
reservoirs and military food depots captured by the enemy

Above: Yamashita inspects the worst-hit areas on Singapore Island after the British surrender.   *Below:* Yamashita's victorious troops march through the almost undamaged City
centre. *(Photos by courtesy of the Imperial War Museum)*

Left: Chinese men, women and children, herded together by the Japanese troops, awaiting their fate after the fall of Singapore. *Below:* British prisoners of war are forced by their captors to sweep the streets under the eyes of the Asiatic population. *(Photos by courtesy of the Imperial War Museum)*

at Bukit Timah, or to surrender. His generals were convinced that a counter-attack was out of the question. The troops were too exhausted and dispirited and had insufficient ammunition and supplies.

Gordon Bennett's diary records: "Silently and sadly we decided to surrender."

II

At that conference on Black Sunday the 15th, General Percival produced a letter, addressed to him from the Japanese commander, Yamashita. The Japanese had littered Singapore with leaflets; and on February 10 several small wooden boxes had hit the ground. They contained copies of the following:

Lieut. Gen. Tomoyuki Yamashita
High Com. Nippon Army
Feb 10th 1942

To:—The High Com. of the British Army in Malaya.

Your Excellency,

I, the High Com. of the Nippon Army based on the spirit of the Japanese chivalry have the honour of presenting the note to your Excellency advising you to surrender the whole force in Malaya.

Many sincere respects are due to your army which, true to the traditional spirit of Great Britain is bravely defending Singapore, which now stands isolated and unaided.

Many fierce and gallant fights have been fought by your gallant men and Officers, to the honour of British warriorship.

But the developments of the general war situation has already sealed the fate of Singapore and continuation of futile resistance would not only serve to inflict direct harm and injuries to thousands of non-combatants in the City, throwing them into further miseries and horrors of war, but also would not certainly add anything to the honour of your Army.

I expect that your Excellency, accepting my advice,

will give up this meaningless and desperate resistance and promptly order the entire front to cease hostilities, and will dispatch at the same time your parliamentaire according to the procedure shown at the end of this advice. If, on the contrary, your Excellency should reject my advice and the present resistance continue, I shall be obliged, though reluctantly, from humanitarian considerations, to order my army to make an annihilating attack on Singapore.

In closing this note of advice I pay again my sincere respects to your Excellency.

Tomoyuki Yamashita.

1. The Parliamentaire should proceed to Bukit Timah Road.
2. The Parliamentaire should bear a white flag and the Union Jack.

On Friday the 13th some more Japanese boxes were airdropped. They contained the following:

ADMONITION

I have the honour of presenting to you this Admonition of Peace from the standpoint of the Nippon Samurai Spirit. Nippon Navy, Army and Air Force have conquered the Philippine Islands and Hong Kong and annihilated the British Extreme Oriental Fleet in the Southern Seas. The command of the Pacific Ocean and the Indian Ocean as well as the Aviation power in the Southern and Western Asian Continents is now under the control of the Nippon Forces. India has risen in rebellion. Thai and Malaya have been subjected to Nippon without any remarkable resistance. The war has almost been settled already and Malay is under Nippon Power. Since the 18th Century Singapore has been the starting point of the development of your country and the important juncture of the civilisation of the West and East. Our Army cannot suffer as well as you to see this district burn to ashes by the war. Traditionally when Nippon is at war, when she takes her arms, she is always based upon the loyalty and breaking wrong and helping right and she does not and never aims at the conquest of other nations nor the expansion of her territories.

The War cause, at this time, as you are well aware, originated from this loyalty. We want to establish new order and some of the mutual prosperity in the Eastern Orient. You cannot deny at the bottom of your impartial hearts that this is divine will and humanity to give happiness to millions of East Orientals mourning under the exploitation and persecution. Consequently the Nippon Army, based upon this great loyalty, attacks without reserve those who resist them, but not only the innocent people but also the surrendered to them will be treated kindly according to their Sumaraism. When I imagined the state of mind of you who have so well done your duty, isolated and without rescuer, and now surrounded by our Armies, how much more could I not sincerely sympathise with you. This is why I do advise you to make peace and give you a friendly hand to co-operate for the settlement of the Oriental Peace. Many thousands of wives and children of your Officers and Soldiers are heartily waiting in their Native Land to the coming home of their husbands and fathers and many hundreds of thousands of innocent people are also passionately wishing to avoid the calamities of War.

I expect you to consider upon the eternal honour of British Tradition, and you, be persuaded by this Admonition. Upon my word, we won't kill you, but treat you as Officers and Soldiers if you come to us. But if you resist us we will gibe swords.

Singapore   Nippon Army.
13 Feb. 42

The "parliamentaire" set off from Fort Canning. Lt.-Col. Chamier, Commandant, Intelligence Corps, Malaya, saw them go. "I was coming up to Fort Canning and a car came out and passed me at the gates. It was an open car with the hood down. Brig. Newbiggin (Administrative Branch, Malaya Command) was in the car, with Hugh Fraser, the Colonial Secretary, Capt. Wild, the Japanese-speaking interpreter, and another chap.

"Sticking out of the back of the car were two flagpoles. One was a furled Union Jack. The other was a furled white flag."

This first party was sent off to discuss truce terms. Yamashita would not even see them. "Send Percival," he said, and the party returned for the General Officer Commanding.

The British drove to a point near Bukit Timah and there the G.O.C. and his party alighted from their car and marched under Japanese escort to the enemy Headquarters, set up in the Ford Motor Factory. Percival had to carry the white flag in person to Yamashita.

Their conference lasted fifty minutes and the G.O.C. has not revealed—either in his book or his dispatch to Parliament—what took place there. The Japanese were less reticent—it was not painful and humiliating for them. They published the following account of the final conversation:

Yamashita. We have just received your reply. The Japanese Army will consider nothing but unconditional surrender.

Percival. It is 9.15 p.m. Japanese Time, I fear we shall not be able to submit our final reply before midnight.

Y. (loudly) Reply to us only whether our terms are acceptable to you or not. Things have to be done swiftly as we are ready to resume firing in the evening.

P. Won't you please wait until you formally file into Singapore?

Y. It is impossible. In the first place, why not disarm all the British troops here, leaving only about 1000 armed gendarmes for maintaining peace? In the second place, under no circumstances can we tolerate further British resistance.

P. One of your terms handed to us demanded that we turn over certain representatives of the Chungking régime to you. Their names are not clear to us.

Y. By that we mean that you arrest and turn over to us Ching Kam Ming, one of the Chinese liaison men.

P. I ask that the Nippon Army reciprocate with us in discontinuing attack.

Y. Agreed. What has become of the Nippon citizens in Singapore?

P. They have all been transferred to India. We do not

know exactly where. The British troops would like to cease fire at 11.30 p.m.

Y. That is too late. By 11.00 p.m. we shall place part of our Army in Singapore proper.

P. Unless you allow us to 11.30 p.m. I fear that I shall not be able to transmit the order to all my troops.

Y. Then 11.30 will do.

P. Please do not allow the Nippon Army to enter Singapore until tomorrow.

Y. Why not assemble all your arms immediately in the heart of Singapore so that our Army can check them?

P. Why not let us arrange that tomorrow morning?

Y. It is a matter we can arrange as a side issue.

P. Even 11.00 p.m. is a little late for all troops to cease fire. Why not let them cease hostilities in their present positions?

Y. In that case we shall continue firing until 11.30 p.m. I would advise you to order cessation of hostilities immediately.

P. I shall see that they cease firing immediately I return to my H.Q., and that the firing ceases by 11.30 p.m. In the City area the firing will cease immediately and in the distant areas not later than 11.30 p.m.

Y. As proof of your good faith we shall hold the Highest British Commanders and the Governor of the Straits Settlements in custody at our headquarters. (A look of amazement was noticed on General Percival's face.)

P. Cannot the Nippon Army remain in its present position so that we may resume notifications again tomorrow at 7.00 a.m.?

Y. What! I want the hostilities to cease tonight and I want to remind you that the question is strictly a matter of this. If you can discontinue resistance by 11.30 p.m. we shall hold the Highest Commander and the Governor of the Straits Settlements in our custody. If you cannot do this, the Highest Commander and the Governor must come to our H.Q. by 10.00 p.m.

P. We shall discontinue firing by 10.00 p.m. Nippon Time. Had we better remain in our present positions tonight?

Y. Speaking on the whole, see that your troops remain in

their positions and assemble tonight, disarmed, at the pre-
scribed places. I approve of the cessation of hostilities at
10.00 p.m. After we have finished firing, all the British troops
should disarm themselves save 1000 men whom we shall permit to carry arms to maintain order. You have agreed to the
terms, but you have not yet made yourself clear as to
whether you have agreed to unconditional surrender or not.

(General Percival, with bowed head and in a faint voice,
gave his consent. It was 7.30 p.m.)

Y. If you have accepted our terms I would like to hear
them from your own lips once more.

P. The British troops will cease hostilities not later than
10.00 p.m. Nippon Time.

Y. The British troops shall disarm themselves completely,
except 1000 men whom the Imperial Army will allow to carry
arms in order to obtain peace and order. If your troops in-
fringe upon these terms, the Imperial Japanese Army will
resume hostilities immediately.

P. I agree. I have a request to make. Will the Imperial
Army protect the women and children and the British civilians, men, women and children?

Y. We shall see to it. Please sign this truce agreement.

The Commander of the surrendered British Garrison af-
fixed his signature at 7.50 p.m. on the 15th February 1942.

III

Memories of that day are vivid in many minds. . . .

"We fired continuously from early morning till the eve-
ning," said Gunner-Signaller Ken Marshall. By this time
Marshall was on the guns in between fixing signal lines to the
O.P.

"We shoved the ammo up, and by now our guns were
beginning to feel the strain. The recoil systems were beginning to leak. They were oil-filled normally. We ran out of oil.
They were leaking so badly they were running empty. We
filled the recoil system with condensed milk. It worked well
enough to go on firing.

"We were very tired, everything was very confused, everyone was dazed. For the past five days and nights we'd had continuous artillery fire, and very little sleep.

"In circumstances like these, men will go on automatically until they reach the point of unwinding. We were getting heavy casualties, which wasn't helping. Most of the casualties were from shrapnel, mortar fire, anti-personnel bombs.

"We wondered what was going on the other side of town. We thought: Here we are, retreating into Singapore Town. We've got the Loyals and Manchesters in front of us—who else is in this war? What's going on over the other side of town?

"Things seemed quiet over there. And then we saw the Jap flag go up over Fort Canning. We had no thought of surrender or being prisoners. We'd been told the Japs didn't take prisoners. We went on firing."

Signalman Fred Mutton watched the gunners in the 28th Indian Brigade sector push "one up and one down", then retire to a safe distance behind each gun and jerk the lanyard. The shell in the breech end hit the shell in the mouth, the two shells exploded, and the gun barrel fell open like a peeled banana.

"Well," said Fred, who had never before seen a gun spiked. "Bang goes the bloody fortress."

Every Gurkha broke his *kukri* knife. They felt the shame worse than the British—it was against their code to surrender.

Lt.-Col. F. W. Young, commanding the 2nd Battalion The Malay Regiment, said: "It hadn't occurred to me to surrender but when it came . . . well it's all part of one's occupational risk. With a bit of luck I might have gone to a higher rank, on the other hand, if I hadn't been in Singapore and become a prisoner of war, I might have been killed somewhere else. One must be philosophical about these matters."

Gordon Bennett said: "One Australian officer broke down completely. 'We can't surrender to the Japs,'" he said.

Mrs. Jill Dawson, wife of the civil secretary of defence and secretary of the War Council, remembers odd unimportant things, "like the canary freed when its owners left. It flew into my house and adopted me.

"When the fighting stopped, the most uncanny thing was the silence. And the most uncanny thing about the silence was the complete silence of all animal life on the island for a day or two."

Mr. H. Miller, one of the non-European *Straits Times* reporters, described the last hours before surrender: "I was at the headquarters of Military Intelligence in the Cathay Building, waiting for Major H. to return with a communiqué. I stood with a brigadier on the balcony of the fifth floor watching the pulverizing of Singapore. He was bare-footed, taking it easy. He had fought all the way down from the Thai border. In the room behind us, a colonel of a Highland Regiment chatted with an Intelligence captain. A subaltern sat on the edge of a table, a worried look in his eyes. Another subaltern, helped by an Indian batman, was preparing his 'prisoner of war' kit—rations, drink, books, clothes. The Colonel was saying, 'Naturally, the terms will be unconditional surrender.'

"Out on our right, Tank Road and its environs were being methodically broken up by Japanese shells. It was a fascinating spectacle but for the fact that we were witnessing the last agonies of a great city. Dust and earth rose in volumes, and branches snapped from trees. Shells whined over the Cathay and whistled into terrific explosions opposite us. From the room behind came the question, 'Did that one get Fort Canning?' and the brigadier called back, 'Yes, but not G.H.Q.' As we stood there, the brigadier traced the Malayan campaign, recounting the reverses all the way down. The chatter of the guns heralded the reappearance of the Japanese bombers. Involuntarily we crouched, straightened as we heard the rumble of explosions. Clouds of dust and debris flew into the air, and tongues of flame leaped skyward. It was the Jap Air Force's final curtain call. The shelling ceased suddenly. An uncanny silence fell over Singapore."

Mr. Charles Martine, of the Kuching Branch of the Borneo Company, also remembers the silence. "It was eerie to a degree. When you'd been running helter-skelter in all directions, with a pandemonium of bangs and explosions and all sorts of noise all around you, the silence after the surrender was sudden and dramatic and eerie."

Mr. Jack Bennett stood on the Mercantile Bank roof that evening and looked over the surroundings.

"From that height one got a general impression of a vast ring of fire," he said, "almost as if there were four walls of fire and Singapore burning in the middle . . . whereas in fact, Singapore as a whole was not very badly knocked about. Fort Canning, for example, was right in the middle, and that was almost untouched. There were no vast areas in the town devastated. Damage was local. Chinatown suffered a lot, probably heavier than the rest.

"Fires, too, were local, though they gave an impression of much more, especially at night. What looked like a holocaust at night—which seems to be the impression carried away by people who departed in rather a hurry—was not really so bad in the calm light of day."

Fighting on Singapore Island ceased at 8:30 P.M. local time. That same evening a message from General Percival was received at General Wavell's Supreme Headquarters in the Dutch East Indies. The message said:

"Owing to losses from enemy action, water, petrol, food and ammunition practically finished. Unable therefore to continue the fight any longer. All ranks have done their best and grateful for your help."

Then silence fell on Singapore.

Postscript

Half an hour after the surrender, at 8:30 P.M. on Sunday the 15th, the Bishop of Singapore, the Rt. Rev. John Leonard Wilson, took a service in St. Andrew's Cathedral. The great building in the heart of the city, marked by a tall spire and exposed in a broad green compound, had miraculously survived the bombardment and for four days had been used by the R.A.M.C. as an emergency hospital for civilians of all nationalities as well as for troops. The pews had been removed and the huge nave was now crowded with wounded people. Some lay on camp-beds but the majority were bedded on blankets and lay in rows on the stone floor.

Several hundred civilians, mostly Asiatic, came in from nearby municipal offices and houses. They crowded into the choir stalls which, like the sanctuary, were kept free for services. There was general relief that the nerve-racking din of bombardment had stopped but relief was mixed with anxiety about tomorrow.

St. Andrew's had witnessed many impressive services but none so intensely moving as when Bishop Wilson, standing below the choir stalls, led that gathering of torn and troubled humanity in a shortened evensong. A major of the R.A.M.C. played the organ, and people quietly sang the hymn, "Praise, my soul, the King of Heaven."

Maj.-Gen. Gordon Bennett, Commander of the Australian Imperial Forces, escaped that night with his A.D.C., Lt. Gordon Walker, and Maj. Moses, A.I.F., in a sampan. He reached Australia by air on February 27. All the other senior officers, except those ordered away, stayed and surrendered.

Immediately after the war ended in 1945, an official inquiry was held in Melbourne into the circumstances of Gordon Bennett's escape, following a letter written by General Percival to the Australian Army Board. The judicial finding was that Gordon Bennett should have stayed with his troops "until surrender was complete" but that his escape was "inspired by high patriotism and by the belief that he was acting in Australia's best interests. He did in fact bring back valuable information which was used in jungle warfare." Gordon Bennett maintained that he did in fact escape after the surrender was complete.

"One thing I shall always remember," said Mr. Jack Bennett of the Borneo Company. "When we all had to go out and be paraded in front of the Japs—I think it was the next day or the day after—there were crowds of us going downstairs in the Mercantile Bank Building. As Martine and I went down to the entrance hall, we noticed the crowd separating and walking round, as if there was some obstruction. It was a British Army revolver, lying there all by itself. Someone had got hold of it and now had thrown it away. It wouldn't do to be found with a revolver in your possession.

"We stood there and watched everybody carefully avoiding it as if it was a red-hot poker. I daresay it stayed there until a Jap came and picked it up."

After the Massacre on Black Friday of British wounded and medical staff at Alexandra Military Hospital, the unknown author concluded his story:

"At about 1800 hours the Japs took a party including Sgt. Anderson and some others away. Their hands were tied behind them and they were led to a drain near the Sergeants' Mess. Here they remained all night but were given some cigarettes and raisins. At about 0800 hours Jap looters arrived. At about 1000 hours a Jap Medical Officer of the rank of D.D.M.S. arrived. On entering the hospital he saluted the dead, complimented the staff on the way the patients had been looked after, and provided a guard against looters."

Later that Monday "the Jap G.O.C. called and expressed his regrets for what had happened and assured the staff that they had nothing further to fear. He told the O.C. that he was to be regarded as the direct representative of the Emperor and that no higher honour could be paid to the hospital."

Lt.-Col. Chamier was given the job of collecting maps. The Japs said all maps must be handed to them. "Most people had destroyed their maps as they were told to do, before the end. I got very few—and collected a rocket from the Jap officer to whom I had to hand them because there were so few. He seemed to blame it on me."

Signalman Mutton watched Sikhs pull out little Japanese flags and pin them on their breasts after the surrender. He also remembers being ordered to burn all his English money. His mate asked: "Are you going to burn yours, Fred?" He answered: "Not ruddy likely!" He and his mates were also told that night when they handed their rifles in, "Nobody is to attempt to escape. The G.O.C. has given his word that we will be handed over intact and the camp is surrounded by Manchesters who have been given instructions to fire if anybody is caught trying to escape." Mutton and some of the others were all for trying to make a break for it, but after that message they resigned themselves to surrender. "Some of the officers seemed dead scared of doing anything to upset the Japs," said Mutton.

Ted Fozard and about a hundred other European volunteers, now disarmed and carrying sticks, acted as auxiliary police during the two or three days before the final handover to the Japanese. "There was no one in power in Singapore during these few days," he said. "The Japs were on the outskirts of the city but had not been allowed to come in, otherwise there would have been wholesale slaughter. The

people in the town were sheltering and trying to feed themselves as best they could—including a lot of deserters. Our
job was to go out on patrol at night and during the day to
try to prevent looting. There was a lot of looting going on—
and rape and other things, too. Some of the Chinese and
Malays, and little Indian boys, showed no great respect for
us. They thought nothing of picking up pieces of glass and
hurling them at anyone who tried to stop them looting.

"One morning, though the Japanese orders were that
Europeans had to keep off the streets, I was walking down
High Street smoking a cigarette when a car camouflaged with
palm leaves came down the road. Behind it was an open
truck full of Japanese troops. I stood still. The car stopped
and a Japanese officer jumped out. In perfect English he
asked me: 'Who are you? What are you doing here?' I told
him: 'We're civilians. We've been helping the police to
prevent looting.' I wasn't sure what to say because we'd been
given no instructions on what we should say to the Japs, and
these were the first I'd seen.

"The Jap said: 'You are not armed?' I told him no, no
arms. He said: 'Throw that cigarette away—I am a Japanese
staff colonel.' He indicated the car. 'And this is the General.'
I was nonplussed. I knew I ought to show some sign of respect but I couldn't salute now I was a civilian again. The
Jap saved me by saying 'You must bow to the General.' I gave
a polite nod. The Colonel showed me how to bow properly.
I bowed.

"Then the Colonel wanted to know what my stick was.
I lifted it up to show him and the Japs in the truck swung
their machine-gun on to me. The Colonel ignored the stick
and asked where my C.O. was. I told him he was being questioned by the Japanese at Headquarters and the Colonel
said: 'Tell your commanding officer and second in command
to report at four o'clock at my office at Fort Canning, where
we have just pulled your flag down.' Then he got back into

the car, which turned round and went back the way it had come.

"Next day I had to parade along with all the others—and they paraded us, of all places, on my favourite bowling green. They kept us—including the Governor—standing in the blazing sun for five hours before a Japanese officer, standing on a box, interrogated us. Next day we walked six miles to our first internment camp. One of my best friends died on the march. He was the first of many who died."

Lt. Alastair Mackenzie, of Headquarters Company, 1st Battalion, The Malay Regiment, decided to get married. He was engaged to Sybil Osborn, who was working at the Singapore General Hospital, and on the morning after the surrender—after handing in his personal weapons—he got permission to go in search of her, and borrowed the Colonel's car.

Near the General Hospital he was stopped by a party of Japs. They took the car. They were quite peaceful about it. He pointed to the Hospital, and the senior Jap nodded. Alastair walked the rest of the way.

In the Matron's office, he was greeted by the Matron herself. Katie Stewart was a woman of about fifty with white hair, pretty eyes and a sweet smile—a charming and warmhearted Scotswoman—who told him that Sybil was alive and well. Katie added: "I suppose you've heard the rumour?"

"Which one? The place is alive with them."

"They say the only condition the General was able to lay down with the Japanese is that European women and children are to be evacuated."

"No, I hadn't heard that one. Where to?"

Matron shrugged. "Your guess is as good as mine. India, perhaps. . . ."

Alastair Mackenzie's mind worked fast. He reckoned if Sybil went to India, he could give her an introduction to the Commercial Union Branch manager in Madras, who would help get her settled. Then what? Alastair's firm would have no obligation to a girl called Miss Sybil Osborn, but, if she were Mrs. Mackenzie, it would make all the difference.

"I want to get married," he told Matron Stewart.

"You're not the only one," said Katie. "This place has been like Caxton Hall since the capitulation." She told how all ten of the British Matrons in Singapore were concerned for the safety of their young nurses. There were 48 British nursing Sisters still there, and the Matrons were going to endless trouble to find the young men of those who were engaged.

"Where's Sybil?" Alastair asked.

"She's helping Dr. Mekie's unit in Theatre D. You find the girl—I'll find the Bishop."

The General Hospital was a shambles. Alastair wondered how Matron, Sisters and nurses could stand it, moving about among this human butchery.

It took Alastair nearly two hours to find Sybil. She was in a sterilizing room when she looked round through the ovals of clear glass in the swing doors and saw him in the corridor. She ran to him and they shared the joy of discovering each other alive and well. Sybil had been working as Dr. Eric Mekie's assistant at the Tan Tock Seng hospital, some miles away from where Alastair was fighting. They had lost touch since the Japs had landed on the Island.

"Will you marry me straight away?" he asked.

"Of course."

On the way to Matron Katie's office, Alastair explained about the rumour he had heard. "If you get to Madras, all you need is to show some sort of marriage lines and you'll be well looked after. I'm jolly glad Katie told me."

Bishop Wilson had left to visit other hospitals, but Archdeacon Grahame-White was waiting to marry them. He was a man of sixty. He leaned over the Matron's desk writing out a marriage certificate on a sheet of hospital paper. He wrote it from memory.

Sybil was dressed in hospital white, stained with the blood of many air-raid victims. She wore moccasins for comfort. She'd cast off her stockings days ago. She was straight from the operating theatre, so wore the regulation white cotton square with drawstring to enclose every wisp of her dark brown hair.

She glanced at herself in Matron's mirror. No make-up, no hair! She made a typically feminine remark; "I look awful!"

"We all do," said Matron Katie, "so there's no need to worry about that."

Captain Peter Tomlinson, a friend of Alastair's, was there as best man. "Where's the ring?" he demanded. "You can't get married without one, and I bet you haven't even thought about it."

"You're dead right," said Alastair. "Where on earth would I get a wedding ring in this shambles?"

Sybil had been given a ring several days earlier by Dr. Mekie, to keep as a memento. It had been his Scottish grandmother's ring. It was too big for her small fingers, and she had slipped it into her apron pocket. Now she remembered it and fished it out.

"What a girl!" admired the Archdeacon. "I've heard about you from Mekie."

The wedding was brief, without fuss or emotion. Archdeacon Grahame-White completed the certificate, which was witnessed by Kathleen Stewart, Matron N.C.S.

"Congratulations, Mrs. Mackenzie. I hope you'll both be happy," said the Archdeacon.

"Thank you," said Sybil. She folded the certificate and put it in her apron pocket.

Outside, she swung her arm like any girl just married, and the gold ring flew off her finger and clinked away down the corridor. She screamed, and they both ran after it. They found it, and went up to Theatre D.

There was no champagne. They drank coffee with Eric Mekie, who snatched a moment to offer congratulations. Then Alastair had to return to his Regiment. "I'll call back this evening if I can."

He was able to call back again that evening for a few moments. By then they knew that the rumour about evacuation had no foundation in fact. He was to go to Changi P.O.W. camp, she to Changi gaol for internment.

They spent their honeymoon in England late in 1945, three and a half years after they were married. As soon as possible, they returned to Singapore and built their home there. It was the home for which they had fought. For their services in the Singapore campaign Alastair was awarded the Military Cross, Sybil was given the M.B.E.

* * *

The fires of Singapore burned low and went out. The smoke blew away across the ocean. The sun shone once more as the curtain fell on "the worst disaster and largest capitulation in British history."

We who surrendered suffered the hardship of prison camps and the torment of the infamous Burma Railway. Figures underline the magnitude of the disaster. Japanese casualties in the Malayan campaign were given as 9,824. The British lost 138,708—fourteen to one—including more than 130,000 prisoners. In addition there were the countless civilian dead, wounded and enslaved.

The victims lived or died, according to their treatment and their strength. They died of neglect, of brutality, of starvation, of overwork, of disease. Unknown thousands were executed. All living or dead suffered wounds, indignity, humiliation and terror. Those who lived bear the scars on minds or bodies to this day.

And though the Japanese were victorious for a time, the

first causes must never be forgotten. The dead and the
wounded were killed and injured by lies, by complacency, by
folly, greed, stupidity and petty jealousy, indulged in by men
who should have known better. All those people who were in
authority at that time must bear a greater or lesser degree of
the blame. One or two, notably Mr. Churchill, have publicly shouldered their share: "I ought to have known," he
afterward recorded. "My advisers ought to have known and
I ought to have been told, and I ought to have asked."
Blame falls on military and civilian alike, and on men in
London and Canberra as well as people in Singapore and
elsewhere. That is why there has never been—and, it is safe
to say, never will be—a public inquiry. Too much dirty linen
would need to be washed in the sight of the world, too
many skeletons would rattle in too many respectable cup-
boards, too many awkward searching questions would need
to be answered. It would not be an edifying spectacle—but
it would be enlightening.

It is a story that should not be forgotten, because it
marked the turn of a tremendous tide in human affairs. Before 1942, Singapore was a place where East and West peace-
fully met and traded. After 1942, Singapore marked the spot
where East and West violently separated.

Have the errors which led to that surrender been corrected? Have all the petty suspicions and jealousies been removed? Do international politicians, the military and
civilian Services now pool their resources for the common
good without thought of this man's authority or that man's
reward? Are there no fake fortresses waiting to fool us? Are
wise men telling the truth and being heard? Has the lesson
of Singapore—disunity—been studied and learned? Or were
the blood and tears which fell and mingled on that faraway
green island shed in vain?

THE END

Index

A page number in italic type indicates that there is a photograph of the subject opposite that page.

Airpower in Malaya, British and Japanese. *See under* British *and* Japanese

Air-raids on Singapore, 24–26, 69–70, 81–83, 163

Alexandra Hospital
Japanese atrocities, 213–16, 231–32

Allen, Dr. G. V., 69

Anderson, Lt.-Col. C. G. W., 86–87, 102–3

Andre, Mrs., 95

Andre, Lt.-Col. J. R. G., 204–5

Argylls. *See* Second Argyll and Sutherland Highlanders

Asherton, Col., 174

Atkins, Sqdn. Ldr. Pat, 196, 218–19

Atrocities, Japanese, 89–90, 214–16, 231–32

Australian Army
and defence of Singapore, 143, 144
and invasion of Singapore, 172–73
antagonism against Argylls, 125–26
deserters, 203–4
Second/Nineteenth and Second/Twenty-Ninth, 86–87, 102–3
strength, 138, 139, 140
Twenty-Second Brigade, 101, 165, 166
Twenty-Seventh Brigade, 182–83
See also Bennett, Maj.-Gen. H. Gordon

Bardwell, Capt. Mike, 105

Beckwith-Smith, Maj.-Gen. H., 86, 87, 206

Behind Bamboo (Rivett) quoted, 177–78

Benedict, Lance-Havildar, 103

Bennett, Maj.-Gen. H. Gordon, 28, 105–6, 140, 159–60, 165, 202–3
and defence of Singapore, 144–45
and invasion of Singapore, 177–78, 182, 206–7, 219–20
and Naval Base, 127
and shelling of hospital, 159–60
escape, 230–31
mentioned, 50, 206, 228

Bennett, Jack, 94–95, 194–95, 205–6, 210–11, 229, 231

Bisseker, A., 208

Blelloch, Ian, 156–57

Bowden, Mr. (Commonwealth Commissioner in Singapore), 64, 203–4, 210

Brewster Buffalo, 19–20, 22, 25, 34, 40

British airpower
in Singapore, 154, 155, 164, 185–87, 199
loss of, 34–36, 79–80
strength of, 42

British Army
bravery of, 101–5
Dallforce, 145–50
defeats in Malayan peninsula, 45–46, 48–51
differences in High Command, 124–25

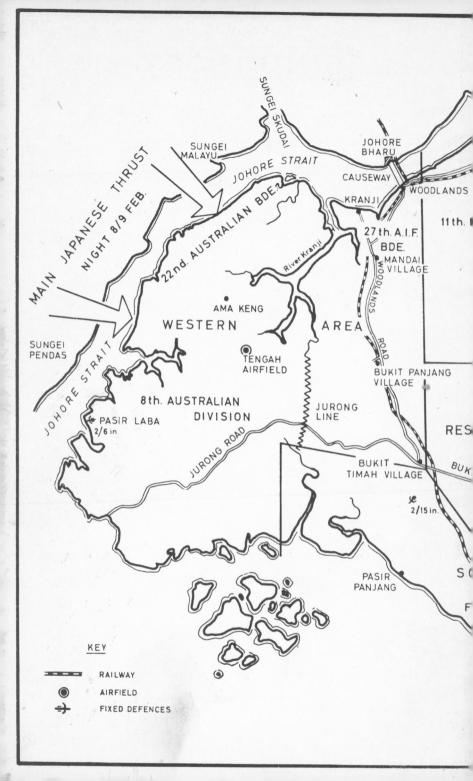